VOLUME NINETY EIGHT

ADVANCES IN MARINE BIOLOGY

Advances in Rhizostomeae Jellyfish Research

ADVANCES IN MARINE BIOLOGY

VOLUME NINETY EIGHT

ADVANCES IN MARINE BIOLOGY

Advances in Rhizostomeae Jellyfish Research

Edited by

ANDRÉ C. MORANDINI

Centro de Biologia Marinha, Universidade de São Paulo, São Sebastião; Departamento de Zoologia, Instituto de Biociências, Universidade de São Paulo, São Paulo, SP, Brazil

SABINE HOLST

Senckenberg am Meer, German Center for Marine Biodiversity Research, Hamburg, Germany

GÖTZ B. REINICKE

Ocean Museum Germany, Stralsund, Germany

ELSEVIER

ACADEMIC PRESS

An imprint of Elsevier

Academic Press is an imprint of Elsevier
50 Hampshire Street, 5th Floor, Cambridge, MA 02139, United States
525 B Street, Suite 1650, San Diego, CA 92101, United States
125 London Wall, London, EC2Y 5AS, United Kingdom

First edition 2024

ISBN: 978-0-443-29646-8
ISSN: 0065-2881

For information on all Academic Press publications
visit our website at https://www.elsevier.com/books-and-journals

Publisher: Zoe Kruze
Acquisitions Editor: Jason Mitchell
Editorial Project Manager: Palash Sharma
Production Project Manager: Maria Shalini
Cover Designer: Gopalakrishnan Venkatraman

Typeset by MPS Limited, India

Contents

Contributors to Volume 98

Dror L. Angel
Department of Maritime Civilizations & Recanati Institute for Maritime Studies, University of Haifa, Mt. Carmel, Haifa, Israel

Lucas Brotz
Sea Around Us, Institute for the Oceans and Fisheries, University of British Columbia, Vancouver, BC, Canada; SeaLifeBase, Quantitative Aquatics, Khush Hall, International Rice Research Institute (IRRI), Los Baños, Philippines

Michael Brown
Department of Biodiversity and Conservation Biology, University of the Western Cape, Bellville, Republic of South Africa

Sean P. Colin
Marine Biology and Environmental Science, Roger Williams University, Bristol, RI; Whitman Center, Marine Biological Laboratory, Woods Hole, MA, United States

John H. Costello
Whitman Center, Marine Biological Laboratory, Woods Hole, MA; Biology Department, Providence College, Providence, RI, United States

Isabella D'Ambra
Integrative Marine Ecology Department, Stazione Zoologica Anton Dohrn, Villa Comunale, Napoli; National Center for Future Biodiversity (NFBC), Palermo, Italy

Nicolas Djeghri
The Marine Biological Association, Continuous Plankton Recorder Survey; The Marine Biological Association, The Laboratory, Citadel Hill, Plymouth, United Kingdom; University of Brest (UBO), Institut Universitaire Européen de la Mer (IUEM), Laboratoire des sciences de l'environnement marin (LEMAR, UMR 6539), Plouzané, France

Tom Doyle
School of Biological, Earth and Environmental Sciences; Science Foundation Ireland Research Centre for Energy, Climate and Marine, Environmental Research Centre, University College Cork, Cork, Ireland

Angelica Enrique-Navarro
Department of Ecology and Coastal Management, Instituto de Ciencias Marinas de Andalucía (ICMAN), Consejo Superior de Investigaciones Científicas (CSIC), Puerto Real, Cádiz, Spain

William Fitt
Odum School of Ecology, University of Georgia, Athens, GA, United States

Edgar Gamero-Mora
Laboratorio de Bioenergética y Genética Molecular, Centro de Investigación en Alimentación y Desarrollo, Hermosillo, Sonora, Mexico

Mark Gibbons
Department of Biodiversity and Conservation Biology, University of the Western Cape, Bellville, Republic of South Africa

Sabine Holst
Senckenberg am Meer, German Centre for Marine Biodiversity Research (DZMB), Hamburg, Germany

Mayara A. Jordano
Departamento de Zoologia, Instituto de Biociências, Universidade de São Paulo, São Paulo, SP, Brazil

Zafrir Kuplik
The Steinhardt Museum of Natural History, Tel Aviv University, Tel Aviv, Israel

Chiara Lauritano
Ecosustainable Marine Biotechnology Department, Stazione Zoologica Anton Dohrn, Napoli, Italy

Jonathan W. Lawley
School of Environment and Science, Centre for Planetary Health and Food Security, Griffith University, Gold Coast, QLD, Australia

Valentina Leoni
CoNISMa, Consorzio Nazionale Interuniversitario per le Scienze del Mare, Rome, Italy

Maximiliano M. Maronna
Instituto de Oceanografia, Universidade Federal do Rio Grande, Rio Grande, RS, Brazil

Hiroshi Miyake
School of Marine Biosciences, Kitasato University, Sagamihara, OL, Japan

André C. Morandini
Centro de Biologia Marinha, Universidade de São Paulo, São Sebastião; Departamento de Zoologia, Instituto de Biociências, Universidade de São Paulo, São Paulo, SP, Brazil

Adriana Muhlia-Almazan
Laboratorio de Bioenergética y Genética Molecular, Centro de Investigación en Alimentación y Desarrollo, Hermosillo, Sonora, Mexico

Renato M. Nagata
Laboratório de Zooplâncton; Programa de Pós-Graduação em Oceanografia Biológica (PPGOB), Instituto de Oceanografia, Universidade Federal do Rio Grande (FURG), Rio Grande, RS, Brazil

Kylie Pitt
School of Environment and Science, Coastal and Marine Research Centre, Australian Rivers Institute, Griffith University, Australia

Laura Prieto
Department of Ecology and Coastal Management, Instituto de Ciencias Marinas de Andalucía (ICMAN), Consejo Superior de Investigaciones Científicas (CSIC), Puerto Real, Cádiz, Spain

Götz B. Reinicke
Ocean Museum Germany, Stralsund, Germany

Agustín Schiariti
Instituto Nacional de Investigación y Desarrollo Pesquero (INIDEP), Mar del Plata; Consejo Nacional de Investigaciones Científicas y Técnicas (CONICET), Buenos Aires, Argentina

Sérgio N. Stampar
Departamento de Ciências Biológicas, Universidade Estadual Paulista, Bauru, SP, Brazil

Ilka Sötje
University of Hamburg, Institute for Cell and Systems Biology of Animals (IZS), Hamburg, Germany

Ilka Straehler-Pohl
Medusa(')s Nursery—Private Laboratory for Life cycle, Developmental, Evolutionary and Taxonomical Research, Haldensleben, Germany

Delphine Thibault
Aix Marseille University, Université de Toulon, CNRS, IRD, MIO, Marseille, France

Hjalmar Thiel
Ocean Museum Germany, Stralsund, Germany

Gisele R. Tiseo
Departamento de Zoologia, Instituto de Biociências, Universidade de São Paulo, São Paulo, SP, Brazil

Shin Uye
Graduate School of Integrated Sciences for Life, Hiroshima University, Higashi-Hiroshima, Japan

Guilherme M. von Montfort
Laboratório de Zooplâncton; Programa de Pós-Graduação em Oceanografia Biológica (PPGOB), Instituto de Oceanografia, Universidade Federal do Rio Grande (FURG), Rio Grande, RS, Brazil

Amanda A. Wiesenthal
University of Rostock, Institute of Biosciences, Rostock, Germany

Series Contents for Last Fifteen Years*,†

Volume 44, 2003.

Hirst, A. G., Roff, J. C. and Lampitt, R. S. A synthesis of growth rates in epipelagic invertebrate zooplankton. pp. 3–142.

Boletzky, S. von. Biology of early life stages in cephalopod molluscs. pp. 143–203.

Pittman, S. J. and McAlpine, C. A. Movements of marine fish and decapod crustaceans: process, theory and application. pp. 205–294.

Cutts, C. J. Culture of harpacticoid copepods: potential as live feed for rearing marine fish. pp. 295–315.

Volume 45, 2003.

Cumulative Taxonomic and Subject Index.

Volume 46, 2003.

Gooday, A. J. Benthic foraminifera (Protista) as tools in deep-water palaeoceanography: environmental influences on faunal characteristics. pp. 1–90.

Subramoniam, T. and Gunamalai, V. Breeding biology of the intertidal sand crab, Emerita (Decapoda: Anomura). pp. 91–182.

Coles, S. L. and Brown, B. E. Coral bleaching—capacity for acclimatization and adaptation. pp. 183–223.

Dalsgaard J., St. John M., Kattner G., Müller-Navarra D. and Hagen W. Fatty acid trophic markers in the pelagic marine environment. pp. 225–340.

Volume 47, 2004.

Southward, A. J., Langmead, O., Hardman-Mountford, N. J., Aiken, J., Boalch, G. T., Dando, P. R., Genner, M. J., Joint, I., Kendall, M. A., Halliday, N. C., Harris, R. P., Leaper, R., Mieszkowska, N., Pingree, R. D., Richardson, A. J., Sims, D.W., Smith, T., Walne, A. W. and Hawkins, S. J. Long-term oceanographic and ecological research in the western English Channel. pp. 1–105.

Queiroga, H. and Blanton, J. Interactions between behaviour and physical forcing in the control of horizontal transport of decapod crustacean larvae. pp. 107–214.

* The full list of contents for volumes 1–37 can be found in volume 38

† The full list of contents for volumes 38–43 can be found in volume 78

Braithwaite, R. A. and McEvoy, L. A. Marine biofouling on fish farms and its remediation. pp. 215–252.

Frangoulis, C., Christou, E. D. and Hecq, J. H. Comparison of marine copepod outfluxes: nature, rate, fate and role in the carbon and nitrogen cycles. pp. 253–309.

Volume 48, 2005.

Canfield, D. E., Kristensen, E. and Thamdrup, B. Aquatic Geomicrobiology. pp. 1–599.

Volume 49, 2005.

Bell, J. D., Rothlisberg, P. C., Munro, J. L., Loneragan, N. R., Nash, W. J., Ward, R. D. and Andrew, N. L. Restocking and stock enhancement of marine invertebrate fisheries. pp. 1–358.

Volume 50, 2006.

Lewis, J. B. Biology and ecology of the hydrocoral *Millepora* on coral reefs. pp. 1–55.

Harborne, A. R., Mumby, P. J., Micheli, F., Perry, C. T., Dahlgren, C. P., Holmes, K. E., and Brumbaugh, D. R. The functional value of Caribbean coral reef, seagrass and mangrove habitats to ecosystem processes. pp. 57–189.

Collins, M. A. and Rodhouse, P. G. K. Southern ocean cephalopods. pp. 191–265.

Tarasov, V. G. Effects of shallow-water hydrothermal venting on biological communities of coastal marine ecosystems of the western Pacific. pp. 267-410.

Volume 51, 2006.

Elena Guijarro Garcia. The fishery for Iceland scallop (*Chlamys islandica*) in the Northeast Atlantic. pp. 1–55.

Jeffrey, M. Leis. Are larvae of demersal fishes plankton or nekton? pp. 57–141.

John C. Montgomery, Andrew Jeffs, Stephen D. Simpson, Mark Meekan and Chris Tindle. Sound as an orientation cue for the pelagic larvae of reef fishes and decapod crustaceans. pp. 143–196.

Carolin E. Arndt and Kerrie M. Swadling. Crustacea in Arctic and Antarctic sea ice: Distribution, diet and life history strategies. pp. 197–315.

Volume 52, 2007.
Leys, S. P., Mackie, G. O. and Reiswig, H. M. The Biology of Glass Sponges. pp. 1–145.
Garcia E. G. The Northern Shrimp (*Pandalus borealis*) Offshore Fishery in the Northeast Atlantic. pp. 147–266.
Fraser K. P. P. and Rogers A. D. Protein Metabolism in Marine Animals: The Underlying Mechanism of Growth. pp. 267–362.

Volume 53, 2008.
Dustin J. Marshall and Michael J. Keough. The Evolutionary Ecology of Offspring Size in Marine Invertebrates. pp. 1–60.
Kerry A. Naish, Joseph E. Taylor III, Phillip S. Levin, Thomas P. Quinn, James R. Winton, Daniel Huppert, and Ray Hilborn. An Evaluation of the Effects of Conservation and Fishery Enhancement Hatcheries on Wild Populations of Salmon. pp. 61–194.
Shannon Gowans, Bernd Würsig, and Leszek Karczmarski. The Social Structure and Strategies of Delphinids: Predictions Based on an Ecological Framework. pp. 195–294.

Volume 54, 2008.
Bridget S. Green. Maternal Effects in Fish Populations. pp. 1–105.
Victoria J. Wearmouth and David W. Sims. Sexual Segregation in Marine Fish, Reptiles, Birds and Mammals: Behaviour Patterns, Mechanisms and Conservation Implications. pp. 107–170.
David W. Sims. Sieving a Living: A Review of the Biology, Ecology and Conservation Status of the Plankton-Feeding Basking Shark *Cetorhinus maximus*. pp. 171–220.
Charles H. Peterson, Kenneth W. Able, Christin Frieswyk DeJong, Michael F. Piehler, Charles A. Simenstad, and Joy B. Zedler. Practical Proxies for Tidal Marsh Ecosystem Services: Application to Injury and Restoration. pp. 221–266.

Volume 55, 2008.
Annie Mercier and Jean-Francois Hamel. Introduction. pp. 1–6.
Annie Mercier and Jean-Francois Hamel. Gametogenesis. pp. 7–72.
Annie Mercier and Jean-Francois Hamel. Spawning. pp. 73–168.
Annie Mercier and Jean-Francois Hamel. Discussion. pp. 169–194.

Volume 56, 2009.

Philip C. Reid, Astrid C. Fischer, Emily Lewis-Brown, Michael P. Meredith, Mike Sparrow, Andreas J. Andersson, Avan Antia, Nicholas R. Bates, Ulrich Bathmann, Gregory Beaugrand, Holger Brix, Stephen Dye, Martin Edwards, Tore Furevik, Reidun Gangst, Hjalmar Hatun, Russell R. Hopcroft, Mike Kendall, Sabine Kasten, Ralph Keeling, Corinne Le Quere, Fred T. Mackenzie, Gill Malin, Cecilie Mauritzen, Jon Olafsson, Charlie Paull, Eric Rignot, Koji Shimada, Meike Vogt, Craig Wallace, Zhaomin Wang and Richard Washington. Impacts of the Oceans on Climate Change. pp. 1–150.

Elvira S. Poloczanska, Colin J. Limpus and Graeme C. Hays. Vulnerability of Marine Turtles to Climate Change. pp. 151–212.

Nova Mieszkowska, Martin J. Genner, Stephen J. Hawkins and David W. Sims. Effects of Climate Change and Commercial Fishing on Atlantic Cod *Gadus morhua*. pp. 213–274.

Iain C. Field, Mark G. Meekan, Rik C. Buckworth and Corey J. A. Bradshaw. Susceptibility of Sharks, Rays and Chimaeras to Global Extinction. pp. 275–364.

Milagros Penela-Arenaz, Juan Bellas and Elsa Vazquez. Effects of the *Prestige* Oil Spill on the Biota of NW Spain: 5 Years of Learning. pp. 365–396.

Volume 57, 2010.

Geraint A. Tarling, Natalie S. Ensor, Torsten Fregin, William P. Good-all-Copestake and Peter Fretwell. An Introduction to the Biology of Northern Krill (*Meganyctiphanes norvegica* Sars). pp. 1–40.

Tomaso Patarnello, Chiara Papetti and Lorenzo Zane. Genetics of Northern Krill (*Meganyctiphanes norvegica* Sars). pp. 41–58.

Geraint A. Tarling. Population Dynamics of Northern Krill (*Meganyctiphanes norvegica* Sars). pp. 59–90.

John I. Spicer and Reinhard Saborowski. Physiology and Metabolism of Northern Krill (*Meganyctiphanes norvegica* Sars). pp. 91–126.

Katrin Schmidt. Food and Feeding in Northern Krill (*Meganyctiphanes norvegica* Sars). pp. 127–172.

Friedrich Buchholz and Cornelia Buchholz. Growth and Moulting in Northern Krill (*Meganyctiphanes norvegica* Sars). pp. 173–198.

Janine Cuzin-Roudy. Reproduction in Northern Krill. pp. 199–230.

Edward Gaten, Konrad Wiese and Magnus L. Johnson. Laboratory-Based Observations of Behaviour in Northern Krill (*Meganyctiphanes norvegica* Sars). pp. 231–254.

Stein Kaartvedt. Diel Vertical Migration Behaviour of the Northern Krill (*Meganyctiphanes norvegica* Sars). pp. 255–276.

Yvan Simard and Michel Harvey. Predation on Northern Krill (*Meganyctiphanes norvegica* Sars). pp. 277–306.

Volume 58, 2010.

A. G. Glover, A. J. Gooday, D. M. Bailey, D. S. M. Billett, P. Chevaldonné, A. Colaço, J. Copley, D. Cuvelier, D. Desbruyères, V. Kalogeropoulou, M. Klages, N. Lampadariou, C. Lejeusne, N. C. Mestre, G. L. J. Paterson, T. Perez, H. Ruhl, J. Sarrazin, T. Soltwedel, E. H. Soto, S. Thatje, A. Tselepides, S. Van Gaever, and A. Vanreusel. Temporal Change in Deep-Sea Benthic Ecosystems: A Review of the Evidence From Recent Time-Series Studies. pp. 1–96.

Hilario Murua. The Biology and Fisheries of European Hake, *Merluccius merluccius*, in the North-East Atlantic. pp. 97–154.

Jacopo Aguzzi and Joan B. Company. Chronobiology of Deep-Water Decapod Crustaceans on Continental Margins. pp. 155–226.

Martin A. Collins, Paul Brickle, Judith Brown, and Mark Belchier. The Patagonian Toothfish: Biology, Ecology and Fishery. pp. 227–300.

Volume 59, 2011.

Charles W. Walker, Rebecca J. Van Beneden, Annette F. Muttray, S. Anne Böttger, Melissa L. Kelley, Abraham E. Tucker, and W. Kelley Thomas. p53 Superfamily Proteins in Marine Bivalve Cancer and Stress Biology. pp 1–36.

Martin Wahl, Veijo Jormalainen, Britas Klemens Eriksson, James A. Coyer, Markus Molis, Hendrik Schubert, Megan Dethier, Anneli Ehlers, Rolf Karez, Inken Kruse, Mark Lenz, Gareth Pearson, Sven Rohde, Sofia A. Wikström, and Jeanine L. Olsen. Stress Ecology in *Fucus*: Abiotic, Biotic and Genetic Interactions. pp. 37–106.

Steven R. Dudgeon and Janet E. Kübler. Hydrozoans and the Shape of Things to Come. pp. 107–144.

Miles Lamare, David Burritt, and Kathryn Lister. Ultraviolet Radiation and Echinoderms: Past, Present and Future Perspectives. pp. 145–187.

Volume 60, 2011.

Tatiana A. Rynearson and Brian Palenik. Learning to Read the Oceans: Genomics of Marine Phytoplankton. pp. 1–40.

Les Watling, Scott C. France, Eric Pante and Anne Simpson. Biology of Deep-Water Octocorals. pp. 41–122.

Cristián J. Monaco and Brian Helmuth. Tipping Points, Thresholds and the Keystone Role of Physiology in Marine Climate Change Research. pp. 123–160.

David A. Ritz, Alistair J. Hobday, John C. Montgomery and Ashley J.W. Ward. Social Aggregation in the Pelagic Zone with Special Reference to Fish and Invertebrates. pp. 161–228.

Volume 61, 2012.

Gert Wörheide, Martin Dohrmann, Dirk Erpenbeck, Claire Larroux, Manuel Maldonado, Oliver Voigt, Carole Borchiellini and Denis Lavrov. Deep Phylogeny and Evolution of Sponges (Phylum Porifera). pp. 1–78.

Paco Cárdenas, Thierry Pérez and Nicole Boury-Esnault. Sponge Systematics Facing New Challenges. pp. 79–210.

Klaus Rützler. The Role of Sponges in the Mesoamerican Barrier-Reef Ecosystem, Belize. pp. 211–272.

Janie Wulff. Ecological Interactions and the Distribution, Abundance, and Diversity of Sponges. pp. 273–344.

Maria J. Uriz and Xavier Turon. Sponge Ecology in the Molecular Era. pp. 345–410.

Volume 62, 2012.

Sally P. Leys and April Hill. The Physiology and Molecular Biology of Sponge Tissues. pp. 1–56.

Robert W. Thacker and Christopher J. Freeman. Sponge-Microbe Symbioses: Recent Advances and New Directions. pp. 57–112.

Manuel Maldonado, Marta Ribes and Fleur C. van Duyl. Nutrient Fluxes Through Sponges: Biology, Budgets, and Ecological Implications. pp. 113–182.

Grégory Genta-Jouve and Olivier P. Thomas. Sponge Chemical Diversity: From Biosynthetic Pathways to Ecological Roles. pp. 183–230.

Xiaohong Wang, Heinz C. Schröder, Matthias Wiens, Ute Schloßmacher and Werner E. G. Müller. Biosilica: Molecular Biology, Biochemistry and Function in Demosponges as well as its Applied Aspects for Tissue Engineering. pp. 231–272.

Klaske J. Schippers, Detmer Sipkema, Ronald Osinga, Hauke Smidt, Shirley A. Pomponi, Dirk E. Martens and René H. Wijffels. Cultivation of Sponges, Sponge Cells and Symbionts: Achievements and Future Prospects. pp. 273–338.

Volume 63, 2012.

Michael Stat, Andrew C. Baker, David G. Bourne, Adrienne M. S. Correa, Zac Forsman, Megan J. Huggett, Xavier Pochon, Derek Skillings, Robert J. Toonen, Madeleine J. H. van Oppen, and Ruth D. Gates. Molecular Delineation of Species in the Coral Holobiont. pp. 1–66.

Daniel Wagner, Daniel G. Luck, and Robert J. Toonen. The Biology and Ecology of Black Corals (Cnidaria: Anthozoa: Hexacorallia: Antipatharia). pp. 67–132.

Cathy H. Lucas, William M. Graham, and Chad Widmer. Jellyfish Life Histories: Role of Polyps in Forming and Maintaining Scyphomedusa Populations. pp. 133–196.

T. Aran Mooney, Maya Yamato, and Brian K. Branstetter. Hearing in Cetaceans: From Natural History to Experimental Biology. pp. 197–246.

Volume 64, 2013.

Dale Tshudy. Systematics and Position of *Nephrops* Among the Lobsters. pp. 1–26.

Mark P. Johnson, Colm Lordan, and Anne Marie Power. Habitat and Ecology of *Nephrops norvegicus*. pp. 27–64.

Emi Katoh, Valerio Sbragaglia, Jacopo Aguzzi, and Thomas Breithaupt. Sensory Biology and Behaviour of *Nephrops norvegicus*. pp. 65–106.

Edward Gaten, Steve Moss, and Magnus L. Johnson. The Reniform Reflecting Superposition Compound Eyes of *Nephrops norvegicus*: Optics, Susceptibility to Light-Induced Damage, Electrophysiology and a Ray Tracing Model. pp. 107–148.

Susanne P. Eriksson, Bodil Hernroth, and Susanne P. Baden. Stress Biology and Immunology in *Nephrops norvegicus*. pp. 149–200.

Adam Powell and Susanne P. Eriksson. Reproduction: Life Cycle, Larvae and Larviculture. pp. 201–246.

Anette Ungfors, Ewen Bell, Magnus L. Johnson, Daniel Cowing, Nicola C. Dobson, Ralf Bublitz, and Jane Sandell. *Nephrops* Fisheries in European Waters. pp. 247–314.

Volume 65, 2013.

Isobel S.M. Bloor, Martin J. Attrill, and Emma L. Jackson. A Review of the Factors Influencing Spawning, Early Life Stage Survival and Recruitment Variability in the Common Cuttlefish (*Sepia officinalis*). pp. 1–66.

Dianna K. Padilla and Monique M. Savedo. A Systematic Review of Phenotypic Plasticity in Marine Invertebrate and Plant Systems. pp. 67–120.

Leif K. Rasmuson. The Biology, Ecology and Fishery of the Dungeness crab, *Cancer magister*. pp. 121–174.

Volume 66, 2013.

Lisa-ann Gershwin, Anthony J. Richardson, Kenneth D. Winkel, Peter J. Fenner, John Lippmann, Russell Hore, Griselda Avila-Soria, David Brewer, Rudy J. Kloser, Andy Steven, and Scott Condie. Biology and Ecology of Irukandji Jellyfish (Cnidaria: Cubozoa). pp. 1–86.

April M. H. Blakeslee, Amy E. Fowler, and Carolyn L. Keogh. Marine Invasions and Parasite Escape: Updates and New Perspectives. pp. 87–170.

Michael P. Russell. Echinoderm Responses to Variation in Salinity. pp. 171–212.

Daniela M. Ceccarelli, A. David McKinnon, Serge Andŕefouët, Valerie Allain, Jock Young, Daniel C. Gledhill, Adrian Flynn, Nicholas J. Bax, Robin Beaman, Philippe Borsa, Richard Brinkman, Rodrigo H. Bustamante, Robert Campbell, Mike Cappo, Sophie Cravatte, Stéphanie D'Agata, Catherine M. Dichmont, Piers K. Dunstan, Cécile Dupouy, Graham Edgar, Richard Farman, Miles Furnas, Claire Garrigue, Trevor Hutton, Michel Kulbicki, Yves Letourneur, Dhugal Lindsay, Christophe Menkes, David Mouillot, Valeriano Parravicini, Claude Payri, Bernard Pelletier, Bertrand Richer de Forges, Ken Ridgway, Martine Rodier, Sarah Samadi, David Schoeman, Tim Skewes, Steven Swearer, Laurent Vigliola, Laurent Wantiez, Alan Williams, Ashley Williams, and Anthony J. Richardson. The Coral Sea: Physical Environment, Ecosystem Status and Biodiversity Assets. pp. 213–290.

Volume 67, 2014.

Erica A.G. Vidal, Roger Villanueva, José P. Andrade, Ian G. Gleadall, José Iglesias, Noussithé Koueta, Carlos Rosas, Susumu Segawa, Bret Grasse, Rita M. Franco-Santos, Caroline B. Albertin, Claudia Caamal-Monsreal, Maria E. Chimal, Eric Edsinger-Gonzales, Pedro Gallardo, Charles Le Pabic, Cristina Pascual, Katina Roumbedakis, and James Wood. Cephalopod Culture: Current Status of Main Biological Models and Research Priorities. pp. 1–98.

Paul G.K. Rodhouse, Graham J. Pierce, Owen C. Nichols, Warwick H.H. Sauer, Alexander I. Arkhipkin, Vladimir V. Laptikhovsky, Marek R. Lipiński, Jorge E. Ramos, Michaël Gras, Hideaki Kidokoro, Kazuhiro Sadayasu, João Pereira, Evgenia Lefkaditou, Cristina Pita, Maria

Gasalla, Manuel Haimovici, Mitsuo Sakai, and Nicola Downey. Environmental Effects on Cephalopod Population Dynamics: Implications for Management of Fisheries. pp. 99–234.

Henk-Jan T. Hoving, José A.A. Perez, Kathrin Bolstad, Heather Braid, Aaron B. Evans, Dirk Fuchs, Heather Judkins, Jesse T. Kelly, Jośe E.A.R. Marian, Ryuta Nakajima, Uwe Piatkowski, Amanda Reid, Michael Vecchione, and José C.C. Xavier. The Study of Deep-Sea Cephalopods. pp. 235–362.

Jean-Paul Robin, Michael Roberts, Lou Zeidberg, Isobel Bloor, Almendra Rodriguez, Felipe Briceño, Nicola Downey, Maite Mascaró, Mike Navarro, Angel Guerra, Jennifer Hofmeister, Diogo D. Barcellos, Silvia A.P. Lourenço, Clyde F.E. Roper, Natalie A. Moltschaniwskyj, Corey P. Green, and Jennifer Mather. Transitions During Cephalopod Life History: The Role of Habitat, Environment, Functional Morphology and Behaviour. pp. 363–440.

Volume 68, 2014.

Paul K.S. Shin, Siu Gin Cheung, Tsui Yun Tsang, and Ho Yin Wai. Ecology of Artificial Reefs in the Subtropics. pp. 1–64.

Hrafnkell Eiríksson. Reproductive Biology of Female Norway Lobster, *Nephrops norvegicus* (Linnaeus, 1758) Leach, in Icelandic Waters During the Period 1960-2010: Comparative Overview of Distribution Areas in the Northeast Atlantic and the Mediterranean. pp. 65–210.

Volume 69, 2014.

Ray Hilborn. Introduction to Marine Managed Areas. pp. 1–14.

Philip N. Trathan, Martin A. Collins, Susie M. Grant, Mark Belchier, David K.A. Barnes, Judith Brown, and Iain J. Staniland. The South Georgia and the South Sandwich Islands MPA: Protecting A Biodiverse Oceanic Island Chain Situated in the Flow of the Antarctic Circumpolar Current. pp. 15-78.

Richard P. Dunne, Nicholas V.C. Polunin, Peter H. Sand, and Magnus L. Johnson. The Creation of the Chagos Marine Protected Area: A Fisheries Perspective. pp. 79–128.

Michelle T. Schärer-Umpierre, Daniel Mateos-Molina, Richard Appeldoorn, Ivonne Bejarano, Edwin A. Hernández-Delgado, Richard S. Nemeth, Michael I. Nemeth, Manuel Valdés-Pizzini, and Tyler B. Smith. Marine Managed Areas and Associated Fisheries in the US Caribbean. pp. 129–152.

Alan M. Friedlander, Kostantinos A. Stamoulis, John N. Kittinger, Jeffrey C. Drazen, and Brian N. Tissot. Understanding the Scale of

Marine Protection in Hawai'i: From Community-Based Management to the Remote Northwestern Hawaiian Islands. pp. 153–204.

Louis W. Botsford, J. Wilson White, Mark H. Carr, and Jennifer E. Caselle. Marine Protected Area Networks in California, USA. pp. 205–252.

Bob Kearney and Graham Farebrother. Inadequate Evaluation and Management of Threats in Australia's Marine Parks, Including the Great Barrier Reef, Misdirect Marine Conservation. pp. 253–288.

Randi Rotjan, Regen Jamieson, Ben Carr, Les Kaufman, Sangeeta Mangubhai, David Obura, Ray Pierce, Betarim Rimon, Bud Ris, Stuart Sandin, Peter Shelley, U. Rashid Sumaila, Sue Taei, Heather Tausig, Tukabu Teroroko, Simon Thorrold, Brooke Wikgren, Teuea Toatu, and Greg Stone. Establishment, Management, and Maintenance of the Phoenix Islands Protected Area. pp. 289–324.

Alex J. Caveen, Clare Fitzsimmons, Margherita Pieraccini, Euan Dunn, Christopher J. Sweeting, Magnus L. Johnson, Helen Bloomfield, Estelle V. Jones, Paula Lightfoot, Tim S. Gray, Selina M. Stead, and Nicholas V. C. Polunin. Diverging Strategies to Planning an Ecologically Coherent Network of MPAs in the North Sea: The Roles of Advocacy, Evidence and Pragmatism in the Face of Uncertaintya. pp. 325–370.

Carlo Pipitone, Fabio Badalamenti, Tomás Vega Fernández, and Giovanni D'Anna. Spatial Management of Fisheries in the Mediterranean Sea: Problematic Issues and a Few Success Stories. pp. 371–402.

Volume 70, 2015.

Alex D. Rogers, Christopher Yesson, and Pippa Gravestock. A Biophysical and Economic Profile of South Georgia and the South Sandwich Islands as Potential Large-Scale Antarctic Protected Areas. pp. 1–286.

Volume 71, 2015.

Ricardo Calado and Miguel Costa Leal. Trophic Ecology of Benthic Marine Invertebrates with Bi-Phasic Life Cycles: What Are We Still Missing? pp. 1–70.

Jesse M.A. van der Grient and Alex D. Rogers. Body Size Versus Depth: Regional and Taxonomical Variation in Deep-Sea Meio- and Macrofaunal Organisms. pp. 71–108.

Lorena Basso, Maite Vázquez-Luis, José R. García-March, Salud Deudero, Elvira Alvarez, Nardo Vicente, Carlos M. Duarte, and Iris E. Hendriks. The Pen Shell, *Pinna nobilis*: A Review of Population Status and Recommended Research Priorities in the Mediterranean Sea. pp. 109–160.

Volume 72, 2015.

Thomas A. Jefferson and Barbara E. Curry. Humpback Dolphins: A Brief Introduction to the Genus *Sousa*. pp. 1–16.

Sarah Piwetz, David Lundquist, and Bernd Würsig. Humpback Dolphin (Genus *Sousa*) Behavioural Responses to Human Activities. pp. 17–46.

Tim Collins. Re-assessment of the Conservation Status of the Atlantic Humpback Dolphin, *Sousa teuszii* (Kükenthal, 1892) Using the IUCN Red List Criteria. pp. 47–78.

Caroline R. Weir and Tim Collins. A Review of the Geographical Distribution and Habitat of the Atlantic Humpback Dolphin (*Sousa teuszii*). pp. 79–118.

Gill T Braulik, Ken Findlay, Salvatore Cerchio, and Robert Baldwin. Assessment of the Conservation Status of the Indian Ocean Humpback Dolphin (*Sousa plumbea*) Using the IUCN Red List Criteria. pp. 119–142.

Stephanie Plön, Victor G. Cockcroft, and William P. Froneman. The Natural History and Conservation of Indian Ocean Humpback Dolphins (*Sousa plumbea*) in South African Waters. pp. 143–162.

Salvatore Cerchio, Norbert Andrianarivelo, and Boris Andrianantenaina. Ecology and Conservation Status of Indian Ocean Humpback Dolphins (*Sousa plumbea*) in Madagascar. pp. 163–200.

Muhammad Shoaib Kiani and Koen Van Waerebeek. A Review of the Status of the Indian Ocean Humpback Dolphin *Sousa plumbea* in Pakistan. pp. 201–228.

Dipani Sutaria, Divya Panicker, Ketki Jog, Mihir Sule, Rahul Muralidharan, and Isha Bopardikar. Humpback Dolphins (Genus *Sousa*) in India: An Overview of Status and Conservation Issues. pp. 229–256.

Volume 73, 2016.

Thomas A. Jefferson and Brian D. Smith. Re-assessment of the Conservation Status of the Indo-Pacific Humpback Dolphin (*Sousa chinensis*) Using the IUCN Red List Criteria. pp. 1–26.

Leszek Karczmarski, Shiang-Lin Huang, Carmen K. M. Or, Duan Gui, Stephen C. Y. Chan, Wenzhi Lin, Lindsay Porter, Wai-Ho Wong, Ruiqiang Zheng, Yuen-Wa Ho, Scott Y. S. Chui, Angelico Jose C. Tiongson, Yaqian Mo, Wei-Lun Chang, John H. W. Kwok, Ricky W. K. Tang, Andy T. L. Lee, Sze-Wing Yiu, Mark Keith, Glenn Gailey, and Yuping Wu. Humpback Dolphins in Hong Kong and the Pearl River Delta: Status, Threats and Conservation Challenges. pp. 27–64.

Bernd Würsig, E.C.M. Parsons, Sarah Piwetz, and Lindsay Porter. The Behavioural Ecology of Indo-Pacific Humpback Dolphins in Hong Kong. pp. 65–90.

John Y. Wang, Kimberly N. Riehl, Michelle N. Klein, Shiva Javdan, Jordan M. Hoffman, Sarah Z. Dungan, Lauren E. Dares, and Claryana Arau´jo-Wang. Biology and Conservation of the Taiwanese Humpback Dolphin, *Sousa chinensis taiwanensis*. pp. 91–118.

Bingyao Chen, Xinrong Xu, Thomas A. Jefferson, Paula A. Olson, Qiurong Qin, Hongke Zhang, Liwen He, and Guang Yang. Conservation Status of the Indo-Pacific Humpback Dolphin (*Sousa chinensis*) in the Northern Beibu Gulf, China. pp. 119–140.

Gianna Minton, Anna Norliza Zulkifli Poh, Cindy Peter, Lindsay Porter, and Danielle Kreb. Indo-Pacific Humpback Dolphins in Borneo: A Review of Current Knowledge with Emphasis on Sarawak. pp. 141–156.

Guido J. Parra and Daniele Cagnazzi. Conservation Status of the Australian Humpback Dolphin (*Sousa sahulensis*) Using the IUCN Red List Criteria. pp. 157–192.

Daniella M. Hanf, Tim Hunt, and Guido J. Parra. Humpback Dolphins of Western Australia: A Review of Current Knowledge and Recommendations for Future Management. pp. 193–218.

Isabel Beasley, Maria Jedensjö, Gede Mahendra Wijaya, Jim Anamiato, Benjamin Kahn, and Danielle Kreb. Observations on Australian Humpback Dolphins (*Sousa sahulensis*) in Waters of the Pacific Islands and New Guinea. pp. 219–272.

Alexander M. Brown, Lars Bejder, Guido J. Parra, Daniele Cagnazzi, Tim Hunt, Jennifer L. Smith, and Simon J. Allen. Sexual Dimorphism and Geographic Variation in Dorsal Fin Features of Australian Humpback Dolphins, *Sousa sahulensis*. pp. 273–314.

Volume 74, 2016.

J. Salinger, A.J. Hobday, R.J. Matear, T.J. O'Kane, J.S. Risbey, P. Dunstan, J.P. Eveson, E.A. Fulton, M. Feng, É.E. Plagányi, E.S. Poloczanska, A.G. Marshall, and P.A. Thompson. Decadal-Scale Forecasting of Climate Drivers for Marine Applications. pp. 1–68.

S.A. Foo and M. Byrne. Acclimatization and Adaptive Capacity of Marine Species in a Changing Ocean. pp. 69–116.

N.D. Gallo and L.A. Levin. Fish Ecology and Evolution in the World's Oxygen Minimum Zones and Implications of Ocean Deoxygenation. pp. 117–198.

R.J. Olson, J.W. Young, F. Ménard, M. Potier, V. Allain, N. Goñi, J.M. Logan, and F. Galván-Magaña. Bioenergetics, Trophic Ecology, and Niche Separation of Tunas. pp. 199–344.

Volume 75, 2016.

G. Notarbartolo di Sciara. Marine Mammals in the Mediterranean Sea: An Overview. pp. 1–36.

L. Rendell and A. Frantzis. Mediterranean Sperm Whales, *Physeter macrocephalus*: The Precarious State of a Lost Tribe. pp. 37–74.

G. Notarbartolo di Sciara, M. Castellote, J.-N. Druon, and S. Panigada. Fin Whales, *Balaenoptera physalus*: At Home in a Changing Mediterranean Sea?. pp. 75–102.

M. Podestà, A. Azzellino, A. Cañadas, A. Frantzis, A. Moulins, M. Rosso, P. Tepsich, and C. Lanfredi. Cuvier's Beaked Whale, *Ziphius cavirostris*, Distribution and Occurrence in the Mediterranean Sea: High-Use Areas and Conservation Threats. pp. 103–140.

R. Esteban, P. Verborgh, P. Gauffier, D. Alarcón, J.M. Salazar-Sierra, J. Giménez, A.D. Foote, and R. de Stephanis. Conservation Status of Killer Whales, *Orcinus orca*, in the Strait of Gibraltar. pp. 141–172.

P. Verborgh, P. Gauffier, R. Esteban, J. Giménez, A. Cañadas, J.M. Salazar-Sierra, and R. de Stephanis. Conservation Status of Long-Finned Pilot Whales, *Globicephala melas*, in the Mediterranean Sea. pp. 173–204.

A. Azzellino, S. Airoldi, S. Gaspari, C. Lanfredi, A. Moulins, M. Podestà, M. Rosso, and P. Tepsich. Risso's Dolphin, *Grampus griseus*, in the Western Ligurian Sea: Trends in Population Size and Habitat Use. pp. 205–232.

D. Kerem, O. Goffman, M. Elasar, N. Hadar, A. Scheinin, and T. Lewis. The Rough-Toothed Dolphin, *Steno bredanensis*, in the Eastern Mediterranean Sea: A Relict Population?. pp. 233–258.

J. Gonzalvo, G. Lauriano, P.S. Hammond, K.A. Viaud-Martinez, M.C. Fossi, A. Natoli, and L. Marsili. The Gulf of Ambracia's Common Bottlenose Dolphins, *Tursiops truncatus*: A Highly Dense and yet Threatened Population. pp. 259–296.

G. Bearzi, S. Bonizzoni, N.L. Santostasi, N.B. Furey, L. Eddy, V.D. Valavanis, and O. Gimenez. Dolphins in a Scaled-Down Mediterranean: The Gulf of Corinth's Odontocetes. pp. 297–332.

M.C. Fontaine. Harbour Porpoises, *Phocoena phocoena*, in the Mediterranean Sea and Adjacent Regions: Biogeographic Relicts of the Last Glacial Period. pp. 333–358.

G. Notarbartolo di Sciara and S. Kotomatas. Are Mediterranean Monk Seals, *Monachus monachus*, Being Left to Save Themselves from Extinction?. pp. 359–386.

T. Scovazzi. The International Legal Framework for Marine Mammal Conservation in the Mediterranean Sea. pp. 387–416.

Volume 76, 2017.

K.S. Meyer. Islands in a Sea of Mud: Insights From Terrestrial Island Theory for Community Assembly on Insular Marine Substrata. pp. 1–40.

E.M. Montgomery, J.-F. Hamel, and A. Mercier. Patterns and Drivers of Egg Pigment Intensity and Colour Diversity in the Ocean: A Meta-Analysis of Phylum Echinodermata. pp. 41–104.

F. Espinosa and G.A. Rivera-Ingraham. Biological Conservation of Giant Limpets: The Implications of Large Size. pp. 105–156.

L. Yebra, T. Kobari, A.R. Sastri, F. Gusmão, and S. Hernández-León. Advances in Biochemical Indices of Zooplankton Production. pp. 157–240.

Volume 77, 2017.

Dayv Lowry and Shawn E. Larson. Introduction to Northeast Pacific Shark Biology, Ecology, and Conservation. pp. 1–8.

David A. Ebert, Jennifer S. Bigman, and Julia M. Lawson. Biodiversity, Life History, and Conservation of Northeastern Pacific Chondrichthyans pp. 9–78.

Shawn E. Larson, Toby S. Daly-Engel, and Nicole M. Phillips. Review of Current Conservation Genetic Analyses of Northeast Pacific Sharks. pp. 79–110.

Joseph J. Bizzarro, Aaron B. Carlisle, Wade D. Smith, and Enric Cortés. Diet Composition and Trophic Ecology of Northeast Pacific Ocean Sharks. pp. 111–148.

Jonathan C.P. Reum, Gregory D. Williams, and Chris J. Harvey. Stable Isotope Applications for Understanding Shark Ecology in the Northeast Pacific Ocean. pp. 149–178.

Mary E. Matta, Cindy A. Tribuzio, David A. Ebert, Kenneth J. Goldman, and Christopher M. Gburski. Age and Growth of Elasmobranchs and Applications to Fisheries Management and Conservation in the Northeast Pacific Ocean. pp. 179–220.

Volume 78, 2017.

Shawn E. Larson and Dayv Lowry. Introduction to Northeast Pacific Shark Biology, Research, and Conservation, Part B. pp. 1–8.

Jackie King, Gordon A. McFarlane, Vladlena Gertseva, Jason Gasper, Sean Matson, and Cindy A. Tribuzio. Shark Interactions With Directed and Incidental Fisheries in the Northeast Pacific Ocean: Historic and Current Encounters, and Challenges for Shark Conservation. pp. 9–44.

Dovi Kacev, Timothy J. Sippel, Michael J. Kinney, Sebastián A. Pardo, and Christopher G. Mull. An Introduction to Modelling Abundance and Life History Parameters in Shark Populations. pp. 45–88.

Michael Grassmann, Bryan McNeil, and Jim Wharton. Sharks in Captivity: The Role of Husbandry, Breeding, Education, and Citizen Science in Shark Conservation. pp. 89–120.

Peter A. Mieras, Chris Harvey-Clark, Michael Bear, Gina Hodgin, and Boone Hodgin. The Economy of Shark Conservation in the Northeast Pacific: The Role of Ecotourism and Citizen Science. pp. 121–154.

Dayv Lowry. Conclusions: The Future of Shark Management and Conservation in the Northeast Pacific Ocean. pp. 155–164.

Volume 79, 2018.

Patricia L.M. Lee, Gail Schofield, Rebecca I. Haughey, Antonios D. Mazaris, and Graeme C. Hays. A Review of Patterns of Multiple Paternity Across Sea Turtle Rookeries. pp. 1–32.

Johanne Vad, Georgios Kazanidis, Lea-Anne Henry, Daniel O.B. Jones, Ole S. Tendal, Sabine Christiansen, Theodore B. Henry, and J. Murray Roberts. Potential Impacts of Offshore Oil and Gas Activities on Deep-Sea Sponges and the Habitats They Form. pp. 33–60.

Gianmarco Ingrosso, Marco Abbiati, Fabio Badalamenti, Giorgio Bavestrello, Genuario Belmonte, Rita Cannas, Lisandro Benedetti-Cecchi, Marco Bertolino, Stanislao Bevilacqua, Carlo Nike Bianchi, Marzia Bo, Elisa Boscari, Frine Cardone, Riccardo Cattaneo-Vietti, Alessandro Cau, Carlo Cerrano, Renato Chemello, Giovanni Chimienti, Leonardo Congiu, Giuseppe Corriero, Federica Costantini, Francesco De Leo, Luigia Donnarumma, Annalisa Falace, Simonetta Fraschetti, Adriana Giangrande, Maria Flavia Gravina, Giuseppe Guarnieri, Francesco Mastrototaro, Marco Milazzo, Carla Morri, Luigi Musco, Laura Pezzolesi, Stefano Piraino, Fiorella Prada, Massimo Ponti, Fabio Rindi, Giovanni Fulvio Russo, Roberto Sandulli, Adriana Villamor, Lorenzo Zane, and Ferdinando Boero. Mediterranean Bioconstructions Along the Italian Coast. pp. 61–136.

Alex D. Rogers. The Biology of Seamounts: 25 Years on. pp. 137–224.

Volume 80, 2018.

Michael H. Schleyer and Sean N. Porter. Drivers of Soft and Stony Coral Community Distribution on the High-Latitude Coral Reefs of South Africa. pp. 1–56.

Sara González-Delgado and José Carlos Hernández. The Importance of Natural Acidified Systems in the Study of Ocean Acidification: What Have We Learned? pp. 57–100.

Volume 81, 2018.

Melissa C. Cook, Adam May, Lucas Kohl, Geert Van Biesen, Christopher C. Parrish, and Penny L. Morrill. The Potential Impact of Hydrocarbons on Mussels in Port au Port Bay, Newfoundland pp. 1–22.

Yunwen Tao, Bing Chen, Baiyu (Helen) Zhang, Zhiwen (Joy) Zhu, and Qinhong Cai. Occurrence, Impact, Analysis and Treatment of Metformin and Guanylurea in Coastal Aquatic Environments of Canada, USA and Europe pp. 23–58.

Gong Zhang, Chun Yang, Mariam Serhan, Graeme Koivu, Zeyu Yang, Bruce Hollebone, Patrick Lambert, and Carl E. Brown. Characterization of Nitrogen-Containing Polycyclic Aromatic Heterocycles in Crude Oils and Refined Petroleum Products. pp. 59–96.

Arpana Rani Datta, Qiao Kang, Bing Chen, and Xudong Ye. Fate and Transport Modelling of Emerging Pollutants from Watersheds to Oceans: A Review. pp. 97–128.

Meijin Du, Wenwen Gu, Xixi Li, Fuqiang Fan, and Yu Li. Modification of Hexachlorobenzene to Molecules with Lower Long-Range Transport Potentials Using 3D-QSAR Models with a Full Factor Experimental Design pp. 129–166.

Weiyun Lin, Xixi Li, Min Yang, Kenneth Lee, Bing Chen, and Baiyu (Helen) Zhang. Brominated Flame Retardants, Microplastics, and Biocides in the Marine Environment: Recent Updates of Occurrence, Analysis, and Impacts. pp. 167–212.

Xixi Li, Zhenhua Chu, Jiawen Yang, Minghao Li, Meijin Du, Xiaohui Zhao, Zhiwen (Joy) Zhu, and Yu Li. Synthetic Musks: A Class of Commercial Fragrance Additives in Personal Care Products (PCPs) Causing Concern as Emerging Contaminants. pp. 213–280.

Volume 82, 2019.

Junaid S. Khan, Jennifer F. Provencher, Mark R. Forbes, Mark L. Mallory, Camille Lebarbenchon, and Karen D. McCoy. Parasites of seabirds: A survey of effects and ecological implications. pp. 1–50.

Charlotte Havermans, Holger Auel, Wilhelm Hagen, Christoph Held, Natalie S. Ensor, and Geraint A. Tarling. Predatory zooplankton on the move: *Themisto* amphipods in high-latitude marine pelagic food webs. pp. 51–92.

Angel Borja, Guillem Chust, and Iñigo Muxika. Forever young: The successful story of a marine biotic index. pp. 93–128.

Ferdinando Boero, Francesco De Leo, Simonetta Fraschetti, and Gianmarco Ingrosso. The Cells of Ecosystem Functioning: Towards a holistic vision of marine space. pp. 129–154.

Volume 83, 2019.

Dayv Lowry and Shawn Larson. Introduction: The sharks of Pacific Mexico and their conservation: Why should we care?. pp. 1–10.

Luz Erandi Saldaña-Ruiz, Emiliano García-Rodríguez, Juan Carlos Pérez-Jiménez, Javier Tovar-Ávila, and Emmanuel Rivera-Téllez. Biodiversity and Conservation of Sharks in Pacific Mexico. pp. 11–60.

Felipe Galván-Magaña, José Leonardo Castillo-Geniz, Mauricio Hoyos-Padilla, James Ketchum, A. Peter Klimley, Sergio Ramírez-Amaro, Yassir Eden Torres-Rojas, and Javier Tovar-Avila. Shark ecology, the role of the apex predator and current conservation status. pp. 61–114.

Jonathan Sandoval-Castillo. Conservation genetics of elasmobranchs of the Mexican Pacific Coast, trends and perspectives. pp. 115–158.

Volume 84, 2019.

Brian Morton and Fabrizio Marcondes Machado. Predatory marine bivalves: A review. pp. 1–98.

Volume 85, 2020.

James T. Ketchum, Mauricio Hoyos-Padilla, Alejandro Aldana-Moreno, Kathryn Ayres, Felipe Galván-Magaña, Alex Hearn, Frida Lara-Lizardi, Gador Muntaner-López, Miquel Grau, Abel Trejo-Ram´ırez, Darren A. Whitehead, and A. Peter Klimley. Shark movement patterns in the Mexican Pacific: A conservation and management perspective. pp. 1–38.

Oscar Sosa-Nishizaki, Emiliano García-Rodríguez, Christian D. Morales-Portillo, Juan C. Pérez-Jiménez, M. del Carmen Rodŕiguez-Medrano, Joseph J. Bizzarro, and José Leonardo Castillo-Géniz. Fisheries interactions and the challenges for target and nontargeted take on shark conservation in the Mexican Pacific. pp. 39–70.

A.M. Cisneros-Montemayor, E.E. Becerril-García, O. Berdeja-Zavala, and A. Ayala-Bocos. Shark ecotourism in Mexico: Scientific research, conservation, and contribution to a Blue Economy. pp. 71–92.

Oscar Sosa-Nishizaki, Felipe Galván-Magaña, Shawn E. Larson, and Dayv Lowry. Conclusions: Do we eat them or watch them, or both? Challenges for conservation of sharks in Mexico and the NEP. pp. 93–102.

Volume 86, 2020.

David J. Vance and Peter C. Rothlisberg. The biology and ecology of the banana prawns: *Penaeus merguiensis* de Man and *P. indicus* H. Milne Edwards. pp. 1–140.

Jeff A. Eble, Toby S. Daly-Engel, Joseph D. DiBattista, Adam Koziol, and Michelle R. Gaither. Marine environmental DNA: Approaches, applications, and opportunities. pp. 141–170.

Volume 87, 2020.

Bernhard M. Riegl and Peter W. Glynn. Population dynamics of the reef crisis: Consequences of the growing human population. pp. 1–30.

Chiara Pisapia, Peter J. Edmunds, Holly V. Moeller, Bernhard M. Riegl, Mike McWilliam, Christopher D. Wells, and Morgan S. Pratchett. Projected shifts in coral size structure in the Anthropocene. pp. 31–60.

Alex E. Mercado-Molina, Alberto M. Sabat, and Edwin A. Hernández-Delgado. Population dynamics of diseased corals: Effects of a Shut Down Reaction outbreak in Puerto Rican *Acropora cervicornis*. pp. 61–82.

Elizabeth A. Goergen, Kathleen Semon Lunz, and David S. Gilliam. Spatial and temporal differences in *Acropora cervicornis* colony size and health. pp. 83–114.

Rian Prasetia, Zi Wei Lim, Aaron Teo, Tom Shlesinger, Yossi Loya, and Peter A. Todd. Population dynamics and growth rates of free-living mushroom corals (Scleractinia: Fungiidae) in the sediment-stressed reefs of Singapore. pp. 115–140.

Joshua S. Feingold and Brandon Brulé. Population fluctuations of the fungiid coral *Cycloseris curvata*, Galápagos Islands, Ecuador. pp. 141–166.

Stuart A. Sandin, Clinton B. Edwards, Nicole E. Pedersen, Vid Petrovic, Gaia Pavoni, Esmeralda Alcantar, Kendall S. Chancellor, Michael D. Fox, Brenna Stallings, Christopher J. Sullivan, Randi D. Rotjan, Federico Ponchio, and Brian J. Zgliczynski. Considering the rates of growth in two taxa of coral across Pacific islands. pp. 167–192.

Geórgenes Cavalcante, Filipe Vieira, Jonas Mortensen, Radhouane Ben-Hamadou, Pedro Range, Elizabeth A. Goergen, Edmo Campos, and Bernhard M. Riegl. Biophysical model of coral population connectivity in the Arabian/Persian Gulf. pp. 193–222.

S.A. Matthews, C. Mellin, and Morgan S. Pratchett. Larval connectivity and water quality explain spatial distribution of crown-of-thorns starfish outbreaks across the Great Barrier Reef. pp. 223–258.

S.A. Matthews, K. Shoemaker, Morgan S. Pratchett, and C. Mellin. COTSMod: A spatially explicit metacommunity model of outbreaks of crown-of-thorns starfish and coral recovery. pp. 259–290.

Tim R. McClanahan. Coral community life histories and population dynamics driven by seascape bathymetry and temperature variability. pp. 291–330.

William F. Precht, Richard B. Aronson, Toby A. Gardner, Jennifer A. Gill, Julie P. Hawkins, Edwin A. Hernéndez-Delgado, Walter C. Jaap, Tim R. McClanahan, Melanie D. McField, Thaddeus J.T. Murdoch, Maggy M. Nugues, Callum M. Roberts, Christiane K. Schelten, Andrew R. Watkinson, and Isabelle M. Côté. The timing and causality of ecological shifts on Caribbean reefs. pp. 331–360.

Howard R. Lasker, Lorenzo Bramanti, Georgios Tsounis, and Peter J. Edmunds. The rise of octocoral forests on Caribbean reefs. pp. 361–410.

Sascha C.C. Steiner, Priscilla Martínez, Fernando Rivera, Matthew Johnston, and Bernhard M. Riegl. Octocoral populations and connectivity in continental Ecuador and Galápagos, Eastern Pacific. pp. 411–442.

Peter W. Glynn, Renata Alitto, Joshua Dominguez, Ana B. Christensen, Phillip Gillette, Nicolas Martinez, Bernhard M. Riegl, and Kyle Dettloff. A tropical eastern Pacific invasive brittle star species (Echinodermata: Ophiuroidea) reaches southeastern Florida. pp. 443–472.

Volume 88, 2021.

Pat Hutchings. Potential loss of biodiversity and the critical importance of taxonomy—An Australian perspective. pp. 3–16.

Ferdinando Boero. Mission possible: Holistic approaches can heal marine wounds. pp. 19–38.

Iraide Artetxe-Arrate, Igaratza Fraile, Francis Marsac, Jessica H. Farley, Naiara Rodriguez-Ezpeleta, Campbell R. Davies, Naomi P. Clear, Peter Grewe, and Hilario Murua. A review of the fisheries, life history and stock structure of tropical tuna (skipjack *Katsuwonus pelamis*, yellowfin *Thunnus albacares* and bigeye *Thunnus obesus*) in the Indian Ocean. pp. 39–90.

Yan Ji and Kunshan Gao. Effects of climate change factors on marine macroalgae: A review. pp. 91–136.

Brian Morton and Fabrizio Marcondes Machado. The origins, relationships, evolution and conservation of the weirdest marine bivalves: The watering pot shells. A review. pp. 139–220.

Volume 89, 2021.

Stanislao Bevilacqua, Laura Airoldi, Enric Ballesteros, Lisandro Benedetti-Cecchi, Ferdinando Boero, Fabio Bulleri, Emma Cebrian, Carlo Cerrano, Joachim Claudet, Francesco Colloca, Martina Coppari, Antonio Di Franco, Simonetta Fraschetti, Joaquim Garrabou, Giuseppe Guarnieri, Cristiana Guerranti, Paolo Guidetti, Benjamin S. Halpern, Stelios Katsanevakis, Maria Cristina Mangano, Fiorenza Micheli, Marco Milazzo, Antonio Pusceddu, Monia Renzi, Gil Rilov, Gianluca Sarà, and Antonio Terlizzi. Mediterranean rocky reefs in the Anthropocene: Present status and future concerns. pp. 1–52.

Johanne Vad, Kelsey Archer Barnhill, Georgios Kazanidis, and J. Murray Roberts. Human impacts on deep-sea sponge grounds: Applying environmental omics to monitoring. pp. 53–78.

Angel Borja and Michael Elliott. From an economic crisis to a pandemic crisis: The need for accurate marine monitoring data to take informed management decisions. pp. 79–114.

Volume 90, 2021.

Shawn Larson, Dayv Lowry, Nicholas K. Dulvy, Jim Wharton, Felipe Galván-Magaña, Abraham B. Sianipar, Christopher G. Lowe, and Erin Meyer. Current and future considerations for shark conservation in the Northeast and Eastern Central Pacific Ocean. pp. 1–50.

Ferdinando Boero. The future ocean we want. pp. 51–64.

Volume 91, 2022.

Jean-François Hamel, Igor Eeckhaut, Chantal Conand, Jiamin Sun, Guillaume Caulier, and Annie Mercier. Global knowledge on the commercial sea cucumber *Holothuria scabra*. pp. 1–286.

Volume 92, 2022.

Monica Montefalcone, Alice Oprandi, Annalisa Azzola, Carla Morri, and Carlo Nike Bianchi. Serpulid reefs and their role in aquatic ecosystems: A global review. pp. 1–54.

Matthew R. Nitschke, Sabrina L. Rosset, Clinton A. Oakley, Stephanie G. Gardner, Emma F. Camp, David J. Suggett, and Simon K. Davy. The diversity and ecology of Symbiodiniaceae: A traits-based review. pp. 55–128.

David Rodríguez-Rodríguez and Javier Martínez-Vega. Ecological effectiveness of marine protected areas across the globe in the scientific literature. pp. 129–154.

Volume 93, 2022.

Fabian Ritter. Opinion: Marine mammal conservation in the 21st century: A plea for a paradigm shift towards mindful conservation. pp. 1–22.

Alex D. Rogers, Ward Appeltans, Jorge Assis, Lisa T. Ballance, Philippe Cury, Carlos Duarte, Fabio Favoretto, Lisa A. Hynes, Joy A. Kumagai, Catherine E. Lovelock, Patricia Miloslavich, Aidin Niamir, David Obura, Bethan C. O'Leary, Eva Ramirez-Llodra, Gabriel Reygondeau, Callum Roberts, Yvonne Sadovy, Oliver Steeds, Tracey Sutton, Derek P. Tittensor, Enriqueta Velarde, Lucy Woodall, and Octavio Aburto-Oropeza. Discovering marine biodiversity in the 21st century. pp. 23–116.

Kailei Zhu and Jiayu Bai. Review study on governance and international law for coastal and marine ecosystems in response to climate change: Social science perspective. pp. 117–146.

Volume 94, 2023.

Jesse van der Grient, Simon Morley, Alexander Arkhipkin, James Bates, Alastair Baylis, Paul Brewin, Michael Harte, J. Wilson White, and Paul Brickle. The Falkland Islands marine ecosystem: A review of the seasonal dynamics and trophic interactions across the food web. pp. 1–68.

Tamara A. Shiganova, Andrei M. Kamakin, Larisa A. Pautova, Alexander S. Kazmin, Aboulghasem Roohi, and Henri J. Dumont. An impact of non-native species invasions on the Caspian Sea biota. pp. 69–158.

Luís Gabriel A. Barboza, Sara Couto Lourenço, Alexandre Aleluia, Natália Carneiro Lacerda dos Santos, Minrui Huang, Jun Wang, and Lúcia Guilhermino. A global synthesis of microplastic contamination in wild fish species: Challenges for conservation, implications for sustainability of wild fish stocks and future directions. pp. 159–200.

Nikolina Rako-Gospić and Marta Picciulin. Addressing underwater noise: Joint efforts and progress on its global governance. pp. 201–232.

Volume 95, 2023.
Julie A. Lively and Jonathan McKenzie. Discards and bycatch: A review of wasted fishing. pp. 1–26.
James J. Bell, Francesca Strano, Manon Broadribb, Gabriela Wood, Ben Harris, Anna Carolina Resende, Emma Novak, and Valerio Micaroni. Sponge functional roles in a changing world. pp. 27–90.
Ferdinando Boero and Joachim Mergeay. Darwin's feathers: Ecoevolutionary biology, predictions and policy. pp. 91–112.
Mauvis Gore, Ewan Camplisson, and Rupert Ormond. The biology and ecology of the basking shark: A review. pp. 113–257.

Volume 96, 2024.
Ayaka T. Matsuda, Takashi F. Matsuishi, Fumika Noto, Masao Amano, Yuko Tajima, and Tadasu K. Yamada. Notes on stomach contents of pygmy and dwarf sperm whales (*Kogia* spp.) from around Japan. pp. 1–24.
Shin Nishida, Atsushi Uchimura, Yuko Tajima, and Tadasu K. Yamada. Comparative analysis of the genetic structures of *Kogia* spp. populations in the western North Pacific. pp. 25–37.
Akira Shiozaki, Shotaro Nakagun, Yuko Tajima, and Masao Amano. A first record of digenean parasites of the dwarf sperm whale *Kogia sima* with morphological and molecular information. pp. 39–61.
Névia Lamas, Pablo Covelo, Alfredo López, Uxía Vázquez, and Nuria Alemañ. A histological study of the facial hair follicles in the pygmy sperm whale (*Kogia breviceps*). pp. 63–83.
Stephanie Plön, Peter B. Best, Pádraig Duignan, Shane D. Lavery, Ric T.F. Bernard, Koen Van Waerebeek, and C. Scott Baker. Population structure of pygmy (*Kogia breviceps*) and dwarf (*Kogia sima*) sperm whales in the Southern Hemisphere may reflect foraging ecology and dispersal patterns. pp. 85–114.

Volume 97, 2024.
Shawna A. Foo, Pauline M. Ross, and Maria Byrne. The 2024 roadmap for understanding marine species' resilience in a changing ocean. pp. 1–10.
Shawna A. Foo and Maria Byrne. Reprint: Acclimatization and Adaptive Capacity of Marine Species in a Changing Ocean. pp. 11–58.
José Carlos Hernández, Sara González-Delgado, M. Aliende-Hernández, B. Alfonso, A. Rufino-Navarro, and C.A. Hernández. Natural acidified marine systems: Lessons and predictions. pp. 59–78.

Sara González-Delgado and José Carlos Hernández. Reprint: The Importance of Natural Acidified Systems in the Study of Ocean Acidification: What Have We Learned?. pp. 79–122.

Patricia L.M. Lee and Graeme C. Hays. A roadmap for multiple paternity research with sea turtles. pp. 123–134.

Patricia L.M. Lee, Gail Schofield, Rebecca I. Haughey, Antonios D. Mazaris, and Graeme C. Hays. Reprint: A Review of Patterns of Multiple Paternity Across Sea Turtle Rookeries. pp. 135–166.

Ferdinando Boero. A roadmap to knowledge-based maritime spatial planning. pp. 167–190.

Ferdinando Boero, Francesco De Leo, Simonetta Fraschetti, and Gianmarco Ingrosso. Reprint: The Cells of Ecosystem Functioning: Towards a holistic vision of marine space. pp. 191–216.

Preface

An aged unpublished review manuscript covering the jellyfish taxon Rhizostomeae (Cnidaria, Medusozoa, Scyphozoa) drafted by the German zoologist Max Egon Thiel (*1898, †1979) incited this edition. Preserved in the family archive of M.E. Thiel's descendants, a typescript of the final volume on the "Rhizostomeae" had failed to finalise his series of monographs on the class Scyphozoa published by the Akademische Verlagsgesellschaft Geest & Portig in Leipzig (Thiel, 1936a-c, 1938a,b, 1959, 1962, 1977). Along with strong interest voiced from various colleagues in the community of jellyfish scientists, inquiries about options for publishing yielded the request for a comprehensive survey of research on Rhizostomeae during the half century since Thiel's work. The initial plan of the editorial team was to publish a single review article about Rhizostomeae research since 1970 as an update to M.E. Thiel's manuscript. However, it soon became clear that the advances in scientific studies on Rhizostomeae were so extensive that a single publication would not suffice to adequately reflect the progress that occurred in the field since the 1970s. This led to the idea to compile the most important areas of Rhizostomeae research in a book volume, essentially following the different sections of the original typescript by M.E. Thiel. An exposé of such a volume, presented during the 6th International Jellyfish Blooms Symposium in 2019 (Cape Town, South Africa), sparked the interest of a team of co-authors. Over five years, a group of 36 experts in their fields contributed to this endeavour. The concept proffered to Elsevier's series *Advances in Marine Biology* (AMB) was accepted, along with Thiel's original bygone manuscript on the Rhizostomeae, his bibliography (1977), as well as a recent literature screening (1970-2023) as electronic supplements.

Species of the order Rhizostomeae are among the most conspicuous organisms in the coastal ocean plankton communities. Their sizes, sometimes over 200 centimetres in diameter, as well as their occasional appearance in vast numbers – "jellyfish blooms" or "outbreaks" – ignited early mention in Aristoteles' "Natural History of Animals" (384-322 BC). Rhizostomes have had a long history of scientific treatments since then, with first descriptions of species in the 18th century. The unpredictable occurrence of the pelagic medusae as well as the tiny and hardly detectable benthic polyp stages of the metagenetic reproduction cycle, however, always caused methodological limits in quantitative investigations of populations. Present knowledge lists 91 species that still lack countless morphological and ecological details, as well as consistency in wider patterns of distribution and seasonal dynamics. Nonetheless, their ecological plasticity and significance in regional planktonic ecosystems has evoked new attention recently, at least in some prominent species. Their apparent success as top predators and their roles in changing pelagic food webs have generated rising awareness of the taxon, along with their possible economic value. This backdrop highlights the pertinence of a comprehensive survey of scientific achievements over the 50 years since Thiel's work. The team of co-authors represented in this volume comprises senior scientists with numerous years of experience in varied fields of jellyfish research, as well as more junior researchers bringing youthful energy and novel perspectives to the contributions.

The volume contents, organised according to the sections of M.E. Thiel's final manuscript, introduce biographic information and present brief reviews of the earlier parts of the series (Chapter 1: Max Egon Thiel's monographs on Scyphozoa (Cnidaria) and a left-behind typescript on the Rhizostomeae). Since then, a profusion of new knowledge has emerged from various lines of current research. Nevertheless, the vast coverage of Thiel's layout revealed some broad and robust baseline views founding the modern perception of the taxon Scyphozoa.

The chapter on rhizostome morphology provides a review of knowledge about the morphological features of different life stages in rhizostome taxa in a contemporary context (Chapter 2: Morphology of Rhizostomeae jellyfishes: what is known and what we advanced since the 1970s). The contribution on histology and micromorphology gives insights into the knowledge about nervous systems, sensing structures and muscles, histological structures of gonads, symbiotic dinoflagellates, and nematocysts based on methodological advances (Chapter 3: Approaches and findings in histological and micromorphological research on Rhizostomeae). In the chapter on life cycles, the authors review the current knowledge about metagenesis and reproduction of rhizostomes, discussing the changes from the benthic to planktonic (asexual reproduction) lifestyle and vice-versa (sexual reproduction). Environmental cues and triggers inducing strobilation are debated (Chapter 4: Life cycles and reproduction of Rhizostomeae), highlighting that we still have much to learn about jellyfish population dynamics. The contribution about physiology and functional biology presents distinct aspects such as body composition, nematocyst toxins, locomotion, feeding strategies, nutrition, digestion, metabolism, and growth in an evolutionary context (Chapter 5: Physiology and functional biology of Rhizostomeae). The chapter on evolution, population genetics and taxonomy provides an updated systematic classification scheme and discusses how different types of evidence contributed to the current understanding of the evolution of Rhizostomeae (Chapter 6: Morphological and molecular data in the study of the evolution, population genetics and taxonomy of Rhizostomeae). The vast theme of the ecology of some Rhizostomeae species is treated, presenting seven iconic and well-studied species including reviews of seasonality and blooms, as well as the role of medusae in ecosystems. The chapter also compares the ecology of rhizostomes with other large jellyfishes (Chapter 7: Ecology of Rhizostomeae). The final chapter highlights the use of rhizostomes as food, sources of bioactive compounds, as well as techno-medical and biotechnological applications, supporting the notion that these animals can be a valuable resource (Chapter 8: Rhizostomes as a resource: The expanding exploitation of jellyfish by humans).

Literature search

To provide all authors of the present volume a broad basis for literature research on their relevant topics in the field of Rhizostomeae, a comprehensive literature search was conducted for the years 1970 to 2023 using the Web of Science search mask (https://www.webofscience.com/wos), provided by Clarivate. Web of Science and Clarivate are trademarks of their respective owners and used herein with permission. The dataset was compiled with the technical assistance of Antje Fischer (Senckenberg, DZMB Hamburg) and Virginia Castilho (CEBIMar, University of São Paulo) and is available as digital supplement Holst et al. 2024, (*Work in progress, to be published soon*). The search terms used were: Rhizostomeae, Rhizostomae, rhizostomes, rhizostome jelly-fish, Daktyliophorae, Dactyliophorae, Kolpophorae, Cepheida, Rhizostomida and all valid orders, infraorders, species and synonyms according to the World Register of Marine Species (WoRMS, www.marinespecies.org). This resulted in 233 search terms for 1970–2020 (search conducted in 2021), 234 search terms for 2021 (search conducted in 2022), 276 search terms in 2022 (search conducted in 2023), and 277 search terms in 2023 (search conducted in 2024).

The searches in all databases of Web of Science resulted in 8,403, 755, 763, and 633 matches for 1970-2020, 2021, 2022, and 2023, respectively (10,584 in total). References were revised and over 80% were deleted primarily because of multiple appearances under different search terms (6,813 deletions) but also because many matches were clearly not related to Rhizostomeae species (1,704 deletions). Typical examples are the search terms "*Cassiopea*" and "*Polyrhiza*" (rhizostome genera) which often resulted in articles mentioning the astronomical constellation Cassiopeia and the aquatic plant *Spirodela polyrhiza*, respectively.

Of the remaining 2,007 hits, the vast majority were journal articles (1,572, Fig. 1). The analysis of published articles per year shows a considerable increase of publications related to Rhizostomeae since the beginning of the twenty-first century (Fig. 2).

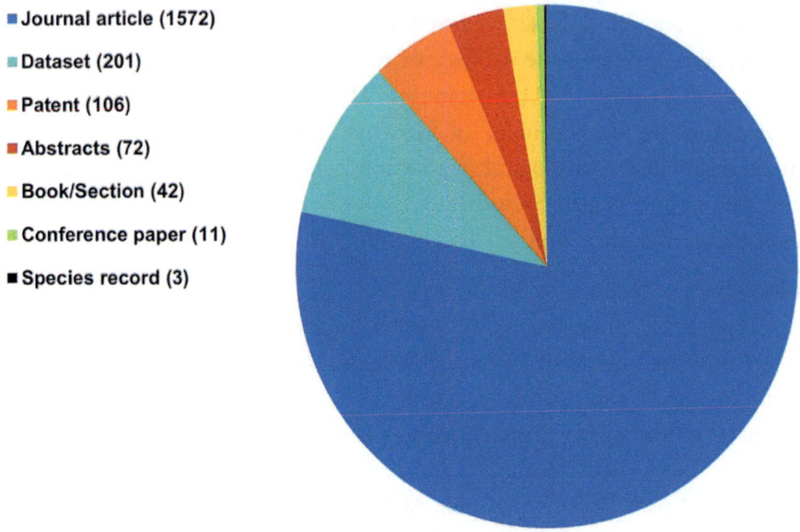

Fig. 1 Numbers (in brackets) and proportions (pie chart) of types of search hits (1970–2023) related to Rhizostomeae. *Source: Data from Web of Science, provided by Clarivate.*

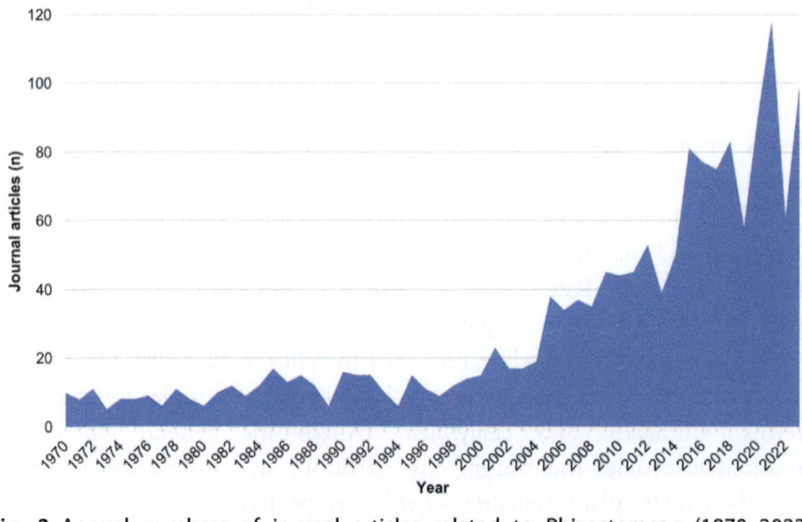

Fig. 2 Annual numbers of journal articles related to Rhizostomeae (1970–2023). *Source: Data from Web of Science, provided by Clarivate.*

Recovery and editing of M.E. Thiel's typescript on the Rhizostomeae

About 840 typed carbon copy pages of the typescript remained after Thiel's death. They rested in files with the family before technology was available to scan and digitise them. M.E. Thiel's grandson and son, Götz B. Reinicke (marine zoologist) and Hjalmar Thiel (marine ecologist) spent extensive time to adjust the scanned reproductions, before diligently revising the original German text[1] to focus and streamline lengthy passages and M.E. Thiel's occasional epic style. Based on the revised German version, the editing team decided to create an English translation of the work to allow wider distribution and access for the international scientific community. After an automatic translation, applying the AI-software *DeepL* (DeepL SE, Maarweg 165, 50825 Cologne, Germany, available at http://www.deepl.com). The resulting text was cross-checked for the trans-literation of scientific terms and language by a team of German-speaking jellyfish researchers. The entire work is published as a digital supplement with this volume (Thiel et al., 2024) at (*Work in progress, to be published soon*).

Acknowledgements

The editors wish to extend their respect and sincere gratitude to all colleagues and friends who contributed and helped with this project. Many researchers from the jellyfish community offered their support in providing advice and granted the use of their figures to assist in creating a comprehensive view of this fascinating taxon. Special thanks are due to Lucas Brotz, Allen Collins, Stefania De Domenico, Liza Gómez-Daglio, Gerhard Jarms, Jamileh Javidpour, Cathy Lucas, Kylie Pitt, Laura Prieto, Agustin Schiariti, Ilka Sötje, Ilka Straehler-Pohl, and Amanda Wiesenthal, who provided comments and insights enriching the different chapters and assisted in the language editing of this volume and the digital supplements. In addition, we are grateful for the assistance in literature searches by Antje Fischer and Virginia Castilho and for granting the copyright of the volume's cover photograph by Alvaro Migotto.

The volume would not have been finalised without the thorough patience and professionalism of the journal's editorial team, namely Palash Sharma, Maria Shalini, and Jason Mitchell, as well as Simon Davy, the series editor of the journal *Advances in Marine Biology*. At the end, an extra warm *Thank you!* is extended to the former editor of the series, Prof. Charles Sheppard OBE of Warwick University, who after more than 50 years of marine scientific research and conservation of tropical coral reefs passed away earlier this year. His inspiration and continuous support of this project gave sympathetic guidance and motivation to the editors and encouraged the cooperation of all scientists involved.

[1] The original and a revised typescript in German are kept in the Ocean Museum Germany in Stralsund and are available as.pdf copies from the library.

Götz B. Reinicke
Ocean Museum Germany, Stralsund, Germany

Sabine Holst
Senckenberg am Meer, German Center for Marine Biodiversity Research, Hamburg, Germany

André C. Morandini
Departamento de Zoologia, Instituto de Biociências and Centro de Biologia Marinha, Universidade de São Paulo, São Sebastião, SP, Brazil

Hjalmar Thiel
Ocean Museum Germany, Stralsund, Germany

References

Holst, S., Fischer, A., Castilho, V., Reinicke, G.B., Morandini, A.C., 2024. Rhizostomeae jellyfish WoS search data 1970-2023. Advances in Marine Biology 98 (electronic supplement) at (*Work in progress, to be published soon*).

Thiel, M.E., 1936a. Stauromedusae. In: Bronns, H.G. (Ed.), Klassen und Ordnungen des Tierreichs. Akademische Verlagsgesellschaft Geest & Portig, Leipzig, Vol. 2, Abt. 2, Buch 2, 1–172.

Thiel, M.E., 1936b. Cubomedusae. In: Bronns, H.G. (Ed.), Klassen und Ordnungen des Tierreichs. Akademische Verlagsgesellschaft Geest & Portig, Leipzig, Vol. 2, Abt. 2, Buch 2, 173–308.

Thiel, M.E., 1936c. Coronatae. In: Bronns, H.G. (Ed.), Klassen und Ordnungen des Tierreichs. Akademische Verlagsgesellschaft Geest & Portig, Leipzig, Vol. 2, Abt. 2, Buch 2, 308–471.

Thiel, M.E., 1938a. Semaeostomae. Morphologie und Histologie. In: Bronns, H.G. (Ed.), Klassen und Ordnungen des Tierreichs. Akademische Verlagsgesellschaft Geest & Portig, Leipzig, Vol. 2, Abt. 2, Buch 2, 471–677.

Thiel, M.E., 1938b. Semaeostomae. Ontogenie. In: Bronns, H.G. (Ed.), Klassen und Ordnungen des Tierreichs. Akademische Verlagsgesellschaft Geest & Portig, Leipzig, Vol. 2, Abt. 2, Buch 2, 677–856.

Thiel, M.E., 1959. Semaeostomae. Physiologie. In: Bronns, H.G. (Ed.), Klassen und Ordnungen des Tierreichs. Akademische Verlagsgesellschaft Geest & Portig, Leipzig, Vol. 2, Abt. 2, Buch 2, 856–1072.

Thiel, M.E., 1962. Scyphomedusae. Ökologie, Geschichte ihrer Kenntnis, ihre Stammesgeschichte und Klassifikation. In: Bronns, H.G. (Ed.), Klassen und Ordnungen des Tierreichs. Akademische Verlagsgesellschaft Geest & Portig, Leipzig, Vol. 2, Abt. 2, Buch 2, 1073–1308.

Thiel, M.E., 1977. Bibliographie. Bronns, H.G. (Ed.), Klassen und Ordnungen des Tierreichs. VEB Gustav Fischer, Jena 2, 1–104. (also available as electronic supplement at https://www.elsevier.com/books-and-journals/book-companion/9780443296468).

Thiel, M.E.(†), Reinicke, G.B., Holst, S., Morandini, A.C., Sötje, I., Straehler-Pohl, I., Wiesenthal, A.A., Thiel, H., 2024. Rhizostomeae-a remaining manuscript to complete the Monograph series on Scyphozoa by M. E. Thiel (published 1936–1977). Advances in Marine Biology 98 (electronic supplement) at (*Work in progress, to be published soon*).

CHAPTER ONE

Max Egon Thiel's monographs on Scyphozoa (Cnidaria) and a left-behind typescript on the Rhizostomeae

Götz B. Reinicke[a,*], Sabine Holst[b], André C. Morandini[c], Ilka Sötje[d], Ilka Straehler-Pohl[e], Amanda A. Wiesenthal[f], and Hjalmar Thiel[a]

[a]Ocean Museum Germany, Stralsund, Germany
[b]Senckenberg am Meer, German Center for Marine Biodiversity Research, Hamburg, Germany
[c]Centro de Biologia Marinha, Universidade de São Paulo, São Sebastião, SP, Brazil
[d]University of Hamburg, Institute for Cell and Systems Biology of Animals (IZS), Hamburg, Germany
[e]Medusa(')s Nursery—Private Laboratory for Life cycle, Developmental, Evolutionary and Taxonomical Research, Haldensleben, Germany
[f]University of Rostock, Institute of Biosciences, Rostock, Germany
*Corresponding author: e-mail address: goetz.reinicke@meeresmuseum.de

Contents

Abstract

Max Egon Thiel worked as curator of the aquatic invertebrates collection at the Zoological Museum in Hamburg until 1963. Specialising in marine planktonic megafauna, he compiled a broad review of the research history on the Scyphozoa (Coronatae, Cubomedusae, Semaeostomeae) including the Staurozoa (as Stauromedusae), written in German. After publishing major parts in 1936 and 1938, World War II delayed further chapters until 1959 and 1962. A complete bibliography covering references up to 1970 was not printed until 1977. The final section on the taxon Rhizostomeae was completed as a typescript before his death, but was never published. In the present paper, the

Advances in Marine Biology, Volume 98
ISSN 0065-2881, https://doi.org/10.1016/bs.amb.2024.07.004

authors provide a synopsis of the published volumes in English. Following Thiel's original outline, the research history, as well as reviews of the current knowledge at the time about morphology, histology, ontogeny (life cycle), physiology, ecology, and phylogeny of the taxa are presented. The paper is complemented by two electronic supplements: A translated and revised version of the left-behind typescript of Max Egon Thiel about the taxon Rhizostomeae, and the revised digital list of references published in Thiel (1977).

1. Introduction

Max Egon Thiel worked as curator of the aquatic invertebrates collection (Niedere Tiere) at the Zoological Museum in Hamburg until 1963 (part of the *Leibniz Institute for the Analysis of Biodiversity Change* (LIB) since 2021). During the years 1936–1977 he published seven volumes of comprehensive monographs within the German series of *BRONN's Klassen und Ordnungen des Tierreiches* of the Akademische Verlagsgesellschaft (Geest and Portig) (in Leipzig), and an extensive bibliography on Scyphozoa (Volume 8) published by the successor, Gustav Fischer Verlag in Jena (see reference list). Research commenced in the early 1930s and the first three parts were published in 1936, and two more in 1938. World War II interrupted this series of publications. Thiel resumed and continued the work after 1956 until his death in 1979. His scientific heritage included over 840 pages of typescript in German on the remaining order Rhizostomeae – the 9th volume, with numerous tables and over 500 figures. This typescript, kept by his family, has recently been digitised, reviewed, and revised (condensed) by the authors HT and GBR (Thiel's son and grandson, respectively), yielding a text file including tables and figure captions of about 207,000 words. Yet, this last volume of the series remained unpublished. Like the other volumes of the series, the typescript provides an extensive review of the literature on the Rhizostomeae scyphozoans up to the early 1970s.

More than 50 years have passed since Thiel's work on the typescript; meanwhile scyphozoan research has progressed in various aspects including advances in systematics, biogeography, and ecology. However, the monograph has sustained its value as an extensive reference to fundamental scyphozoan research from the 18th into the second half of the 20th century, including discussions of various transient hypotheses and scientific disputes. This paper includes chapter summaries of Thiel's earlier publications (1962, 1959b, 1938a, 1938b, 1936a, 1936b, 1936c) for quick orientation.

1.1 Biographic remarks about Max Egon Thiel

Born in 1898, Max Egon Thiel (Fig. 1) was the fourth of seven sons of the mayor of Merzig, Germany (all information based on Dzwillo, 1980; Meise, 1980, and personal memories of HT and GBR). After participating in WWI, he took up studies of philosophy and chemistry at the University of Heidelberg and later concentrated on biology at the University of Hamburg, where he presented his PhD dissertation on the bivalve genus *Sphaerium* from the River Elbe. As a member of the hydro-biological section of the Zoological State Institute in Hamburg, he continued working on this subject, performing growth experiments in the field related to pollution from the town's sewage discharge into the river. In 1923, he started working as a lecturer assistant in the Zoological Museum, which then belonged to the University of Hamburg, founded in 1919. He became responsible for the collection of aquatic invertebrates in 1926. Through strong determination, he qualified as a profound expert on the scyphomedusae (Scyphozoa) and

Fig. 1 Max-Egon Thiel, 1957 in his lab at the Zoological Museum in Hamburg. *Photo: Author unknown, printed from the family archive.*

also published about Arctic and Antarctic hydromedusae (Hydrozoa), Scleractinia (Anthozoa) from West Africa, SE-Australia and the Philippines, and about Chaetognatha collected in the South Atlantic Ocean during the "METEOR" Expedition (1925–1927). He extended his cnidarian research as a guest researcher in Brussels and during a half-year collection of Scleractinia in Santo Domingo, Dominican Republic (1938–1939). The first monographs on Scyphozoa were published between 1936 and 1938 (Thiel, 1938a, 1938b, 1936a, 1936b, 1936c), before WWII interrupted his professional career for 17 years. He resumed his work on the Semaeostomeae, with the remaining sections on physiology (Thiel, 1959b), as well as ecology and phylogeny (Thiel, 1962). Additionally, to research on e.g. *Rhizostoma, Cyanea, Cassiopea,* and the evolution of jellyfishes (Thiel, 1970, 1976), he published the comprehensive bibliography of the Volumes 1–7 in Volume 8 (Thiel, 1977), because the editor had realised that the manuscript on the Rhizostomeae would probably remain unfinished. Gradually, forced by severe illness, he also retired from working at home, before his death in 1979. The final typescript (Volume 9) on the Rhizostomeae remained unpublished.

1.2 The monograph on Scyphomedusae by M. E. Thiel

With his series of publications, Thiel (1936a, 1936b, 1936c, 1938a, 1938b, 1959b, 1962, with consecutive pagination) undertook to review the contemporary natural history knowledge of the taxon Scyphozoa, based on his own observations and comprehensive research of literature, often including publications in various languages such as Chinese, Danish, Dutch, Finnish, Italian, Japanese, Latin, Norwegian, Portuguese, Russian (which he spoke himself), Spanish, and Swedish. The reviews about historic developments of various subjects reflect the stepwise improvement in scientific instruments (e.g. microscopes) and advancement of methodologies. Disagreeing speculations and interpretations on new zoological discoveries occasionally sparked robust conflicts between some leading proponents, or between their scholars. Often, only time brought about new observations and enlightenments to understand the morphological structures and their functions. Thiel carefully reported the gradual stages of progress in knowledge.

Scyphozoan systematics, as represented in the series by Thiel, progressively developed during the following decades. His system, in relation to the state of the World Register of Marine species (WoRMS) classification as of 2023, is represented in Table 1. Several species names used by Thiel (1936a, 1936b, 1936c, 1938a, 1938b, 1959b, 1962) have later been systematically revised in subsequent publications. In this paper, asterisks (*) indicate these

Table 1 Scyphozoa systematic classification as applied by Thiel (1936a, 1936b, 1936c, 1938a, 1938b, 1959b, 1962), according to the World Register of Marine Species (WoRMS 2023).

Systematic rank	Taxon, Author/s in Thiel	Systematic rank	Taxon, Author/s in WoRMS 2023
Phylum	**CNIDARIA**	Phylum	**CNIDARIA Verrill, 1865**
Scyphomedusae			
Class		**Class**	**Staurozoa** Marques & Collins, 2004
Order	Stauromedusae	Order	Stauromedusae Haeckel, 1879
		+ Suborder	**Amyostaurida** Miranda, Hirano, Mills, Falconer, Marques and Collins, 2016
Family	Cleistocarpidae James-Clark, 1863 *(partim)*	Family	Craterolophidae Uchida, 1929
Family	Kishinouyeidae	Family	Kishinouyeidae Uchida, 1929
		+ Suborder	**Myostaurida** Miranda, Hirano, Mills, Falconer, Marques & Collins, 2016
		Family	Lucernariidae Johnston, 1847
Family	Haliclystidae Haeckel 1879 *(partim)*	Family	Haliclystidae Haeckel, 1879
Family	Lipkeidae	Family	Lipkeidae Vogt, 1887
		+ Family	Kyopodiidae Larson, 1988
		Class	**Cubozoa** Werner, 1973
Order	Cubomedusae Haeckel, 1880		

(continued)

Table 1 Scyphozoa systematic classification as applied by Thiel (1936a, 1936b, 1936c, 1938a, 1938b, 1959b, 1962), according to the World Register of Marine Species (WoRMS 2023). (*cont'd*)

Systematic rank	Taxon, Author/s in Thiel	Systematic rank	Taxon, Author/s in WoRMS 2023
		Order	**Carybdaeida** Gegenbaur, 1857
Family	Charybdaeidae *s.str.*	Family	Carybdeidae Gegenbaur, 1857
		Family	Tamoyidae Haeckel, 1880
Family	Tripedaliidae Conant, 1897	Family	Tripedaliidae Conant, 1897
		+ Family	Alatinidae Gershwin, 2005
		+ Family	Carukiidae Bentlage, Cartwright, Yanagihara, Lewis, Richards & Collins, 2010
		Order	**Chirodropida** Haeckel, 1880
		Family	Chirodropidae Haeckel, 1880
Family	Chiropsalmidae	Family	Chiropsalmidae Thiel, 1936
		+ Family	Chiropsellidae Toshino, Miyake and Shibata, 2015
		Class	**Scyphozoa** Goette, 1887
		Subclass	**Coronamedusae** Calder, 2009
Order	Coronatae	Order	Coronatae Vanhöffen, 1892
Family	Atollidae Bigelow, 1909	Family	Atollidae Hickson, 1906

Family	Atorellidae	Family	Atorellidae Vanhöffen, 1902
Family	Linuchidae	Family	Linuchidae Haeckel, 1880
Family	Nausithoidae	Family	Nausithoidae Haeckel, 1880
Family	Paraphyllinidae	Family	Paraphyllinidae Maas, 1903
Family	Periphyllidae	Family	Periphyllidae Haeckel, 1880
		Subclass	**Discomedusae** Haeckel, 1880
Order	Semaeostomae	Order	Semaeostomeae Agassiz, 1862
Family	Cyaneidae	Family	Cyaneidae Agassiz, 1862
	Pelagiidae	Family	Pelagiidae Gegenbaur, 1856
	Aurelidae Agassiz, 1862	Family	Ulmaridae Haeckel, 1880
		Family	Drymonematidae Haeckel, 1880
		+ Family	Phacellophoridae Straehler-Pohl, Widmer and Morandini, 2011
Superorder	Rhizostomoidea Thiel subordo nov.	Order	Rhizostomeae Cuvier, 1800
		Suborder	**Dactyliophorae** Stiasny, 1920
Family	Rhizostomidae	Family	Rhizostomatidae Cuvier, 1800
Family	Rhopilemidae		as part of Rhizostomatidae

(continued)

Table 1 Scyphozoa systematic classification as applied by Thiel (1936a, 1936b, 1936c, 1938a, 1938b, 1959b, 1962), according to the World Register of Marine Species (WoRMS 2023). (*cont'd*)

Systematic rank	Taxon, Author/s in Thiel	Systematic rank	Taxon, Author/s in WoRMS 2023
Family	Stomolophidae	Family	Stomolophidae Haeckel, 1880
Family	Lychnorhizidae	Family	Lychnorhizidae Haeckel, 1880
Family	Catostylidae	Family	Catostylidae Claus, 1883
Family	Lobonematidae	Family	Lobonemidae Stiasny, 1920
		Suborder	**Kolpophorae** Stiasny, 1920
Family	Cassiopeidae	Family	Cassiopeidae Tilesius, 1831
Family	Cepheidae	Family	Cepheidae Agassiz, 1862
Family	Thysanostomidae Gegenbaur, (1857)	Family	Leptobrachidae Agassiz, 1862
Family	Mastigiadidae Stiasny, (1920)	Family	Mastigiidae Stiasny, 1920
Family	Versurigidae Gegenbaur, (1857)	Family	Versurigidae Kramp, 1961

"+" indicates taxa newly established after Thiel's work.

names, referring to a list of species names in Table 2, which also provides the respective modified nomenclature according to WoRMS: https://www.marinespecies.org/aphia.php?p=taxdetails&id=135220 on 2023-11-17.

In many cases, Thiel provided numerous detailed figures published by the various earlier researchers, with the respective features described. Occasionally, original drawings are used to exemplify general morphological patterns. For the present publication, Thiel's German text passages (e.g. taxon diagnoses) were translated by the AI-software *DeepL*, before applying thorough transliteration and language reviews by the authors.

Table 2 Species names used by Thiel (1936a, 1936b, 1936c, 1938a, 1938b, 1959b, 1962), and the respective modified nomenclature according to World Register of Marine Species at: https://www.marinespecies.org/aphia.php?p=taxdetails&id=135220 on 2023-11-17.

Species names used by M.E. Thiel	Original reference	Accepted species names in WoRMS (2023)
CUBOMEDUSAE		
Charybdea marsupialis (sic!)	(Linnaeus, 1758)	*Carybdea marsupialis* (Linnaeus, 1758)
Charybdea xaymacana (sic!)	Conant 1897	*Carybdea xaymacana* Conant 1897
Carybdea rastonii	(in Uchida, 1926)	*Carybdea prototypus* (Haeckel, 1880)
Charybdea alata (sic!)	Reynaud, 1830	*Alatina alata* (Reynaud, 1830)
Tripedalia cystophora	Conant 1897	dto.
Chiropsalmus quadrigatus	Haeckel, 1880	*Chiropsoides quadrigatus* (Haeckel, 1880)
Chiropsalmus zygonema	Haeckel, 1880	dto.
Chiropsalmus quadrumanus	(F. Müller, 1859)	dto.
Chiropsalmus buitendijki	van der Horst, 1907	*Chiropsoides buitendijki* (van der Horst, 1907)
Chirodropus gorilla	Haeckel, 1880	*Chirodropus gorilla* Haeckel, 1880
Chirodropus palmata	Haeckel, 1880	*Chimaerus palmatus* (Haeckel, 1880)
CORONATAE		
Stephanoscyphus mirabilis	Allmann, 1874	*Nausithoe punctata* Kölliker, 1853
Stephanoscyphus racemosus	Komai, 1936	*Nausithoe racemosa* (Komai, 1936)
Spongicola fistularis	F.E. Schultze (1877)	*Nausithoe punctata* Kölliker, 1853
Periphylla hyazinthina	Steenstrup, 1837	*Periphylla periphylla* (Péron & Lesueur, 1810)
Pulmo marinus	Pliny the Elder (AD 23/24–79)	*Periphylla periphylla* (Péron & Lesueur, 1810)

(continued)

Table 2 Species names used by Thiel (1936a, 1936b, 1936c, 1938a, 1938b, 1959b, 1962), and the respective modified nomenclature according to World Register of Marine Species at: https://www.marinespecies.org/aphia.php?p=taxdetailsid=135220 on 2023-11-17. (*cont'd*)

Species names used by M.E. Thiel	Original reference	Accepted species names in WoRMS (2023)
SEMAEOSTOMEAE		
Aurelidae		**Ulmariidae** Haeckel 1880 **(Aureliinae Agassiz 1862)**
Aurelia aurita	(Linnaeus, 1758)	dto.
Aurelia limbata	Brandt, 1835	dto.
Discomedusa lobata	Claus, 1877	dto.
Undosa undulata	Haeckel, 1879	[uncertain status]
Poralia rufescens	Vanhöffen, 1902	dto.
Sthenonia albida	Eschscholtz, 1829	dto.
Ulmaropsis drygalskii	Vanhöffen, 1908	*Diplulmaris antarctica* Maas 1908
Stygiomedusa fabulosa	Russell, 1959	*Stygiomedusa gigantea* (Browne, 1910)
Pelagidae (sic!)	Gegenbaur, 1856	**Pelagiidae**
Chrysaora hysoscella	(Linnaeus, 1767)	dto.
Chrysaora helvola	Brandt, 1835	*Chrysaora fuscescens* Brandt, 1835
Chrysaora pacifica	(Goette, 1886)	dto.
Dactylometra quinquecirrha	Agassiz, 1862	*Chrysaora quinquecirrha* (Desor, 1848)
Dactylometra africana	Vanhöffen, 1902	*Chrysaora africana* (Vanhöffen, 1902)
Dactylometra pacifica	(Goette 1886)	*Chrysaora pacifica* (Goette 1886)
Dactylometra lactea	Agassiz	*Chrysaora lactea* Eschscholtz, 1829
Pelagia noctiluca	(Forsskål, 1775)	dto.

Pelagia panopyra	Péron & Lesueur, 1810	dto.
Sanderia malayensis	Goette, 1886	dto.
Cyaneidae	Agassiz 1862	dto.
Cyanea annaskala	Lendenfeld, 1882	dto.
Cyanea capillata	(Linnaeus, 1758)	dto.
Cyanea var. *lamarckii*	Péron & Lesueur, 1810	*Cyanea lamarckii* Péron & Lesueur, 1810
Cyanea nozakii	Kishinouye, 1891	dto.
Desmonema chierchiana	Vanhöffen, 1888	*Desmonema gaudichaudi* (Lesson, 1830)
Desmonema gaudichaudi	(Lesson, 1830)	dto.
Phacellophoridae		Straehler-Pohl, Widmer & Morandini, 2011
Phacellophora sicula	Haeckel, 1880	*Ph. camtschatica* Brandt, 1835
Drymonematidae	Haeckel, 1880	
Drymonema dalmatina	Haeckel, 1880	*Drymonema dalmatinum* Haeckel, 1880

2. Brief summaries of the published Volumes 1–8 of M. E. Thiel

2.1 Volume 1: Introduction to the Scyphomedusae and Stauromedusae

The series commenced with a general introduction to five orders of the class **Scyphomedusae** [Thiel, 1936a, pp. 1–41 of the series, accepted as the class Scyphozoa by WoRMS (2023)]. Besides the order Coronatae (1936c, pp. 308–471), Thiel also included the order Stauromedusae (Thiel, 1936a, pp. 42–172) [accepted as the only order in the class Staurozoa by

WoRMS (2023)] and the order Cubomedusae (Thiel, 1936b, pp. 173–308) [accepted as the class Cubozoa by WoRMS (2023)] within the taxon (see Fig. 2). Within the order Discomedusae he differentiated the Semaeostomeae (1938a, 1938b, 1959b, 1962, pp. 471–1308) and the Rhizostomeae [Thiel's unpublished German typescript translated and revised as a digital supplement herein: Thiel et al., 2024: (*Work in progress, to be published soon*)] [Semeostomeae and Rhizostomeae are today accepted as separate orders in the subclass Discomedusae by WoRMS (2023)].

Thiel presented this classification as the outcome of several centuries of research, beginning with the first mention of free-floating sea nettles by Aristoteles (384–322 BC). Later descriptions by e.g. Aldrovandi (1618) and reviews of comprehensive systematic "attempts" from the 18th - 20th centuries were compared and discussed: Gegenbaur (1856), Agassiz (1862), Haeckel (1880), Maas (1897), Delage and Hérourd (1901), Mayer (1910) and Krumbach (1925); for the comprehensive references before 1970 compare the complete bibliography of Thiel (1977) or https://www.elsevier.com/books-and-journals/book-companion/9780443296468). The section ends with a **diagnosis** of the Scyphomedusae (p. 30): The Scyphomedusae are wineglass-, bell- or umbrella-shaped organisms of gelatinous or cartilaginous consistency with a structured or unstructured surface, consisting of only two germ layers equipped with cnidocysts. On the clefted (= lobed) or non-clefted margin, there are bean-shaped adhesive bodies or piston-shaped sensory organs, as well as marginal tentacles either solitary or in clusters. The animals are benthic or planktonic, but a large

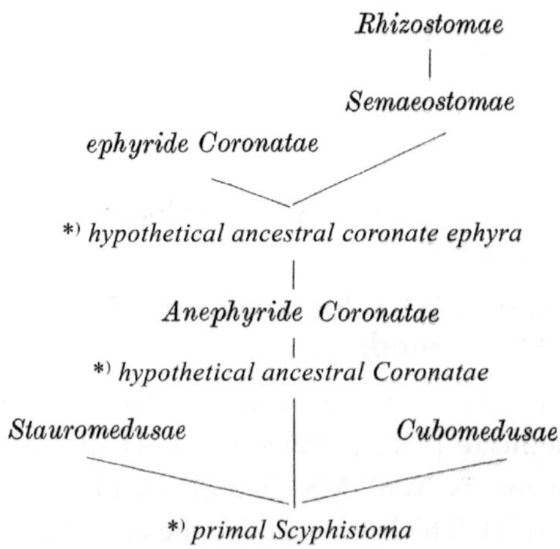

Fig. 2 Thiel's scheme of the phylogenetic development in the Scyphozoa (Thiel, 1936a, p. 35); (*) german wording translated.

proportion of the latter develop through a benthic stage. In the medusae, there is a mouth tube of variable length in the body centre, which may be extended at its end into four oral lobes (= oral arms). In the order Rhizostomeae, these have a large number of small funnel-shaped openings [Trichterkrausen] and various club- and finger-shaped appendages, while a central mouth opening is only present in early juveniles and closes later. The oral openings lead into a gastric cavity. This is divided into a central stomach and peripheral gastric pouches [Gastraltaschen] of variable numbers, formed by four adhesions of the body walls (gastric septa) – which, however, may be absent. In some forms, a large number of branched or unbranched radial canals extend from the gastric cavity to the periphery, where a circular canal may connect them all. Finger-shaped gastric cirri arranged in clusters or rows arise from the septa while the sexual organs, which always arise from a bilateral endodermal fold of the septa, are located between the septa. Externally, the position of the septa is characterised by a more or less deep depression (= taenioles) of the subumbrellar surface. The nervous system consists of a diffuse network spread over the whole body, as well as nerve centres located near the sensory bulbs, which are only connected by a nerve cord in the Cubomedusae. – *("Die Scyphomedusae sind aus nur zwei Keimblättern bestehende, mit Nesselorganen versehene weinglas-, glocken- oder schirmförmige Organismen von gallertiger oder knorpeliger Beschaffenheit mit gegliederter oder ungegliederter Oberfläche. Am geteilten oder ungeteilten Rand, befinden sich bohnenförmige Haft- oder kolbenförmige Sinnesorgane, sowie einzeln oder in Büscheln stehende Tentakel. Die Tiere leben benthonisch oder planktonisch, ein großer Teil der letzteren durchläuft in seiner Entwicklung jedoch ein benthonisches Stadium. Im Zentrum des [Medusen-] Körpers befindet sich ein mehr oder weniger langes Mundrohr, das an seinem Ende in vier mehr oder weniger lange Mundzipfel ausgezogen sein kann, die als Mundarme bezeichnet werden. In der Ordnung Rhizostomeae sind sie mit einer großen Zahl von kleinen trichterförmigen Öffnungen (sog. Trichterkrausen) und verschiedenartigen kolben- und fingerförmigen Anhängen versehen, während die zentrale Mundöffnung nur in der frühen Jugend vorhanden, später geschlossen ist. Die Mundöffnung führt in einen Magenraum, der durch vier Verwachsungen (Septen) der Körperwände – die allerdings fehlen können – in einen Zentralteil und verschieden zahlreiche periphere Taschen (Gastraltaschen, Gastrobursen) gegliedert wird. Vom Magenraum ziehen bei einem Teil der Formen eine große Zahl von verzweigten oder unverzweigten Kanälen nach der Peripherie hin, wo sie durch einen ringförmigen Kanal, den Ringkanal, miteinander verbunden sein können. An den Septen befinden sich, in der Entwicklung zuweilen erst später auftretende, in Büscheln oder in Reihen angeordnete fingerförmige Fortsätze*

(Gastralcirren) und zwischen den Septen liegen die Geschlechtsorgane, die stets aus einer beiderseitigen entodermalen Falte der Septen hervorgehen. Äußerlich ist die Lage der Septen durch eine mehr oder weniger tiefe Einsenkung (Taeniolen) der Oberfläche gekennzeichnet. Das Nervensystem besteht aus einem diffusen Netzwerk, das über den ganzen Körper verbreitet ist, sowie örtlich begrenzten Nervenzentren in der Nähe der Sinneskolben, die nur bei den Cubomedusae durch einen Nervenstrang miteinander verbunden sind.").

Further sections reviewed the phylogeny of the taxon and added remarks on collection, preservation, and dissection methods. A comment on the impact of medusae on humans briefly described the toxic effects of stinging cells of some species and recommended treatment options. The introductory chapter ends with a note on edible jellyfish species and their consumption, primarily in East Asian countries.

2.1.1 Stauromedusae

The subsequent chapter (Thiel, 1936a, pp. 42–172) introduces the Stauromedusae [i.e. stalked jellyfish, accepted as the only order in the class Staurozoa by WoRMS (2023)], with a general characterisation of the organisms. Detailed morphological descriptions exemplify various structures as observed in the different species: the exumbrella, stalk and basal discs, animal sizes, coloration, the inner septae of the calyx, the septal funnel and gastric cavity, the umbrella margins with tentacles, the musculature, reproductive tissues, and the nervous systems. The following section on histology describes examples of ecto- and endodermal tissues, mesoglea, nervous and sensory cells, secretory, cnidocysts, and reproductive cells. Many detailed figures reprinted from the original publications (e.g. from Antipa, Carlgren, Clark, Hertwig, Kassianow, Kling, Leuschel, Mayer, Uchida) illustrate and complement the texts.

Descriptions about the **ontogenetic development** include historical remarks on early studies, the development from the zygote to the planula larva, and the complex changes in body structures during the metamorphosis of the stauropolyp, also including the formation of the tentacles and anchors [Randanker]. Again, many original illustrations show details of the histological sections described.

The **physiology** section comprises information about the metabolism, neuro- and sensory physiology as well as locomotion, metamorphosis, growth variations, and regenerative capacities. A short review about the **ecology** and geographic records of 13 genera proposed at the time includes the genus *Haliclystus* with the most widespread known distribution.

Recorded stauromedusae were usually attached on hard substrata or rigid algal rhizoid structures (e.g. *Fucus*, *Macrocystis*) in shallow coastal zones, but reported vertical distributions extended down to 1000 m depth.

The final section covers the **classification and phylogeny** of the Stauromedusae. Discovered in 1776 by Müller (1776), several authors of the early 19th century discussed the relation of this taxon with Actiniaria (Cuvier, 1817; Lamouroux, 1815), and with Ctenophora and Siphonophorae (Lamarck, 1816). Agassiz (1862) and Clark (1863) finally proposed the relation with Scyphozoa, which was also followed by Haeckel (1880). Thiel provides an overview to the history of the different genera and discusses their possible relations according to e.g. Mayer (1910), Krumbach (1925), Uchida (1929) and Carlgren (1935). He gives the following **diagnosis** of the taxon (p. 154): The Stauromedusae are scyphomedusae with cup- or wineglass-shaped body, their stalk is provided with an adhesive disc and the margin can be smooth, or more or less clefted and bears variously shaped tentacle-like or transformed appendages which derived from tentacles. *("Die Stauromedusae sind Scyphomedusen von becher- oder weinglasförmiger Gestalt, deren Stiel mit einer Haftscheibe versehen ist und deren Rand teils glatt, teils mehr oder weniger stark gebuchtet ist und verschieden gestaltete, tentakelähnliche oder durch Umbildung von Tentakeln entstandene Anhänge trägt").* Together with the description of characters used for classification (and a table of species characters), Thiel presented a detailed reasoning of his scheme of generic relations within the taxon, including a key to families and genera.

[Today, the Staurozoa – along with the Scyphozoa – are classified separately within Cnidaria, and hold about 50 valid species in 6 families and 11 genera (WoRMS, 2023; Daly et al., 2007; Mills, 2012; Collins and Jarms, 2018; Miranda et al., 2018).].

2.2 Volume 2: Cubomedusae

Thiel (1936b, pp. 173–308) referred to the Cubomedusae as an order within the class Scyphomedusae [today accepted as class Cubozoa by WoRMS (2023)]. After a general characterisation of overall body features, he provided detailed descriptions of the morphological structures such as the exumbrella with specific accessory organs, the subumbrella, gastric cavity, reproductive organs, musculature, the nervous system, and the sensory organs. Again, details were visualised by many original figures from various other researchers such as e.g. Berger, Claus, Conant, Haeckel, Schewiakoff, Stiasny, Yatsu, along with his own drawings. Thiel thoroughly reported and discussed other authors' views and strived to collect the state of knowledge at the time.

The **histology** section only referred to Claus' (1878) work on *Carybdea marsupialis* (Linnaeus, 1758) and Berger's (1898, 1900) studies of *Carybdea xaymacana* Conant, 1897 in the Caribbean. They described and depicted examples from ecto- and endodermal tissues, the gelatinous layers, cnidocyst and cnidocytes, nervous and sensory cells, secretory and muscle cells, as well as reproductive cells. Detailed figures reprinted from the original publications illustrate and complement Thiel's texts. Observations on the **ontogenetic development** of Cubomedusae were scarce at that time. Thiel reported on embryo and planula descriptions by Okada (1927) and observations on a few young medusae of *Carybdea rastonii*[(*1)] by Uchida (1926, 1929) [=*C. prototypus* (Haeckel, 1880)], and included some original figures. The section on **physiology** reported fish as being the preferred food resource and described prey capture, including ingestion and metabolism. Further data on nerves and senses largely referred to Conant (1898), Berger (1900), and especially to Yatsu (1917). Data on the physiology of movements, metamorphosis, growth variations, polyp metamorphosis, and regenerative capacities were scarce or absent during that time. Thiel reported and valued existing information, also referring to growth measurements related to the development of gonads, as tabled by Mayer (1910).

With regard to the **ecology**, Thiel reported on diverse early assumptions about Cubomedusae being deep-sea creatures. Stiasny (1919) was the first to establish their occurrence in tropical coastal habitats, often near nutrient-rich estuaries or ports. Sessile cubopolyps and juvenile medusae were always observed in shallow waters, never in deep waters. Tropical water temperatures appeared to limit their distribution, and light seemed to influence daily patterns. Based on data of that time, the latitudinal distribution between 30° N and S occasionally extended due to warmer seas (in the Mediterranean) or current patterns (e.g. near the Bermuda Islands, in Japanese and south Australian coastal waters). The known geographic distributions were tabled (p. 277) for *Carybdea marsupialis* (Linnaeus, 1758), *Carybdea alata*[(*1)] [=*Alatina alata* (Reynaud, 1830)], *Tripedalia cystophora* Conant, 1897, *Chiropsalmus quadrigatus*[(*)] [=*Chiropsoides quadrigatus* (Haeckel, 1880)],

[1] Several species names used by Thiel (1936a, 1936b, 1936c, 1938a, 1938b, 1959b, 1962) referring to other authors have later been systematically revised in subsequent publications. In this paper, asterisks (*) indicate these names, referring to a list of species names in Table 2, which also provide the respective modified nomenclature according to WoRMS.

Chiropsalmus zygonema Haeckel, 1880, *Chiropsalmus quadrumanus* (Müller, 1859), *Chiropsalmus buitendijki*(*) [=*Chiropsoides buitendijki* (van der Horst, 1907)], *Chirodropus gorilla* Haeckel, 1880 and *Chirodropus palmata*(*) [=*Chimaerus palmatus* (Haeckel, 1880)].

Meanwhile, the systematic knowledge about the taxon has increased. A recent taxonomic revision (ISP in 2023, this paper) of the specimen records for the nine species in Thiel's list yielded a total count of 21 species. Based on these findings, an updated survey of the geographic distribution is included here as Table 3.

The last section covered the **classification and phylogeny** of Cubomedusae. Following the first description of a single specimen in 1739 by Plancus, early works were limited due to only anecdotal records. This changed when Gegenbaur described *Carybdea marsupialis* from the Mediterranean in 1856, and detailed drawings by Claus (1878) and Haeckel (1882) reported about *Carybdea murrayana* Haeckel, 1880 as part of the Challenger Expedition collection, providing the first descriptions of the taxon's general morphology. Further studies on sense organs by Schewiakoff (1889) and Conant (1898) on anatomy and histology of newly described species provided numerous plates with detailed figures. Conant's work, which was continued after his death by Berger (1900), still merits reference today. Based on further comprehensive works of Yatsu (1917), Uchida (1926, 1929) and Okada (1931), Thiel gave the following **diagnosis** of the taxon (p. 280): Cubomedusae are scyphomedusae of hemispherical, truncated conical, or cubical shape with a deep umbrella cavity and four or twelve branched or unbranched leaf-shaped gelatinous appendages (pedalia). These are prolongations of the interradial edges of the bell, each bearing one or, depending on the number of branches, several tentacles. (*"Die Cubomedusae sind Scyphomedusen von halbkugeliger, abgestutzt kegelförmiger oder kubischer Gestalt mit tiefer Schirmhöhle und mit vier oder zwölf verzweigten oder unverzweigten blattförmigen Gallertanhängen (Pedalien), die in der Verlängerung der interradialen Kanten der Glocke angeordnet sind und je einen oder, je nach Zahl der Verzweigungen, mehrere Tentakel tragen").*

Assumptions about the phylogeny of Cubomedusae grew with the increasing knowledge of the taxon. Thiel provided a historic protocol of genera descriptions and evaluated their phylogenetic relationship, as suggested in the complex statements and discussions of various authors. Based on Uchida (1929), he reviewed relevant characteristics and identified the pedalia, sensory niches with rhopalia, velarial-lobes and -channels, and the gastric pouches as key-features for

Table 3 Taxonomic revision of Cubomedusae represented in the material reported by Thiel (1936b, p. 277) – revised by I. Straehler-Pohl (this paper) with updated species list and geographic records.

Species (Thiel 1936b)	Revised species (Straehler-Pohl)	Atlantic Ocean	Indian Ocean and Malayan Archipelago	Pazific Ocean
***Charybea marsupialis* (Linnaeus, 1758)**				
	Carybdea marsupialis (Linnaeus, 1758)	Mediterranean Sea		
	Carybdea murrayana Haeckel 1880	Sierra Leone (West Africa)		
	Carybdea xaymacana Conant, 1897	Bermudas, Southern East coast of North America, South of Cape Cod, West Indies		
	Carybdea rastonii Haacke, 1887			South coast of Australia
	Alatina tetraptera (Haeckel, 1880)		Malayan Archipelago	
	Meteorona kishinouyei Toshino, Miyake & Shibata, 2015			Japan
	Carybdea prototypus (Haeckel, 1880)			Chinese Sea

	Alatina philippina (Haeckel, 1880)			Philippines
	Carybdea cuboides (Haeckel, 1880)			Hawaiian Islands
***Charybdea alata* Reynaud, 1830**				
	Alatina alata (Reynaud, 1830)	Atlantic coasts of North and South America, between 30°N u. 30°S, West Indies, Cabo Verde	Gulf of Mannar, Borneo, Sumatra, Malay Archipelago, Southern Indian Ocean, Seychelles, Bay of Bengal	Hawaii, Philippines
	Alatina madraspatana Menon, 1930		Coast of Madras	
	Carybdea murrayana Haeckel, 1880	South West Africa		
	Species uncertain		Mergui Archipelago [Thailand]	
	Morbakka virulenta Bentlage & Lewis, 2012			Japan
	Alatina grandis Agassiz & Mayer, 1902			Marianas

(continued)

Table 3 Taxonomic revision of Cubomedusae represented in the material reported by Thiel (1936b, p. 277) – revised by I. Straehler-Pohl (this paper) with updated species list and geographic records. *(cont'd)*

Species (Thiel 1936b)	Revised species (Straehler-Pohl)	Atlantic Ocean	Indian Ocean and Malayan Archipelago	Pazific Ocean
	Species uncertain			New Guinea
	Alatina grandis Agassiz & Mayer, 1902			[French Polynesia] Tahiti, Society Inseln, Paumotus [Fakarava Islands]
	Carybdea irregularis Straehler-Pohl, 2019			Marquesas
***Tripedalia cystophora* Conant, 1897**				
	Tripedalia cystophora Conant, 1897	West Indies (Jamaica)	-	Philippines
	Species uncertain			Ecuador* (West coast South America)
Chiromedusa debrachiata* Thiel = *Chiropsalmus quadrumanus				
	Chiropsella bart Gershwin & Alderslade, 2006			New Pommern [= New Britain, Papua New Guinea]
Chiropsalmus quadrigatus Haeckel 1880				
	Chiropsoides quadrigatus Haeckel, 1880		Yangoon (Gulf of Bengal) [Myanmar]	

	Species uncertain, [*Stiasny 1922: guess only*]		Johore (Str. of Malacca)	
	Chiropsoides buitendijki Horst 1907		East coast of the Indian subcontinent	
	Chironex yamaguchii Lewis & Bentlage 2009			Philippines
Chiropsalmus zygonema Haeckel 1880				
	Chiropsalmus zygonema Haeckel, 1880	Coast of Argentina		
***Chiropsalmus quadrumanus* F. Müller 1859**				
	Chiropsalmus quadrumanus	Coast of Brazil		
	?*Chironex indrasaksajiae* Sucharitakul et al. 2017		Mergui Achipelago, Padang [Thailand], Sumatra	
	Chiropsella bart Gershwin & Alderslade, 2006			Port Darwin (N.-Australia)

(continued)

Table 3 Taxonomic revision of Cubomedusae represented in the material reported by Thiel (1936b, p. 277) – revised by I. Straehler-Pohl (this paper) with updated species list and geographic records. (*cont'd*)

Species (Thiel 1936b)	Revised species (Straehler-Pohl)	Atlantic Ocean	Indian Ocean and Malayan Archipelago	Pazific Ocean
***Chiropsalmus buitendijki* Horst 1907**				
	Chiropsoides buitendijki Horst 1907		Java, Tijmera, Madras coast	
	Species uncertain			Bowen Harbour (Queensland)
	Species uncertain			Port Denison [Western Australia]
***Chirodropus gorilla* Haeckel, 1880**				
	Chirodropus gorilla Haeckel, 1880	West Africa		
	Chimaerus palmatus (Haeckel, 1880)	Loanga [Gabun], Muculla?		
***Chirodropus palmata* Haeckel, 1880**				
	Chimaerus palmatus (Haeckel, 1880)	St. Helena		

the classification of the three accepted genera at that time: *C(h*[2]*)arybdea, Tripedalia and Chiropsalmus*. He established and diagnosed the family Chiropsalmidae within the Cubomedusae to accommodate the latter genus.

[Later, the taxon diversified considerably due to an increase in observations and new species discoveries. Today, the Cubomedusae are classified either as Cubozoa, an independent class within Cnidaria, alongside the Scyphozoa, or as the suborder Cubomedusae within the subclass Discomedusae within the Scyphozoa. The taxon holds 50 valid species within 2 orders, split in 8 families with 19 genera (WoRMS, 2023; Marques and Collins, 2004; Daly et al., 2007; Jarms and Morandini, 2019; Collins and Jarms, 2021; Straehler-Pohl and Jarms, 2022; Straehler-Pohl et al., 2022).]

2.3 Volume 3: Coronatae

In this chapter, Thiel (1936c, pp. 308–471) introduced the order with a general characterisation of the coronate medusa, describing overall features like the central umbrellar disc and the corona with its diverse accessory structures (e.g. marginal lappets, rhopalia, tentacles, oral tube, subumbrellar pouches). Different body shapes accounted for the generic characteristics. Significant features like the subumbrellar cavity, the oral canal and lips, the taeniols, and gastric cirri were described, as well as the general organisation of gonads. Thiel added remarks on overall sizes and colour patterns in e.g. *Atolla* Haeckel, 1880 and *Periphylla* F. Müller, 1861, mentioning completely transparent species and the greenish-brown coloration caused by symbiotic algae in *Linuche* Eschscholtz, 1829. Additionally, some morphological and ontogenetic information regarding other stages (polyp, strobila, planula) were addressed in the genus *Nausithoe* Kölliker, 1853.

The **morphology** section provided detailed descriptions of the inner and outer epithelium in various body parts. It included observations on the exumbrella and subumbrella, the gastric cavity, gonads, musculature, the nervous system and sensory structures in selected medusa specimens. Detailed figures reproduced from Haeckel (1880), Claus (1883), Vanhöffen (1902) and Maas (1897, 1903) illustrate various details. Further, Thiel collected occasional remarks and examinations about **histological** features, which he found distributed among the former descriptions of the ecto- and endoderm, as well as the gastrodermal lamellae [Gastrodermis] and the supportive lamellae [Mesogloea]. Glands, secretory cells, and cnidocytes (stinging cells) were included, along with the nervous and sensory

[2] Thiel (1936b) erroneously referred to the genus as *Charybdea* sp., which is regarded here as a synonym.

complex, muscles and reproductive tissues, also referring to the authors mentioned above. Most observations, however, remained preliminary or fragmentary because of the limited data available at that time.

For a long time, the available information about the **ontogeny** was limited, since studies on coronate taxa were rare. But early observations on a single polyp found in a sponge noted by Kowalewski (1873, describing strobilation, in Russian) gave rise to cautious speculation. Allmann (1874) briefly reported on *Stephanoscyphus mirabilis*(*) [=*Nausithoe punctata* Kölliker, 1853] and Schultze (1877) described *Spongicola fistularis*(*) [=*N. punctata*] in greater detail. Despite neither of them drawing a conclusion on the systematic relation, F.E. Schultze noted a possible link to the acraspedote medusae [a taxon that comprised Stauromedusae, Coronatae, Cubomedusae and Discomedusae]. Only Komai (1936) reported another polyp (*Stephanoscyphus racemosus*(*) [=*Nausithoe racemosa* (Komai, 1836)]) from Japanese waters, but knowledge about the entire reproductive cycle remained incomplete. Developments of oocytes and sperm was histologically shown by Metschnikoff (1886) in *N. punctata* and by Maas (1897) in *Periphylla* and *Atolla*, while Conklin (1908) described details of oocyte development in *Linuche unguiculata* (Swartz, 1788). Conklin also gave detailed accounts of cleavage stages and planula development, and Thiel (1936c) included his figures.

The metamorphosis of the planula into a polyp was shown by Metschnikoff (1886 for *N. punctata*). Based on several studies, Thiel gave a detailed account about settlement, growth, and morphology, as well as strobilation, including the metamorphosis of *N. punctata* and *L. unguiculata* stages and medusa maturation. Most information was available on coronate polyps studied by Schultze (1877, on *S. fistularis*(*)) and Komai (1936, on *Stephanoscyphus racemosus*(*)) (=*N. racemosa*). This paragraph also included further information on morphological and histological details on the soft body, periderm tube and cnidocysts.

The section on **physiology** of Coronatae remained brief on data about digestion, excretion, and respiration due to the lack of knowledge. Also, due to the dearth of experiments, information on the sensory structures is limited to a brief description of the equilibrium organ consisting of the statocyst, touch-plates [Tastpolster], and epidermal hoods [Deckschuppen]. A central neuronal organisation unit was assumed to be located in the rhopalia (Vanhöffen, 1902). Stimulus reception by the sensory structures provokes the main ganglia and stimulates the supposed motor ganglia of the ring muscle into controlled contractions. In the experiment, a cutout bell margin of *Linuche* produced alternate contractions of opposite sides of the umbrella, which narrows its opening to a slot in altering directions. Available experiments, however, did not provide a consistent picture of the topic. Information about the physiology of movement refers to brief comments of Kiefenstein and Ehlers (1861) on *Nausithoe* and *Periphylla*, and includes observations of Conklin (1908) on *L. unguiculata*, describing three different swimming modes of the latter.

In addition, the reproductive physiology of the dioecious Coronatae still lacked observations. The oocytes are released with a jelly cover, drawn and described (after Fewkes, 1882 and Conklin, 1908) as a characteristic of *Nausithoe* and *Linuche* and (after Vanhöffen 1892) possibly also of *Atolla*. Their origin from the endodermal ovarial lamellae and the formation of jelly [Gallerte] with pigments and cnidocysts were discussed. Further reports on observations of fertilisation in *Nausithoe, Linuche* and *Atolla* describe early developmental stages. Fertilisation seemed to be ascertained through timed aggregations in *Nausithoe* (after Metschnikoff, 1886) and *Linuche* (after Conklin, 1908). Nothing was known about the fertilisation processes in deep-sea species like *Atolla* (after Maas, 1897), but maturity graduations of eggs led to the assumption of similar spawning processes. Data on the duration of ontogenetic processes and stages were lacking. Reproductive periods observed in Mediterranean and Japanese waters seemed to be related to summer seasonal changes of the northern hemisphere. This was confirmed by observations of Mayer (1910) on ephyrae of *Linuche* in Fiji waters during November and December, which developed into mature individuals in the following months. According to Conklin (1908) these specimens died immediately after spawning. Very few available data on growth rates and alternation of generations allowed only preliminary considerations. General age estimates were given to be one year, and in larger species like *Periphylla* at least two years were expected. A paragraph describing growth variations and aberrations (number of tentacles, height and width of umbrella, pigmentation, subumbrellar pouches) completed this section.

A section on the **ecology** of Coronatae provides observations on the distribution of neritic species like *N. punctata* and *L. unguiculata*, as well as oceanic species of Periphyllidae and Atollidae. Their occurrence in coastal areas and different water bodies was discussed. While mean annual temperature ranges of 22–28 °C allowed a wide distribution of *N. punctata* at 35–36.5 ‰ salinity, regional presence of the species coincided with elevated salinity values (37.5–38 ‰) in the Mediterranean (after Krumbach, 1925). *Atolla* species caught below 600 m depth during the German Deep

Sea Expedition (1898–1899) from 0 to 10 °C (Vanhöffen, 1902), and specimens of *Atolla wyvillei* Haeckel, 1880, collected by the British Discovery-Expedition (1901–1904) to the Antarctic at 1.25–4 °C demonstrated much tighter ranges (Stiasny, 1934), suggesting the significance of light on vertical distribution. These early ideas of biological oceanography, relating species distribution patterns to physical data, indicated differences in ecological ranges of salinity tolerance and a general dependency on environmental conditions. Further remarks discuss the seasonality of the occurrence of *Atolla, Nausithoe*, and *Linuche* species. Available data on horizontal and bathymetrical distributions were summarised in two tables and discussed. Size (diameter) vs. depth distribution and abundance data are given for *Periphylla hyacinthina*(*) [=*Periphylla periphylla* (Péron and Lesueur, 1810)], the Mediterranean *P. periphylla, Atolla bairdii* Fewkes, 1886, and *Atolla verrillii* Fewkes, 1886, and demonstrated a general increase of size with depth. Further attention is given to colour variations with depth in *Periphylla* and *Atolla* (after Broch, 1913), reflecting an increase in darker coloration (pigmentation) in deeper habitats.

The section on the **classification and phylogeny** of Coronatae gives a detailed account of the research history of the taxon. Thiel introduces the first descriptions of a species by Swartz (1788), *Medusa unguiculata*(*) [=*L. unguiculata*] and *Carybdea periphylla*(*) [=*P. periphylla*]. Several early ideas to define systematic relations with the Semaeostomeae remained preliminary and disputed, while more genera were described, e.g. *Nausithoe* by Kölliker (1853). Initial attempts of systematic classifications by Gegenbaur (1856) and Agassiz (1862) remained preliminary due to still limited information. Based on descriptions of all 29 known species, Haeckel (1880) was the first to realise the uniqueness of the group. He hypothesised a close relation with Stauromedusae and defined two subgroups (Coronatae and Cannostomae, the latter also including some Semaeostomeae and Rhizostomeae) based on the gastrovascular systems and the marginal bodies = rhopalia [Randkörper]. However, his arrangement sparked discussions and criticism due to inconsistencies in character interpretation (Claus, 1883, 1886; Fewkes, 1882). It was ultimately Vanhöffen (1891, revised in 1902) who established the core classification of Coronatae "used" today. A few alternative models were rebutted by Maas (1903, 1907), based on his works about medusae in the collections of the Dutch Siboga-Expedition to the Indonesian Archipelago (1899–1900). The ongoing period of intense studies in the Coronatae resulted in Mayer's (1910) system and account of 5 families, 10 genera, and 24 species of the taxon, including a revision of the system proposed by Haeckel (1880).

The ongoing period of extensive international marine expeditions yielded numerous further studies and newly described species (collected in Thiel, 1936c, p. 451 ff). At the time of publication, it was Krumbach (1925) who had incorporated all these findings in his most recent classification system, which had been further supplemented by Bigelow's (1928) thorough discussion of accepted species in *Periphylla*, *Nausithoe*, *Atolla*, and *Linuche*.

The section closes with a **diagnosis** of the taxon (p. 456), also including a brief account of features used for distinction of the order: "Coronatae are scyphomedusae of hemispherical, thimble-like, or discoid shape, the body of which is divided by an circular furrow into a peripheral part divided into marginal lappets and pedalia, provided with tentacles and sensory organs, and a central disc, which may also be more or less conical, giving the whole body a conical shape. The gastric cavity of the central disc is divided into a central stomach and a marginal area [Kranzmagen] by four interradial, node- and ridge-shaped adhesions (septal nodes or septal ridges) of the exumbrella and subumbrella." (in German: *"Die Coronatae sind Scyphomedusae von halbkugeliger, fingerhutförmiger oder scheibenförmiger Gestalt, deren Körper durch eine Ringfurche in einen peripheren, in Randlappen und Pedalien gegliederten, mit Tentakeln und Sinneskolben versehen Kranzteil und eine Zentralscheibe gegliedert ist, welch' letztere auch mehr oder weniger kegelförmig sein und dem ganzen Körper eine kegelförmige Gestalt verleihen kann. Der gastrale Hohlraum der Zentralscheibe ist durch 4 interradiale, knoten- und leistenförmige Verwachsungen (Septalknoten oder Septalleisten) der Ex- und Subumbrella in einen Zentralmagen und einen Kranzmagen gegliedert."*).

Concluding the chapter, Thiel explained his thoughts on the phylogeny of Coronatae. Based on the origin from a stauromedusoid ancestor, he described the hypothetical steps of differentiation from *Pericolpa-Periphylla* to *Periphyllopsis-Atolla*, to *Atorella* and *Palephyra-Nausithoe-Linuche*. Alternatively, he explicitly did not exclude a separate parallel development of *Paraphyllina-Nausithoë-Linuche*. Finally, Thiel (1936c) presented his system of Coronatae, consisting of 6 families, 10 recent genera (+1 fossil), and 18 species of the taxon, also giving his diagnoses of six families and eleven genera. The phylogenetic development of the different families was deduced from potential transformation processes of rhopalia and tentacles. The families were determined by the position of rhopalia, number of tentacles, the location of the gonads, and/or presence of sub-umbrellar vesicles. Genera were distinguished by the number of tentacles and/or location of the gonads. An identification key to the families, genera, and species completed the chapter.

[Thiel's (1936c) overall classification of the taxon is still accepted today. During the 20th century it further diversified due to the increase in observations and new species descriptions. The Coronatae are Scyphozoa, and hold 51 valid species in 6 families and 13 genera (WoRMS, 2023; Daly et al., 2007; Jarms and Morandini, 2019; Collins et al., 2021)].

2.4 Volumes 4–7: Semaeostomeae

The most comprehensive chapter is dedicated to the Semaeostomeae. Thiel (1938a, 1938b, 1959b, 1962, pp. 471–1308) introduced this taxon with a brief characterisation, describing the general body pattern and morphological features of scyphomedusae, including figures of well-known species from northern and southern seas. Possible sizes are mentioned for *Cyanea capillata* (Linnaeus, 1758) with bell diameters > 200 cm from boreal and cold seas, *Desmonema chierchiana* Vanhöffen, 1888[(*)] [= *D. gaudichaudi* (Lesson, 1829)] (≤75 cm), *Phacellophora camtschatica* Brandt, 1835 (50–60 cm) and *Aurelia aurita* (Linnaeus, 1758) (up to 40 cm). Colour patterns are described, including remarks on the chemical nature of some pigments (pp. 471–483).

2.4.1 Morphology

The subsequent section on **morphology** (pp. 483–587) provides detailed accounts of the various body structures and organ systems in the Semaeostomeae. The exumbrella with marginal structures is described, as well as the subumbrella, muscular system, gastric cavities, and reproductive organs, in addition to the nervous system and sense organs. Diverse figures reproduced from Agassiz (1860), Bigelow (1928), Claus (1877, 1883), Eimer (1878), Goody (1908), Hertwig and Hertwig (1879), Krasinska (1914), Lendenfeld (1882), Mayer (1910), Schäfer (1878), Schewiakow (1889), and Vanhöffen (1902) illustrate the text descriptions. For the reproductive systems, Thiel (1938a) included a brief survey of early research on these structures, especially highlighting male gonads in *Chrysaora*. Similar historical remarks introduce the descriptions of the nervous systems and sense organs. The accounts given for the diverse structures concentrate on those frequent and conspicuous species that were in the focus of research zoologists of the 19th and early 20th century. They were occasionally referenced for the respective family taxon: *A. aurita, Sthenonia albida* Eschscholtz, 1829, *Poralia rufescens* Vanhöffen, 1902, *Discomedusa lobata* Claus, 1877 (Ulmariidae); *Pelagia noctiluca* (Forsskål, 1775), *Chrysaora hysoscella* (Linnaeus, 1767), *Dactylometra quinquecirrha* Agassiz, 1862 [= *Chrysaora quinquecirrha* (Desor, 1848)], *Sanderia malayensis* Goette, 1886

(Pelagiidae); *C. capillata, Cyanea lamarckii* Péron and Lesueur, 1810, *C. annaskala* von Lendenfeld, 1882, *Desmonema gaudichaudi* (Lesson, 1829) (Cyaneidae); and *Phacellophora sicula* Haeckel, 1880$^{(*)}$ [=*Ph. camtschatica* Brandt, 1835] (Phacellophoridae).

2.4.2 Histology

The section on **histology** (pp. 587–677) again focuses on a small selection of frequently observed species sampled and studied by 19th century zoologists, namely *Cyanea annaskala* von Lendenfeld 1882 (e.g. in von Lendenfeld, 1882), *A. aurita* (M. Schultze, 1856), *P. noctiluca* (Krasinska, 1914), and *C. hysoscella* (Claus, 1877). Accounts of observations are given for the ectoderm, endoderm, the endodermal lamella, and the mesoglea, with the latter including experimental results about physical and chemical properties. In this respect, Thiel (1938a) carefully reviewed data on inorganic ions content (Na^+, K^+, Ca^{2+}, Mg^{2+}) of medusae in relation with ambient water salinity values and other organic and inorganic components.

The account of ectodermal cells, endodermal gland, and secretory cells is followed by descriptions of the nervous and sensory complex, muscles and reproductive tissues, referring to the authors mentioned above. An extended section on cnidocysts and cnidocytes (pp. 632–665) collects available results for the better-known genera *Pelagia, Chrysaora, Dactylometra, Cyanea,* and *Aurelia* as reviewed by Weill (1934), including the history of cnidocyst research. Detailed figures (e.g. from Chuin, 1930; Krasinska, 1914; Papenfuss, 1936; Toppe, 1910) illustrate structural details. Additional remarks cover the development of cnidocysts and anecdotal information about their distribution in semaeostome planulae, ephyrae, scyphistomae, and medusae. Concluding the chapter, endodermal migrating trophocytes [Nährzellen] are described from male gonads in *A. aurita* (with figures from Aders, 1902). The development of oocytes (e.g. Claus, 1877; von Lendenfeld, 1882) and sperm cells of *C. capillata* and *A. aurita* (after Retzius, 1909, 1910) are depicted and described.

2.4.3 Ontogeny

A historical account of ontogenetic research in the taxon opens the chapter on Semaeostomeae **ontogeny** (Volume 5, pp. 677–856). Thiel (1938b) describes five periods, beginning with the discovery of Semaeostomeae development during 1816–1842. While Baster (1762) had described and drawn an ephyra, as Müller (1775) did for a polyp stage of semaeostomes, they had not at all guessed about connections of the developmental stages.

Gäde (1816) presented the oocytes of *A. aurita* medusae. After Sars (1829) described "*Scyphistoma filicorne*", and "*Strobila octoradiata*" producing medusa-like free-swimming animals, von Siebold (1838, 1839) reared planulae into polyps without connecting them to Sars' *Scyphistoma* and *Strobila*. This work, and his observations on *C. capillata*, led Sars (1841) to publish his findings on the complete developmental cycle of the Semaeostomeae. Steenstrup (1842), based on similar observations found in other taxa, coined the term "Generationswechsel" (= alternation of generations). The following period (1842–1866) produced further results, especially variations from the general reproduction pattern, like the formation of new polyps by stolons or other various modes of polyp budding in scyphistomae. Also, the observation of hermaphroditism and viviparity in *Chrysaora* species by Derbès (1850) and the direct development of planktonic larvae into ephyrae ("hypogenesis") in *P. noctiluca* (Krohn, 1855) fell into this period. Subsequently (1866–1886), more authors carried out studies on details of the development of early life stages in the most common species. Zoologists like Claus (1877, 1883) and Haeckel (1880) published comprehensive works, which in some respects reflected differing views, e.g. on the nature of strobilation (terminal budding vs. constriction of the distal body wall), and the phyletic significance of hypogenesis. Those studies and ongoing discussions (1886–1908) lead to further robust controversy, especially between Claus (1883, 1891) and Götte (1885, 1887), mainly about three subjects: (1) the mode of gastrulation, (2) the origin of septal muscles in the gastric septa [Magenwülsten] and (3) the ecto- or endodermal derivation of the manubrium in scyphistomae. While the former two problems were solved, the latter led Götte to conclude a close relation of Scyphozoa with Anthozoa. While his student Hyde (1894) assisted his arguments, other researchers rebutted Götte's results, not least as being artefacts of presumed transport and maintenance complications (Hein, 1900, 1902; Friedeman, 1902). A number of other, also rather popular studies (e.g. Haacke, 1896; von Lendenfeld, 1896), and a review of recent research advances by Maas (1907) closed this period. The following phase (1908–1930) of research, mainly on asexual propagation, focused on the budding phenomena of semaeostomes, especially in the scyphistoma stage. The entire range of these studies on development, podocyst formation, and strobilation in *Chrysaora* scyphistomae was summarised by Chuin (1930).

The following extended sections focus on (1) the development from sperm and eggs to the planulae (maturation of eggs, fertilisation and cleavage, the formation of germinal sheets and the planulae), (2) the hypogenetic development in *P. noctiluca* (formation of planulae into ephyrae) and (3) the metagenesis in Semaeostomeae. The development of eggs in *P. noctiluca* is carefully illustrated using the figures of Schaxel (1910a,b, 1912), the gastrulation and formation of germinal sheets are described and drawn according to Hein (1900). The hypogenetic transition of planulae into ephyrae in this species is drawn according to Götte (1893) and Delap (1905b).

For all other Semaeostomeae known at that time the alternation of asexual and sexual generations (metagenesis) is described. For the **planulae**, this includes histological studies on the process of substrate attachment (from Chuin, 1930) and the transformation into scyphistomae (e.g. after Hein, 1900), the development of tentacles, septa, longitudinal muscles, and the peristome. Occasional observations of planula budding (from Busch, 1851 in Krumbach, 1925) and encystation (drawn after Hyde, 1894) are described. The general morphology of **scyphistomae** is described according to Friedemann (1902), also explaining the mode of subsequent tentacle formation. Further morphological details are reviewed for *A. aurita* (from Hein, 1900; Friedemann, 1902) and *C. hysoscella* (Chuin, 1930): the ectoderm, longitudinal muscles, peristome, endoderm, and the development of the mesoglea. For the asexual phase of the life cycle, Thiel (1938b) reviews the history of discovery and various modes of propagation (Fig. 3): budding in scyphistomae and stolons with the possible liberation of buds (after Hérouard, e.g. 1909; Perez, 1922; Hadzi, 1910, 1912; Chuin, 1930; Thiel 1938c), the formation of podocysts, pseudoplanulae (Hérouard, 1909), and the formation of ephyrae by strobilation (Percival, 1923).

For the early development of the sexual generation in the metagenetic life cycle, observations on the **strobilation**[3] process are available from Sars (1829 ff.) and Dalyell (1836). Van Beneden (1860) discovered the general nature of the process based on detailed description and figures by Agassiz (1862). Initial modifications of the scyphistoma indicate the beginning of ephyra formation. Rhopalia, lappet pouches, and septal ostia form, before the ex- and internal differentiation precedes the release of the ephyra from the scyphistoma: the oral tube [Mundrohr], gastric filaments, septal and peristome funnels develop (figures given from Percival, 1923; Chuin, 1930). Thiel (1938b) gives a general characterisation of the eight-armed **ephyra** morphology, before these stages of various semaeostome species are thoroughly described and diagrammed (e.g. from Delap and Delap, 1906): *P. noctiluca*, *C. hysoscella* (figures also from Claus, 1883), *C. capillata*, *A. aurita*, *D. quinquecirrha*[(*)] (figures from Mayer, 1910), *D. gaudichaudi*

[3] M.E. Thiel used the term 'Strobil**is**ation', while 'strobilation' has become generally used.

and *Ulmaropsis drygalskii* Vanhöffen, 1908[(*)] [=*Diplulmaris antarctica* Maas, 1908], figures from Vanhöffen (1908). Direct comparison of the stages, however, may be obscured by different ages of the diagrammed specimen.

With growing numbers of gastric filaments and the development of marginal tentacles, along with modifications of the bell rim [Schirmrand] and the gastrovascular system, the ephyra transforms into the **metephyra** and subsequently into the mature medusa. An ongoing metamorphosis is therefore part of the nature of this series of developmental stages. Thiel reviews available descriptions and figures of metephyrae of the following species: *P. noctiluca, C. hysoscella, A. aurita, D. lobata* (after Claus, 1877), *Pelagia panopyra* Péron and Lesueur, 1810 (after Uchida, 1934), *C. annaskala* (after von Lendenfeld, 1882), and *C. capillata* (after Delap, 1905a).

Finally, looking at the **medusa** stage, Thiel (1938b) described metamorphosis in Semaeostomeae: modifications of the bell rim with the formation of marginal lappets by intercalation (Aurelidae*) or fission (Pelagidae*, Cyaneidae), marginal tentacles, and the differentiation of gastrovascular canal systems with the multiplication of gastric filaments. Further details are reviewed for the tentacle development, the gastric lamella, the formation of mouth arms, muscles, and sensory organs based on descriptions from several researchers (e.g. von Lendenfeld, 1882). A survey of gonad differentiation in *P. noctiluca* (from Hamann, 1883), *C. hysoscella, A. aurita,* and *D. lobata* (after Claus, 1877) closes the chapter.

2.4.4 Physiology

After releasing his earlier chapters, a 20-year break ensued before Thiel (1959b) was able to publish the chapter on the physiology of the Semaeostomeae (Volume 6, pp. 856–1072). Earliest attention to jellyfish, e.g. of free-floating sea-nettles by Aristoteles (384–322 BC), focused on the pulsating movements of the medusae, the burning stings of their cnidocysts, and the occasional mass occurrence of these animals in often kilometre-long swarms, the latter also interfering with the nets set out by fishermen. With the increased observations, the metagenetic mode of reproduction – especially the processes of strobilation and budding – attracted scientific and public curiosity.

Thiel reviewed data on the diet of scyphistomae (chaetognaths, ctenophores, medusae), ephyrae ("small fish"), and metephyrae ("usually fish", Thiel 1959b). Preferred prey items change with growth and developmental stage. The capture and ingestion of food is described with respect to the ciliary surface cover in juveniles, as well as tentacles, mouth arm folds, and cnidocysts in adults for *A. aurita*, Adriatic *Drymonema dalmatinum* Haeckel

1880 (from Krumbach, 1924) and *C. annaskala* (after von Lendenfeld, 1882). The circulation of food particles in the gastrovascular system, including transport speed measurements investigated by colour-particle injections, revealed cilia-driven movements in two different modes with or without ring canals described for *A. aurita* and *C. capillata* (from Widmark, 1911, 1913). Extensive experiments to elucidate the site (tissues vs. gastrovascular cavity) and mode of digestion are described, including the amoebocytes in scyphistomae of *A. aurita, C. hysoscella*, and *C. capillata* (after Wetochin, 1930). Extra- and intracellular processing of nutrients in vacuoles is reviewed for *C. hysoscella* scyphistomae after Chuin (1929).

All available data on Semaeostomeae **excretion** refer to *A. aurita*. Early observers had described eight anal pores at the ends of the radial canals that Agassiz (1862) had already denied, before Widmark (1913) described the canal pathway from the mouth opening through the adradial into the ring canals, and returning via interradial and radial canals through the egestion pores at the end of the four mouth arms.

For obvious reasons, the pulsing bells of medusae had long been related to the inhaling and exhaling movement of **respiration** in humans by early researchers, but no mention was made about gas exchange in jellyfish bodies. The diverse interpretations of water movement and gas exchange sparked variable controversies between researchers for more than 100 years. All attempts to relate respiratory processes with pulsation or morphological structures remained inconclusive. Only Thill (1937) linked observations of Baltic *A. aurita* medusa activities with animal sizes, temperature, and salinity of the experimental tank to oxygen consumption. His results showed that oxygen consumption was negatively related to the animals' sizes. Consumption increased with temperature up to 25 °C, before it decreased in warmer waters, and increased with salinity.

The history of knowledge about **sense organs** and the **nervous system** in general began with the description of marginal rhopalia. These were speculated to represent anal openings due to some extruding particulate matter observed by Müller (1779–1784), but they remained "mysterious bodies" (Gäde, 1816). Ehrenberg (1835) first referred to them as sense organs, based on the brown-red spot pigment in *A. aurita*, which he interpreted as "eyes". With more detailed observations available, interpretations among zoologists included sensory mechanisms for light or acoustics. It took more than 50 years of various attempts and reasoning, before Delage (1887), based on experimental evidence in cephalopods and crustaceans, correctly concluded rhopalia to be the static sense that in medusae also controls the animal's equilibrium. The following period falsified the visual or acoustic perception for many species. The role of rhopalia in orientation and locomotion was complemented by von Uexküll (1901) and others who proposed that they also produce tonus stimuli inducing muscle contractions throughout the bell, resulting in the pulsating movement. The complex interactions of the rhopalia with the nervous system, as well as the chemical, tactile [Tastplatte], and the light sensing reception, became successively established by further experimental work and analyses of morphological structures.

Subsequently, Thiel (pp. 903 ff.) describes the **physiology of the nervous system** in Semaeostomeae. He summarises the functioning for **the senses** of taste, of light perception – referring to *A. aurita* investigated by Horstmann (1934), of the sensory pads, hypothesised to carry the sensory epithelia related to the rhopalia, and of the rhopalia detecting the position of the medusa in the water column. While at first rhopalia were suspected to be acoustic sensory organs, early studies by Bethe (1910) and Lehmann (1923) discussed their role for the tonus control of umbrella contractions. In a lengthy discussion about the "normal" physiological and physical balance position of swimming medusae (and ephyrae) in the water; it was then concluded to be mainly directed by gravity. Further observations revealed a dominant inclination of the radial medusa axis towards the vertical gravity vector and the role of rhopalia possibly controlling it. Observations on the mode of movement are discussed below.

The basic structure of the **nervous system** in the Semaeostomeae is presented by Thiel (pp. 922–948), primarily based on the works of Horridge (1953, and following studies) conducted on the ephyra of *A. aurita*. These studies examined the interactions of a large and small nerve fibre network and their connections with the rhopalia. In addition, the structure of different nerve cell types is shown. Furthermore, experimental approaches investigating the functions of the nerves by means of incisions in the umbrella are described (e.g. Eimer, 1876, 1878; Romanes, 1876). These experiments demonstrated the interaction of the nerve plexus on the sensory bulbs with the rhythmic pulsating swimming movements. It was further shown that the inward folding of the umbrella – performed during feeding and during the sinking reaction – functions independently of the contraction movement, since contraction is stimulated through the large-fibre network, and inward folding of the umbrella is innervated by the small-fibre network. The studies of Bozler (1926, and following studies) and Horridge (1956a, 1956b) demonstrate the conduction of stimuli from the tentacles to the mouthparts in *P. noctiluca* and *A. aurita*.

Including the studies of von Uexküll (1901) and Bethe (1943), Thiel (1959b) explained the expansion of the rhythmic contraction of the ring muscles: the stimulus originating from the nerve centres at the base of the rhopalia and the contraction occurs as soon as a certain stimulus threshold is reached. The nerve centres work as uptake and switching points for stimuli and thus are similar in function to ganglia. However, the medusa ganglia are not as highly developed as the ganglia of other animals. The contraction is initiated from one rhopalium and then continues to both sides. Interaction of the rhopalia is further demonstrated by the experiments of Horstmann (1934), who studied the effects of rhopalia being removed from the umbrella of *C. capillata*. The function of the rhopalia as sensory organs that perceive the 3-dimensional position of the medusa is explained with reference to the studies of Fränkel (1925), Bozler (1926, 1927), Bauer (1927), Horstmann (1934), and Horridge (1956): The position of a rhopalium generates pressure on certain areas of the sensory pad and the stimulus is transmitted to the nerve centre connecting the umbrella muscles, which induces a change in the position of the medusa.

The section "Physiology of **movement**" explains the recoil principle of the swimming movement. This is followed by descriptions of the rhythm of the movement and a discussion of its possible discretionary control. Subsequently, the significance of the rhopalia in locomotion as well as the frequency of the pulsating motion of the umbrella are described, summarising the works of various authors (e.g. Bozler, 1926; Eimer, 1878; Fränkel, 1924; Horridge, 1956, Horstmann, 1934, von Skramlik, 1945). The results of experimental studies on the effects of various factors on the movement of medusae are presented. These studies include the effects of temperature (e.g. Schaefer, 1921; Bozler, 1926; von Skramlik 1945), light (e.g. Horstman 1934), oxygen, hydrogen ions, and salinity (e.g. Thill, 1937), the effects of salts and other chemical substances (e.g. Hykes, 1928; Heyman and Moore, 1923), the effects of the toxins strychnine (Romanes, 1876), alcohol (Bethe, 1903), adrenaline, nicotine, and curare (Schaefer, 1921), and the effects of electrical and mechanical stimuli (Horstmann, 1934, von Uexküll 1904, Bethe 1909).

The subsequent section dedicated to the "Physiology of **shape changes**" (pp. 976–1061) is divided into four parts about the **reproduction**, the physiology of **growth**, as well as morphological **variations and abnormalities** and the ability for **regeneration** of the

Semaeostomeae. Until the year 1934, the physiology of reproduction and shape change was described as part of the ontogenetic development of Semaeostomeae (Thiel, 1938b, p. 677–856). Because WWII had delayed further publications of Thiel, and due to the advancements made in reproduction research of Semaeostomeae during that time, a separate chapter describing and discussing the 6th period of the history of the knowledge of reproduction was needed. This period was defined by many extensive reviews on the development of Semaeostomeae (e.g. Korschelt and Heider, 1936; Wilson, 1947; Kramp, 1950), as well as on the effects of environmental factors on the development (and thus seasonal appearance) of different developmental stages (e.g. Moore, 1937; Künne, 1952). Littleford (1939) published a detailed description of the developmental stages of *Dactylometra quinquercirrha*[(*)] and their association with seasonality, bringing a non-European species into focus. Work by Verweys (1942) extensively investigated the geographical and periodic appearance of the various developmental stages of European species: *C. hysoscella*, *C. lamarckii*, *C. capillata*, *A. aurita*, and *Rhizostoma pulmo* (Macri, 1778), and how they are influenced by environmental parameters. For the first time, a coherent picture connecting the appearance of jellyfish with environmental conditions in western European waters was generated. Studies on various species concerning budding, cell cleavage, rearing in the laboratory, cannibalism, and the nervous system were published, as well as an extensive study by Kramp (1944), discussing the metagenesis of Semaeostomeae. Based on the existing literature and the new insights on the behaviour of scyphistomae up to 1958, Thiel published a paper concluding that the development of Semaeostomeae could not be viewed as metagenesis (Thiel 1959a).

Thiel (1959b) briefly described the development from the planula larvae to a sessile stage and the release of ephyrae via a strobila that develops into a medusa for most Semaeostomeae. The hypogenetic development of the holoplanktonic *P. noctiluca* and the metagenetic development of other Semaeostomeae had previously been described in detail, including various forms of asexual reproduction (Thiel 1938b, pp. 716, 724 ff). He stated that the various forms of asexual propagation had been observed in *C. hysoscella*, *C. capillata*, *C. lamarckii*, *A. aurita and D. quinquecirrha*[(*)], which are native to the North American Atlantic and/or German waters. Up to that point, not every developmental stage had been observed in every mentioned species, but as reproduction depends on environmental conditions, it remained possible that these will trigger the respective asexual reproductive modes. Since reports had largely been based on observations made in the laboratory, it was not clear if all forms could also be found in the wild. Yet, it allowed the certainty that the polyps were capable of various modes of reproduction and likely initiated them in the wild under certain environmental conditions.

Both planula and scyphistoma stages can multiply via budding (Thiel, 1938b, p. 729, 757, Fig. 3). They can also produce dormant stages known

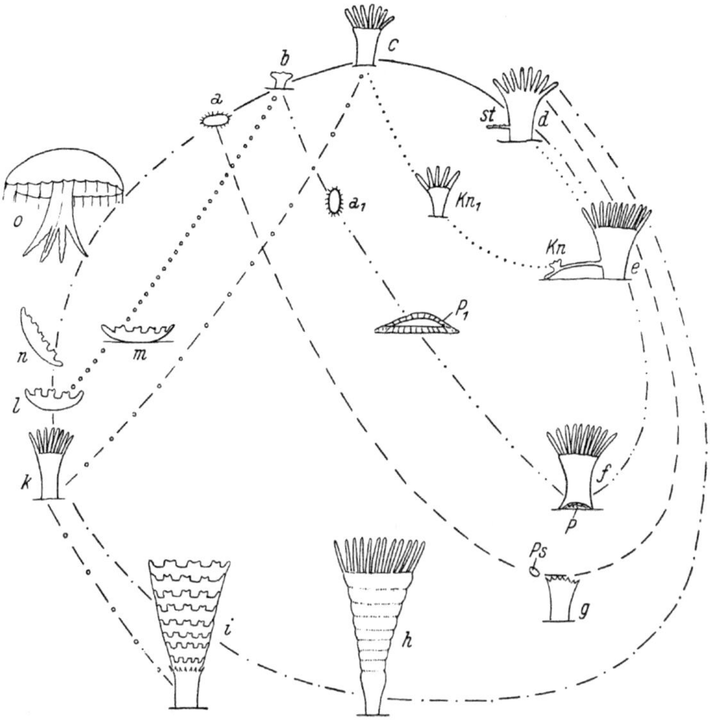

Fig. 3 Scheme of observed reproductive cycles in Semaeostomeae (Thiel, 1938b, Fig. 369, p. 754, modified after Hérouard, 1909, p. 322, legend slightly amended, read clockwise): *a, a1* planula larvae, *b, c* young scyphistoma buds, *d, e* scyphistoma with stolon *(St)* and bud *(Kn)*, *f* scyphistoma with podocyst (*p*), *g* scyphistoma in depression with reduced tentacles and pseudoplanula (*ps*), *h* scyphistoma with initial strobilation, *i* strobila, *k* remaining polyp, *l* loose ephyra, *m* attaching ephyra, *n* free swimming ephyra, *o* medusa. *P1* free podocyst, *Kn* young scyphistoma by budding [german term: Knospung].

···············	Asexual reproduction by budding on stolons
–··–··–··–··	Asexual reproduction by podocyst development
– – – – – –	Asexual reproduction by pseudoplanula development
oooooooooooo	Asexual reproduction by scyphistoma development from strobilation
o–o–o–o–o	Recovery of scyphistoma after strobilation
–·–·–·–·–·–·	Sexual reproduction by strobilation and ephyra development

as podocysts by encysting (Thiel, 1938b, p. 730, 765). Additionally, released tentacles of degenerating scyphistomae (living under environmentally challenging conditions) can turn into pseudo-planulae, while the degenerated polyp may recover and develop new tentacles (Thiel, 1938b, p. 771). In addition, observations on abnormal strobilae (Percival, 1923) revealed that released ephyra discs can develop into polyps which was considered to be a form of asexual reproduction by Thiel. He points to the important role of this fact for the understanding of the evolutionary history of the development of ephyrae and medusae.

Observations of *P. noctiluca* indicated that this species reproduces in autumn and reaches sexual maturity with a diameter size of 80–90 mm (Hunt, 1952; Kramp, 1924; Legendre, 1945; Stiasny, 1935). Thiel (1959b) continued to discuss findings of sexual maturity in four other species: highlighting the hermaphroditism in *C. hysoscella*, the smaller size of *C. lamarckii* when reaching sexual maturity as compared to *C. capillata*, and the varying sizes of mature *A. aurita*, possibly related to different temperature conditions (Claus, 1877; Derbès, 1850; Künne, 1952; Östergren, 1909; Thiel, 1958; Wright, 1861). Most likely, however, they reach maturity with a diameter size of 60–65 mm. Thiel (1959b) lists findings from the literature of various species – where they can be found and at what time of the year they reach maturity.

Concerning the release of reproductive cells, Thiel (1959b) briefly described the tissue in which male reproductive cells were found and from which they are released. He also describes the tissue holding the eggs of the Semaeostomeae. He compared contradictory observations in *Cyanea,* i.e. eggs found in folds of the oral arms as opposed to eggs being released into the water under laboratory conditions (Delap, 1905a). He considered the possibility of egg release into the water in the wild, based on the vast number of eggs produced while few larvae are found in oral arm folds. In contrast, *A. aurita* is described to likely brood the fertilised in egg breeding sacs in the oral arms, however, tissues have been found suggesting that eggs can both reach the gastrovascular system and be released (Claus, 1883; Widmark, 1911, 1913).

Parental care in Semaeostomeae is not reported in *P. noctiluca*, as fertilisation and development take place entirely in the open water. While fertilisation in *D. quinquecirrha*[(*)] takes place in the gastric cavity, thus ensuring reproduction to a certain extent, there is no further care of the offspring (Littleford, 1939). *Cyanea* sp. and *A. aurita* brood fertilised eggs and developing planula larvae in crevices of their oral arms and in designated breeding sacs, respectively.

The brooding pockets of *A. aurita* were first described by Ehrenberg (1835), and increase in size when they are filled with embryos, yet begin to disintegrate when the developed planula larvae are released. Nonetheless, Thiel (1959b) reports the lack of data concerning parental care of the offspring during this developmental stage of *Aurelia* sp. In *C. hysoscella*, however, Claus (1877, 1883) and Teissier (1925) had published these observations. They report fertilisation and embryo development into planula larvae to take place in the ovaries of the mature female, thereby supplying the offspring with nutrients during this time.

Thiel (1959b) summarised the findings of several authors on early development, from cleavage, through gastrulation, to planula larvae in various European and North American species (Berrill, 1949a; Hollowday, 1951; Krohn, 1855; Lambert, 1936; Teissier, 1925; among others). He explicitly stressed the difference in egg sizes and the correlation with the subsequent developmental path. For example, while large eggs follow a path that leads to direct development of ephyrae in *P. noctiluca*, planula larvae from small eggs go through planula budding [Planulaknospung] before developing into scyphistomae. The planula encystment in *C. capillata* supports this hypothesis (McMurrich, 1891). So, based on various observations, the developmental path depends on the egg size and not on environmental conditions, and may explain the developmental diversity. On the contrary, the time in which planulae are free-swimming differs between species and appears to depend on environmental cues (von Siebold, 1839). Since planula budding and planula encystment do not regularly occur, they seem to represent intermediate developmental stages. Conventional development then follows, theretofore acknowledged, from free-swimming planula, to attachment, to scyphistoma. The settlement of planula larvae has extensively been described in the section 'Ontogeny' (Thiel, 1938b), including the anatomical rearrangement from planula to scyphistoma. Thiel (1959b) nonetheless stressed the importance of Galigher (1925) observations in the field as well as Thill (1937) work on the effect of salinity on planula settlement.

Despite scyphistomae of various Semeaostomeae hardly being distinguishable because of intraspecific variations in size and colour, in combination with very similar interspecific morphology, Thiel (1959b) described species-specific behaviour towards light, as well as characteristic budding behaviour in *C. capillata*, *C. lamarckii*, *C. hysoscella*, and *A. aurita*. The time of planula transformation into scyphistomae varies both within and between species, and is driven by environmental conditions (Hollowday, 1951; Delap,

1905a, among others). Once tentacle cnidocytes are developed, the animals are able to catch and paralyse larger prey. Until this developmental state, Thiel (1959b) noted that it is not certain whether or not scyphistomae can feed. He described feeding of animals held in the laboratory and stated that they cannot digest plants and are selective in their choice of food. According to Lambert (1937), Gilchrist (1937), and Berrill (1949a) the type and amount of food affects budding, strobilation, and ephyra production. Thiel (1959b) revisited the five different types of budding in *Chrysaora*, *Cyanea*, and *Aurelia* described by Berrill (1949a), as well as a sixth type of multiple circular budding, that is only found in *Nausithoe punctata* Kölliker, 1853 (Komai, 1935). He followed up by giving a detailed description of Gilchrist (1937) 'lateral budding', 'lateral budding by means of stolon', 'stolonic budding' and 'hydra-type budding'. Based on these budding types and his other experiments, Gilchrist tried to deduce developmental principles concerning plasticity of scyphistomae paired with a certain stability of the body morphology and the role of ectoderm and endoderm during budding. He also hypothesised on the mechanisms underlying pedal stolon production.

Another type of asexual reproduction, or autotomic regeneration termed "Depression" by Thiel (1959b), is also considered and has been observed in large *C. hysoscella* scyphistoma after periods of excessive food intake (i.e. "planuloids originating from tentacle tips/tentacle pieces", Adler and Jarms (2009), compare Schiariti et al. (2024), see chapter 4 in this volume). This mode is characterised by a degeneration of the animal and the release of tentacles. These free-drifting tentacles, named pseudo-planula, settle and develop into scyphistomae within two weeks. Often, the tentacles of the 'mother polyp' regenerate. After periods of starvation, however, degeneration and shape changes were observed in all developmental stages of several species. The degeneration included body volume shrinkage, dedifferentiation of gonad tissue cells, and a decrease in oocytes and sperm cells, among other changes.

Temperature was considered the most influential parameter triggering budding and strobilation. While Semaeostomeae scyphistomae bud when temperatures rise above a certain threshold, they strobilate and release ephyrae when temperatures fall below this threshold. There was, however, also a report of scyphistomae that were held at a constant temperature in an aquarium and began strobilating. Strobilation and ephyra release are considered as a single physiological process rather than separate events. The number of ephyrae produced varies between species, and also ranges within species, possibly might be affected by the amount and quality of available food, as observed in *A. aurita*. Verwey (1942) compiled a list of the appearance of ephyrae in western European waters for *C. lamarckii*, *C. capillata*, *A. aurita*, and *C. hysoscella*, thereby adding to previously published literature, including documentation of the appearance and size of *D. chierchiana*(*) and *U. drygalskii*(*) throughout the year. The initial assumption that the entire strobila divides into ephyrae was falsified in 1848 by Dalyell and other authors, who found that a part of the polyp remained after strobilation and regenerated into a complete scyphistoma.

In order to investigate the **physiology of growth,** the process was defined to begin after metamorphosis has ended and all organs are developed. Thiel (1958) was the first to investigate the actual growth rates of *A. aurita* in more detail. Previous publications thus far had merely been on the size ranges of species during a given month, which does not give any information on the growth rate of individuals. Yet to use these findings, Thiel (1959b) included a table with the size of various ephyrae immediately after release, as well as the size of the fully-grown medusae. He also described findings of Delap (1905a, 1905b), who tracked the growth rates of *C. lamarckii*, *C. hysoscella*, and *A. aurita* in laboratory aquaria. Both the amount and quality of food, as well as the temperature, influenced the growth rates in these species, with growth and temperature being positively correlated. Thiel then compared Delap's results to Browne's (1901) and his own (1958) observations, and found Delap's deducted growth rates to be almost twice as high as Browne's. Since Browne's and Thiel's observations took place in the field, he regarded these growth rates to be much more representative of free-living medusae. However, they only represent the mean growth rate of a population, and not individual growth rates, as the authors did not track single individuals during their observations. Thiel also noted that because the diameter of the umbrella was used to determine the growth rates, the increase in body volume was not considered and was much larger.

The growth rate gives some notion on the correlation between size and age of the medusae, yet Thiel (1959b) mentioned that there is no data on the specific age that can be reached by the various species. From single sightings of unusually large individuals of *A. aurita*, *Phacellophora ornata* Haeckel (species uncertain), or *Cyanea* sp., Thiel concluded that Semaeostomeae medusae can grow very old if they are carried out to the open sea where they are not subjected to the risks of being washed ashore or touching the sea floor. This contradicts the observations in the western European waters theretofore, where medusae life span was assumed to be ½–¾ of a year.

The German word for metagenesis ['Generationenwechsel'] which literally means an alternation of asexually and sexually reproducing generations, led to the false conception of a polyp 'generation' and a medusa 'generation' appearing alternately, because actually the scyphistoma stage does not disappear when medusae begin to be present. The scyphistomae are a long-lived stage in the life cycle and persist after the release of ephyrae, and they continue to asexually reproduce. Thus, Thiel (1959b) explained that in Semaeostomeae 'Generationenwechsel' is rather an alternation of two types of reproduction, i.e. the asexual and sexual modes, hence metagenesis. The scyphistomae were found to practise more than one mode of asexual reproduction, live for many years, and are able to release ephyrae in several consecutive years, while medusae merely represent a temporary stage of reproduction. Thiel (1959b) argued that the medusae should not be considered as fully developed adult individuals, but rather as sexually reproducing stages within the life cycle. He also argued that scyphistomae and medusae do not alternate as generations, but polyps – based on their longer lifetime – determine the actual length of a generation, which may be several years.

Variations and abnormalities in the body shapes and organs of Semaeostomeae were reported by a large number of authors very early on (e.g. Müller, 1788; Agassiz, 1862; Haeckel, 1880, 1881; Vanhöffen, 1888; Browne, 1901; Littleford and Truitt, 1937; Berrill, 1949b). Because of the large number of reports on anatomical anomalies, Thiel (1959b) divided them into groups, first discussing variations and anomalies in medusae, followed by those found in ephyrae, and lastly in scyphistomae.

Most publications report on variations in *A. aurita*. In this species, an irregular number of gonads can often be found, both decreased and increased. They usually correlate with the number of oral arms, yet Ballowitz (1899) also observed some irregularities with mismatching numbers of gonads, oral arms, and radial canals, as well as anomalies with up to three gonads sharing a gastric pouch.

Information on the frequency of variations in the anatomy of *A. aurita* vary greatly and range from 0.1% to 50%. This may be due to anomalies reported from animals in waters with low salinities, where the stressful environmental conditions may contribute to variability. Also, the type of variation plays a role in the frequency. Variations in the number of gonads affect the whole organism and reproductive success much more than irregular numbers of rhopalia, which is likely why abnormalities in gonad numbers were found with a lower frequency than those of the rhopalia. Browne (1901) studied the correlation between the number of rhopalia and the variation of other body parts. He also examined how these variations in ephyrae affect variations in the medusa stage. The results showed that rhopalia variations in ephyrae prevail into the medusa stage and that they correlate with the number of lappets and radial canals. Anomalies with increased numbers of rhopalia were observed more often than decreased numbers. Thiel (1959b) included a table with Browne (1901) recorded anomalies that shows a mean of 21.3%, yet these anomalies do not seem to be a selective disadvantage, as the percentage does not change with size of the medusae. The number of gonads also correlates with the number of oral arms.

Detailed research of anomalies as was done for *A. aurita* unfortunately did not exist for other Semaeostomeae species. Thiel (1959b), therefore, described the occasionally mentioned sightings of anomalies concerning e.g. length of the oral tube [Mundrohr], patterning, colour, number of rhopalia, number of tentacles, number of marginal lappets, and the number of gonads in *P. noctiluca, Chrysaora melanaster* Brandt, 1835, *C. hysoscella, C. capillata, D. quinquecirrha*[(*)], *S. malayensis, Poralia* sp., *Diplulmaris malayensis* Stiasny 1935, and *Ph. camtschatica* medusae.

Anomalies found in ephyrae were the same as for medusae and, as mentioned before, persist into the medusa stage. Scyphistomae also often show anomalies and are most likely based already on anomalies in the egg. Thus, any variation from the basic pattern probably runs through all life cycle stages.

Regeneration has been observed in the medusa, ephyra, and scyphistoma stages, yet the latter shows a much higher potential for it. Medusae only manage to close their wounds, contract back to a near bell shape and continue pulsation as long as the rhopalia are not missing. Similarly, ephyrae regenerate the lappets, but not rhopalia. In contrast, every tissue part of the scyphistoma has the potential to regenerate, but each to a different extent.

The first mention of **luminescence** in Semaeostomeae (Thiel, 1959b, p. 1061 ff.) was conveyed from Pliny the Elder (AD 23/24–79), referring to a *Pulmo marinus*[(*)]. Subsequent descriptions were reviewed by Ehrenberg (1834). A number of reports on other scyphozoan species emitting light could not be verified reliably. Early investigations about the origin of this light started with Spallanzani (1793, 1796), who first identified the mucus as the luminescent carrier material and tested some media for their influence on the emitted light. With rising interest in animal luminescence in the early 20th century, Dahlgren (1916, p. 247) described the mucus release from the exumbrella epithelium cells containing the substance luciferin. In contact with oxygen in the water, the enzyme luciferase induces the light emission, which can persist for several seconds to dozens of minutes depending on conditions such as ion concentrations, temperature, and electric polarisation. Details of the species-specific chemical mechanism were only elucidated during the 1950s.

2.4.5 Ecology, research history, phylogeny, and classification

In the section on the **ecology** of Semaeostomeae, Thiel (1962, Volume 7, pp. 1073–1121) reviewed the distributional data available at that time. Most meroplanktonic species are typically confined to near surface waters in coastal regions due to the necessity of substrates for their reproductive stages. Few exceptions referred to the holoplanktonic *P. noctiluca* and the viviparous *Stygiomedusa fabulosa* Russel 1959[(*)] from the deep Biscaya Sea. Information available on the role of salinity, temperature, and light on distribution were limited to a few species, namely *A. aurita*'s capacity and dependency on osmoregulation in the North and Baltic Seas, and *C. hysoscella* and *D. quinquecirrha*[(*)] from brackish water habitats. Occasional observations of daily vertical migrations were attributed to the light cycle or migration of prey organisms, but were also subject to other local effects such as shallow nearshore concentrations of younger individuals in spring avoiding direct sunlight, or coastal shading.

Factors inducing jellyfish swarms and "**blooms**" had not been investigated at that time. Incidental reports hinted at external effects such as local tide patterns, autumn storms driving surface currents, concentrations along frontal systems of ocean current convergences, and the possible aggregations for synchronous spawning of eggs and sperm. Active "gregarious habits" (Agassiz, 1862) were not considered. Further, effects of food availability in upwelling areas and convergent sinking of water were known to concentrate organisms, as was shown for hydrozoan medusae and Chaetognatha (Thiel, 1936d, 1938d) collected by the Deutsche Atlantische Expedition (German Atlantic Expedition) on the RV "METEOR" I during 1925–1927. [Early ideas linking biological plankton distributions with physical oceanography observations (e.g. salinity and temperature patterns, nutrient chemistry, currents and meteorological conditions, etc.) to establish biological oceanography as an interdisciplinary research field had only originated during that expedition (Reinicke and Thiel, 2019)].

Distribution patterns (p. 1093 ff.) observed from the sea surface on larger scales exhibit different species dominating in warmer (tropic-subtropic) and colder (boreal-arctic) conditions. Thiel describes an example from Uchida (1954) in Japanese waters over a distance of 8000 km between Sachalin and Formosa (Taiwan). There was a northern group of *Chrysaora helvola* Brandt, 1835, *C. capillata*, and *Aurelia limbata* Brandt, 1835, and the southern species *Dactylometra pacifica*[(*)] (Goette, 1886), *Cyanea nozakii* Kishinouye, 1891, and *A. aurita*, which Uchida related to predominant temperature patterns. Similar situations were described for *A. limbata* and *A. aurita* in the North Atlantic (Kramp, 1943), as well as *C. lamarckii* and *C. capillata* (Kramp, 1955). Other information on horizontal distributional patterns was still patchy, reflecting observer patterns rather than a consistent picture of the dynamic situation. Data on vertical distributions are restricted to a few common species, mostly from shallower depths of 0–150 m, as well as some individual reports. Nevertheless, differing depth preferences had been described for different age stages (in *C. capillata*, Agassiz, 1862), and were known to be affected by wind conditions as well as diurnal and seasonal changes. In general, available observations were related to only a small number of species and limited data. The picture was blurred further due to provisional species definitions and synonymies under review. The overall knowledge about biogeographic, distributional, and taxonomic data has considerably increased since 1962.

Reports of **seasonal changes** relating to the presence of developmental stages (planulae, scyphistomae, strobila, ephyrae, and adult medusae) focus on a few common species. These had been studied in European and North American coastal areas as well as few southern hemisphere seas, and reflect variation in latitude, annual temperature regimes, and reproductive cycle dynamics (*C. hysoscella* and *D. quinquecirrha*[(*)], *C. capillata* and *D. chierchiana*[(*)], *A. aurita* and *D. antarctica*).

Early observations of **species associations** with semaeostome medusae (Thiel, 1962, p. 1118 ff.) date back to Malm (1852) and Sars (1879) regarding *C. capillata* and juvenile *Caranx trachurus* [= *Trachurus trachurus* (Linnaeus, 1758)] feeding on the medusa's eggs. Later, Malm also observed young *Gadus merlangus* [= *Merlangius merlangus* (Linnaeus, 1758)] with the medusae, and Sars found young cod *Gadus morhua* Linnaeus, 1758 inside *A. aurita.* Heincke (1905) described the general co-occurrence of young gadids and jellyfish, which are otherwise absent when the medusae are not present. Similar observations were reported from Japanese waters (Kishinouye, 1910). Experiments revealed a parasitic relationship of the fish feeding on the medusae's tissues and ovaries. The regular association of *Hyperia galba* (Montagu, 1813) (Amphipoda) with medusae receives no further mention by Thiel.

The section on the **classification and phylogeny** of Semaeostomeae gives a detailed account of the research history of the taxon (Thiel, 1962, pp. 1122–1210). The first mention of a Semaeostomeae dates back to the XVI century, before significant knowledge expansion during the second half of the XVIII century. Thiel introduced the first descriptions of five species by Linnaeus (1747), four of which represent genera still valid today, followed by a series of descriptions with two figures from other authors. Péron and Lesueur (1809) were the first to present an early system, establishing the genera *Aurelia, Chrysaora, Cyanea,* and *Pelagia,* all holding several "species," which are mostly synonyms or currently revised into other taxa. While highly recognised at the time, the work exhibited only minor progress, as the diagnoses were often vague, mostly referring to external features, and missing true understanding of the inner organisation of the animals. Further species diagnoses and detailed descriptions followed during the 1820s. Many new anatomical features (e.g. reproductive organs, gastrovascular cavities, and swimming modes) were observed and described during the voyages by Eschscholtz (1829). He established the Discomedusae based on a single stomach cavity and the disc-like shape, including a basic separation of the "Medusidae" and Rhizostomidae. The discovery and description of the scyphistoma and strobila featured as a next major step. The metagenetic process was further studied and deciphered by Sars (1835) and Steenstrup (1842), thus eliminating the generic names *Strobila* and *Ephyra* from the system. This context was then supplemented by unveiling the connection between adult medusae and the scyphistomae by von Siebold (1839) and Sars (1841). Around that time, Ehrenberg (1836) described the location and processes of digestion in the radial canals, as well as new observations about the nature of rhopalia in *A. aurita.* His studies included a survey of knowledge on luminescence in Semaeostomeae and several figures about morphological variability in *A. aurita.* During this period, a number of authors added descriptions of more species, most of which later turned out to be synonyms of earlier known species. Important new distributional data on Semaeostomeae were contributed by Lesson (1829) and Brandt (1835, 1838).

The following period generated both general and a number of more specific works. These included descriptions of new species (Lesson, 1843) and faunistic studies, as well as specific research on morphological details of individual organs in Semaeostomeae by various authors. Thiel reviewed the primary results of these works, leading to Agassiz's (1862) comprehensive treatise on anatomy, ontology, and systematics (among other subjects) of North American medusae. This also included the first comprehensive systematic revision of the known species diversity of the taxon. The establishment of the families Aurelidae(*), Sthenonidae, Cyaneidae, and Pelagidae(*) – with further adjustments – largely became the basis of all subsequent classifications. The subsequent extended attempt of a general revision by Haeckel (1880) provided a structured and consistent character evaluation of all known Semaeostomeae.

"The main merit of Haeckel's work therefore lies, on the one hand, in the summary of the entire knowledge of the Semaeostomeae [...] at that time under logical nexuses of the individual systematic units arising from the phylogenetic view, which led to a completely different arrangement of the families and genera. In contrast to Agassiz, who placed the genera and families next to each other without distinct relation, Haeckel arranged an unbroken series from simple (Pelagidae)(*) to more complicated forms (Aurelidae), and his system has essentially been maintained to today." (Translated from Thiel, 1962, p. 1141).

The following period included complex revisions of some of Haeckel's species, as well as the addition of 35 new species, 11 of which are still accepted (e.g. von Lendenfeld, 1882; Vanhöffen, 1908), and the remaining later described as varieties of other species. Many more studies of that time dealt with anatomy and ontogeny, distribution and seasonality of occurrence, as well as physiology and systematics. Especially Vanhöffen (1888) and Maas (1906) worked towards clear definitions of species and genera to consolidate the system. Based on a variety of new material coming in from numerous expeditions, they verified or revised existing species and described the new findings [e.g. from the Italian "Vettor Pisani" world circumnavigation 1882–1885, the "Valdivia" German Deep Sea Expedition 1898–1899, the Dutch "Siboga" Expedition 1899–1900 to Indonesian waters, and the RV "Gauss" 1st German South Polar Expedition 1901–1903 (among others)]. With his large monograph of Semaeostomeae, Mayer (1910) concluded this period of research, also presenting an updated system of the known taxa. Compared to Haeckel's view, it was largely restricted with only 17 genera and 30 species remaining. This was confirmed, and further condensed to 17 species by Cockerell (1911).

In the following period, a larger number of records on **occurrence and distribution** of individual species were collected. Authors like Browne, Bigelow, Stiasny, Mayer, and Light evaluated available data and evidence to confirm or adjust the contemporary base of species distributional knowledge.

In a vast treatment of Semaeostomeae, Krumbach (1925) assembled contemporary research on the taxon, focusing on comparative morphology, histology, and ontogeny, though not explicitly including the systematics. Special mention was made to ovulation (Schaxel, 1912), developmental processes, budding, strobilation, and feeding. Further investigations included sensory physiology and effects of temperature, toxins, and electrical currents on the pulsation rate of medusae. For the first time, this period gave special attention to medical reports about the effects of cnidocyst venoms on humans (e.g. Castellani and Chalmers, 1913; Light, 1914; Lord and Wills, 1918; Stitt, 1923).

For the period following Krumbach (1925), Thiel reported various studies from mostly European authors, but also from zoologists from the United States, Japan, or India, revolving around regional observations of several well-established species during the 1930s. *P. noctiluca, C. hysoscella, A. aurita*, and *C. capillata* (i.a., *Dactylometra*[(*)] sp., *Drymonema* sp.) from various locations were investigated for their identification as valid species or regional varieties. During this period, Weill (1934) and Papenfuss (1936) also introduced the use of different types of cnidocysts for the differentiation of several *Cyanea* species. Ranson (1945) published a summary of the extensive discussions on *Pelagia, Cyanea*, and *Aurelia* species. While more and more material arrived in the museum collections from ongoing expeditions (e.g. Lund University's Chile Expedition 1948–1949, the Galathea Deep Sea Expedition 1950–1952) the validation of species status between different oceanic regions remained the subject of intensive debate. Regional surveys like Uchida's (1954) review of scyphomedusae in Japanese waters was questioned for consistency with other reports by Kramp (1955).

Thiel (1962) wrote: *"All these works on Semaeostomeae from different parts of the world give a much more definite picture of their distribution, even if this is hardly mentioned. The fact that the more recent findings of the various expeditions have confirmed the existing picture of the occurrences indicates that a certain conclusion has probably been reached* [leaving aside the taxonomic status of species]. *Only in the case of rare forms, such as Haeckel's Floresca parthenia and the more common Poralia rufescens and Undosa undulata*[(*)]*, which are often found as poorly preserved specimens from only a few places, the picture of their distribution is [still] uncertain."* (Translated from Thiel, 1962, p. 1177).

Thiel compiled the available records of semaeostome species throughout all oceans: Arctic and northern Atlantic Ocean, the Baltic Sea, European coastal waters of the North Sea and the Mediterranean (including the Balearic Islands, Rovinj, the Bosporus and the Black Sea). In the northern Atlantic, regions included seas around the Canary Islands, the

Azores, and Bermuda, as well as Canadian and U.S. coastal waters south to the Gulf of Mexico. Reports from the comparably understudied southern tropical Atlantic Ocean came from the Guinea coast, Martinique, and Brazil. Reports from the Indian Ocean included the Maldives, Nicobar, and Seychelles archipelagos, further the Islands of Lakshadweep (Laccadives), Cocos (Keeling) Islands and the Sumatra coast. In the north, reports came from the Arabian and Red Seas, the Persian Gulf, and the coastal waters off Calcutta (India). Across the vast Pacific Ocean, only a few locations had been sampled for Semaeostomeae, namely the Malay Archipelago including the Timor area, the Sulu-Sea and Philippine waters, the South China Sea, as well as waters around Japan. Two species were reported from Kamchatka and the Sea of Okhotsk. In the far offshore Pacific Ocean, reports were published from the Truk and Salomon Islands, east of New Caledonia, and from New Guinea. Along the eastern Pacific shores, reports came from the Central American coasts, Galapagos, off Bogota (Colombia), and Tocopilla (Chile). Australian and New Zealand reports included New South Wales and Port Jackson.

In the final debate on the **phylogenetic relations** of the Semaeostomeae, Thiel (1962, p. 1207 ff.) reported the diverse early attempts of assorting the taxon before Eschscholtz's (1829) initial classification as *Discophorae phanerocarpae*, that also included the Rhizostomeae. This positioning was largely agreed upon – with some alterations of naming – by Ehrenberg (1835), Haeckel (1880), Claus (1889), and others. Eschscholtz's taxon already included many of the valid genera accepted today, defining characteristic differences, while his genus *Ephyra* later became obsolete. Definitions of families were proposed by Agassiz (1862): Cyaneidae, Sthenonidae, and Aurelidae[(*)]. Haeckel (1880) was the first to investigate phylogenetic relations. Thiel (1962) gave an account of phylogenetic character analyses for the known genera at the time, also presenting a phylogenetic tree (Fig. 4). Many of his considerations have meanwhile been further detailed or revised in the light of molecular data (e.g. Daly et al., 2007; Bayha et al., 2010).

Further, Thiel gave an account and evaluation of semaeostome families, subfamilies, genera, and species (1962, p. 1222 ff.), carefully discussing and partly diagramming specific characters (e.g. from Haeckel, 1880; Vanhöffen, 1888; Mayer, 1910; Kramp, 1955). The evaluation of historic interpretations provides detailed reasoning of the known species, and also considers possible questions for subsequent revision. While some species appear well founded, others were not conclusively resolved in the light of morphological character

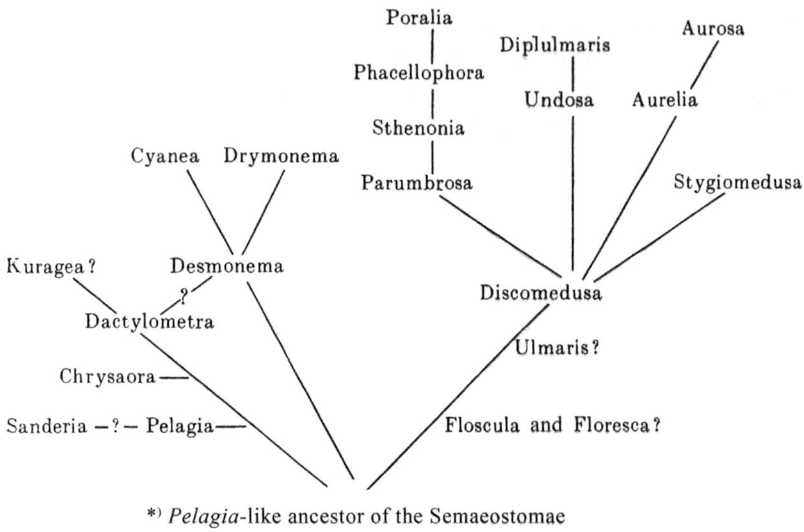

Fig. 4 Thiel's schemes of the phylogenetic development in the Semaeostomae (Thiel, 1962, p. 1218; (*) german wording translated).

variability. Based on these reflections, Thiel listed three families, 18 genera, and 30 species, nine of which he marked as uncertain.

The conclusive pages (1962, pp. 1301 ff) provide a **systematic account** of families, subfamilies, and genera, with diagnoses for all taxa and a key to the recognised species at the time of publication (Thiel, 1962): *P. noctiluca, C. melanaster, C. hysoscella, C. helvola*(*), *Dactylometra lactea* L. Agassiz(*) (species uncertain), *D. quinquecirrha*(*), *D. africana*(*), *S. malayensis, D. gaudichaudi, D. chierchiana*(*), *C. capillata, C. var. lamarkii*(*), *Drymonema dalmatina* Haeckel, 1880(*), *Ulmaris prototypus* Haeckel, 1880, *D. lobata, Undosa undulata*(*), *D. antarctica, Aurelia solida* Browne, 1905, *A. limbata, A. maldivensis, A. aurita, Aurosa furcata* Haeckel, 1880, *Parumbrosa polylobata* Kishinouye, 1910, *Sthenonia albida* Eschscholtz, 1829, *P. rufescens, P. camtschatica, S. fabulosa*(*) (asterisks refer to Table 2 listing the accepted species names according to WoRMS 2023).

2.5 Volume 8: Bibliography

The printed edition of Thiel's (1977, Volume 8) **bibliography** for all seven volumes has been scanned, text processed, and subsequently revised for misreading, typos, bugs, and other inconsistencies. Errors (e.g. in language) were corrected or verified using diverse internet sources. Addenda and occasional comments by the authors were included in brackets […]. Doubtful or incomplete references were checked with the help of Virginia Castilho

(Centre for Marine Biology of the University of São Paulo, CEBIMar) and Antje Fischer (Senckenberg am Meer, DZMB) to clarify and complete the missing information. A handful of references remained annotated as incomplete or inaccessible for verification (**). In some cases, web links to the papers were included, when available. Preliminary notes by the author M. E. Thiel are also provided in English. – Elsevier, as the legal successor of the original publisher VEB Gustav Fischer in Jena, agreed to include the revised version as a digital supplement to accompany this volume at https://www.elsevier.com/books-and-journals/book-companion/9780443296468.

The early volumes on Stauromedusae, Cubomedusae, and Coronatae, as well as on Semaeostomeae morphology, histology, and ontogeny, were published by Thiel during the period 1936–1938. Therefore, the comprehensive bibliography (Volume 8) published in 1977 might turn out to be incomplete regarding those subject-related references available after WWII. In addition, multiple references to authors published in the same year are indicated using letters a, b, c, etc. However, as the editing of the bibliography occurred about 40 years after the first editions, the letters were often not included in Thiel (1936a, 1936b, 1938a, 1938b, 1938c). Furthermore, a few works on the above taxa printed after 1938 had escaped the author's attention, when Volume 8 (the Bibliography) was published in 1977. Thiel worked on the Rhizostomeae typescript until close to his death in 1979, but gave up regular literature searches after 1971. As such, some references may have been overlooked. In this review project, additional references were included with the bibliographies in the respective chapters of this volume and in the new literature search carried out for the period 1970–2023.

3. The remaining typescript on the Rhizostomeae

Max Egon Thiel's magnum opus offers a comprehensive overview of the history of research on Scyphomedusae, from the first mentions in antiquity to the early 1960s. In addition to his own smaller studies, he undertook to assemble the available knowledge about the taxon in one place. To finalise this project, he worked on the remaining manuscript on the Rhizostomeae until the late 1970s. About 840 typed carbon copy pages of the typescript had remained after Thiel's death. They rested in files with the family before technology was available to scan and digitise them. Extensive time was used to adjust the scans, before the authors diligently revised the original German text to focus and streamline some lengthy passages as well as the author's occasionally epic style. (The original and a revised typescript in German language are kept in the Ocean Museum Germany in Stralsund and are available as. pdf copies from the library). Based on the revised German version, the co-authors' team of this paper decided to publish this final part of Thiel's work as a revised version in English language to allow wider distribution and access for the international scientific readership (Thiel et al., 2024). It is provided as an electronic supplement with this edition at (*Work in progress, to be published soon*). After an automatic translation, applying the AI-software *DeepL* (DeepL SE, Maarweg 165, 50825 Cologne, Germany. Available at http://www.deepl.com), the resulting text was cross-checked for the trans-literation of scientific terms and language by the co-authors. References relating to the Rhizostomeae and published after 1970, were intensively searched and databased by Antje Fischer and Virginia Castilho to compile an updated list of publications for this taxon, including up to the year 2023, (*Work in progress, to be published soon*).

Acknowledgements

The authors extend their gratitude to the many helpers who supported the transfer of the hand-written original to typescript processing and final publication, including H. Reinicke, F. Bitschofsky and K. Weber. T. Korth took on the extensive task to re-edit many of the original figures as prepared by M.E. Thiel. Thanks are owed to Virginia Castilho (Center for Marine Biology of the University of São Paulo, CEBIMar) and Antje Fischer (Senckenberg am Meer, German Centre for Marine Biodiversity Research, DZMB) for their amazing help on recent literature and to clarify and complete the missing reference information. The manuscript substantially profited from careful revisions of language and content by L. Brotz and G. Jarms.

References

For the complete bibliography of Thiel (1977) see digital attachment in https://www.elsevier.com/books-and-journals/book-companion/9780443296468.

Aders, W.M., 1902. Beiträge zur Kenntnis der Spermatogenese bei den Coelenteraten. Z. wiss. Zool 74, 81–108.

Adler, L., Jarms, G., 2009. New insights into reproductive traits of scyphozoans: special methods of propagation in *Sanderia malayensis* Goette, 1886 (Pelagiidae, Semaeostomeae) enable establishing a new classification of asexual reproduction in the class Scyphozoa. Mar. Biol. 156, 1411–1420. https://doi.org/10.1007/s00227-009-1181-6.

Agassiz, L., 1860/62. Monograph of the Acalephae. In: Contributions to the Natural History of North America, Boston, Vol. III, pp. 3–301 and Vol. IV, 1–380.

Ballowitz, E., 1899. Über Hypomerie und Hypermerie von *Aurelia aurita* LAM. Arch. Entwickl. Mech. Org. 8, 239–252.

Baster, J., 1762. Natuurkundige Uitspanningen behelzende einige waarnemingen over sommige Zeeplanten en Zeeinsecten. (translation of original publication). Harlem 1762.

Bauer, V., 1927. Die Schwimmbewegungen der Quallen und ihre reflektorische Regulierung. Z. wiss. Biol. Abt. C 5 (1), 37–69.

Bayha, K.M., Dawson, M.N., Collins, A.G., Barbeitos, M.S., Haddock, S.H.D., 2010. Evolutionary relationships among scyphozoan jellyfish families based on complete taxon sampling and phylogenetic analyses of 18S and 28S ribosomal DNA. Integr. Comp. Biol. 50 (3), 436–455.

Beneden, P.J. van, 1860. On the strobilisation of the Scyphistoma. Ann. Mag. Nat. Hist. 5, 504, and Ann. Sci. Nat. 4, 154–159.

Berger, E.W., 1898. The histological structure of the eyes of Cubomedusae. J. Comp. Neurol. 8 (3), 223–230.

Berger, E.W., 1900. Physiology and histology of the Cubomedusae, including Dr. F. S. Conant's notes on physiology. Mem. Biol. Lab. Johns Hopkins Univ. 4, 1–84.

Berrill, N.I., 1949a. Developmental analysis of Scyphomedusae. Biol. Rev. Camb. 24, 393–410.

Berrill, N.I., 1949b. Form and growth in the development of a scyphomedusa. Biol. Bull. Woods Hole 96 (3), 283–292.

Bethe, A., 1910. Notizen über die Erhaltung des Körpergleichgewichtes. In: Festschr. Richard K. W. T. Hertwig 3, Jena, Gustav Fischer, Jena, 129–138.

Bethe, A., 1943. Die periodische Unterbrechung rhythmischer Vorgänge (Cheyne Stokesches Phaenomen und Lucianische Perioden) im Vergleich zu ähnlichen Vorgängen an elektrischen Kippsystemen. Arch. ges. Physiol. 246 (4), 485–519.

Bigelow, H.B., 1928. Scyphomedusae from the Arcturus Oceanographie Expedition. Zool. New York 8 (10), 495–524.

Bozler, E., 1926. Sinnes- und nervenphysiologische Untersuchungen an Scyphomedusen. Z. vergl. Physiol. 4 (1), 37–80.

Brandt, J.F., 1835. Prodromus descriptiones animalium ab Mertensio in orbis terrarum circumnavigationi observatum, Fasc. I: Polypos, Acalephas, Discophoras nec non Echinodermata continens. Petropoli 1835, sumptibus Academiac Petropoli apud Graeff, Lipsiae apud Leop. Voss, pp. 203–275.

Brandt, J.F., 1838. Ausführliche Beschreibung der von C. H. MERTENS auf seiner Weltumseglung beobachteten Schirmquallen. Mém. Sci. St. Petersburg, Sci. Nat. (Ser. 6) 2, 237–411.

Broch, H., 1913. Coelentérés du fond. In: Duc D'Orléans Campagne Arctic de 1907, Brüssel, 1–129.

Browne, E.T., 1901. Variation in *Aurelia aurita*. Biometrica 1, 90–108.

Carlgren, O., 1935. Über eine neue südafrikanische Lucernariide, *Depastromorpha africana* n. g., n. sp., nebst Bemerkungen über den Bau und die Systematik dieser Tiergruppe. Kgl. Svensk. Vetensk. Akad. Handl. 3. Ser. 15 (1), 1–24.

Castellani, A., Chalmers, A.J., 1913. Manuel of Tropical Medicine, second ed. Bailleri, Tindall & Cox, London. i-xxxii + 1-1747.

Chuin, T.T., 1929. Sur les vacuoles digestives chez le scyphistome. C. R. Soc. Biol. Paris 102, 825–827.

Chuin, T.T., 1930. Le cycle evolutif du scyphistome de *Chrysaora*, étude histo-physiologique. Trav. Stat. Biol. Roscoff. 48, 1–179.

Clark, H.J., 1863. Prodromus of the history, structure and physiology of the order Lucernariae. J. Boston Soc. Nat. Hist. 7, 531–567.

Claus, C., 1877. Studien über Polypen und Quallen der Adria. Denkschr. Akad, Wiss. Wien 38, 1–64.

Claus, C., 1878. Untersuchungen über *Charybdea marsupialis*. Arb. Zool. Inst. Wien 1 (2), 221–276.

Claus, C., 1883. Untersuchungen über die Organisation und Entwicklung der Medusen. Haase, Prag and Leipzig, ii + 1–96.

Claus, C., 1886. Über die Classification der Medusen mit Rücksicht auf die Stellung der sog. Peromedusae, der Periphyllidae und Pericolpidae. Arb. Zool. Inst. Wien 7, 97–110.

Cockerell, T.D., 1911. The nomenclature of the Scyphomedusae. Proc. Biol. Soc. Washington 24, 1–12.

Collins, A.G., Jarms, G., 2018. WoRMS Staurozoa, World list of Staurozoa (version 2018–04-01). In: Roskov, Y., Abucay, L., Orrell, T., Nicolson, D., Bailly, N., Kirk, P. M., Bourgoin, T., DeWalt, R.E., Decock, W., De Wever, A., van Nieukerken, E., Zarucchi, J., Penev, L. (Eds.), Species 2000 & ITIS Catalogue of Life, 2018 Annual Checklist. Digital resource at www.catalogueoflife.org/annual-checklist/2018. Species 2000, Naturalis, Leiden, Netherlands. ISSN 2405–884X.

Collins, A.G., Jarms, G., Morandini, A.C., 2021. World List of Scyphozoa. Coronatae. Accessed through World Register of Marine Species at http://www.marinespecies.org/aphia.php?p=taxdetails&id=851649 on 2023-06-30.

Collins, A., Jarms, G., 2021. World list of Cubozoa. In: Bánki, O., Roskov, Y., Vandepitte, L., DeWalt, R.E., Remsen, D., Schalk, P., Orrell, T., Keping, M., Miller, J., Aalbu, R., Adlard, R., Adriaenssens, E., Aedo, C., Aescht, E., Akkari, N., Alonso-Zarazaga, M.A., Alvarez, B., Alvarez, F., Anderson, G. (Eds.), Catalogue of Life Checklist (Oct. 2021). https://doi.org/10.48580/d4sl-3fr.

Conant, F.S., 1898. The Cubomedusae. Mem. Biol. Lab. Johns Hopkins Univ. 4 (1), 1–61.

Conklin, E.G., 1908. The habits and early development of *Linerges mercurius*. Carneg. Inst. Washington Publ. 103 (2), 153–170.

Cuvier, G.D., 1817. Le Règne animal, T. IV, Zoophytes, Paris.

Dahlgren, U., 1916. The production of light by animals. Porifera and Coelenterata. J. Franklin Inst. 181, 243–261.

Daly, M., Brugler, M.R., Paulyn, C., Collin, A.G., Dawson, M.N., Fautin, D.G., France, S.C., McFadden, C.S., Opresko, D.M., Rodriguez, E., Romano, S.L., Stake, J.L., 2007. The phylum Cnidaria: a review of phylogenetic patterns and diversity 300 years after Linnaeus. Zootaxa 1668, 127–182.

Dalyell, J.G., 1836. Further illustrations to the propagation of Scottish Zoophytes. Edinb. New. Philos. J. 21, 88–94.

Dalyell, J.G., 1848. Rare and remarkable animals of Scotland, represented from living subjects with practical observations on their nature. J. van Voorst, London Vol. I, i–xii + 1–268, Vol. II, i–iv + 1–322.

Delage, Y., 1887. Sur une fonction nouvelle des otocystes comme organes d'orientation locomotrice. Arch. Zool. Expér. Gén. (Ser. 2) 5 (1), 1–226.

Delage, Y., Hérourd, E., 1901. Traité de Zoologie concrete. T. II. Les Coelentérés, Scyphozoaires. Schleicher, Frères, Paris, 294–366.

Delap, M.J., 1905a. Notes on the rearing in an aquarium of *Cyanea lamarcki* Pér. et Les. Ann. Rep. Fish. Sci. Invest. Ireland 1902–1903. II, 20–22.

Delap, M.J., 1905b. Notes on the rearing in an aquarium of *Aurelia aurita* L. and *Pelagia perla* (SLABB.). Fish. Ireland Sci. Invest. 1905 (7), 160–164.

Delap, M.J., Delap, C., 1906. Notes on the plancton of Valencia Harbour 1902–1905. Fish. Ireland Sci. Invest. (1905) Pt. III, 141–159.

Derbès, A., 1850. Note sur les organes réproducteurs et l'embryogénie du *Cyanea chrysaora*. Ann. Sci. Nat. (3. Ser.) 13, 377–382.

Dzwillo, M., 1980. Max Egon Thiel zum Gedächtnis. Mitt. Hamb. Zool. Mus. Inst. 77, 11–17.

Ehrenberg, C.G., 1835. Das Leuchten des Meeres. Abh. Kgl. Akad. Wiss. Berlin 1835, 411–575.

Ehrenberg, C.G., 1835. Über die Acalephen des Roten Meeres und den Organismus der Medusen der Ostsee. Abh. Kgl. Akad. Wiss. Berlin 18 Juni 1836, 181–259. (published in 1837).

Eimer, Th, 1878. Die Medusen, physiologisch und morphologisch auf ihr Nervensystem untersucht. H. Laupp'sche Buchhdl. Tübingen 1–278.

Eschscholtz, Fr, 1829. System der Acalephen, eine ausführliche Beschreibung aller medusenartigen Strahltiere. Ferd. Dümmler, Berlin 1–190.

Fewkes, J.W., 1882. Notes on Acalephs from Tortugas with description of new genera and species. Bull. Mus. Comp. Zool. Harv. Coll. 9 (7), 251–289.

Friedemann, O., 1902. Untersuchungen über die postembryonale Entwicklung von *Aurelia aurita*. Z. wiss. Zool. 71, 227–275.

Gäde, H.M., 1816. Beiträge zur Anatomie und Physiologie der Medusen nebst einem Versuch einer Einleitung über das, was den älteren Naturforschern in Hinsicht dieser Tiere bekannt war, Maurer'sche Buchhandlung, Berlin, 1–28, 2 pls.

Galigher, A.E., 1925. On the occurrence of the larval stages of Scyphozoa in the Elkhorn slough, Monterey Bay. Amer. Naturalist 59 (660), 94–96.

Gegenbaur, C., 1856. Bemerkungen über die Randkörper der Medusen. Arch. Anat. Physiol. 1856, 230–250.

Gegenbaur, C., 1857. Versuch eines Systems der Medusen mit Beschreibung neuer oder wenig bekannter Formen, zugleich ein Beitrag zur Kenntnis der Fauna des Mittelmeeres. Z. wiss. Zool. 8, 202–273.

Gilchrist, F.G., 1937. Budding and locomotion in the Scyphistomae of *Aurelia*. Biol. Bull. Woods Hole 72, 99–124.

Goody, T., 1908. On the presence of gonadial grooves in a medusa, *Aurelia aurita*. Proc. Zool. Soc. London 1908, 55–59.

Götte, A., 1893. Vergleichende Entwicklungsgeschichte von *Pelagia noctiluca*. Z. wiss. Zool. 55, 645–712.

Haacke, W., 1896. Ein Beitrag zur Lebensgeschichte der Ohrenqualle (*Aurelia aurita*). Zool. Garten 37, 344–346.

Hadzi, J., 1912. Über die Podocysten der Scyphopolypen. Biol. Zbl. 32, 52–60.

Haeckel, E., 1879. System der Craspedoten: Erste Hälfte des Systems der Medusen. Denkschr. Med. Naturw. Ges. Jena 2, 1–360.

Haeckel, E., 1880. System der Acraspeden: Zweite Hälfte des Systems der Medusen. Denkschr. Med. Naturw. Ges. Jena 2, 361–672.

Haeckel, E., 1882. Report on the deep-sea medusae, dredged by HMS "Challenger" during the years 1873–1876. In: Thomson, F.T., Murray, J. (Eds.), Report Sci. Res. Challenger (Zool.) 4, pp. 1–154, figs. A–Q, pl. 1–32.

Hamann, O., 1883. Beiträge zur Kenntnis der Medusen. Z. wiss. Zool. 38, 419–429.

Hein, W., 1900. Untersuchungen über die Entwicklung von *Aurelia aurita*. Z. wiss. Zool. 67, 401–438.

Heincke, F., 1905. Die Eier und Jugendformen der Nutzfische der Nord- und Ostsee und die Altersbestimmungen der Nutzfische. Rapp. Proc. Verb. Cons. Intern. III, Gesamtber. Anl. E, 1–39.

Hérouard, E., 1909. Sur les cycles évolutifs d'un Scyphistoma. C. R. Acad. Sci. Paris 148, 320–323.

Hertwig, O., Hertwig, R., 1879. Die Aktinien, anatomisch und histologisch, mit besonderer Berücksichtigung des Nervensystems untersucht (including reference to the oogenesis of Pelagia noctiluca and gametes of Charybdaea marsupialis). Gustav Fischer, Jena, viii + 1-224 + 224.

Hollowday, E.D., 1951. Planula, Hydra tuba and Ephyra stages of the common jellyfish *Aurelia aurita*. Microscope 8 (8), 193–198.

Horstmann, E., 1934. Untersuchungen zur Physiologie der Schwimmbewegungen der Scyphomedusen. Pflügers Arch. ges. Physiol. 234, 406–420.

Hunt, O.D., 1952. Occurrence of *Pelagia* in the River Yealm estuary (South Devon). Nature 169 (4309), 934.

Hyde, I.H., 1894. Entwicklungsgeschichte einiger Scyphomedusen. Z. wiss. Zool. 58, 531–565.

Jarms, G., Morandini, A.C., 2019. World Atlas of Jellyfish. Abh. Naturw. Ver. Hamburg (N.F.), special volume, Dölling und Galitz, Hamburg 1–816.

Kölliker, A.V., 1853. II. Über Quallen. In: Gegenbaur, G.A., Kölliker, A.V., Müller, H. (Eds.), Bericht über einige im Herbst 1852 in Messina angestellte vergleichend-anatomische Untersuchungen, pp. 315–329 Z. wiss. Zool. 4 (3–4), 299–370.

Komai, T., 1935. On *Stephanoscyphus* and *Nausithoë*. Mem. Coll. Sci. Kyoto (Ser. B) 10. 289–339.

Komai, T., 1936. On another form of *Stephanoscyphus* found in the waters of Japan. Mem. Coll. Sci. Kyoto (B) 11 (13), 175–183.

Korschelt, E., Heider, K., 1936. Vergleichende Entwicklungsgeschichte Vol. I. Gustav Fischer, Jena, xx + 1-536, 158–220 (Scyphozoa).

Kowalewski, A.O., 1873. Untersuchungen über die Entwicklung der Coelenteraten. Nachr. Kais. Ges. Freunde der Naturerkenntnis. Anthropologie und Ethnographie (russ.) 10. 1–36 (2), Moskau.

Kramp, P.L., 1943. The Godthaab Expedition 1928: Medusae. Medd. om Grønland 81 (1), 1–168.

Kramp, P.L., 1944. On development through alternating generations, especially on coelenterates. Vid. Medd. Nath. Fören 107, 13–32.

Kramp, P.L., 1955. The Medusae of the tropical west coast of Africa. Atlantide Rep. Nr. 3, Sci. Res. Dan. Exp. Coasts Trop. West Africa 1945/46 239–324 pls. 1-3.

Krasinska, S., 1914. Beiträge zur Histologie der Medusen. Z. wiss. Zool. 109, 256–348.

Krohn, A., 1855. Über die frühesten Entwicklungsstufen der *Pelagia noctiluca*. Müll. Arch. f. Anat. Physiol. 1855, 491–497.

Krumbach, T., 1925. Scyphozoa. In: W. Kükenthals Handbuch der Zoologie, Walter de Gruyter, Berlin Bd. I, 522–686.

Künne, C., 1952. Untersuchungen über das Großplankton der Deutschen Bucht. Wiss. Meeresunters. Helgoland 4, 1–54.

Lamarck, J.B.P. de, 1801–1815–1827. Histoire naturelle des animaux sans vertebres. 2 Vol., Paris.

Lambert, F.I., 1936. Jelly fish, the difficulties of the study of their life history and other problems. Essex Naturalist 25 (2), 70–86.

Lambert, F.I., 1937. Observations on scyphomedusae of the Thames estuary and their metamorphosis. Trav. Stat. Zool. Wimereux 12, 281–307.

Lamouroux, I.V.F., 1815. Mémoire sur la Lucernaire campanulée. Mém. Mus. Hist. Nat. Paris 2, 460–473.

Legendre, R., 1945. Coelenterées. In: La faune pelagique de l'Atlantique au large du Golfe de Gascogne recueillie dans les estomacs de Germons: 3ème partie Invertébrés (Cephalopodes exclus). Ann. Inst. Océanogr. Paris (NS) 20. 40–148.

Lendenfeld, R.von, 1882. Über Coelenteraten der Südsee. I. *Cyanea annaskala* n. sp. Z. wiss. Zool. 37, 465–552.

Lendenfeld, R.von, 1896. Die Lebensgeschichte der Schirmquallen. Der Zoologische Garten 37, 196–203, contd.

Lesson, R.P., 1829. Voyage autour du monde exécuté par ordre du roi sur la corvette "La Coquille" pendant les années 1822–1825. Zoologie 2 Vol., 1-135, Paris.

Lesson, R.P., 1843. Histoire naturelle des Zoophytes Acalèphes, Livre II: Les médusaires. Librairie Encyclopédique de Roret, Paris 159–425.

Light, S.F., 1914. Some Philippine Scyphomedusae including two new genera, five new species and one new variety. Phil. J. Sci. Manila (Sect. D) 9, 195–231 (Manila).

Linnaeus, C., 1747. Wästgöta Resa förratted ar 1746. Lars Salvii, Stockholm 1–284.

Littleford, R.A., 1939. The life cycle of *Dactylometra quinquecirrha* in the Chesapeake Bay. Biol. Bull. Woods Hole 77, 368–381.

Littleford, R.A., Truitt, R.V., 1937. Variation of *Dactylometra quinquecirrha*. Science New York 86 (2236), 426.

Lord, R.E., Wills, S.L.B., 1918. A case of jellyfish sting simulating an acute abdomen. Lancet 2, 390.

Maas, O., 1897. Reports on an exploration off the west coast of Mexico, Central and South America and off the Galapagos Islands, in Charge of Alexander Agassiz by the U. S. fish commission steamer "Albatross" during 1891. Lieut. Comm. z. L. Tanner USN, Commanding, Vol. XXI. Medusae. Mem. Mus. Comp. Zool. Harv. Coll 23 (1), 7–92.

Maas, O., 1903. Die Scyphomedusen der Siboga-Expedition. Siboga Expeditie, 11, 1-91, + 12 plates. 1–91.

Maas, O., 1907. Die Scyphomedusen. Ergebn. Fortschr. Zool. 1 (1), 189–238.

Marques, A.C., Collins, A.G., 2004. Cladistic analysis of medusozoa and cnidarian evolution. Invertebrate Biol. 123 (1), 23–42.

Mayer, A.G., 1910. Medusae of the world. Publ. Carneg. Inst. Wash. Nr 109, 1–735 (III).

McMurrich, I.P., 1891. The development of *Cyanea arctica*. Amer. Naturalist 25, 287–289.

Meise, W., 1980. Max Egon Thiel 1898–1979. Verh. Naturwiss. Ver. Hamburg, NF 23, 365–368.

Metschnikoff, E., 1886. Embryologische Studien an Medusen, ein Beitrag zur Genealogie der Primitivorgane. A. Hölder, Wien, vi + 1-159.

Mills, C.E., 2012. Stauromedusae: List of all valid species names. Electronic document http://faculty.washington.edu/cemills/Staurolist.html. Published on web page established October 1999, most recent update 2012 (accessed on January 17, 2024).

Miranda, L.S., Mills, C.E., Hirano, Y.M., Collins, A.G., Marques, A.C., 2018. A review of the global diversity and natural history of stalked jellyfishes (Cnidaria, Staurozoa). Mar. Biodiv. 48, 1695–1714. https://doi.org/10.1007/s12526-017-0721-4.

Moore, H.B., 1937. Annual Report of the Liverpool Marine Biology Committee and their Biological Station at Port Erin, Nr. 12. Proc. Trans. Liverpool Biol. Soc. 50, 1–293.

Müller, O.F., 1776 (1788). Zoologicae danicae prodromus seu animalium Danise et Norvegiae rariorum ac minus notorum icones. Fasc. II, Havniae, typis Hallageriis, xxxii + 1–282.

Okada, Y.K., 1927. Note sur l'ontogénie de *Carybdaea rastonii* Haacke. Bull. Biol. France Belg. 61, 241–249.

Okada, Y.K., 1931. Studies on regeneration of Coelenterata, 3rd Part: Hydromedusae, *Charybdaea* and *Coeloplana*. Mem. Coll. Sci. Kyoto Imp. Univ. (Ser. B) 7 (5), 205–221.

Östergren, Hj, 1909. *Cyanea palmstruchii* (Swartz), eine verkannte Qualle aus dem Skagerrak. Zool. Anz. 35 (15), 464–474.

Papenfuss, E.J., 1936. The utility of the nematocysts in the classification of certain Scyphomedusae 1. *Cyanea capillata, Cyanea palmstruchii, Dactylometra quinquecirrha, Dactylometra quinquecirrha var. chesapeakei* and *Chrysaora hysoscella*. Acta Univ. Lund. (NS) 31, 19–26.

Percival, E., 1923. On the strobilisation of *Aurelia*. Quart. J. Microsc. Sci. (NS) 67, 85–100.

Péron, F., Lesueur, C.A., 1809. Histoire générale et particulière de tous les animaux que composent la familie des méduses. Ann. Mus. d'Histoire Nat. 14, 218–228.

Reinicke, G.B., Thiel, H., 2019. Early multidisciplinary cooperation in oceanography. Hist. Oceanogr. Yearbook 24, 59–78.

Retzius, G., 1909. Die Spermien von *Aurelia aurita* L. Biol. Unters. (NF) 14, 67–68.

Retzius, G., 1910. Die Spermien von *Cyanea capillata* (L.). Biol. Unters. (NF) 15, 39.

Romanes, G.I., 1876. Preliminary observations on the locomotor system of medusae. Philos. Trans. Roy. Soc. London 166, 269–313.

Sars, M., 1829. Strobila octoradiata. Bidrag til Sødyrenes Naturhistorie af M. Sars, cand. theol. Förste Haefte, med sex illuminerede Steentryktafler, Bergen 17–26.

Sars, M., 1841. Mémoire sur le dévelopment de la *Medusa aurita* et de la *Cyanea capillata*, translated from German by Dr. Young (Arch. Naturgesch. 7). Ann. Sci. Nat. Ser. 16(2), 321–348.

Schäfer, E.A., 1878. On the nervous system of *Aurelia aurita*. Philos. Trans. Roy. Soc London 169 II, 563–575.

Schaxel, J., 1910a. Die Oogenese von *Pelagia noctiluca* Pér. & Les. mit besonderer Berücksichtigung der Chromidien und Nucleolen. Zool. Anz. 35, 407–414.
Schaxel, J., 1910b. Die Eibildung der Meduse *Pelagia noctiluca* Pér. & Les., Untersuchungen über die morphologischen Beziehungen der Kernsubstanzen untereinander und zum Cytoplasma. Arch. mikrosk. Anat. 94, 167–212.
Schaxel, J., 1912. Weitere Untersuchungen über die Eibildung der Meduse *Pelagia*. Jen. Z. Naturw. 48 (4), 479–491.
Schewiakoff, W.L., 1889. Beiträge zur Kenntnis des Acalephenauges. Morph. Jb 10, 21–60.
Schiariti, A., Holst, S., Tiseo, G.R., Miyake, H., Morandini, A.C., 2024. Life cycles and reproduction of Rhizostomeae. Adv. Mar. Biol. 98, 193–254.
Schultze, F.E., 1877. *Spongicola fistularis*, ein in Spongien wohnendes Hydrozoon. Arch. mikrosk. Anat. 13, 795–817.
Schultze, M., 1856. Über den Bau der Gallertscheibe der Medusen. Arch. Path. Anat. Physiol. 1856 (3), 311–320.
Siebold, T.von, 1839. Beiträge zur Naturgeschichte der wirbellosen Tiere I. Über *Medusa aurita*. Neueste Schriften Danz. Naturf. Ges 3 (2), 1–35.
Steenstrup, J.J., 1842. Om Fortplanting og udvikling gjennem vexlende Generationsraekker. Kopenhagen, 1–76.
Stiasny, G.A., 1919. Die Scyphomedusensammlung des Naturhistorischen Reichsmuseums in Leiden. I. Carybdeiden (Cubomedusen). Zool. Mededel. Deel 5, 35–58.
Stiasny, G.A., 1934. Scyphomedusae. Discov. Rep. 8, 329–396.
Stiasny, G.A., 1935. Die Scyphomedusen der Snellius Expedition. Verh. Akad. Wet. Amst. (Sect. 2) Afd. Natuurwiss 34 (6), 1–44.
Stitt, E., 1923/1946. Diagnosis, Prevention and Treatment of Tropical Diseases 2, Coelenterata, eighth ed., The Blakiston Company, Philadelphia, 1549–1550.
Straehler-Pohl, I., Jarms, G., 2022. Back to the roots, Part 2—Rhopaliophora (Scyphozoa, Cubozoa and Staurozoa) reborn based on early life cycle data. Plankton Benthos Res. 17 (2), 1–22.
Straehler-Pohl, I., Ghepdeu, G.F.Y., Chougong, D.T., Tchoumboungnang, F., 2022. Ernst Haeckel's mysterious species, Part II: African Chirodropida (Cnidaria, Cubomedusae). Plankton Benthos Res. 17 (4), 406–429.
Swartz, O., 1788. *Medusa unguiculata* und *Actinea pusilla*. Kgl. Vet. Acad. Handl. 9, 188–202.
Teissier, G., 1925. Sur la croissance embryonnaire de *Chrysaora hysoscella* (L.) (Meduse Acalephe). C. R. Acad. Sci. Paris 181, 530–532.
Thiel, M.E., 1936a. Stauromedusae. In: Bronns, H.G. (Ed.), Klassen und Ordnungen des Tierreichs Vol. 2. Akademische Verlagsgesellschaft Geest & Portig, Leipzig, pp. 1–172 Abt. 2, Buch 2.
Thiel, M.E., 1936b. Cubomedusae. In: Bronns, H.G. (Ed.), Klassen und Ordnungen des Tierreichs Vol. 2. Akademische Verlagsgesellschaft Geest & Portig, Leipzig, pp. 173–308 Abt. 2, Buch 2.
Thiel, M.E., 1936c. Coronatae. In: Bronns, H.G. (Ed.), Klassen und Ordnungen des Tierreichs Vol. 2. Akademische Verlagsgesellschaft Geest & Portig, Leipzig, pp. 308–471 Abt. 2, Buch 2.
Thiel, M.E., 1936d. Systematische Studien zu den Trachylinae der Meteorexpedition, zugleich ein Beitrag zu einer Revision der Trachylinae. Zoologische Jahrbücher, Abteilung für Systematik, Ökologie und Geographie der Tiere 69, 1–92 Heft 1.
Thiel, M.E., 1938a. Semaeostomae. Morphologie und Histologie. In: Bronns, H.G. (Ed.), Klassen und Ordnungen des Tierreichs Vol. 2 Akademische Verlagsgesellschaft Geest & Portig, Leipzig, 471–677 Abt. 2, Buch 2.
Thiel, M.E., 1938b. Semaeostomae. Ontogenie. In: Bronns, H.G. (Ed.), Klassen und Ordnungen des Tierreichs Vol. 2 Akademische Verlagsgesellschaft Geest & Portig, Leipzig, 677–856 Abt. 2, Buch 2.

Thiel, M.E., 1938c. Über eine Scyphistomabildung durch Strobilisation und ihre phylogenetische und eidonomische Bedeutung. Zool. Anz. 121 (5/6), 97–107.

Thiel, M.E., 1938d. Die Chaetognathen-Bevölkerung des südatlantischen Ozeans. Wissenschaftliche Ergebnisse der Deutschen Atlantischen Expedition auf dem Forschungs- und Vermessungsschiff, Meteor 1925-1927 Vol. 13. 1–100 Biologische Sonderuntersuchungen, 1. Lfg.

Thiel, M.E., 1958. Beiträge zur Kenntnis der Wachstums- und Fortpflanzungsverhältnisse von *Aurelia aurita* L. Abh. Verh. Naturw. Ver. Hamburg, N. F 3 (1959), 13–26.

Thiel, M.E., 1959a. Gibt es einen Generationswechsel bei den Scyphomedusen? Ein Beitrag zum Begriff der Generation, der Generationenfolge und des Generationenwechsels. Abh. und Verh. Naturw. Ver. Hamburg. N. F 3 (1958), 37–54.

Thiel, M.E., 1959b. Semaeostomae. Physiologie. In: Bronns Klassen und Ordnungen des Tierreichs, H.G. (Ed.), Akademische Verlagsgesellschaft Geest & Portig, Leipzig Vol. 2, 856–1072 Abt. 2, Buch 2.

Thiel, M.E., 1959c. Beobachtungen über Wachstum, Variationen und Abnormitaten bei *Cyanea capillata* der Ostsee. Abh. und Verh. Naturw. Ver. Hamburg, N.F, 4. 89–108.

Thiel, M.E., 1962. Scyphomedusae. Ökologie, Geschichte ihrer Kenntnis, ihre Stammesgeschichte und Klassifikation. In: Bronns Klassen, H.G., des Tierreichs, O. (Eds.), Akademische Verlagsgesellschaft Geest & Portig, Leipzig. Vol. 2 Akademische Verlagsgesellschaft Geest & Portig, Leipzig, 1073–1308.

Thiel, M.E., 1970. Über den zweifachen stammesgeschichtlichen ("biphyletischen") Ursprung der Rhizostomeae (Scyphomedusae) und ihre Aufteilung in die zwei neuen Ordnungen Cepheida und Rhizostomida. Abh. Verh. Naturwiss. Ver. Hamburg 14, 145–168.

Thiel, M.E., 1976. Die Cassiopeida (Arcodomyariae Maas, 1903, Kampylomyariae Stiasny, 1921) als dritte Ordnung der Rhizostomae (Scyphomedusae) und die Erhebung der letzteren zu der Überordnung Rhizostomoidea (Wurzelmundartige) Supraord. nov. Mitt. Hamburg. Zool. Mus. Inst. 73, 1–16.

Thiel, M.E., 1977. Bibliographie. In: Bronns, H.G. (Ed.), Klassen und Ordnungen des Tierreichs. VEB Gustav Fischer, Jena Vol. 2. 1–104 Abt. 2, Buch 2 (Electronic supplement).

Thiel, M.E. (†), Reinicke, G.B., Holst, S., Morandini, A.C., Sötje, I., Straehler-Pohl, I., Wiesenthal, A.A., Thiel, H., 2024. Rhizostomeae—a remaining manuscript to complete the Monograph series on Scyphozoa by M. E. Thiel (published 1936–1977). Adv. Mar. Biol. 98, (Electronic supplement) at (*Work in progress, to be published soon*).

Thill, H., 1937. Beiträge zur Kenntnis der *Aurelia aurita*. Z. wiss. Zool. 150, 51–96.

Uchida, T., 1926. Notes on a young specimen of Cubomedusae. Proc. Imp. Acad. 2 (6), 295–298.

Uchida, T., 1929. Studies on the Stauromedusae and Cubomedusae with special reference to their metamorphosis. Jap. J. Zool. 2 (2), 103–192.

Uchida, T., 1934. Metamorphosis of a scyphomedusa (*Pelagia panopyra*). Proc. Imp. Acad. 10 (7), 428–430.

Uchida, T., 1954. Distribution of Scyphomedusae in Japanese and its adjacent waters. J. Fac. Sci. Hokk. Univ. Ser. 6 12 (1–2), 209–219.

Uexküll, J.von, 1901. Die Schwimmbewegungen von *Rhizostoma pulmo*. Mitt. Zool. Stat. Neapel 14, 620–626.

Uexküll, J.von, 1904. Die ersten Ursachen der Rhythmen in der Tierreihe. Ergebn. Physiol. (Abt. IV) 3 (2), 1–11.

Vanhöffen, E., 1888. Untersuchungen über semaeostome und rhizostome Medusen. Bibliotheca Zoologica 3, 1–54.

Vanhöffen, E., 1891. *Periphylla* und *Nausithoë*. Zool. Anz. 14, 38–42.

Vanhöffen, E. 1892. Die Akalephen der Plankton-Expedition. Plankton Expedition Vol. II Kd., Lipsius & Tischer, Kiel u. Leipzig, pp. 1–30.

Vanhöffen, E., 1902. Die acraspeden Medusen der Deutschen Tiefsee-Expedition 1898–1899. Ergebn. dtsch. Tiefsee-Exped. Valdivia 3, 1–52.

Vanhöffen, E., 1908. Die Lucernariden und Skyphomedusen der Deutschen Südpolar-Expedition 1901-1903. Deutsche Südpolar-Expedition 10, 25–49.

Verwey, I., 1942. Die Periodizität im Auftreten und die aktiven und passiven Bewegungen der Quallen. Arch. Néerl. Zool. 6, 363–468.

Weill, R., 1934. Contribution à l'étude des cnidaires et de leurs nématocysts. Trav. Statn. Zool Wimereux 10 + 11, 1–347 + 351–701.

Wetochin, I., 1930. Über das Ergreifen der Nahrung und die Ernährung der Meduse *Aurelia aurita* in Verbindung mit der Tätigkeit des Flimmerepithels. Russ. Physiol. J. 13 (4–5), 585–600.

Widmark, E.M.P., 1911. Über die Gastrovaskularströmungen bei *Aurelia aurita* L. und *Cyanea capillata* Eschsch. Zool. Anz. 38, 378–382.

Widmark, E.M.P., 1913. Über die Wasserströmungen in dem Gastrovaskularapparat von *Aurelia aurita* L. Z. Allg. Physiol. 15, 33–48.

WoRMS Editorial Board, 2023. World Register of Marine Species. Available from https://www.marinespecies.org at VLIZ. Accessed on 2023-11-17. https://doi.org/10.14284/170.

Wright, T.Str, 1861. On hermaphroditic reproduction in *Chrysaora hysoscella*. Ann. Mag. Nat. Hist. (Ser. 3) 7, 357–359.

Yatsu, N., 1917. Notes on the physiology of *Charybdaea rastoni*. J. Coll. Sci. Tokyo 40 (3), 1–12.

CHAPTER TWO

Morphology of Rhizostomeae jellyfishes: What is known and what we advanced since the 1970s

André C. Morandini[a,b,*]
[a]Departamento de Zoologia, Instituto de Biociências, Universidade de São Paulo, São Paulo, SP, Brazil
[b]Centro de Biologia Marinha, Universidade de São Paulo, São Sebastião, SP, Brazil
*Corresponding author. e-mail address: acmorand@ib.usp.br

Contents

Abstract

The morphology of members of the order Rhizostomeae is revisited considering all life cycle stages, but with emphasis on the medusa. The current classification of the group is presented, and some aspects of species diversity are discussed. The main issues investigated since the 1970s are briefly presented by decade.

1. Introduction

Scyphomedusae (class Scyphozoa Goette, 1887) – also known as true jellyfishes or macromedusae – are conspicuous members of the gelatinous zooplankton (Cornelius, 1997a). They form a group of cnidarians that vary in size from the minute members of the order Coronatae Vanhöffen, 1892 (*e.g.* genera *Atorella* Vanhöffen, 1902 and *Nausithoe* Kölliker, 1853) to gigantic

Advances in Marine Biology, Volume 98
ISSN 0065-2881, https://doi.org/10.1016/bs.amb.2024.09.004

specimens of the order Semaeostomeae Agassiz, 1862 (*e.g.* genus *Cyanea* Péron and Lesueur, 1810). Among the scyphomedusae, the jellyfishes of the order Rhizostomeae Cuvier, 1800 comprise an amazing clade of 91 species (Collins and Morandini, 2024) in which the majority of them inhabit shallow and tropical waters (Kramp, 1970). Although their medusae are large-sized animals and live close to human populations, knowledge about the rhizostomes is poor. Besides some historical studies on the morphology (*e.g.* Agassiz, 1862; Grenacher and Noll, 1876), life cycle (*e.g.* Paspaleff, 1938; Uchida, 1926) and biology (*e.g.* Gohar and Eisawy, 1960; Thiel, 1969) only in the last decades this group gained importance due to the more noticed jellyfish blooms and the subsequent negative impacts (*e.g.* Daryanabard and Dawson, 2008; Edelist et al., 2020; Keesing et al., 2016; Rottini Sandrini and Avian, 1986; UNEP, 1991; Uye, 2008).

The description of the general morphology of rhizostome jellyfishes did not change since Thiel's time (1970s) (compare Thiel et al., 2024). So, what can we add to about the morphological aspects of Rhizostomeae? And what is new in the last 50 years? In fact, their morphology has not changed since Cambrian times, which is the estimated dating of the first recognized fossils of this group (Adler, 2019; Haeckel, 1865, 1869, 1874; Young and Hagadorn, 2010). Although M.E. Thiel´s manuscript is only being published as a supplement of this volume (Thiel et al., 2024), the main morphological features of the group are long known and can be consulted by browsing ancient authors' monographs (Agassiz, 1862; Haeckel, 1880; Mayer, 1910) as well as in recent ones (Gibbons et al., 2022; Jarms and Morandini, 2019, 2023; Russell, 1970). But it is worth spending some time reading Thiel's text, especially the section on morphology, which is surprisingly full of details.

This chapter aims to contribute to revisiting the morphology of the rhizostome jellyfishes but based on the current knowledge about the group, highlighting new approaches and concluding what is still missing for a better understanding of these graceful and enchanting marine animals.

2. Definition of a rhizostome jellyfish

Following Thiel et al. (2024) in the first sections of his review, this chapter will also start with a definition of the rhizostomes. Rhizostome (plural rhizostomes) is a noun used to refer to any member of the order Rhizostomeae. The medusa of the order Rhizostomeae (Fig. 1) can be clearly recognized by its unique morphological external features as: absence

Fig. 1 A typical specimen of a rhizostome jellyfish species, *Lychnorhiza lucerna* Haeckel, 1880. The individual was collected from the São Sebastião Channel (São Sebastião, São Paulo state, SE Brazil) in April 2013, and was photographed in an aquarium just after being sampled. Bell diameter 16–18 cm. *Photo credit: Alvaro E. Migotto, CEBIMar-USP.*

of marginal/subumbrellar tentacles; eight oral arms (= mouth arms) bearing tiny apertures (mouthlets); and absence of a central mouth opening. Note that these are general characters, and some variation might be expected especially in the number of structures, and presence of a central mouth in the genus *Stomolophus* Agassiz, 1860. These general features were already recognized and used to distinguish the group by previous researchers like Johann Friedrich Eschscholtz, Ernst Haeckel, Otto Maas, Alfred G. Mayer, and Gustav Stiasny.

Evolution works at features of a given organism at the same time. When we try to understand the evolutionary pathways of rhizostome jellyfishes, however, a striking macroevolutionary trait should be highlighted: the feeding habits and its associated structures (see details of this trait in Nagata et al., 2024). As the name of the group proposed by Cuvier (1800) cleverly suggests and indicates "*...multiplicité de ses bouches...*" (a free translation of the French could be: multitude of mouth openings) and can be understood as a ramified mouth (or "root mouth" as Thiel et al., 2024 mentioned). Thus, the order Rhizostomeae evolved from this main *Bauplan* (body pattern/architecture), which is a jellyfish without tentacles, with a ramified mouth and a complex gastrovascular cavity with many canals. The diversification of the group occurred as a complex combination of the evolution of different features like the umbrella and oral arms shape and size, and the differentiation of the gastrovascular canal system. Thiel also recognized the importance of the oral arms in the following part: "The mouth arms and their appendages are probably the most prominent organs of the "Rhizostomae", which not only distinguish them from the Semaeostomae, but also determine their anatomy and way of life (Thiel et al, 2024)"

Recognizing different species is not an easy task, even using identification keys. Most of the keys are constructed in a way that requires a good knowledge of the morphological features in order to advance through every step of the identification keys. But recognizing the general diversity of rhizostome jellyfishes is relatively easy. As highlighted by Jarms and Morandini (2019, 2023) in their pictorial key, only considering a few external characters allow to broadly differentiate the main families and even genera of Rhizostomeae medusae – namely the umbrella shape and ornamentation, the structure of the oral arms, and the types of oral arm appendages, if present.

3. Phylogenetic position and origin of the group

It is not the subject of this chapter to discuss the relationship of the rhizostomes (see Gamero-Mora et al., 2024), but some insights and suggestions need to be mentioned. There is no doubt in the literature and among researchers of the group that the most traditional orders of the class Scyphozoa are all related (Coronatae, Semaeostomeae and Rhizostomeae). They all share the asexual process of strobilation, ephyrae, and presence of rhopalia as common features (*e.g.* Marques and Collins, 2004; Schuchert, 1993). The Cubozoa Werner, 1973 and Staurozoa Marques and Collins, 2004 do share some characters with the Scyphozoa and, depending on the authors considered, the interrelationships with them can vary (*e.g.* Kayal et al., 2018; Straehler-Pohl and Jarms, 2022a; Van Iten et al., 2006).

But the focus of this chapter is the order Rhizostomeae, which has also has an interesting taxonomic perspective. Due to the general morphological aspects, most authors consider the group as a single taxonomic unit (monophyletic group). Several classification schemes were proposed considering the unity of the Rhizostomeae (see Gamero-Mora et al., 2024; Haeckel, 1882; Mayer, 1910; Stiasny, 1921; and the historical section in Thiel et al., 2024). But earlier, Thiel (1970, 1976) had a different view, with a non-monophyletic origin of the rhizostomes – and he was followed in part by Jarms and Morandini (2019, 2023) with the classification presented by these authors. Straehler-Pohl and Jarms (2022a) suggested a different classification, but with some resemblances with the idea of Thiel (1970). But even Thiel (1976) was not completely convinced of the non-monophyletic status and proposed to group all the rhizostomes under the superorder Rhizostomoidea Thiel, 1976, with three orders (Cassiopeida Thiel, 1976, Cepheida Thiel, 1970, and Rhizostomida Thiel, 1970) – as highlighted by Gamero-Mora et al. (2024) and Thiel et al. (2024).

The resolution of the internal relationships of the group has not yet come to a consensus. There are some phylogenies available in the literature, in which different groups of rhizostomes were used as terminals (Bayha et al., 2010; Dawson and Hamner, 2009; Gómez- Daglio and Dawson, 2017; Gamero–Mora et al., 2024; Neethling et al., 2011). And that is the main reason for many unresolved questions and contradictory findings: different authors used specimens from different groups! Because different individual studies used different taxa for their analyses, there are insufficient robust data to substantiate the links between families and genera. So, there is plenty of space to advance in this topic: a robust phylogeny with as many terminals as possible and embracing all the different families and genera of rhizostomes. However, that is an ambitious task to accomplish by one person, the whole community of researchers studying jellyfishes should work in a collaborative manner.

The origin of the rhizostomes seems to be a delicate topic. As the oldest fossils already presented the key features of the group, it is not trivial to explore how they evolved and differentiated. There are, indeed, some resemblances among members of the semaeostome family Ulmaridae Haeckel, 1880 and rhizostomes, which were noted and mentioned by different researchers (*e.g.* Mayer, 1910; Thiel, 1970; Uchida, 1926). We can mention the gastrovascular system with ramified and anastomosed canals, a tendency to reduce the marginal tentacles, oral arms with digitata, and a microphagous diet. However, no research study actually concentrated on elucidating the group's origins so far. Thus, this is another line of research to be explored.

4. Classification of Rhizostomeae and diagnostic features of the main groups

The current accepted classification of the order Rhizostomeae is shown below. It is the same used in WoRMS (Collins and Morandini, 2024) and also presented by Gamero-Mora et al. (2024). Additionally, the morphological diagnostic features of each taxon are provided, compiled from Calder (2009), Cornelius (1997b), Jarms and Morandini (2019, 2023), Kramp (1961), and Russell (1970).

- Order Rhizostomeae Cuvier, 1800
  - Suborder Kolpophorae Stiasny, 1920
    - Family Cassiopeidae L. Agassiz, 1862
    - Family Cepheidae L. Agassiz, 1862
    - Family Leptobrachidae L. Agassiz, 1862
    - Family Mastigiidae Stiasny, 1920
    - Family Versurigidae Kramp, 1961
  - Suborder Dactyliophorae Stiasny, 1920
    - Family Catostylidae Vanhöffen, 1888
    - Family Lobonemidae Stiasny, 1920
    - Family Lychnorhizidae Haeckel, 1880
  - Superfamily Rhizostomatoidea Cuvier, 1800
    - Family Rhizostomatidae Cuvier, 1800
    - Family Stomolophidae Haeckel, 1880
  - Suborder Ptycophorae Gershwin and Davie, 2013
    - Family Bazingidae Gershwin and Davie, 2013

Order Rhizostomeae – Scyphomedusae with eight adradial oral arms possessing small apertures (mouthlets); absence of a single central mouth opening in adult medusae; and absence of marginal/subumbrellar tentacles.

- Suborder Kolpophorae – Rhizostome medusae with dichotomous branching oral arms; subumbrellar musculature vary depending on the family; network of anastomosing canals directly connected with central stomach; smooth rhopalia pits; and subgenital ostia without papillae.
  - Family Cassiopeidae – Rhizostome kolpophore medusae with feather-like subumbrellar musculature; typically 16 rhopalia and 16 main radial canals; the arm disc usually round.
  - Family Cepheidae – Rhizostome kolpophore medusae with radial subumbrellar musculature; absence of ring canal; 8 main radial canals and several inter-rhopaliar canals; the arm disc octogonal.
  - Family Leptobrachidae – Rhizostome kolpophore medusae with circular subumbrellar musculature; ring canal present; 8 main radial canals and several inter-rhopaliar canals; arm disc without appendages; no appendages on oral arms, but each arm with a naked mouthless

terminal portion (may be confused with a terminal appendage); extremely elongated and slender oral arms (2–3 times bell diameter).
  - Family Mastigiidae – Rhizostome kolpophore medusae with circular subumbrellar musculature; ring canal present; 8 main radial canals and several inter-rhopaliar canals; arm disc with filamentous appendages; most species with single terminal appendage (terminal club) on each arm.
  - Family Versurigidae – Rhizostome kolpophore medusae with circular subumbrellar musculature; ring canal present; 8 main radial canals and several inter-rhopaliar ones; arm disc with filamentous appendages; oral arms without terminal appendages; characteristic exumbrella with a network of anastomosing furrows.
- Suborder Dactyliophorae – Rhizostome medusae with three-winged oral arms; subumbrellar musculature circular; peripheral network of anastomosing canals mostly separated from central stomach by a ring canal, connection with stomach through radial canals; furrowed rhopalia pits; in most species subgenital ostia with papillae.
  - Family Catostylidae – Rhizostome dactyliophore medusae with or without appendages on oral arms depending on the genus; rhopaliar and inter-rhopaliar radial canals reach the stomach; internal network of anastomosing canals with or without connections to rhopaliar and/or inter-rhopaliar canals, in some species may reach the central stomach.
  - Family Lobonemidae – Rhizostome dactyliophore medusae with appendages on oral arms (mostly filamentous); rhopaliar and inter-rhopaliar radial canals reach the stomach; internal network of anastomosing canals with connections to rhopaliar and/or inter-rhopaliar canals, but never reaching stomach; characteristic elongated marginal lappets, and window-like openings in the connecting membranes of the oral arms.
- Superfamily Rhizostomatoidea – Rhizostome medusae with scapulets (outgrowths of the oral arms, located near the arm disc).
  - Family Rhizostomatidae – Rhizostome rhizostomatoid medusae without a central mouth opening; with or without appendages on oral arms.
  - Family Stomolophidae – Rhizostome rhizostomatoid medusae with a central mouth opening; without appendages on oral arms.
- Suborder Ptycophorae – Rhizostome medusae with coalescent short oral arms; subumbrellar musculature circular; hooded rhopalia with no pits; small subgenital ostia without papillae.

- Family Stomolophidae – Rhizostome rhizostomatoid medusae with a central mouth opening; without appendages on oral arms.
- Family Bazingidae – with characters of the suborder.

Images with selected representatives of each family are provided in Figs. 2 and 3.

5. Species diversity among the rhizostomes

The first rhizostome species to be described was *Cassiopea frondosa* (Pallas, 1774) from the Caribbean Sea, and the 8 oral arms were already mentioned. Cumulative number of species described in the order Rhizostomeae is shown in Fig. 4 (only considering the valid species according to WoRMS). The graphic reflects some periods with gaps (more than 20 years) without any descriptions of rhizostome species (1792–1820, 1839–1859, 1946–1965, 1967–1989). The time frame in which there was an increasing

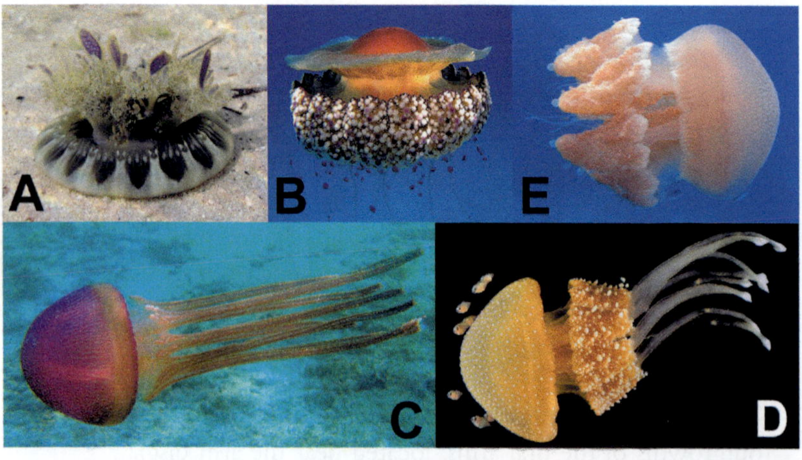

Fig. 2 Selected species representing all families of the Order Rhizostomeae, suborder Kolpophorae. (A) Cassiopeidae, *Cassiopea xamachana*, bell diameter ~6 cm. (B) Cepheidae, *Cotylorhiza tuberculata*, bell diameter ~25 cm. (C) Leptobrachidae, *Thysanostoma loriferum*, bell diameter ~13 cm. (D) Mastigiidae, *Phyllorhiza punctata*, bell diameter ~13 cm. (E) Versurigidae, *Versuriga anadyomene*, bell diameter ~25 cm. *Photo credits: (A) André C. Morandini. (B) Fredski 2013, licensed under Creative Commons 3.0, https://commons.wikimedia.org/w/index.php?curid=29319370. (C) David Burdick/NOAA, https://commons.wikimedia.org/wiki/File:Jellyfish_on_Guam_(reef113319074).jpg. (D) Alvaro E. Migotto, CEBIMar USP. (E) Anne Hoggett, Lizar Island Research Station.*

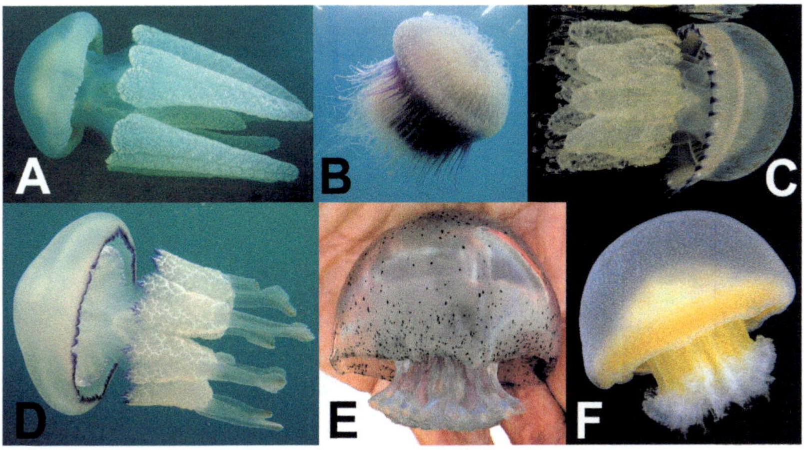

Fig. 3 Selected species representing all families of the Order Rhizostomeae, suborders Dactyliophorae and Ptychophorae. (A) Catostylidae, *Catostylus mosaicus*, bell diameter ~25 cm. (B) Lobonemidae, *Lobonema smithii*, bell diameter ~30 cm. (C) Lychnorhizidae, *Lychnorhiza lucerna*, bell diameter ~16 cm. (D) Rhizostomatidae, *Rhizostoma pulmo*, bell diameter ~15 cm. (E) Stomolophidae, *Stomolophus* sp. 2?, bell diameter ~6 cm. (F) Bazingidae, *Bazinga rieki*, bell diameter ~2 cm. *Photo credits: (A) Mark Norman, Licensed under Creative Commons Attribution 4.0, https://collections.museumsvictoria.com.au/species/8681. (B) Lee Goldman, Coral Triangle Adventures Inc. (C) Alvaro E. Migotto, CEBIMar USP. (D) André C. Morandini. (E) John Snow, Mexican Marine Life.org. (F) Denis Riek.*

number of species discoveries (pulse of activity) was from the publication of the System der Medusen by E. Haeckel in 1880 until the "Discovery expedition report on scyphomedusae" by Stiasny (1934). Given that only one species was described between 1946 and 1989, I tend to say that this represents the largest gap (*Catostylus ouwensi* Moestafa and McConnaughey, 1966 – based on only 5 specimens and never found again).

Ernst Haeckel was the author who described the highest number of rhizostome valid species – 14. Gustav Stiasny and Alfred G. Mayer followed him with 8 species each, S.F. Light described 7 species, and then Ernst Vanhöffen with 5 species described.

Since the 1970s, only a few new species of rhizostomes were described (see Table 1). From the 91 species considered as valid nowadays in WoRMS (Collins and Morandini, 2024), 8 species were described since 1970, accounting ~9% of the known diversity of the order. Finding new species of conspicuous and large animals, which are usually colourful and living in tropical shallow waters, is quite surprising because these are intensively

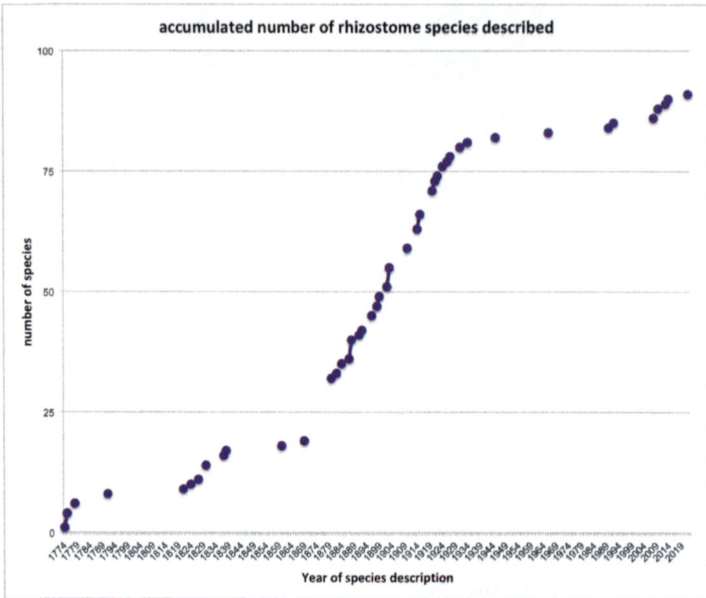

Fig. 4 Species accumulation curve of the currently valid rhizostomes (n = 91) from 1774–2022. *Data compiled from WoRMS (Collins, A.G., Morandini, A.C., 2024. World List of Scyphozoa. Accessed through: World Register of Marine Species at: https://www.marinespecies.org/aphia.php?p=taxdetails&id=135220 (accessed on August 29, 2024)).*

studied areas. The reason why new species are being found and described is perhaps related to misidentification (*e.g.* Kienberger and Prieto, 2018) and also cryptic diversity (Gómez Daglio and Dawson, 2017). Additionally, Gómez Daglio and Dawson (2019) further discussed the topic on the context of a renaissance of the taxonomy (integrative taxonomy) using scyphozoan medusae as an example.

6. The planula and polyp (scyphistoma) stages

It is essential to state that in all rhizostome species with known life cycle (see data summarized by Schiariti et al., 2024) – comprising 28 out of 91 valid species – they all exhibit the metagenesis or metagenetic life cycle.

The early stages of the development of the rhizostome medusae do not show a special morphology, nor it is distinct from other scyphozoans. General aspects concerning the histology, life cycle and morphological

Table 1 Eight species of the order Rhizostomeae described since 1970. The list is derived from WoRMS (Collins and Morandini, 2024).

Year	Species	Authority	Locality
1990	*Rhopilema nomadica*	Galil, Spanier and Ferguson, 1990	Israel
1992	*Eupilema inexpectata*	Pagès, Gili and Bouillon, 1992	South Africa
2008	*Netrostoma nuda*	Gershwin and Zeidler, 2008	E Australia
2010	*Cassiopea maremetens*	Gershwin, Zeidler and Davie, 2010	E Australia
2010	*Marivagia stellata*	Galil and Gershwin, 2010	Israel/India
2013	*Bazinga rieki*	Gershwin and Davie, 2013	E Australia
2014	*Crambionella helmbiru*	Nishikawa, Mulyadi and Ohtsuka, 2014	Indonesia
2022	*Cassiopea mayeri*	Gamero-Mora et al. (2022)	Philippines

features about the planula and polyp are to be found and summarized in Holst et al. (2024), Schiariti et al. (2024), and Thiel et al. (2024). The planula larva derived from gametes fertilization does not possess a mouth and consists of a slightly elongated mass of externally epidermal ciliated cells and an internal core of gastrodermis.

Especially concerning the morphology, there is no difference in the polyps of rhizostomes if compared to the other orders, except the general shape and exoskeleton covering. The polyps have a general funnel-shaped body (sac-like or cone-shaped), with the slender part attached to the substrate. The polyps have a single terminal opening – the mouth. This opening is surrounded by tentacles, which vary in number but usually count around 16 – relating to the polyps tetramerous symmetry [although there are some variation, as seen in semaeostomes Thiel (1963), and also documented for rhizostomes by Bigelow (1900) and Gohar and Eisawy (1960)]. From the oral view of the polyp it is possible to see the insertion of the four gastric septae that connect the oral membrane to the basal part of the polyp. Following Gamero-Mora et al. (2024) and WoRMS (Collins and Morandini, 2024), we can recognize differences in the polyps of Dactyliophorae and Kolpophorae. Note that the status of the suborder Ptychophorae (with its single species *Bazinga rieki* Gershwin and Davie, 2013) is considered uncertain until further morphological details of the canal system are properly described.

The Dactyliophorae comprise the families Catostylidae, Lobonemidae, Lychnorhizidae, Rhizostomatidae, and Stomolophidae. In the species with known life cycle of this group (see Schiariti et al., 2024), the polyps have a relatively short peduncle (or stalk) compared to the main column (or calyx) (Fig. 5A). The Kolpophorae comprise the families Cassiopeidae, Cepheidae, Leptobrachidae, Mastigiidae, and Versurigidae. In the species with known life cycle of this group (see Schiariti et al., 2024), the polyps have a pronounced peduncle, sometimes being much longer than the main column (Fig. 5B). In the polyps of Dactyliophorae there is a small exoskeleton covering just the basal part. On the contrary, part of the polyps of the Kolpophorae (family Cepheidae) lacks an exoskeleton; but the rest of the group have a more conspicuous exoskeleton (also called periderm; see Mendoza-Becerril et al., 2016) surrounding the most basal third of the peduncle. Additional images of the different polyps can be found in Straehler-Pohl and Jarms (2022b).

Interestingly, those two taxa (Dactyliophorae and Kolpophorae) differ significantly in their asexual reproductive types or strategies in addition to having minor variations in polyp morphology (see Schiariti et al., 2014, 2024). The Dactyliophorae asexually propagate by means of podocysts (Fig. 6A), while the Kolpophorae only produce free-swimming buds (Fig. 6B). If we consider the strobilation, there is also some distinction between the groups (summarized by Schiariti et al., 2024): in general, the Dactyliophorae present polydisc strobilation (but see the definition of

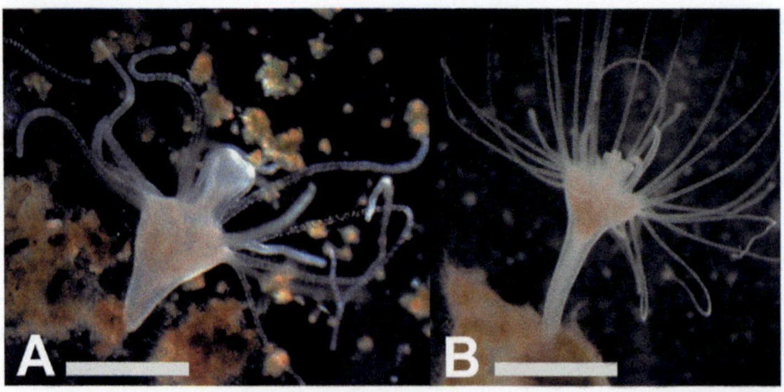

Fig. 5 The polyps of the rhizostome suborders Dactyliophorae and Kolpophorae. (A) Dactyliophorae, a polyp of *Lychnorhiza lucerna*. Scale bar = 1 mm. (B) Kolpophorae, a polyp of *Cassiopea xamachana*. Scale bar = 2 mm. Note the differences in general shape. *Photo credits: André C. Morandini.*

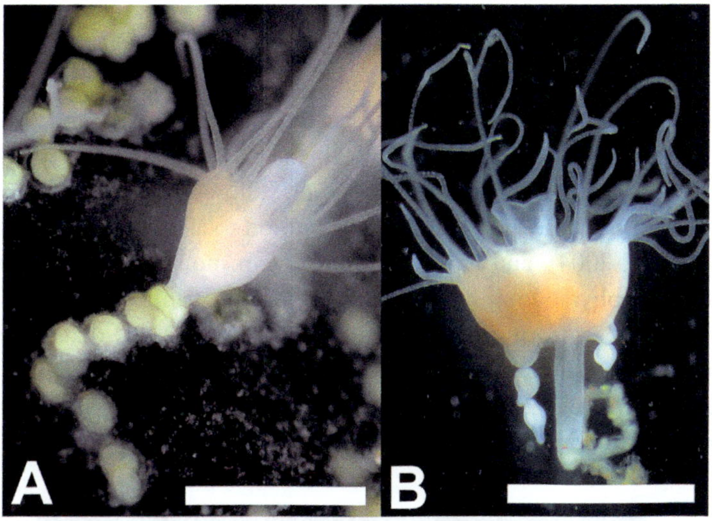

Fig. 6 Asexual reproduction in polyps of the rhizostome suborders Dactyliophorae and Kolpophorae. (A) Dactyliophorae, a polyp of *Rhopilema asamushi*. Scale bar = 3 mm. Note the production of a series of podocysts. (B) Kolpophorae, a polyp of *Cassiopea andromeda*. Scale bar = 5 mm. Note the production of many free-swimming buds forming a chain. *Photo credits: André C. Morandini.*

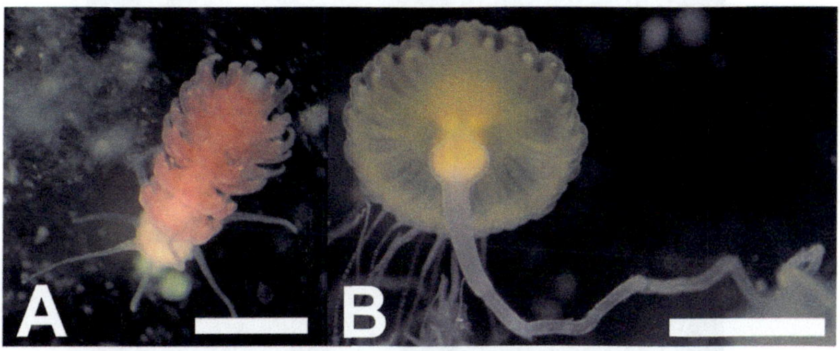

Fig. 7 Strobilation in the polyps of the rhizostome suborders Dactyliophorae and Kolpophorae. (A) Dactyliophorae, a polydisc strobilating polyp of *Stomolophus* sp. 2. Scale bar = 2 mm. Note the 5 ephyrae in the chain. (B) Kolpophorae, a monodisc strobilating polyp of *Cassiopea xamachana*. Scale bar = 4 mm. Note the single ephyra. *Photo credits: André C. Morandini.*

oligodisc strobilation by Straehler-Pohl and Jarms, 2022b) (Fig. 7A); and the Kolpophorae reproduce by monodisc strobilation (Fig. 7B).

Regarding the Ptychophorae (*Bazinga rieki*) there is not yet evidence available about early development and life cycle.

7. The ephyra stage

During the progression of strobilation, the tissues of the polyps are reorganized, and a free-swimming stage is formed: the ephyra (= young/juvenile jellyfish) (Fig. 8A and B), which grows and develops into adult medusa during the gonadal development. The ephyra is a flat 8-pointed star-shaped early medusa stage that is common to all scyphozoans (Marques and Collins, 2004), and has a single cruciform mouth opening. In rhizostomes, the ephyrae look similar having 8 rhopalia flanked by two ocular or rhopaliar lappets (forming one lobe) (Jordano et al., 2022) that persist until the medusa is fully grown (Thiel et al., 2024). The only exception to this general ephyra *Bauplan* is the genus *Cassiopea* Péron and Lesueur, 1810, in which the number of rhopalia is higher (12, 16 or more) and the ephyrae

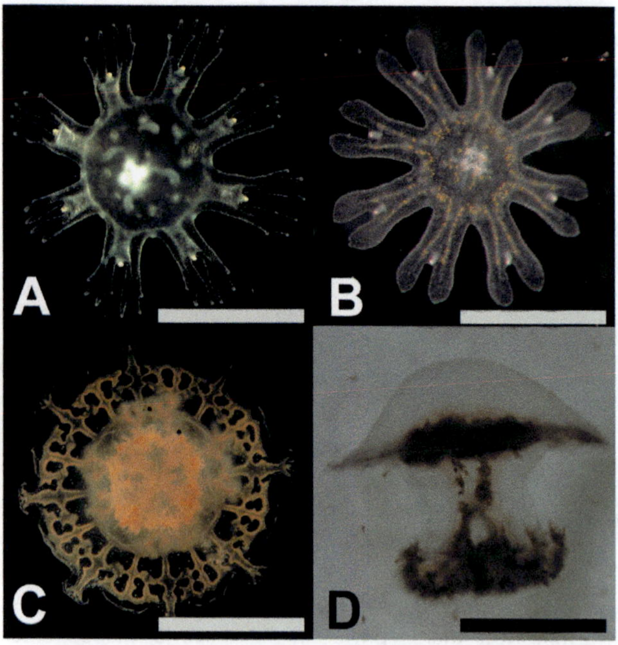

Fig. 8 Ephyrae of the rhizostome suborders Dactyliophorae and Kolpophorae. (A) Dactyliophorae, a recently released ephyra of *Lychnorhiza lucerna*. Scale bar = 1 mm. (B) Kolpophorae, a recently released ephyra of *Cotylorhiza tuberculata*. Scale bar = 0.8 mm. Note the differences in the contour of the ocular/rhopaliar lappets. (C) Dactyliophorae, metaephyra of *Lychnorhiza lucerna* in oral view, 10 days old. Scale bar = 3 mm. Note the growing of the canal system in the umbrella. (D) Dactyliophorae, metaephyra of *Lychnorhiza lucerna* in side view, 15 days old. Scale bar = 1 cm. Note the convex shape of the umbrella. *Photo credits: André C. Morandini.*

already resemble a small adult medusa (*e.g.* Gohar and Eisawy, 1960; Jordano et al., 2022). There are differentcharacters that might be used to distinguish ephyrae, like the presence and pattern of nematocyst warts or clusters (see Holst et al., 2024), gastric filaments, colour/pigment spots and shape of marginal lappets (see Straehler-Pohl and Jarms, 2010); but they are species-specific without any pattern among higher taxonomical ranks. However, known recently released ephyrae of the Dactyliophorae have a serrated contour of the ocular/rhopaliar lappets (Fig. 8A) (noting that there might be some exceptions, *e.g. Rhizostoma* spp.), while the ones of the Kolpophorae have a smooth contour of the ocular/rhopaliar lappets (as stated by Jordano et al., 2023) (Fig. 8B).

As soon as the ephyra grows and morphological features develop (velar lappets, gastrovascular canals, oral arms), it starts to increase its volume due to the thickening of the mesoglea and the bell becomes slightly convex (tending to be hemispherical in most species) (Fig. 8C and D). The term metaephyra (sometimes written as metephyra) can be found in the literature to describe an intermediate stage between the ephyra and adult medusa. But no clear definition is proposed or suggested, except by Kienberger et al. (2018), who states: "The mesogloea thickened and the flat exumbrella of the young ephyra transformed into a typically bell-shaped medusa". Another possible definition is that "the metaephyra is an intermediate developmental stage of discomedusae ephyrae in which the marginal bell rim reaches the rhopaliar lappets before developing velar lappets" (Straehler-Pohl *pers. comm.,* 2018). Depending on the study considered, it could be interpreted differently from the images indicating a metaephyra (Ballesteros et al., 2021; López-Martínez et al., 2022; Miyake et al., 2021; Nakanishi et al., 2009; Prieto et al., 2010; Schiariti et al., 2008) sometimes being more developed in one study than in the other. What can be concluded is that a metaephyra is the stage in between a flat star-shaped form (the ephyra) and a young bell-shaped medusa. A proposed definition of a metaephyra for rhizostomes is: an intermediate developmental stage in which the marginal bell rim starts to develop (forming or not distinct velar lappets) and the flat umbrella thickens and starts to become bell or dome-shaped (convex); canal system complexity varies depending on the species. The closure of the central mouth is a crucial developmental stage that coincides with the growth of the ephyra. Depending on the species and environmental conditions, this can occur in different timing, but they all pass through this crucial change. Although the development of body structures is quite well described, there are still some unknown details about the steps from a single mouth to the ramified oral arms and all the changes in the gastrovascular system connections of the umbrella and with the oral arms (but see Thiel et al., 2024).

8. The structure of the medusa stage

The manuscript of Thiel that appeared with this volume (Thiel et al., 2024) is a wonderful and comprehensive summary of the knowledge about the morphology of the order Rhizostomeae, and it is unique because it is focused exclusively on this group of scyphomedusae. Nevertheless, good descriptions can be found in old monographs and revisionary studies by Agassiz (1862), Haeckel (1880, 1882) and Mayer (1910) – of course considering the knowledge at that time. Some general aspects of the rhizostomes can also be read in general textbooks (*e.g.* Franc, 1993; Hyman, 1940; Werner, 1984). In addition, there are morphological descriptions in more recent texts like Russell (1970), Jarms and Morandini (2019, 2023), and Gibbons et al. (2022). General definitions used for other jellyfish groups are also valid for the rhizostomes (see Fig. 9); but a description will be presented hereafter pointing some specificities. The aim of this review is to combine information that are fairly spread in different publications, and to bring all information together in a contemporary context.

The adult medusa has a gelatinous body resembling a bell-like structure (also named umbrella), usually dome-shaped. The umbrella can vary significantly in shape depending on the genus, from a flat disc/saucer-like structure to a hemispherical bell (Figs. 2 and 3). The most eye-catching features are the larger bell sizes compared to other medusae (hydromedusae, cubomedusae, coronamedusae), and the lack of marginal tentacles (the synapomorphy of the order) (Fig. 1). The outer part of the bell is called exumbrella and might be smooth or ornate with warts or different types of projections and grooves (Figs. 2 and 3). The underneath part of the body, named subumbrella has 8 feeding structures hanging from the central part, the oral or mouth arms (Figs. 9 and 10). Such oral arms are attached to the subumbrella by 4 connections named pillars (Fig. 11). Additionally, between these pillars we find holes called subgenital ostia (Fig. 11). Thus, the umbrella is composed of two sides: the convex exumbrella and the concave or flat subumbrella with the oral arms. The thickness and flexibility of these umbrella parts also vary depending on the species and is

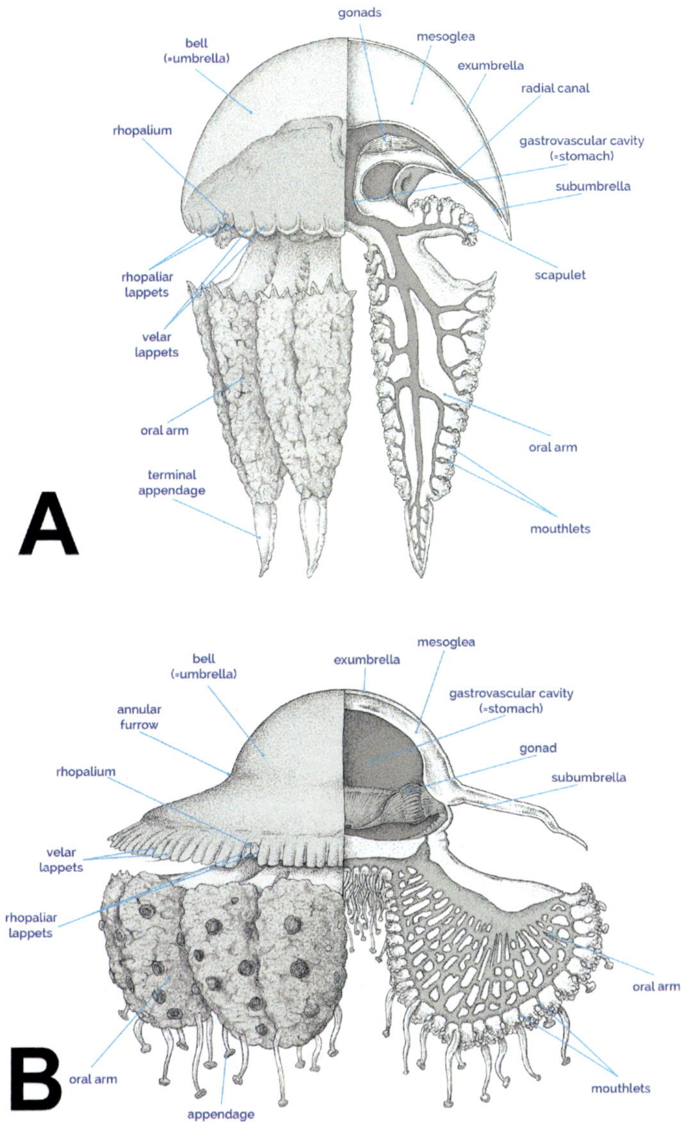

Fig. 9 Schematic view of the rhizostome suborders Dactyliophorae and Kolpophorae. (A) Dactyliophorae, *Rhizostoma* in lateral view. Left side, external view and right side, internal organization. (B) Kolpophorae, *Cotylorhiza* in lateral view. Left side, external view and right side, internal organization. *Figures adapted from Gibbons et al. (2022 :55). Food and Agriculture Organization of the United Nations (FAO). Reproduced with permission.*

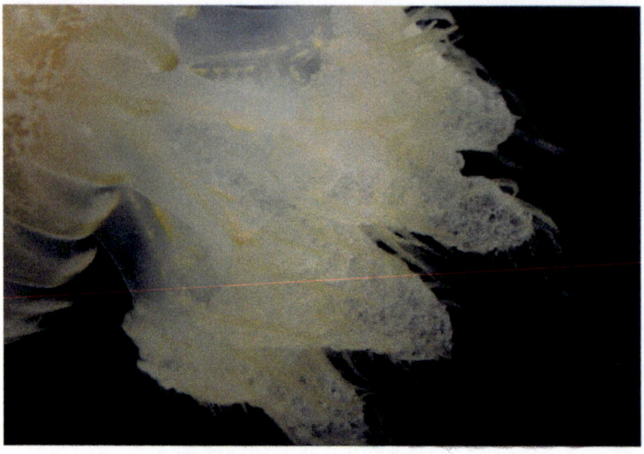

Fig. 10 Lateral view of the oral arms of a swimming adult specimen of *Lychnorhiza lucerna* (size ~15 cm bell diameter). Note the downward directed oral arms with filamentous appendages. *Photo credit: André C. Morandini.*

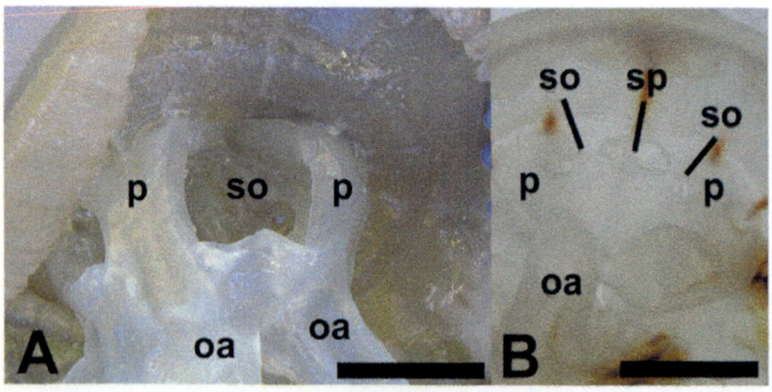

Fig. 11 Pillars and subgenital ostia of rhizostome medusae. (A) Oral view of an adult specimen of *Catostylus tagi* showing the pillars (p), subgenital ostia (so), and part of the oral arms (oa). Scale bar = 15 cm. (B) Oral view of a young specimen of *Acromitus flagellatus* showing the pillars (p), subgenital ostia (so), a subgenital papilla (sp), and part of an oral arm (oa). Scale bar = 5 cm. *Photo credits: André C. Morandini.*

another feature that distinguishes the rhizostomes from other jellyfish groups (quite important in edible species, see more in Brotz et al., 2024). The positions of the body structures are relevant and must be considered following the locations of the body axes (symmetrical planes): considering an ephyra, the axis lines passing through the mouth tips are considered the perradia, alternating with the interradia and the adradia between these two; considering an adult medusa, care should be taken to identify the basal part of the branching oral arms which are perradial in position (Fig. 12A).

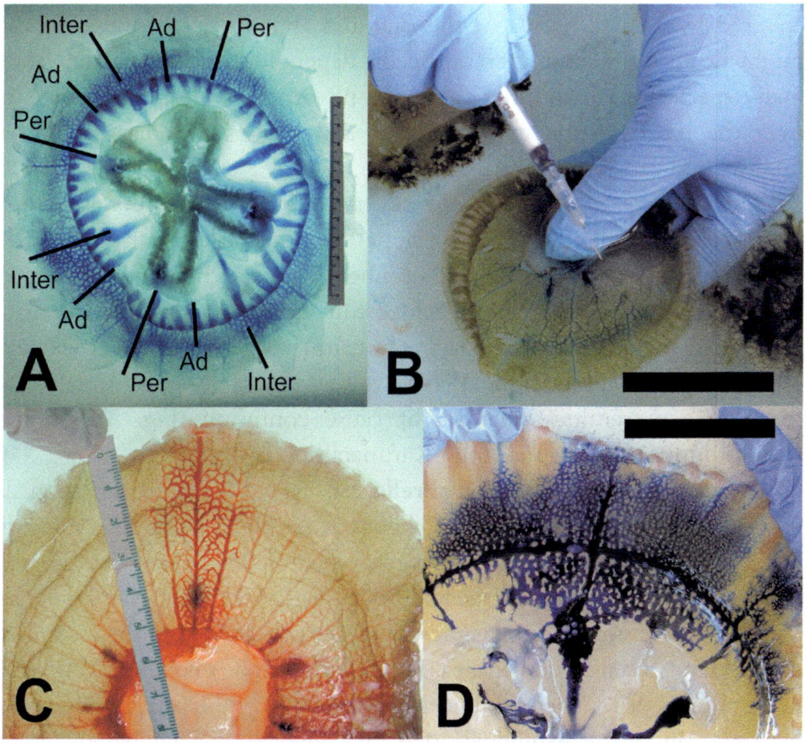

Fig. 12 Symmetry axes and gastrovascular canal system in the umbrella of rhizostome medusae. (A) Oral view of an adult specimen of *Lychnorhiza lucerna* (oral arms removed) in which the gastrovascular canal system was injected with Methylene blue dye solution. Scale bar in cm. (B) Umbrella of a *Cassiopea xamachana* specimen (oral arms removed) being injected with Methylene blue solution. Scale bar = 5 cm. (C) Umbrella of a *Cassiopea xamachana* specimen (oral arms removed) injected with Mayer's Paracarmine solution. Scale bar in cm. (D) Umbrella of an adult *Catostylus tagi* specimen (oral arms removed) injected with Methylene blue solution. Scale bar = 5 cm. *Photo credit: (A) Renato M. Nagata, FURG. (B) Edgar Gamero-Mora, Mexico. (C and D) André C. Morandini.*

The rhopalia (grouped sensing structures) are located in the **margin of the umbrella**. They can vary in structure, number and shape, but in most species they are 8 of adradial position (see Holst et al., 2024 for further details). Sötje and Henne (2019) gave important details about the typical shape and functioning of the rhopalia; but here only general aspects are mentioned. Each rhopalium protrudes from the bell margin and are flanked by, usually, small lappets. The rhopalium consist of a bulbous projection with a gravity sense organ at the tip (statocyst), which is in many cases covered by a small hood. At the oral and aboral parts of the bulb there might occur light sensing areas (ocelli) depending on the species. And in many species of rhizostomes there is a pit on the exumbrellar side – just over the basal portion of the bulb – and this exumbrellar pit presents furrows and grooves (Fig. 13).

The margin is scalloped into several **marginal lappets**, which are called velar (if flanked by another lappet) or ocular/rhopaliar (if flanked by a rhopalium) – the size, shape and number of velar and rhopaliar lappets can be significant for identification. Rhizostome scyphomedusae lack tentacles in the bell margin – that is a fact – but, in two genera (*Lobonema* Mayer, 1910 and *Lobonemoides* Light, 1914), we find much elongated lappets (with thread-like extremities) that might cause confusion (Fig. 3B) (see the redescription of *Lobonemoides* by Karunarathne et al., 2024).

The general shape of the **umbrella** can be helpful in identification, like proposed by some pictorial keys (Gibbons et al., 2022; Jarms and Morandini, 2019, 2023). It is possible to use the contour of the animal, but also some additional features that ornate the body. Members of the family Cepheidae have a marked depression on the exumbrella at some distance from the margin; and also an elevated central domed area (Fig. 2B) that can be smooth or with varying sized, pointed or rounded protuberances. Jellyfishes of the family Lobonemidae have also an ornate exumbrella (Fig. 3B); they have long bluntly pointed papillae. Specimens of the genus *Versuriga* Kramp, 1961 have a different type of

Fig. 13 Detail of the bell margin of a specimen of *Acromitus maculosus* showing the furrowed exumbrellar pit of the rhopalium. Scale bar = 2 mm. *Photo credit: André C. Morandini.*

bell ornaments: the exumbrella has an irregular pattern of grooves (named network of anastomosing furrows by Kramp, 1961) (Fig. 2E). Several other rhizostome jellyfishes exhibit different colour patterns and warts that vary in size and distribution over the umbrella. The epidermis of the subumbrella can be differentiated into what is called musculature. There are some distinct patterns of musculature arrangements that can additionally help in identification, and they were used to distinguish some groups in the past (Arcadomyaria, Radiomyaria and Cyclomyaria of Maas, 1903).

The downward hanging **oral arms** are another unique structure of the rhizostome medusae. They evolved from the elongation of the ephyra mouth lips, turning into a V-shaped structure, and then into a complexly branched tube-like structure with internal gastrovascular canals. The oral arms can be short like in the genus *Stomolophus* (Fig. 3E) or they can be quite elongated like in the genus *Thysanostoma* Agassiz, 1862 (Fig. 2C). There are different modes of branching and shapes, as dichotomous or irregular branching and J-shaped in side view. But they are highly branched in all species and along these branches there are thousands of miniature mouth openings surrounded by small tentacle-like structures called digitata (Fig. 14). The oral arms can also be distinct in cross-section of the most distal part; they can be V-shaped (as in the family Cepheidae) or Y-shaped (sometimes termed three-winged, as in most other families). A particular group of the rhizostomes with a different structure in the oral arms, the

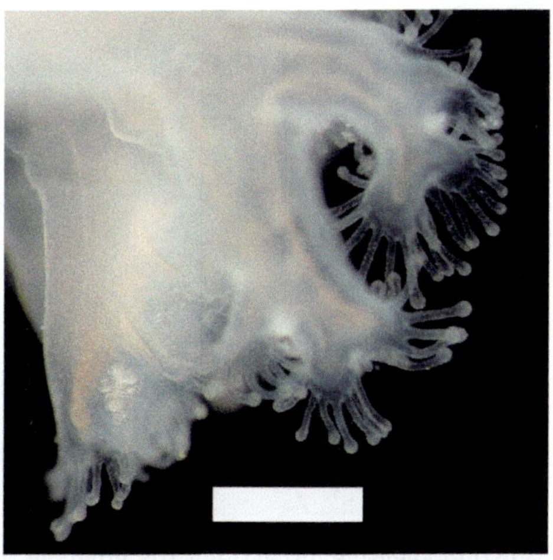

Fig. 14 Detail of an oral arm tip of a young specimen of *Catostylus tagi* showing the digitata. Scale bar = 5 mm. *Photo credit: André C. Morandini.*

scapulets (families Rhizostomatidae and Stomolophidae) (Fig. 15) form a monophyletic clade (superfamily Rhizostomatoidea; see Gamero-Mora et al., 2024). These structures are an additional part of the oral arms, located near the base of the outer side of each arm, which also possess the tiny mouth openings.

Appendages of the oral arms are another remarkable feature of the rhizostome jellyfishes. Despite the colourful bodies and rhythmic swimming pulsations, the different types of structures pendant from the oral arms are noticeable. The appendages are also of taxonomic value. Being present or not, the size, shape, colour and where they are inserted in the arms are helpful to differentiate families and genera (Figs. 2 and 3). There are different shapes of appendages and, unfortunately, there is no standardized naming for the distinct types – and that is an interesting topic to be studied in detail – the terminology describing them in the literature is a mess; but it is important not to use the word "tentacle" to indicate an appendage, to avoid confusion with tentacles of other scyphomedusae. Some appendages are massive and solid; some are somewhat hollow and harbour symbionts (Fig. 16); others have gastrovascular connections to the main system. Thiel et al. (2024) mentioned in his review that some of the appendages might work for excretion of food items, stabilization during swimming, and release of gametes. We know that in the genus *Cassiopea* most of the appendages harbour symbionts (*e.g.* Ohdera et al., 2018) and others produce cassiossomes (defensive structures; Ames et al., 2020; Holst et al., 2024); and in some

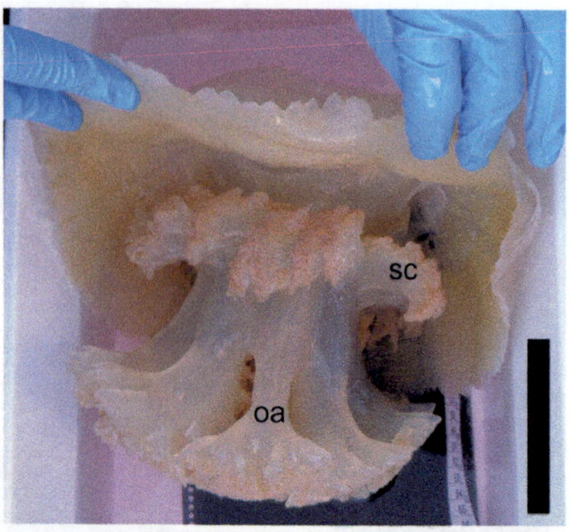

Fig. 15 Adult specimen of *Eupilema inexpectata.* The umbrella is folded to show the structure of the oral arms (oa) and scapulets (sc). Scale bar = 10 cm. *Photo credit: Sergio N. Stampar, UNESP Bauru.*

Fig. 16 Underwater photograph of a mangrove area in Florida (USA) showing different specimens of *Cassiopea xamachana.* Note the variety in shape and colours of the oral arms appendages. Scale bar = 20 cm. *Photo credit: André C. Morandini.*

Mastigiidae there are filamentous appendages inserted in the tissue between the bases of the oral arms (arm disc) that serve as brooding structures (*e.g.* von Lendenfeld, 1888).

The **subgenital cavity** (subgenital porticus) is a space between the umbrella and the insertion of the oral arms (pillars). As mentioned by Thiel et al. (2024), it is possible to see the gonadal tissues in the roof of this cavity through the openings of the subgenital ostia (Fig. 11). These ostia vary in size and shape (*e.g.* slit-like, round, crescentic), and can bear membranes or papillae (Fig. 11B) at the opening that are diagnostic features of certain species.

The **gastrovascular cavity** is divided into a central part (usually called stomach) and a peripheral canal system. The cavity is the space between the exumbrella (upper part) and the subumbrella (lower part), in the central region it is reasonably broad, but in the periphery it is reduced to narrow canals. The central stomach can be somewhat round (*e.g. Cassiopea*) (Fig. 17A), but it is cruciform in the majority of the jellyfishes (Fig. 17B) – visible through the umbrella due to its transparency. Most of the floor lining the stomach is covered by gastric cirri (= gastric filaments) responsible for digestion, and the folded gonadal tissues – which in some cases, can occupy almost the entire cavity if completely mature. The mouthlets, throughout the oral arms, are all connected by a series of small canals, which further connect to each other

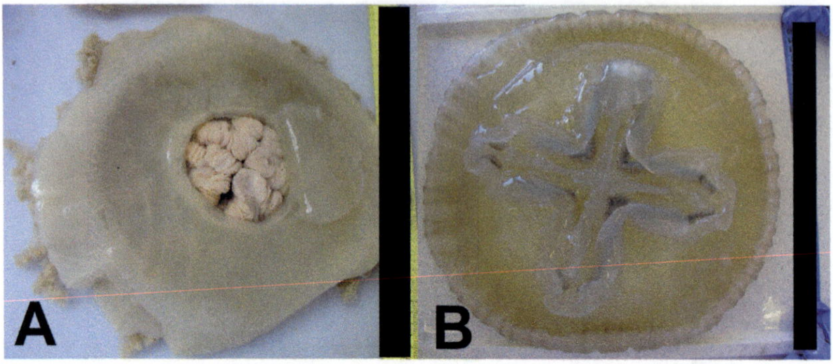

Fig. 17 Central gastrovascular cavity (stomach) of rhizostome medusae. (A) Aboral view of an adult dissected specimen of *Cassiopea xamachana*. Scale bar = 18 cm. Note the round central stomach with folded gonadal tissue. (B) Oral view of an adult dissected specimen of *Catostylus tagi*. Scale bar = 30 cm. Note the cruciform central stomach with parts of the folded gonadal tissue. *Photo credits: André C. Morandini.*

forming a larger duct at the base of each arm. These eight larger ducts open up in the central stomach passing through the pillars that support the arm disc. Once the main cavity is reached, there are peripheral smaller canals extending into the umbrella that branch and anastomose forming different combinations of networks that are essential for definition of subgroups. There are radial and circular canals and the way they are formed and combined are crucial for distinguishing the main lineages of rhizostomes (Dactyliophorae *vs.* Kolpophorae) (Fig. 12). The classical study of Stiasny (1921) established the details and methods to use the organization of the canal system as an important identification feature of rhizostomes, which was followed and adapted by some authors (Kramp, 1961; Thiel, 1970, 1976, 1978; Uchida, 1926).

9. Summary of selected morphological studies in last decades (1970's–2020's)

As previously indicated, morphological observations are frequently linked to different approaches. But there are some recent studies with focus only on the morphology of rhizostomes, and they are mostly related to inventories (*e.g.* Goy, 1979; Mianzan and Cornelius, 1999; Pagès et al., 1992; Rizman-Idid et al., 2016; Yap and Ong, 2012), and systematics and description of species (*e.g.* Cooke, 1984; Das et al., 2023; Galil et al., 2010, 2017; Gershwin and Davie, 2013; Gul and Morandini, 2013; Karunarathne and de

Croos, 2020, 2023; Neethling et al., 2011; Nishikawa et al., 2014; Omori and Kitamura, 2004; Silveira and Cornelius, 2000; Souza and Dawson, 2018; Syazwan et al., 2020; Thiel, 1975). The following chapters of this volume provide details and references regarding the different areas (Brotz et al., 2024; Gamero-Mora et al., 2024; Holst et al., 2024, Nagata et al., 2024; Schiariti et al., 2024; Thibault et al., 2024). Since 1970, special attention must be paid to life cycle studies, which increased the knowledge and also helped to understand the development of the rhizostome morphology in an ontogenetic perspective (data summarized in Schiariti et al., 2024). Here only a few of these studies will be highlighted, especially those with some morphological data involved – the reader should be warned that this is not an exhaustive list; in fact, the decision was to emphasize the main topics of research during the years by selecting some papers to use as examples.

Chronologically just after Thiel finished his manuscript (published as Thiel et al., 2024), there were two studies: one on the systematics (Russell, 1970) and the other on biogeography of rhizostomes (Kramp, 1970). Thiel explored the evolution of rhizostome jellyfishes (Thiel, 1970, 1976) using specific morphological features mostly related to the development of the gastrovascular canal system (Thiel, 1978). In more recent years, the taxonomy and evolution of scyphozoans in general (containing groups of rhizostomes) included, besides morphological observations, also molecular data – mostly DNA barcoding (Bayha et al., 2010; Dawson and Hamner, 2003, 2009; Dawson, 2004, 2005a,b,c,d). Even though, a molecular approach is usually associated with morphological systematics studies as mentioned (*e.g.* Nishikawa et al., 2014; Souza and Dawson, 2018), a recent series of rhizostome species redescriptions were solely based on a strong and carefully detailed morphology (Behera et al., 2020; Das et al., 2023; Karunarathne and de Croos, 2020, 2021, 2023; Karunarathne et al., 2024; Syazwan et al., 2020). To summarize, the 1970s were marked by traditional studies on morphology and evolution, with systematics (*e.g.* Thiel, 1975) and also development works (*e.g.* Hofmann et al., 1978). But, evolutionary aspects of rhizostome jellyfishes gained some spotlight again in early 2000s with redescriptions in the 2020s decade.

As mentioned previously in Section 2 (definition of a rhizostome) and 8 (structure of pelagic stage), morphological identification of rhizostome species requires broad observation of external features and additionally detailed inspections of anatomical characters. Species identification is greatly helped with the use of identification keys. Although there are some keys available, they are mostly restricted to certain areas (*e.g.* Larson, 1976; Morandini et al., 2005; O'Sullivan, 1982). Luckily, there are two studies that provide keys to all genera of scyphomedusae, they are Cornelius (1997b) and Martellos et al. (2016). But there are some papers that presented identification keys to specific groups of Rhizostomeae. Kitamura and Omori (2010) while synopsing the edible jellyfishes, provided a key to all families, genera and species of Dactyliophorae. Gul et al. (2015) while reporting cepheid medusae from Pakistan, provided a key to the family Cepheidae and to species of *Cephea* Péron and Lesueur, 1810 and *Netrostoma* Schultze, 1898.

In the 1980s, there was an increasing attention to the occurrence of the jellyfish blooms, considering the problems for ecosystems and humans that they can cause (see Rottini Sandrini and Avian, 1986). Some authors dedicated their time to study the growth and morphology of jellyfishes (Kikinger, 1986) and also systematics (Mianzan, 1989a,b). It is worth to highlight the study of Avian (1986) in which the author provided valuable and rare data on the development of the subgenital cavity of *Cotylorhiza tuberculata* (Macri, 1778).

The 1990s decade started with some inventories in which species and new species were described (Atta, 1991; Galil et al., 1990; Goy, 1990; Larson, 1990; Pagès et al., 1992). Kikinger (1992) provided a complete study on *C. tuberculata* where morphological details were emphasized. Lu et al. (1992) described the morphology of what they identified as *Stomolophus meleagris* Agassiz, 1860, but probably the authors had *Nemopilema nomurai* Kishinouye, 1922 in their hands. As in the previous decade, an increase in the studies about jellyfish blooms was noted (UNEP, 1991).

In the 2000s decade the concern regarding jellyfish blooms (Mills, 2001) was coming to an unprecedented status due to their impact in energy supply (Masilamoni et al., 2000), on fisheries (Uye, 2008), and also on tourism (Dawson et al., 2001). The researchers community responded to this major concern of the society by organizing two international meetings to discuss the topic (Pitt and Purcell, 2009, Purcell et al., 2001). There were also some studies focusing on morphology. Bolton and Graham (2004) described morphological variation on the invasive *Phyllorhiza punctata* von Lendenfeld (1884) in waters of the Gulf of Mexico. Omori and Kitamura (2004) reviewed three rhizostome species that were important for fisheries. And some redescriptions were provided considering rhizostome jellyfishes as mentioned previously (Dawson, 2005 a, b, c, d); and also new species descriptions (Gershwin and Zeidler, 2008); while

Calder (2009) reviewed the knowledge on scyphomedusae of the SE USA coast. In this decade, several authors also pointed to the biological invasion issue, presenting records of different species (*e.g.* Graham and Bayha, 2007; Graham et al., 2003; Haddad and Nogueira, 2006; Silveira and Cornelius, 2000). While some research topics were focusing on the effects of jellyfish blooms (*e.g.* Billet et al., 2006; Uye, 2008), detection of blooms (Houghton et al., 2006), and summarizing the knowledge (*e.g.* Richardson et al., 2009), some others were interested on enhancing the population of rhizostome species (Dong et al., 2009).

The 2010s decade continued showing concern of the jellyfish researchers' community to the so-called blooms (*e.g.* Angel et al., 2016; Dong et al., 2010; Duarte et al., 2012; Kogovšek et al., 2010; Straehler-Pohl and Jarms, 2010), including the proceedings of four international meetings (Condon et al., 2014; Condon et al., 2018; Purcell and Angel, 2010; Purcell et al., 2012). There were also some invasion or new findings reports (*e.g.* Galil et al., 2017; Gul and Morandini, 2013; Morandini et al., 2017). But the number of studies related to scyphozoans, and specifically rhizostomes, increased a lot in comparison with previous decades (see Preface by Reinicke et al., 2024). The period was also prolific in systematics papers including strong morphological information about rhizostome species (*e.g.* Galil et al., 2010; Gershwin and Davie, 2013; Neethling et al., 2011; Nishikawa et al., 2014; Souza and Dawson, 2018). An interesting approach on the evolution and development of the scyphozoans were published by Helm (2018); as well as the effort of a huge group of researchers to categorize the rhizostome *Cassiopea xamachana* Bigelow, 1892 as a model system (Ohdera et al., 2018). And closing the decade, Gómez Daglio and Dawson (2019) highlighted the integrative approach to study scyphozoans as a general plea. Finally, the late Dale Calder wrote "a comprehensive overview of the current state of scientific knowledge about these unique invertebrate animals" at the preface of the World Atlas of Jellyfish (Jarms and Morandini, 2019). The book contains a lot of morphological information in the different chapters and also on the species descriptions, and is a landmark of this century for scyphozoan as well as rhizostome studies. There are other relevant books that explore morphology issues and should be on the shelves of anyone interested in rhizostome and jellyfishes in general (with amazing photographs): Gershwin (2016); Goy and Toulemont (1997); Heeger (1998); Miyake and Lindsay (2013); Minakuchi and Toshino (2022); Namikawa and Soyama (2000); Purushothaman et al. (2022).

The 2020s, until now, can be highlighted by new morphological techniques that refined the knowledge both on polyps and medusae. Regarding the polyp stage, Khabibulina and Starunov (2019, 2021) used fluorescent microscopy to highlight musculature development and cell proliferation in polyps and buds of *Cassiopea* (see also Holst et al., 2024). An interesting new approach (3D X-ray microtomography) combined with an old methodology (injection of dye) transformed the way we can see and study the gastrovascular canal system of jellyfishes (Avian et al., 2022). Jordano et al. (2020, 2022) while studying the growth of scyphomedusae highlighted the influence of temperature on the development of the bell and feeding structures, which can also affect the cilia efficiency (Jordano et al., 2024). Furthermore, von Montfort et al. (2023) discussed and correlated macroevolutionary patterns of the swimming modes with different morphological outlines of scyphozoan species. Although some efforts involving citizen science monitoring and apps with bloom alert systems were already being used (*e.g.* Jellywatch, MedJelly, Jellyfish App, Jelly Spotter) as well as social media like Facebook and iNaturalist, only recently a paper presenting rhizostome species recorded using citizen science was published (Terenzini and Falkenberg, 2022).

10. Concluding remarks and future perspectives

The comprehension of the evolutionary trends of rhizostome jellyfishes only makes sense if we can build a connection between all the different traits (*e.g.* ecology, reproduction, behaviour) through morphology. Morphological observations are essential to understand life cycle patterns and also to propose relationship hypothesis to be tested, because they provide evidences for comparison and recognition of differences and similarities.

Despite the recent advances promoted by molecular techniques, morphological approach still provides the ground basis for comparative and evolutionary studies in order to comprehend the biodiversity of scyphozoan jellyfishes. The finding of new species of rhizostomes and redescriptions in the last few years reinforce the rationale that there is still a lot to do, not only on morphology, but in all aspects of the biology and the natural history of these marine creatures. The knowledge and evolution of the diversity of body forms in rhizostome jellyfishes are still to be unravelled. And if we take more benefits of new techniques (*e.g.* 3D microtomography) and approaches regarding morphological aspects of rhizostomes, our knowledge of the animals will increase exponentially.

The current perception of climate change, especially in the coastal areas, emphasizes the increasing need of more studies to help humankind understand how our near future will be – and most important, how we can act to make it better. The proliferation of jellyfishes in the oceans will certainly be part of this future, and field- and laboratory-based experiments will advance the knowledge about medusae.

Acknowledgements

Several colleagues must be acknowledged due to fruitful discussions and opportunities to observe different species: F.L. da Silveira, P.F.S. Cornelius, G. Jarms, D.R. Calder, A.C. Marques, H.W. Mianzan, I. Straehler-Pohl, A. Schiariti, S.N. Stampar, A.E. Migotto, R.M. Nagata, G.R. Tiseo, J.W. Lawley, E. Gamero-Mora, M.A. Jordano, M.M. Maronna, L. Gómez Daglio, M.J. Gibbons, A.G. Collins, M.N. Dawson, Z. Morais, W.K. Fitt, D.K. Hofmann, S. Gul, R. Saravanan, A. Ryias, B. Kumar, L. Gershwin. Thanks also goes to an uncountable list of people that received me in amazing museums and lovely aquaria, such visits make the difference. Agustin Schiariti, Götz B. Reinicke, and Sabine Holst provided important insights and comments to the manuscript. Alvaro E. Migotto, Anne Hogget, Denis Riek, Edgar Gamero-Mora, John Snow, Lee Goldman, Renato M. Nagata, and Sergio N. Stampar provided some of the excellent images that illustrated this chapter, besides the ones available for public use. Data used for this contribution was gathered while the author received grants from FAPESP (2010/50174-7, 2013/05510-4, 2015/21007-9, 2019/20042-6, 2023/12416-9) and CNPq (307832/2022-8). This is a NP-BioMar USP contribution.

References

Adler, L., 2019. The fossil record of medusae. In: Jarms, G., Morandini, A.C. (Eds.), 2019. World atlas of jellyfish, first ed. Dölling und Galitz Verlag, Hamburg, pp. 24–27.

Agassiz, L., 1860. Contributions to the natural history of the United States of America. III. Second monograph, in five parts, Acalephs in general, Ctenophorae, Discophorae, Hydroidae, homologies of the Radiata. Little, Brown and Co, Boston.

Agassiz, L., 1862. Contributions to the natural history of the United States of America. IV. Second monograph, in five parts, Acalephs in general, Ctenophorae, Discophorae, Hydroidae, homologies of the Radiata. Little, Brown and Co, Boston.

Ames, C.L., Klompen, A.M.L., Badhiwala, K., Muffett, K., Reft, A.J., Kumar, M., et al., 2020. Cassiosomes are stinging-cell structures in the mucus of the upside-down jellyfish *Cassiopea xamachana*. Nat. Comm. Biol. 3, 67.

Atta, M.M., 1991. The scyphomedusae of the Mediterranean cost of El Arish. J. Egyp. Germ. Soc. Zool. 6, 251–256.

Angel, D.L., Edelist, D., Freeman, S., 2016. Local perspectives on regional challenges: jellyfish proliferation and fish stock management along the Israeli Mediterranean coast. Reg. Environ. Change 16, 315–323.

Avian, M., 1986. Biological cycle of *Cotylorhiza tubercolata* (Macri 1778): morphological aspects of the development from ephyra to young medusa. Nova Thalassia 8, 47–58.

Avian, M., Mancini, L., Voltolini, M., Bonnet, D., Dreossi, D., Macaluso, V., et al., 2022. A novel endocast technique providing a 3D quantitative analysis of the gastrovascular system in *Rhizostoma pulmo*: an unexpected through-gut in Cnidaria. PLoS One 17 (8), e0272023.

Ballesteros, A., Östman, C., Santín, A., Marambio, M., Narda, M., Gili, J.-M., 2021. Cnidome and morphological features of *Pelagia noctiluca* (Cnidaria: Scyphozoa) throughout the different life cycle stages. Front. Mar. Sci. 8 (1059), 714503.

Bayha, K.M., Dawson, M.N., Collins, A.G., Barbeitos, M.S., Haddock, S.H.D., 2010. Evolutionary relationships among scyphozoan jellyfish families based on complete taxon sampling and phylogenetic analyses of 18S and 28S ribosomal DNA. Integr. Comp. Biol. 50, 436–455.

Behera, P.R., Jishnudev, M.A., Saravanan, R., Roul, S.K., Ghosh, S., Uma Mahesh, V., et al., 2020. Redescription of the enigmatic jellyfish, *Crambionella annandalei* (Cnidaria: Scyphozoa) from Indian waters. J. Mar. Biol. Assoc. U. K. 100 (5), 691–699.

Bigelow, R.P., 1892. On a new species of *Cassiopea* from Jamaica. Zool. Anz. 15, 212–214.

Bigelow, R.P., 1900. The anatomy and development of *Cassiopea xamachana*. Mem. Boston Soc. Nat. Hist. 5, 193–237.

Billet, D.S.M., Bett, B.J., Jacobs, C.L., Rouse, I.P., Wigham, B.D., 2006. Mass deposition of jellyfish in the deep Arabian Sea. Limnol. Oceanogr. 51, 2077–2083.

Bolton, T.F., Graham, W.M., 2004. Morphological variation among populations of an invasive jellyfish. Mar. Ecol. Prog. Ser. 278, 125–139.

Brotz, L., Angel, D.L., D'Ambra, I., Enrique-Navarro, A., Lauritano, C., Thibault, D., et al., 2024. Rhizostomes as a resource: the expanding exploitation of jellyfish by humans. Adv. Mar. Biol. 98.

Calder, D.R., 2009. Cubozoan and Scyphozoan jellyfishes of the Carolinian biogeographic province, southeastern USA. R. Ont. Mus. Contr. Sci. 3, 1–58.

Collins, A.G., Morandini, A.C., 2024. World list of Scyphozoa. Accessed through: World Register of Marine Species at: https://www.marinespecies.org/aphia.php?p=taxdetailsandid=135220 (accessed on August 29, 2024).

Cooke, W.J., 1984. New scyphozoan records for Hawaii: *Anomalorhiza shawi* Light, 1921, and *Thysanostoma loriferum* (Ehrenberg, 1835); with notes on several other rhizostomes. Proc. Biol. Soc. Wash. 97, 583–588.

Condon, R.H., Lucas, C.H., Pitt, K.A., Uye, S., 2014. Jellyfish blooms and ecological interactions. Mar. Ecol. Progr. Ser. 510, 107–288.

Condon, R.H., Costello, J.H., Fuentes, V., Lombard, F., Lucas, C., Purcell, J.E., et al., 2018. Jellyfish bloom research: advances and challenges. Mar. Ecol. Progr. Ser. 591, 1–302.

Cornelius, P.F.S., 1997a. Class Scyphozoa – jellyfish. In: Richmond, M.D. (Ed.), A guide to the seashores of Eastern Africa and the Western Indian Ocean Islands. SIDA/Department for Research Cooperation. SAREC, Stockholm, pp. 122–125.

Cornelius, P.F.S., 1997b. Keys to the genera of cubomedusae and scyphomedusae (Cnidaria). In: Den Hartog, J.C. (Ed.), Proceedings of the 6th International Conference on Coelenterate Biology, 1995. Nationaal Natuurhistorisch Museum, Leiden, pp. 109–122.

Cuvier, G., 1800. Mémoire sur l'organisation de quelques méduses. Bull. Sci. Soc. Philom. Paris, Ser. I, Tome 2, 69.

Daryanabard, R., Dawson, M.N., 2008. Jellyfish blooms: *Crambionella orsini* (Scyphozoa: Rhizostomeae) in the Gulf of Oman, Iran, 2002–2003. J. Mar. Biol. Assoc. U. K. 88, 477–483.

Das, Y., Karunarathne, K.D., Roy, M., Chowdhury, M.S.N., Sharifuzzaman, S.M., 2023. Record of *Crambionella annandalei* Rao, 1931 from Bangladesh, with a review of the geographic distribution of the genus *Crambionella* (Cnidaria: Scyphozoa). Mar. Biol. Res. 19, 279–293.

Dawson, M.N., 2004. Some implications of molecular phylogenetics for understanding biodiversity in jellyfishes, with emphasis on Scyphozoa. Hydrobiologia 530/531 249–260.

Dawson, M.N., 2005a. Incipient speciation of *Catostylus mosaicus* (Scyphozoa, Rhizostomeae, Catostylidae), comparative phylogeography and biogeography in south-east Australia. J. Biogeogr. 32, 515–533.

Dawson, M.N., 2005b. Morphological variation and systematics in the Scyphozoa: *Mastigias* (Rhizostomeae, Mastigiidae) – a golden unstandard? Hydrobiologia 537, 185–206.

Dawson, M.N., 2005c. Five new subspecies of *Mastigias* (Scyphozoa: Rhizostomeae: Mastigiidae) from marine lakes, Palau, Micronesia. J. Mar. Biol. Assoc. U. K. 85, 679–694.

Dawson, M.N., 2005d. Morphologic and molecular redescription of *Catostylus mosaicus conservativus* (Scyphozoa: Rhizostomeae: Catostylidae) from south-east Australia. J. Mar. Biol. Assoc. U. K. 85, 723–731.

Dawson, M.N., Hamner, W.M., 2003. Geographic variation and behavioral evolution in marine plankton: the case of *Mastigias* (Scyphozoa, Rhizostomeae). Mar. Biol. 143, 1161–1174.

Dawson, M.N., Hamner, W.M., 2009. A character-based analysis of the evolution of jellyfish blooms: adaptation and exaptation. Hydrobiologia 616, 193–215.

Dawson, M.N., Martin, L.E., Penland, L.K., 2001. Jellyfish swarms, tourists, and the Christ-child. Hydrobiologia 451, 131–144.

Dong, J., Jiang, L., Tan, K., Liu, H., Purcell, J.E., Li, P., et al., 2009. Stock enhancement of the edible jellyfish (*Rhopilema esculentum* Kishinouye) in Liaodong Bay, China: a review. Hydrobiologia 616, 113–118.

Dong, Z., Liu, D., Keesing, J.K., 2010. Jellyfish blooms in China: dominant species, causes and consequences. Mar. Pollut. Bull. 60, 954–963.

Duarte, C.M., Pitt, K.A., Lucas, C.H., Purcell, J.E., Uye, S., Robinson, K.L., et al., 2012. Is global ocean sprawl a cause of jellyfish blooms? Front. Ecol. Environ. 11, 91–97.

Edelist, D., Guy-Haim, T., Kuplik, Z., Zuckerman, N., Nemoy, P., Angel, D.L., 2020. Phenological shift in swarming patterns of *Rhopilema nomadica* in the Eastern Mediterranean Sea. J. Plankton Res. 42, 211–219.

Franc, A., 1993. Classe des Scyphozoaires. In: Doumenc, D. (Ed.), Traité de Zoologie, Tome III, Cnidaires – Cténaires, fascicule 2. Masson, Paris, pp. 597–884.

Galil, B.S., Spanier, E., Ferguson, W.W., 1990. The scyphomedusae of the Mediterranean coast of Israel, including two lessepsian migrants new to the Mediterranean. Zool. Med. 64, 95–105.

Galil, B.S., Gershwin, L., Douek, J., Rinkevich, B., 2010. *Marivagia stellata* gen. et sp. nov. (Scyphozoa: Rhizostomeae: Cepheidae), another alien jellyfish from the Mediterranean coast of Israel. Aquat. Inv. 5, 331–340.

Galil, B.S., Gershwin, L.-A., Zorea, M., Rahav, A., Rothman, S.B.-S., Fine, M., et al., 2017. *Cotylorhiza erythraea* Stiasny, 1920 (Scyphozoa: Rhizostomeae: Cepheidae), yet another erythraean jellyfish from the Mediterranean coast of Israel. Mar. Biodiv. 47, 229–235.

Gamero-Mora, E., Collins, A.G., Boco, S.R., Geson III, S.M., Morandini, A.C., 2022. Revealing hidden diversity among upside-down jellyfishes (Cnidaria: Scyphozoa: Rhizostomeae: *Cassiopea*): distinct evidence allows the change of status of a neglected variety and the description of a new species. Inv. Syst. 36, 63–89.

Gamero-Mora, E., Lawley, J.W., Marona, M.M., Stampar, S.N., Morandini, A.C., 2024. Morphological and molecular data in the study of the evolution, population genetics and taxonomy of Rhizostomeae. Adv. Mar. Biol. 98.

Gershwin, L., 2016. Jellyfish: A Natural History. The University of Chicago Press, Chicago.

Gershwin, L., Davie, P.J.F., 2013. A remarkable new jellyfish (Cnidaria: Scyphozoa) from coastal Australia, representing a new suborder within the Rhizostomeae. Mem. Qld. Mus. 56, 625–630.

Gershwin, L., Zeidler, W., 2008. Two new jellyfishes (Cnidaria: Scyphozoa) from tropical Australian waters. Zootaxa 1764, 41–52.

Gershwin, L., Zeidler, W., Davie, P.J.F., 2010. Medusae (Cnidaria) of Moreton Bay, Queensland, Australia. Mem. Qld. Mus. 54, 47–108.

Gibbons, M.J., Morandini, A.C., Straehler-Pohl, I., Bezio, N., 2022. Identification Guide to Macro Jellyfishes of West Africa. FAO, Rome.

Gohar, H.A.F., Eisawy, A.M., 1960. The biology of *Cassiopea andromeda* (from the Red Sea) (With a note on the species problem. Publ. Mar. Biol. Stat., Al-Ghardaqa 11, 5–42.

Gómez Daglio, L., Dawson, M.N., 2017. Species richness of jellyfishes (Scyphozoa: Discomedusae) in the Tropical Eastern Pacific: missed taxa, molecules, and morphology match in a biodiversity hotspot. Invert. Syst. 31, 635–663.

Gómez Daglio, L., Dawson, M.N., 2019. Integrative taxonomy: ghosts of past, present and future. J. Mar. Biol. Assoc. U. K. 99, 1237–1246.

Goy, J., 1979. Campagne de la Calypso au large des côtes atlantiques de l'Amérique du Sud (1961–1962) – 35. Méduses. Ann. Inst. Océanogr Paris 55 (Suppl), 263–296.

Goy, J., 1990. Preliminary observations on the zooplankton of Shark Bay, Western Australia, with emphasis on medusae and description of a new rhizostome jellyfish (Cnidaria, Scyphozoa). In: Berry, P.F., Bradshaw, S.D., Wilson, B.R. (Eds.), Research in Shark Bay. Report of the France-Australe Bicentenary Expedition Committee. Western Australian Museum, Perth, pp. 107–113.

Goy, J., Toulemont, A., 1997. Méduses. Musée Océanographique Monaco (Monaco).

Graham, W.M., Bayha, K.M., 2007. Biological invasions by marine jellyfish. In: Nentwig, W. (Ed.), Biological Invasions. Springer, Berlin, pp. 239–255.

Graham, W.M., Martin, D.L., Felder, D.L., Asper, V.L., Perry, H.M., 2003. Ecological and economic implications of a tropical jellyfish invader in the Gulf of Mexico. Biol. Inv. 5, 53–69.

Grenacher, H., Noll, F.C., 1876. Beiträge zur Anatomie und Systematik der Rhizostomeen. Abh. Senckenb. Naturf. Ges. 10, 119–179.

Gul, S., Morandini, A.C., 2013. New records of scyphomedusae from Pakistan coast: *Catostylus perezi* and *Pelagia* cf. *noctiluca* (Cnidaria: Scyphozoa). Mar. Biodiv. Rec. 6, e86.

Gul, S., Moazzam, M., Morandini, A.C., 2015. Crowned jellyfish (Cnidaria: Scyphozoa: Rhizostomeae: Cepheidae) from waters off the coast of Pakistan, northern Arabian Sea. Check List. 11, 1551.

Haddad, M.A., Nogueira Jr., M., 2006. Reappearance and seasonality of *Phyllorhiza punctata* von Lendenfeld (Cnidaria, Scyphozoa, Rhizostomeae) medusae in southern Brazil. Rev. Bras. Zool. 23, 824–831.

Haeckel, E., 1865. Ueber fossile Medusen. Z. Wiss. Zool. 15 (4), 504–514.

Haeckel, E., 1869. Ueber die fossilen Medusen der Jura-Zeit. Z. Wiss. Zool. 19 (4), 538–562.

Haeckel, E., 1874. Ueber eine sechszählige fossile Rhizostomee und eine vierzählige fossile Semaeostomee. Vierter Beitrag zur Kenutniss der fossilen Medusen. Jen. Zeits. Naturwiss. 8, 308–330.

Haeckel, E., 1880. Das System der Medusen. I, 2: System der Acraspeden. Gustav Fischer, Jena.

Haeckel, E., 1882. Report on the deep-sea medusae dredged by H.M.S. Challenger during the years of 1873-1876. Report on the scientific results of the voyage of H.M.S. Challenger during the years 1873–1876. Zoology IV, 1–154 (i–cv).

Heeger, T., 1998. Quallen. Gefährliche Schönheiten. Wissenschaftliche Verlagsgesellschaft mbH, Stuttgart.

Helm, R.R., 2018. Evolution and development of scyphozoan jellyfish. Biol. Rev. 93, 1228–1250.

Hofmann, D.K., Neumann, R., Henne, K., 1978. Strobilation, budding and initiation of scyphistoma morphogenesis in the rhizostome *Cassiopea andromeda* (Cnidaria: Scyphozoa). Mar. Biol. 47, 161–176.

Holst, S., Tiseo, G.R., Djeghri, N., Sötje, I., 2024. Approaches and findings in histological and micromorphological research on Rhizostomeae. Adv. Mar. Biol. 98.

Houghton, J.D.R., Doyle, T.K., Davenport, J., Hays, G.C., 2006. Developing a simple, rapid method for identifying and monitoring jellyfish aggregations from the air. Mar. Ecol. Progr. Ser. 314, 159–170.

Hyman, L.H., 1940. The Invertebrates: Protozoa through Ctenophora. McGraw-Hill Book Company, Inc, New York.

Jarms, G., Morandini, A.C., 2019. World atlas of jellyfish, first ed. Dölling und Galitz Verlag, Hamburg.

Jarms, G., Morandini, A.C., 2023. (e-Book). World atlas of jellyfish. Dölling und Galitz Verlag, Hamburg.

Jordano, M.A., Morandini, A.C., Nagata, R.M., 2020. Is phenotypic plasticity determined by temperature and fluid regime in filter-feeding gelatinous organisms? J. Exp. Mar. Biol. Ecol. 522, 151238.

Jordano, M.A., Miyake, H., Nagata, R.M., Morandini, A.C., 2022. Patterns of morphological development in Scyphozoa ephyrae (Cnidaria, Medusozoa). Mar. Biodiv. 52 (3), 31.

Jordano, M.A., Nagata, R.M., Morandini, A.C., 2024. A review of the role played by cilia in medusozoan feeding mechanics. Biol. Rev. 99, 950–964.

Karunarathne, K.D., de Croos, M.D.S.T., 2020. First records of three cepheid jellyfish species from Sri Lanka with redescription of the genus *Marivagia* Galil and Gershwin, 2010 (Cnidaria: Scyphozoa: Rhizostomeae: Cepheidae). Sri Lanka J. Aquat. Sci. 25, 45–55.

Karunarathne, K.D., de, Croos, M.D.S.T., 2021. Jellyfish species used as live baits in traditional trap fishery of Sri Lanka: *Acromitus flagellatus* and *Lychnorhiza malayensis* (Cnidaria: Scyphozoa). J. Indian. Soc. Coast. Agric. Res. 39 (2), 172–182.

Karunarathne, K.D., de Croos, M.D.S.T., 2023. Untapped edible jellyfish of Sri Lanka: *Crambione mastigophora* and *Rhopilema hispidum* (Cnidaria: Scyphozoa). Thalassas Int. J. Mar. Sci. 39, 657–670.

Karunarathne, K.D., Amaya, H.K.T., Digamadulla, D.S., De Croos, M.D.S.T., 2024. Redescription of the commercially harvested jellyfish, *Crambionella orsini* and *Lobonemoides gracilis* (Cnidaria: Scyphozoa) from Sri Lanka. Mar. Biol. Res.

Kayal, E., Bentlage, B., Sabrina Pankey, M., Ohdera, A.H., Medina, M., Plachetzki, D.C., et al., 2018. Phylogenomics provides a robust topology of the major cnidarian lineages and insights on the origins of key organismal traits. BMC Evol. Biol. 18, 68.

Keesing, J.K., Gershwin, L., Trew, T., Strzelecki, J., Bearham, D., Liu, D., et al., 2016. Role of winds and tides in timing of beach strandings, occurrence, and significance of swarms of the jellyfish *Crambione mastigophora* Maas 1903 (Scyphozoa: Rhizostomeae: Catostylidae) in north-western Australia. Hydrobiologia 768, 19–36.

Khabibulina, V., Starunov, V., 2019. Musculature development in planuloids of *Cassiopeia xamachana* (Cnidaria: Scyphozoa). Zoomorphology 138, 297–306.

Khabibulina, V., Starunov, V., 2021. Proliferation activity in the polyps of *Cassiopea xamachana* where the planuloid buds grow. Biol. Comm. 66, 333–340.

Kienberger, K., Prieto, L., 2018. The jellyfish *Rhizostoma luteum* (Quoy and Gaimard, 1827): not such a rare species after all. Mar. Biodiv. 48, 1455–1462.

Kienberger, K., Riera-Buch, M., Schönemann, A.M., Bartsch, V., Halbauer, R., Prieto, L., 2018. First description of the life cycle of the jellyfish *Rhizostoma luteum* (Scyphozoa: Rhizostomeae). PLoS One 13, e0202093.

Kikinger, R., 1992. *Cotylorhiza tuberculata* (Cnidaria: Scyphozoa) – life history of a stationary population. P.S.Z.N.I: Mar. Ecol. 13, 333–362.

Kishinouye, K., 1922. Echizen kurage. Zool. Mag. (Dobutsugaku zasshi) 34, 343–346 (in Japanese).

Kitamura, M., Omori, M., 2010. Synopsis of edible jellyfishes collected from Southeast Asia, with notes on jellyfish fisheries. Plankton Benthos Res. 5, 106–118.

Kogovšek, T., Bogunović, B., Malej, A., 2010. Recurrence of bloom-forming scyphomedusae: wavelet analysis of a 200-year time series. Hydrobiologia 645, 81–96.

Kramp, P.L., 1961. Synopsis of the medusae of the world. J. Mar. Biol. Assoc. U. K. 40, 7–469.

Kramp, P.L., 1970. Zoogeographical studies on Rhizostomeae (Scyphozoa). Vidensk. Meddel. Dan. Naturhist. Foren. 133, 7–30.

Larson, R.J., 1976. Marine flora and fauna of the northeastern United States. Cnidaria: Scyphozoa. NOAA Tech. Rep. NMFS Circular 397, 1–17.

Larson, R.J., 1990. Scyphomedusae and cubomedusae from the Eastern Pacific. Bull. Mar. Sci. 47, 546–556.

López-Martínez, J., Álvarez-Tello, F.J., Porchas-Cornejo, M.A., Nevárez-López, C.A., Muhlia-Almazán, A., Urías-Padilla, K.V., 2022. Multiple reproduction forms in the polyps of the cannonball jellyfish *Stomolophus* sp. 2: Probable life-cycle reversal. J. Exp. Zool. Part A: Ecol. Integr. Physiol. 339 (3), 239–252.

Lu, N., Zhao, Y., Jiang, S., 1992. The morphology and structure of *Stomolophus meleagris* L. Agassiz. Fish. Sci. (Liaoning) 11 (1), 5–8 (in Chinese).

Maas, O., 1903. Die Scyphomedusen der Siboga-Expedition. Siboga-Exped 11, 1–91.

Macri, S., 1778. Nuove osservazioni intorno la. Stor. naturale del. polmone marino degli antichi. Napoli, pp. 1–36.

Marques, A.C., Collins, A.G., 2004. Cladistic analysis of Medusozoa and cnidarian evolution. Invert. Biol. 123, 23–42.

Martellos, S., Ukosich, L., Avian, M., 2016. JellyWeb: an interactive information system on Scyphozoa, Cubozoa and Staurozoa. ZooKeys 554, 1–26.

Masilamoni, J.G., Jesudoss, K.S., Nandakumar, K., Satpathy, K.K., Nair, K.V.K., Azariah, J., 2000. Jellyfish ingress: a threat to the smooth operation of coastal power plants. Curr. Sci. 79, 567–569.

Mayer, A.G., 1910. The medusae of the world. The Scyphomedusae. Carnegie Inst. Wash., Publ. 109 III, pp. 499–735.

Mendoza-Becerril, M.A., Maronna, M.M., Pacheco, M.L.A.F., Simões, M.G., Leme, J.M., Miranda, L.S., et al., 2016. An evolutionary comparative analysis of the medusozoan (Cnidaria) exoskeleton. Zool. J. Linn. Soc. 178, 206–225.

Mianzan, H.W., 1989a. Sistematica y zoogeografia de scyphomedusae en aguas neriticas argentinas. Invest. Mar. Cicimar 4, 15–34.

Mianzan, H.W., 1989b. Las medusas Scyphozoa de la Bahía Blanca, Argentina. Bolm. Inst. Oceanogr. 37, 29–32.

Mianzan, H.W., Cornelius, P.F.S., 1999. Cubomedusae and Scyphomedusae. In: Boltovskoy, D. (Ed.), South Atlantic zooplankton, vol. 1. Backhuys Publishers, Leiden, pp. 513–559.

Mills, C.E., 2001. Jellyfish blooms: are populations increasing globally in response to changing ocean conditions? Hydrobiologia 451, 55–68.

Minakuchi, H., Toshino, S., 2022. The most beautiful photographs of jellyfish. Seibundo Shinkosha Tokyo. (in Japanese).

Miyake, H., Honda, S., Nishikawa, J., Yusoff, F.M., 2021. Life cycle of edible jellyfish *Acromitus hardenbergi* Stiasny, 1934 (Scyphozoa: Rhizostomeae) inhabiting a brackish-water environment. Animals 11 (7), 2138.

Miyake, H., Lindsay, D., 2013. Encyclopedia of jellyfish. Seibundo Shinkosha, Tokyo (in Japanese).

Moestafa, S.H., McConnaughey, B.H., 1966. *Catostylus ouwensi* (Rhizostomeae, Catostylidae), a new jellyfish from Irian (New Guinea) and *Ouwensia catostyli* n. gen., n. sp., parasitic in *C. ouwensi*. Treubia 27, 1–9.

Morandini, A.C., Ascher, D., Stampar, S.N., Ferreira, J.F.V., 2005. Cubozoa e Scyphozoa (Cnidaria: Medusozoa) de águas costeiras do Brasil. Iheringia, Sér. Zool. 95, 281–294.

Morandini, A.C., Stampar, S.N., Maronna, M.M., Silveira, F.L., 2017. All non-indigenous species were introduced recently? The case study of *Cassiopea* (Cnidaria: Scyphozoa) in Brazilian waters. J. Mar. Biol. Assoc. U. K. 97, 321–328.

Nagata, R.M., D'Ambra, I., Lauritano, C., von Montfort, G.M., Djeghri, N., Jordano, M.A., et al., 2024. Physiology and functional biology of Rhizostomeae jellyfish. Adv. Mar. Biol. 98.

Nakanishi, N., Hartenstein, V., Jacobs, D.K., 2009. Development of the rhopalial nervous system in *Aurelia* sp.1 (Cnidaria, Scyphozoa). Dev. Genes. Evol. 219, 301–317.

Namikawa, H., Soyama, I., 2000. Jellyfish Japanese waters. Hankyu, TBS Britanica Tokyo (in Japanese).

Neethling, S., Channing, A., Gershwin, L., Gibbons, M.J., 2011. A modern description of *Crambionella stuhlmanni* (Scyphozoa: Rhizostomeae) from St Lucia Estuary, South Africa. J. Mar. Biol. Assoc. U. K. 91, 357–367.

Nishikawa, J., Ohtsuka, S., Mulyadi, Mujiono, N., Lindsay, D.J., Miyamoto, H., et al., 2014. A new species of the commercially harvested jellyfish *Crambionella* (Scyphozoa) from central Java, Indonesia with remarks on the fisheries. J. Mar. Biol. Assoc. U. K. 95, 471–481.

O'Sullivan, D., 1982. A guide to the scyphomedusae of the Southern ocean and adjacent waters. ANARE Res. Notes 4, 1–43.

Ohdera, A.H., Abrams, M.J., Ames, C.L., Baker, D.M., Suescún-Bolívar, L.P., Collins, A.G., et al., 2018. Upside-down but headed in the right direction: review of the highly versatile *Cassiopea xamachana* system. Front. Ecol. Evol. 6, 35.

Omori, M., Kitamura, M., 2004. Taxonomic review of three Japanese species of edible jellyfish (Scyphozoa: Rhizostomeae). Plankton Biol. Ecol. 51, 36–51.

Pagès, F., Gili, J.-M., Bouillon, J., 1992. Medusae (Hydrozoa, Scyphozoa, Cubozoa) of the Benguela Current (southeastern Atlantic). Sci. Mar. 56, 1–64.

Pallas, P.S., 1774. Spicilegia Zoologica quibus novae imprimis et obscurae animalium species iconibus, descriptionibus atque commentariis illustrantur. Fasciculus 10. Berolini, Prostant Apud Gottl 1–41.

Paspaleff, G.W., 1938. Über die Entwicklung von *Rhizostoma pulmo* Agass. Arb. biol. Meeresst. Varna 7, 1–17.

Pitt, K.A., Purcell, J.E., 2009. Jellyfish blooms: causes, consequences, and recent advances. Hydrobiologia 616, 1–289.

Prieto, L., Astorga, D., Navarro, G., Ruiz, J., 2010. Environmental control of phase transition and polyp survival of a massive-outbreaker jellyfish. PLoS One 5 (11), e13793.

Purcell, J.E., Angel, D.L., 2010. Jellyfish blooms: new problems and solutions. Hydrobiologia 645, 1–234.

Purcell, J.E., Graham, W.M., Dumont, H.J., 2001. Jellyfish blooms: ecological and societal importance. Hydrobiologia 451, 1–334.

Purcell, J.E., Mianzan, H., Frost, J.R., 2012. Jellyfish blooms: interactions with humans and fisheries. Hydrobiologia 690, 1–290.

Purushothaman, J., Praveen Raj, C., Prasad, H., Bhowal, A., Siddique, A., Raghunathan, C., et al., 2022. Handbook on Indian Scyphozoa. Handb. Pict. Guide, Zool. Surv. India 74, 9–155.

Reinicke, G.B., Holst, S., Morandini, A.C., Thiel, H., 2024. Preface. Adv. Mar. Biol. 98.

Richardson, A.J., Bakun, A., Hays, G.C., Gibbons, M.J., 2009. The jellyfish joyride: causes, consequences and management responses to a more gelatinous future. TREE 24, 312–322.

Rizman-Idid, M., Farrah-Azwa, A.B., Chong, V.C., 2016. Preliminary taxonomic survey and molecular documentation of jellyfish species (Cnidaria: Scyphozoa and Cubozoa) in Malaysia. Zool. Stud. 55, 1–19.

Rottini Sandrini, L., Avian, M. (Eds.), 1986. Workshop on jellyfish in the Mediterranean Sea, 3–4 September 1984. UNEP-MAP, Trieste, Italy, pp. 191.

Russell, F.S., 1970. The medusae of the British Isles II. Pelagic Scyphozoa with a supplement to the first volume on hydromedusae. Cambridge University Press, London.

Schiariti, A., Kawahara, M., Uye, S., Mianzan, H., 2008. Life cycle of the jellyfish *Lychnorhiza lucerna* (Scyphozoa: Rhizostomeae). Mar. Biol. 156, 1–12.

Schiariti, A., Morandini, A.C., Jarms, G., Paes, R.G., Franke, S., Mianzan, H., 2014. Asexual reproduction strategies and blooming potential in Scyphozoa. Mar. Ecol. Prog. Ser. 510, 241–253.

Schiariti, A., Holst, S., Tiseo, G.R., Miyake, H., Morandini, A.C., 2024. Life cycles and reproduction of Rhizostomeae. Adv. Mar. Biol. 98.

Schuchert, P., 1993. Phylogenetic analysis of the Cnidaria. Z. Zool. Syst. Evol. forsch. 31, 161–173.

Schultze, L.S., 1898. Rhizostomen von Ambon. Denkschr. Med. Naturw. Ges. Jena. 8, 443–466.

Silveira, F.L., Cornelius, P.F.S., 2000. Novas observações sobre medusas (Cnidaria, Scyphozoa, Rhizostomeae) no nordeste e sul do Brasil. Acta Biol. Leopold. 22, 9–18.

Sötje, I., Henne, S., 2019. The senses of Scyphozoa. In: Jarms, G., Morandini, A.C. (Eds.), World atlas of jellyfish, first ed. Dölling und Galitz Verlag, Hamburg, pp. 46–53.

Souza, M.R.D., Dawson, M.N., 2018. Redescription of *Mastigias papua* (Scyphozoa, Rhizostomeae) with designation of a neotype and recognition of two additional species. Zootaxa 4457, 520–536.

Stiasny, G., 1920. Die Scyphomedusen-Sammlung des Naturhistorischen Reichsmuseums in Leiden. III. Rhizostomeae. Zool. Meded. 5, 213–230.

Stiasny, G., 1921. Studien über Rhizostomeen mit besonderer Berücksichtigung der Fauna des Malaiischen Archipels nebst einer Revision des Systems. Capita Zool. 1, 1–179.

Stiasny, G., 1934. Scyphomedusae. Discovery Rep. 8, 329–396.

Straehler-Pohl, I., Jarms, G., 2010. Identification key for young ephyrae: a first step for early detection of jellyfish blooms. Hydrobiologia 645, 3–21.

Straehler-Pohl, I., Jarms, G., 2022a. Back to the roots, Part 2 – Rhopaliophora (Scyphozoa, Cubozoa and Staurozoa) reborn based on early life cycle data. Plankton Benthos Res. 17, 105–126.

Straehler-Pohl, I., Jarms, G., 2022b. Back to the roots, Part 1 – Early life cycle data of Rhopaliophora (Scyphozoa, Cubozoa and Staurozoa). Plankton Benthos Res. 17, 1–33.

Syazwan, W.M., Low, L.B., Rizman-Idid, M., 2020. First record in peninsular Malaysia and morphological redescription of *Lychnorhiza malayensis* (Scyphozoa: Rhizostomeae: Lychnorhizidae). Raffles Bull. Zool. 68, 32–49.

Terenzini, J., Falkenberg, L.J., 2022. First records of the jellyfishes *Thysanostoma loriferum* (Ehrenberg, 1837) and *Netrostoma setouchianum* (Kishinouye, 1902) in Hong Kong waters. Check List. 18, 357–362.

Thibault, D., Kuplik, Z., Prieto, L., Enrique-Navarro, A., Brown, M.K., Uye, S.-I., et al., 2024. Ecology of Rhizostomeae. Adv. Mar. Biol. 98.

Thiel, H., 1963. Untersuchungen über die Entstehung abnormer Scyphistomae, Strobilae und Ephyrae von *Aurelia aurita* Lam. und ihre theoretische Bedeutung. Zool. Jb. Anat. 81, 301–358.

Thiel, M.E., 1969. Beobachtungen über Variationen und Aberrationen an der *Rhizostoma octopus* L.Ag. im Elbmündungsgebiet. Abh. Verh. Naturwiss. Ver. Hambg. 13, 119–142.

Thiel, M.E., 1970. Über den zweifachen stammesgeschichtlichen ("biphyletischen") Ursprung der Rhizostomae (Scyphomedusae) und ihre Aufteilung in die zwei neuen Ordnungen Cepheida und Rhizostomida. Abh. Verh. Naturwiss. Ver. Hambg. 14, 145–168.

Thiel, M.E., 1975. Bemerkungen zur Systematik der Gattung *Cassiopea* (Cepheida, Scyphomedusae). Mitt. Hamb. Zool. Mus. Inst. 72, 25–46.

Thiel, M.E., 1976. Die Cassiopeida (Arcadomyariae MAAS, 1903, Kampylomyariae STIASNY, 1921) als dritte Ordnung der Rhizostomae (Scyphomedusae) und die Erhebung der letzteren zu der Überordnung Rhizostomoidea (Wurzelmundartige) Supraord. Nov. Mitt. Hamb. Zool. Mus. Inst. 73, 1–16.

Thiel, M.E., 1978. Die postephyrale Entwicklung des Gastrovascularsystems der Rhizostomida nebst Ergänzungen und Berichtigungen zu den Stiasnyschen Typen dieser Entwicklung, zugleich ein Zeugnis für das Haeckelsche biogenetische Grundgesetz. Z. Zool. Syst. Evol. Forsch. 16, 267–289.

Thiel (†), M.E., Reinicke, G.B., Holst, S., Morandini, A.C., Sötje, I., Straehler-Pohl, I., et al., 2024. Rhizostomeae – a remaining manuscript to complete the monograph on Scyphozoa by M.E. Thiel (published 1936–1977). Adv. Mar. Biol. 98 (Electronic supplement).

Uchida, T., 1926. The anatomy and development of a rhizostome medusa, *Mastigias papua* L. Agassiz, with observations on the phylogeny of Rhizostomae. J. Fac. Sci. Imp. Univ. Tokyo Sect. IV 1, 45–95.

UNEP, 1991. Jellyfish blooms in the Mediterranean. Proceedings of the II workshop on jellyfish in the Mediterranean Sea. Athens, UNEP. MAP Tech. Rep. Ser. 47, 1–320.

Uye, S., 2008. Blooms of the giant jellyfish *Nemopilema nomurai*: a threat to the fisheries sustainability of the East Asian marginal seas. Plankton Benthos Res. 3, 125–131.

Van Iten, H., Leme, J.M., Simões, M.G., Marques, A.C., Collins, A.G., 2006. Reassessment of the phylogenetic position of conulariids (?Ediacaran-Triassic) within the subphylum Medusozoa (phylum Cnidaria). J. Syst. Palaeont 4, 109–118.

Vanhöffen, E., 1888. Untersuchungen über semäostome und rhizostome Medusen. Bibl. Zool. 1, 5–52.

von Lendenfeld, R., 1884. The scyphomedusae of the southern hemisphere. Part. III. Proc. Linn. Soc. NSW 9, 259–306.

von Lendenfeld, R., 1888. Über Coelenteraten der Südsee. VII. Die australischen rhizostomen Medusen. Z. Wiss. Zool. 47, 201–324.

von Montfort, G.M., Costello, J.H., Colin, S.P., Morandini, A.C., Migotto, A.E., Maronna, M.M., et al., 2023. Ontogenetic transitions, biomechanical trade-offs and macroevolution of scyphozoan medusae swimming patterns. Sci. Rep. 13, 9760.

Werner, B., 1984. Stamm Cnidaria, Nesseltiere. Lehrbuch der Speziellen Zoologie (co-ordinated by A. Kaestner) In: Gruner, H.-E. (Ed.), Band I, Wirbellose Tiere, 2 Teil, Cnidaria, Ctenophora, Mesozoa, Plathelminthes, Nemertini, Entoprocta, Nemathelminthes. Priapulida. Fisher, Stuttgart, pp. 11–133 Lehrbuch der Speziellen Zoologie (co-ordinated by A. Kaestner).

Yap, N.W.L., Ong, J.Y., 2012. A survey of jellyfish (Cnidaria) around St John's Island in the Singapore straits. Contr. Mar. Sci. 2012, 57–74.

Young, G.A., Hagadorn, J.W., 2010. The fossil record of cnidarian medusae. Palaeoworld 19, 212–221.

Further reading

Minemizu, R., Kubota, S., Hirano, Y., Lindsay, D., 2015. A Photographic Guide to the Jellyfishes of Japan. Heibonsha, Tokyo [in Japanese].

CHAPTER THREE

Approaches and findings in histological and micromorphological research on Rhizostomeae

Sabine Holst[a,*], Gisele R. Tiseo[b], Nicolas Djeghri[c,d], and Ilka Sötje[e]
[a]Senckenberg am Meer, German Centre for Marine Biodiversity Research (DZMB), Hamburg, Germany
[b]Departamento de Zoologia, Instituto de Biociências, Universidade de São Paulo, São Paulo, SP, Brazil
[c]The Marine Biological Association, Continuous Plankton Recorder Survey, Plymouth, United Kingdom
[d]University of Brest (UBO), Institut Universitaire Européen de la Mer (IUEM), Laboratoire des sciences de l'environnement marin (LEMAR, UMR 6539), Plouzané, France
[e]University of Hamburg, Institute for Cell and Systems Biology of Animals (IZS), Hamburg, Germany
*Corresponding author. e-mail address: sabine.holst@senckenberg.de

Contents

Advances in Marine Biology, Volume 98
ISSN 0065-2881, https://doi.org/10.1016/bs.amb.2024.07.005

Abstract

The substantial development of microscopic techniques and histological examination methods during the past five decades allowed for many new insights into the histology and microanatomy of Rhizostomeae. The present review focuses on new findings about histologically important structures: nerves, senses, muscles, gonads, zooxanthellae and nematocysts. Different ontogenetic stages of rhizostome species were included in the literature research, supplemented with the authors' unpublished data and figures. The overview of the research results reveals that the application of chemo- and immunohistochemical techniques have provided deeper insights into neuronal and sensory structures and their interconnections. Modern microscopic methods led to new findings on the histological gonadal organization and details of the processes of gametogenesis, fertilization, cleavage, gastrulation, and brooding. Advanced optical methods also allowed for a better understanding of Rhizostomeae-zooxanthellae associations and the morphology and function of nematocysts. Improvements in molecular biology allowed for more precise identification of zooxanthellae associated with rhizostome species. Although there has been significant progress in all of the research subjects covered here, we identify several knowledge gaps and conclude with some recommendations for future research.

1. Introduction

The optimization of microscopic techniques and staining methods as well as the introduction of immunohistochemistry and transmission electron microscopy (TEM) in the past 50 years revolutionized investigations at the cellular and ultrastructural level (Breton, 1999; Friedrich et al., 2014, Richter et al., 2010). In this review we present an overview on new findings in histology and microanatomy including different life stages of Rhizostomeae with the focus on histologically interesting and functionally relevant structures, namely nerves, senses, muscles, gonads, zooxanthellae and nematocysts, which have been also comprehensively addressed by Thiel in his review on Rhizostomeae up to 1970 (hereafter referred to as Thiel et al., 2024). To avoid possibly inappropriate generalizations, we have indicated the species names of the investigated specimens in the text, which in turn allows the readers to search for specific information on the species of their interest. In addition to literature data, our review includes some unpublished data and figures provided by the authors.

The application of modern techniques has greatly enhanced the knowledge on senses and nervous systems in Rhizostomeae in recent decades (e.g. Hofmann and Honegger, 1990; Satterlie, 2002) but still only few data are available for Scyphozoa compared to Hydrozoa (Garm and Ekström, 2010). Within the Scyphozoa, information about Rhizostomeae is typically even more scarce than for the Semaeostomeae for which most references refer to the species *Aurelia aurita* (Linnaeus, 1758) (Sötje and Henne, 2019). In the earlier review by Thiel (Thiel et al., 2024) the author stated that the rhopalia of Semaeostomeae and Rhizostomeae are similar regarding their location and structure and thus he provided examples from both taxa. Later authors agreed that the nervous systems of Rhizostomeae species are generally comparable to nervous systems in other scyphozoan taxa (Passano, 1988). With the development of various progressive techniques, the neuronal structures of rhizostomes have been explored to a much higher degree over the last five decades.

As mentioned by Thiel (Thiel et al., 2024), studies on the gonads (or reproductive organs) of rhizostome species were only conducted by a few authors before 1970. Even in the past five decades, studies on the histology and ultrastructure of rhizostome gonads and germ cells remained surprisingly rare, although they do provide histological details on gonadal location and arrangement, and on the development of male and female gametes. Most of the published reviews tended to generalize the findings of semaeostome species for all scyphozoans. Historically, Widersten (1965) is one of the pioneers in describing the histology of the gonads of the rhizostome jellyfish *Rhizostoma pulmo* (Macri, 1778). In later studies, the aspects of general cnidarian reproduction were reviewed (Arai, 1997; Campbell, 1974; Fautin, 1992; Larson, 1976). Lesh-Laurie and Suchy (1991) reviewed and detailed the morphology of the gonadal tissues and cell types. Harrison and Jamieson (1999) and Miller (1983) summarized the findings in spermatogenesis and sperm morphology. Recent studies have described details of the gonadal tissue organization, and processes like gametogenesis, germ cell release, fertilization, cleavage, gastrulation, and brooding in several rhizostome species (Eckelbarger and Larson, 1992; Hofmann and Hadfield, 2002; Ikeda et al., 2011a; Kikinger, 1992; Ohtsu et al., 2007; Pitt and Kingsford, 2000; Schiariti et al., 2012; Tiseo, 2022).

Some species of Rhizostomeae host dinoflagellates from the family Symbiodiniaceae as photosynthetic endosymbionts (historically called zooxanthellae). This kind of relationship was earlier described by Thiel

(Thiel et al., 2024) who provided a census of Rhizostomeae species believed to be associated with zooxanthellae at that time. Thiel also mentioned the significance of zooxanthellae for Rhizostomeae nutrition and physiology (Thiel et al., 2024) which has been investigated further over the past 50 years (see Nagata et al., 2024). Recent advances have also been made in the histology of zooxanthellae, particularly due to the emergence of the jellyfish *Cassiopea* spp. as a model organism for cnidarian-dinoflagellates associations (Ohdera et al., 2018). In addition, there are new insights into the diversity of rhizostome-dinoflagellates symbioses (Djeghri et al., 2019).

Few investigations on the occurrence of specific nematocyst categories in Rhizostomeae were conducted prior to 1970, for example, on the medusae of *Cotylorhiza tuberculata* (Macri, 1778) and *Rhizostoma octopus* (Gmelin, 1791), and polyps of *R. pulmo* (see Calder, 1983; Thiel et al., 2024). Improved light microscopic (LM) techniques, scanning electron microscopy (SEM), TEM, confocal laser scanning microscopy (cLSM) and the use of high-resolution microscope cameras have significantly improved the possibilities for morphological and histological nematocyst investigations in the past fifty years (e.g. Mariscal, 1984; Mariscal and Bigger, 1976; Östman, 2000; Szczepanek et al., 2002). Moreover, immunocytochemical approaches have uncovered the localization of the toxins (Lotan et al., 1996) and high-speed cameras have visualized the nematocyst discharge process (Holstein and Tardent, 1984). Nematocysts have been the subject of several review articles and book chapters (e.g. Arai, 1997; Hessinger and Lenhoff, 1988; Killi and Mariottini, 2018; Lesh-Laurie and Suchy, 1991; Sötje and Holst, 2019).

The present review presents selected details of research on rhizostome nerves, senses and muscles from the past five decades. Assuming that the neuronal and sensory structures do not differ significantly in the two taxa, information on Rhizostomeae have been supplemented by findings on Semaeostomeae to provide a complete overview of the current state of knowledge. Regarding the descriptions of rhizostome gonads, our review includes studies that used different microscopy techniques to understand different issues of the reproduction. The association of Rhizostomeae and dinoflagellates is presented in light of a more recent account of their diversity and histology to update and critically assess Thiel's descriptions (Thiel et al., 2024). Finally, we focus on nematocyst studies conducted on rhizostome species over the past 50 years to provide an overview on the most important aspects of nematocyst research by using various investigation techniques.

2. Nervous system, sensing and effector structures

Studies up to 1970 cited in the review by Thiel (Thiel et al., 2024), were mainly descriptive and fine neuronal structures were difficult to detect by earlier researchers because of the limited technical methods available. Many studies conducted during that time were based on physiological test methods instead of examinations on the structure itself (Thiel et al., 2024), thus limiting the level of knowledge about the nervous system in Rhizostomeae.

All stages in the scyphozoan life cycle need well-developed senses, nerve systems and epitheliomuscular cells to receive stimuli from their environment, to conduct the information and to perform movements (Arai, 1997; Sötje and Henne, 2019).

The medusa's nervous system is the most complex in comparison to all other life-stages, enabling the specimens to survive in a pelagic environment. Since the medusa is also the most conspicuous stage in the scyphozoan life cycle, the majority of older studies were conducted on medusae whereas investigations on the nervous systems in other life cycle forms (e.g. polyps) were rare (Arai, 1997). During the 20th century, several rhizostome life cycles have been described (see Schiariti et al., 2024), and an increasing number of studies on planulae, polyps and asexual reproduction led to crucial new findings about Rhizostomeae histology.

The first part of the following review (2.1.) summarizes general basic knowledge about the nervous system, the sensing and the effector, while the second part (2.2.) highlights the knowledge about these structures in specific life stages. In addition to the references about neuronal structures published in the past 50 years we included selected earlier references which are not mentioned in the review on Rhizostomeae up to 1970 (Thiel et al., 2024).

2.1 General knowledge

2.1.1 Nervous system

The nervous system, which can be developed as a diffuse nerve-net or as a more centralized system interacting with several diffuse nerve nets, has always a double function: conducting and integrating sensory inputs (Satterlie, 2011). The anatomy of the nervous system in Cnidaria is rather diverse, but all taxa share several distinct characters (Galliot, 2010). Neurosensory cells, ganglionic neurons and nematocytes (see Section 5) for mechano-chemo-sensing and venom secretion, as well as nerve-nets with chemo-conduction, can be differentiated in all cnidarian taxa. In planula larvae and swimming buds an apical sensory organ, also referred to as aboral knob, serves for the initiation of the metamorphosis

(Galliot, 2010; Hofmann and Honegger, 1990; Martin and Chia, 1982; Van Lieshout and Martin, 1992). As demonstrated for the rhizostome *Cassiopea xamachana* Bigelow, 1892, nerve nets are found in both epithelia (epidermis and gastrodermis), but are most abundant in the epidermis (Holland, 2003; Passano, 1988). The developmental level of nerve nets and sensory structures depends on the ontogenetic stage and on the life cycle stage (Chapman, 1974). Generally, the nervous system is located at the base of the epithelium or mesoglea and it can be differentiated in basoepithelial ganglion cells with long processes (including different neurons and also pacemaker neurons) and neurosecretory cells (without specialized neurosecretory sensory cells) (Chapman, 1974; Passano, 2004; Westfall, 1996). The neuron cells are of the neurosensory type, the sensory-motor type or the ganglionic type (Westfall, 1973).

Affiliated conducting cells and presumed photoreceptors are additional important parts of the nervous system (Chapman, 1974; Passano, 2004; Westfall, 1996). The affiliated conducting structures can be subdivided in (1) nervous and epithelial conducting cells, including receptor cells or sensory cells with an apical cilium for the perception of different stimuli (e.g. vibration, touch, chemicals, light), (2) cnidocysts which represent very specialized mechano-receptor or rather receptor-effector cells (Galliot, 2010), (3) statocyst mechanoreceptors, and (4) photoreceptors. Nerve-nets are connected to both, sensing-structures like sense organs, primary sense cells, and nematocysts and effector structures represented by the muscles (Chapman, 1974; Katsuki and Greenspan, 2013; Westfall, 1996). Each nerve-net, has its specific function and is composed of bipolar and multipolar nerve cells (Bozler, 1926; Horridge, 1956; reviewed in Thiel et al., 2024; Passano, 2004) that are branching and interconnecting and have axons at their basal ends (Passano and Passano, 1971). Only two of the several networks are conducting through with impulses or contraction waves that invade the whole region without declining (Satterlie, 2011, 2002). The giant fibre nerve net or motor nerve net (MNN) is a specialized and quick conducting pathway only present in medusae like the rhizostome *C. xamachana* but also in the semaeostomes for example, *Cyanea capillata* (Linnaeus, 1758), *A. aurita* (Anderson et al., 1986; Horridge, 1956; Passano, 1988). The diffuse nerve net (DNN) is a slower and more generalized pathway in medusae and in polyps, as described for *C. xamachana* and *A. aurita* (Horridge, 1956; reviewed in Thiel, 1959 see Reinicke et al., 2024, Passano, 2004, 1988). The MNN neurons activate muscle contractions for scyphomedusa swimming movements whereas the DNN serves as MNN in the medusa stage as well as in the polyp stage and has a sensory function (Horridge, 1956; Katsuki and Greenspan, 2013; Passano, 2004; Satterlie and Eichinger, 2014). In medusae, the rhopalia (complex marginal sense organs that contain a statocyst and eyes) are neuronal centres (rudimentary ganglia, mostly ectodermal), which are of crucial importance for the nervous system (Jacobs et al., 2007; Rentzsch et al., 2016; Satterlie, 2011). They are the integrative area in the MNN serving for the motor activity of swimming in *Nemopilema nomurai* Kishinouye, 1922, as also shown for other scyphozoans (Ohtsu and Uye, 2011; Passano, 1965). Both nerve nets are composed of bipolar and multipolar neurons, but the neurites, which are long processes of the basoepithelial nerve cells, are larger in the MNN of the rhizostome *C. xamachana* (Passano, 1988) and in the semaeostome *Chrysaora quinquecirrha* (Desor, 1848) (Westfall, 1996). In both taxa, the neurites of the DNN are morphologically similar to the nerve-types in the nerve nets of tentacles, oral arms, and gastrodermis (Katsuki and Greenspan, 2013; Passano, 1988). The gastrodermal nerve net cells (DNN) are strongly connected to each other, but are generally not connected to primary sense cells and free nerve-endings are rare (Passano and Passano, 1971). Cnidarian neurites contain mitochondria, ribosomes, glycogen, granular and agranular vesicles and smooth endoplasmic reticulum like other cells and they do not have wrapping cells, which makes it difficult to distinguish them histologically (Chapman, 1974). As shown for Semaeostomeae (*C. capillata*), neurites have a typical shape with axons and branching and form a layer between the cell bodies and the mesoglea (Horridge and Mackay, 1962). Generally, chemical synapses serve as neuronal connections in Cnidaria with ultrastructural features morphologically similar to synapses in higher taxa (Mackie et al., 1984; Satterlie, 2002; Westfall, 1996). The synapses in Scyphozoa can be unidirectional or bidirectional with peptidergic neurotransmitters (Chia et al., 1984; Grimmelikhuijzen, 1983; Westfall, 1996). MNN neurons are connected through bidirectional synapses in Rhizostomeae (*C. xamachana*) and in Semaeostomeae (*C. capillata*) (Anderson, 1985; Anderson and Schwab, 1981; Katsuki and Greenspan, 2013; Passano, 1988). Therefore, the excitation proceeds nonpolarized, conducts through the complete nerve-net and results in a synchronized contraction of the swimming musculature (Katsuki and Greenspan, 2013).

Neurite cells can have synapses to other neurites (interneurite or interneuronal connections) to nematocytes or to sensing structures (sensory-neuronal connections) and to the musculature (neuronal-motor connections) (Westfall, 1996). The synaptic gap is 11–18 nm in width in *Cassiopea andromeda* (Forskål, 1775) (Fig. 1A), which is similar to other

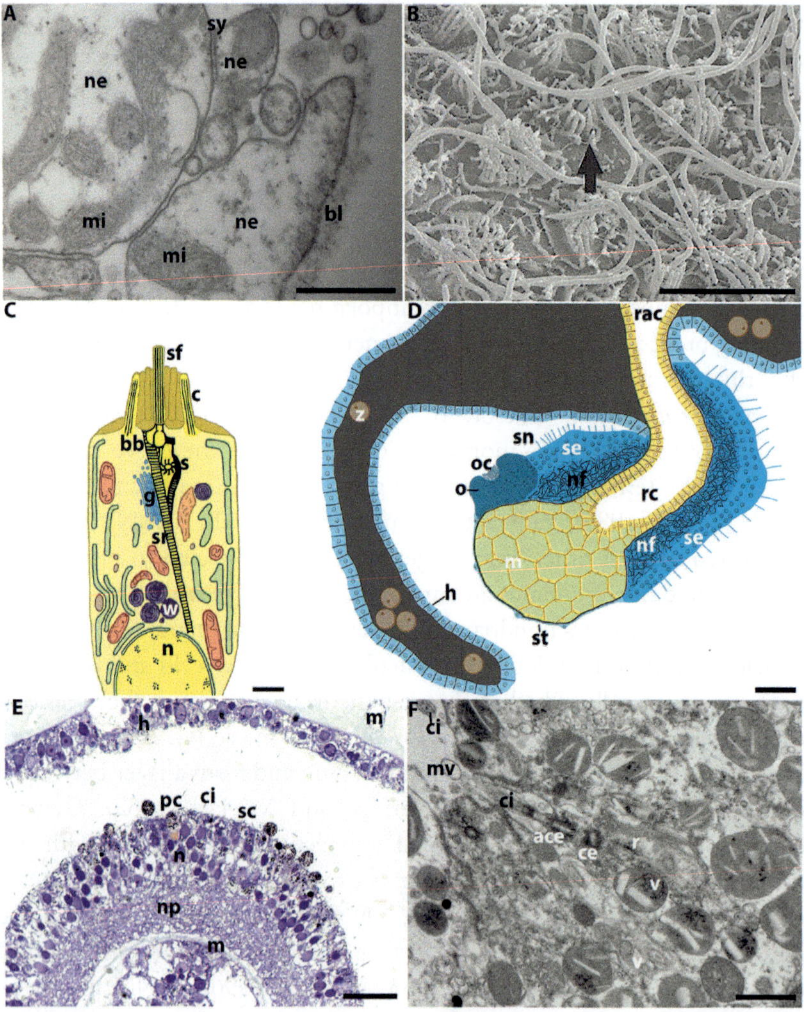

Fig. 1 Nerves and senses (A) *Cassiopea andromeda,* TEM transversal section through the rhopalial neuropil with neurons and a synapse. (B) *Cotylorhiza tuberculata,* SEM, hair cells bearing nonmotile kinocilia surrounded by a crown of stereocilia (arrow) (C) *Aurelia aurita,* schematic drawing of a mechanoreceptor with collar and static flagellum. (D) *Cassiopea xamachana,* schematic drawing of longitudinal section through the median area of the rhopalium. (E) *Cassiopea andromeda*, transversal section through rhopalium with section of ocellus, LM toluidine-blue stained. (F) *Cassiopea andromeda*, TEM, longitudinal section through ocellus with crystalline pigment vesicles and cilum. Abbreviations: *ace,* axial centriole; *bb,* basal body; *bl,* basallamina; *c,* collar; *ce,* centriole; *ci,* cilium; *f,* flagellum; *g,* golgi body; *h,* hood; *m*, mesoglea; *mi,* mitochondrium; *mv,* microvilli; *n,* nucleus; *ne,* neuron; *nf,* nervous fibres; *np,* neuropil; *o,* ocellus; *oc,* ocellus cavity; *pc,* pigment cell; *r,* rootlet; *rc,* rhopalium channel; *s,* satellite; *sc,* sense cell; *se,* sensoric epithelium; *sf,* static flagellum;

(continued)

cnidarian synapses (13–25 nm, Westfall, 1996) but *C. andromeda* synapses lack recognizable intracleft filaments (Hess and Sötje, unpublished) which are present in other cindarians (Westfall, 1996).

Three vesicle types, namely dense-cored, clear, and opaque, occur in Cnidaria synapses (Westfall, 1996). Only unidirectional synapses were identified in *C. andromeda* (Fig. 1A) with electron dense (diameter 55–90 nm) and electron transmitting vesicles (diameter 40–55 nm), leading to the assumption that different neurotransmitters may be involved (Hess and Sötje, unpublished). Neuromuscular synapses have been described in scyphozoan species, most of them are interneural, but some have terminal-like junctions with synaptic loci that innervate two or more effector structures (Westfall, 1996). Neuromuscular and neuron-nematocyte synapses have clear, dense-cored or opaque vesicles, as shown for semaeostome *A. aurita* medusae (Chia et al., 1984, Westfall, 1996). The vesicles appear larger and are less abundant compared to higher taxa, and postjunctional folds are lacking (Westfall, 1996). Scyphozoa lack gap junctions like Cubozoa, and studies using immunohistochemical staining led to the assumption that each exumbrella muscle cell is innervated by a motor-neuron with synaptic sites along the length of their neurites allowing synaptic interactions with other muscle cells and crossing neurons (Satterlie, 2015). The neuro-cnidocyte contact is similar to the interneural and the neuromuscular synapse (junction) (Chapman, 1974).

The nervous structures of Rhizostomeae (*C. xamachana*) and Semaeostomeae (*C. capillata*, *A. aurita*) have been examined, with several antibodies revealing the importance of several substances as neurotransmitters and differences regarding nervous structures. The MNN of the subumbrella is reactive to tubulin molecules, whereas the DNN is reactive to FMRFamide (Phe-Met-Arg-Phe-NH_2), RF-amide (arginine-phenylalanin terminal ending peptides) and taurine-like molecules

Fig. 1—cont'd *sn*, sensoric niche; *sr*, striated rootlet; *st*, statocyst; *sy*, synapse; *v*, vacuole with pigment and crystalline structure; *w*, whorled myelin figures; *z*, zooxanthelle. Scale bars: *A* = 0.5 μm, *B* = 100 μm, *C* = 10 μm, *D* = 1 μm, *E* = 10 μm, *F* = 20 μm. *(A) Photo: C. Hess, with permission. (B) From Solé, M., Lenoir, M., Vortuno, J.M., Durfort, M., Schar, vd. M., André, M., 2016. Evidence of cnidarians sensitivity to sound after exposure to low frequency under water sources. Sci. Rep. 6, 37979; SpringerNature, CC BY license. (C) Redrawn from Chapman, D.M., 1974. Cnidarian histology. In: Muscatine, L., Lenhoff, H.M. (Eds.), Coelenterate Biology: Reviews and New Perspectives. Academic Press, New York, pp. 1–92. Coelenterate Biology, Elsevier. (D) Redrawn from Bouillon, J., Nielsen, M., 1974. Etude de quelques organs sensoriels de cnidaires. Arch. Biol. 85, 307–328; (E) Photo: C. Hess, with permission (F) Photo: C. Hess, with permission.*

(Carlberg et al., 1995, Katsuki and Greenspan, 2013, Satterlie and Eichinger, 2014). FMRF-markers recognize sensory cells and neurons in sensory cell areas (Jacobs et al., 2007, Satterlie and Eichinger, 2014). These structures are denser in the polyp tentacles than in the polyp body wall and in the medusa bell (Jacobs et al., 2007). Sulfonated amino acid taurine depolarizes neurons in some nerve-nets (epidermal and gastrodermal) in *C. capillata*, revealing its function as an excitatory neurotransmitter (Carlberg et al., 1995).

2.1.2 Sensory structures

Sensory structures (impulse-conducting cells) can be of several different structural types (Jacobs et al., 2011). Scyphozoans have sensory structures in both epithelia, most of them located in the epidermis especially on the tentacles and around the mouth, with fewer present in the gastrodermis (Kass-Simon and Hufnagel, 1992; Werner, 1993). Sensory structures can be differentiated by their function in light-, mechano- and chemoreceptors, and appear already in the planula stage (Werner, 1993). The simplest sensory structure (primary sense cell, sense nerve cell) has a sensory neuron with a dendrite that appears as a cilium, whereas secondary sensory cells are connected to a conducting nerve cell (Jacobs et al., 2011; Werner, 1993). Primary sense cells have either a long and stiff cilium or a shorter cilium, located in a depression of the surface (Werner, 1993). Primary receptor cells include mechanoreceptors, specialized photoreceptors and cnidocytes (Chapman, 1974), which are receiving single or several stimuli from the environment like vibration, touch, light and chemical triggers. Epidermal sensory structures (sensory neurons, sensory organs) reach the epidermal surface, have cilia on their apical surface, which can be motile (e.g. exumbrellar sensory pit) or static (rhopalium), and have a basal axon (Chapman, 1974; Jacobs et al., 2007). Other sensory neuron cell bodies are located below the epidermis and can be even integrated within the nervous system (Jacobs et al., 2007). Sensory cells can also be connected with epitheliomuscular cells (e.g. in *A. aurita*) (Westfall, 1973). Specific sensory cells, which are immunoreactive for FMRF neuropeptide antibodies, occur at the tentacle surface and are connected to peptinergic nerve nets (Anderson et al., 2004). The stimulation of the apical cilia generates an electrical potential that runs along the axon of the sensory cell or is transmitted to a neural cell via chemical synapses (Jacobs et al., 2007). Further there are intraepithelial or intramesogleal sensory cells, which do not reach the epithel surface (Chapman, 1974), like ganglion cells (bipolar neurons, small and large isopolar neurons), neurosecretory cells (including some neurosecretory sensory cells) and presumed photoreceptors (e.g. endodermal eyes in *A. aurita*) (Chapman, 1974). In sensory structures for light reception, mechanoreception (e.g. acoustic, equilibrium, touch, stretch) and chemoreception (e.g. taste, smell, morphogens), sensory neurons are often part of specialized nonneuronal cells and in specific sense organs, large groups of sensory neurons are combined with nonneuronal cells forming macroscopic structures (rhopalia) (Jacobs et al., 2011). Sensory organs are more sophisticated in having individual or grouped sensory neurons associated with non-neuronal sensory cells, which are specialized for light, mechano- or chemoreception (e.g. vibration, touch, taste) and often bear cilia and microvilli on their apical surfaces (Jacobs et al., 2011). The rhopalium is a complex multifunctional neuro-sensory organ in the ephyra and the medusa stage (Rentzsch et al., 2016). Special receptors (GPCRs-G protein-coupled) for photo- and chemoreception or membrane-bound ion channels for reception of mechanical stimuli are involved (Jacobs et al., 2011).

Cnidocytes are specialized mechano- (vibration receptor) and chemoreceptor cells (Kass-Simon and Hufnagel, 1992). The morphology of cnidocytes (see also Section 5) is similar to a sensory cell, with a cilium, 18–22 surrounding stereocilia and a basal body-rootlet complex (Chapman, 1974). They respond to sensorial stimuli and are innervated (Anderson et al., 2004; Jacobs et al., 2011). Cnidocytes in the tentacles of scyphozoans are associated with FMRF- and RF-amid neuropeptide containing neurons and nerve nets indicating that the chemosensory reception for the cnidocysts discharge (see Section 5.4) may be regulated by a peptinergic innervation (Anderson et al., 2004). Another mechanoreceptor cell (hair cell) occurring in the rhopalial touch-plates in Semaeostomeae (e.g. *A. aurita*) and in Rhizostomeae (*R. pulmo*, *C. tuberculata*) (Chapman, 1974, Solé et al., 2016) bears apically a static cilium (kinocilium) with a long rootlet surrounded by a microvilli collar (stereocilia) and is very similar to epithelial cells without sensory function (Fig. 1B and C). Photoreceptor structures are divided in three categories with different specialization levels (Garm and Ekström, 2010): Extraocular reception, specialized photoreceptors (single-celled or multicellular ocelli) and proper eyes. Extraocular reception, the simplest photoreception structure, is represented by any cellular structure that responds to light but does not have screening pigments. Extraocular photoreception has been described in Hydrozoa and Anthozoa but has not yet been documented in Rhizostomeae. Ocelli occur in ephyrae and medusae of some rhizostome species of the genus *Cassiopea* (Fig. 1D–F) (Bouillon and Nielsen, 1974; Hess and Sötje, unpublished), and in *Mastigias papua* (Lesson, 1830) (Uchida, 1926), whereas proper eyes with screening pigments and optic structures like lenses have not been reported in Rhizostomeae (Garm and Ekström, 2010).

2.1.3 Effector structures (muscles)

The effector structures, represented by the muscles, enable movements by contractions in medusae and polyps. These effector structures appear mostly in form of epitheliomuscular cells with thinner and thicker myofibrils in their basal extensions and sometimes as myocytes without epithelial components (Blanquet and Riordan, 1981; Chapman, 1974). The ultrastructure of epitheliomuscular fibrils and filaments (Fig. 2A–F) is similar to higher taxa. Different types of myofibrils with different structural details have been classified (Chapman, 1974). Muscle extensions are coated individually by their plasma membranes, always attached to the subjacent mesoglea, and can form larger muscle fields (Blanquet and Riordan, 1981; Chapman, 1974). Therefore, the mechanical performance properties of the muscle are also dependent on features of the connected mesoglea. Epitheliomuscular cells in *A. aurita* are innervated through neuromuscular junctions that are specific in having an endoplasmatic reticulum with subsurface saccular or ballooned cisterna of endoplasmic reticulum. The adjacent cytoplasmic gap is 180–300 Å wide and the opposing epitheliomuscular cell is coated with filaments (Westfall, 1973).

2.2 Planula

Planulae are mouthless, with slightly broader anterior end, an inner solid gastrodermis area and an outer fully ciliated epidermis with nematocysts (Calder, 1982, 1973; Holst et al., 2007; Martin and Chia, 1982). Ultrastructural studies on planulae of the rhizostome *C. xamachana* suggest the lack of septal muscle strands, neural elements and an aboral knob (Martin and Chia, 1982), but later studies using methylene blue staining, have revealed sensory and ganglion type structures at least at the anterior end of *C. xamachana* planulae (Hofmann and Hellmann, 1995). Light sensing structures have not been detected in *Cassiopea* sp. planulae (Müller and Leitz, 2002; Widersten, 1968), but they have numerous highly immunoreactive nerve cell processes and perikarya, which can be detected with immunohistological methods (Grimmelikhuijzen, 1983). However, nervous structures have not been detected with electron microscopy in *Rhopilema verrilli* (Fewkes, 1887) and *C. xamachana* planulae (Lesh-Laury and Suchy, 1991; Martin and Chia, 1982).

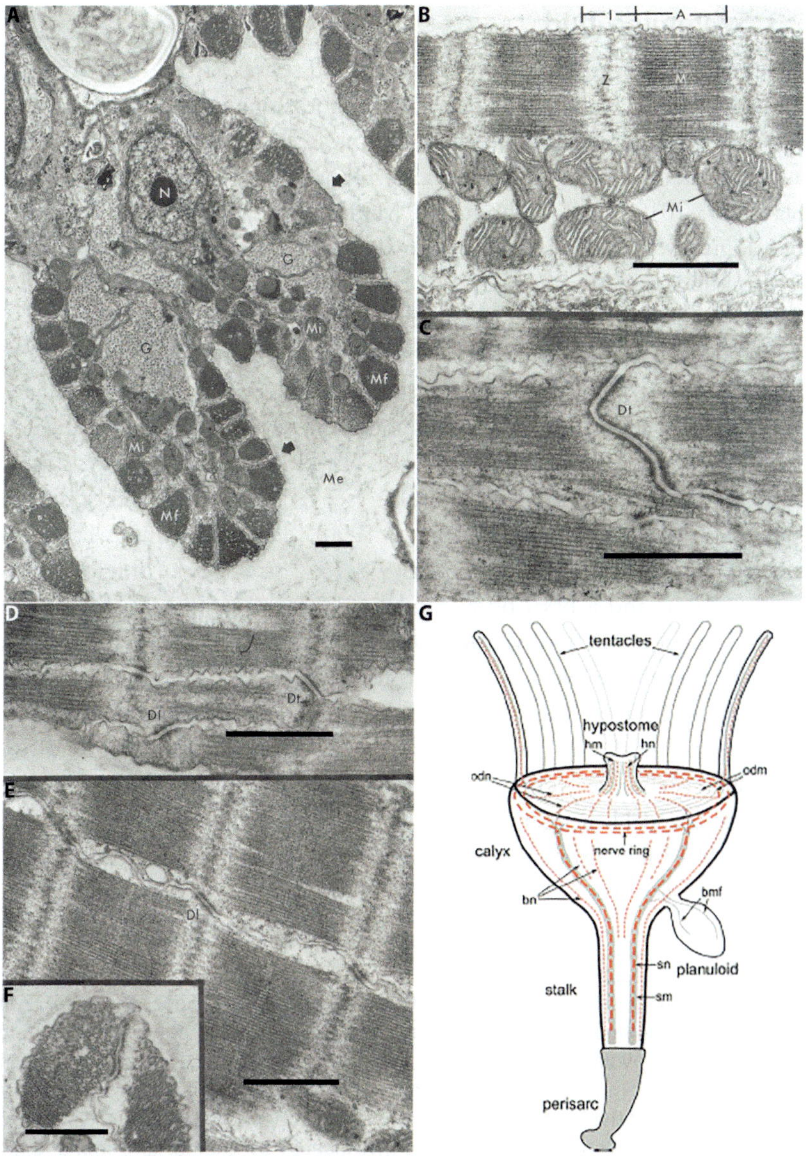

Fig. 2 *Cassiopea xamachana*, medusa musculature (A–F), and polyp musculature (G). (A) Transversal section (TEM) through the folded epithelial musculature. Each epithelio-muscular cell contains a myofibril, numerous mitochondria, and extensive glycogen deposits. A condensed zone of mesogleal fibres (arrows) occurs along the point of contact between the muscle fibres and the mesoglea. (B, C) longitudinal section through a muscle fibre (B) with single myofibrils and parallel rows of mitochondria, striated muscle fibril exhibit distinct I and A-bands, Z-lines and M-bands, (C) terminal desmosomal junction connecting the ends of two muscle fibres. (D) Integrated muscle fibres joined by

(continued)

2.3 Swimming bud

External and internal histological aspects of *Cassiopea* spp. free-swimming buds have been examined in detail with light, scanning and TEM as well as laser scanning confocal microscopy and intracellular vital staining (Hofmann and Gottlieb, 1991; Hofmann and Honegger, 1990; Khabibulina and Starunov, 2019; Van Lieshout and Martin, 1992). The swimming bud is histologically similar but also shows some differences to the planula (Khabibulina and Starunov, 2019; Martin and Chia, 1982; Van Lieshout and Martin, 1992). In *C. xamachana* swimming buds are larger in size (450–500 μm) than planulae (120–220 μm), have more cell types (15–20 instead of just 4), and have muscle strands in their septa as well as an aboral knob at their aboral end (Hofmann and Honegger, 1990; Martin and Chia, 1982; Van Lieshout and Martin, 1992). The buds contain cnidocysts in both epithelia and, in contrast to planulae (see Section 4.2), also some zooxanthellae (Hofmann and Honegger, 1990). The epidermis is completely ciliated with cilia that have a typical $9 \times 2 + 2$ structure of microtubules in cross-sections and a basal body-rootlet complex (Hofmann and Honegger, 1990, Van Lieshout and Martin, 1992). The mesoglea is very thin, whereas the gastrodermis develops as a thick layer. Gastrodermis cells in the knob-like area of the bud are also ciliated and extend into the large gastrovascular cavity. Epithelial cells are vacuolated and contain abundant lysosomes. Glandular cells are very abundant in the attachment area of the bud (Hofmann and Honegger, 1990).

FMRF-amide immunoreactive nerve structures have been examined in *C. andromeda* and *C. xamachana* buds (Hofmann and Hellmann, 1995;

Fig. 2—cont'd terminal and lateral (Z-line) desmosomes. (E) Series of lateral (Z-line) desmosomes joining adjacent muscle fibres with sarcomeres in phase. Note continuity of the I-band myofilaments with the desmosomal plaque. (F) Transverse section through adjacent muscle fibres showing the elongated area of desmosomal contact. Note continuity of the I-band myofilaments with the desmosomal plaque. (G) Overview on muscle and nervous system of polyp investigated by chemo- and immune-histological dyed samples. Abbreviations: *Dl*, lateral Desmosome; *Dt*, terminal desmosome; *G*, glycogen; *Me*, mesoglea; *Mf*, myofibril; *Mi*, mitochondria. Scale bars = 1 μm. *(A–F) TEM from Blanquet, R.S., Riordan, G.P., 1981. An ultrastructural study of the subumbrellarmusculature and desmosomal complexes of Cassiopea xamachana (Cnidaria: Scyphozoa). Trans. Am. Microscopical Soc. 100, 109–119. (G) Schematic drawing from Khabibulina, V., Starunov, V., 2020. FMRFamide immunoreactive nervous system in the adult Cassiopeia xamachana scyphopolyp and at the early stages of planuloid formation. Invertebr. Zool. 17, 371–384, reproduced with permission.*

Khabibulina and Starunov, 2020). Developing buds still connected to parental polyps show just a reaction to RF-amid neuropeptides in a small number of cells at the anterior end. Nervous structures of the parental polyp are not involved in the development of the bud nervous system. Only released buds have nerve elements in the ecto- and endoderm located close to the mesoglea (Hofmann and Honegger, 1990; Khabibulina and Starunov, 2020). Immunoreactive neurons have been detected in buds during the metamorphosis into polyps (Hofmann and Honegger, 1990). The bud to polyp metamorphosis depends on exogenous inducers (Fitt et al., 1987; Hofmann and Honegger, 1990) indicating that the bud has sensing receptors for morphogenetic chemicals. The cells at the anterior end of *C. xamachana* buds are putative sensory cells (Khabibulina and Starunov, 2020). Muscle structures of *C. xamachana* buds have been examined with modern fluorescent histological methods (TRITC-conjugated phalloidin labelling) (Khabibulina and Starunov, 2020; Fig. 2G) and with light- and electron microscopy (*C. andromeda*) (Hofmann and Honegger, 1990). In the early budding stage, the parental polyp septal muscle bands are connected to the developing bud muscle, whereas additional muscle structures develop in the bud after the detachment from the polyp (Khabibulina and Starunov, 2019). Newly released elongated buds have four longitudinal muscle groups with short orderless fibres located within the mesoglea, each composed of 3–18 elongated myocytes, which are connected by desmosomes and each containing a nucleus in their centre (Hofman and Honegger, 1990; Khabibulina and Starunov, 2019). They represent muscle structures of the prospective polyp (Khabibulina and Starunov, 2019) and become more prominent with the onset of the metamorphosis into the polyp stage (Hofmann and Honegger, 1990). Vesicles without or with electron-dense inclusions similar to dense-cored vesicles in neurosecretory material in other scyphozoans (Westfall, 1973) appear in some muscle processes, but there is no evidence for nervous structures in these muscle strands so far (e.g. *C. andromeda*, Hofmann and Honegger, 1990). During the early metamorphosis that starts before settlement, the bud changes into a cone-shaped stage and additional muscle structures appear in the hypostome and tentacle area of the developing polyp. The four muscle groups transform into muscle bands before the bud settles (Khabibulina and Starunov, 2019).

2.4 Podocyst

Podocysts (also termed as pedal cysts or pedalocysts) are produced beneath the pedal discs or at the end of stolons (see Schiariti et al., 2024). Podocysts are filled with a milky white mass of irregular shaped-cells containing a nucleus with nucleolus, few mitochondria, and granules in various sizes constructed of nutrients like proteins, carbohydrates, and lipids; nervous, sensory and muscle structures have not been detected with light or electron microscopic methods (Fu et al., 2019; Ikeda et al., 2011b).

2.5 Polyp

Rhizostome planulae and swimming buds transform into 8–24 tentacled polyps after settlement (see Schiariti et al. 2024). *C. xamachana* polyps have several nerve elements that are reactive to FMRFamide. Longitudinal muscles of the septa are aligned with neurites, a nerve ring in the marginal area of the oral disc contains circular neurites, and also the tentacles have neurites (Hofmann and Hellmann, 1995; Khabibulina and Starunov, 2020; Fig. 2G). The FMRF-immunoreactive nerve nets of the tentacles in *Cyanea* spp. are connected to the margin of the hypostome and to a prominent nerve ring at the tentacle bases (Hofmann and Hellmann, 1995). The dense nerve net of the hypostome and the oral disc forms a diffuse network with numerous processes, which consists mostly of three- and multipolar neurons, whereas the tentacles have mostly bipolar neurons (Hofmann and Hellmann, 1995; Khabibulina and Starunov, 2020). The oral area is connected to the surrounding part of the calyx and to the stalk only by a few delicate neurons (Hofmann and Hellmann, 1995). The density of neuronal processes is low in scyphozoan polyps (e.g. *A. aurita, C. xamachana*) in comparison to cubozoan and anthozoan polyps and they never form concentrated cords (Khabibulina and Starunov, 2020). Polyps lack obvious light sense organs, and chemical receptors have not yet been described. Primary sensory receptor cells with a single cilium and a single axon that are connected through synapses with a subepithelial neuron have been detected in the tentacles of the Semaeostomeae *A. aurita* (Chia et al., 1984). The arrangement of *C. xamachana* polyp muscle structures have been examined with chemo-histological methods (phalloidin-labelling) (Khabibulina and Starunov, 2020). As previously shown for *C. andromeda* (Hofmann and Gottlieb, 1991) each of the four polyp septa has an internal longitudinal muscle strand extending from the foot to the oral pole where it connects with the numerous thin tentacle muscles, which run along the tentacle (Khabilbulina and Starunov, 2020). Polyp muscle fibres are generally smooth, the septal muscle strands are ectodermal, located below the epidermis and are separated from the epidermis by a thin mesoglea layer. Scyphozoan polyp muscle cells are myocytes without an epithelial cell part (Werner, 1993) and can have thick and thin smooth myofibrils (Chapman, 1974). After strobilation, a tentacular nerve net is already present in the early stage of regenerating polyp tentacles in the polyp residuum (see Schiariti et al., 2024 for polyp regeneration after strobilation), whereas a nerve ring at the tentacle base develops later (Hofmann and Hellmann, 1995).

2.6 Ephyra and medusa

Young medusae (ephyrae) are produced by the polyp during strobilation (see Schiariti et al., 2024 for details). During strobilation, a dense nerve net develops at the oral side of the marginal lobes in each ephyra segment, a wide-meshed net of three- and multipolar nerves appears at the subumbrella, and the manubrium bears a dense aggregation of neurites along the longitudinal axis (Hofmann and Hellmann, 1995). Conduction takes place within nerve-nets in all scyphomedusae but without epithelial conduction because of lacking gap junctions (Mackie et al., 1984; Passano, 2004, 1982). The MNN (FMRF-IR stainable) and the DNN (tubulin-IR stainable) occur in the exumbrella and in the subumbrella but both nerve nets are less dense in the exumbrella (Satterlie and Eichinger, 2014). The DNN reaches all areas of the subumbrellar epidermis, transports information to the rhopalia and innervates some muscles including the oral arm musculature, whereas the MNN is responsible for conducting stimuli between the rhopalia and from rhopalia to swimming muscles and to sense organs, resulting in medusa movements (Passano, 1982). The MNN (tubulin reactive) of the subumbrella is restricted to the circular and radial muscle bands of the medusa (e.g. in *A. aurita*) (Satterlie and Eichinger, 2014). The DNN (FRMF-reactive) network is overlapping with the MNN in the subumbrellar area in the rhizostome *C. xamachana* and in semaeostome medusae and is extending from the margin into the manubrium (Satterlie and Eichinger, 2014). As already described in the review by Thiel (Thiel et al., 2024), *R. pulmo* medusae have two ectodermal nerve nets located in the subumbrella, connected to the rhopalia and composed of bipolar and multipolar nerve cells. In *C. xamachana* and other rhizostome medusae, the MNN is restricted to that part of the subumbrella, which bears the striated swimming musculature and the rhopalia (Passano, 1988) connected to each other only through the rhopalial ganglia. Additionally, local and specialized pathways within the rhopalia conduct light and gravity sensory information to the ganglia without involvement of the DNN (Passano, 2004). In contrast to semaeostome medusae, the MNN does not reach in the mouth-arm collar and the oral apparatus in rhizostome medusae (Passano, 1988). The MNN neurons receive signals from the pacemaker neurons in the rhopalia and activate the muscle contractions directly as a response (Katsuki and Greenspan, 2013). The DNN neurons activate marginal tentacle contractions, but do not induce muscle contractions directly. It is assumed that these neurons affect the pacemakers with passing sensory information towards them and use this pathway to affect the musculature indirectly. In addition, they influence the musculature with modulation (Katsuki and Greenspan, 2013). DNN pulses (in *Cassiopea* medusae) pass around the umbrella margin, from subumbrella to exumbrella or vice versa and the oral arm movement is controlled by these DNN pulses while the rhopalia seem not to trigger swimming spasms, but influence their intensity and period (Passano, 2004). The endodermal nerve-net (in *N. nomurai* and *C. capillata*) is generally structured like the ectodermal DNN, but lacks a primary sense cell (Passano and Passano, 1971). The rhizostome medusa *C. xamachana* has two large ectodermal subumbrellar nerve-nets which are conducting through, one in the gastrodermis and the other one in the oral apparatus (Horridge, 1956; Passano, 1988). As shown for Semaeostomeae (*Phacellophora camtschatica* Brandt, 1835, and *C. capillata*), the gastrodermal nerve-net is mainly composed of bipolar and a few tripolar neurons with unbranched neurites (Passano and Passano, 1971). In addition to the nerve nets in the manubrium and in the subumbrella, as well as nerve cell aggregations close to the rhopalia in rhizostome medusae, studies on rhizostome ephyrae (*C. andromeda*, C. *xamachana*) revealed a nerve net in the exumbrella and a circular cell aggregation structure at the umbrella margin consisting of bipolar neurons connected to the rhopalial sensory complex (Hofmann and Hellmann, 1995).

The most important sensing-structure in scyphozoan ephyrae and medusae is the rhopalium, a multifunctional neuro-sensory organ which is a highly modified polyp tentacle (Rentzsch et al., 2016) developing during strobilation (Fig. 3A–E). Newly released ephyrae of most species in the rhizostome taxa Kolpophorae (e.g. *C. tuberculata, M. papua*) and Dactyliophorae (e.g. *R. octopus*, *R. verilli*, *Stomolophus meleagris* Agassiz, 1860) have eight free rhopalia (without a surrounding subumbrellar niche) lined with the rhopalar lappets. Some species have more than eight rhopalia as for example *C. andromeda* with usually 16 but exceptionally more or less than 16 (11–23) rhopalia (Calder, 1982, 1973; Straehler-Pohl and Jarms, 2010).

In medusae, the rhopalia are located at the bell margin flanked by the rhopalar lappets (Fig. 3C). Each rhopalium accumulates morphological and molecular distinct groups of neuron ganglia and sensory structures, including contraction initiation for the swimming rhythm (pacemaker)

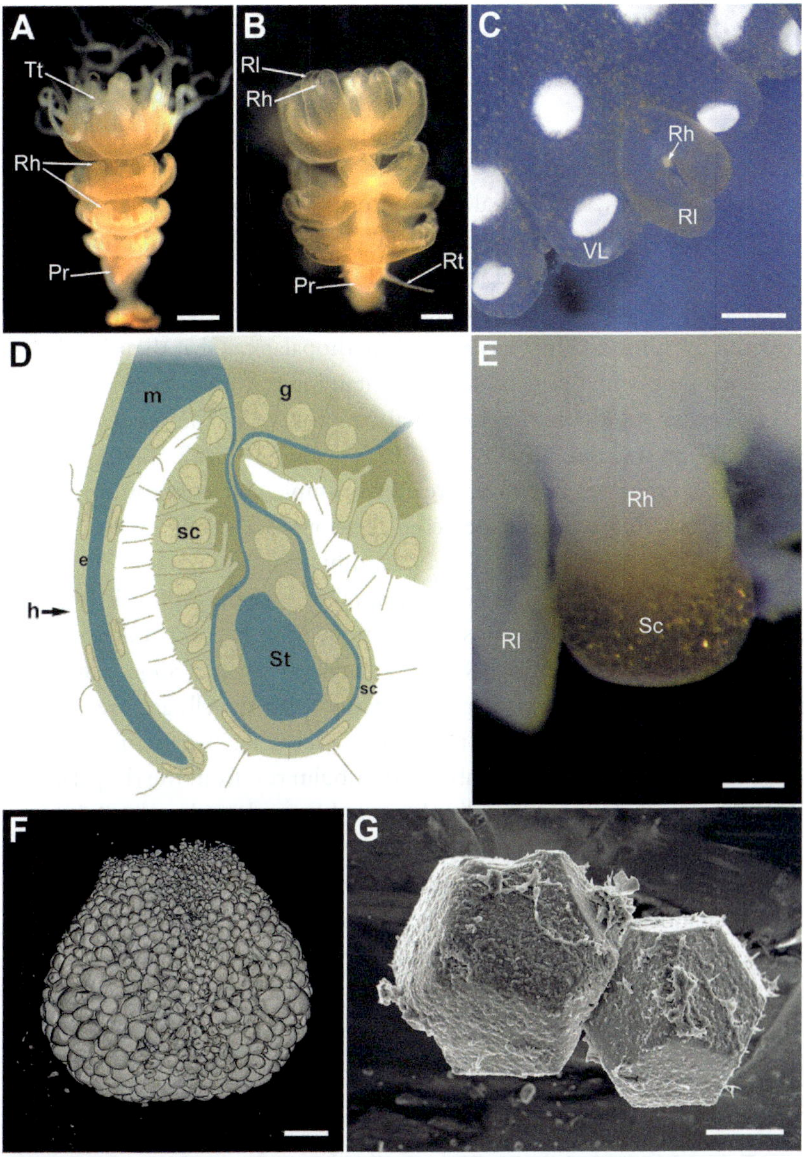

Fig. 3 Rhopalium, statocyst and statoliths. (A, B) *Rhizostoma octopus* living specimens. (A) Early strobila with tentacles transforming into rhopalia in the first (distal) ephyra. (B) Fully developed rhopalia in the distal ephyra. (C) *Phyllorhiza punctata*, living specimen. Bell rim of young medusa with rhopalium flanked by rhopalar lappets. Note: brownish colouring due to zooxanthellae. (D) Schematic drawing of a longitudinal section through a scyphozoan medusa rhopalium. (E, F, G) *Rhizostoma octopus* medusa rhopalium and statoliths. (E) Ethanol preserved rhopalium with statocyst containing reflective statolith crystals. (F) Micro-CT 3D reconstruction of statoliths

(continued)

(Arai, 1997; Horridge, 1959; Katsuki and Greenspan, 2013; Passano, 1973; Russel, 1970) and integration of sensory input and motor output interaction (Satterlie, 2011).

The rhizostome medusa rhopalium is morphologically similar to the semaeostome rhopalium regarding the hood-like extension between the lappets that roofs the rhopalium or forms the roof of a sensorial niche (rhopalar cleft) (Figs. 1D and 3D) (Arai, 1997; Bouillon and Nielsen, 1974; Russel, 1970; Satterlie and Eichinger, 2014; Solé et al., 2016). Rhopalia are often club-shaped and are located in a protective and sensorial niche covered by a hood. They have a hollow rhopalar canal at their basal stem, a proximal bulb and a distal statocyst containing statoliths (Solé et al., 2016).

Rhopalia are a complex neuronal component of the medusa's nervous system (Katsuki and Greenspan, 2013; Passano, 1982; Satterlie, 2011). Several different sensing structures for example, for equilibrium (gravity), light and touch have been detected in rhopalia (Ax, 1995; Bouillon and Nielsen, 1974; Chapman, 1974; Nakanishi et al., 2009).

The statolith area of the *C. andromeda* medusa rhopalium is surrounded by a multi-row single layered sensory epithelium. It consists of sensory cells and a basally located neuropil without ganglia (Hess and Sötje, unpublished, Fig. 1E). In *C. xamachana* rhopalia, putative sensory cells are present on the rhopalium and the adjacent niche area (Satterlie and Eichinger, 2014). Additionally, nerve fibres stainable with tubulin run from the rhopalar base in the subumbrella, where they distribute and form the subumbrella nerve-net (DNN) (Satterlie and Eichinger, 2014). The sensory cells in rhopalial sensory epithelia of the rhizostome species *C. tuberculata* and *R. pulmo* are crucial for the mechanoreception in interaction with the statocyst (Solé et al., 2016).

Gastrodermal cells of the statocyst area contain statoliths within a vacuole in scyphozoans like *Periphylla periphylla* (Péron and Lesueur, 1810), *C. tuberculata, C. xamachana* (Fig. 1D) and *R. pulmo* (Solé et al., 2016; Sötje et al., 2011). Statoliths in Rhizostomeae (e.g. *R. octopus, Cassiopea* sp.) increase in number and size with the growth of medusae (Holst et al., 2007; Hopf and Kingsford, 2013),

---

Fig. 3—cont'd arrangement in the statocyst. (G) SEM micrograph of compact statoliths with pentagonal surfaces. Abbreviations: *e*, epidermis; *g*, gastrodermis; *h*, hood; *Pr*, polyp residuum; *Rh*, rhopalium; *Rl*, rhopalar lappet; *Rt*, regenerating tentacle; *Tt*, transforming tentacle, *Sc*, statocyst, *St*, statolith; *Vl*, velar lappet. Scale bars: *A*, *B* = 500 μm, *C* = 1 mm, *D* = 20 μm, *E* = 100 μm, *F* = 50 μm, *G* = 10 μm. *(A, B) Photo: S. Holst; (C) Photo: I. Sötje; (D) Solé, M., Lenoir, M., Vortuno, J.M., Durfort, M., Schar, vd. M., André, M., 2016. Evidence of cnidarians sensitivity to sound after exposure to low frequency under water sources. Sci. Rep. 6, 37979. (E) Photo: S. Holst. (F) Photo: S. Holst (G) Photo: I. Sötje.*

as also shown for Semeostomeae (Heins et al., 2018; Sötje et al., 2017, 2011), and therefore the potential for medusa age determination in Scyphozoa by statolith parameters like number, size and calcein-binding properties is discussed. In Scyphozoa, the smallest statoliths are located in the proximal area of the statocyst where they develop first and the oldest statoliths are located distally (Boßelmann et al., 2007; Heins et al., 2018; Sötje et al., 2011). Micro-computed tomography examinations allow a detailed three-dimensional visualization of the statolith arrangement within the statocyst (Fig. 3F) and the analysis of species-specific statocyst shapes (Heins et al., 2018; Holst et al., 2016). The chemical shape of statoliths is described as trigonal needles which have two times three head-faces and six side-faces in the investigated Rhizostomeae *R. octopus, C. tuberculata* and *C. andromeda* as well as in other Scyphozoa (Becker et al., 2005; Boßelmann et al., 2007; Holst et al., 2007; Tiemann et al., 2002). Even though the chemical crystal shape is the same in all rhopaliophoran statoliths the morphological shape of the statoliths tends to be taxon-specific. For example, *R. octopus* has very compact statoliths with all faces similar in size (Fig. 3G), whereas *Cassiopea spp.* show hexagonal disc-like crystals in all stages (Boßelmann et al., 2007; Holst et al., 2007; Hopf and Kingsford, 2013). In living specimens, statoliths appear as bright refractive crystals with colour variations: for example, orange-brown (*R. octopus*, Fig. 3E), bright orange [*Catostylus mosaicus* (Quoy and Gaimard, 1824) in Pitt, 2000], conspicuous, shiny (*Rhopilema nomadica* Galil, Spanier and Ferguson, 1990 in Lotan et al., 1992) or bright yellow gold (*R. verrilli* in Cargo, 1971). Statoliths have been termed as otoliths (= Otholithen, German) or Konkrementkörner (German) by previous authors (Gohar and Eisawy, 1960a; Passano, 1982; Reinicke et al., 2024) and the statolith material of *C. andromeda* was described as a dissolvable chalk compound (Keller, 1883 as *C. polypoides* reviewed in Thiel et al., 2024). Later, the material of semaeostome statoliths was identified as gypsum (calcium sulfate dihydrate, $CaSO_4 \cdot 2H_2O$) by using x-ray microanalysis or spectrometry (Chapman, 1985; Spangenberg and Beck, 1968). More recent studies using sophisticated methods (energy dispersive x-ray spectroscopy, high-resolution x-ray powder diffractometry with synchrotron beamline) on different species with rhopalia (Rhopaliophora) including the Rhizostomeae *R. octopus*, as well as representatives of Semaeostomeae, Coronatae and Cubozoa revealed that the statoliths of these taxa are composed of calcium sulfate hemihydrate ($CaSO_4 \cdot 0.5\ H_20$, basanite) (Becker et al., 2005; Boßelmann et al., 2007). The higher density of this material is more effective regarding the gravity sensing compared to gypsum (Tiemann et al., 2002). Later studies with energy dispersive X-ray spectroscopy confirmed calcium sulfate in the statoliths of *Catostylus tagi* (Haeckel, 1869) (Santos et al., 2018) and

Phyllorhiza punctata Von Lendenfeld (1884) (Sötje, unpublished). The significance of statoliths for gravity sensing and the interaction with epidermal sensory areas has been investigated in the semaeostome *A. aurita* (Pollmans and Hündgen, 1981; Spangenberg et al., 1994).

Ocelli located at the rhopalia are found only in the pelagic ephyra and medusa stages of some rhizostome species like *C. xamachana* (Fig. 1D), *C. andromeda* (Fig. 1E) and *Mastigias* sp. (Garm and Ekström, 2010; Eimer, 1878; Russel, 1970, Bouillon and Nielsen, 1974). In other species, ocelli are lacking, e.g in *R. pulmo* (Eimer, 1878; Russel, 1970), *S. meleagris* (Hyman, 1940; Shanks and Graham, 1987), *N. nomurai* (Kawahara et al., 2006; Ohtsu and Uye, 2011) or are not mentioned in the descriptions of the medusa stage as in *S. meleagris* (Calder, 1982). Ocelli are nonvisual photosystems that enable the animals to perform controlled horizontal migrations (see also Nagata et al., 2024) directed by light conditions as observed in *Mastigias* sp. (Hamner and Hauri, 1981). Horizontal migrations of species (e.g. *S. meleagris*) lacking ocelli are correlated to wind and surface waves instead (Shanks and Graham, 1987), indicating that they may have sensory structures for the detection of surrounding-mechanical stimuli. The ocellus is composed of an accumulation of pigment granules on the aboral epidermal surface of the rhopalium (Gohar and Eisawy, 1960a). Cnidaria have ocelli with ciliary receptor cells which cellular projections (cilia) consist of microtubules (Eakin, 1979; Garm and Ekström, 2010). These receptor cells are neurons, which are specialized for photoreception, containing screening pigments (opsins) in their apical ends but lack optic structures like lenses (Albert, 2011; Martin, 2002; Plachetzki et al., 2005). The histological structure of the ocelli in medusae has been described in detail for *C. xamachana* (Bouillon and Nielson, 1974, in French): Two histologically identical rust-red ocelli are located at the base of the statocyst of each rhopalium, directed towards the ceiling of the sensory niche and connected to each other by a wide band of cubic and pigmented cells. They are part of the ectodermal sensory epithelium, have a central cavity, and measure 20 μm in height and 30 μm in diameter. The ocellus cells are sensory cells, with cilia at their apical surface having a typical $9 \times 2 + 2$ structure. Similar unspecialized photosensory cilia have been described in the *C. andromeda* (Fig. 1E and F) and the semaeostome *A. aurita* (Yamasu and Yoshida, 1973). These cilia project into the cavity of the ocellus, but *C. xamachana* and *C. andromeda* ocellus cilia do not have rootlets like *A. aurita* ocellus cilia (Bouillon and Nielson, 1974; Hess and Sötje, unpublished; Yamasu and Yoshida, 1973).

Visual pigments (photopigments) are important for the function of photoreceptors. The visual pigment molecule has a protein part (opsin) that is bound to an aldehyde derived from vitamin A. Opsins belong to G-protein coupled-receptors with signalling functions. Opsins (termed as Cnidorps in Cnidaria), covalently bind to a chromophore, are photosensitive and exist outside of visual photoreceptors. Ciliary photoreceptors contain c-type opsins that bleach after photoactivation through losing their chromophore (Porter et al., 2016). In the light sensory area with intraepithelial cilia of the *C. andromeda* rhopalium, rhodopsin is detectable using antibodies against the c-terminus of rhodopsin. The pigment cells of the ocellus show a weaker signal than the rest of the sensory area, leading to the assumption that the ocellus produces a shady area, which is detected by the photopigment cells (Hess and Sötje, unpublished). It has previously been assumed that the pigment cells in the ocellus are non-sensorial (Arai, 1997). The specimens are able to distinguish shady and sunny areas, and thus to find the optimal light intensity for the function of their zooxanthellae (Hess and Sötje, unpublished) (see also Section 4).

Regarding the musculature of rhizostome medusae, muscle cells are not present in all body parts and appear in varying extents. The exumbrella and the margin of the subumbrella ectoderm of the *C. andromeda* medusa lack muscles, whereas other parts of the subumbrella ectoderm are underlaid by an undulating and circular sheet of muscles, and muscle fibres extend along the oral arms and appendages (Gohar and Eisawy, 1960a). The subumbrellar muscles are arranged circular in rhizostome Dactyliophorae whereas in rhizostome Kolpophorae the arrangement of the musculature varies depending on the family (see Morandini, 2024). The ultrastructure of *C. xamachana* medusa muscles (Blanquet and Riordan, 1981) differ in different body regions (Blanquet and Riordan, 1981; Fig. 2A). The subumbrellar epitheliomuscular cells are striated with ortho-striated fibrils (Blanquet and Riordan, 1981; Passano, 1988), and a row of mitochondria on the side directed to the apical side of the epithelium (Chapman, 1974) (Fig. 2B–E). Muscle cells have terminal and z-line desmosomes with dense fibrous plaques and an extracellular space of 28.7 and 30.3 nm respectively. The ends of two muscle fibres are connected by a terminal desmosomal junction (Blanquet and Riordan, 1981; Fig. 2C–F). Adjacent muscles fibres are distanced variably (16–60 nm) and do not have nexal or tight junctions. Striated myofibrils exhibit distinct I and A-bands, Z-lines, and M-bands (Blanquet and Riordan, 1981; Fig. 2B–E).

3. Gonads

3.1 History of research on cnidarian gonads

Widersten (1965) conducted a histological comparative study on scyphozoan gonads. He described the gonad formation and gametogenesis in four scyphozoan species including the Rhizostomeae *R. pulmo*. The author was the first to point out the presence of accessory cells during the vitellogenesis process and referred to it as "nurse cells". The review of Campbell (1974) addressed the complex issue on reproduction in cnidarians focusing mainly on representatives of Hydrozoa and Anthozoa. The author mentioned colour differences in gonads allowing distinguishing between oocytes and sperms, and points on the size and shape differences between "ovaries" and "testicles". Campbell (1974) stated that germ cells originate from interstitial cells of the gastrodermis, cells that seem to have a wide distribution within the body and that are difficult to differentiate from each other in histological sections. These germ cells, which start the processes of spermatogenesis and oogenesis, respectively, were recognized for the first time by Campbell (1974) and were termed as "gonian" cells (spermatogonia and oogonia).

Miller (1983) presented general data on spermatogenesis and sperm ultrastructure, mainly for species of Hydrozoa, commenting on the origin of the gonads, and noting that there are few descriptive studies detailing with the process. Lesh-Laurie and Suchy (1991) compiled and reviewed the knowledge of the microscopic anatomy of Scyphozoa. The authors claimed that the anatomical organization of jellyfish is usually separated into a marginal or sensory "system" and a vegetative or central "system". The latter concept considers the gonads along with the manubrium, mouth, and oral appendages as part of the gastric cavity.

Arai (1997) presented the most comprehensive review of reproduction in Scyphozoa compiling data on the structure of the gonads for each order of Scyphozoa (Coronatae, Semaeostomeae and Rhizostomeae) and described the processes of gametogenesis, fertilization, and embryogenesis in a general way. The gonads are usually located close to the gastrovascular cavity adjacent to the gastric cirri, and each gonad is associated to a subgenital sinus with a total of eight gonads in Coronatae and four gonads in Semaeostomeae and Rhizostomeae (Arai, 1997).

Hinsch (1974) described the general ultrastructural sperm morphology with specific details for the rhizostome *Cassiopea* sp. and Hedwig and Schäfer (1986) investigate the sperm morphology of *R. pulmo*. In a study on the development of *C. tuberculata* from the ephyra to the young medusa, Avian (1986) observed the formation and fusion of the subgenital cavity. The study of Avian and Rottini-Sandrini (1991) is one of the first describing the detailed oocyte development in scyphomedusae and pointing on similarities in oocyte maturation process in *R. pulmo* compared to three semaeostome species.

Considering the total number of species of Scyphozoa (222 morphospecies *sensu* Jarms and Morandini, 2019) gametogenesis has been investigated for only a few of them (15 in total, nine being rhizostomes). Several studies on the gonadal morphology, oogenesis and/or spermatogenesis and brooding in Rhizostomeae species (e.g. *Cassiopea* sp., *R. pulmo*, *S. meleagris*, *C. tuberculata*, *C. mosaicus*, *N. nomurai* and *Lychnorhiza lucerna* Haeckel, 1880) include ultrastructural investigations in addition to light microscopy (Avian and Rottini-Sandrini, 1991; Eckelbarger and Larson,1992; Eckelbarger and Hodgson, 2021; Hofmann and Hadfield, 2002; Ikeda et al., 2011a; Kikinger, 1992; Ohtsu et al., 2007; Pitt and Kingsford, 2000; Rouse and Pitt, 2000; Schiariti et al., 2012; Tiseo, 2022, 2016) (Table 1). Together, these have resulted in a better understanding and deeper insights into these processes. Additional studies provide data on seasonal changes in oocyte sizes in *N. nomurai* (Toyokawa et al., 2010), lipid variation in oocytes in *S. meleagris* (Carvalho-Saucedo et al., 2010), biochemical composition of "ovaries" in *R. pulmo* (Stabili et al., 2019) and metabolic profile in female gonads in *R. pulmo* (Angile et al., 2020) completing the general understanding of female sexual maturation and development.

3.2 Formation, location, and arrangement of Rhizostomeae gonads

In Rhizostomeae, as in the majority of all scyphozoans, the sexes are separate and gonochorism is a commonly observed feature (Arai, 1997; Campbell, 1974; Eckelbarger, 1994; Fautin, 1992; Holst et al., 2007; Kikinger, 1992; Rouse and Pitt, 2000; Widersten, 1965). Hermaphroditism has been observed in *Pseudorhiza haeckeli* Haake, 1884 (Haacke, 1884) and *C. andromeda* (Fig. 4A and B) from the Red Sea in the Indo-Pacific (Hofmann and Hadfield, 2002) with female and male parts intermingling gonads. The gonad has its origin from interstitial cells of gastrodermis and cavities (Widersten, 1965) that appear in segregated cell masses near the genital sinus. These cavities fuse giving rise to the genital sinus first reported by Widersten (1965), and later by Lesh-Laurie and Suchy (1991) and Arai (1997). The fully developed gonad is usually located interradially on the floor of the

Table 1 Published studies addressing gametogenesis in Rhizostomeae species. − = no information, + = information available.

Species	Spermatogenesis	Sperm morphology	Oogenesis	References
Cotylorhiza tuberculata	+	−	+	Kikinger (1992)
Cassiopea andromeda	+	−	+	Gohar and Eisawy (1960b), Hofmann et al. (1996), Mammone et al. (2023), Tiseo (2022)
Cassiopea frondosa	+	−	+	Smith (1936), Tiseo (2022)
Cassiopea maremetens	+	+	−	Tiseo (2022)
Cassiopea xamachana	+	+	+	Tiseo (2022)
Phyllorhiza punctata	−	+	−	Rippingale and Kelly (1995), Rouse and Pitt (2000)
Versuriga anadyomene	−	−	+	Tiseo (2022)
Catostylus mosaicus	+	+	+	Pitt and Kingsford (2000), Rouse and Pitt (2000)
Catostylus tagi	+	−	+	Lisboa et al. (2018)
Lychnorhiza lucerna	+	+	+	Schiariti et al. (2008), Schiariti et al. (2012), Tiseo (2022)
Nemopilema nomurai	+	+	+	Ohtsu et al. (2007), Iguchi et al. (2010), Ikeda et al. (2011a)
Rhizostoma pulmo	+	−	+	Widersten (1965), Avian and Rottini-Sandrini (1991), Paspaleff (1938)
Rhopilema nomadica	+	−	−	Lotan et al. (1992), Lotan et al. (1994), Tiseo (2022)
Stomolophus meleagris	−	−	+	Eckelbarger and Larson (1992), Rouse and Pitt (2000)

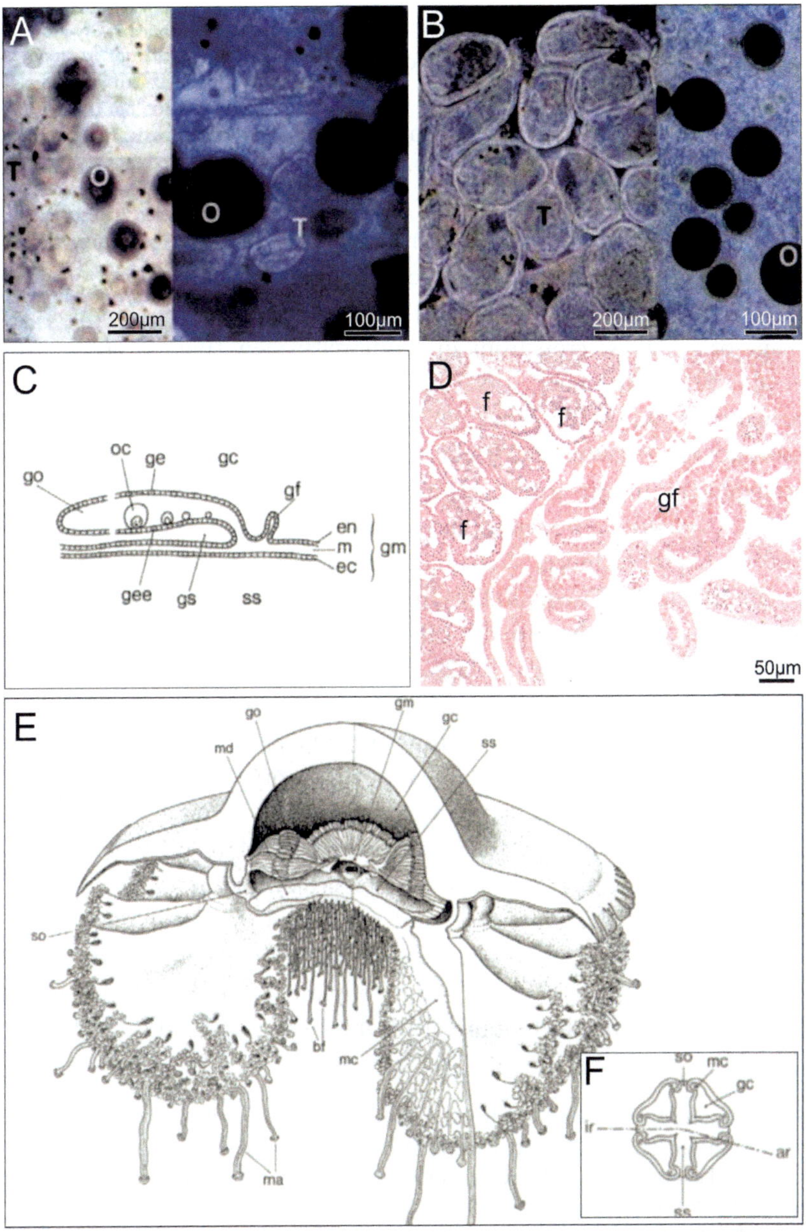

Fig. 4 (A) *Cassiopea andromeda*. Hermaphroditic gonads. Male (left) and female (right). (B) *Cassiopea andromeda* sperm follicles (left) and oocytes (right). (C) Longitdutinal section through gonad of *Cotylorhiza tuberculata*. (D) *Lychnorhiza lucerna*. Longitudinal section of male gonad of near the gastric cirri stained with (hematoxylin and eosin stained). Note: male follicles filled with sperm. (E) *Cotylorhiza tuberculata* longitudinal

(continued)

gastrovascular cavity, adjacent to the gastric cirri (Fig. 4C and D) (Arai, 1997; Kikinger, 1992; Larson, 1976; Tiseo, 2022) and above the subgenital cavity (Eckelbarger, 1994).

In contrast to the Semaeostomeae, the Rhizostomeae do not have a central mouth (see Morandini, 2024), and the gonads are oriented towards the mouth arms (Fig. 4E) forming bands or sacs that may hang down into the subumbrellar cavity (Eckelbarger, 1994). Rhizostome species have gonads with a distinct L-shaped morphology that increases in size with the growth of the medusa, resulting in a cruciform appearance resembling the shape of a Maltese cross from an exumbrellar view (Fig. 4E and F) (Ikeda et al., 2011a; Kikinger, 1992; Lisboa, 2018; Pitt and Kingsford, 2000; Schiariti et al., 2012; Tiseo, 2022). The main difference in the gonad morphology between the eight families of Rhizostomeae is the gonad arrangement in the gastric cavity, which is narrower in specimens of Cepheidae and Cassiopeidae and wider in species of the families Mastigiidae, Leptobrachidae, Versurigidae, Catostylidae, Lobonemidae, and Lychnorhizidae (Tiseo, 2022). In Rhizostomatidae and Stomolophidae the "L" is wider than in the other families giving the impression of a "V-shaped" gonad (Tiseo, 2022).

In the rhizostomes *C. tuberculata, C. mosaicus, R. pulmo* and *R. octopus,* the gonads are described as a tissue band of gametic cells with different stages of development (Kikinger, 1992; Pitt and Kingsford, 2000; Widersten, 1965). In *L. lucerna* each arm of the cross-shaped gonad is a band-like evagination of the gastrodermis within the gastric pouch, forming a fold which contains digitate prolongations of the epithelium (gastric cirri) at the inner side, and reproductive cells at the outer side (Schiariti et al., 2012). *N. nomurai* is a gonochoristic jellyfish with a ribbon-like "ovary" and "testis" at the bottom of pleated interradial gastric pouches that protrude beneath the umbrella (Ohtsu et al., 2007; Uye, 2014). The

Fig. 4—cont'd section through medusa bell with gonad. (F) *Cotylorhiza tuberculata* transversal section showing cross shape of gonad. Abbreviations: *a*, adrradius; *bf*, brood-carrying filaments; *ect*, ectoderm; *en*, endoderm; *f*, male follicle; *gc*, gastric cavity; *gf*, gastric filaments; *gm*, gastrogenital membrane; *go*, gonad; *gs*, genital sinus; *ir*, interradius; *m*, mesoglea; *ma*, moutharm appendage; *mc*, moutharm canal; *md*, moutharm disk; *O*, Ovary; *so*, subgenital ostium; *ss*, subgenital space; *T*, testis. *(A, B) from Hofmann, D.K., Hadfield, M.G., 2002. Hermaphroditism, gonochorism, and asexual reproduction in Cassiopea sp.—an immigrant in theislands of Hawaii. Invertebr. Reprod. Dev. 41, 215–221; (C, E, F) from Kikinger, R., 1992. Cotylorhiza tuberculata (Cnidaria: Scyphozoa)—life history of astationary population. Mar. Ecol. 13, 333–362 (D) Photo: G. Tiseo.*

mature gonad of *N. nomurai* can hang down from the subgenital ostia in sac-like structures termed as "large sacs" (Ohtsu et al., 2007). In *S. meleagris,* the bluish-coloured female gonads occur on four pleated membranes located on the floor of the stomach and numerous oocytes in all stages of development arise in each ovary from a 1 cm wide band just distal to the gastric cirri (Eckelbarger, 1994; Eckelbarger and Larson, 1992).

3.3 Histological structure of Rhizostomeae gonads

Similar to Semaeostomeae species, the gonad in Rhizostomeae is divided into three tissue layers: an outer cylindrical, ciliated endodermal layer, an inner genital epithelium and the germ cells immersed in the mesoglea (Fig. 5A and B) (Miller, 1983; Tiseo, 2022, 2016; Widersten, 1965). The histology of the gonadal structures is similar in all Rhizostomeae species but different authors use different nomenclatures and terms to describe the gonadal epithelia. Avian and Rottini-Sandrini (1991) described the inner genital epithelium of *R. pulmo* gonads as a "secondary endoderm", which mainly consists of cuboid cells with no maturation gradient of oocytes immersed in the mesoglea. Pitt and Kingsford (2000) used the term "gastrodermis" referring to the gonadal epithelia of *C. mosaicus* whereas earlier authors (reviewed in Thiel et al., 2024) used the term "germinal layer" to refer to the inner genital epithelium of *Cassiopea frondosa* (Pallas, 1774) and *C. andromeda* (Gohar and Eisawy, 1960b; Smith, 1936). Kikinger (1992) refers to the gonad of *C. tuberculata* as a band-like endodermal epithelium, which separates the gastric cavity from the subgenital space, but also mentions a "genital epithelium" which is located near the mesoglea and contains zooxanthellae. The gonad of *N. nomurai* is described as an epithelium of endodermal origin, containing ovarian epithelial cells in the female gonad and testicular epithelial cells in the male gonad (Ikeda et al., 2011a). In *L. lucerna* gonads are two-layered in cross-section. The gastrodermis surrounds the gastrovascular cavity with gonadal mesoglea lying beneath. The genital epithelium is located on the subumbrellar side interior to the mesoglea separating it from the genital sinus. The four subgenital cavities lay between the subumbrella and the oral disc each with one subgenital ostium located between the four oral pillars (Schiariti et al., 2012; Tiseo, 2022). In cross section the female gonad of *S. meleagris* is described to be of "gastrodermis" with the developing oocytes immersed in mesoglea and maintaining intimate contact with the gastrodermal trophocytes (Eckelbarger and Larson, 1992).

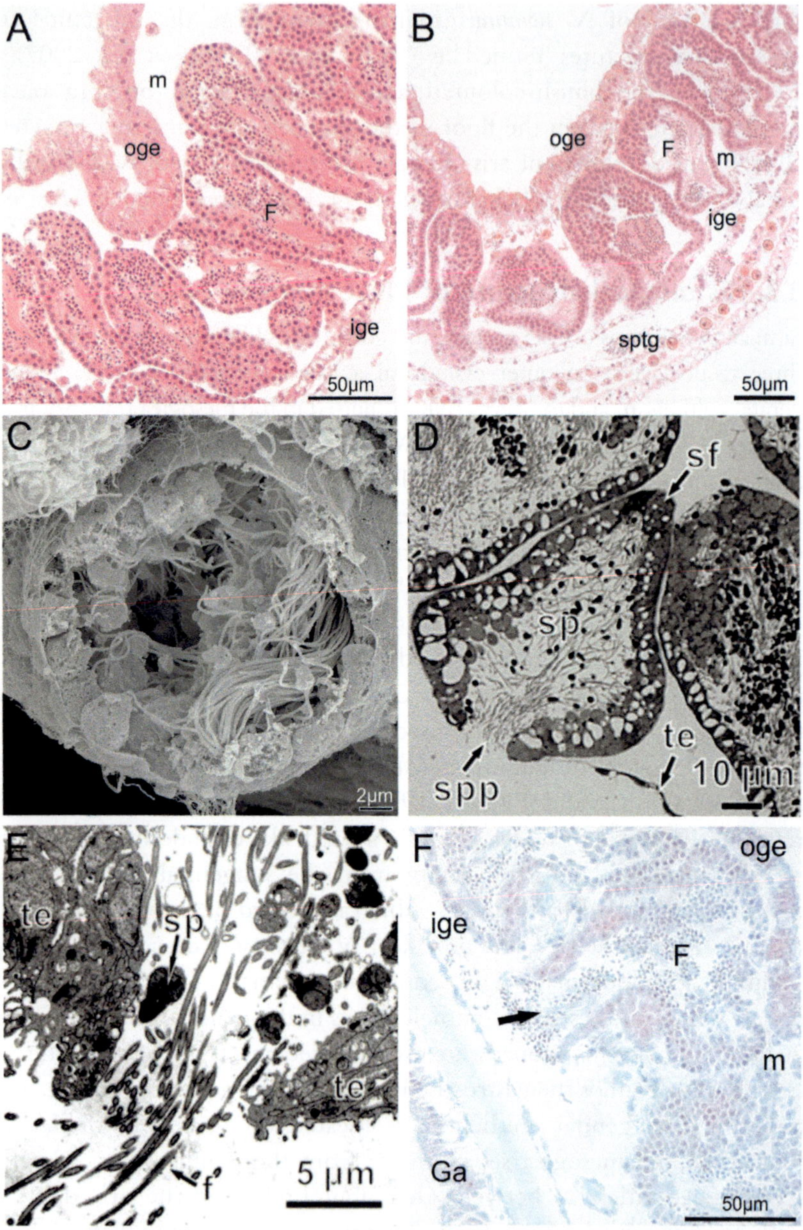

Fig. 5 (A) *Lychnorhiza lucerna*. Longitudinal section of the male gonad (hematoxylin and eosin). The gonad is divided in inner and outer gonadal epithelium and between them, the male follicles immersed in the mesoglea. (B) Longitudinal section of the male gonad of *Cassiopea maremetens* stained with hematoxylin and eosin. Note the gonadal epithelia and the spermatozeugmata in the genital sinus. (C) Scanning electron *(continued)*

3.4 Development and release of germ-cells

In the Rhizostomeae species investigated so far, the germs cells usually go through a series of morphological transformations resulting in the final stages of the two gamete types, namely sperm and oocyte. These are released (or not) from the gonad into the gastrovascular cavity (Avian and Rottini-Sandrini, 1991; Eckelbarger and Larson, 1992; Ikeda et al., 2011a; Schiariti et al., 2012; Tiseo, 2022; Widersten, 1965). Widersten (1965) described the gametogenesis of *R. pulmo* as being similar to other Discomedusae species and Avian and Rottini-Sandrini (1991) detailed the oogenesis of the species using ultrastructural investigations. Eckelbarger and Larson (1992) focused only on the oocyte development of *S. meleagris* and Eckelbarger (1994) reviewed the knowledge of scyphozoan ovarian structure and oocyte nutrition, including information of *C. xamachana* and *S. meleagris* oogenesis. Pitt and Kingsford (2000) mentioned the gametogenesis of *C. mosaicus* and Rouse and Pitt (2000) described the sperm ultrastructure. Ohtsu et al. (2007) and Ikeda et al. (2011a) examined the gametogenesis of *N. nomurai* detailing the oocyte and sperm histology and ultrastructure including details on the fertilization induction by light exposure. Schiariti et al. (2012) reported on the processes of gametogenesis in *L. lucerna*. Tiseo (2022) described the histochemistry and sperm ultrastructure of *Cassiopea maremetens* Gershwin, Zeidler and Davie, 2010, the spermatogenesis of *R. nomadica* (histology and histochemistry), the oogenesis of *Veruriga anadyomene* (Maas, 1903) (histology, histochemistry and ultrastructure) and the gametogenesis (histology and histochemistry) of three species of the genus *Cassiopea* (*C. andromeda*, *C. frondosa* and *C. xamachana*).

Fig. 5—cont'd microscopy of the sperm of *Lychnorhiza lucerna* being released by dissociation of the sperm follicles cells forming a pore in the follicular wall. (D) Longitudinal section of male gonad of *Nemopilema nomurai* with sperm being released by the spermiation pit. (E) Transmission electron microscopy of the sperm of *Nemopilema nomurai* being released by the spermiation pit. (F) Longitudinal section of male gonad of *Rhopilema nomadica* (stained with Gomori Trichomic). Note the sperm being released free in the genital sinus (arrow). Abbreviations: *F*, male follicles; *f*, flagellum; *Ga*, gastrodermis; *ige*, inner gonadal epithelium; *m*, mesoglea; *oge*, outer gonadal epithelium; *sf*, sperm follicles; *sp*, sperm; *spgt*, spermatozeugmata; *spp*, spermiation pit; *te*, testicular epithelium. *(A, B, C) Photo: G. Tiseo (D, E) Modified from Ikeda, H., Ohtsu, K., Uye, S.-I., 2011a. Structural changes of gonads duringartificially induced gametogenesis and spawning in the giant jellyfish Nemopilema nomurai (Scyphozoa: Rhizostomeae). J. Mar. Biol. Assoc. U.K. 91, 215–227, Cambridge Universtiy Press, reproduced with permission. (F) from Tiseo, G.R., 2022. Sexual reproduction in jellyfish species (Scyphozoa,Cnidaria) from the southeastern Brazilian coast. PhD Thesis, Universidade de São Paulo, São Paulo, Brazil, pp. 257.*

The spermatogonia of the investigated Rhizostomeae species usually have a large basophil nucleus with prominent visible nucleolus while the spermatocytes are smaller with a strongly basophil nucleus. In ultrastructural examinations, meiotic phases as prophase with synaptonemal complex and metaphase are visible (Tiseo, 2022). The spermatids are interconnected by cellular bridges with the strongly basophil nucleus (Ikeda et al., 2011a; Tiseo, 2022, 2016). The sperm have an ovoid basophil head as observed in *L. lucerna* and *C. mosaicus* (Rouse and Pitt, 2000; Tiseo, 2022), a long conical head as reported in *N. nomurai* (Ikeda et al., 2011a) or an elongated head found in *R. pulmo*, *C. maremetens* and *P. punctata* (Hedwig and Schäfer, 1986; Hinsch, 1974; Rouse and Pitt, 2000; Tiseo, 2022). The sperm have small pre-acrosomal vesicles above the nucleus and laterally in the midpiece, proximal and distal centrioles, a pericentriolar complex that anchor the flagellum and always four mitochondria in the midpiece (Hedwig and Schäfer, 1986; Hinsch, 1974; Ikeda et al., 2011a; Rouse and Pitt, 2000; Tiseo, 2022).

Two different sperm release strategies are observed in rhizostomes: either single separated or groups of several sperm cells are released into the genital sinus. The rhizostome species *C. andromeda, C. frondosa, C. maremetens, C. xamachana* (Mammone et al., 2023; Tiseo, 2022), *M. papua* (Uchida, 1926) and *C. tuberculata* (Kikinger, 1992) release the sperm in groups called "spermatozeugmata", whereas other investigated rhizostomes (*C. mosaicus, L. lucerna, N. nomurai* and *R. pulmo*) release isolated cells into the genital sinus. In *L. lucerna* sperm release (Fig. 5C) occurs by dissociation of the sperm follicle cells forming a pore in the follicular wall (Schiariti et al., 2012; Tiseo, 2022). In *N. nomurai* a spermiation pit has been observed (Fig. 5D and E) where the sperm cells shed through into the subgenital sinus before they enter the gastrovascular cavity and probably reach the water column (Hedwig and Schäfer, 1986; Ikeda et al., 2011a; Pitt and Kingsford, 2000; Schiariti et al., 2012; Tiseo, 2022). Rouse and Pitt (2000) report that both strategies—free sperm and "spermatozeugmata" occur in *P. punctata* and *C. mosaicus*. Tiseo (2022) reported that the sperm cells in *R. nomadica* are released separately in the genital sinus like in other scyphozoan species (Fig. 5F).

For rhizostome species, there is no consensus in the literature about how many stages of oocyte development exist; depending on the species, three, four or five stages of development have been distinguished (Avian and Rottini-Sandrini, 1991; Eckelbarger and Larson, 1992; Kikinger, 1992; Ikeda et al., 2011a; Pitt and Kingsford, 2000; Schiariti et al., 2012; Tiseo, 2022). Kikinger (1992) mentions fully developed oocytes with nucleus, nucleolus, and nurse cell (Fig. 6A). Using light microscopy, Pitt and Kingsford (2000),

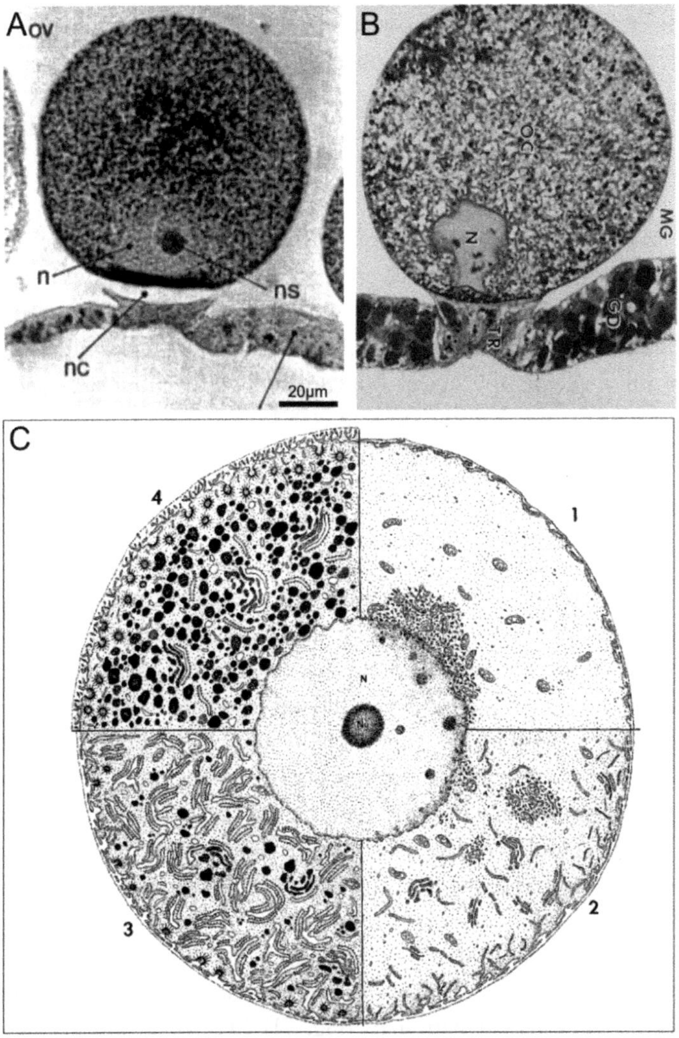

Fig. 6 (A) *Cotylorhiza tuberculata*. Vitellogenic oocyte protruding into the mesoglea. (B) *Stomolophus meleagris*. Vitellogenic oocyte in cross section. Note: oocyte in close contact with the trophocyte. (C) *Stomolophus meleagris*. Diagrammatic synopsis of four oocyte stages of maturation. 1 = previtellogenic stage showing fibro- granular material (nuage) and several mitochondria in the perinuclear ooplasm; 2 = Previtellogenic stage in which nuage has dispersed into central ooplasma, Golgi and rough endoplasmic reticulum (RER) appears and the oolemma is invaginatlng to form intraooplasmic channels; 3 = early vitellogenic stage in which yolk is being synthesized by RER-Golgi. The surface of oocyte is covered with endocytotic pits which migrate into the cortical ooplasm; 4 = vitellogenic oocyte filled with yolk bodies, Golgi complexes and RER. Although endocytotic activity is still apparent along the surface of the cell, the ooplasmic channels have disappeared, *N*, nucleus; *n*, nucleus; *nc*, nurse cell; *ns*, nucleolus. *(A) from Kikinger, R., 1992. Cotylorhiza*

(continued)

recognized three categories of oocytes in *C. mosaicus*: previtellogenic (absence of yolk granules), early-mid vitellogenic (some yolk granules present) and late vitellogenic (oocytes containing large quantities of yolk granules). Eckelbarger and Larson (1992) describe the ultrastructure of oogenesis in *S. meleagris* and differentiate four oocyte stages of development: previtellogenic, early vitellogenic, middle vitellogenic, and late vitellogenic. In the early stage, the yolk is synthesized by Golgi-rough endoplasmic reticulum (RER), ooplasmic channels closely associated with RER cisternae and endocytotic pits are present in the cortical ooplasm. The vitellogenic (middle and late stage) (Fig. 6B and C) is characterized by the presence of ooplasm completely filled with yolk bodies, shorter microvilli in the apical ooplasm and with endocytotic activity still apparent. Avian and Rottini-Sandrini (1991) described the mature oocytes of *R. pulmo* stating that the development of these germ cells is similar to that described in the semaeostome *Pelagia noctiluca*, which have four different stages of oocytes.

Schiariti et al. (2012) and Tiseo (2022) found four stages of oocyte development in *L. lucerna*: pre-, early-, mid- (or intermediate) and late-vitellogenic oocytes (Fig. 7A). Histologically, the pre-vitellogenic oocytes are inside the genital epithelium and have a round shape with prominent nucleolus. When they maturate, they become larger and protrude into the gonadal mesoglea maintaining contact with the genital epithelium by the trophocytes. The early stage is distinguished from the other two stages by the ooplasm basophilia. Mid (or intermediate) and late oocytes have ooplasm in different levels of acidophily being more acidophil in late than middle oocytes (Fig. 7A). Four stages of oocytes (pre, early, intermediate and late vitellogenic) were also found during the oogenesis of *C. andromeda* (Mammone et al., 2023), *C. frondosa*, *C. xamachana* and *V. anadyomene* (Tiseo, 2022). In light microscopy, late oocytes of *C. andromeda*, *C. frondosa*, *C. xamachana*, and *V. anadyomene* have acidophil yolk granules in the ooplasm periphery (Fig. 7B–D), which are positive to total (Fig. 7E) and neutral proteins (Fig. 7F).

N. nomurai is the only species for which five stages of oocyte maturation have been reported (Ikeda et al., 2011a) (Fig. 8A). In the first stage, the yolk bodies are formed by nutrients transported from the subgenital sinus

Fig. 6—cont'd *tuberculata (Cnidaria: Scyphozoa)—Life history of a stationary population. Mar. Ecol. 13,333–362, John Wiley and Sons, reproduced with permission; (B, C) from Eckelbarger, K.J., Larson, R., 1992. Ultrastructure of the ovary and oogenesis in the jellyfish Linuche unguiculata and Stomolophus meleagris, with a review of the ovarian structure in the Scyphozoa. Mar. Biol. 114, 633–643, Springer Nature, reproduced with permission from SNCSC.*

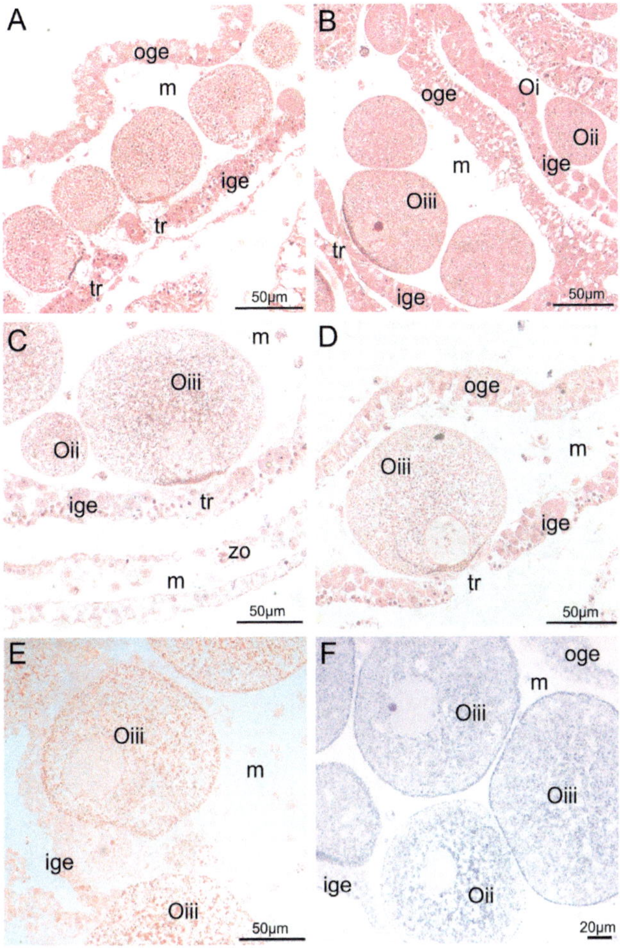

Fig. 7 (A) *Lychnorhiza lucerna*. Pre-, early, mid and late-oocytes in longitudinal section (hematoxylin and eosin). (B) *Cassiopea andromeda*. Pre-, early, intermediate, and late oocytes (hematoxylin and eosin). (C) *Cassiopea frondosa*. Longitudinal section of late oocyte in close association with trophocyte. (D) *Cassiopea xamachana*. Female gonad with late oocyte immersed in the mesoglea and with contact with trophocyte (longitudinal section, hematoxylin and eosin). (E) *Versuriga anadyomene* female gonad with yolk positive to total proteins (longitudinal section, stained with Xylidine Ponceau). (F) *Versuriga anadyomene* oocyte with yolk positive for neutral proteins (stained with Mercuric bromophenol blue). *ige*, inner gonadal epithelium; *m*, mesoglea; *oge*, outer gonadal epithelium; *Oi*, early oocyte; *Oii*, intermediate oocyte; *Oiii*, late oocyte; *tr*, trophocytes; *zo*, zooxanthelleae. Photos: G. Tiseo.

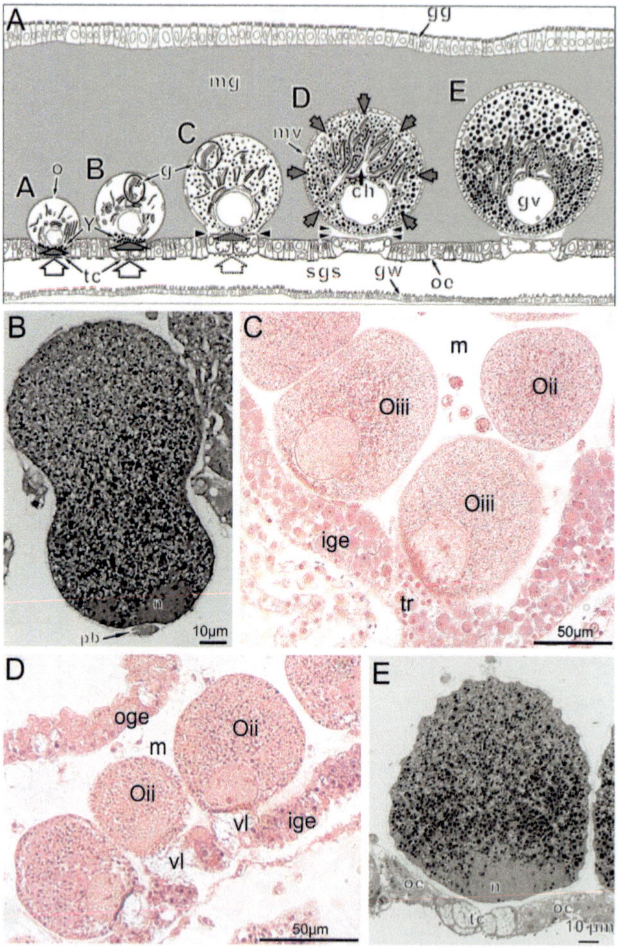

Fig. 8 (A) *Nemopilema nomurai*. Schematic drawing of five stages of oocytes (see A–E letters in image). (B) *Nemopilema nomurai*. Peanut-shaped oocyte being released in the genital sinus. (C) *Cassiopea frondosa*. Trophocyte type 1 in close association with vitellogenic oocytes. (D) *Lychnorhiza lucerna*. Trophocytes type 2 in close contact with intermediate and late oocytes. (E) *Nemopilema nomurai*. Trophocytes type 2. *Ch*, intraooplasmic channel; *g*, Golgi complex; *gg*, gonadal gastrodermis; *gv*, germinal vesicle; *gw*, genital wall; *ige*, inner gonadal epithelium; *m*, mesoglea; *mg*, mesoglea; *mv*, microvilli; *n*, nucleus; *O*, oocyte; *oe*, ovarian epithelium; *oge*, outer gonadal epithelium; *Oi*, early oocyte; *Oii*, intermediate oocyte; *Oiii*, late oocyte; *pb*, polar body; *sgs*, subgenital sinus; *tc*, trophocytes; *tr*, trophocytes; *vl*, vacuoles; *Y*, yolk bodies. *(A, B, E) from Ikeda, H., Ohtsu, K., Uye, S.-I., 2011a. Structural changes of gonads during artificially induced gametogenesis and spawning in the giant jellyfish Nemopilema nomurai (Scyphozoa: Rhizostomeae). J. Mar. Biol. Assoc. U.K. 91, 215–227, Cambridge Universtiy Press, reproduced with permission (C, D) Photo: G. Tiseo.*

by the trophocytes. The second stage is characterized by the presence of yolk deposition in the ooplasm adjacent to the trophocytes and near the Golgi complexes. When the oocyte begins to separate from the trophocytes it passes to the third stage characterized by the decreasing transfer of nutrients by trophocytes. During the fourth stage, the oocyte completely disconnects from the trophocytes taking nutrients from the surrounding mesoglea. Finally, the yolk-filled oocyte becomes mature. and is enveloped by the vitelline membrane in the fifth stage (Ikeda et al., 2011a). Despite the contrasting classifications of developmental stages (due the different views of the authors) different histological and ultrastructural techniques expanded the knowledge on oocyte development. Histological investigations allow for the differentiation of the vitellogenic oocytes stages through the acidophilic or basophilic reactions depending on the stain used, and TEM can detect the presence of yolk bodies in the ooplasm and endocytotic pits in the cortical ooplasm with a clearly vitelline membrane surrounding the mature oocyte.

Regarding ovulation, the rule seems to be that matured oocytes are released into the genital sinus, as observed in the majority of the rhizostome species (Gohar and Eisawy, 1960b; Ikeda et al., 2011a; Kikinger, 1992; Lotan et al., 1992; Ohtsu et al., 2007; Schiariti et al., 2012; Smith, 1936) (see Table 2 for details). There are only two known exceptions (*R. pulmo* and *C. frondosa*), in which the oocytes are kept inside the female gonad (Tiseo, 2022; Widersten, 1965). In *L. lucerna* and *N. nomurai* the release of oocytes occurs through an ovulation pit located in the centre of the trophocytes vacuoles. The oocyte projects towards the pit, taking on a peanut-like shape while passing through the pit (Fig. 8B) and is released in the genital sinus (Ikeda et al., 2011a; Schiariti et al., 2012; Tiseo, 2022). In *R. nomadica* the fully developed oocytes migrate to the scapulae from where they are released into the seawater (Lotan et al., 1992). Gohar and Eisawy (1960b) reported that the oocyte in *C. andromeda* is released ("discharged") into the genital sinus, as similarly described for *C. frondosa* (Smith, 1936). Widersten (1965) described that *R. pulmo* maintained the oocyte inside the female gonad whereas Avian and Rottini-Sandrini (1991) remarked that the *R. pulmo* oocytes were released into the genital sinus. Kikinger (1992) and Tiseo (2022) observed that oocytes in *C. tuberculata* and *C. frondosa* (Fig. 8C), respectively, were kept inside the female gonad (see Tables 2 and 3 for details). The release of mature oocyte in the genital sinus with posterior occurrence of the gametes in the gastrovascular cavity, as documented for *L. lucerna* (Schiariti et al., 2012; Tiseo, 2022) and *N nomurai* (Ikeda et al., 2011a), suggests external

Table 2 Sexual reproductive features in Kolpophorae species. * not described in the text, but observed from the figures.

Family	Species	Gonad colour	Trophocytes	Release of sperm	Vitelline membrane	Late oocyte size (μm)	Ovulation	Fertilization	Gastrulation	Brooding	References
Cepheidae	*Cotylorhiza tuberculata*	No data	Type 1*	Spermato-zeugmata	Present	140–160	Maintain inside the female gonad	Takes place inside the female gonad	Ingression	Carrying planulae in mouth arms	Berrill (1949), Kikinger (1992)
Cassiopeidae	*Cassiopea andromeda*	No data	Type 1	Spermato-zeugmata	Present	140–170	Release into the genital sinus	Takes place in the genital sinus	No data	Carrying planulae in the gastric cavity or mouth arms	Gohar and Eisawy (1960b), Hofmann et al. (1996), Tiseo (2022)
	Cassiopea culionensis	No data	No data	No data	No data	No data	No data	No data	No data	Carrying planulae in mouth arms	Tiseo (2022)
	Cassiopea frondosa	No data	Type 1	Spermato-zeugmata	Present	140	Release into the genital sinus/ Maintain inside the female gonad	Takes place in the genital sinus or inside the female gonad	No data	Carrying planulae in mouth arms	Smith (1936), Tiseo (2022)
	Cassiopea maremetens	Whitish (male)	No data	Spermato-zeugmata	Present	No data	No data	No data	No data	Carrying planulae in mouth arms	Tiseo (2022)
	Cassiopea xamachana	No data	Type 1	Spermato-zeugmata	Present	190	No data	Takes place in the genital sinus	No data	Carrying planulae in mouth arms	Eckelbarger and Larson (1992), Tiseo (2022)
Mastigiidae	*Mastigias papua*	No data	No data	Spermato-zeugmata	No data	100	No data	No data	Ingression	Carrying planulae in mouth arms	Uchida (1926), Berrill (1949)
	Phyllorhiza punctata	No data	No data	Free sperm—spermato-zeugmata	No data	No data	No data	Seems to be internal	No data	Carrying planulae in mouth arms	Rippingale and Kelly (1995), Rouse and Pitt (2000)
Versurigidae	*Versuriga anadyomene*	Brownish-orange	Type 1	No data	Present	170	No data	No data	No data	No data	Tiseo (2022)

Table 3 Sexual reproductive features in Dactyliophorae species. *not described but observed in the figures.

Family	Species	Gonad colour	Trophocytes	Release of sperm	Vitelline membrane	Late oocyte size (μm)	Ovulation	Fertilization	Gastrulation	Brooding	References
Catostylidae	*Catostylus mosaicus*	Most frequently brown (male) most frequently green (female)	Type 2*	Free sperm	No data	120	No data	No data	No data	Brood larvae in mouth arms	Pitt and Kingsford (2000), Rouse and Pitt (2000)
	Catostylus tagi	Greenwish (male) browish-greenwish (female)	No data	No data	No data	average 31	No data	No data	No data	No data	Lisboa et al. (2018)
	Pseudorhiza aurosa	No data	No data	No data	No data	No data	No data	No data	No data	Embryos in "brood pouches"	von Lendenfeld (1882)
	Pseudorhiza haeckeli	No data	No data	No data	No data	No data	No data	No data	No data	"Spermaries" in mouth arms	von Lendenlfeld (1888), Jarms and Morandini (2019)
Rhizostomatidae	*Nemopilema nomurai*	Milky white to pinkish (male) milky white to dark brown (female)	Type 2*	Free sperm through a spermiation pit	Present	170–180	Release of oocytes into the genital sinus through ovulation pit	Supposed to be external or takes place in the gastrovascular cavity	No data	Non-brooder	Ohtsu et al. (2007), Iguchi et al. (2010), Ikeda et al. (2011a)
	Rhizostoma octopus	Violet or blue (male) reddish-brown (female)	No data	No data	No data	180	No data	No data	Invagination	No data	Berril (1949)
	Rhizostoma pulmo	Yellowish	Type 1*	Free sperm	Present	150–180	Maintain inside the female gonad/ Release of oocytes into the genital sinus (the endodermal cells of which fenestrate)	Takes place in the genital sinus or inside the female gonad	Invagination	Carrying planulae in mouth arms	Avian and Rottini-Sandrini (1991), Paspalev (1938), Widersten (1965)

(continued)

Table 3 *(cont'd)* Sexual reproductive features in Dactyliophorae species. *not described but observed in the figures.

Family	Species	Gonad colour	Trophocytes	Release of sperm	Vitelline membrane	Late oocyte size (μm)	Ovulation	Fertilization	Gastrulation	Brooding	References
	Rhopilema nomadica	No data	No data	Sperm imersed in mucus	No data	100-110	Oocytes migrate to the scapulae and are released into the seawater	External	Invagination	Non-brooder	Lotan et al. (1992), Lotan et al. (1994), Tiseo (2022)
	Rhopilema verrilli	No data	No data	No data	No data	No data	No data	Supposed to be internal	No data	Planulae retained inside the gonad	Calder (1973)
Stomolophidae	*Stomolophus meleagris*	Bluish-colored (female)	Type 2*	No data	Present (as basal lamina)	No data	No data	No data	No data	Non-brooder	Calder (1983), Eckelbarger and Larson (1992), Rouse and Pitt (2000)

fertilization. The maintenance of oocytes inside the genital sinus or inside the female gonad suggests internal fertilization, as documented for *Cassiopea* spp. (Smith, 1936; Gohar and Eisawy, 1960b; Tiseo, 2022), *R. pulmo* (Widersten, 1965) and *C. tuberculata* (Kikinger, 1992).

3.5 Accessory cells

Rhizostomeae species (as well as Semaeostomeae) have specific accessory cells, called trophocytes. These cells surround the oocytes, separating them from the mesoglea but maintaining contact with the inner genital epithelium also named "gastrodermal epithelium" (Eckelbarger and Larson, 1988). These cells have been reported for *S. meleagris*, *N. nomurai*, *R. pulmo* and *C. tuberculata*, in the latter two referred to as "nurse cells" (Avian and Rottini-Sandrini, 1991; Eckelbarger and Larson, 1992; Kikinger, 1992; Ikeda et al., 2011a; Widersten, 1965). Eckelbarger and Larson (1988) first proposed the term "trophocytes" while studying oogenesis in the semaeostome species *A. aurita*. They demonstrated that the previously proposed term "nurse cells" (used by Kikinger, 1992; Widersten, 1965) was inappropriate because the trophocytes do not maintain cytoplasmic connection with the developing oocyte by intercellular bridges which is the classical definition of nurse cells (*sensu* Anderson, 1974; Huebner and Anderson, 1976), whereas the origin of the trophocytes is uncertain.

The term "accessory cells" was suggested to substitute the previously used terms (nurse cells, paraovular bodies and follicular cells) (Eckelbarger and Hodgson, 2021) and we suggest the same for the trophocyte. Two types of trophocytes can be distinguished (Tiseo, 2022): type 1 (Fig. 8C) previously described by Eckelbarger and Larson (1988) with electrodense, slightly flattened cells that form a thin plate just over the internal genital epithelium, and type 2 (Fig. 8D and E) first described by Avian and Rottini-Sandrini (1991) with columnar cells rich in rough and smooth endoplasmic reticulum, and with several large vacuoles (rich in mucopolysaccharides) near the nucleus. Based on the available published images on trophocytes of ten rhizostome species, we were able to assign the trophocyte type to each different species (Tables 2 and 3) (after Tiseo, 2022).

3.6 Fertilization

Earlier reviews reported that fertilization in jellyfishes (including rhizostomes) is usually external (Arai, 1997; Campbell, 1974). However, this seems to be a generalization since: (1) data on fertilization are available for only 27 of 241 scyphozoan species (18 Semaeostomeae and 9 Rhizostomeae, see Tables 2 and 3); (2) in most of these species, fertilization has not yet been observed under natural conditions; and (3) in some cases the conclusion of external fertilization is just based on the absence of evidence of internal fertilization (e.g. presence of planulae in the oral arms). In the case of external fertilization, sperm and mature oocytes are first released into the genital sinus before they reach the gastrovascular cavity and seawater where fertilization occurs (Arai, 1997). Internal fertilization can take place inside the female gonad, in the genital sinus, in the gastric cavity or in the canals of the mouth arms (Calder, 1973; Gohar and Eisawy, 1960b; Rippingale and Kelly, 1995; Rouse and Pitt, 2000; Smith, 1936; Tiseo, 2022). Kikinger (1992) described sperm clustered in "*spermatozeugmata*" were released in seawater and made their way from the open water into the female genital sinus through the canals of the mouth arms. After the degeneration of the "*spermatozeugmata*" single sperm were found in the female gonad (Kikinger, 1992).

In the Rhizostomeae, only a few data on fertilization are derived from life cycle studies and "in vivo" fertilization conducted in laboratory (Ikeda et al., 2011a; Schiariti et al., 2008; Ohtsu et al., 2007). However, some authors have described external or internal fertilization based on in situ observations (Kikinger, 1992; Lotan et al., 1992; Widersten, 1965). Characteristic external fertilization with oocytes released in the seawater were observed in *R. nomadica* (Lotan et al., 1992), and probably also occurs in other rhizostome species, for example, *L. lucerna* and *N. nomurai* (Ikeda et al., 2011a; Schiariti et al., 2012; Tiseo, 2022). In *R. pulmo, C. tuberculata* and *C. frondosa*, several sperms have been observed inside the genital epithelium near the trophocytes, indicating that fertilization is internal and inside the female gonad (Kikinger, 1992; Tiseo, 2022; Widersten, 1965). In other species (e.g. *C. andromeda, C. xamachana, C. maremetens, P. punctata, R. verrilii*) there is a hint of internal fertilization due to oocytes released in the genital sinus and planulae observed in the mouth arms (Calder, 1973; Gohar and Eisawy, 1960b; Rippingale and Kelly, 1995; Rouse and Pitt, 2000; Smith, 1936; Tiseo, 2022) (Tables 2 and 3).

Ohtsu et al. (2007) stated that some enzymatic activities (like the acrosome reaction in other metazoans) might be involved in the penetration of the sperm and light exposure triggered the processes of gonadal maturation, spawning and fertilization in *N. nomurai* (see also Thibault et al., 2024). Male gonads exposed to continuous light for 30 min release sperm into the seawater and female gonads exposed to light release peanut-shaped oocytes through a pit in the genital epithelium (Ohtsu et al., 2007). A clear vitelline membrane forms just after fertilization and the eggs are transported through the subgenital sinus into the gastrovascular cavity by ciliary movements.

3.7 Cleavage and gastrulation processes

In Rhizosomeae species cleavage is found to be holoblastic (total), equal and radial (Kikinger, 1992; Ohtsu et al., 2007; Paspaleff, 1938; Smith, 1936) with a gastrula formed by ingression (Uchida, 1926; Berrill, 1949) or invagination (Berrill, 1949; Lotan et al., 1992; Paspaleff, 1938). In *C. frondosa* cleavage occurs in the gastric cavity. As already reviewed by Thiel (Thiel et al., 2024), the egg passes through the brachial canal to the mouth arms during the cleavage phase and remains there until the end of the development into the planula (Smith, 1936). Paspaleff (1938) reported total and equal cleavage of *R. pulmo* in the same way as other Rhizostomeae with the gastrula arising by invagination. Early cleavage stages in *C. tuberculata* females have been observed inside the canals of the mouth arms (Kikinger, 1992) with vegetative and animal blastomeres of the same size. The whole development until the planula occurs inside the mouth arms canals, and the ciliated planula is released still inside the thin egg membrane (Kikinger, 1992). Ohtsu et al. (2007) also observed holoblastic and radial cleavage of *N. nomurai* with a clearly seen vitelline membrane that begins to disappear at the vegetative pole before the embryos develop into 200 μm long planulae.

3.8 Brooding

Brooding, the presence of planulae or embryos in the mouth arms or in the gastric cavity, is reported for 11 of the 19 rhizostome species studied to date (Tables 2, 3). In *C. tuberculata,* planulae were observed inside the mouth arms canals until the optimum time for release (Kikinger, 1992). In *Cassiopea* species planulae were observed in the gastric cavity (*C. andromeda*) or in the mouth arms of other species (Gohar and Eisawy, 1960b; Hofmann et al., 1996; Ohdera et al. 2018; Tiseo, 2022). In Mastigiidae, *M. papua* and *P. punctata* brood planulae in mouth arms (Rippingale and Kelly, 1995; Rouse and Pitt, 2000; Uchida, 1926). Rouse and Pitt (2000) discussed the brooding in scyphozoans stating that *C. mosaicus* brood larvae in the mouth arms in contrast to *S. meleagris*. Brooding has also been reported for *Pseudorhiza aurosa* Von Lendenfeld (1882), *R. pulmo* and *R. verrilli* (Calder, 1973; Paspaleff, 1938; Widersten, 1965). See Schiariti et al. (2024) for more details about sexual reproductive strategies.

4. Zooxanthellae

A particular feature of rhizostomes compared to semaeostomes is that some of them host photosynthetic dinoflagellates symbionts – "zooxanthellae"—in their tissues (Fig. 9). This was an active research field in 1970's

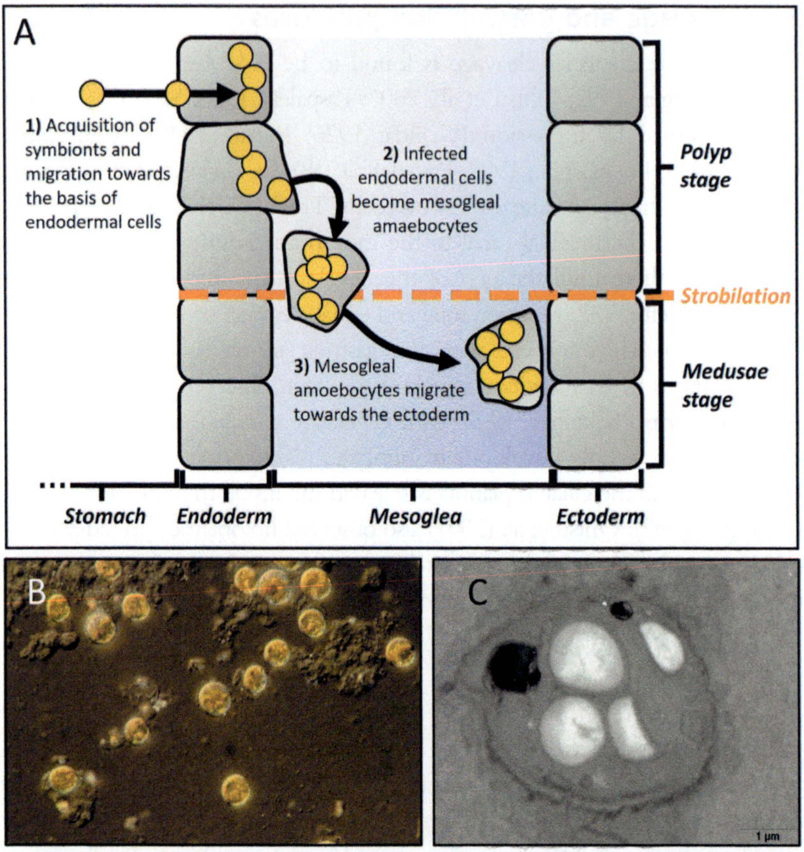

Fig. 9 (A) Summary of the location of dinoflagellate symbionts (zooxanthellae) in kolpophore rhizostome tissues during ontogeny. (B) Zooxanthellae (diameter approximately 10 µm) isolated from *Cassiopea xamachana* in optic microscopy. (C) Close-up of a zooxanthella isolated from *Versuriga anadyomene* in transmission electron microscopy. *(A) Image: N. Djeghri. (B) Photo: Cheryl Lewis Ames, Smithsonian NMNH, with permission. (C) Photo: G. Tiseo.*

when Thiel (Thiel et al., 2024) wrote his review and continues nowadays in particular with the upside-down jellyfish, *Cassiopea* spp., considered as a new model organism (Ohdera et al. 2018). The rhizostome-zooxanthellae symbiosis is considered mutualistic, the jellyfish host providing excretion products and a refuge from predators, while the dinoflagellate symbiont provides photosynthates. The holobiont (host + symbionts considered as one organism) can be considered a mixotroph opening peculiar ecological niches to these jellyfishes (Djeghri et al., 2019). Here, we provide an update and critical review of Thiel's account on the diversity of rhizostomes-zooxanthellae associations and on their histology. In addition, since the 1970's our understanding of the role of the symbionts in host nutrition and potential ecological consequences greatly progressed. This aspect is addressed in detail in the chapter dedicated to the physiology of Rhizostomeae (Nagata et al., 2024).

4.1 Diversity of associations

4.1.1 Hosts

The review by Thiel (Thiel et al., 2024) gives a census of the Rhizostomeae species reported to live in association with zooxanthellae. Thiel states that zooxanthellae "occur in representatives of the various families of both the 'Kolpophorae' and the 'Dactyliophorae'". This contradicts a more recent review stating that, within the Rhizostomeae, zooxanthellate species are—except for *Bazinga rieki* Gershwin and Davie, 2013—restricted to the monophyletic clade Kolpophorae (Djeghri et al., 2019). Compared with other rhizostomes *B. rieiki* phylogenetic relationships are not yet resolved, making the origin(s) of symbiosis between Rhizostomeae and dinoflagellates difficult to pin-point. We review the evidence for the symbiosis with zooxanthellae in Kolpophorae and Dactyliophorae species.

Symbiosis with zooxanthellae is widespread in Kolpophorae (Table 4), which was already noted by Thiel (Thiel et al., 2024). From similar evidence, Djeghri et al. (2019) parsimoniously inferred that all Kolpophorae might be able to engage in symbiosis albeit some (mostly in Cepheidae) appear to display a facultative symbiosis. Within Dactyliophorae Thiel reports symbiosis with zooxanthellae in *C. mosaicus*, *P. haeckeli*, *S. meleagris* and *S. fritillaria* (Thiel et al., 2024). The sources of Thiel for these reports are mostly based on the work of Von Lendenfeld (1884, 1885) on *C. mosaicus* (as *Crambessa mosaica*) in Australia (Thiel et al., 2024). Von Lendenfeld (1885) described blue and brown colour morphs of *C. mosaicus* and ascribed the brown colour morph to the presence of zooxanthellae (Von Lendenfeld, 1885 mentions "yellow cells"). The brown coloration or the existence of different colour morphs was used as evidence for the presence of zooxanthellae in *P. haeckeli* (Von Lendenfeld, 1888) and later in *Stomolophus* sp. (Bigelow, 1914). However, all these records predate largely the modern understanding of the nature of zooxanthellae that started with the seminal paper of Freudenthal (1962), calling caution on what was called "zooxanthellae" at that time. Meanwhile, *C. mosaicus* is a well-studied species without a proof of zooxanthellae and has been used as the non-symbiotic example in comparative studies on jellyfish species with and

Table 4 Evidences of symbiosis with zooxanthellae in Kolpophorae.

Family	Genus	Documentation of symbiosis with zooxanthellae	Reported by Thiel? (see Thiel et al., 2024)	References
Cassiopeidae	*Cassiopea*	Abundant(direct observations, genetics, measures of chlorophyll, measures of photosynthesis)	Yes (in *C. andromeda* and *C. xamachana*)	e.g. Freudenthal (1962), Lampert (2016), Mellas et al. (2014), Ohdera et al. (2018), Verde and McCloskey (1998)
Cepheidae	*Cephea*	Rare direct observations of zooxanthellae, most likely because of: 1) facultative symbiosis in this genus, and 2) ontogenic loss of zooxanthellae	Yes (in *C. cephea*)	Sugiura (1969)
	Cotylorhiza	Abundant (direct observation, measure of photosynthesis, genetics)	Yes (in *C. tuberculata*)	Kikinger (1992), LaJeunesse et al. (2021)
	Marivagia	No data	No	None
	Netrostoma	Rare direct observations of zooxanthellae perhaps for the same reasons as in the closely related *Cephea*	No	Straehler-Pohl and Jarms (2010)
Mastigidae	*Mastigias*	Abundant (direct observations, genetics, measures of chlorophyll and photosynthesis)	Yes (in *M. andersoni*, *M. papua* and *M. siderea*)	McCloskey et al. (1994), Muscatine et al. (1986), Sugiura (1963), Vega de Luna et al. (2019)
	Phyllorhiza	Abundant (direct observations, measures of nutrient absorptions and of photosynthesis)	Yes (in *P. punctata* and *P. pacifica*)	Djeghri (2019), Pitt et al. (2005)
Thysanostomatidae	Thysanostoma	No update	Yes (in *T. thysanura* as *Lorifera lorifera*)	None
Versurigidae	*Versuriga*	Rare (few direct observations)	Yes (in *V. palmata*)	Tiseo (2022)

without zooxanthellae (Pitt et al., 2005; West et al., 2009). The blue and brown colour morphs of some Rhizostomeae are also a research subject. A family of blue pigments (the rhizostomins) has been recently characterized and is found in rhizostomes with and without zooxanthellae (Lawley et al., 2021). A brown pigment is currently under investigation but does not appear to be related to zooxanthellae in *C. mosaicus* (Lawley, pers. com.). Hence, the sole argument of colour to ascribe the presence of zooxanthellae in jellyfish appears weak at best.

Another report of Dactyliophorae with zooxanthellae is given by Trench (1971) for *Rhizostoma* sp. However, the record by Trench (1971) is doubtful as the author describes a *Rhizostoma* species from Hawaii settled "with oral surface upwards". However, species of the genus *Rhizostoma* do not occur in Hawaii but only in Atlantic and Mediterranean waters (Jarms and Morandini, 2019) and the behaviour described by Trench (1971) is typical of the genus *Cassiopea.*

To conclude, the evidence of the existence of Dactyliophorae symbiotic with zooxanthellae seems incorrect. In contrast, most, if not all, Kolpophorae are photosymbiotic, albeit potentially facultative (Table 4). These elements agree with the conclusion that symbiotic rhizostomes are, with the potential exception of *B. rieki*, all found in Kolpophorae (Djeghri et al., 2019).

4.1.2 Symbionts

When Thiel wrote his review in the 1970s (Thiel et al., 2024), the nature of zooxanthellae had only recently been determined as dinoflagellates mutualistic endosymbionts. Their diversity was yet to be explored, as pointed out by Thiel himself (Thiel et al., 2024). The true diversity has been revealed only recently, largely through molecular methods that were not yet available. The zooxanthellae found in rhizostomes are a group of dinoflagellates belonging to the family Symbiodiniaceae, which are also the symbionts of many metazoans including scleractinian corals and giant clams (LaJeunesse et al., 2018). Until recently, the family was comprised of a single genus: *Symbiodinium* (see Freudenthal, 1962), to which different clades were progressively recognized. However, the main clades have now been elevated to genera (LaJeunesse et al., 2018). The symbionts are best studied for the zooxanthellate jellyfish species of the genus *Cassiopea.* In this genus, symbionts of the genera *Symbiodinium* LaJeunesse, 2017 and *Cladocopium* LaJeunesse and H.J. Jeong, 2018 appear as the most common ones (Dall'Olio et al., 2022; LaJeunesse, 2001; Mellas et al., 2014; Santos et al., 2003; Thornhill et al., 2006). The zooxanthellae-jellyfish association presents some plasticity, as a given population of jellyfish can take up species of different genera of the family Symbiodiniaceae, which sometimes occur together within the same host at the polyp stage (Dall'Olio et al., 2022; Lampert, 2016; Mellas et al., 2014; Thornhill et al., 2006). However, the establishment of symbiosis may be facilitated by some specific host-symbiont pairs (Trench et al. 1981). At the medusae stage, only one symbiont genus tends to be present in each individual host suggesting that selection or competitive exclusion occurs among symbionts within the host (Dall'Olio et al., 2022; Thornhill et al., 2006). Symbionts in other genera of Kolpophorae jellyfish are poorly known, but *Cladocopium* has been reported for *M. papua* from marine lakes in Palau, Vietnam and Indonesia (Vega de Luna et al., 2019; Ferreira et al., 2022), and *Symbiodinium* from Mediterranean and Australian *P. punctata* (Dall'Olio et al., 2022). *Philozoon medusarum* Geddes, 1882 has been described from Mediterranean *C. tuberculata* (LaJeunesse et al., 2021) which can also be associated with *Breviolum* spp. (Dall'Olio et al., 2022; LaJeunesse et al., 2022).

4.2 Histology of the zooxanthellae-Kolpophorae association

The histology of the Kolpophorae-zooxanthellae association is tightly coupled with their life-cycle. This is due to the timing of the acquisition of the symbionts and to changing location of zooxanthellae with the changing life-stages in the life-cycle of Kolpophorae jellyfishes (summarized in Fig. 9A). This was already largely known by Thiel (Thiel et al., 2024), and so we only give an update and synthesis. Zooxanthellae are not transmitted during sexual reproduction. Planulae are therefore non-symbiotic and zooxanthellae must be acquired at the polyp stage (see for *Cassiopea* spp., Colley and Trench, 1985; Fitt, 1984; Ludwig, 1969; Newkirk et al., 2018; Trench et al., 1981; for *Cephea cephea* (Forskål, 1775), Sugiura, 1969; for *C. tuberculata*, Astorga et al., 2012; Kikinger, 1992; and for *M. papua*, Sugiura, 1963). The histological processes of zooxanthellae acquisition are described in detail for *C. xamachana* polyps by Colley and Trench (1985). Zooxanthellae enter the coelenteron of polyps (generally in their free-living motile stage, see Freudenthal, 1962) and are uptaken phagocytotically by endodermal cells (Fig. 9A). However, phagosome-lysosome fusion occurs at low frequency preventing the digestion of zooxanthellae cells. Zooxanthellae-filled phagosomes (symbiosomes) are transported towards the basis of endodermal cells, far from the apical region where digestion occurs (Colley and Trench, 1985, Fig. 9A). In *C. xamachana* polyps, a zooxanthellae-containing endodermal cell may contain one to three (and up to about eight) symbiotic cells (Trench et al., 1981). Within a few days after infection, most of the endodermal cells infected with zooxanthellae cease phagocytotic activities and migrate towards the mesoglea, becoming mesogleal amoebocytes within which zooxanthellae population can increase (Colley and Trench, 1985; Fig. 9A). Such increase in symbiont numbers is a critical aspect of strobilation induction in Kolpophorae although the mechanism is unknown and other factors such as temperature and food are also at play (reviewed in Djeghri et al., 2019).

In the ephyrae and medusae stage, most zooxanthellae remain in mesogleal amoebocytes with small percentages in endodermal or ectodermal cells as shown for *C. xamachana* and *M. papua* (Colley and Trench, 1985; Muscatine et al., 1986). However, the precise location of zooxanthellae in the mesoglea is apparently dependent on the developmental stage and the species. In ephyrae, symbionts tend to line the endoderm (Kikinger, 1992; Straehler-Pohl and Jarms, 2010), suggesting that zooxanthellae-filled amoebocytes stay close to where they originate. However, in most medusae amoebocytes migrate towards the ectoderm with growth (see in *M. papua*, Muscatine et al., 1986; *P. punctata*, De Souza et al., 2007; in *Cassiopea* spp., Blanquet and Riordan, 1981; Blanquet and Phelan, 1987; Estes et al., 2003; Lyndby et al., 2020; Fig. 9A). In Cepheidae however, observations on *C. tuberculata* suggest that most symbionts remain closely associated with the endoderm (Kikinger, 1992) but this requires further observations. It has been proposed that amoebocytes could act as "cargo-ships" for zooxanthellae, bringing them closer to nutrient sources and/or to medusae tissue in need of autotrophically derived molecules (e.g. glucids, lipids, Lyndby et al., 2020) (see also below for more information on organ-dependent variation in zooxanthellae density).

4.3 Density and distribution of zooxanthellae in Kolpophorae body

The density of zooxanthellae in Rhizostomeae specimens can vary as a function of life-stage, state of the symbiosis, species, or environmental factors. In the polyp it can vary from none to tens of thousands of cells per polyp (Enrique-Navarro et al., 2021; Newkirk et al., 2018). In the medusa, density of zooxanthellae can be fairly stable (in the order of 10^6–10^7 cells.g^{-1} of wet mass, Enrique-Navarro et al., 2022; Muscatine et al., 1986; Verde and McCloskey, 1998). However, jellyfish may lose their zooxanthellae ontogenetically (e.g. in *C. cephea*, Sugiura, 1969; and perhaps many other Cepheidae, Djeghri et al., 2019) or in response to environmental factors such as heat stress (bleaching, see e.g. Dawson et al., 2001; Klein et al., 2019; McGill and Pomoroy, 2008).

Zooxanthellae density also varies as a function of the medusae organ. Precise mapping of symbiont distribution within medusae bodies has recently been published for *Cassiopea* sp. (Lyndby et al., 2023). It largely confirms and completes previous studies indicating that zooxanthellae are present in particularly high concentrations in the subumbrella, associated with the coronal muscle, and in the appendages of the oral arms (Blanquet and Riordan, 1981; Blanquet and Phelan, 1987; Enrique-Navarro et al., 2022; Estes et al., 2003; De Souza et al., 2007; Lyndby et al., 2020; Muscatine et al., 1986). Some other organs host less symbiotic cells (e.g. exumbrella, Enrique-Navarro et al., 2022; Lyndby et al., 2020). This suggests that there is a degree of body region specialization in symbionts hosting, possibly linked to access to light, nutrients, or to the proximity with an area of the medusae in need of autotrophy products (Lyndby et al., 2020). Furthermore, pigmentation in some medusa body regions can modulate advantageously the light environment of the symbionts, for example, the pigmentation of the central dome of the umbrella of *C. tuberculata* may provide protection against ultraviolet radiation (Enrique-Navarro et al., 2022), and white granules within *Cassiopea* sp. umbrella may scatter light which may lead to increased photosynthesis (Lyndby et al., 2023). Finally, zooxanthellae are found in the recently described cassiosomes (see Section 5.6) from *C. xamachana*, and the authors suggest they could lengthen the viability of the structure by providing autotrophic products to it (Ames et al., 2020).

Zooxanthellae populations within their hosts are controlled through cellular processes such as expulsion, digestion, or limitation in nutrient access (see Davy et al., 2012 for a review focused on scleractinian corals). In zooxanthellate Kolpophorae, recent results from *C. xamachana* suggest the control of symbiont populations could be conducted primarily via a limitation by the host to the access to dissolved nitrogen (Röthig et al., 2021).

5. Nematocysts

5.1 Introduction to nematocyst morphology and function

Cnidocysts (cnidae) are intracellular capsulated organelles produced by the cnidocytes (stinging cells), which occur in all life stages of all cnidarian taxa. The cnidocyst is a round, oval or elongate capsule containing toxins and a coiled tubule (Mariscal, 1974; Sötje and Holst, 2019). Nematocysts, varying from 3 to ~100 μm in size (Mariscal, 1974), are secretory products of the Golgi apparatus. During the development of the nematocyst, a stage termed nematoblast, the tubule forms as an appendix of the capsule, which is invaginated in a later developmental state (Berking and Herrmann, 2006). Different categories of nematocysts have been described according to morphological differences of the capsule and the tubule inside the capsule, which everts during discharge (reviewed by Sötje and Holst, 2019). All cnidocyst types known from the Rhizostomeae and other scyphozoans belong to the category nematocyst, while the other main categories, spirocysts and ptychocysts, differ in morphology and function and only occur in anthozoans (Shick, 1991; Östman, 2000). More than 30 morphological types of nematocysts were proposed in various nomenclature systems (reviewed in Östman, 2000). A standardization of the terminology describing cnidae has been aspired (Arai, 1997; Mariscal, 1984; Östman, 2000; Watson and Wood, 1989). The classification proposed by Weill (1934) was used as a baseline study until the 1970s and was further modified by later authors (Calder, 1974; Mariscal, 1974; Östman, 2000). The fully developed nematocyst is located close to the cell membrane. Its discharge is induced by a cnidocil apparatus and driven by a high internal pressure which causes the fast evagination of the tubule (Berking and Herrmann, 2006; Szczepanek et al., 2002; Tardent, 1995). The tubule is used to penetrate potential prey, to defend against predators or to facilitate attachment (Sötje and Holst, 2019). To paralyze prey, most nematocyst capsules contain microdrops of toxins (see Nagata et al., 2024) but only a few rhizostome species are harmful to humans (reviewed in Haddad and Morandini, 2019; Kondo et al., 2020).

In contrast to the Semaeostomeae, tentacles at the bell margin and a central mouth opening are lacking in Rhizostomeae medusae (see Morandini, 2024), and instead their oral arms have numerous mouthlets and are occasionally equipped with oral tentacles (digitata) and oral appendages. Rhizostome medusae capture smaller prey (primarily microplankton) than semaeostome medusae, but similarly high capture efficiency was observed in both taxa (Bezio et al., 2018). It has therefore been suggested that the differences in capture surface morphology probably explains the difference in prey selection patterns between Semaeostomeae and Rhizostomeae rather than bell shape and nematocyst complement (Bezio et al., 2018). Recent results indicate that rhizostomes need higher flow speeds around their bell to sieve fluid through their dense feeding surfaces, while the less dense feeding structures of semeostomes do not require such high velocity flow (Von Montfort et al., 2023) (compare Nagata et al., 2024 for mechanics of food capture).

5.2 Nematocysts categories, types and varieties in Rhizostomeae

Light microscopic studies in the late 19th and at the beginning of the 20th centuries have already revealed the occurrence of two different capsule types in rhizostomes: one type with a shaft and another type without a shaft visible in the undischarged capsule. These capsules have been categorized as euryteles and atrichs as already reviewed by Thiel (Thiel et al., 2024). However, later studies showed that the so-called atrichs in some cases were armed with tiny spines, which had been overlooked in previous studies (Calder, 1974). According to current knowledge, the two main nematocyst categories occurring in all Rhizostomeae are haploneme isorhizas (holotrich or atrich) and heteroneme rhopaloids (euryteles and/or birhopaloids) (Tables 5 and 6).

5.2.1 Haplonemes

Haplonemes are nematocysts without a prominent, rod-shaped shaft visible inside the undischarged capsule (Fig. 10). Different subtypes of haploneme isorhizas have been distinguished based on the capsule shape and the windings (coils) of the tubule inside the undischarged capsule (Östman, 2000; Sötje and Holst, 2019). Of these isorhiza types, a-isorhizas (capsule pyriform, tubule regularly coiled), α-isorhiza (capsule ellipsoidal to reniform, tubule regularly coiled) and O-isorhizas (capsule nearly spherical, tubule loosely and irregularly coiled) occur in Rhizostomeae. Based on SEM observations on *R. nomadica*, the nematocyst variety heterotrichous isorhiza haploneme has been introduced, a nematocyst type with a tubule with spines throughout, which are differently shaped in the proximal and distal portions of the tubule (Avian et al., 1995). This variety differs from the previously described heterotrichous anisorhiza haploneme, which exhibits a fairly evident and/or constant diameter reduction and spinal size reduction all over the everted tubule (Avian et al., 1995). The A-isorhizas (capsule ovate, tubule irregularly coiled) and polyspiras (capsule linguliform, tubule irregularly coiled) have been reported for the Semaeostomeae only (Calder, 1974, 1983; Lesh-Laurie and Suchy, 1991). Recent reports on these capsule types in the rhizostome species *C. tuberculata* and *P. punctata* (Cengiz and Killi, 2021) are not in accordance with previous descriptions of the latter species (Peach and Pitt, 2005) (Table 6) and should be considered with some caution.

The occurrence of anisorhizas in Rhizostomeae (Tables 5 and 6), defined as haplonemes with a tubule of unequal diameter according to Calder (1974), has been described by LM investigations on *R. verilli* for the first time (Calder, 1977). When present in a rhizostome species, anisorhizas appear to be largely restricted to the ephyra and medusa stages and their occasional occurrence in

Table 5 Rhizostomeae, Dactyliophorae. Tissues, analysis methods and nematocyst types/varieties in various studies (see references). Measurements have been conducted on undischarged capsules unless otherwise specified under measurement parameters. N.d. = no data, ref. = reference.

Species	Reference	Life stage(s)	Body part(s)	Nematocyst terminology	Nematocyst types /verities	Measurements (n) per type	Measurement parameters	Measurement data (in ref.)
Catostylus mosaicus	Peach and Pitt (2005)	Medusa	Oral arms and bell margins	Östmann (2000)	Pear-shaped holotrichous isorhiza, oval-shaped holotrichous isorhizae, rhopaloid, birhopaloids	6	Length, width (mean)	Table 1, p. 101
Catostylus tagi	Parracho and Morais (2015)	Medusa	Oral arms and bell margins	n.d.	eurytele, birhopaloid, isorhiza	n.d.	Estimates only	Text, p. 7
Lobonemoides robustus	Kondo et al. (2020)	Medusa	Mouth arm	Östmann (2000)	Ellipsoid isorhiza, round isorhiza, ellipsoid rhopaloid (small and large)	40	Length, width, tubule length (min, max, mean)	Table 2, p. 705
Nemopilema nomurai	Kubota et al. (2006)	Medusa	Mouth cirri	n.d.	Holotrichous anisorhiza, microbasic masigophore, atrichous isorhizas	n.d.	Length, width (min, max., mean)	Text, p. 58
Nemopilema nomurai	Pyo et al. (2016)	Medusa	Oral tentacle	Östmann (2000)	Isorhiza, birhopaloid, rhopaloid, mastigophore	n.d.	Length, width (mean)	Table 1, p. 28
Rhizostoma octopus	Holst et al. (2007)	Planula, polyp, ephyra, medusa	Planula: entire. Polyp: body-wall and tentacle. Ephyra: marginal lappet, mouth lip, exumbrella, gastric filament. Medusa: marginal lappet, mouth lip, exumbrella, gastric filament	Werner (1993), Calder (1974)	Atrichous haplonneme a-isorhiza, heterotrichous microbasic eurytele	10–64	Length, width (min, max, mean)	Table 4, p. 1700

(continued)

Table 5 (*cont'd*) Rhizostomeae, Dactyliophorae. Tissues, analysis methods and nematocyst types/varieties in various studies (see references). Measurements have been conducted on undischarged capsules unless otherwise specified under measurement parameters. N.d. = no data, ref. = reference.

Species	Reference	Life stage(s)	Body part(s)	Nematocyst terminology	Nematocyst types /verities	Measurements (n) per type	Measurement parameters	Measurement data (in ref.)
Rhizostoma pulmo	Avian et al. (1991)	Medusa	Marginal lobes, oral arms	Mariscal (1974)	Heterotrichous microbasic eurytele, holotrichous isorhizas, atrichous a-isorhizas, atrichous α-isorhizas	Approx. 10	Length, width (min, max)	Text, p. 617
Rhopilema esculentum	Chen and Ding (1981)	Planula, polyp, strobila, ephyra, young medusa, matured medusa	Planula: entire. Polyp: tentacle. Ephyra: marginal lappet, mouth. Medusa: marginal lappet, mouth lip, gastric filament	Calder (1974)	Heterotrichous microbasic eurytele, a-isorhiza, o-anisorhiza, e-anisorhiza	20–150	Length, width (min, max, mean)	Table 2, p. 312
Rhopilema hispidum	Othman et al. (1996)	Medusa	Oral appendage	n.d.	Atrichous isorhiza, holotrichous isorhiza, microbasic eurytele	n.d.	Diameter discharged (SEM) and undischarged (TEM)	Text, p. 365-368
Rhopilema nomadica	Avian et al. (1995)	Medusa	Appendages, gastric cirri, tentacular processes surrounding the mouths	Weill (1934)	Small heterotrichous isorhiza haploneme, large holotrichous isorhiza haploneme, heterotrichous microbasic eurytele	100–200	Length, width (min, max, mean) also discharged	Table 1, p. 225
Rhopilema verilli	Calder (1972)	Medusa	Appendages, gastric cirri, tentacular processes surrounding the mouths	Weill (1934)	Atrichous isorhiza ("a" atrich), holotrichous haploneme (isorhiza or anisorhiza), heterotrichous microbasic eurytele	30	Length, width (min, max, mean)	Table II

Rhopilema verilli	Calder (1973)	Planula, Young polyp, Fully-developed polyp, Early srobila, Young ephyra	n.d.	Weill (1934)	Atrichous isorhiza ("a" atrich), microbasic heterotrichous eurytele, holotrichous haploneme	25	Length, width (min, max)	Text, p. 110-111
Rhopilema verilli	Calder (1977)	Ephyra	Entire	Weill (1934), Calder (1974)	a-isorhiza, heterotrichous ansorhiza, heterotrichous microbasic eurytele	25	Length, width (min, max, mean, ratio	Table I, p. 15
Stomolophus meleagris	Calder (1983)	Planula, polyp, ephyra, medusa	Planula: entire; polyp: proboscis, calyx, and tentacles; ephyra: manubrium, exumbrella, lappets; medusa: manubrium, scapulets, gastric cirri	Weill (1934), Calder (1974)	Holotrichous a-isorhiza, holotrichous α-isorhiza, heterotrichous microbasic eurytele (small, medium, medium-large, large)	30–100	Length, width (min, max, mean, ratio),	Table 1, p. 1187

Table 6 Rhizostomeae, Kolpophorae. Tissues, analysis methods and nematocyst types/varieties in various studies (see references). Measurements have been conducted on undischarged capsules unless otherwise specified under measurement parameters. N.d. = no data, ref. = reference.

Species	Reference	Life stage(s)	Body part(s)	Nematocyst terminology	Nematocyst types /verities	Measurements (n) per type	Measurement parameters	Measurement data (in ref.)
Cassiopea andromeda	Gülsahin (2016)	Medusa	Oral arm	Calder (1974), Östmann (2000)	a-isorhiza, birhopaloid medium, birhopaloid small	100	Length, width (mean) discharged and undischarged	Table 1, table 2, p. 210
Cassiopea andromeda	Heins et al. (2015)	Bud, polyp, strobila, ephyra, medusa	Bud: entire; polyp: tentacle and oral lip; strobila, ephyra, medusa: marginal lappet and oral lip	Östman and Hydman (1997), Östmann (2000)	Holotrichous a-isorhiza, holotrichous O-isorhiza, oval heterotrichous microbasic birhopaloids, heterotrichous microbasic euryteles, lemon-shaped microbasic birhopaloids	1–304	Length, width (min, max, mean, ratio, volume), discharged and undischarged	Table 1, p. 355
Cassiopea andromeda	Jensch and Hofmann (1997)	Planula, polyp, bud, ephyra, medusa	Planula: entire; polyp: tentacle, hypostome, calyx, stalk; bud: entire; ephyra: exumbrella, subumbrella, manubrium, digitellae, oral lobes; medusa: exumbrella, subumbrella, manubrium, digitellae, oral lobes	Calder (1974)	Heterotrichous microbasic eurytele, a-isorhiza, heterotrichous anisorhiza	n.d.	Length, width (min, max)	Text, p. 280
Cassiopea culionensis	Gamero-Mora et al. (2022)	Medusa	Oral arms	Heins et al. (2015), Morandini and Marques (2010)	Holotrichous O-isorhiza, small holotrichous a-isorhiza, large holotrichous a-isorhiza, heterotrichous microbasic rhopaloid, two unidentified capsules	30 (3 if rare)	Length, width (min, max, mean)	Text, p. 74

Cassiopea mayeri	Gamero-Mora et al. (2022)	Medusa	Oral arms	Heins et al. (2015), Morandini and Marques (2010)	Holotrichous O-isorhiza, small holotrichous a-isorhiza, large holotrichous a-isorhiza, heterotrichous microbasic rhopaloid, two unidentified capsules	30 (3 if rare)	Length, width (min, max, mean)	Text, p. 81
Cassiopea ornata	Anthony et al. (2022)	Medusa	Digitate cirri, oral arm, cassiosome, mucus	Östmann (2000)	Small a-isorhizas, medium birhapoloids, large birhapoloids, O-isorhizas, long eurytele (rare)	Up to 130	Length, width (mean)	Table 1, p. 11
Cassiopea xamachana	Ames et al. (2020)	Polyp, strobila, ephyra, medusa	Tentacle, bell margin, oral arm, cassiosome, mucus	Östmann (2000)	O-isorhiza, a-isorhiza, rhopaloid (eurytele or birhopaloid), lemon shaped rhopaloid (eurytele or birhopaloid)	1–158	Length, width (min, max, mean)	Supplement table, p. 8
Cassiopea xamachana	Jensch and Hofmann (1997)	Planula, polyp, bud, ephyra, medusa	Planula: entire; polyp: tentacle, hypostome, calyx, stalk; bud: entire; ephyra: exumbrella, subumbrella, manubrium, digitellae, oral lobes; medusa: exumbrella, subumbrella, manubrium, digitellae, oral lobes	Calder (1974)	Heterotrichous microbasic eurytele, a-isorhiza, heterotrichous anisorhiza	n.d.	Length, width (min, max)	Text p. 280

(continued)

Table 6 (*cont'd*) Rhizostomeae, Kolpophorae. Tissues, analysis methods and nematocyst types/varieties in various studies (see references). Measurements have been conducted on undischarged capsules unless otherwise specified under measurement parameters. N.d. = no data, ref. = reference.

Species	Reference	Life stage(s)	Body part(s)	Nematocyst terminology	Nematocyst types /verities	Measurements (n) per type	Measurement parameters	Measurement data (in ref.)
Cotylorhiza tuberculata	Cengiz and Killi (2021)	Medusa	Bell margin and oral arms	Calder (1974), Östmann (2000)	Eurytele (small, medium), O-isorhiza, a-isohrhiza, A-isorhiza	n.d.	Length, width (min, max, mean)	Table 1, p. 36
Phyllorhiza punctata	Cengiz and Killi (2021)	Medusa	Bell margin and oral arms	Calder (1974), Östmann (2000)	Eurytele (small, medium, large), birhopaloid, stenotele, O-isorhiza, a-isorhiza, A-isorhiza, polyspira	n.d.	Length, width (min, max, mean)	Table 1, p. 36
Phyllorhiza punctata	Peach and Pitt (2005)	Medusa	Oral arm, bell margin	Östmann (2000)	Round holotrichous isorhiza, oval holotrichous isorhiza, rhopaloid medium sized, rhopaloid large, birhopaloid	1–6	Length and width (mean)	Table 2, p. 101

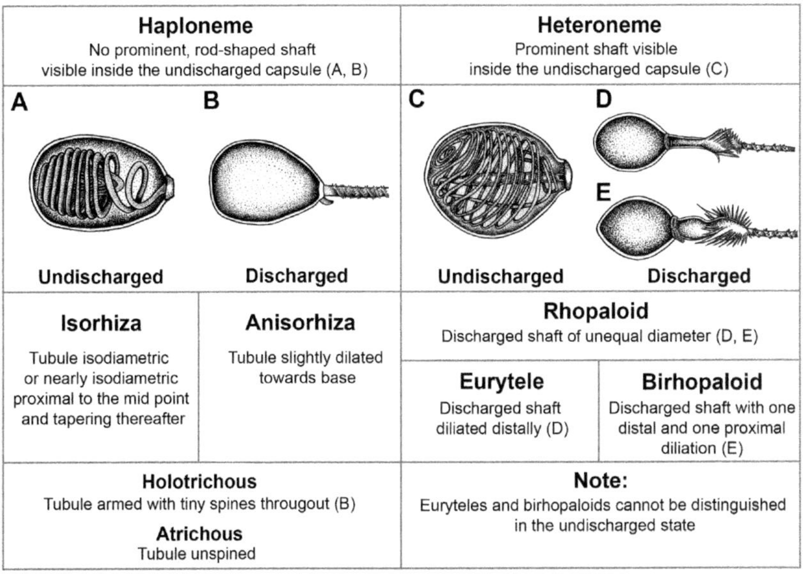

Fig. 10 Nematocyst categories in Rhizostomeae (nomenclature according to Östman, 2000). (A) Undischarged haploneme. (B) Discharged haploneme. (C) Undischarged heteroneme rhopaloid. (D) Discharged eurytele. (E) Discharged birhopaloid. *Drawings from Heins, A., Glatzel, T., Holst, S., 2015. Revised descriptions of the nematocysts and the asexual reproduction modes of the scyphozoan jellyfish Cassiopea andromeda (Forskål, 1775). Zoomorphology 134, 351–366, Springer, reproduced with permission from SNCSC.*

the polyp is probably a signal for the onset of strobilation in *R. verrilli*, *S. meleagris* (Calder, 1973, 1983), and in *C. andromeda* (Jensch and Hofmann, 1997). The occurrence of anisorhizas in the strobila, ephyra and medusa stage but not in the planula and polyp has also been reported for *Rhopilema esculentum* Kishinouye, 1891 (Chen and Ding, 1981). Later, the definition of anisorhizas was under debate and according to the new classification introduced by Östman (2000) the tubule of an anisorhiza has a broader base (close to the capsule) and tapers gradually, whereas the tubule of an isorhiza is isodiametric or nearly isodiametric proximal to the mid-point and tapering thereafter (Fig. 10). This change of definition can lead to confusion since descriptions of anisorhizas, which are based on older nomenclature may match the category isorhizas after the new definition by Östman (2000). For example, round nematocysts with a tapering tubule were identified as anisorhizas in *C. andromeda* (Jensch and Hofmann, 1997). However, according to the new classification (Östman, 2000), the round nematocysts occurring in *C. andromeda* have been classified as O-isorhizas (Heins et al., 2015).

5.2.2 Rhopaloids

The euryteles have a shaft visible in the undischarged capsule, which is distally dilated in the everted stage (Fig. 10). This capsule type was mentioned as the only heteroneme nematocyst found in Rhizostomeae until the end of the 20th century (Tables 5 and 6). After the first description of birhopaloids in Scyphozoa, including the rhizostomes *C. tuberculata* and *C. xamachana* (Östman 2000), this rhopaloid type has later also been documented for the species *C. mosaicus*, *P. punctata* (Peach and Pitt, 2005), *C. tagi* (Parracho and Morais, 2015), *C. andromeda* (Gülsahin, 2016; Heins et al., 2015), and *N. nomurai* (Pyo et al., 2016). It has to be highlighted, that birhopaloids were overlooked in earlier descriptions of rhizostome nematocysts, since they cannot be distinguished from euryteles when undischarged (Ames et al., 2020; Morandini and Marques, 2010) (Fig. 10). For example, different subtypes of euryteles in *S. meleagris,* which differ in shape as well as in size (Calder, 1983) may also include birhopaloids. Light microscopy was successfully used to identify the two dilations of discharged birhopaloids in rhizostomes (Gülsahin, 2016; Heins et al., 2015; Peach and Pitt, 2005). However, even in the discharged stage the distinction between euryteles and birhopaloids can be difficult since the dilations of the shaft can be almost imperceptible (Peach and Pitt, 2005). Recent descriptions of stenoteles in *P. punctata* medusae (Cengiz and Killi, 2021) should be regarded with caution since these rhopaloids with extraordinary long spines have not been described in any Scyphozoa before (Östman, 2000; Sötje and Holst, 2019). Similarly, the exceptional observation of "mastigophore-type nematocysts" in *N. nomurai* (Kubota et al., 2006; Pyo et al., 2016) needs further verification since mastigophores usually do not occur in Scyphozoa (Ballesteros et al., 2021; Östman, 2000).

In addition to the subtypes and categories of haplonemes and rhopaloids mentioned above, several varieties of these main nematocysts categories have been described based on their particular capsule shape, as for example the "lemon-shaped birhopaloid" (Ames et al. 2020; Heins et al., 2015), the "ellipsoid isorhiza" (Kondo et al., 2020), or the "pear-shaped isorhiza" (Peach and Pitt, 2005). Such descriptive terms for nematocyst varieties are useful to highlight the diversity of nematocysts, intraspecific variations and interspecific differences in cnidomes but should not be misinterpreted, as new nematocyst categories. In addition, it has to be ensured that developing and atypical forms of nematocyst should not be considered as distinct categories (Calder, 1977). Also, incompletely discharged nematocysts and capsules with malformations or unusual appearance should be excluded from the analyses (Heins et al., 2015).

5.3 Microscopic techniques to explore the nematocysts

LM techniques are useful for descriptions of the capsule shape, for studies on the spatial position and the windings of the tubule within the capsule as well as for observations and measurements on the everted tubules (e.g. Kondo et al., 2020). The magnification range of LM can be insufficient to depict tubule details, such as tiny spines or variations in the tubule diameter (Avian et al., 1991). However, it would be rather impractical to base nematocyst identification and analyses on ultrastructural characteristics (Mariscal, 1974), and the differentiation and documentation of rhizostome capsule categories and their subtypes have successfully been conducted with high quality light microscopes (e.g. Heins et al., 2015; Kondo et al., 2020) (Fig. 11A–I). Various staining techniques (hematoxylin-eosin, Masson's trichrome, alcian blue, von Kossa) applied on *C. tagi* nematocysts in combination with LM demonstrated that this approach can provide important contributions to clarify the chemical nature of the components of nematocytes, for example a basic protein with cationic nature in the tubules and collagenous material inside the capsule (Parracho and Morais, 2015). As demonstrated for small and large *R. nomadica* isorhizas, different nematocyst types differ in their dielectric properties, which can be useful for dielectrophoretic characterization and isolation (Park et al., 2017).

SEM is an excellent tool for the investigation and depiction of capsule and tubule details (Fig. 11J and K) including descriptions and measurements on the tiny spines as conducted for the rhizostome species *C. andromeda* (Jensch and Hofmann, 1997), *Rhopilema hispidum* (Vanhöffen, 1888) (Ohtman et al., 1996), *P. punctata, and C. mosaicus (*Peach and Pitt, 2005). In addition to heterotrichous microbasic euryteles and atrichous isorhizas, SEM investigations revealed holotrichous isorhizas and atrichous α-Isorhizas, with fusiform, compressed capsules in *R. pulmo* (Avian et al., 1991). A disadvantage of SEM is, that during the preparation of the sample, some of the discharged nematocysts may lose their spines in part or completely, and thus, a type that is spined when observed under a light microscope, can occasionally appear to be atrichous in SEM micrographs (Avian et al., 1991). Moreover, studies on *P. punctata* and *C. mosaicus* demonstrated that nematocysts appeared smaller when investigated with SEM than with LM, and dilations of the shaft in rhopaloids and birhopaloids were less distinct than in LM observations, due to the shrinkage during SEM preparations (Peach and Pitt, 2005). SEM micrographs of *R. octopus* nematocysts revealed that the operculum of the haploneme capsule was split up during discharge, whereas it was bent up but remained attached to the discharged eurytele capsule (Holst et al. 2007).

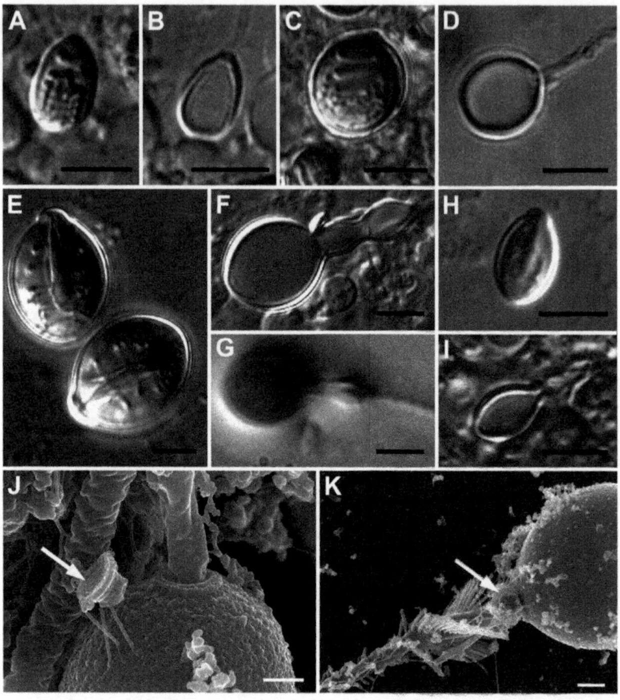

Fig. 11 (A–I) *Cassiopea andromeda*. Light microscopic images of nematocysts. (A, B) a-isorhiza without shaft and with isodiametric tubule, (A) undischarged, (B) discharged. (C, D) O-isorhizas without shaft and with isodiametric tubule, (C) undischarged, (D) discharged. (E) Euryteles or birhopaloids (indistinguishable in the undischarged state). (F) Oval birhopaloid, discharged. (G) Eurytele, discharged. (H, I) Lemon-shaped birhopaloids, (H) undischarged, (I) discharged. (J, K) *Rhizostoma octopus*. SEM micrographs of nematocysts. (J) Haploneme with split-off operculum (arrow). (K) Eurytele with bent-up operculum (arrow). Scale bars (A–I) = 5 μm, (J) = 500 nm, (K) = 1 μm. *(A-I) from Heins, A., Glatzel, T., Holst, S., 2015. Revised descriptions of the nematocysts and the asexual reproduction modes of the scyphozoan jellyfish Cassiopea andromeda (Forskål, 1775). Zoomorphology 134, 351–366, reproduced with permission from SNCSC. (J, K) Photo: I. Sötje.*

TEM depicted the arrangement of the nematocyte in the ectoderm of the planula of *C. xamachana* embedded within the supportive cells (Martin and Chia, 1982). The large double-layered nematocyst capsule occupied the upper two-thirds of the cell while a small nucleus and a well-developed Golgi apparatus were present in the basal region of the cell, and above the cyst a modified cilium gave rise to the cnidocil (Martin and Chia, 1982). TEM sections of *R. hispidum* nematocysts found similarities between different capsule types (two isorhizas and one eurytele) with an electron-dense outer layer and an electron transparent inner layer surrounding the electron-dense inner matrix of the capsule (Othman et al., 1996). Combined SEM and TEM observations on *C. andromeda* revealed that the cnidocil is a multicellular flagellum-stereocilia complex: the central cnidocyte is surrounded by peripheral cells contributing to the stereocilium which is associated to a much longer flagellum in the eurytele complex (15 μm) than in the isorhiza complex (4 μm) (Mariscal and Bigger, 1976; Fig. 12A). Immunocytochemistry approaches in combination with LM and TEM have been used to visualize the toxin localization in undischarged and discharged euryteles and isorhizas of *R. nomadica*, revealing that the toxin is limited to the tubular segments of the nematocysts but absent from the capsular lumen and walls (Lotan et al., 1996) (Fig. 12B).

5.4 Nematocyst discharge and nematoblasts

The immunocytochemical approach on *R. nomadica* nematocysts suggested that the cnidarian venom is stored on the outer surface of the folded, undischarged tubule while the toxin (see Nagata et al., 2024) is translocated into the lumen of the tubule, and when the spines fold out, the toxin passes from the everted tubule through the hollow barbs during the discharge process (Lotan et al., 1996; Fig. 12B). The discharged tubule with the spirally arranged array of the extended hollow spines resembles a multi-headed poisonous arrow (Lotan et al., 1996). Another study on *R. nomadica* tentacle isorhizas confirmed that the increase of osmotic pressure affecting the tubule discharge is related to a concentration change of cations in the capsule (Park et al., 2017).

The nematocyst discharge can only be induced in living animals, but also stranded jellyfish can potentially sting if their tissues are still intact (e.g. reported for the siphonophore *Physalia* sp. in Mulyadi and Sianturi, 2021). Discharged nematocysts are essential for detailed investigations on the morphological details of the nematocyst tubule, its shaft and spines, which are needed for the distinction of nematocyst categories and subtypes (see Section 5.2). Thus, the distinct differentiation between certain capsule categories as euryteles and birhopaloids is usually not possible if fresh tissues are unavailable (Heins et al., 2015). For this reason, most authors who described discharged nematocysts of rhizostomes used fresh tissues from living specimens (Tables 7 and 8). But even in preserved or frozen samples, rhizostome nematocyst capsules that were discharged before or during preservation or freezing were documented (e.g. Cengiz and Killi, 2021; Gamero-Mora et al., 2022; Gülsahin, 2016; Kondo et al., 2020). However,

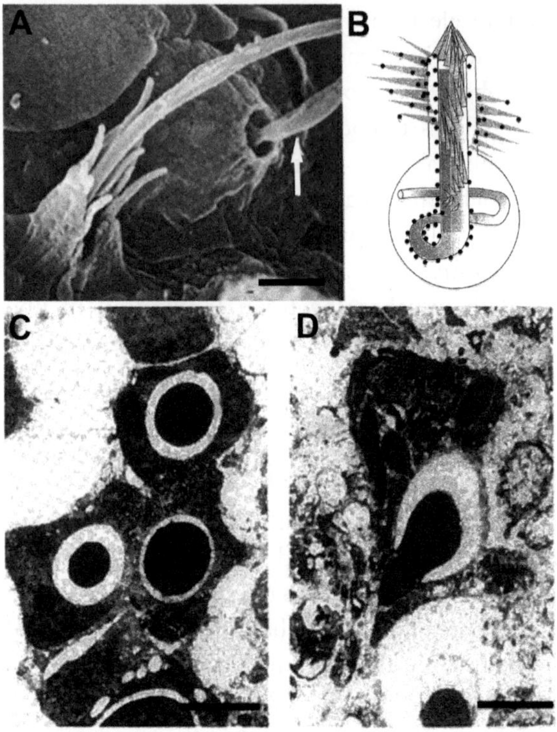

Fig. 12 (A) *Cassiopea xamachana*. SEM micrograph of F-S complex (left) and single flagellum (arrow). (B) Schematic drawing of toxin compartmentation and delivery in the nematocyst system. Central part: resting tubule whose lumen is filled with spines. The toxin (dots) is located on the external side of the tubule's membrane. During the discharge toxin is translocated into the tubule's lumen while the barbs emerge and extend. Toxin is delivered from the everted-extended tubule through the hollow barbs. (C, D) *Rhopilema hispidum*. Transmission electron micrographs of nematocytes of with unencapsulated threads. (C) Transverse section of three nematocytes containing expanded terminus. The expanded bulbs are enclosed within an electron-transparent capsule of various thickness. The inner core is basophilic with electron-dense matrix. (D) The electron-transparent capsule is interrupted to allow the extrusion of secreting mass from the bulb. Scale bars: A = 1 μm, C = 2 μm, D = 3 μm. *(A) from Mariscal, R.N., Bigger, C.H., 1976. A comparison of putative sensory receptors associatedwith nematocysts in an anthozoan and a scyphozoan, In: Mackie, G.O. (Ed.), Coelenterate Ecology and Behavior. Plenum Press, New York, pp. 559–568, reproduced with permission from SNCSC; (B) from Lotan, A., Fishmann, L., Zlotkin E., 1996. Toxin compartmentation and delivery in the cnidarian: the nematocyts tuble as a multiheaded poisonous arrow. J. Exp. Zool. 275, 444–451, Wiley, reproduced with permission; (C, D) from Othman, I., Fathilah, A.R., Mohd Saad, J., Mohd Yusof, M., Mustaffa, M.R., Nor, A., 1996. Studies on the venomous coelenterate: Rhopilema hispidum. J. Nat. Toxins 5, 361–375, reproduced with permission by the author.*

after preservation, only parts of the thread and the shaft can be squeezed out mechanically from the undischarged nematocyst (Heins et al., 2015).

Squash preparations without additional application of any discharge inducers have been successfully conducted to investigate discharged capsules in fresh or preserved tissue samples (Tables 7 and 8). For the SEM documentation of discharged nematocysts of *C. andromeda*, *Artemia* nauplii and human stratum corneum penetrated by nematocysts have been used (Jensch and Hofmann, 1997). Homogenization and centrifugation of samples can be advantageous over squash preparations for discharge induction and for the separation of nematocyst capsules from the tissues (e.g. Cengiz and Killi, 2021; Lotan et al., 1996). Acetic acid has been used in several studies on rhizostomes to induce the capsule discharge (DiMatteo, 1981; Fitt and Costley, 1998; Kondo et al., 2020) (Table 7). Nematocysts from polyp tentacles of *C. xamachana* polyps maintained at temperature ranges from 6 °C to 38 °C discharged when subjected to 5% acetic acid, but at temperatures higher than 36 °C, the tentacles began to disintegrate (Fitt and Costley, 1998). Other studies used lyotropic ions, potassium iodide, sodium citrate, or sodium isothiocyanate as discharge inducers for rhizostome nematocysts (Avian et al. 1991, 1995). Experimental results on the rhizostome *N. nomurai* and the cubozoan *Carybdea brevipedalia* Kishinouye, 1891 (as *C. mora)* clearly demonstrated that the effect of different chemicals and reagents on nematocyst discharge is taxon specific; for example, acetic acid did not induce discharge in the cubozoan but had a strong effect on nematocyst discharge in *N. nomurai*. Ethanol strongly affected the discharge in *C. brevipedalia* but had only slight effects in *N. nomurai*; isopropanol caused severe discharge, whereas no discharge was observed after lidocaine treatment in both species (Pyo et al., 2016). A significant reduction of the nematocyst capsule volumes after discharge was demonstrated for *R. nomadica* (Avian et al., 1995) and *C. andromeda* with volume reduction of up to 51% (Heins et al., 2015).

Since nematocysts are released from the nematocytes after use (Mariscal, 1974), the nematocyst repertoire has to be constantly renewed. In the process of nematocyst morphogenesis, the tubule forms at the apical end of the nematocyst capsule and is later invaginated into the capsule matrix (Beckmann and Özbeck, 2012). Such developmental stages of nematocysts (nematoblasts) (Mariscal, 1984) are shown by TEM micrographs on oral appendages of *R. hispidum* (Othman et al. 1996; Fig. 12C and D). TEM investigations on *C. xamachana* planulae demonstrated that the nematoblasts were located at the base of the epidermis without contact to the ectoderm surface (Martin and Chia, 1982).

Table 7 Rhizostomeae, Dactyliophorae. Collection data, fixation and preparation methods used for nematocyst analyses in various studies (references). LM = Light Microscopy, n.d. = no data, SEM = Scanning electron microscopy, TEM = Transmission electron microscopy, *Figure number and page in reference.

Species	Reference	Locality	Collection date (year)	Fixation	Preparation	Discharge induction	Microscopic technique	Nematocyst photos*
Catostylus mosaicus	Peach and Pitt (2005)	Lake Illawarra, Australia	2002, 2003	Formalin 10%	Homogenization, squash preparation	Tissues rubbed against small pieces of prawn carapace	LM, SEM	1 and 3, p. 99 and 103
Catostylus tagi	Parracho and Morais (2015)	Tagus and Sado, Portugal	2010-2013	Formaldehyde 4%, fresh tissues	5 μm sections, various stains	Smashing, centrifugation of fresh tissues	LM, histological sections, SEM	Figs. 2–6, p. 5-7
Lobonemoides robustus	Kondo et al. (2020)	Nathung, Thailand	2014	Formalin in seawater 5%	Vinegar added before preservation	Acidic acid (vinegar)	LM	Fig. 5, p. 704
Nemopilema nomurai	Kubota et al. (2006)	Tanabe Bay, Japan	2005	Fresh tissues	squash preparation	no	LM	Fig. 3, p. 53
Nemopilema nomurai	Pyo et al. (2016)	Tongyeong coast, Korea	2013	Fresh tissues	Tentacles on glass slides	Distilled water, ethanol, acetic acid, isopropanol, lidocaine	LM	Fig. 1, p. 27
Rhizostoma octopus	Holst et al. (2007)	North Sea, Germany	2003, 2004	Fresh tissues, formalin 4%	Squash preparations	No	LM, SEM	Fig. 6, p. 1702
Rhizostoma pulmo	Avian et al. (1991)	Gulf of Trieste, Italy	1985-1986	Fresh tissues	Tissues immersed in distilled water, Homogenization, centrifugation	Lyotropic ions, mechanical pressure	LM, SEM	1–8, p. 616

Rhopilema esculentum	Chen and Ding (1981)	Bohai Sea, China	1978-1980	Formalin	Squash preparations with distilled water	Acetic acid	LM	Fig. I, p. 317
Rhopilema hispidum	Othman et al (1996)	n.d.	n.d.	n.d.	Homogenization for SEM	n.d.	LM, SEM, TEM	Figs. 3–5, p. 164-167
Rhopilema hispidum	Kondo et al. (2020)	Khampuan, Thailand	2014	Formalin in seawater 5%	Vinegar added before preservation	Acidic acid (vinegar)	LM	Fig. 3, p. 702
Rhopilema nomadica	Avian et al. (1995)	Haifa Bay, Israel	1992-1993	Formaldehyde in seawater 4%, Fresh tissues	For SEM: tissues immersed in distilled water, Homogenization, centrifugation	Potassium iodide, sodium citrate, sodium isotiocyamte	LM, SEM, TEM	2–18, p. 224–229
Rhopilema verilli	Calder (1972)	Virginia waters, USA	1971, 1972	Fresh or frozen tissues	Squash preparations	No	LM	Figs. 1–3, p. 215
Rhopilema verilli	Calder (1973)	Virginia waters, USA	1970	n.d.	n.d.	n.d.	LM	no
Rhopilema verilli	Calder (1977)	Virginia waters, USA	1970	Fresh tissues	Squash preparations	No	LM	Figs. 13–15, p. 16
Stomolophus meleagris	Calder (1983)	South Carolina, USA	1974	Formalin	Squash preparations	No	LM	Fig. 1–13, p. 1188

Table 8 Rhizostomeae, Kolpophorae. Collection data, fixation and preparation methods used for nematocyst analyses in various studies (references). LM = Light Microscopy, n.d. = no data, SEM = Scanning electron microscopy, *Figure number and page in reference.

Species	Reference	Locality	Collection date (year)	Fixation	Preparation	Discharge induction	Microscopic technique	Nematocyst photos*
Cassiopea andromeda	Gülsahin (2016)	Güllük Bay, Turkey	2015	Frozen (–20°C)	Squash preparations	No	LM	Fig. 3, p. 357
Cassiopea andromeda	Heins et al. (2015)	Red Sea	n.d.	Fresh tissues	Squash preparations	No	LM	Fig. 3, p. 357
Cassiopea andromeda	Jensch and Hofmann (1997)	Red Sea, Israel	n.d.	Fresh tissues	Squash preparations, semithin sections	*Artemia* nauplii, human stratum corneum	LM, histological sections, SEM	No
Cassiopea culionensis	Gamero-Mora et al. (2022)	Lapu-Lapu, Philippines	2019	Formalin 4%	Squash preparations	No	LM	Fig. 7, p. 75
Cassiopea mayeri	Gamero-Mora et al. (2022)	Lapu-Lapu, Philippines	2019	Formalin 4%	Squash preparations	No	LM	Fig. 7, p. 75
Cassiopea ornata	Anthony et al. (2022)	Guam, Micronesia	2021	Fresh tissues	Squash preparations	No	LM	Fig. 7, p. 10
Cassiopea xamachana	Ames et al. (2020)	Florida Keys, USA	n.d.	Fresh tissues	Squash preparations	Dehydration	LM	6b–d, p. 8
Cassiopea xamachana	Jensch and Hofmann (1997)	Grassy Key, USA	n.d.	Fresh tissues	Squash preparations, semithin sections	*Artemia* nauplii, human stratum corneum	LM, histological sections, SEM	1–6, p. 282

Cotylorhiza tuberculata	Cengiz and Killi (2021)	Gökova Bay, Turkey	2017	Frozen (−18°C)	Homogenization, centrifugation	No	LM	Fig. 7, p. 36
Phyllorhiza punctata	Cengiz and Killi (2021)	Sülüngür Bay, Turkey	2017	Frozen (−18°C)	Homogenization, centrifugation	No	LM	Fig. 7 and 8, p. 36
Phyllorhiza punctata	Peach and Pitt (2005)	Lake Illawarra, Australia	2003	Formalin 10%	Homogenization, squash preparation	Tissues rubbed against small pieces of prawn carapace	LM, SEM	Figs. 2 and 4, p. 102 and 104

In semithin sections on *C. andromeda* stained with "routine methods" nematoblasts could be distinguished from nematocytes due to their size and different, uniform staining of the capsule rudiment (Jensch and Hofmann, 1997). In hematoxylin-eosin stained tissues from deep layers of oral arms of *C. tagi*, early nematocyst stages appeared as small circular eosinophilic (orange) structures (Parracho and Morais, 2015). Distinct patterns of distributions of nematoblasts and nematocytes in different developmental stages (bud, polyp, ephyra, medusa, planula) have been described using LM for *C. andromeda* and it was suggested that discharged cnidae are replaced by cells recruited from neighbouring areas (Jensch and Hofmann, 1997). In *C. andromeda* polyps, nematoblasts were limited to the ectoderm and endoderm of the calyx and the hypostome, whereas in ephyrae and medusae, nematoblasts were more widely distributed (Jensch and Hofmann, 1997).

5.5 Quantitative measurements on nematocysts

Quantitative measurements are usually conducted on undischarged nematocyst capsules, which is also possible in preserved material (see Section 5.4). An overview on nematocyst capsule sizes of various rhizostome medusae by Kondo et al. (2020) (Table 9) demonstrates that both, rhopaloids and isorhizas, are very variable in size. Thiel (Thiel et al., 2024) has reviewed nematocyst measurements on several rhizostome species without mentioning the investigated life stages or measurement replicates. Calder (1972, 1973, 1977) measured 25–30 capsules of the same category (Table 5). Approximately ten measurements on nematocysts of *R. pulmo* medusae were regarded as provisional (Avian et al., 1991) and the authors measured 100–200 capsules in a later study on *R. nomadica* medusae (Avian et al., 1995). In other studies, the number of measured capsules varied with their abundance (e.g. Ames et al., 2020; Calder, 1983; Chen and Ding, 1981; Gamero-Mora et al., 2022; Heins et al., 2015) (Tables 5 and 6).

Although undischarged euryteles and birhopaloids are often morphologically indistinguishable under LM (see Section 5.2), size measurements on an appropriate number of discharged capsules identified as euryteles or birhopaloids may allow the conclusive evaluation and identification of undischarged rhopaloid capsules (Heins et al., 2015). Most authors used ranges and mean values of capsule lengths and widths as measurement parameters for rhizostome nematocysts (Tables 5 and 6). Occasionally also the length/width ratio and/or the capsule volume have been calculated as an additional feature for size comparisons (Calder, 1983; Heins et al., 2015; Kondo et al., 2020). Light microscopic measurements on discharged

Table 9 Capsule sizes of nematocysts from rhizostome species *(modified from Kondo et al., 2020, Journal of the Marine Biological Association of the United Kingdom 100, Cambridge Universtiy Press, reproduced with permission).*

Species	Reference	Nematocyst type	Length (μm)	Width (μm)
Cassiopea andromeda	Heins et al. (2015)	a-isorhiza	4.15–8.29	2.68–4.88
		O-isorhiza	5.37–9.02	4.88–8.54
		Rhopaloid	5.97–17.80	4.52–15.85
		Lemon-shaped microbasic birhopaloid	4.63–9.27	2.93–5.85
Catostylus mosaicus	Peach and Pitt (2005)	Oval-shaped holotrichous isorhiza	3.9–4.8	2.9–3.4
		Pear-shaped holotrichous isorhiza	5.8–8.7	2.4–3.0
		Rhopaloid	7.1–9.3	4.9–6.6
		Birhopaloid	14.0–15.9	9.5–10.2
Lobonemoides robstus	Kondo et al. (2020)	Small ellipsoid isorhiza	3.1–6.9	2.3–4.2
		Small ellipsoid rhopaloid	5.3–8.1	4.4–6.3
		Large ellipsoid rhopaloid	9.0–15.5	7.0–11.9
Nemopilema nomurai	Kubota et al. (2006)	Holotrichous anisorhiza	25.0–30.0	24.0–26.0
		Microbasic mastigophore	17.0–22.0	11.0–14.0
		Large atrichous isorhiza	13.7–15.8	3.74–4.15
		Small atrichous isorhiza	8.47–10.8	5.40–7.06
Phyllorhiza punctata		Round holotrichous isorhiza	5.6–5.8	4.9
		Oval holotrichous isorhiza	5.1–5.7	3.4–3.8
		Medium rhopaloid	7.2–7.7	4.7–5.5
		Large rhopaloid	12.5–14.7	8.4–10.9
		Birhopaloid	13.5–14.6	10.0–10.5
Rhizostoma octopus	Holst et al. (2007)	Atrichous haploneme	3.5–6.5	2.3–5.3
		Heterotrichous microbasic eurytele	5.8–10.2	4.1–7.8

(continued)

Table 9 *(cont'd)* Capsule sizes of nematocysts from rhizostome species *(modified from Kondo et al., 2020, Journal of the Marine Biological Association of the United Kingdom 100, Cambridge Universtiy Press, reproduced with permission).*

Species	Reference	Nematocyst type	Length (μm)	Width (μm)
Rhizostoma plumo	Avian et al. (1991)	Heterotrichous microbasic eurytele	5.9–8.0	3.5–6.0
		Holotrichous isorhiza	3.0–4.0	2.0–2.5
		Atrichous a-isorhiza	3.0–3.5	2.0–3.0
		Atrichous α-isorhiza	3.3–4.0	1.1–1.4
Rhopilema esculentum	Chen and Ding (1981)	a-isorhiza	2.9–7.5	2.0–5.0
		o-anisorhiza	5.9–15.0	3.9–12.5
		e-anisorhiza	4.5–12.8	2.5–6.7
		Eurytele	4.4–16.0	2.9–11.5
Rhopilema hispidum	Kondo et al. (2020)	Small ellipsoid isorhiza	3.5–6.1	2.1–3.5
		Medium round isorhiza	5.0–7.6	4.7–5.0
		Large round isorhiza	12.2–13.3	10.4–15.0
		Rhopaloid	5.3–7.5	3.8–5.0
Rhopilema nomadica	Avian et al. (1995)	Small holotrichous isorhiza	4.20–8.50	2.90–5.56
		Large holotrichous isorhiza	10.50–18.30	10.10–16.80
		Heterotrichous microbasic eurytele	6.10–12.70	4.10–10.50
Rhopilema verrilli	Calder (1972)	a-atrichous isorhiza	5.3–6.9	3.3–4.5
		Holotrichous haploneme (isorhiza or anisorhiza)	6.9–8.9	5.4–7.4
		Microbasic heterotrichous eurytele	7.6–10.1	5.0–7.1
Stomolophus meleagris	Calder (1983)	a-isorhiza	3.8–4.8	2.4–3.1
		Small eurytele	6.4–10.1	4.5–6.5
		Medium-large eurytele	10.5–12.6	7.3–9.0

capsules and everted tubules of *R. hispidum* and *Lobonemoides robustus* Stiasny, 1920 mouth arm nematocysts revealed a positive correlation of tubule lengths and capsule volumes (Kondo et al., 2020). Detailed measurements on the tiny spines covering the everted tubule are only possible with high-resolution techniques such as SEM (see Section 5.3).

Until the 1970s different nematocyst types and sizes had not been analysed with regard to different medusa sizes or various developmental stages. Nematocyst studies on *R. verrilli* medusae (Calder, 1972; reviewed in Thiel et al., 2024) comprised five length-width measurements on nematocysts (a-atrichs, holotrichs, euryteles) of six medusae in different sizes (3–45 cm) and stages (4 juveniles, 1 mature female, 1 mature male), but for the analyses of nematocyst size variations the data from different medusae were pooled for each capsule type.

Recent studies on nematocyst dimensions of *L. robustus* (40–48 cm, $n = 4$) and *R. hispidum* (39–60 cm, $n = 3$) provide size measurements on individual specimens suggesting that the capsule dimensions do not significantly increase with size but that the composition of nematocyst types may slightly change with medusa growth (Kondo et al., 2020).

Similarly, a change of the nematocyst composition but no clear correlation of medusa sizes and nematocyst sizes was observed in recent studies on *C. tuberculata* and *P. punctata* (Cengiz and Killi, 2021). In earlier studies on *P. punctata* medusae (Peach and Pitt, 2005), a significant correlation was found between the bell diameter and length of oval isorhizas but the analysed sample size used in this study was low (six measurements in each of six specimens).

Investigations on different ontogenetic stages of *R. verrilli* (Calder, 1973) and *S. meleagris* (Calder, 1983) revealed stage-specific differences in nematocyst types and sizes. It was suggested that changes in the nematocyst complement occur during strobilation (Calder, 1977, 1983) and studies on *R. esculenta* (Chen and Ding, 1981) and *C. andromeda* (Heins et al., 2015; Jensch and Hofmann, 1997) confirmed the increase of nematocyst types from the polyp stage to the strobila stage. Nematocyst sizes tend to increase with the progressing development and increasing size of developmental stages as for example shown for *C. andromeda* (Heins et al., 2015). A changing diet with changing preferences for certain prey items (different trophic niche) could probably explain the differences in nematocyst types and sizes in different life stages (Heins et al., 2015).

It has been suggested that size measurements on cnidocysts are especially valuable for distinguishing species with the same cnidome (Östman, 2000). However, it has to be considered that nematocyst sizes increase with ontogenetic development and perhaps with medusa growth, which requires further verification. Moreover, as documented for *P. punctata*, sizes (and abundances) of certain nematocyst types can vary between specimens of the same species even if the same body regions are analysed (Peach and Pitt, 2005). Measurements on at least 40 capsules of each type per tissue were recommended for nematocyst analyses in soft-bodied hexacorals (Häussermann, 2004). Replicate measurements in similar extent (30–40 size measurements per nematocyst type and body region in at least five specimens) should be targeted for quantitative studies on rhizostome nematocyst but may not be achievable for rare capsule types.

5.6 Nematocyst abundances in different body regions

Nematocysts are particularly numerous on the polyp tentacles and often occur in batteries [see e.g. Calder, 1983 for *S. meleagris*, Lotan et al.,1992 for *R. nomadica*, Schiariti et al., 2008 for *L. lucerna*, Kienberger et al., 2018 for *Rhizostoma luteum* (Quoy and Gaimard, 1827)] (Fig. 13A). The same distribution pattern of cnidocytes in planulae have been found in Rhizostomeae (e.g. *R. octopus*) and Semaeostomeae (e.g. *A. aurita*) with abundant cnidocysts in the epidermis of the one end and rare or absent cnidocysts at the other end (Holst et al., 2007; Nakanishi et al., 2008; Yuan et al., 2008). In rhizostome medusae, the oral arms are equipped with higher numbers of nematocysts compared to other body regions (see Calder, 1983 for *S. meleagris*; Cengiz and Killi, 2021 for *C. tuberculata* and *P. punctata*). Oral appendages and oral tentacles armed with nematocysts batteries are documented for Kolpophorae, e.g. *Cassiopea* spp. (Jensch and Hofmann, 1997; Larson, 1997) (Fig. 13B) as well as for Dactyliophorae, for example, *N. nomurai* (Kawahara et al., 2006; Pyo et al., 2016) (Fig. 13C). Several studies include descriptions of intraspecific variations of nematocyst abundances in different life stages or body parts, however, a standardized format for the analysis and presentation of such data is lacking (Table 10). Dense aggregations of nematocysts appearing as refractive patterns termed as nematocyst clusters, warts or batteries occur in several rhizostome ephyrae and can be scattered over the exumbrella (Fig. 13D) or appear as specific patterns in certain species (Fig. 13E and F). The pattern of nematocyst clusters may change or disappear during the development from the young ephyra to the metaephyra (Kawahara et al., 2006; Schiariti et al., 2008; Tronolone et al., 2002). Distinct patterns of nematocyst clusters or warts are also present in several adult rhizostome medusae (Jarms and Morandini, 2019).

Investigations on the feeding behaviour of *C. frondosa* and *C. xamachana* (Larson, 1997) described nematocysts covering polygonal-shaped bodies,

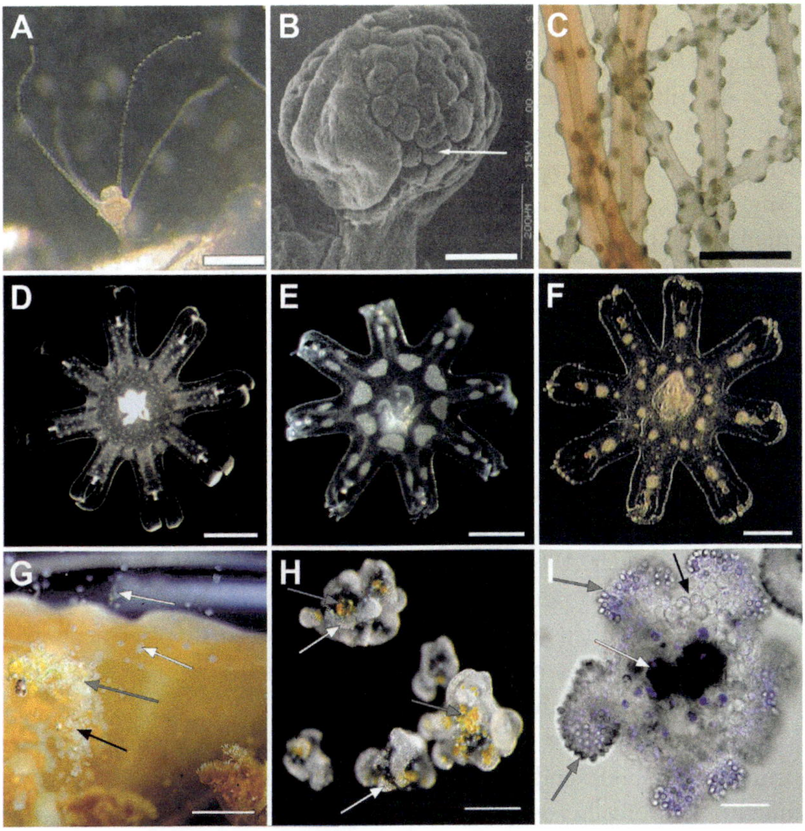

Fig. 13 Nematocyst batteries and clusters. (A) *Lychnorhiza lucerna* polyp with nematocyst batteries along the tentacles. (B) *Cassiopea andromeda* medusa, oral lobe with nematocyte bags (arrow). (C) *Nemopilema nomurai* metaephyra, oral arm appendages with nematocyst batteries. (D) *Rhizostoma octopus* ephyra, scattered nematocyst clusters. (E) *Rhopilema esculenta* ephyra, pattern of nematocyst clusters. (F) *Lychnorhiza lucerna* ephyra, pattern of nematocyst clusters. (G–I) *Cassiopea xamachana*. (G) Cassiosome nest (black arrow) at the oral appendages, with mucus (grey arrow) released into the water, and small white flecks correspond to live cassiosomes (white arrows). (H) Cassiosomes with peripheral nematocystes (white arrows), and centralized zooxanthellae (grey arrows). (I) Confocal image of cassiosome with peripheral O-isorihzas (grey arrows), non-nematocyte cells (black arrow) and central zooxanthellae (white arrow). Scale bars: A = 200 µm, B, F = 100 µm, C, E = 500 µm, D = 1 mm, G = 3 mm, H = 300 µm, I = 50 µm. *(A, F) From Schiariti, A., Kawahara, M., Uye,S.-I., Mianzan, H.W., 2008. Life cycle of the jellyfish Lychnorhiza lucerna (Scyphozoa: Rhizostomeae). Mar. Biol. 156, 1–12, Springer, reproduced with permission from SNCSC. (B) SEM micrograph from Jensch, F., Hofmann D.K., 1997. The cnidomes of Cassiopea andromeda Forskaal, 1775, and Cassiopea xamachana Bigelow, 1882 (Cnidaria: Scyphozoa). Proceedings of the 6th International Conference on Coelenterate Biology, Het Nationaal Natuurhistorisch Museum (today Naturalis Biodiversity Center), Leiden, pp. 279–285, reproduced with permission by Naturalis Biodiversity Center. (C) from Kawahara, M., Uye, S.-I., Ohtsu, K., Iizumi, H.,*

(continued)

which are released when contacted. These structures, released by *C. xamachana* medusae and termed "cassiosomes" (see also Nagata et al., 2024), were described as irregularly-shaped cellular masses of 100–550 μm in diameter, composed of an outer epithelial layer containing high numbers of nematocytes and a core filled by zooxanthellae and presumptive mesoglea (Ames et al., 2020) (Fig. 13G–I). In two *Cassiopea ornata* Haeckel, 1880 populations from Guam (Micronesia), cassiosomes with and without zooxanthaellae were found (Anthony et al., 2022). Cassiosome-like structures have also been reported in other rhizostome species including Kolpophorae [*M. papua, P. punctata, Netrostoma setouchianum* (Kishinouye, 1902)] and Dactyliophorae (*C. mosaicus*) (Ames et al., 2020). Prey items (*Artemia* nauplii) were immediately immobilized and killed after being added to dishes containing isolated *C. xamachana* cassiosomes (Ames et al., 2020). Similarly, disturbed *S. meleagris* medusae released a sticky mucus containing large numbers of nematocysts, which can cause death of fish and crustaceans (Calder, 1983; Shanks and Graham, 1988) and probably serves as a defence against potential predators (Camacho-Pacheco et al., 2022). Cassiosomes or mucus released to the water by rhizostome medusae can cause varying degrees of pain and irritation of the human skin without physical contact with the medusa (Muffett et al., 2021).

Higher abundances of nematocysts in specific body regions, as for example exumbrella and oral arms, and the occurrence of nematocyst warts and clusters were already noticed in early studies of rhizostomes (reviewed in Thiel et al., 2024), but without differentiation of different nematocyst types or categories. Later, distinct variability in abundances of nematocyst types in different body parts within the same life stage were detected in several species including *S. meleagris* (Calder, 1983), *C. andromeda* (Heins et al., 2015; Jensch and Hofmann, 1997), *P. punctata* (Cengiz and Killi, 2021), and *C. tuberculata* (Cengiz and Killi, 2021). The abundances of certain nematocyst types in a sample have often been estimated from squash preparations, and given in categories as frequent, rare or absent (Heins et al., 2015; Jensch and Hofmann, 1997). For the determination of relative

Fig. 13—cont'd *2006. Unusual population explosion of the giant jellyfish Nemopilema nomurai (Scyphozoa: Rhizostomeae) in East Asian waters. Mar. Ecol. Prog. Ser. 307, 161–173, Inter-Research, reproduced with permission; (D, E) Photo: S. Holst. (G–I) Ames, C.L., Klompen, A.M.L., Badhiwala, K., Muffett, K., Reft, A.J., Kumar, M., et al., 2020. Cassiosomes are stinging-cell structures in the mucus of the upside-down jellyfish Cassiopea xamachana. Commun. Biol. 3, 67, SpringerNature, CC BY licence.*

Table 10 Nematocyst abundances in rhizostome species. SO = suborder; DP = Dactyliophorae; KP = Kolpophorae, n.d. = no data.

SO	Species	Reference	Description
DP	*Catostylus mosaicus*	Peach and Pitt (2005)	Oral arm: oval isorhizas 74%, rhopaloids 19% birhopaloids 6%, Pear-shaped isorhizae 1%. Bell margin: birhopaloids 56%, oval isorhizae 21%, pear-shaped isorhizae 18%, rhopaloids 5%
	Catostylus tagi	Parracho and Morais (2015)	Higher nematocyte abundance in oral arms than in bell margins. Euryteles 55 %, isorhizae 30 % and birhopaloids 15 %, both in oral arms and bells
	Lobonemoides robustus	Kondo et al. (2020)	Irrespective of medusa bell diameter small ellipsoid rhopaloids were most dominant (>70%), small ellipsoid isorhizas and large round isorhizas comprised 6.3–12.8% and 5.9–17.0% of the total cnidome, respectively. Large ellipsoid rhopaloids comprised < 1%
	Nemopilema nomurai	Kubota et al. (2006)	The four types of nematocysts were concentrated on knobbed cirri around mouthlets
	Nemopilema nomurai	Pyo et al. (2016)	Nematocyst batteries on oral tentacles composed of various kinds of nematocysts
	Rhizostoma octopus	Holst et al. (2007)	Planula: Haplonemes present at the broader end of the planula only. Polyp: euryteles more abundant than haplonemes in the body wall, both types with similar abundance in the tentacles
	Rhizostoma pulmo	Avian et al. (1991)	Nematocysts more abundant in marginal lobes and oral arms
	Rhopilema esculentum	Chen and Ding (1981)	n.d.
	Rhopilema hispidum	Othman et al (1996)	n.d.
	Rhopilema hispidum	Kondo et al. (2020)	Irrespective of medusa bell diameter rhopaloids and medium round isorhizas were more abundant (55.2–60.5% and 28.6–30.9%, respectively) than small ellipsoid and large round isorhizas (8.0–13.4% and 0.8–0.9%, respectively)
	Rhopilema nomadica	Avian et al. (1995)	Scapular tentacles: small holotrichous isorhiza, 92.6%; large holotrichous isorhiza, 6.4%; eurytele, 1.0%. Mouth-arm tentacles: small holotrichous isorhiza, 84.5%; large holotrichous isorhiza, 14.0%; eurytele, 1.5%. Similar frequencies were observed in the ectoderm of oral lobes and exumbrellar marginal lappets. The endoderm of the gastric cirri and of the gonads contains a higher concentration of the eurytele type
	Rhopilema verilli	Calder (1972)	The relative abundance of the three nematocyst categories was not constant from region to region on a given medusa. Euryteles were the most conspicuous type on the gastric cirri. On the appendages, the euryteles and "a" atrichs were common only at the tips; elsewhere on these structures the nematocyst batteries consisted largely or entirely of holotrichs. All three categories were abundant the tiny tentacular processes surrounding the mouths
	Rhopilema verilli	Calder (1973)	n.d.

(continued)

Table 10 (*cont'd*) Nematocyst abundances in rhizostome species. SO = suborder; DP = Dactyliophorae; KP = Kolpophorae, n.d. = no data.

SO	Species	Reference	Description
DP (*cont'd*)	*Rhopilema verilli*	Calder (1977)	All nematocyst types well represented in the examined ephyrae
	Stomolophus meleagris	Calder (1983)	Planula: nematocysts scattered throughout the ectoderm, a-isorhizas were more numerous than euryteles. Polyp: nematocysts particularly numerous on the tentacles. A-isorhizas and small euryteles common, α-isorhizas only in the proboscis region. Ephyra: nematocysts abundant on the lappets, a-isorhizas outnumbering small euryteles. Prominent nematocyst batteries, consisting largely or entirely of medium euryteles, present on the exumbrella. The exumbrella also bores scattered a-isorhizas and small euryteles. On the manubrium, nematocysts were most common along the margins of the lips. Euryteles in dense aggregations within pockets on both the scapulets and at the distal end of them manubrium
KP	*Cassiopea andromeda*	Gülsahin (2016)	Birhopaloids (38.89%), a-isorhizas (61.11%). Medium birhopaloids more abundant in 190-220 mm bell sized medusae than in smaller medusae
	Cassiopea andromeda	Heins et al. (2015)	The a-isorhizas often appeared in batteries and were frequent in all developmental stages and body regions. O-isorhizas on the other hand were mainly found in marginal lappets of strobilae and ephyrae, but were less abundant in the oral lips of these stages and completely absent in buds and polyps. Rhopaloids were frequent in all body parts except the marginal lappets of medusae (rare) and ephyrae (absent). The lemon-shaped birhopaloids were present in all developmental stages but often in few numbers except in the oral lips of the medusa stage
	Cassiopea andromeda	Jensch and Hofmann (1997)	Euryteles present in the ectoderm of all stages and occur in the ectodermis and endodermis while a-isorhizas and anisrohizas are restricted to the ectodermis of some stages
	Cassiopea culionensis	Gamero-Mora et al. (2022)	Holotrichous O-isorhizas and two unidentified capsule types rare
	Cassiopea mayeri	Gamero-Mora et al. (2022)	Holotrichous O-isorhizas and two unidentified capsule types rare
	Cassiopea ornata	Anthony et al. (2022)	Cassiosomes harbored O-isorhizas exclusively. The abundance of O-isorhizas in cassiosomes differed in two populations
	Cassiopea xamachana	Jensch and Hofmann (1997)	Euryteles present in the ectoderm of all stages and occur in the ectodermis and endodermis while a-isorhizas and anisrohizas are restricted to the ectodermis of some stages
	Cotylorhiza tuberculata	Cengiz and Killi (2021)	No significant difference between numbers of nematocysts per milliliter in samples from medusae with different bell size. O-isorhizas dominant in the bell margin samples. In oral arm samples eurytele dominant in medusae with 30-35 cm diameter, O-isorhiza dominant in smaller medusae
	Phyllorhiza punctata	Cengiz and Killi (2021)	No significant difference between numbers of nematocysts per milliliter in samples from medusae with different bell size. Euryteles dominant in the bell margin and oral arm samples up to 25 cm medusa diameter. O- isorhizas dominant in 31-34 cm medusa diameter. Birhopaloids more abundant in oral arms than in bell margins
	Phyllorhiza punctata	Peach and Pitt (2005)	Abundances of round holotrichous isorhizas, larger rhopaloids and birhopaloids varied between the specimens. Bell margin: isorhizae 93%, medium rhopaloids 5%, large rhopaloids (including birhopaloids) up to 3%. Oral arm: isorhizae 75%, medium rhopaloids 25%, large rhopaloids (including birhopaloids) up to 3%

abundances, the number of nematocysts in each category in several randomly chosen fields of view in a squash preparation is useful (Peach and Pitt, 2005). Calculating percentages of nematocyst types present in one millilitre of homogenized, centrifuged samples from specific body regions may allow for a more precise description of nematocyst abundances and distribution, however, in zooxanthellate species, high densities of the zooxanthellae can make the isolation and counting of nematocysts difficult (Cengiz and Killi, 2021).

5.7 Nematocysts as a taxonomic feature

The use of the cnidom for taxonomic purposes has been discussed by several authors and is still under debate. Scyphozoans are generally reported to have a rather uniform complement of nematocysts and this uniformity in cnidome reflects the close phylogenetic relationship of the orders Coronatae, Semaeostomeae, and Rhizostomeae (Calder, 1974; Jensch and Hofmann, 1997; Sötje and Holst, 2019). It is suggested that the nematocyst complement is rather consistent for a specific stage of a species and thus can be used as a taxonomic feature (Lesh-Laurie and Suchy, 1991). However, nematocyst descriptions and measurements are rarely included as part of species descriptions of rhizostomes (but see Gamero-Mora et al., 2022; Yap and Ong, 2012).

Nematocysts types of scyphozoans from the US Atlantic coast, including the rhizostomes *R. verilli* (Calder, 1972, 1973, 1977) and *S. meleagris* (Calder, 1983), provided a useful morphological feature which has been included in identification keys on polyps and ephyrae from this region (Calder, 1977, 1983). Species-specific patterns of nematocyst clusters on the exumbrella of the ephyra were described for some rhizostome species (see Section 5.6). Although the arrangement of the clusters may be similar in different species, this conspicuous feature can be useful to differentiate between early ephyra stages of co-occurring species (Calder, 1977; Fuentes et al., 2011; Holst, 2012; Lotan et al., 1992; Schiariti et al., 2008; Straehler-Pohl and Jarms 2010; Tronolone, 2002). However, the application of this feature for taxonomic purposes is limited since the clusters usually only occur in the youngest ephyra stages and can change or disappear during ephyra growth (see Section 5.6).

Several studies included various body regions in nematocyst descriptions of investigated specimens, but interspecific comparison can be difficult since different authors often used different body parts for their observations (Tables 5 and 6). Moreover, cnidome descriptions of the same species can vary depending on the methods used and nomenclature adopted as demonstrated on the example of *C. andromeda* (Table 6). Nematocyst complements and abundances often show interspecific differences as well as similarities, for example reported for *P. punctata* and *C. mosaicus* medusae (Peach and Pitt, 2005). No species-specific differences of taxonomic value were detected in comparisons between *C. andromeda* and *C. xamachana* nematocysts (Jensch and Hofmann, 1997). Nevertheless, differences between *Cassiopea* species could be detected by detailed investigations focusing on nematocyst varieties (Gamero-Mora et al., 2022; Heins et al., 2015). Indeed, statistical analyses of *Cassiopea* spp. medusae nematocysts separated in six classes by type and size (small a-isorhiza, large a-isorhiza, O-isorhiza, medium birhapoloid, large birhapoloid, small lemon-shaped birhapoloid) demonstrate the potential of nematocyst analyses as a species delineation and identification tool (Anthony et al., 2022). This example highlights the need to consider varieties of main nematocysts categories to detect interspecific differences in nematocyst complements.

6. Conclusions and outlook

Although the revolutionary development of histological methods and optical techniques in the last 50 years has brought many new insights, there are still many gaps in the knowledge concerning the histology of the Rhizostomeae. This provides a wide field for future exciting research and we would like to point out some interesting research questions and to give some methodological recommendations.

Sophisticated chemo- and immune-histological methods have been shown to be very effective tools for histological investigations of nervous systems, sense- and muscle cells in Cnidaria, which should be used for further studies in Rhizostomeae. Modern comprehensive comparative immune-histological studies of neuronal and sensing structures and chemo-histological studies of muscles in species of different rhizostome taxa including different life stages could unravel details of their development and histological taxonomic features. The existence of specialized receptors for environmental stimuli and their important role for developmental and ecological processes like planula settlement, planula and polyp metamorphosis have been confirmed by several authors but the histological structure of these receptors remains unknown to date and needs further investigation. It has been shown that semaeostome planulae lack well-developed muscle structures, but comparable examinations on rhizostome planulae and swimming buds would be an interesting field for future research.

Regarding the reproductive structures, the present review highlights new findings on the histology of gonadal organization and details of the processes of gametogenesis, fertilization, cleavage, gastrulation, and brooding for Rhizostomeae species in the past fifty years. Despite the advances in the microscopic techniques there are still many uncertainties regarding sexual reproduction in Rhizostomeae species. For example, it remains largely unclear how and where internal and external fertilization occurs, and what kind of triggers the sperms need to recognize and fertilize the oocytes. Some studies have shown that various different cell types, of which some were recently discovered, are involved in the rhizostome sexual reproduction, but detailed descriptions of these cell types are lacking. Intensifying light and electron microscopic studies is necessary to broaden the understanding of the cellular and tissue processes of the taxon Rhizostomeae.

Much progress has been made since 1970 regarding Rhizostomeae-zooxanthellae associations and the knowledge of the diversity of associations, particularly on the symbiont side due to the emergence of molecular methods. For hosts, it appears now quite clear that the symbiosis with zooxanthellae is a strong feature of Kolpophorae but is absent in nearly all other Rhizostomeae (with perhaps the exception of *B. reiki*). Nonetheless, some host species remain very scarcely studied particularly in the families Cepheidae, Thysanostomatidae and Versurigidae. Further studies focused on the histology of the association in those families, and in particular a precise mapping of symbionts in hosts bodies (as conducted for *Cassiopea* sp. Lyndby et al., 2023) would be invaluable for a better understanding of the evolution of rhizostomes-zooxanthellae associations.

Methodological differences in nematocyst studies (optical devices, preparation methods, classifications, nomenclatures, developmental stages or sizes of specimens, investigated body regions) often hinder the comparison of nematocyst data from different studies. Future investigations on nematocyst descriptions should clearly indicate all these methodological aspects. On the other hand, intra- or interspecific comparisons are only reasonable if potential differences in methodologies are considered. Further improvements of optical and imaging methods or the use of advanced techniques may allow for more detailed descriptions of nematocyst and tubule features in future studies. This technical progress may in turn lead to changes in nematocyst classification, and consequently species-specific nematocyst descriptions will need to be re-examined and updated. For further investigations on nematocysts, it is highly recommended to use fresh living tissues, when possible, because it is preferable for the observation of morphological details of undischarged and discharged capsules. Since the nematocyst complement varies between different life stages, live cultures are useful for investigations on the entire cnidome of a species. Although the cnidome of different rhizostome species is similar, interspecific comparisons could be of taxonomic value if sizes and abundances of different nematocyst categories and their varieties in different developmental stages and body regions are considered and appropriate replicate measurements (see Section 5.5) are conducted. Nematocyst analyses including all these aspects are very time consuming, but the main problem is the lack of consistent methods. Therefore, a standardization of methods for nematocyst analyses should be aspired.

Acknowledgements

We thank our colleagues for fruitful exchanges and we are grateful for the valuable comments and suggestions by the reviewers C. Lucas and A.C. Morandini. Many thanks to A. Fischer (Senckenberg am Meer, DZMB Hamburg) for her support in the literature search and to Virginia Castilho (CEBIMar/USP) for her assistance in obtaining the permissions for published images. We greatly acknowledge the photos provided by C. Hess and C.L. Ames.

References

Albert, D.J., 2011. Adaptive behaviours of the jellyfish *Aurelia labiata* in Roscoe Bay on the west coast of Canada. J. Sea Res. 59, 198–201.

Ames, C.L., Klompen, A.M.L., Badhiwala, K., Muffett, K., Reft, A.J., Kumar, M., et al., 2020. Cassiosomes are stinging-cell structures in the mucus of the upside-down jellyfish *Cassiopea xamachana*. Commun. Biol. 3, 67.

Anderson, E., 1974. Comparative aspects of the ultrastructure of the female gamete. In: Bourne, G.H., Danielli, J.E., Jean, K.W. (Eds.), Aspects of Nuclear Structure and Function. Academic Press, New York, pp. 1–70.

Anderson, P.A.V., 1985. Physiology of a bidirectional, excitatory, chemical synapse. J. Neurophysiol. 53, 821–835.

Anderson, P.A.V., Schwab, W.E., 1981. The organization and structure of nerve and muscle in the jellyfish *Cyanea capillata* (Coelenterata; Scyphozoa). J. Morphol. 170, 383–329.

Anderson, P.A.V., Schwab, W.E., Gilbert, S., Allen, R.D., 1986. Fast axonal transport by neurons from the jellyfish *Cyanea capillata*. J. Neurobiol. 17, 29–37.

Anderson, P.A.V., Thompson, L.F., Moneypenny, C., 2004. Evidence for a common pattern of peptidergic innervation of cnidocytes. Biol. Bull. 207, 141–146.

Angile, F., Coco, L.D., Girelli, C.R., Basso, L., Rizzo, L., Piraino, S., et al., 2020. H-1 NMR metabolic profile of Scyphomedusa *Rhizostoma pulmo* (Scyphozoa, Cnidaria) in female gonads and somatic tissues: preliminary results. Molecules 25, 806.

Anthony, C.J., Heagy, M., Bentlage, B., 2022. Phenotypic plasticity in *Cassiopea ornata* (Cnidaria: Scyphozoa: Rhizostomeae) suggests environmentally driven morphology. Zoomorphology 141, 115–131.

Arai, M.N., 1997. A Functional Biology of Scyphozoa. Chapman and Hall, London.

Astorga, D., Ruiz, J., Prieto, L., 2012. Ecological aspects of early life stages of *Cotylorhiza tuberculata* (Scyphozoa: Rhizostomae) affecting its pelagic population success. Hydrobiologia 690, 141–155.

Avian, M., 1986. Biological cycle of *Cotylorhiza tuberculata* (Macri 1778): morphological aspects of the development from ephyra to young medusa. In: Rottini-Sandrini, L., Avian, M. (Eds.), Workshop on jellyfish in the Mediterranean Sea. Nova Thalassia, pp. 47–58.

Avian, M., Del Negro, P., Rottini Sandrini, L., 1991. A comparative analysis of nematocysts in *Pelagia noctiluca* and *Rhizostoma pulmo* from the north Adriatic Sea. Hydrobiologia 216/217. pp. 615–621.

Avian, M., Rottini-Sandrini, L., 1991. Oocyte development in four species of scyphomedusa in the northern Adriatic Sea. Hydrobiologia 216, 189–195.

Avian, M., Spanier, E., Galil, B., 1995. Nematocysts of *Rhopilema nomadica* (Scyphozoa: Rhizostomeae), an immigrant jellyfish in the eastern Mediterranean. J. Morphol. 224, 221–231.

Ax, P., 1995. Das System der Metazoa I: Ein Lehrbuch der phylogenetischen Systematik. Gustav Fischer Verlag, Stuttgart.

Ballesteros, A., Östman, C., Santín, A., Marambio, M., Narda, M., Gili, J.-M., 2021. Cnidome and morphological features of *Pelagia noctiluca* (Cnidaria: Scyphozoa) throughout the different life cycle stages. Front. Mar. Sci. 8, 714503.

Becker, A., Sötje, I., Paulmann, C., Beckmann, F., Donath, T., Boese, R., et al., 2005. Calcium sulfate hemihydrate is the inorganic mineral in statoliths of scyphozoan medusae (Cnidaria). Dalton Trans. 2005, 1545–1550.

Beckmann, A., Özbek, S., 2012. The nematocyst: a molecular map of the cnidarian stinging organelle. Int. J. Dev. Biol. 56, 577–582.

Berking, S., Herrmann, K., 2006. Formation and discharge of nematocysts is controlled by a proton gradient across the cyst membrane. Helgol. Mar. Res. 60, 180–188.

Berrill, N.J., 1949. Developmental analysis of scyphomedusae. Biol. Rev. 24, 393–409.

Bezio, N., Costello, J., Perry, E., Colin, S., 2018. Effects of capture surface morphology on feeding success of scyphomedusae: a comparative study. Mar. Ecol. Prog. Ser. 596, 83–93.

Bigelow, H.B., 1914. Note on the medusan genus *Stomolophus* from San Diego. Univ. Calif. Public. Zool. 13, 239–241.

Blanquet, R.S., Phelan, M.A., 1987. An unusual blue mesogleal protein from the mangrove jellyfish *Cassiopea xamachana*. Mar. Biol. 94, 423–430.

Blanquet, R.S., Riordan, G.P., 1981. An ultrastructural study of the subumbrellar musculature and desmosomal complexes of *Cassiopea xamachana* (Cnidaria: Scyphozoa). Trans. Am. Microsc. Soc. 100, 109–119.

Boßelmann, F., Epple, M., Sötje, I., Tiemann, H., 2007. Statoliths of calcium sulfate hemihydrate are used for gravity sensing in rhopaliophoran medusae (Cnidaria). In: Bäuerlein, E. (Ed.), Handbook of Biomineralization. Wiley-VCH Verlag, Weinheim, pp. 261–272.

Bouillon, J., Nielsen, M., 1974. Etude de quelques organs sensoriels de cnidaires. Arch. Biol. 85, 307–328.

Bozler, E., 1926. Sinnes- und nervenphysiologische Untersuchungen an Scyphomedusen. Z. f. vergl. Physiologie 4, 37–80.

Breton, P.J., 1999. From microns to nanometers: early landmarks in the science of scanning electron microscope imaging. Scan. Microsc. 13, 1–6.

Calder, D.R., 1972. Development of the sea nettle *Chrysaora quinquecirrha* (Scyphozoa, Semaeostomeae). Chesap. Sci. 13, 40–44.

Calder, D.R., 1973. Laboratory observations on the life history of *Rhopilema verrilli* (Scyphozoa: Rhizostomeae). Mar. Biol. 21, 109–114.

Calder, D.R., 1974. Nematocysts of the coronatae scyphomedusa, *Linuche unguiculata*, with a brief reexamination of Scyphozoa nematocyst classification. Chesap. Sci. 15, 170–173.

Calder, D.R., 1977. Nematocysts of the ephyra stages of *Aurelia*, *Chrysaora*, *Cyanea*, and *Rhopilema* (Cnidaria, Scyphozoa). T. Am. Micros. Soc. 96, 13–19.

Calder, D.R., 1982. Life history of the cannonball jellyfish *Stomolophus meleagris* L. Agassiz, 1860 (Scyphozoa, Rhizostomida). Biol. Bull. 162, 149–162.

Calder, D.R., 1983. Nematocysts of stages in the life cycle of *Stomolophus meleagris*, with keys to scyphistomae and ephyrae of some western Atlantic Scyphozoa. Can. J. Zool. 61, 1185–1192.

Camacho-Pacheco, A.V., Gómez-Salinas, L.C., Cisneros-Mata, M.Á., Rodríguez-Félix, D., Díaz-Tenorio, L.M., Unzueta-Bustamante, M.L., 2022. Feeding behavior, shrinking, and the role of mucus in the cannonball jellyfish *Stomolophus* sp. 2 in captivity. Diversity 14, 103.

Campbell, R.D., 1974. Cnidaria. In: Giese, A.C., Pearse, J.S. (Eds.), Reproduction of marine invertebrates, Volume 1: Acoelomate and pseudocoelomate metazoans. Academic Press, Cambridge, pp. 133–199.

Cargo, D.G., 1971. The sessile stages of a scyphozoan identified as *Rhopilema verrilli*. Tulane Stud. Zool. Bot. 17 (2), 31–34.

Carlberg, M., Alfredsson, K., Nielsen, S.-O., Anderson, P.A.V., 1995. Taurine-like immunoreactivity in the motor nerve net of the jellyfish *Cyanea capillata*. Biol. Bull. 1888, 78–82.

Carvalho-Saucedo, L., Garcia-Dominguez, F., Rodriguez-Jaramillo, C., Lopez-Martinez, J., 2010. Lipid variation in oocytes of the jellyfish *Stomolophus meleagris* (Scyphozoa: Rhizostomeae) from Las Guasimas Lagoon, Mexico, during gonadal development. Rev. Biol. Trop. 58, 119–130.

Cengiz, S., Killi, N., 2021. Nematocysts types and morphological features of some Scyphozoa species in the southwest Turkey. Fresenius Environ. Bull. 30, 32–40.

Chapman, D.M., 1974. Cnidarian histology. In: Muscatine, L., Lenhoff, H.M. (Eds.), Coelenterate biology: reviews and new perspectives. Academic Press, New York, pp. 1–92.

Chapman, D.M., 1985. X-ray microanalysis of selected coelenterate statoliths. J. Mar. Biol. Ass. UK 65, 617–627.

Chen, J., Ding, G., 1981. On the nematocysts of the various stages of development of *Rhopilema esculenta* Kishinouye (Cnidaria, Scyphozoa). Acta Zool. Sin. 27, 310–317.

Chia, F.-S., Amerongen, H.M., Peteya, D.J., 1984. Ultrastructure of the neuromuscular system of the polyp of *Aurelia aurita* L., 1758 (Cnidaria, Scyphozoa). J. Morph. 180, 69–79.

Colley, N.J., Trench, R.K., 1985. Cellular events in the reestablishment of a symbiosis between a marine dinoflagellate and a coelenterate. Cell Tissue Res. 239, 93–103.

Dall'Olio, L.R., Beran, A., Flander-Putrle, V., Malej, A., Ramšak, A., 2022. Diversity of dinoflagellate symbionts in scyphozoan hosts from shallow environments: the Mediterranean Sea and Cabo Frio (Rio de Janeiro, Brazil). Front. Mar. Sci. 9, 867554.

Davy, S.K., Allemand, D., Weis, V.M., 2012. Cell biology of cnidarian-dinoflagellate symbiosis. Microbiol. Mol. Biol. Rev. 76, 229–261.

Dawson, M.N., Martin, L.E., Penland, L.K., 2001. Jellyfish swarms, tourists, and the Christ-child. Hydrobiologia. 451, 131–144.

De Souza, L.M., Iacomini, M., Gorin, P.A.J., Sari, R.S., Haddad, M.A., Sassaki, G.L., 2007. Glyco- and sphingophoslipids from the medusa *Phyllorhiza punctata*: NMR and ESI-MS/MS fingerprints. Chem. Phys. Lipids. 145, 85–96.

DiMatteo, T., 1981. Investigation into interspecific encounters of the sea hare *Aplysia dactylomela* Rang, 1828 (Gastropoda: Opisthobranchia). Veliger 24, 72–75.

Djeghri, N., Pondaven, P., Stibor, H., Dawson, M.N., 2019. Review of the diversity, traits, and ecology of zooxanthellate jellyfishes. Mar. Biol. 166, 147.

Djeghri, N., 2019. Variability and plasticity of the nutrition of zooxanthellate jellyfishes: insights from experimental and field studies. PhD Thesis, Université de Bretagne Occidentale, Brest, France.

Eakin, R.M., 1979. Evolutionary significance of photoreceptors. Retrospect. Am. Zool. 19, 647–653.

Eckelbarger, K.J., 1994. Oocyte nutrition in the lower metazoa: the Scyphozoa. In: Wilson, W.H., Stricker, S.A., Shinn, G.L. (Eds.), Reproduction and development of marine invertebrates. John Hopkins University Press, Baltimore, pp. 15–28.

Eckelbarger, K.J., Hodgson, A.N., 2021. Invertebrate oogenesis: a review and synthesis: comparative ovarian morphology, accessory cell function and the origins of yolk precursors. Invertebr. Reprod. Dev. 65, 71–140.

Eckelbarger, K.J., Larson, R., 1988. Ovarian morphology and oogenesis in *Aurelia aurita* (Scyphozoa: Semaeostomae): ultrastructural evidence of heterosynthetic yolk formation in a primitive metazoan. Mar. Biol. 100, 103–115.

Eckelbarger, K.J., Larson, R., 1992. Ultrastructure of the ovary and oogenesis in the jellyfish *Linuche unguiculata* and *Stomolophus meleagris*, with a review of the ovarian structure in the Scyphozoa. Mar. Biol. 114, 633–643.

Eimer, T., 1878. In: Die Medusen physiologisch und morphologisch auf ihr Nervensystem untersucht. Verlag der H. Lauppschen Buchhandlung, Tübingen.

Enrique-Navarro, A., Huertas, I.E., Cobo, M.J.L., Prieto, L., 2021. Impact of ocean warming and ocean acidification on asexual reproduction and statolith formation of the symbiotic jellyfish *Cotylorhiza tuberculata*. PLoS One 16, e0254983.

Enrique-Navarro, A., Huertas, E., Flander-Putrle, V., Bartual, A., Navarro, G., Ruiz, J., et al., 2022. Living inside a jellyfish: the symbiosis case study of host-specialized dinoflagellates, "zooxanthellae", and the scyphozoan *Cotylorhiza tuberculata*. Front. Mar. Sci. 9, 817312.

Estes, A.M., Kempf, S.C., Henry, R.P., 2003. Localization and quantification of carbonic anhydrase activity in the symbiotic scyphozoan *Cassiopea xamachana*. Biol. Bull. 204, 278–289.

Fautin, D.G., 1992. Cnidaria. In: Adiyodi, K.G., Adiyodi, R.G. (Eds.), Reproductive biology of invertebrates (sexual differentiation and behaviour). John Wiley and Sons, New York, pp. 31–52.

Ferreira, M., Cleary, D.F.R., Bat, N.K., Polónia, A.R.M., Gomes, N.C.M., 2022. Assessing the core microbial symbionts of jellyfish in Indonesian and Vietnamese marine lakes. J. Mar. Biol. Ass. U. K 102, 486–495.

Fitt, W.K., 1984. The role of chemosensory behavior of *Symbiodinium microadriaticum*, intermediate hosts, and host behavior in the infection of coelenterates and molluscs with zooxanthellae. Mar. Biol. 81, 9–17.

Fitt, W.K., Costley, K., 1998. The role of temperature in survival of the polyp stage of the tropical rhizostome jellyfish *Cassiopea xamachana*. J. Exp. Mar. Biol. Ecol. 222, 79–91.

Fitt, W.K., Hofmann, D.K., Wok, M., Rahat, M., 1987. Requirement of exogenous inducers for metamorphosis of axenic larvae and buds of *Cassiopea andromeda* (Cnidaria: Scyphozoa). Mar. Biol. 94, 415–422.

Freudenthal, H.D., 1962. *Symbiodinium* gen. nov. and *Symbiodinium microadriaticum* sp. nov., a zooxanthella: taxonomy, life cycle, and morphology. J. Protozool. 9, 45–52.

Friedrich, F., Matsumura, Y., Pohl, H., Bai, M., Hoernschemeyer, T., Beutel, R.G., 2014. Insect morphology in the age of phylogenomics: innovative techniques and its future role in systematics. Entomol. Sci. 17, 1e24.

Fu, Z., Xing, Y., Gu, Z., Liu, C., Chen, S., 2019. Effects of temperature, salinity and dissolved oxygen on excystment of podocysts in the edible jellyfish *Rhopilema esculentum* Kishinouye, 1891. Indian. J. Fish. 66, 69–77.

Fuentes, V., Straehler-Pohl, I., Dacha, A., Franco, I., Tilves, U., Gentile, M., et al., 2011. Life cycle of the jellyfish *Rhizostoma pulmo* (Scyphozoa: Rhizostomeae) and its distribution, seasonality and inter-annual variability along the Catalan coast and the Mar Menor (Spain, NW Mediterranean). Mar. Biol. 158, 2247–2266.

Galliot, B., 2010. A key innovation in animal evolution, the emergence of neurogenesis: cellular and molecular cues from cnidarian nervous systems. In: Schierwater, B., De Salle, R. (Eds.), Key transitions in animal evolution. Science Publishers and CRC Press, New York, pp. 127–147.

Gamero-Mora, E., Collins, A.G., Boco, S.R., Geson, S.M., Morandini, A.C., 2022. Revealing hidden diversity among upside-down jellyfishes (Cnidaria: Scyphozoa: Rhizostomeae). Invertebr. Syst. 36, 63–89.

Garm, A., Ekström, P., 2010. Evidence for multiple photosystems in jellyfish. Int. Rev. Cell. Mol. Bio. 280, 41–78.

Gohar, H.A.F., Eisawy, A.M., 1960a. The development of *Cassiopea andromeda* (Schyphomedusae) Publ. Mar. Biol. Sta. Ghardaga, Red. Sea 11, 147–190.

Gohar, H.A.F., Eisawy, A.M., 1960b. The biology of *Cassiopea andromeda* (with note on the species problem). Publ. Mar. Biol. Sta. Ghardaga, Red. Sea 11, 3–42.

Grimmelikhuijzen, C.J.P., 1983. FMRFamide immunoreactivity is generally occurring in the nervous systems of coelenterates. Histochemistry 78, 361–381.

Gülsahin, N., 2016. Preliminary study on nematocyst types and venom isolation of *Cassiopea andromeda* Forskal, 1775 (Scyphozoa, Cnidaria) from Turkey. Cent. Nerv. Syst. Agents Med. Chem. 16, 208–212.

Haacke, W., 1884. *Pseudorhiza haeckelii*, spec. nov., der Endspross des Discomedusenstammes. Biol. Centralblatt 4, 291–294.

Haddad, V., Morandini, A.C., 2019. Jellyfish stings. In: Jarms, G., Morandini, A.C. (Eds.), World atlas of jellyfish. Dölling und Galitz Verlag, München, pp. 64–69.

Hamner, W.M., Hauri, I.R., 1981. Long-distance horizontal migrations of zooplankton (Scyphomedusae: *Mastigias*). Limnol. Oceanogr. 26, 414–423.

Harrison, P.L., Jamieson, B.G.M., 1999. Cnidaria and ctenophora. In: Jamieson, B.G.M. (Ed.), Reproductive biology of the invertebrates, Volume 9 Part A, Progress in male gamete ultrastructure and phylogeny. John Wiley and Sons Ltd, Chichester, pp. 21–95.

Häussermann, V., 2004. Identification and taxonomy of soft-bodied hexacorals exemplified by Chilean sea anemones; including guidelines for sampling, preservation and examination. J. Mar. Biol. Assoc. UK 84, 931–936.

Hedwig, M., Schäfer, W., 1986. Vergleichende Untersuchungen zur Ultrastruktur und zur phylogenetischen Bedeutung der Spermien der Scyphozoa. Z. Zool. Syst. Evolt. -forsch. 24, 109–122.

Heins, A., Glatzel, T., Holst, S., 2015. Revised descriptions of the nematocysts and the asexual reproduction modes of the scyphozoan jellyfish *Cassiopea andromeda* (Forskål, 1775). Zoomorphology 134, 351–366.

Heins, A., Sötje, I., Holst, S., 2018. Assessment of investigation techniques for scyphozoan statoliths, with focus on early development of the jellyfish *Sanderia malayensis*. Mar. Ecol. Progr. Ser. 591, 37–56.

Hessinger, D.A., Lenhoff, H.M., 1988. The biology of nematocysts. Academic Press, Inc, San Diego.

Hinsch, G.W., 1974. Comparative ultrastructure of cnidarian sperm. Am. Zool. 14, 457–465.

Hofmann, D., Fleck, J., Fitt, W., 1996. Checkpoints in the life-cycle of *Cassiopea* spp; control of metagenesis and metamorphosis in a tropical jellyfish. Int. J. Bio. 40, 331–338.

Hofmann, D.K., Gottlieb, M., 1991. Bud formation in the scyphozoan *Cassiopea andromeda*: epithelia dynamics and fate map. Hydrobiologia 216/217, 53–59.

Hofmann, D.K., Hadfield, M.G., 2002. Hermaphroditism, gonochorism, and asexual reproduction in *Cassiopea sp.*—an immigrant in the islands of Hawaii. Invertebr. Reprod. Dev. 41, 215–221.

Hofmann, D.K., Hellmann, M., 1995. Studies in the reef-dwelling cnidarian *Cassiopea* spp.: RF-amide positive elements of the nervous system at different stages of development. Beitr. Paläont 20, 21–29.

Hofmann, D.K., Honegger, T.G., 1990. Bud formation and metamorphosis in *Cassiopea andromeda* (Cnidaria: Scyphozoa): a development and ultrastructural study. Mar. Biol. 105, 509–518.

Holland, N.D., 2003. Early central nervous system evolution: an era of skin brains? Nat. Rev. 4, 1–11.

Holst, S., 2012. Morphology and development of benthic and pelagic life stages of North Sea jellyfish (Scyphozoa, Cnidaria) with special emphasis on the identification of ephyra stages. Mar. Biol. 159, 2707–2722.

Holst, S., Michalik, P., Noske, M., Krieger, J., Sötje, I., 2016. Potential of X-ray micro-computed tomography for soft-bodied and gelatinous cnidarians with special emphasis on scyphozoan and cubozoan statoliths. J. Plankton Res. 38, 1225–1242.

Holst, S., Sötje, I., Tiemann, H., Jarms, G., 2007. Life cycle of the rhizostome jellyfish *Rhizostoma octopus* (L.) (Scyphozoa, Rhizostomeae), with studies on cnidocysts and statoliths. Mar. Biol. 151, 1695–1710.

Holstein, T., Tardent, P., 1984. An ultrahigh-speed analysis of exocytosis: nematocyst discharge. Science 223, 830–833.

Hopf, J.K., Kingsford, M.J., 2013. The utility of statoliths and bell size to elucidate age and condition of a scyphomedusa (*Cassiopea* sp.). Mar. Biol. 160, 951–960.

Horridge, A., 1956. The nervous system of the ephyra larva of *Aurellia aurita*. Quart. J. Micr. Sci. 97, 59–74.

Horridge, G.A., 1959. The nerves and muscles of medusae VI. The rhythm. J. Exp. Biol. 36, 72–91.

Horridge, G.A., Mackay, B., 1962. Naked axons and symmetrical synapses in coelenterates. Quart. J. Micr. Sci. 103, 531–541.

Huebner, E., Anderson, E., 1976. Comparative spiralian oogenesis—ultrastructural aspects: an overview. Am. Zool. 16, 315–343.

Hyman, L.H., 1940. The Invertebrates: Protozoa through Ctenophora, 1, McGraw-Hill Book Company, New York.

Iguchi, N., Lee, H.E., Yoon, W.D., Kim, S., 2010. Reproduction of the giant jellyfish, *Nemopilema nomurai* (Scyphozoa: Rhizostomeae), in 2006–2008 as peripherally-transported populations. Ocean. Sci. 45, 129–138.

Ikeda, H., Ohtsu, K., Uye, S.-I., 2011a. Structural changes of gonads during artificially induced gametogenesis and spawning in the giant jellyfish *Nemopilema nomurai* (Scyphozoa: Rhizostomeae). J. Mar. Biol. Assoc. U. K. 91, 215–227.

Ikeda, H., Ohtsu, K., Uye, S.-I., 2011b. Fine structure, histochemistry, and morphogenesis during excystment of the podocysts of the giant jellyfish *Nemopilema nomurai* (Scyphozoa, Rhizostomeae). Biol. Bull. 221, 248–260.

Jacobs, D.K., Gold, D.A., Nakanishi, N., Yuan, D., Camara, A., Nichola, S.A., et al., 2011. Basal metazoan sensory evolution. In: DeSalle, R., Schierwater, B. (Eds.), Key transitions in animal evolution. Science Publishers and CRC Press, New York, pp. 175–196.

Jacobs, D.K., Nakanishi, N., Yuan, D., Camara, A., Nichols, S.A., Harenstein, V., 2007. Evolution of sensory structures in basal metazoan. Integr. Comp. Biol. 47, 712–723.

Jarms, G., Morandini, A.C., 2019. World atlas of jellyfish. Abhandlungen des Naturwissenschaftlichen Vereins in Hamburg, Hamburg.

Jensch, F., Hofmann, D.K., 1997. The cnidomes of *Cassiopea andromeda* Forskaal, 1775, and *Cassiopea xamachana* Bigelow, 1882 (Cnidaria: Scyphozoa). Proceedings of the 6th international conference on coelenterate biology. Het Nationaal Natuurhistorisch Museum (today Naturalis Biodiversity Center), Leiden, pp. 279–285.

Kass-Simon, G., Hufnagel, L.A., 1992. Suspected chemoreceptors in coelenterates and ctenophores. Microsc. Res. Tech. 22, 265–284.

Katsuki, T., Greenspan, R.J., 2013. Jellyfish nervous systems. Curr. Biol. 23, R592.

Kawahara, M., Uye, S.-I., Ohtsu, K., Iizumi, H., 2006. Unusual population explosion of the giant jellyfish *Nemopilema nomurai* (Scyphozoa: Rhizostomeae) in East Asian waters. Mar. Ecol. Prog. Ser. 307, 161–173.

Keller, C., 1883. Untersuchungen über neue Medusen aus dem Rothen Meere. Z. Wiss. Zool. Abt. A Z 38, 621–670.
Khabibulina, V., Starunov, V., 2019. Musculature development in planuloids of *Cassiopeia xamachana* (Cnidaria: Scyphozoa). Zoomorphology 138, 297–306.
Khabibulina, V., Starunov, V., 2020. FMRFamide immunoreactive nervous system in the adult *Cassiopeia xamachana* scyphopolyp and at the early stages of planuloid formation. Invertebr. Zool. 17, 371–384.
Kienberger, K., Riera-Buch, M., Schönemann, A.M., Bartsch, V., Halbauer, R., Prieto, L., 2018. First description of the life cycle of the jellyfish *Rhizostoma luteum* (Scyphozoa: Rhizostomeae). PLoS One 13, e0202093.
Kikinger, R., 1992. *Cotylorhiza tuberculata* (Cnidaria: Scyphozoa) – Life history of a stationary population. Mar. Ecol. 13, 333–362.
Killi, N., Mariottini, G., 2018. Cnidarian jellyfish: ecological aspects, nematocyst isolation, and treatment methods of sting. In: Kloc, M., Kubiak, J.Z. (Eds.), Marine organisms as model systems in biology and medicine. Spinger, Cham, pp. 477–513.
Klein, S.G., Pitt, K.A., Lucas, C.H., Hung, S.-H., Schmidt-Roach, S., Aranda, M., et al., 2019. Night-time temperature reprieves enhance the thermal tolerance of a symbiotic cnidarian. Front. Mar. Sci. 6, 453.
Kondo, Y., Suzuki, Y., Ohtsuka, S., Nagai, H., Tanaka, H., Srinui, K., et al., 2020. Differences in the cnidomes and toxicities of the oral arms of two commercially harvested rhizostome jellyfish species in Thailand. J. Mar. Biol. Assoc. UK 100, 701–711.
Kubota, S., Kawamura, M., Ueno, S., 2006. First occurrence of a mature medusa of *Nemopilema nomurai* (Cnidaria, Scyphozoa, Rhizostomae) in Tanabe Bay, Wakayama Prefecture, Japan. Nanki Seibutu 48, 57–59.
LaJeunesse, T.C., 2001. Investigating the biodiversity, ecology, and phylogeny of endosymbiotic dinoflagellates in the genus *Symbiodinium* using the ITS region: in search of a "species" level marker. J. Phycol. 37, 866–880.
LaJeunesse, T.C., Casado-Amezúa, P., Turnham, K.E., Hume, B.C.C., Butler, C.C., Mordret, S., et al., 2022. Mutualistic dinoflagellates with big disparities in ribosomal DNA variation may confound estimates of symbiont diversity and ecology in the jellyfish *Cotylorhiza tuberculata*. Symbiosis 88, 1–10.
LaJeunesse, T.C., Parkinson, J.E., Gabrielson, P.W., Jeong, H.J., Reimer, J.D., Voolstra, C.R., et al., 2018. Systematic revision of Symbiodiniaceae highlights the antiquity and diversity of coral endosymbionts. Curr. Biol. 28, 2570–2580.
LaJeunesse, T.C., Wiedenmann, J., Casado-Amezúa, P., D'Ambra, I., Turnham, K.E., Nitschke, M.R., et al., 2021. Revival of *Philozoon* Geddes for host-specialized dinoflagellates, 'zooxanthllae', in animals from coastal temperates zones of northern and southern hemispheres. Eur. J. Phycol. 57, 166–180.
Lampert, K.P., 2016. *Cassiopea* and its zooxanthellae. In: Goffredo, S., Dubinsky, Z. (Eds.), The Cnidaria, past, present and future. Springer, Cham, pp. 415–423.
Larson, R.J., 1976. Marine flora and fauna of the northeastern United States. Cnidaria: Scyphozoa. NOAATech Rep NMFS 397, pp. 1–19.
Larson, R., 1997. Feeding behaviour of Caribbean scyphomedusae: *Cassiopea frondosa* (Pallas) and *Cassiopea xamachana* (Bigelow). Stud. Nat. History Caribb. Reg. 73, 43–54.
Lawley, J.W., Carroll, A.R., McDougall, C., 2021. Rhizostomins: a novel pigment family from rhizostome jellyfish (Cnidaria, Scyphozoa). Front. Mar. Sci. 8, 752949.
Lesh-Laurie, G.E., Suchy, P.E., 1991. Cnidaria: scyphozoa and cubozoa. In: Harrison, F., Westfall, J. (Eds.), Microscopy anatomy of invertebrates, Volume 1: Placozoa, Porifera, Cnidaria and Ctenophora. Wiley-Liss, Inc., New York, pp. 185–266.

Lisboa, R., Mascarenhas, P. Gil, F., Ruano, F., Ladeira, C., Morais, Z., 2018. Preliminary studies on the lifecycle of *Catostylus tagi*: differentiation between female and male at morphological level. Second international congress of CiiEM: Translational research and Innovation in human in health sciences. Egas Moniz, Caparica. (Poster).

Lotan, A., Ben-Hillel, R., Loya, Y., 1992. Life cycle of *Rhopilema nomadica*: a new immigrant scyphomedusan in the Mediterranean. Mar. Biol. 112, 237–242.

Lotan, A., Fine, M., Ben-Hillel, R., 1994. Synchronization of the life cycle and dispersal pattern of the tropical invader scyphomedusan *Rhopilema nomadica* is temperature dependent. Mar. Ecol. Prog. Ser. 109, 59–65.

Lotan, A., Fishmann, L., Zlotkin, E., 1996. Toxin compartmentation and delivery in the cnidarian: the nematocyts tuble as a multiheaded poisonous arrow. J. Exp. Zool. 275, 444–451.

Ludwig, F.-D., 1969. Die Zooxanthellan bei *Cassiopea andromeda* Eschscholtz 1829 (Polyp-Stadium) und ihre Bedeutung für die Strobilation. Zool. Jahrbücher. Abt. Anat. Ontog. Tiere 86, 238–277.

Lyndby, N.H., Murthy, S., Bessestte, S., Jakobsen, S.L., Meibom, A., Kühl, M., 2023. Non-invasive investigation of the morphology and optical properties of the upside-down jellyfish *Cassiopea* with optical coherence tomography. Proc. R. Soc. B. 290, 20230127.

Lyndby, N.H., Rädecker, N., Bessette, S., Søgaard Jensen, L.H., Escrig, S., Trampe, E., et al., 2020. Amoebocytes facilitate efficient carbon and nitrogen assimilation in the *Cassiopea*-Symbiodiniaceae symbiosis. Proc. R. Soc. B. 287, 20202393.

Mackie, G.O., Anderson, P.A.V., Singla, C.L., 1984. Apparent absence of gap junctions in two classes of Cnidaria. Biol. Bull. 167, 120–123.

Mammone, M., Bosch-Belmar, M., Milisenda, G., Castriota, L., Sinopoli, M., Allegra, A., et al., 2023. Reproductive cycle and gonadal output of the Lessepsian jellyfish *Cassiopea andromeda* in NW Sicily (Central Mediterranean Sea). PLoS One 18, 1–14.

Mariscal, R.N., 1974. III. Nematocysts. In: Muscatine, L., Lenhoff, H.M. (Eds.), Coelenterate biology: reviews and new perspectives. Academic Press, New York, pp. 129–178.

Mariscal, R.N., 1984. Cnidaria: cnidae. In: Bereiter-Hahn, J., Matoltsy, A.G., Richards, K.S. (Eds.), Biology of the integument. Springer, Berlin, pp. 57–68.

Mariscal, R.N., Bigger, C.H., 1976. A comparison of putative sensory receptors associated with nematocysts in an anthozoan and a scyphozoan. In: Mackie, G.O. (Ed.), Coelenterate ecology and behavior. Plenum Press, New York, pp. 559–568.

Martin, V.J., 2002. Photoreceptors of cnidarians. Can. J. Zool. 80, 1703–1722.

Martin, V.J., Chia, F.-S., 1982. Fine structure of a scyphozoan planula, *Cassiopea xamachana*. Biol. Bull. 163, 320–328.

McCloskey, L.R., Muscatine, L., Wilkerson, F.P., 1994. Daily photosynthesis, respiration, and carbon budgets in a tropical marine jellyfish (*Mastigias* sp.). Mar. Biol. 119, 13–22.

McGill, C.J., Pomoroy, C.M., 2008. Effects of bleaching and nutrient supplementation on wet weight in the jellyfish *Cassiopea xamachana* (Bigelow) (Cnidaria: Scyphozoa). Mar. Freshw. Behav. Physiol. 41, 179–189.

Mellas, R.E., McIlroy, S.E., Fitt, W.K., Coffroth, M.A., 2014. Variation in symbiont uptake in the early ontogeny of the upside-down jellyfish, *Cassiopea* spp. J. Exp. Mar. Biol. Ecol. 459, 38–44.

Miller, R.L., 1983. Cnidaria. In: Adiyodi, K.G., Adiyodi, R.G. (Eds.), Reproductive biology of invertebrates. Vol II Spermatogenesis and sperm function. Wiley and Sons, New York, pp. 23–69.

Morandini, A.C., 2024. Morphology of Rhizostomeae jellyfishes: what is known and what we advanced since the 1970s. Adv. Mar. Biol. 98.

Morandini, A.C., Marques, A.C., 2010. Revision of the genus *Chrysaora* Péron and Lesueur, 1810 (Cnidaria: Scyphozoa). Zootaxa 2464, 1–97.

Muffett, K.M., Klompen, A.M.L., Collins, A.G., Lewis Ames, C., 2021. Raising awareness of the severity of "contactless stings" by *Cassiopea* jellyfish and kin. Animals 11, 3375.
Müller, W.A., Leitz, T., 2002. Metamorphosis in cnidaria. Can. J. Zool. 80, 1755–1771.
Mulyadi, H.A., Sianturi, O.R., 2021. The occurrence of harmful jellyfish outbreaks and human stung reported at recreational beaches in special region of Yogoyakarta, Indonesia. IOP Conf. Ser. Earth Environ. Sci. 789, 1–9.
Muscatine, L., Wilkerson, F.P., McCloskey, L.R., 1986. Regulation of population density of symbiotic algae in a tropical marine jellyfish (*Mastigias* sp.). Mar. Ecol. Prog. Ser. 32, 279–290.
Nagata, R.M., D'Ambra, I., Lauritano, C., von Montfort, G.M., Djeghri, N., Jordano, M.A., et al., 2024. Physiology and functional biology of Rhizostomeae jellyfish. Adv. Mar. Biol. 98.
Nakanishi, N., Hartenstein, V., Jacobs, D.K., 2009. Development of the rhopalial nervous system in *Aurelia* sp. 1 (Cnidaria, Scyphozoa). Dev. Genes. Evol. 219, 301–317.
Nakanishi, N., Yuan, D., Jacobs, D.K., Hartenstein, V., 2008. Early development, pattern, and reorganization of the planula nervous system in *Aurelia* (Cnidaria, Scyphozoa). Dev. Genes. Evol. 218, 511–524.
Newkirk, C.R., Frazer, T.K., Martindale, M.Q., 2018. Acquisition and proliferation of algal symbionts in bleached polyps of the upside-down jellyfish, *Cassiopea xamachana*. J. Exp. Mar. Biol. Ecol. 508, 44–51.
Ohdera, A.H., Abrams, M.J., Ames, C.L., Baker, D.M., Suescún-Bolivar, L.P., Collins, A.G., et al., 2018. Upside-down but headed in the right direction: review of the highly versatile *Cassiopea xamachana* system. Front. Ecol. Evol. 6, 35.
Ohtsu, K., Kawahara, M., Ikeda, H., Uye, S.-I., 2007. Experimental induction of gonadal maturation and spawning in the giant jellyfish *Nemopilema nomurai* (Scyphozoa: Rhizostomeae). Mar. Biol. 152, 667–676.
Ohtsu, K., Uye, S.-I., 2011. Anomalous infrared taxis of an aquatic animal, the giant jellyfish *Nemopilema nomurai* (Scyphozoa, Rhizostomeae). Biol. Bull. 221, 243–247.
Östman, C., 2000. A guideline to nematocysts nomenclature and classification, and some notes on the systematic value of nematocysts. Sci. Mar. 64 (Supl. 1), 31–46.
Östman, C., Hydman, J., 2007. Nematocyst analysis of *Cyanea capillata* and *Cyanea lamarckii* (Scyphozoa, Cnidaria). Sci. Mar. 61, 313–344.
Othman, I., Fathilah, A.R., Mohd Saad, J., Mohd Yusof, M., Mustaffa, M.R., Nor, A., 1996. Studies on the venomous coelenterate: *Rhopilema hispidum*. J. Nat. Toxins 5, 361–375.
Park, S., Capelin, D., Piriatinskiy, G., Lotan, T., Yossifon, G., 2017. Dielectrophoretic characterization and isolation of jellyfish stinging capsules. Electrophoresis 38, 1996–2003.
Parracho, T., Morais, Z., 2015. *Catostylus tagi*: partial rDNA sequencing and characterization of nematocyte structures using two improvements in jellyfish sample preparation. J. Venom. Anim. And. Toxins 21, 40.
Paspaleff, G.W., 1938. Über die Entwicklung von *Rhizostoma pulmo* Agass. Arb. Biol. Meeresst. Varna 7, 1–17.
Passano, L.M., 1965. Pacemakers and activity patterns in medusae: homage to Romanes. Am. Zool. 5, 465–481.
Passano, L.M., 1973. Behavioral control systems in medusae. A comparison between hydro- and scyphomedusae. Seto. Mar. Biol. Lab. 20, 615–645.
Passano, L.M., 1982. Scyphozoa and Cubozoa. In: Shelton, G.A.B. (Ed.), Electrical conduction and behavior in *simple* invertebrates. Clarendon Press, Oxford, pp. 149–2020.
Passano, K.N., 1988. Variability in the initiation of diffuse nerve-net impulses in the mangrove jellyfish *Cassiopea xamachana* (Coelenterate: Scyphozoa). Comp. Biochem. Physiol. 91C, 273–279.
Passano, K.N., 2004. Spasm behavior and the diffuse nerve-net in *Cassiopea xamachana* (Scyphozoa: Coelenterata). Hydrobiologia 510/511, 91–96.

Passano, K.N., Passano, L.M., 1971. The endodermal nerve net of Scyphozoa. J. Morph. 133, 105–124.
Peach, M.B., Pitt, K.A., 2005. Morphology of the nematocysts of the medusae of two scyphozoans, *Catostylus mosaicus* and *Phyllorhiza punctata* (Rhizostomeaea): implication for capture of prey. Invertebr. Biol. 124, 98–108.
Pitt, K.A., 2000. Life history and settlement preferences of the edible jellyfish *Catostylus mosaicus* (Scyphozoa: Rhizostomeae). Mar. Biol. 136, 269–279.
Pitt, K.A., Kingsford, M.J., 2000. Reproductive biology of the edible jellyfish *Catostylus mosaicus* (Rhizostomeae). Mar. Biol. 137, 791–799.
Pitt, K.A., Koop, K., Rissik, D., 2005. Contrasting contributions to inorganic nutrient recycling by the co-occuring jellyfishes, *Catostylus mosaicus* and *Phyllorhiza punctata* (Scyphozoa, Rhizostomeae). J. Exp. Mar. Biol. Ecol. 315, 71–86.
Plachetzki, D.C., Serb, J.M., Oakley, T.H., 2005. New insights into the evolutionary history of photoreceptor cells. Trends Ecol. Evol. 20, 465–467.
Pollmans, D., Hündgen, M., 1981. Licht- und elektronenmickroskopische Untersuchungen der Rhopalien von *Aurelia aurita* (Scyphozoa, Semaeostomeae) Zool. Jahrb. Abt. Anat. Ontol. Tiere 105, 508–525.
Porter, M.L., Blasic, J.R., Bok, M.J., Cameron, E.G., Pringle, T., Cronin, T.W., et al., 2016. Shedding new light on opsin evolution. Proc. R Soc. B 279, 3–14.
Pyo, M.J., Lee, H., Bae, S.K., Heo, Y., Choudhary, I., Yoon, W.D., et al., 2016. Modulation of jellyfish nematocyst discharges and management of human skin stings in *Nemopilema nomurai* and *Carybdea mora*. Toxicon 109, 26–32.
Reinicke, G.B., Holst, S., Morandini, A.C., Sötje, I., Straehler-Pohl, I., Wiesenthal, A.A., et al., 2024. Max Egon Thiel's monographs on Scyphozoa (Cnidaria) and a left-behind typescript on the Rhizostomeae. Adv. Mar. Biol. 98.
Rentzsch, F., Layden, M., Manuel, M., 2016. The cellular and molecular basis of cnidarian neurogenesis. WIREs Dev. Biol. 6, 1–19.
Richter, S., Loesel, R., Purschke, G., Schmidt-Rhaesa, A., Scholtz, G., Stach, T., et al., 2010. Invertebrate neurophylogeny: suggested terms and definitions for a neuroanatomical glossary. Front. Zool. 7, 29.
Rippingale, R.J., Kelly, S.J., 1995. Reproduction and survival of *Phyllorhiza punctata* (Cnidaria: Rhizostomeae) in a seasonally fluctuating salinity regime in Western Australia. Mar. Freshw. Res. 46, 1145–1151.
Röthig, T., Puntin, G., Wong, J.C.Y., Burian, A., McLeod, W., Baker, D., 2021. Holobiont nitrogen control and its potential for eutrophication resistance in an obligate photosymbiotic jellyfish. Microbiome 9, 127.
Rouse, G.W., Pitt, K., 2000. Ultrastructure of the sperm of *Catostylus mosaicus* and *Phyllorhiza punctata* (Scyphozoa, Cnidaria): implications for sperm terminology and the inference of reproductive mechanisms. Invertebr. Reprod. Dev. 38, 23–34.
Russel, F.S., 1970. The medusae of the British Isles. Cambridge University Press, Cambridge.
Santos, G., Nogueira, I., Morais, Z., 2018. The statoliths of *Catostylus tagi*, chemical characterization by energy dispersive X-ray spectroscopy. Ann. Med. Res. 511, 11–12.
Santos, S.R., Taylor, D.J., Kinzie, R.A., III, Hidaka, M., Sakai, K., et al., 2003. Molecular phylogeny of symbiotic dinoflagellates inferred from partial chloroplast large subunit (23S)-rDNA sequences. Mol. Phylogenet. Evol. 23, 97–111.
Satterlie, R.A., 2002. Neuronal control of swimming in jellyfish: a comparative story. Can. J. Zool. 80, 1654–1669.
Satterlie, R.A., 2011. Do jellyfish have central nervous systems? J. Exp. Biol. 214, 1215–1223.
Satterlie, R.A., 2015. The search for ancestral nervous systems: an integrative and comparative approach. J. Exp. Biol. 218, 612–617.

Satterlie, R.A., Eichinger, J.M., 2014. Organization of the ectodermal nervous structures in jellyfish: scyphomedusae. Biol. Bull. 226, 29–40.

Schiariti, A., Christiansen, E., Morandini, A.C., da Silveira, F.L., Agustin, G., Walter, D., et al., 2012. Reproductive biology of *Lychnorhiza lucerna* (Cnidaria: Scyphozoa: Rhizostomeae): individual traits related to sexual reproduction. Mar. Biol. Res. 8, 255–264.

Schiariti, A., Holst, S., Straehler-Pohl, I., Tiseo, G.R., Miyake, H., Morandini, A.C., 2024. Life cycles and reproduction of Rhizostomeae. Adv. Mar. Biol. 98.

Schiariti, A., Kawahara, M., Uye, S.-I., Mianzan, H.W., 2008. Life cycle of the jellyfish *Lychnorhiza lucerna* (Scyphozoa: Rhizostomeae). Mar. Biol. 156, 1–12.

Shanks, A.L., Graham, W.M., 1987. Orientated swimming in the jellyfish *Stomolophus meleagris* L. Agassiz (Scyphozoan: Rhizostomida). J. Exp. Mar. Biol. Ecol. 108, 159–169.

Shanks, A.L., Graham, W.M., 1988. Chemical defense in a scyphomedusa. Mar. Ecol. Prog. Ser. 45, 81–86.

Shick, J.M., 1991. A functional biology of sea anemones. Chapman and Hall, London.

Smith, H.G., 1936. Contribution to the anatomy and physiology of *Cassiopea frondosa*. 31. Carnegie Institution of Washington, Washington, pp. 18–52.

Solé, M., Lenoir, M., Vortuno, J.M., Durfort, M., Schar, vd, M., André, M., 2016. Evidence of cnidarians sensitivity to sound after exposure to low frequency underwater sources. Sci. Rep. 6, 37979.

Sötje, I., Dishon, T., Hoffmann, F., Holst, S., 2017. New methods of morphometric analyses on scyphozoan jellyfish statoliths including the first direct evidence for statolith growth using calcein as a fluorescent marker. Microsc. Microanal. 23, 553–568.

Sötje, I., Henne, S., 2019. The senses of Scyphozoa. In: Jarms, G., Morandini, A.C. (Eds.), World atlas of jellyfish. Dölling und Galitz Verlag, München, pp. 46–53.

Sötje, I., Holst, S., 2019. The stinging cell of Scyphozoa. In: Jarms, G., Morandini, A.C. (Eds.), World atlas of jellyfish. Dölling und Galitz Verlag, München, pp. 54–63.

Sötje, I., Neues, F., Epple, M., Ludwig, W., Rack, A., Gordon, M., et al., 2011. Comparison of the statolith structures of *Chironex fleckeri* (Cnidaria, Cubozoa) and *Periphylla periphylla* (Cnidaria, Scyphozoa): a phylogenetic approach. Mar. Biol. 158, 1149–1161.

Spangenberg, D.B., Beck, C.W., 1968. Calcium sulfate dihydrate statoliths in *Aurelia*. Trans. of Am. Microsc. Soc. 87, 329–335.

Spangenberg, D.B., Jernigan, T., McCombs, R., Lowe, B.T., Sampson, M., Slusser, J., 1994. Development studies of *Aurelia* (jellyfish) ephyrae which developed during the SLS-1 mission. Adv. Space Res. 14, 239–247.

Stabili, L., Rizzo, L., Fanizzi, F.P., Angile, F., Coco, L.D., Girelli, C.R., et al., 2019. The jellyfish *Rhizostoma pulmo* (Cnidaria): Biochemical composition of ovaries and antibacterial lysozyme-like activity of the oocyte lysate. Mar. Drugs. 17, 1–15.

Straehler-Pohl, I., Jarms, G., 2010. Identification key for young ephyrae: a first step for early detection of jellyfish blooms. Hydrobiologia 645, 3–21.

Sugiura, Y., 1963. On the life-history of *Rhizostome medusae* I. *Mastigia papua* L. Agassiz. Annot. Zool. Jpn. 36, 194–202.

Sugiura, Y., 1969. On the life-history of rhizostome medusae V. On the relation between zooxanthellae and the strobilation of *Cephea cephea*. Bull. Mar. Biol. Stn. Asamushi 8, 227–233.

Szczepanek, S., Cikala, M., David, C.N., 2002. Poly-g-glutamate synthesis during formation of nematocyst capsules in hydra. J. Cell Sci. 115, 745–751.

Tardent, P., 1995. The cnidarian cnidocyte, a hightech cellular weaponry. BioEssays 17, 351–362.

Thibault, D., Kuplik, Z., Prieto, L., Enrique-Navarro, A., Brown, M.K., Uye, S.-I., Doyle, T.K., Pitt, K., Fitt, W.K., Gibbons, M.J., 2024. Ecology of Rhizostomeae. Adv. Mar. Biol. 98.

Thiel, M.E., 1959. Scyphomedusae: Semaeostomae, Physiologie. In: Bronns, H.G. (Ed.), Klassen und Ordnungen des Tierreichs. Akademische Verlagssgesellschaft Geest und Portig K.-G., Leipzig, pp. 856–1072.

Thiel, M.E., Reinicke, G.B., Holst, S., Morandini, A.C., Sötje, I., Straehler-Pohl, I., et al., 2024. Rhizostomeae—a remaining manuscript to complete the monograph on Scyphozoa by M.E. Thiel (published 1936-1977). Adv. Mar. Biol. 98 (Electronic supplement).

Thornhill, D.J., Daniel, M.W., LaJeunesse, T.C., Schmidt, G.W., Fitt, W.K., 2006. Natural infections of aposymbiotic *Cassiopea xamachana* scyphistomae from environmental pools of *Symbiodinium*. J. Exp. Mar. Biol. Ecol. 338, 50–56.

Tiemann, H., Sötje, I., Jarms, G., Paulmann, C., Epple, M., Hasse, B., 2002. Calcium sulfate hemihydrate in statoliths of deep-sea medusae. J. Chem. Soc. Dalton Trans. 2002, 1266–1268.

Tiseo, G.R., 2016. Conformação gonadal, caracterização histoquímica e ultraestrutural da gônada masculina e espermatozoides em espécies de águas-vivas (Cubozoa e Scyphozoa, Medusozoa, Cnidaria). Master Thesis, Universidade de São Paulo, São Paulo, Brazil, pp.148.

Tiseo, G.R., 2022. Sexual reproduction in jellyfish species (Scyphozoa, Cnidaria) from the southeastern Brazilian coast. PhD Thesis, Universidade de São Paulo, São Paulo, Brazil, pp. 257.

Toyokawa, M., Shimizu, A., Sugimoto, K., Nishiuchi, K., Yasuda, T., 2010. Seasonal changes in oocyte size and maturity of the giant jellyfish, *Nemopilema nomurai*. Fish. Sci. 76, 55–62.

Trench, R.K., 1971. The physiology and biochemistry of zooxanthellae symbiotic with marine coelenterates II. Liberation of fixed 14C by zooxanthellae in vitro. Proc. R. Soc. Lond. B. 177, 237–250.

Trench, R.K., Colley, N.J., Fitt, W.K., 1981. Recognition phenomena in symbioses between marine invertebrates and "zooxanthellae"; uptake, sequestration and persistence. Ber. Deutsch. Bot. Ges. Bd. 94, 529–545.

Tronolone, V.B., Morandini, A.C., Bigotto, A.E., 2002. On the occurrence of scyphozoan ephyrae (Cnidaria, Scyphozoa, Semestomae and Rhizostomae) in the southeastern Brazilian coast. Biota Neotrop. 2, 1–18.

Uchida, T., 1926. The anatomy and development of a rhizostome medusa, *Mastigias papua* L. Agassiz, with observations on the phylogeny of Rhizostomeae. J. Fac. Sci. Uni. Tokyo Sect. 4, Zool. 1, 45–95.

Uye, S.-I., 2014. The giant jellyfish *Nemopilema nomurai* in East Asian marginal seas. In: Pitt, K.A., Lucas, C.H. (Eds.), Jellyfish blooms. Springer Science and Business Media, Dordrecht, pp. 185–205.

Van Lieshout, J.S., Martin, V.J., 1992. Development of planuloid buds of *Cassiopea xamachana* (Cnidaria: Scyphozoa). Trans. Am. Microsc. Soc. 111, 89–110.

Vega de Luna, F., Dang, K.V., Cardol, M., Roberty, S., Cardol, P., 2019. Photosynthetic capacity of the endosymbiotic dinoflagellate *Cladocopium* sp. Is preserved during digestion of its jellyfish host *Mastigias papua* by the anemone *Entacmaea medusivora.* FEMS Microbiol. Ecol. 95, fiz141.

Verde, E.A., McCloskey, L.R., 1998. Production, respiration, and photophysiology of the mangrove jellyfish *Cassiopea xamachana* symbiotic with zooxanthellae: effect of jellyfish size and season. Mar. Ecol. Prog. Ser. 168, 147–162.

Von Lendenfeld, R., 1882. Über eine Übergangsform zwischen Semostomen und Rhizostomen. Zool. Anz. 5, 380–383.

Von Lendenfeld, R., 1884. . The Scyphomedusae of the southern hemisphere. Proc. Linn. Soc. N. South. Wales 9 (pt 1), 155–169 (pt 2): 242–249, (pt 3): 259–306, Sydney.

Von Lendenfeld, R., 1885. Local colour variety of Scyphomedusae, a new species produced in forty years. Ann. Mag. Nat. Hist. 14 (5), 409–412 (London).

Von Lendenfeld, R., 1888. Der Charakter der australischen Coelenteratenfauna. Biol. Zbl 7. pp. 641–646 (Erlangen).

Von Montfort, G.M., Costello, J.H., Colin, S.P., Morandini, A.C., Migotto, A.E., Maronna, M.M., et al., 2023. Ontogenetic transitions, biomechanical trade-offs and macroevolution of scyphozoan medusae swimming patterns. Sci. Rep. 13, 9760.

Watson, G.M., Wood, R.L., 1989. Colloquium on terminology. In: Hessinger, D.A., Lenhoff, H.M. (Eds.), The biology of nematocysts. Academic Press, San Diego, pp. 21–23.

Weill, R., 1934. Contribution à l'étude des cnidaires et de leurs nématocystes. Trav. Sta. Zool. Wimereux 10, 1–701.

Werner, B., 1993. Stamm Cnidaria, Nesseltiere. In: Gruner, H.E. (Ed.), Lehrbuch der speziellen Zoologie, Vol. 1: Cnidaria, Ctenophora, Mesozoa, Plathelminthes, Nemertini, Entoprocta, Nemathelminthes, Priapulida. Gustav Fischer Verlag, Jena, pp. 11–305.

West, E.J., Pitt, K.A., Welsh, D.T., Koop, K., Rissik, D., 2009. Top-down and bottom-up influences of jellyfish on primary productivity and planktonic assemblages. Limnol. Oceanogr. 54, 2058–2071.

Westfall, J.A., 1973. Ultrastructural evidence for neuromuscular systems in coelenterates. Amer. Zool. 13, 237–246.

Westfall, J.A., 1996. Ultrastructure of synapses in the first-evolved nervous systems. J. Neurocytol. 25, 735–746.

Widersten, B., 1965. Genital organs and fertilization in some Scyphozoa. Zool. Bidr. Upps. 37, 45–58.

Widersten, B., 1968. On the morphology and development in some cnidarian larvae. Zool. Bidr. Upps. 37, 139–182.

Yamasu, T., Yoshida, M., 1973. Electron microscopy on the photoreceptors of an anthomedusan and a scyphomedusan. Publ. Seto. Mar. Lab. 20, 757–778.

Yap, N.W.L., Ong, Y.J., 2012. A survey of jellyfish (Cnidaria) around St John's Island in the Singapore Straits. Contrib. Mar. Sci. 57–74.

Yuan, D., Nakanishi, N., Jacobs, D.K., Hartenstein, V., 2008. Embryonic development and metamorphosis of the scyphozoan *Aurelia*. Dev. Genes. Evol. 218, 525539.

CHAPTER FOUR

Life cycles and reproduction of Rhizostomeae

Agustín Schiariti[a,b,*], Sabine Holst[c], Gisele R. Tiseo[d], Hiroshi Miyake[e], and André C. Morandini[d,f]

[a]Instituto Nacional de Investigación y Desarrollo Pesquero (INIDEP), Mar del Plata, Argentina
[b]Consejo Nacional de Investigaciones Científicas y Técnicas (CONICET), Buenos Aires, Argentina
[c]Senckenberg am Meer, German Center for Marine Biodiversity Research, Hamburg, Germany
[d]Departamento de Zoologia, Instituto de Biociências, Universidade de São Paulo, São Paulo, SP, Brazil
[e]School of Marine Biosciences, Kitasato University, Sagamihara, OL, Japan
[f]Centro de Biologia Marinha, Universidade de São Paulo, São Sebastião, SP, Brazil
*Corresponding author. e-mail address: agustin@inidep.edu.ar

Contents

Abstract

In the present study we reviewed the life cycles and reproduction strategies of the order Rhizostomeae. We found 28 species with described life cycles representing ~30% of the valid species. The metagenetic life cycle of most scyphozoans, which includes the benthic asexually-reproducing polyp and the pelagic sexually-reproducing medusa, is exhibited by all rhizostome species. Rhizostomeae are dioecious with only two exceptions described as hermaphroditic. Sexual dimorphism can be found in species with special external structures utilised for brooding but others show no sexual dimorphism despite the colour of mature gonads. Six asexual reproduction modes have been described for the production of new polyps but rhizostome polyps propagate through a main mode that differs among taxa. Species belonging to Dactyliophorae produce new polyps by podocysts whereas the Kolpophorae new polyps develop from free-swimming buds. The number of ephyrae formed per

Advances in Marine Biology, Volume 98
ISSN 0065-2881, https://doi.org/10.1016/bs.amb.2024.07.006

strobila differs between taxa with monodisc and polydisc strobilation in the Kolpophorae and Dactyliophorae, respectively. Given the low number of studied species it is expectable that new reproductive strategies will be discovered when additional species are investigated. We recommend increasing (1) descriptions on life cycles and reproductive strategies for a greater number of species, (2) attempts to locate the polyps in the field, (3) the study of species in their natural environment, to understand the population dynamics of Rhizostomeae and to clarify the potential of artificial structures to increase medusa populations. In addition, experimental studies are needed to improve our understanding of the factors affecting transitions between life cycle stages and medusa production rates.

1. Introduction

The species of the scyphozoan order Rhizostomeae Cuvier, 1800 (hereafter, the rhizostomes) are significant from the ecological (see Thibault et al., 2024 for seasonality of selected species), evolutionary (Gamero-Mora et al., 2024), and human (Brotz et al., 2024) perspectives. Interest in these species has increased over the past few decades as a result of the growing number of reports on the socio-economic issues caused by their frequent "blooms" (for definitions of "blooms", see Lucas and Dawson, 2014) on various industries (fisheries, tourism, and aquaculture; Richardson et al., 2009), the fishery scientists' recognition of the significance of including these species in ecosystem models (Pauly et al., 2009), and the expansion of the jellyfish fisheries from its origin in Asia abroad (Brotz et al., 2017). All of these issues highlighted the need for a deeper comprehension of the medusae role in marine ecosystems as well as the variables influencing their abundance (i.e., population dynamics; see Box 1 in Schiariti et al. (2018) for a definition of medusozoan populations). Furthermore, they pose interesting conceptual subjects, not only because of the key position they occupy in the evolutionary tree (Gamero-Mora et al., 2024; Morandini, 2024) but also due to distinctive aspects of their biology and development (Mackie, 2002; Technau et al., 2015).

The development of long-term data sets that can be used to describe patterns of abundance and distribution of scyphomedusae has gained more attention in recent years, primarily to test the plausibility of increasing medusae abundance through an increase in the size and frequency of its blooms (Condon et al., 2012). The ecology of jellyfish has also received more attention as a means of better comprehending their function in ecosystem dynamics (e.g., Chiaverano et al., 2018; Robinson et al., 2014).

However, frustration with explaining a given situation in terms of trophic ecology and physical factors has directed the attention on recruitment, which necessarily demands basic knowledge of life history and, hence, of life cycles and reproductive strategies.

Life history explains the broad features of a life cycle—how fast the organism will grow, when it will mature, how long it will live, how frequently it will reproduce, how many offspring it will have, and so forth (Stearns, 1992). Population dynamics is determined by life history features because they play a direct role in reproduction and survival (Stearns, 1992). Because animals have developed so many ways of combining these traits to affect fitness, studying life history strategies is complex and fascinating. Rhizostomes are especially intriguing since they go through a particularly complex life cycle and exhibit a wide range of sexual and asexual strategies, living in benthic and pelagic environments, and having sessile and free-living stages.

While aiming to describe the importance of a certain species in a given environment (which is expected to be proportional to its abundance), a common mistake is to consider species only based on morphological and functional features of the adult stage, ignoring the fact that, for instance, several species present complex life cycles including the alternation of sessile-benthic and free-living pelagic stages, sexual and asexual reproductive strategies as part of the same life cycle (Giangrande et al., 1994).

According to Stearns (1992), the life cycle is the continual series of changes that an organism goes through from one initial form, such as a gamete, through to the creation of the same form again. In the typical rhizostome life cycle, fertilisation of gametes results in a planula larva that settles onto the substrate and metamorphoses into a sessile polyp (i.e., scyphistoma). The polyps reproduce asexually generating more polyps. When certain environmental conditions are met, polyps begin producing ephyrae (i.e., young/juvenile medusae), which are released into the water column and become adult medusae which produce the gametes (Fig. 1). This basic scheme has been described as a *metagenetic* life cycle and is characterised by the alternation (or succession) of generations (or stages) (see Morandini et al., 2016). Unmatched diversity and plasticity have been found within the medusozoans (Scyphozoa Goette, 1887, Cubozoa Werner, 1973, Staurozoa Marques and Collins, 2004, Hydrozoa Owen, 1843) from this general life cycle, possibly because cnidarians have simple anatomical and physiological structures and are hence evolutionarily plastic (Boero et al., 1997; Fautin, 2002). Yet, the rhizostomes present the typical metagenetic life cycle formerly described with no exceptions reported so far (Jarms and Morandini, 2019).

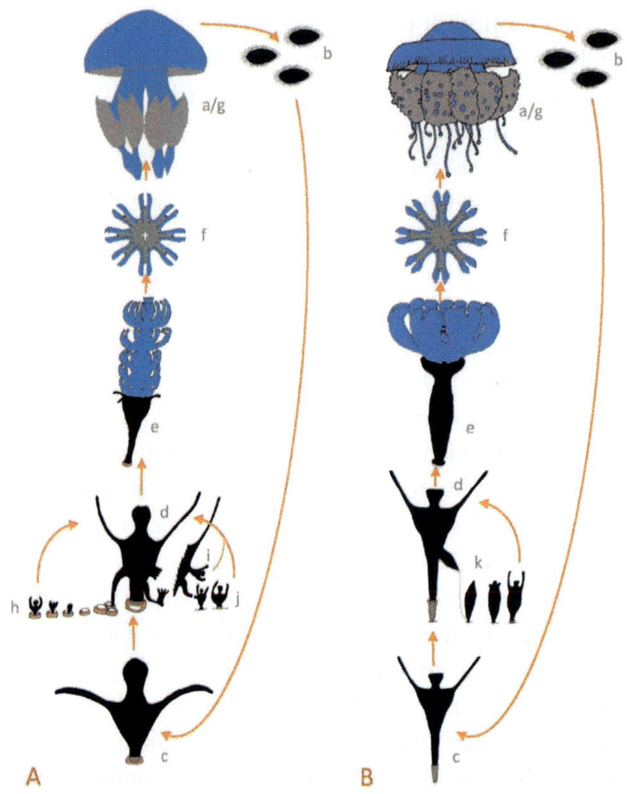

Fig. 1 Schematic representation of the life cycle types of Rhizostomeae: (A) Dactyliophorae, (B) Kolpophorae; a/g: adult medusa, b: planulae, c: polyp; d: asexually reproducing polyp; e: polydisk (A) and monodisc (B) strobila; f: ephyra; h: secondary polyps produced by podocyst; i: secondary polyp production by lateral budding (distal end in front); j: secondary polyp production by lateral budding by means of stolons; k: secondary polyp production by free-swimming buds (proximal end in front). Blue: medusoid tissue/stage; Black: polypoid or planula tissue/stage. Drawing by I. Straehler-Pohl.

Thiel et al. (2024) review of the knowledge about rhizostomes chronologically stopped in the 1970s. At that time, the developmental stages of only a handful of species were known. The section dedicated to the "ontogeny" focused mostly on presenting and describing the different stages of the life cycle. Due to the restricted knowledge coverage, it was not possible to M.E. Thiel to have insights and convert the literature data into general patterns for the rhizostomes. The section has a historical part, fertilisation and early development, scyphistoma and asexual reproduction, strobilation modes, and finally ephyrae growth and gastrovascular cavity differentiation.

Science aims at simple explanations of complex things. In the present study, the diversity of life cycles and reproduction strategies (sexual and asexual) of the order Rhizostomeae are reviewed. We aim to analyse the knowledge on life cycles and reproduction strategies of rhizostomes and to give some recommendations for future research in this field.

2. Life cycles

Our literature search reported 28 rhizostome species with formally described life cycles or completed from laboratory observations, representing 31% of the currently valid rhizostome species (Table 1). Nineteen of these described life cycles correspond to species belonging to the suborder Dactyliophorae and 9 to the Kolpophorae. Rhizostomatidae is the best-studied family with 9 life cycles described over 12 species (9/12 = 75%), followed by Stomolophidae (2/3 = 66%), Catostylidae (6/24 = 25%), Cassiopeidae (3/12 = 25%), Leptobrachidae (1/4 = 25%), Lobonemidae (1/4 = 25%), Cepheidae (3/13 = 23%), Mastigiidae (2/12 = 17%) and Lychnorhizidae (1/6 = 17%). No life cycles of Versurigidae and Bazingidae have been described so far.

All the rhizostomes present the typical metagenetic life cycle described for most scyphozoans (see Jarms, 2010; Morandini et al., 2016) including three stages: the polyp stage, which reproduces asexually in the benthos, the medusoid stage, which reproduces sexually in the plankton, with the planula larvae as a link in between them (Fig. 1). Sexually mature medusae produce and release gametes that become ready for fertilisation either in the surrounding water or within the female body (e.g. Arai, 1997; Jarms, 2010). The zygote, which results from fertilisation, develops into the planula larva. After a short planktonic period, the planula settles on a suitable substrate and transforms into a benthic polyp. The benthic polyps then reproduce and propagate forming new polyps via different asexual reproduction modes and then, under specific environmental conditions, produce and release young/juvenile medusae (ephyrae) back to the plankton through strobilation. The ephyrae then grow and reach sexual maturity closing the life cycle.

All the rhizostome species described so far present the same general features of the life cycle with differences only in the polyp-to-polyp reproduction modes and strobilation (see below). Exceptions to this pattern have been described in other scyphozoan taxa reaching the

Table 1 Described or completed* (not formally described but observed in laboratory cultures) life cycles of Rhizostomeae and applied methods to collect fertilised eggs (FE), planulae (PL) or polyps (PO) for cultivation. Culture tanks containing medusae are abbreviated with "tank". # *LC*, number of described life cycles; # *SP*, number of currently accepted species.

Suborder (# LC/# SP)	Family (# LC/# SP)	Species	Methodology	Source
Dactyliophorae Stiasny, 1920 (19/48)	Catostylidae Claus, 1883 (6/24)	*Acromitus flagellatus* (Maas, 1903)*	FE from tank	Miyake et al. (2018)
		Acromitus hardenbergi Stiasny, 1934	FE from tank	Miyake et al. (2021)
		Acromitus maculosus Light, 1914*	FE from tank	Miyake et al. (2018)
		Catostylus mosaicus (Quoy and Gaimard, 1824)	PL from female medusae	Pitt (2000)
		Catostylus tagi (Haeckel, 1869)	Artificial fertilisation	Gueroun et al. (2021)
		Catostylus townsendi Mayer, 1915	PL from tank	Phuangsanthia et al. (2018)
	Lobonemidae Stiasny, 1920 (1/4)	*Lobonemoides robustus* Stiasny, 1920	FE from tank	Miyake et al. (2018)

Lychnorhizidae Haeckel, 1880 (1/6)	*Lychnorhiza lucerna* Haeckel, 1880	Artificial fertilisation	Schiariti et al. (2008)
Rhizostomatidae Cuvier, 1800 (9/12)	*Rhizostoma luteum* (Quoy and Gaimard, 1827)	PL from female medusae	Kienberger et al. (2018)
	Rhizostoma octopus (Gmelin, 1791)	PL from tank	Holst et al. (2007)
	Rhizostoma pulmo (Macri, 1778)	PO from tank	Paspaleff (1938); Fuentes et al. (2011)
	Rhopilema asamushi Uchida, 1927*	Not described	Jarms and Morandini (2019)
	Rhopilema esculentum Kishinouye, 1891	FE, PL, PO from tank	Ding and Cheng (1981)
	Rhopilema hispidum (Vanhöffen, 1888)	Artificial fertilisation	Miyake et al. (2018); Phuangsanthia et al. (2020)
	Rhopilema nomadica Galil, Spanier, Ferguson, 1990	PL from tank	Lotan et al. (1992)

(continued)

Table 1 Described or completed* (not formally described but observed in laboratory cultures) life cycles of Rhizostomeae and applied methods to collect fertilised eggs (FE), planulae (PL) or polyps (PO) for cultivation. Culture tanks containing medusae are abbreviated with "tank". # *LC*, number of described life cycles; # *SP*, number of currently accepted species. (*cont'd*)

Suborder (# LC/# SP)	Family (# LC/# SP)	Species	Methodology	Source
		Rhopilema verrilli (Fewkes, 1887)	PL from medusae	Cargo (1971); Calder (1973)
	Stomolophidae Haeckel, 1880 (2/3)	*Nemopilema nomurai* Kishinouye, 1922	Artificial fertilisation	Kawahara et al. (2006)
		Stomolophus meleagris Agassiz, 1860	PL from tank	Calder (1982)
		Stomolophus sp.2*	PL from tank	López-Martínez et al. (2022)
Kolpophorae Stiasny, 1920 (9/42)	Cepheidae Agassiz, 1862 (3/13)	*Cephea cephea* (Forskål, 1775)*	PL from female medusae	Sugiura (1966); Straehler-Pohl (2009)
		Cotylorhiza tuberculata (Macri, 1778)	PL from female medusae	Kikinger (1992)
		Netrostoma setouchianum (Kishinouye, 1902)*	PL from female medusae	Straehler-Pohl (2009); present study
	Cassiopeidae Tilesius, 1831 (3/12)	*Cassiopea andromeda* (Forskål, 1775)	Artificial fertilisation	Gohar and Eisawy (1960)

	Cassiopea frondosa (Pallas, 1774)	PL from female medusae	Bigelow (1893)
	Cassiopea xamachana Bigelow, 1892	PL found on natural substrates	Bigelow (1900)
Leptobrachidae Agassiz, 1862 (1/4)	*Thysanostoma thysanura* Haeckel, 1880*	PL from tank	Yamamori et al. (2017); Jarms and Morandini (2019)
Mastigiidae Stiasny, 1920[a] (2/12)	*Mastigias papua* (Lesson, 1830)	PL from female medusae	Uchida (1926); Sugiura (1963)
	Phyllorhiza punctata von Lendenfeld, 1884	PL from female medusae	Rippingale and Kelly (1995)

[a]Lange and Kaiser (1995) presented a report on polyps of *Mastigias roseus* (Reynaud, 1830) from the tropical Atlantic that we considered as doubtful because the picture presented resembles *Phyllorhiza punctata* (Plate 2 in Lange and Kaiser, 1995). Furthermore, the species *M. roseus* has been considered as doubtful by Jarms and Morandini (2019). Therefore, we did not include the record by Lange and Kaiser (1995) in our study.

greatest diversity of reproductive strategies in the order Coronatae Vanhöffen, 1892 which contain species lacking an entire half of the life cycle such as *Periphylla periphylla* (Péron and Lesueur, 1810) (lacking the polyp stage) or *Thecoscyphus zibrowii* Werner, 1984 (lacking pelagic medusae) (Jarms, 2010). The polyp stage can be absent in Scyphozoa, as in the semaeostome *Pelagia noctiluca* (Forskål, 1775) (Rottini Sandrini and Avian, 1983). Other interesting deviations from the metagenetic pattern have been described for the semaeostome *Aurelia coerulea* von Lendenfeld, 1884 (identified as *Aurelia* sp.1; He et al., 2015), where the normal direction of the life cycle was reversed when the ephyrae, and medusa tissues, are back transformed into polyps (He et al., 2015). In this species, planula strobilation was also observed and named "direct development" or "planula-strobilated ephyrae" (Takauchi et al., 2021; Yasuda, 1975). So far, the only exception to the typical metagenetic life cycle described within the rhizostomes was proposed by López-Martínez et al. (2022) for *Stomolophus* sp.2. The authors reported that "reversed ephyrae" were observed (*always at "hot" environmental temperatures*), and that these ephyrae "*settled on both aboral and oral parts, forming a stem attached to the substrate*" (Figure 8 and Supporting Information Material in López-Martínez et al., 2022). They also described "*the autophagy of the umbrella tissue producing a polyp*". However, the images and videos provided by the authors may instead show (abnormal) ephyrae which were not released but retained in a possibly abortive strobilation. The autophagy of tissues developed in abortive reproductive processes has been previously observed for some species on different occasions (Schiariti and Morandini *pers. obs.*). Thus, the description made by López-Martínez et al. (2022) probably does not refer to the reversion of the life cycle but to the reabsorption (autophagy) of an unreleased ephyra due to an interrupted strobilation process. The specified "hot" temperatures might have been the environmental factor that interrupted strobilation. Furthermore, whereas *Stomolophus* sp.2 normally produce several ephyrae per strobilation (González-Valdovinos, 2010; Jacobson, 2016; Mendoza, 2015; Schiariti unpubl. data), the above-mentioned cases produced only one ephyra suggesting that these polyps could have been too small to produce more than one (see strobilation section below). The evolutionary and ecological plasticity among cnidarians is overwhelming, and life cycle is a good example of these diverse expressions (Boero, 1990; Fautin, 2002; Jarms, 2010) but new evidence is needed to confirm this exceptional observation.

3. From plankton/sexual to benthos/asexual

3.1 Sexual reproduction

Like in most scyphozoans, conspicuous pelagic medusae are the sexually reproducing stage in the rhizostomes life cycle with no exceptions described so far. Different aspects of sexual reproduction have been studied for 20 rhizostomes achieving 22% of the known described species (Table 2). Despite a few studies on selected species, data on scyphozoan sexual reproduction are scattered and not studied using an evolutionary approach (Tiseo, 2022).

Rhizostomeae species have separate sexes (i.e., dioecious) with only two exceptions, represented by *Pseudorhiza haeckeli* from Australian waters (Haacke, 1887) and *Cassiopea andromeda* from the Red Sea but found in Hawaii (Hofmann and Hadfield, 2002) (Table 2). These species were described as hermaphroditic and develop both eggs and sperm in the same individual (see Hofmann and Hadfield, 2002; Thiel et al., 2024 for details). Sexual maturity has been interpreted at the tissue level by some authors (Mammone et al., 2023; Pitt and Kingsford, 2000; Schiariti et al., 2012; Tiseo, 2022). Although specimens may appear to have fully grown gonads, they sometimes are not ready to spawn and might require some special stimulus as light exposure (Ohtsu et al., 2007). Like in some semaeostomes, exposure to light in *N. nomurai* is also a spawning trigger; the fully developed oocytes were spawned 80–100 min after females were exposed to light and males released sperm after 30 min of light exposure (Ohtsu et al., 2007; Thibault et al., 2024). It has been reported that some peptides released from hydrozoan female gonads may play a role in fertilisation (Ikegami et al., 1978). Although there is no documentation of oocyte and sperm identification for rhizostomes, Ohtsu et al. (2007) postulated that these compounds might be involved in the activation of the sperm in *N. nomurai*.

Gametogenesis is known for a few species with no significant differences among the rhizostomes (but see Holst et al., 2024a for further details). In fact, oogenesis and spermatogenesis do not differ between Discomedusae species in general with the gonia cells (oogonia and spermatogonia) passing through several morphological changes before, finally becoming late oocyte and sperm (Avian and Rottini-Sandrini, 1991; Campbell, 1974; Eckelbarger and Larson, 1992; Ikeda et al., 2011; Kikinger, 1992; Pitt and Kingsford, 2000; Schiariti et al., 2012; Tiseo, 2022; Uchida, 1926; Widersten, 1965). Discomedusae oocytes maintain a connection with the genital epithelium by the trophocytes (Eckelbarger, 1994; Ikeda et al., 2011; Mammone et al., 2023; Pitt and Kingsford, 2000; Schiariti et al., 2012; Tiseo, 2022), specialised cells

Table 2 Sexual reproduction features in Rhizostomeae species.

Species	Monoic vs dioic	sexual dimorphism	Fertilisation	Brooding	Gametogenesis	Female spawning	Male spawning	Sources
Dactyliophorae								
Catostylidae								
Acromitus hardenbergi	Dioic	Not observed	Presumed to be external	Non-brooding	Not described	Not described	Not described	Miyake et al. (2021)
Catostylus mosaicus	Dioic	Not observed	Internal	Brooding	Oogenesis and spermatogenesis	Not described	Free sperm	Pitt and Kingsford (2000); Rouse and Pitt (2000)
Catostylus tagi	Dioic	Not observed	Presumed to be external	Non-brooding	Not described	Not described	Not described	Gueroun et al. (2021)
Lychnorhizidae								
Lychnorhiza lucerna	Dioic	Gonadal colour	Presumed to be external	Non-brooding	Oogenesis and spermatogenesis	Oocytes released into the genital sinus through ovulation pit	Free sperm through Follicular pore	Schiariti et al. (2012); Tiseo (2022)

Pseudorhiza haeckeli Haackel, 1884	Monoic	Not described	Presumed to be internal	Brooding	Not described	Not described	Not described	Haacke (1887)
Rhizostomatidae								
Rhizostoma luteum	Dioic	Not described	Internal	Brooding	Not described	Oocytes (might be) kept inside female gonad; eggs released into genital sinus	Not described	Kienberger et al. (2018)
Rhizostoma octopus	Dioic	Gonadal colour	Presumed to be external	Non-brooding	Not described	Not described	Not described	Holst et al. (2007)
Rhizostoma pulmo	Dioic	Not observed	Internal	Brooding	Oogenesis and spermatogenesis	Oocytes kept inside female gonad; eggs released into genital sinus	Free sperm in the genital sinus	Widersten (1965); Avian and Rottini-Sandrini (1991); Holst et al. (2007)

(continued)

Table 2 Sexual reproduction features in Rhizostomeae species. (*cont'd*)

Species	Monoic vs dioic	sexual dimorphism	Fertilisation	Brooding	Gametogenesis	Female spawning	Male spawning	Sources
Rhopilema nomadica	Dioic	Not described	External	Non-brooding	Spermatogenesis	Oocytes migrate to the scapulet before being released into the seawater	Free in the genital sinus	Lotan et al. (1992); Tiseo (2022)
Rhopilema verrilli	Dioic	Not described	Presumed to be internal	Brooding	Not described	Not described	Not described	Calder (1973)
Nemopilema nomurai	Dioic	Gonadal colour	Presumed to be external	Non-brooding	Oogenesis and spermatogenesis	Oocytes released into the genital sinus through ovulation pit	Free sperm through a spermia-tion pit	Kawahara et al. (2006); Ohtsu et al. (2007); Ikeda et al. (2011)
Stomolophidae								
Stomolophus meleagris	Dioic	Not described	Presumed to be external	Non-brooding	Not described	Not described	Not described	Calder (1982)

Kolpophorae								
Cepheidae								
Cotylorhiza tuberculata	Dioic	Brooding filaments	Internal	Brooding	Oogenesis and spermatogenesis	Eggs released from the ovary to the genital sinus	*Spermatozeugmata*	Kikinger (1992)
Cassiopeidae								
Cassiopea andromeda	Monoic and dioic	Brooding filaments	Internal	Brooding	Oogenesis and spermatogenesis	Eggs released into the genital sinus	*Spermatozeugmata*	Neumann (1979); Gohar and Eisawy (1960); Hofmann et al. (1996); Tiseo (2022)
Cassiopea frondosa	Dioic	Planulae in mouth arms	Internal	Brooding	Oogenesis and spermatogenesis	Eggs released into the genital sinus and kept inside the female gonad	*Spermatozeugmata*	Smith (1936); Tiseo (2022); Mammone et al. (2023)

(continued)

Table 2 Sexual reproduction features in Rhizostomeae species. (*cont'd*)

Species	Monoic vs dioic	sexual dimorphism	Fertilisation	Brooding	Gametogenesis	Female spawning	Male spawning	Sources
Cassiopea maremetens Gershwin, Zeidler and Davie, 2010	Dioic	Not described	Not described	Not described	Spermatogenesis	Not described	*Spermatoz-eugmata*	Tiseo (2022)
Cassiopea xamachana	Dioic	Planulae in mouth arms	Presumed to be internal	Brooding	Oogenesis and spermatogenesis	Not described	*Spermatoz-eugmata*	Eckelbarger (1994)Tiseo (2022)
Mastigiidae								
Mastigias papua	Dioic	Gonadal colour and planulae attached to the ventral side of oral disk	Presumed to be internal	Brooding	Oogenesis and spermatogenesis	Eggs released into the genital sinus	*Spermatoz-eugmata*	Uchida (1926); Berrill (1949)

Phyllorhiza punctata	Dioic	Not described	Presumed to be internal	Brooding	Not described	Not described	Free sperm/ *spermatozeugmata*	Rippingale and Kelly (1995); Rouse and Pitt (2000)
Versurigidae								
Versuriga anadyomene (Maas, 1903)	Dioic	Not described	Not described	Not described	Oogenesis	Not described	Not described	Tiseo (2022)

which may play a role in the transfer of nutrients from the gastrovascular cavity to the oocyte (Eckelbarger and Larson, 1992). Tiseo (2022) recognised two probable types of trophocytes: type one, forming a cup around the lower half of the egg (sensu Widersten, 1965; Eckelbarger and Larson, 1988) and type two, with vacuoles in the cytoplasm (Avian and Rottini-Sandrini, 1991; Ikeda et al., 2011; Tiseo, 2022). The oocyte may stay inside the female gonad (Kikinger, 1992; Tiseo, 2022; Widersten, 1965), be released through a pit to the genital sinus (Ikeda et al., 2011; Schiariti et al., 2012) and, in some cases, being further released to the gastrovascular cavity (Ikeda et al., 2011; Ohtsu et al., 2007; Schiariti et al., 2012; Tiseo, 2022) and later to the water (Lotan et al., 1992). The mature spermatozoa can be released freely or in groups immersed in a mucus secretion called *spermatozeugmata* (Table 2; see Holst et al., 2024 chapter for details).

Sexual dimorphism can be found within the species *Cotylorhiza tuberculata*, *C. andromeda*, *C. frondosa*, *C. xamachana* and *M. papua* (Table 2). In these species, fertilisation occurs within the female body, which presents special external structures (brood filaments) for brooding embryos and planulae (Table 2) (Gohar and Eisawy, 1960; Kikinger, 1992; Smith, 1936; Tiseo, 2022; Uchida, 1926; Widersten, 1965; Table 2). Internal fertilisation may take place inside the female gonad, as described for *R. pulmo* (Widersten, 1965), *C. tuberculata* (Kikinger, 1992) and *C. frondosa* (Tiseo, 2022) or inside the genital sinus, as seen in *C. andromeda* (Gohar and Eisawy, 1960), *C. frondosa* (Smith, 1936) and *C. xamachana* (Tiseo, 2022). Conversely, *L. lucerna*, *M. papua*, *Rhizostoma octopus* and *N. nomurai* appear to show no sexual dimorphism despite the colour of mature gonads (Table 2) (Uchida, 1926; Schiariti et al., 2012). In other species (except for *R. nomadica*, in which Lotan et al. (1992) mentioned observing the fully-developed oocytes being released into the seawater), fertilisation is presumed to occur externally (Table 2) (Arai, 1997; Schiariti et al., 2012). This is the case of *A. hardenbergi* (Miyake et al., 2021), *C. tagi* (Gueroun et al., 2021), *L. lucerna* (Schiariti et al., 2012; Tiseo, 2022), *R. octopus* (Holst et al., 2007); *N. nomurai* (Ohtsu et al., 2007) and *S. meleagris* (Calder, 1982).

In general, there is a trend in which internal fertilisation may be inferred by the presence of planulae or embryos inside the female gastrovascular cavity or mouth arms (brooding). Generally, species with internal fertilisation and brooding usually have trophocytes type 1 and release sperm in *spermatozeugmata* (Table 2) (Tiseo, 2022). This is the case of *Cassiopea* spp. (Eckelbarger, 1994; Tiseo, 2022), *C. tuberculata* (Kikinger, 1992) and

R. pulmo (Widersten, 1965). In contrast, species with external fertilisation do not incubate planulae (non-brooding), have trophocytes type 2 (with vacuoles) and release free spermatozoa instead of in groups (Tiseo, 2022) like in *L. lucerna* (Schiariti et al., 2012; Tiseo, 2022), *N. nomurai* (Ikeda et al., 2011) and *S. meleagris* (Eckelbarger, 1994) (Table 2).

3.2 Planula development and settlement

Planulae emerge from the zygote and start swimming in the water column rotating around their axis due to ciliary movements until they find a suitable substrate to metamorphose into a new polyp. This is a critical stage of the life cycle switching from the pelagic sexual and free-living stages to the benthic asexual and sessile stages.

Rhizostome planulae are similar to planulae of Semaeostomeae Agassiz, 1862, being oval to elongate in shape with a broader anterior end (Calder, 1973; Holst and Jarms, 2007; Kikinger, 1992; Schiariti et al., 2008). Colouration is slightly variable from translucent-whitish to brownish and reddish depending on the species (e.g., Kikinger, 1992; Kienberger et al., 2018). No mouth aperture has been observed suggesting that they are lecithotrophic (e.g., Pitt, 2000; Phuangsanthia et al., 2018). In laboratory conditions, the developmental time from the fertilised oocyte to the planula larva varied between 7 and 48 h (Ding and Chen, 1981; Wolk et al., 1985). Variable planula sizes (length) were reported in Dactyliophorae and Kolpophorae with no important differences among the suborders. The smallest planulae (<100 μm) were documented for some non-brooding Dactyliophorae species (Table 3).

The settlement process and the subsequent transformation into polyps was described for planulae of a few rhizostomes (Table 3). Like other scyphozoans, rhizostome planulae were observed moving along the substrate, orienting perpendicular to the substrate, attaching with its anterior end and metamorphosing directly (and irreversibly) into a polyp. In the semaeostome genus *Cyanea* Péron and Lesueur, 1810 planulae can form a cyst (called planulocyst) instead of a polyp (Holst et al., 2024b) but this was not observed for any rhizostome so far. The fixation of planulae to the substrate was described to be mediated by gland cells at the anterior end (Hofmann et al., 1992). A gregarious settlement has been reported in some species (Kikinger, 1992; Pitt, 2000) but was not confirmed in others, suggesting an artefact of restricted space or laboratory conditions.

Planula settlement depends on temperature and salinity (Prieto et al., 2010; Takao and Uye, 2018); and usually, the first settlement occurs within 1–7 days. Some authors observed that most planulae settled from 1 to

Table 3 Literature review on planula size, time until first settlement, duration of planula-to-polyp metamorphosis, preferences for substrate types and for orientation to the substrate.

Species	Planula length (μm)	Settlement (days)	Metamorphosis (days)	Substrate preference	Orientation preference	Source
Kolpophorae						
Cepheidae						
Cephea cephea	130 −230	Not described	2	Not mentioned, but glass avoided	Not described	Sugiura (1966)
Cotylorhiza tuberculata	270 −330	5 - >15	Not described	None	None	Kikinger (1992); Prieto et al. (2010); Astorga et al. (2012)
Mastigiidae						
Mastigias papua	120 −140	1	1 −2	Mangrove leaves	Underside	Sugiura (1963); Hofmann and Crow (2002)
Phyllorhiza punctata	300 −500	1	2 −3	Not described	Underside	Rippingale and Kelly (1995)
Cassiopeidae						

Cassiopea andromeda	160 −210	1 −3	2	Not described	Not described	Gohar and Eisawy (1960)Neumann (1979)
Cassiopea frondosa	220 −300	10	11	Plastic	Underside	Morandini (unpubl. data)
Cassiopea xamachana	120 −200	1 −2	Not described	Mangrove leaves (degrading), sand, rock	Underside	Martin and Chia (1982); Fleck and Fitt (1999); Fitt et al. (2021)
Dactyliophorae						
Catostylidae						
Acromitus hardenbergi	168.8 ± 25.3	2	Not described	Not described	Not described	Miyake et al. (2021)
Catostylus mosaicus	100 −130	4 −5	Not described	Glass	Underside	Pitt (2000)
Catostylus tagi	Not described	1	4	Not described	Not described	Gueroun et al. (2021)
Catostylus townsendi	80 −130	3 −4	Not described	Not described	Not described	Phuangsanthia et al. (2018)

(continued)

Table 3 Literature review on planula size, time until first settlement, duration of planula-to-polyp metamorphosis, preferences for substrate types and for orientation to the substrate. (*cont'd*)

Species	Planula length (μm)	Settlement (days)	Metamorphosis (days)	Substrate preference	Orientation preference	Source
Lychnorhizidae						
Lychnorhiza lucerna	95 −207	1	1	Air-water interface, styrene glass and shells	Underside	Schiariti et al. (2008)
Stomolophidae						
Stomolophus meleagris	120 −390	2 −5	Not described	Not described	Not described	Calder (1982)
Stomolophus sp.2	161 ± 3.1	Not described	5	Not described	Not described	López-Martínez et al. (2022)
Rhizostomatidae						
Rhizostoma luteum	126 −139	3 −5	3	Not described	None	Kienberger et al. (2018)
Rhizostoma octopus	107 −150	2, 5	2	Plastic	Underside	Holst et al. (2007); Holst and Jarms (2007); Holst (2012b)
Rhizostoma pulmo	c. 500	Not described	Not described	Not described.	Not described	Paspaleff (1938)

Rhopilema esculentum	95–150	1	1–4	Not described	Underside	Ding and Chen (1981); Takao and Uye (2018); You et al. (2007)
Rhopilema hispidum	100–140	Not described	4–5	Not described	Not described	Phuangsanthia et al. (2020)
Rhopilema nomadica	c. 200	3–4	Not described	Not mentioned, but stones avoided	Not described	Lotan et al. (1992)
Rhopilema verrilli	165–310	7	7–10	Not described	Not described	Calder (1973); Cargo (1971)
Nemopilema nomurai	159–200	2	2–7	Not described	Underside	Kawahara et al. (2006); Ohtsu et al. (2007); Takao and Uye (2018)

> 30 days. The planktonic larval duration (PD) is considered the time until no more active planulae were present or no increase in settlement numbers was observed (Conley and Uye, 2015; Takao and Uye, 2018). The PD can be substantially longer than the time until most planulae have settled (Kienberger et al., 2018) and is temperature-dependent (Prieto et al., 2010; Gueroun et al., 2021) but has only been documented by a few authors. Some features of the planula (swimming speed, body swelling) can be affected by environmental factors like salinity and temperature which may also affect the PD but further data are still needed to precisely evaluate such effects (Takao and Uye, 2018).

The location of polyps in the natural environment is unknown for most species but settlement preferences have been assessed through laboratory experiments to help in the identification of potential settlement areas in the wild. Although laboratory observations demonstrated that planulae settled on sand grains, swimming or floating particles, or even at the air-water interface (e.g. Miyake et al., 2021; Schiariti et al., 2008), most authors state that rhizostome polyps usually colonise hard substrates in the wild (Fitt and Costley, 1998; Kikinger, 1992; Müller and Leitz, 2002). Different experimental setups were used to investigate preferences for different substrate types and positions (bottom, sides, undersides). In general, planulae settled on various natural and artificial substrates often without significant differences in settlement numbers on different substrate types (Astorga et al., 2012; Pitt, 2000) while others demonstrated preferences for particular substrates (e.g. Holst and Jarms, 2007; Schiariti et al., 2008). The evidence that artificial substrates are suitable settling surfaces for polyps suggests that increasing marine constructions and marine litter may lead to an increase of new habitats for planula settlement, which may cause increasing polyp populations (Astorga et al., 2012; Holst and Jarms, 2007). Settlement on the air-water interface was often observed in laboratory experiments (e.g., Hofmann and Crow, 2002; Kawahara et al., 2006; Rippingale and Kelly, 1995; Schiariti et al., 2008). It was suggested that planulae tended to preferentially settle on hydrophobic surfaces (Tomaru et al., 2014) and that this settlement behaviour was only possible because of the small water volumes and the calm water surface under laboratory conditions but may be impeded in the field due to water turbulence (Astorga et al. 2012; Holst and Jarms, 2007; Schiariti et al., 2008).

Several rhizostome species preferred to colonise the underside of substrates (Table 3). The upside-down hanging position of polyps was suggested to be advantageous for survival and reproduction (Fleck and Fitt, 1999; Holst and Jarms, 2007). However, the settlement preference for substrate undersides was studied in very few cases (Astorga et al., 2012; Kienberger et al., 2018). Several studies focused on optimal conditions for planula development, stating that the water quality, salinity, temperature and suitable substrates are the main factors for a successful planula development. Temperature and salinity affect the survival rates and settlement success of planulae (Gueroun et al., 2021; Takao and Uye, 2018). Compounds originating from bacteria appear to induce planula settlement and metamorphosis, whereas settlement was significantly reduced in sterile artificial seawater and after treatments with antibiotics (Fleck and Fitt, 1999; Hofmann et al., 1996; Neumann, 1979; Wolk et al., 1985). Thus, several experimental studies preconditioned substrates by exposing them to seawater to allow bacterial films to develop which were suggested to facilitate planula settlement (Rippingale and Kelly, 1995; Müller and Leitz, 2002).

Detailed descriptions of the morphogenesis from planula to polyp are rare but general aspects of the transformation are similar in Dactyliophorae and Kolpophorae. The attached planula elongates, forms a foot, a stalk and a calyx, and subsequently a hypostome and an increasing number of tentacles appear (Calder, 1982; Hofmann et al., 1992; Hofmann and Crow, 2002; Neumann, 1979; Wolk et al., 1985). The early tentacle development can start with four primary tentacles but early two-tentacled polyp stages were documented (Feng et al., 2015a; Miyake et al., 2021; Rippingale and Kelly, 1995; Pitt, 2000). Durations from the settled planula into the 4-tentacle polyp stage varied from 1 to 11 days (Table 3). The duration of metamorphosis can be affected by variations in abiotic factors (Ohtsu et al., 2007; Takao and Uye, 2018). Generally, planulae need external triggers to start metamorphosis, but this is only well-studied for *Cassiopea* species (reviewed in Leitz, 1997; Müller and Leitz, 2002). In *C. xamachana*, an inducing peptide is synthesised after fertilisation of the oocytes. This endogenous inducer is localised in the basal part of the bud (and probably also in the planula) and remains inactive until external stimulation by an exogenous metamorphosis inducer (Thieme and Hofmann, 2003).

4. From benthos/asexual to benthos/asexual

4.1 Polyp-to-polyp reproduction modes

Information about polyp-to-polyp reproduction modes was found for 26 species totalising 29% of the described species so far (Table 4). The absence of the solid chitinous periderm that covers the polyps of coronate scyphozoans allows rhizostome polyps to develop a variety of asexual

Table 4 List of the asexual reproduction modes observed for the Rhizostomeae polyps. *LB*, typical lateral budding; *LBst*, lateral budding by means of stolons; *POD*, podocysts; *LF*, longitudinal fission; *FSB*, typical free-swimming buds; *MP*, motile bud-like tissue particles; *#E*, number of ephyrae produced per strobilation event per strobila. (-): not observed; (+): observed sporadically; (++): main asexual mode to produce new polyps. See Adler and Jarms (2009) for the definition of polyp-to-polyp reproduction modes. Some information about the asexual reproductive modes was derived from the authors' personal observations in their laboratories and referred to as the present work.

Species	LB	LBst	POD	LF	FSB	MP	#E	Source
Dactyliophorae								
Catostylidae								
Acromitus hardenbergi	++	-	-	-	-	-	1	Miyake et al. (2021)
Acromitus maculosus	++	-	-	-	-	-	1	Miyake (*pers. obs.*)
Catostylus tagi	-	-	++	-	-	-	>1	Gueroun et al. (2021)
Catostylus mosaicus	+	-	++	+	-	-	>1	Pitt (2000); Straehler-Pohl (2009); Straehler-Pohl and Jarms (2022)
Catostylus townsendi	-	-	++	-	-	-	>1	Phuangsanthia et al. (2018)
Lychnorhizidae								
Lychnorhiza lucerna	-	-	++	-	-	-	>1	Schiariti et al. (2008, 2014)
Rhizostomatidae								
Rhizostoma luteum	-	-	++	-	-	-	>1	Kienberger et al. (2018)

Rhizostoma octopus	+	+	++	+	-	-	>1	e.g., Holst and Jarms (2006); Holst et al. (2007); Holst (2012a, b); Straehler-Pohl and Jarms (2022)
Rhizostoma pulmo	+	+	++	+	-	+	>1	Paspaleff (1938); Fuentes et al. (2011); Schiariti et al. (2014); Straehler-Pohl and Jarms (2022)
Rhopilema asamushi	-	-	++	-	-	-	>1	Jarms and Morandini (2019)
Rhopilema esculentum	-	-	++	-	-	-	>1	e.g., Ding and Chen (1981); Guo (1990); You et al. (2007); Schiariti et al. (2014)
Rhopilema hispidum	-	-	++	-	-	-	>1	Phuangsanthia et al. (2020)
Rhopilema nomadica	-	-	++	-	-	-	>1	Lotan et al. (1992); Dror and Angel (2023)
Rhopilema verrilli	-	-	++	-	-	-	>1	Cargo (1971); Calder (1973)
Nemopilema nomurai	-	-	++	-	-	-	>1	e.g., Kawahara et al. (2006); Ikeda et al. (2011); Kawahara et al. (2013); Dong et al. (2014); Feng et al. (2015a, b)
Stomolophidae								
Stomolophus meleagris	-	-	++	-	-	-	>1	Calder (1982)
Stomolophus sp.2	-	-	++	-	-	-	>1	López-Martínez et al. (2022)

(continued)

Table 4 List of the asexual reproduction modes observed for the Rhizostomeae polyps. *LB*, typical lateral budding; *LBst*, lateral budding by means of stolons; *POD*, podocysts; *LF*, longitudinal fission; *FSB*, typical free-swimming buds; *MP*, motile bud-like tissue particles; *#E*, number of ephyrae produced per strobilation event per strobila. (-): not observed; (+): observed sporadically; (++): main asexual mode to produce new polyps. See Adler and Jarms (2009) for the definition of polyp-to-polyp reproduction modes. Some information about the asexual reproductive modes was derived from the authors' personal observations in their laboratories and referred to as the present work. (*cont'd*)

Species	LB	LBst	POD	LF	FSB	MP	#E	Source
Kolpophorae								
Cepheidae								
Cephea cephea	+	-	-	-	++	-	1	Sugiura (1966); Schiariti et al. (2014); Straehler-Pohl and Jarms (2022)
Cotylorhiza tuberculata	+	-	-	-	++	-	1	Claus (1883, 1884); Kikinger (1992); Straehler-Pohl and Jarms (2022)
Netrostoma setouchianum	-	-	-	-	++	-	1	Hashimoto et al. (2016); Straehler-Pohl and Jarms (2022)
Cassiopeidae								
Cassiopea andromeda	-	-	-	+	++	-	1 (rarely 2)	e.g., Bigelow (1900); Gohar and Eisawy (1960); Hofmann et al. (1978); Straehler-Pohl and Jarms (2022)
Cassiopea frondosa	-	-	-	-	++	-	1	Fitt et al. (2021)
Cassiopea xamachana	-	-	-	-	++	-	1	Bigelow (1900); Fitt et al. (2021)

Leptobrachidae								
Thysanostoma thysanura	–	–	–	–	+	–	1	Yamamori et al. (2017); present study
Mastigiidae								
Mastigias papua	–	–	–	–	+ +	–	1	e.g., Sugiura (1963, 1964, 1965); Schiariti et al. (2014); Straehler-Pohl and Jarms (2022)
Phyllorhiza punctata	–	–	–	–	+ +	–	1	Rippingale and Kelly (1995); Lange and Kaiser (1995); Schiariti et al. (2014); Rato et al. (2021); Straehler-Pohl and Jarms (2022); present study

reproduction modes through which they produce new polyps; ten different modes were stated by Adler and Jarms (2009), six of which have been observed in the rhizostomes:

1. *Typical lateral budding* (LB) (Fig. 2A): Buds growing at the junction point of calyx and stalk; the new polyp develops calyx, tentacles, and hypostome before detachment from the parental polyp representing a kind of "double-head polyp stage". The new polyp develops a stolon that attaches near the parental polyp where the new polyp remains settled.
2. *Lateral budding by means of stolons* (LBst) (Fig. 2B): Buds growing on stolons formed by the parental polyp. They are either situated at the tip of the stolon (near the attachment point) or midway along the stolon. The bud completes its development before detachment from the parental polyp and remains nearby.
3. *Podocysts* (POD) (Fig. 2E–F): Formation of cysts covered by a chitinous protein sheath occurs beneath the basal region of the polyp, or at the attachment area of stolons (or both). Regardless of its origin, POD formation always involved the development of stolons. Even if POD are formed beneath the polyp stalk, the polyp forms a stolon to leave the cyst and to form a new stalk close to the cyst.
4. *Longitudinal fission* (LF): Fission or "rupture" of the polyp between two attachment areas; the polyps just split into two halves and lacking body parts regenerate in each halve during (or after) the fission, being two polyps at the end of the process.
5. *Typical free-swimming buds* (FSB) (Fig. 2G–I): formation of the so-called "true free-swimming buds" occurs from the body wall of the parental polyp; the buds are formed at the lower part of the calyx or the junction point of the calyx and stalk. In contrast to the LB, detachment of these buds from the parental polyp takes place before further development of the new polyp. A FSB rotates around its own axis in a very similar way to what has been described for planulae. These buds are highly motile and consistent in appearance and size. Sometimes also named planuloids.
6. *Motile bud-like tissue particles* (MP) (Fig. 2J–K): Bud-like tissue particles are formed by a constriction from the junction point of the calyx to the stalk. Detachment of these particles also occurs before the development of the new polyp. They have no discernible body axis and present a rotating movement but are not directed in any way; their motility is much lower than that of the FSB. Their appearance and size are highly variable.

Fig. 2 Types of asexual reproduction modes found among Rhizostomeae species. (A) typical lateral budding = LB on *Cassiopea frondosa* (Florida, USA); (B) lateral budding by means of stolons = LBst on *Aurelia coerulea* (Roskoff, France); (C) stolon = ST on *Catostylus townsendi* (Thailand); (D) stolon = ST on *Rhizostoma luteum* (La Herradura, Spain); (E) podocysts = POD being formed at the stolon tip of a *Catostylus tagi* polyp (Lisbon, Portugal); (F) chain of podocysts = POD formed around *Lychnorhiza lucerna* polyp (Argentina); (G) typical free-swimming bud = FSB formed at the junction of calyx and stalk of a *Cassiopea* sp. polyp (Singapore); (H) typical free-swimming bud = FSB formed at the junction of calyx and stalk of a *Cephea cephea* polyp (Japan), note that the polyp is starting to strobilate; (I) chain of FSBs formed at the junction of calyx and stalk of a *Cassiopea andromeda* (Cabo Frio Brazil), note that there are more than one FSB; (J) motile bud-like tissue particle = MP formed at the junction of calyx and stalk of a *Sanderia malayensis* polyp (Japan); (K) motile bud-like tissue particles = MPs released from an *Aurelia coerulea* polyp (Roskoff, France). For a definition of asexual reproduction modes see text and Adler and Jarms (2009). Note that: images on B, J, K are examples of asexual reproduction modes from Semaeostomeae species. Scales: *A*, *B* = 2 mm; *C*, *D*, *I* = 1.5 mm; *E*, *K* = 0.1 mm; *F* = 0.2 mm; *G* = 3 mm; *H*, *J* = 1 mm. Photos: A. Schiariti.

Four more asexual reproduction modes were described by Adler and Jarms (2009) including reproduction from parts of stolons/stalks, planuloids originating from tentacle tips/tentacle pieces, internally produced propagules, and *Sanderia*-type budding (see Adler and Jarms, 2009 for details), but none of these have been reported for any rhizostome species so far. A previously uncharacterised reproduction mode was described for *Stomolophus* sp.2 by López-Martínez et al. (2022). These authors reported that this species "*produce stolons that produce free-swimming propagules that become new scyphistomae*". The mention of free-swimming propagules might refer to the production of MP, reported also for *R. pulmo* (Fuentes et al., 2011). However, the production of this kind of free-swimming particles does not involve the formation of stolons but a budding process that starts at the junction of the calyx and the stalk (see Adler and Jarms, 2009; Schiariti et al., 2014, 2015). On the other hand, the development of stolons is associated with different reproduction modes (POD, ST, LBst), or can also be formed for locomotion, but never have been described as part of the development of free-swimming particles. Unfortunately, the authors did not provide pictures, drawings, or a more detailed description to allow proper identification of the observed mode and assess if this species displayed a previously undescribed mode for polyp production or a mode not previously described for this species.

4.1.1 Asexual reproductive strategies in the Rhizostomeae polyps

The reproductive strategy has been defined as the number and type of asexual reproduction modes displayed by species and the relative importance of each of these modes to the polyp density (Schiariti et al., 2014). It has been observed that the semaeostomes *Aurelia* spp. and *Sanderia malayensis* Goette, 1886 produce new polyps through different asexual modes, even simultaneously (multi-mode strategy, Schiariti et al., 2014). Furthermore, it has been observed that *Aurelia solida* Browne, 1905 polyps (reported as *A. aurita* s.l.) respond to environmental clues like the room for expansion by switching their reproductive strategy (Schiariti et al., 2015). This capacity of producing new polyps through different modes, adapting the strategy in response to environmental conditions allows *Aurelia* to colonise the available substrate rapidly when conditions are favourable (by LB, LBst and ST), to encyst and withstand starvation and other adverse conditions (e.g., the presence of predators, see Lucas et al., 2012) (by POD formation) and to develop MP or to detach to facilitate dispersion and avoid the disadvantages of extremely dense areas. Several rhizostomes can also reproduce asexually using multiple modes (Table 4). However, they produce new polyps through a main asexual mode (mono-mode strategy, see Schiariti et al., 2014) that differs among taxa. Species belonging to Dactyliophorae produce new polyps by POD with the only exception of *A. hardenbergi* and *A. maculosus* which only reproduce through LB (Table 4). Moreover, these rhizostomes seem to develop such cysts as the only way of producing new polyps, or at least as the main one (Table 4). In contrast, no PODs were observed in the Kolpophorae families Cepheidae, Cassiopeidae and Mastigiidae that carry out asexual reproduction of the polyps by production of FSB with no exceptions described so far.

4.1.2 Environmental control of the production of new polyps

There are numerous reports of POD and FSB production among rhizostome polyps (Adler and Jarms, 2009; Jarms and Morandini, 2019 and references therein). Quantitative information on the variables affecting the rates and duration of their development and survival, however, is still limited.

The development of POD (i.e., encystment) has been described for several species utilising different terms to refer to the same structures (see Cargo and Rabenold, 1980; Magnusen, 1980; Arai, 1997; Pitt, 2000). To avoid confusion, we followed the definition proposed by Blanquet (1972), where POD are defined as small bits of tissue containing epidermal and mesenchymal cells separated from the pedal disc and covered by a thin chitinous cuticle. The POD are usually formed either beneath the pedal disc of the polyp (*L. lucerna*, *R. nomadica*), at the tip of extended stolons (*Catostylus mosaicus*), or both (*R. verrilli*) (Calder, 1973; Lotan et al., 1992; Pitt, 2000; Schiariti et al., 2008).

POD can survive extreme conditions, protecting against unfavourable physicochemical conditions (Blanquet, 1972), starvation (*N. nomurai*, Kawahara et al., 2013) and predation (Cargo and Schultz, 1966; Hernroth and Gröndahl, 1985). Therefore, it is likely that the ability to encyst has evolved to withstand conditions in which polyps would not otherwise survive (Arai, 2009). However, more recent papers showed that POD may indeed protect against adverse conditions, but the rate of their production is usually positively correlated with the availability of food and temperatures within optimal ranges that vary according to the species and populations. For example, encystment rates increased with food supply and warmer conditions in *L. lucerna*, *R. pulmo* and *Rhopilema esculentum* (Schiariti et al., 2008, 2014) and reached higher rates at warmer temperatures also in

N. nomurai and *Stomolophus* sp.2 (Feng et al. 2015a, b; Jiang et al., 1993; Kawahara et al., 2013; Lu et al., 1997; Purcell et al., 2012; Schiariti et al., 2008, 2014). Therefore, the analysis of the literature suggests that favourable environmental conditions represented by increased food availability and temperatures within optimal ranges, enhance encystment rates in the species of the suborder Dactyliophorae.

Other factors like salinity, light, and bacterial fouling have been also reported to induce POD production but only a few references include rhizostome species. The encystment rate of *N. nomurai* is likely facilitated by hyposaline seawater and potentially enhanced around the estuaries that comprise the major nurseries of medusae (Feng et al., 2018). These findings confirm that estuarine areas with salinities from 11 to 20 are appropriate for *N. nomurai* polyp colonisation. An asexual reproduction strategy also corresponded with autumn salinity fluctuations. The intensity of diluted water and monsoonal rainfall, as well as plankton supply around the estuaries in autumn, may fundamentally affect polyp abundance and size, determining the population size of medusae in the following spring (Feng et al., 2018). Light conditions also seem to affect encystment in *R. esculentum*, with more POD produced in darker conditions (Dong et al. 2014; Jiang et al., 1993). These results indicated that *N. nomurai* polyps are adapted to survive, grow, and reproduce in estuaries and near-shore environments (Dong et al., 2014). On the other hand, bacterial fouling has been mentioned as a condition triggering encystment in semaeostomes (Blanquet, 1972; Cargo and Schultz, 1966). Schiariti et al. (2008) attributed the high POD production rate observed for *L. lucerna* to a combination of conspicuous bacterial fouling in culture dishes combined with the favourable experimental temperature, salinity and feeding conditions. However, observations from our laboratory cultures on several POD-producers [Rhizostomeae: *L. lucerna, Rhopilema asamushi* Uchida, 1927, *R. pulmo, R. octopus, Stomolophus* sp.2, *R. esculentum*; Semaeostomeae: *Chrysaora lactea* Eschscholtz, 1829, *Chrysaora achlyos* Martin, Gershwin, Burnett, Cargo and Bloom, 1997, *Chrysaora colorata* (Russell, 1964), *Chrysaora fuscescens* Brandt (1835), *Chrysaora pacifica* (Goette, 1886), *C. plocamia* (Lesson, 1830), *Chrysaora quinquecirrha* (Desor, 1848), *Chrysaora chesapeakei* (Papenfuss, 1936)] indicate that polyps detached from the substrate cannot produce PODs. As described above, the production of PODs requires the formation and attachment of a stolon. Hence, special substrate properties, and probably the presence of bacterial fouling, influence the attachment rate of stolons (Schmahl, 1985). Therefore, differences in the substrate conditions (material and presence of bacterial fouling) may affect encystment rates through their influence on stolon attachment.

The emergence of new polyps from the cysts (i.e., excystment) can occur after an indefinite dormant period that could last up to 6 years (Kawahara et al., 2013). This great longevity has been attributed to the extremely low metabolism of the dormant cells (Ikeda et al., 2011), in addition to their physical robustness against bacterial attack (even in organic-rich silt) and predators (Arai, 2009 and references therein). The environmental control over excystment (induction and rates) has been even less studied than encystment. A few studies dealing with rhizostomes demonstrated that POD of *N. nomurai* can remain dormant for 1–6 years without significant changes in internal structure or nutrient reserves (Ikeda et al., 2011; Kawahara et al., 2013) and podocyst dormancy up to ten years have been reported (Thibault et al., 2024). Different processes occur during excystment including nematocyst formation in the internal cell mass, layering of the cell mass into endoderm and ectoderm, extrusion of the cell mass through a gradual opening of the cyst, and formation of a 4-tentacled tiny polyp (Ikeda et al., 2011).

Histological observations revealed that newly-formed POD contained rich organic reserves (e.g., carbohydrate, protein, and lipid) that were gradually consumed while encysted. Excystment has been stimulated by artificial removal of their covering (Kawahara et al., 2013). Spontaneous excystment was also observed in the laboratory in response to stressful conditions like exposure to abnormally high temperatures, low salinities, hypoxia, and burial in organic-rich mud (Ikeda et al., 2011; Kawahara et al., 2013). Low excystment rates were also reported for *L. lucerna* but with no significant correlation with the temperatures tested (Schiariti et al., 2014). However, the later experiment considered temperature ranges assumed to be favourable for *L. lucerna* polyps and abnormal and stressful values for this species need to be tested to corroborate the hypothesis. The warmer temperature and lower salinity facilitated podocyst excystment in *R. esculentum* (Sun et al., 2022).

Although the contribution of POD to the polyp density and further medusa populations′ size has been little studied (but Blanquet, 1972; Thein et al., 2012), their production and excystment may play important roles (Ikeda et al., 2011; Kawahara et al., 2013). From the literature examination and laboratory observations, a temporal gap exists between encystment and excystment, generating a kind of disconnection between the formation of POD and the increase in polyp density. Our laboratory observations on several POD-producers (Rhizostomeae and Semaeostomeae) demonstrate that the encystment rate always exceeds the excystment rate and that PODs accumulated in the experimental dishes with limited formation of new polyps. Preliminary quantitative laboratory estimations indicated that the percentage of excystment ranged between 0% and 15% of the total POD formed by *L. lucerna* over 3 years under stable conditions of temperature (20 °C–22 °C), salinity (20–23), darkness, and feeding with rotifers (*Brachionus* sp.) once weekly (Schiariti unpubl. data). This temporal gap between POD production and polyp density has been utilised by Kawahara et al. (2013) to propose an explanation for the intermittent nature of *N. nomurai* medusae blooms in the East Asian Marginal Seas. These authors proposed that in non-bloom years, a few POD may excyst, whereas in bloom years, a large number of POD may excyst (see also (Thibault et al., 2024)). This hypothesis could explain at least partially the strong inter-annual fluctuation commonly observed for Dactyliophorae species worldwide. The different set of environmental stimuli that seems to be required for encystment and excystment would support this hypothesis.

As stated before, the asexual production of new polyps in the Kolpophorae occurs through a special type of buds named FSB. FSBs have been also described as planuloids because of their similarities with the planula larvae. Formation of these swimming buds occurs from the body wall of the polyp, at a point situated at the lower part of the calyx, right at the junction of the calyx and the stalk (Adler and Jarms, 2009). Upon detachment, the bud is propelled by the cilia of the ectodermal cells and rotates slowly around its longitudinal axis with the knob-like, originally distal, pole directed forward. Detailed descriptions of morphology and ultrastructure can be consulted in Hofmann and Honegger (1990). FSB are produced at one or more perradial sites of the calyx close to the stalk (Fig. 2). Some major stages have been defined by Neumann et al. (1980); first, ectoderm, mesoglea, and endoderm of the parental polyp bulge to form a roundish protrusion which quickly expands; then the developing bud starts constricting at its base; and finally, it elongates and assumes its definitive spindle-shape. The connection between bud and polyp becomes narrow and is reduced to a thin ectodermal tubular strand which is eventually obliterated. Upon release, the bud, which is devoid of tentacles and an oral opening, is propelled by the cilia of the ectoderm and moves with the previously distal, knob-shaped end in front rotating around its longitudinal axis (Hofmann and Gottlieb, 1991). Under optimal feeding and temperature conditions, new buds frequently emerge before the previous ones are released forming chains of buds (Fig. 2I). Moreover, polyps could simultaneously produce chains of buds at 2 to 4 budding regions.

Budding rate is affected by several rearing conditions like the availability of food, the number of individuals kept per culture dish, the size and age of the polyps and the temperature (Hofmann et al., 1978). According to most of the studies, budding rates increased with food supply and raised temperatures within specific ranges (Hofmann et al., 1978; Hofmann and Hadfield, 2002; Schiariti et al., 2014). Underlying this pattern, species respond differently to experimental settings; for example, no budding was observed under starvation for *C. andromeda* (identified as *Cassiopea* sp.) and *C. cephea* (Schiariti et al., 2014). Conversely, *M. papua* and *Phyllorhiza punctata* produced buds after some days without food, although at lower rates compared to fed polyps (Schiariti et al., 2014). Some species require temperatures above specific thresholds to start budding. For example, in *M. papua*, budding was observed only at 20 and 25°C but not at 15°C regardless of food availability. In contrast, upon reaching a critical temperature of 20°C, this species started budding at rates that were elevated by food availability, but not affected by higher temperatures (Schiariti et al., 2014). In the same experiments, *P. punctata* showed the lowest budding rates but produced more buds than the other species at 15 °C and when food was available (Schiariti et al., 2014). Salinity was also observed to affect the budding of *P. punctata* with higher rates upon lower salinities but always within specific ranges (below 10, 100% mortality was reported) (Rato et al., 2021; Rippingale and Kelly, 1995). Low salinity regimes caused by rainwater might likely act as an environmental threshold for polyp reproduction (Rato et al., 2021; Rippingale and Kelly, 1995).

The time required to develop and release a bud is influenced by various endogenous and exogenous factors, as is the case with all reproductive processes. FSB-producers are usually symbiotic species associated with zooxanthellae that are crucial for metabolic processes. However, according to Hofmann et al. (1978), FSB production appears to be unrelated to such associations. The periodic production of FSB observed in several species kept under dark conditions for more than 15 years supports this hypothesis (Schiariti and Morandini *unpubl. data.*).

FSB and POD are both reproduction modes used by polyps to propagate but remarkable differences exist between them in terms of their ecological and developmental implications. In addition to the lack of chitinous cover in FSB, that allows it to withstand theoretically stressful conditions, the planula-like buds detach from the parental polyp before further development. Therefore, in contrast to the sessile POD, they drift away minimising the density-dependent effects of increased local densities

(Schiariti et al., 2015). This not only diminishes the negative effects of intraspecific competition for space and food but also allows for the potential colonisation of new habitats, and a further increase in polyp density away from the original source (Schiariti et al., 2015). Despite the differences, in both reproductive strategies, the reproduction rates are increased by "favourable" environmental conditions. Like the species producing POD, the FSB-producers also have a mono-mode strategy and have no other alternatives for propagating polyps. Besides, these species seem to be mostly tropical and more sensitive to cold temperatures. Therefore, polyps that are not capable of encystment occur throughout the mid-to-late summer but are completely absent during winter and early spring (*C. xamachana*; Fitt and Costley, 1998). In these cases, not only polyp density but also the continuity of the species strongly depends on the success of sexual reproduction and its pelagic stages (Schiariti et al., 2014).

Once the buds are released, the production of new polyps depends on the settlement rate and further development of the buds. The metamorphic development of a polyp from the bud of *C. andromeda* has been described by Hofmann et al. (1992); the first step is the settlement and attachment of the bud to a substratum, mediated by gland cells at the knoblike anterior pole; later on, a pedal disk is formed and irreversibly attaches to the settlement site; once attached, individuals elongate and segregate a pedal disk, a stalk area and a broader calyx anlage which then bears the rudiments of tentacles and a hypostome. When the basic body plan of the polyp is established, a perisarc covers the foot region, a slender contractile stalk, and a calyx rimmed by the full complement of 32 filiform tentacles, which encircle the tetragonal expandable hypostome (Hofmann et al., 1992).

The settlement success of FSB seems to be like the settlement of planulae. Larvae and buds sporadically explore their habitat with this aboral knob, searching for a suitable solid substrate to which attachment will be made (see also Holst et al., 2024a for histology of swimming buds). The "touch-and-go" swimming behaviour of buds observed in the laboratory leads to frequent contact with the wall and the bottom of the dish.

Settlement of FSB and the irreversible morphogenetic transition to a primary polyp do not occur autonomously but strictly require exogenous inducers (Fitt et al., 1987; Wolk et al., 1985). Laboratory experiments demonstrated that such inducers could be released by species of marine bacteria (Hofmann et al., 1978; Hofmann et al., 1992). It has been shown that growing *Vibrio alginolyticus* bacteria produce peptides which induce bud metamorphosis (Hofmann et al., 1978). Settlement of *C. andromeda* buds can be observed in less than 24 h by exposing the animals to oligopeptides that are likely produced by marine bacteria (e.g., Fitt and Hofmann, 1985; Fleck and Hofmann, 1990; Fleck and Bischoff, 1993; Hofmann and Brand, 1987; Hofmann et al., 1996). Marine bacteria could influence larvae and buds in three fundamental ways: (1) by establishing microbial films with particular surface properties to which the buds could be exposed upon contact, (2) by secreting biologically active, metabolic products into the surrounding medium and (3) by modifying or degrading of organic substrates through exogenous enzymes, thereby leading to the release of soluble compounds (Hofmann et al., 1996). In the presence of metamorphosis-inducing substances secreted by bacteria, buds start secreting mucus from their presumptive basal portion. They thereby "glue to all sorts of materials" (e.g., glass, plastics, cellulose fibres, other organic matter, etc.) and then complete metamorphosis (Hofmann and Brand, 1987). Some buds can start metamorphosing prior to attachment to a substratum. These observations show that in solutions of inducing compounds, even in the absence of specific environmental cues, buds can attach to a variety of substrate types and transform into polyps (Hofmann and Brand, 1987). On the other hand, these processes are suppressed in sterilised seawater and sterilised seawater containing antibiotics proving that the bacteria are involved in releasing the metamorphosis and attachment trigger substances (Hofmann et al., 1978). Since a number of these inducers might be present in the sea, it is assumed that these phenomena could also occur in the natural environment (Hofmann et al., 1996).

5. From benthos/asexual to plankton/sexual

5.1 Polyp-to-medusa = strobilation

The term strobilation describes the entire process involved in the transition of a polyp into one or more ephyrae (Percival, 1923; Lambert, 1935; Hyman, 1940) with no exceptions among the rhizostomes described so far. Although there is a common agreement about the former concept, different definitions of strobilation are found in the literature; the simplest criterion considered strobilation as a process of transverse fission (Fautin, 2002; Mills and Larson, 2007; Adler and Jarms, 2009; Holst, 2012b) or "a special type of transverse division" (Holst, 2012a). Different authors noted that strobilation is not restricted to the Scyphozoa but also occurs in other cnidarians like anthozoans and hydrozoans, or even in other groups of animals (flatworms, annelids, tunicates)

(Bavestrello et al., 2000; Karande and Nakauchi, 1981; Hand and Uhlinger, 1995; Straehler-Pohl and Jarms, 2005). A more complete definition considers strobilation as a process of transverse segmentation (=transverse fission), formation, and release of ephyrae (Straehler-Pohl and Jarms, 2005; Sötje and Jarms, 1999). The "formation of ephyrae" includes metamorphosis in which structures of the polyp are lost and replaced in each disc/segment with those of the developing ephyrae (da Silveira and Morandini, 1998; Spangenberg, 1965; Thiel, 1938). As Spangenberg (1965) wrote: "*this is because the use of the term as a fission process alone is inadequate to describe the production of new form which occurs in strobilation. If strobilation in jellyfish were merely a fission process, it would result in the production of many new polyps. Since organisms of new form are produced, it is evident that strobilation involves an actual metamorphosis*". Therefore, we followed Spangenberg (1965), using the term strobilation in this manner, emphasising that two separate processes are involved: segmentation (i.e., transversal fission = formation of the segments/discs) and metamorphosis.

Beyond specific definitions, the entire strobilation process has been described in detail by different authors (see Berrill, 1949; Custance, 1967; Helm, 2018; Spangenberg, 1968; Straehler-Pohl and Jarms, 2022) and can be summarised into four main phases: body elongation, segmentation, differentiation of the primordia and ephyra detachment. Recent and very detailed descriptions of phases can be consulted in Helm (2018) and Straehler-Pohl and Jarms (2022). The most unambiguous sign of strobilation initiation is the formation of a strobilation furrow directly under the polyp tentacle crown; this is the beginning of the segmentation phase which can include the appearance of one or more constrictions leading to the formation of one or more segments (Fig. 3). This furrow separates the topmost part of the polyp calyx and oral disc (which are destined to become the topmost, or 'terminal' ephyra) from the main body column (Helm, 2018). Each of the segments then metamorphoses by forming structures that are absent in the polyp stage, successively developing into the ephyra that finally starts pulsing while still being attached to the strobila. During strobilation, only a part of the polyp is transformed to released ephyrae while the remaining part (residuum) remains attached to the substrate, usually regenerating a complete polyp at the end of the strobilation process. However, this regeneration process seems to be associated with high mortality that likely varies between species, the size of the polyp at the beginning of strobilation, and the environmental conditions (Gröndahl, 1988; Hernroth and Gröndahl, 1983; Watanabe and Ishii, 2001; Wilcox et al., 2007, 2008). If the residuum does not die and the polyp successfully regenerates all its structures (tentacles and mouth opening), strobilation can be repeated several times if the environmental conditions are favourable and the nutritional condition of the polyp is sufficient.

There are two distinct types of strobilation in the Rhizostomeae: polydisc (more than one ephyra is regularly produced per strobilation) and monodisc (a single ephyra regularly produced per strobilation) (Fig. 3). The number of segments/ephyrae formed per strobila seems to be evolutionarily conserved and differs distinctly in different taxa of Rhizostomeae but it is also regulated by environmental conditions. Strobilation is always monodisc in the studied Kolpophorae (Table 4),

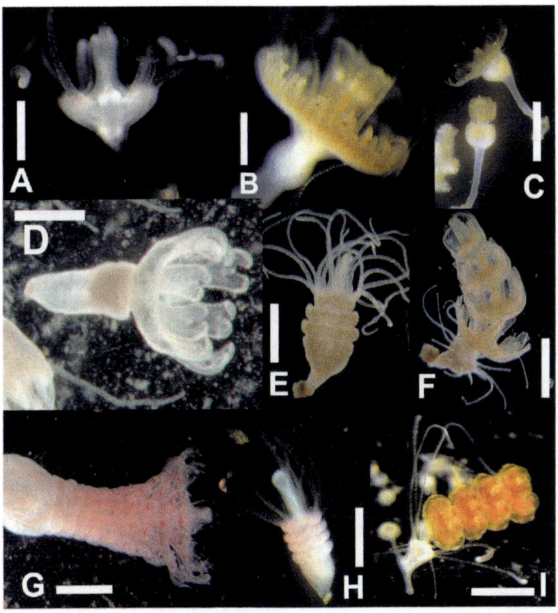

Fig. 3 Strobilation among Rhizostomeae species. (A) early strobilation stage of *Lychnorhiza lucerna* (Argentina); (B) strobilation of *Cassiopea andromeda* (Cabo Frio, Brazil), note the advanced stage of ephyra development with rhopalia almost completely formed; (C) monodisc strobilation of *Cephea cephea* (Japan); (D) monodisc strobilation of *Cotylorhiza tuberculata* (Spain); (E) early polydisk strobilation stage of *Catostylus mosaicus* (Australia); (F) advanced polydisk strobilation stage of *Catostylus mosaicus* (Australia); (G) advanced polydisk strobilation stage of *Catostylus tagi* (distal portion of the polyp) (Lisbon, Portugal); (H) early polydisk strobilation stage of *Stomolophus* sp.2 (La Paz, Mexico); (I) advanced polydisk strobilation stage of *Stomolophus* sp.2. Scales: *A, B, H, I* = 2 mm; *C* = 3 mm; *D* = 1 mm; *E, F* = 1.5 mm; *G* = 4 mm. Photos: A.C. Morandini.

with an exceptional case in *C. andromeda*, where two ephyrae were observed per strobila on only one occasion (Gohar and Eisawy, 1960). In all the Dactyliophorae, however, strobila can develop more than one ephyra per strobilation with the only exception of *Acromitus flagellatus* and *A. maculosus*, forming one large ephyra by monodisc strobilation (Table 4) without a regenerating polyp residuum, resembling the direct development described for *A. coerulea* (see Takauchi et al., 2021; Yasuda, 1975). Since *Acromitus* spp. Inhabit low-salinity and sediment-rich areas such as mangroves and estuaries with drastic salinity changes, the regular monodisc strobilation of the genus *Acromitus* would be an adaptation to "extreme" environmental conditions (Takauchi et al., 2021). Repeated observations of a single ephyra in the strobila have been observed for many Dactyliophorae species and were referred to as monodisc strobilation (e.g. Calder, 1973; Cargo, 1971; Kienberger et al., 2018; Pitt, 2000). However, these observations are special cases probably caused by certain environmental factors or polyp conditions in species which are capable of polydisc strobilation. Conversely, the Kolpophorae always form a single ephyra per strobilation regardless of the environmental conditions. A third type of strobilation, where an intermediate number of ephyrae is produced per strobilation (i.e., the oligodisc), has been proposed by Straehler-Pohl and Jarms (2022) and observed among the Dactyliophorae. These oligodisc species produce more than one ephyra but rarely less than 20, in contrast to the so-called polydisc taxa (Semaeostomeae and Coronatae) where a higher number of ephyrae are regularly produced (see Table S2b in Straehler-Pohl and Jarms, 2022) when the nutritional condition of polyps is within optimal ranges. Following this definition, all Rhizostomeae could be classified into the oligodisc or monodisc strobilation types.

5.2 Environmental control of strobilation

External environmental conditions often play a key role in triggering strobilation and controlling the number of ephyrae produced within the strobilation window. About 360 studies have been published about strobilation, 93 of which provide information about 17 rhizostome species (Table 5). This means that our knowledge about the environmental control on the number of ephyrae produced by rhizostome species comes from only 18% of the species described so far.

Although strobilation in jellyfish is controlled by endogenous factors, the initiation and rate of ephyrae production are also influenced by

Table 5 Compiled factors affecting strobilation induction (triggering factors), number of ephyrae produced per strobila over time (strobilation rates), and inhibiting strobilation.

Species	Triggering factor	Strobilation rates*	Inhibiting factor	Source
Catostylus mosaicus	Not described	Not described	Inadequate nutritional state	Pitt (2000)
Thysanostoma thysanura	Indole compounds	Not described	Not described	Yamamori et al. (2017)
Rhopilema verrilli	Raising temperature; salinity increase	Not described	Not described	Calder (1973); Cargo (1975)
Rhopilema nomadica	Not described	Enhanced by higher nutritional state	Not described	Lotan et al. (1992)
Rhopilema esculentum	Raising temperature; low preconditioning temperature + raising temperature; indole compounds	Diminished with the number of repeated strobilations, increased with polyp size and nutritional state	Darkness for "prolonged" periods, inadequate nutritional state	e.g., Chen and Ding (1983, 1984); Chen et al. (1984, 1985); Wu et al. (2009); You et al. (2010); Yamamori et al. (2017)
Rhizostoma octopus	Raising temperature; temperature decrease	Not described	Not described	Holst et al. (2007); Holst (2012b)

(continued)

Table 5 Compiled factors affecting strobilation induction (triggering factors), number of ephyrae produced per strobila over time (strobilation rates), and inhibiting strobilation. (*cont'd*)

Species	Triggering factor	Strobilation rates*	Inhibiting factor	Source
Rhizostoma pulmo	Addition of potassium iodide	Not described	Not described	Paspaleff (1938)
Nemopilema nomurai	Raising temperature	Enhanced by higher temperatures	Temperature above a critical threshold	Kawahara et al. (2006); Feng et al. (2015a,b); Lee et al. (2017)
Stomolophus sp.2	Raising temperature	Increased by higher nutritional state	Temperature below a critical threshold; temperature decrease	Schiariti (unpubl. data)
Lychnorhiza lucerna	Rising temperature, low preconditioning temperature + raising temperature	Decrease with the "age" of polyp cultures, increased by a higher nutritional state	Temperature decrease, poor nutritional state	Schiariti et al. (2008, 2014); Pereyra (2015); Schiariti (unpubl. data)
Cassiopea andromeda	Rising temperature	Enhanced by zooxanthellae	Temperature below a critical value	Hoffmann et al. (1978); Rahat and Adar (1980); Hofmann and Kremer (1981)
Cassiopea xamachana	Indomethacin	Enhanced by higher temperatures	Lack of endosymbiotic algae (*Symbiodinium*)	Colley and Trench (1985); Sicarello et al. (1979); Fitt and Costley (1998); Cabrales-Arellano et al. (2017)

Mastigias papua	Indole compounds	Not described	Lack of endosymbiotic algae (*Symbiodinium*); temperature below a critical value	Sugiura (1962, 1964, 1965); Yamamori et al. (2017); Helm and Dunn (2017)
Mastigias sp.	Not described	Not described	Temperature above a critical value	Dawson et al. (2001)
Phyllorhiza punctata	Rising temperature; increase in length of daylight; indole compounds	Not described	Lack of endosymbiotic algae (*Symbiodinium*)	Rippingale and Kelly (1995); Jufri (2001); Helm and Dunn (2017)
Netrostoma setouchianum	Indole compounds	Not described	Not described	Hashimoto et al. (2016)
Cephea cephea	Raising temperature; indole compounds	Enhanced by zooxanthellae	Temperature below a critical value	Sugiura (1962, 1966, 1969); Helm and Dunn (2017); Schiariti (unpubl. data)
Cotylorhiza tuberculata	Raising temperature; indole compounds	Not described	Not described	Kikinger (1986, 1992); Yamamori et al. (2017); Prieto et al. (2010)

chemical and physical environmental factors. In all the species studied to date, temperature plays a critical role in regulating the number of ephyrae produced. Temperature effects on strobilation can be seen in both variations (increases and decreases) and absolute values, and the way that these factors affect strobilation differs between species; raising the temperature acts to trigger strobilation in *C. andromeda*, *P. punctata*, *C. cephea*, *C. tuberculata*, *R. verrilli*, *R. octopus*, *R. esculentum*, *N. nomurai*, *Stomolophus* sp.2 and *L. lucerna* (Table 5). Decreasing the temperature triggered strobilation only in *R. octopus* (Table 5). A combination of a low preconditioning temperature followed by a further increase induced the formation of ephyrae in *L. lucerna* and *R. esculentum* (Table 5). The magnitude of these temperature variations (range and time from initial to final temperature) may vary with species and populations (see references in Table 5).

Despite the critical role of temperature changes in initiating strobilation, the maintenance of polyps at a constant temperature (Calder, 1982; López-Martínez et al., 2022; Schiariti unpubl. data.) shows that strobilation is regulated by environmental factors other than temperature changes. Among the rhizostomes, the addition of different chemical compounds (indomethacin, indoles, potassium iodide), salinity increase, and the duration of daylight have been also reported as triggering strobilation in some species (see Table 5) but the limited detail regarding the experimental conditions provided in these studies prevent us from identifying specific patterns. The biochemical processes involved in strobilation are not well understood and the limited information available is mostly restricted to *Aurelia* species and other semaeostomes like *Chrysaora* spp. Among rhizostomes, the addition of indole compounds, indomethacin and potassium iodide triggers the formation of ephyrae (Table 5). Although all these observations were conducted in the laboratory, iodine is a trace element in seawater and, therefore, it is assumed that mechanisms involving these substances also regulate the strobilation in the wild.

In several zooxanthellate species (commonly found in Kolpophorae, see Holst et al., 2024a; Nagata et al., 2024), strobilation can only be initiated once the polyp has acquired *Symbiodinium* spp. cells (Fitt and Costley, 1998; Sugiura, 1965), although this is not a universal requirement in all jellyfish species containing zooxanthellae (Hofmann and Kremer, 1981; Sugiura, 1969). In *C. andromeda*, symbionts favour the production of ephyrae and enhance strobilation rates but are not essential to start the process (Hofmann and Kremer, 1981; Rahat and Adar, 1980). The nutritional state of the polyps (commonly studied through different levels of food availability) acts similarly for all species studied so far. It seems that a "relatively long" nutritive preparation is a necessary condition for strobilation (Berrill, 1949). It has been reported that starved polyps of *C. mosaicus* and *L. lucerna* did not strobilate (Pitt, 2000; Schiariti et al., 2008), and that scyphistomae of *R. esculentum* only strobilated when "well fed" (Chen et al., 1985). Therefore, although the nutritional state of polyps would not trigger the strobilation, it may need to exceed a critical value for ephyrae to be produced. Once the strobilation has been triggered, the better nutritional state, the more ephyrae produced.

Analysis of the literature suggests that the onset of strobilation and the number of ephyrae produced during a certain period are controlled by different environmental clues. Once strobilation is triggered, the number of ephyrae produced per strobilation within a specific period (i.e., strobilation rates) is determined by the absolute values of temperature, the presence of zooxanthellae, and the food availability as a proxy of the nutritional state of polyps. Almost universally, the time it takes for polyps to develop strobila decreases with increasing temperature (Lotan et al., 1994; Liu et al., 2009; Purcell et al., 1999) and the same occurs for the polyp to regenerate the lost structures and become ready to repeat strobilation. On the other hand, the number of ephyrae produced per strobilation, and the number of repeated strobilations, are reduced by lower temperature values, lower food availability, and by repeated strobilations. Laboratory observations indicate that typical polydisc species initiate strobilation in response to the mentioned inducers but produce only one ephyra when they are poorly fed or after several years of being kept in artificial cultures (Schiariti *pers. obs.*). Instead, fewer ephyrae are produced by smaller polyps than by larger polyps of the same species (Holst, 2012b; Kakinuma, 1975; Kroiher et al., 2000; Willcox et al., 2007) and there is a general trend, both in the field and in the lab, for more ephyrae to be produced per polyp when more food is available because polyps reach larger sizes. In the Güllmar fjord, Hernroth and Gröndahl (1983, 1985) reported that strobilating polyps of *A. aurita* (Linnaeus, 1758) were polydisc in autumn (higher zooplankton abundance, higher temperature) and monodisc in winter/spring (lower zooplankton abundance, lower temperature). The metabolic rate and nutritional condition of the polyp are influenced by factors like temperature, food, and polyp size; larger, better-fed polyps produce more ephyrae.

In general, most body parts of the polyp transformed into ephyrae except for the lower part of the stalk. The tiny residuum can disappear without regenerating into a polyp after liberating an ephyra with the concomitant increase of polyp mortality and density reduction. High mortality of polyps after strobilation was observed in the laboratory for some species like *C. tuberculata* and *L. lucerna* (Heins et al., 2015; Prieto et al., 2010; Schiariti, *pers. obs.*). In addition to the increased mortality, the importance of the nutritional state of polyps in regulating strobilation rates suggests that this reproductive process is highly energy-consuming. Despite this, polyps can regenerate the lost parts and re-start the strobilation cycle while the environmental conditions remain favourable defining what is called the strobilation window (see Lucas et al., 2012 and references therein). In reviewing natural cycles of strobilation in the wild (as indicated by the presence of strobilating polyps or newly released ephyrae in the water column), the length of the strobilation window varies from species to species, from location to location and from year to year (Lucas, 2001; Purcell et al., 2009). Discrete seasonal periods of strobilation have been noted for the rhizostomes *C. andromeda*, *C. cephea*, *C. tuberculata*, *M. papua*, *N. nomurai*, *R. esculentum* and *R. nomadica* with the onset of strobilation corresponding with warming temperatures (Chen and Ding, 1983; Kawahara et al., 2006; Lotan et al., 1994; Prieto et al., 2010; Rahat and Adar, 1980; Sugiura, 1965, 1966), with tropical species requiring warmer temperatures (Lucas et al., 2012). Tropical marine lakes are characterised by pulsed strobilation throughout the year, possibly associated with changing weather patterns (Dawson and Martin, 2001). Similarly, protracted strobilation periods also occur in enclosed temperate lakes (*Aurelia* Lamarck, 1816; Lucas, 2001) most likely due to relaxed selection for synchronised breeding (Dawson and Martin, 2001). Multiple (2–5) strobilation events per polyp per season have also been observed in *R. nomadica* (Lotan et al., 1994) and *C. mosaicus* (Pitt and Kingsford, 2003).

All these studies have focused on the description of the factors that define the onset of the strobilation and control the number of ephyrae produced but surprisingly, the identification of the environmental signals indicating the end of the strobilation window has not received much attention. A few studies described specific conditions that "inhibit" strobilation in different rhizostomes and semaeostomes species (see Table 5); for example, the lack of symbionts in *C. andromeda* (Hoffmann et al., 1978), the lack of food in *Aurelia* (Hyman, 1940) and *L. lucerna* (Pereyra, 2015; Schiariti et al., 2008), "high-intensity" light conditions in *Aurelia* sp. (Custance, 1964) and temperature or salinity values out of the physiological boundaries of different species (see Lucas et al., 2012 and references therein). Each species has its physiological optimum range of environmental parameters above and below which strobilation declines, stops, or never occurs. For the present review, we define that the factors that inhibit strobilation are those reported as interrupting an ongoing process and, therefore, which would favour the end of the strobilation window. The addition of some chemical compounds can inhibit strobilation. For example, the addition of tyrosine or iodine concurrently with goitrogens (compounds that interfere with thyroid function in mammals by inhibiting iodine uptake) to *A. aurita* polyp cultures inhibited strobilation, and these compounds may be natural strobilation inhibitors (Berking et al., 2005; Spangenberg, 1974). On the other hand, strobilation ceased in response to changing environmental conditions; it has been suggested that a salinity decrease inhibited strobilation of *A. coerulea* polyps Tokyo Bay (Watanabe and Ishii, 2001), a temperature decrease stopped strobilation of polyp cultures of *L. lucerna* and *Stomolophus* sp.2 (from 20 °C to 15 °C and from 30 °C to 25 °C, respectively: Pereyra, 2015; Schiariti, *unpubl. data*).

6. Conclusions

Understanding population dynamics requires knowledge of life cycles and reproductive features. These two topics are also crucial to comprehend the ecological role that species play in the functioning of marine ecosystems. Ecological modelling provides the best alternative to understand the role of jellyfish in large fisheries-based ecosystems (Pauly et al., 2009). However, considering the number of investigated species, our understanding of the life cycle and reproduction of Rhizostomeae is still far from complete despite being an increasing issue in jellyfish studies for the past 20 years. Only one-third of the rhizostome species' life cycles have been documented and sexual or asexual reproduction has been described for even fewer species. It is reasonable to expect that when additional species are investigated, new reproductive strategies will be discovered given the enormous diversity of reproductive strategies typical of cnidarians. The broad generalisations needed in ecology necessitate knowledge of a greater number of species.

Without the information included in these "descriptive papers", ecologists and modellers lack the information and data necessary to conduct such broad analyses. We therefore encourage authors and editors to publish this kind of descriptive studies. We recommend describing life cycles and reproductive strategies of a greater number of species, to locate the polyps in the field, to increase the study of species in their natural environment, and to understand their population dynamics and the potential for artificial structures to enhance medusa populations. Experimental studies are also needed to improve understanding of the conditions that influence transitions between life cycle stages and rates of production of medusae.

Acknowledgements

This work was partially funded by Grants PICT 2018–03872 (Argentina), and ACM was supported by CNPq grant (307832/2022–8). We greatly appreciate Ilka Straehler-Pohl's contribution to Fig. 1. We thank the Editors, Kylie Pitt and Laura Prieto for their helpful comments which greatly improved the quality of this manuscript. This is INIDEP Contribution N°2307 and a NP-BioMar USP contribution.

References

Adler, L., Jarms, G., 2009. New insights into reproductive traits of scyphozoans: special methods of propagation in *Sanderia malayensis* GOETTE, 1886 (Pelagiidae, Semaeostomeae) enable establishing a new classification of asexual reproduction in the class Scyphozoa. Mar. Biol. 156, 1411–1420.

Agassiz, L., 1860. Contributions to the natural history of the United States of America. III. Second monograph, in five parts, Acalephs in general, Ctenophorae, Discophorae, Hydroidae, homologies of the Radiata. Little, Brown and Co., Boston.

Arai, M.N., 1997. A Functional Biology of Scyphozoa. Chapman and Hall, London.

Arai, M.N., 2009. The potential importance of podocyst to the formation of scyphozoan blooms: a review. Hydrobiologia 616, 241–246.

Astorga, D., Ruiz, J., Prieto, L., 2012. Ecological aspects of early life stages of *Cotylorhiza tuberculata* (Scyphozoa: Rhizostomeae) affecting its pelagic population success. Hydrobiologia 690 (1), 141–155.

Avian, M., Rottini-Sandrini, L., 1991. Oocyte development in four species of scyphomedusa in the northern Adriatic Sea. Hydrobiologia. 216, 189–195.

Bavestrello, G., Puce, S., Cerrano, C., Senes, L., 2000. Strobilation in a species of *Bougainvilliodea* (Cnidaria, Hydrozoa). Scient. Mar. 61, 147–150.

Berking, S., Czech, N., Gerharz, M., Herrmann, K., Hoffmann, U., Raifer, H., et al., 2005. A newly discovered oxidant defence system and its involvement in the development of *Aurelia aurita* (Scyphozoa, Cnidaria): reactive oxygen species and elemental iodine control medusa formation. Int. J. Dev. Biol. 49, 969–976.

Berrill, N., 1949. Developmental analysis of *Scyphomedusae*. Biol. Rev. Camb. Philos. Soc. 24, 393–410.

Bigelow, R.P., 1892. On a new species of Cassiopea from Jamaica. Zool. Anz. 15, 212–214.

Bigelow, R.P., 1893. Some observations on *Polyclonia frondosa*. Johns. Hopkins Univ. Circulars 12 (106), 106.

Bigelow, R.P., 1900. The anatomy and development of *Cassiopea xamachana*. Mem. Boston Soc. Nat. History 5, 190–244.

Blanquet, R.S., 1972. Structural and chemical aspects of the podocyst cuticle of the scyphozoan medusa, *Chrysaora quinquecirrha*. Biol. Bull. 142, 1–10.

Boero, F., 1990. Life cycles, life histories, and recruitment. TREE 5, 200.

Boero, F., Bouillon, J., Piraino, S., Schmid, V., 1997. Diversity of hydroidomedusan life cycles: ecological implications and evolutionary patterns. Nationaal Natuuhistorisch Museum, Leiden, The Netherlands.

Brandt, J.F., 1835. Prodomus descriptionis animalium ab H. *Mertensio observatorum*. Fasc. I. Polypos, Acalephas Discophoras et Siphonophoras, nec non Echinodermata continens. *Recueil des Actes de la Séance Publique de l'Académie Impériale des Sciences de St.* Pétersbourg 1834, 201–275.

Brotz, L., Schiariti, A., López-Martínez, J., Álvarez-Tello, J., Hsieh, Y.-H.P., Jones, R.P., et al., 2017. Jellyfish fisheries in the Americas: origin, state of the art, and perspectives on new fishing grounds. Rev. Fish. Biol. Fish. 27, 1–29.

Brotz, L., Angel, D.L., D'Ambra, I., Enrique-Navarro, A., Lauritano, C., Thibault, D., Prieto, L., 2024. Rhizostomes as a resource: The expanding exploitation of jellyfish by humans. Adv. Mar. Biol. 98.

Browne, E.T., 1905. Scyphomedusae. In: Gardiner, I.S. (Ed.), The fauna and geography of the Maldive and Laccadive archipelagoes Vol. 2. Cambridge University Press, Cambridge, pp. 960–962 Suppl. 1.

Cabrales-Arellano, P., Islas-Flores, T., Thomé, P.E., Villanueva, M.A., 2017. Indomethacin reproducibly induces metamorphosis in *Cassiopea xamachana* scyphistomae. PeerJ 5, e2979.

Calder, D.R., 1973. Laboratory observations on the life history of *Rhopilema verrilli* (Scyphozoa: Rhizostomeae). Mar. Biol. 21, 109–114.

Calder, D.R., 1982. Life history of the cannonball jellyfish, *Stomolophus meleagris* L. Agassiz, 1860 (Scyphozoa, Rhizostomida). Biol. Bull. (Woods Hole). 162, 149–162.

Campbell, R.D., 1974. Cnidaria. In: Giese, A.C., Pearse, J.S. (Eds.), Reproduction of marine invertebrates, Volume 1: Acoelomate and Pseudocoelomate Metazoans. Academic Press, Cambridge, pp. 133–199.

Cargo, D.G., 1971. The sessile stages of a scyphozoan identified as *Rhopilema verrilli*. Tulane Stud. Zool. Botany 17, 31–34.

Cargo, D.G., 1975. Comments on the laboratory culture of Scyphozoa. In: Smith, W.L., Chanley, M.H. (Eds.), Culture of marine invertebrate animals. Plenum Press, New York.

Cargo, D.G., Rabenold, G.E., 1980. Observations on the asexual reproductive activities of the sessile stages of the sea nettle *Chrysaora quinquecirrha* (Scyphozoa). Estuaries 3, 20–27.

Cargo, D.G., Schultz, L.P., 1966. Notes on the biology of the sea nettle, *Chrysaora quinquecirrha*, in Chesapeake Bay. Chesap. Sci. 7, 95–100.

Chen, J., Ding, G., 1983. Effect of temperature on the strobilation of jellyfish (*Rhopilema esculenta* Kishinouye—Scyphozoa, Rhizostomeae). Acta Zool. Sin. 29, 195–206.

Chen, J., Ding, G., Liu, C.-Y., 1984. Effect of light on the strobilation of edible medusa, *Rhopilema esculenta* Kishinouye (Cnidaria, Scyphozoa). Oceanologia Limnol. sin. 15, 310–316.

Chen, J., Ding, B., Liu, C., 1985. Effect of nutritional conditions on the strobilation of edible medusa, *Rhopilema esculenta* Kishinouye. Shuichan Xuebao (J. Fish. China) 9, 321–329.

Chen, J., Ding, G., 1984. On the seasonal regularity of strobilation of edible medusa. J. Fish. China 8, 55–68.

Chiaverano, L.M., Robinson, K.L., Tam, J., Ruzicka, J.J., Quiñones, J., Aleksa, K.T., et al., 2018. Evaluating the role of large jellyfish and forage fishes as energy pathways, and their interplay with fisheries, in the Northern Humboldt current system. Prog. Oceanogr. 164, 28–36.

Claus, C., 1883. Untersuchungen über die Organisation und Entwicklung der Medusen. Prag, Leipzig, pp. 1–96.

Claus, C., 1884. Die Ephyren von *Cotylorhiza* und *Rhizostoma* und ihre Entwicklung zu achtarmigen Medusen. Arb. aus dem Zoologischen Inst. der Universität Wien. und der Zoologischen Stn. Triest. 5 (2), 169–178.

Colley, N.J., Trench, R.K., 1985. Cellular events in the reestablishment of a symbiosis between a marine dinoflagellate and a coelenterate. Cell tissue Res. 239, 93–103.

Condon, R.H., Graham, W.M., Duarte, C.M., Pitt, K.A., Lucas, C.H., Haddock, S.H.D., et al., 2012. Questioning the rise of gelatinous zooplankton in the world's oceans. BioScience 62, 160–169.

Conley, K., Uye, S.-i., 2015. Effects of hyposalinity on survival and settlement of moon jellyfish (*Aurelia aurita*) planulae. J. Exp. Mar. Biol. Ecol. 462, 14–19.

Custance, D.R.N., 1964. Light as an inhibitor of strobilation in *Aurelia aurita*. Nature 204, 1219–1220.

Custance, D.R.N., 1967. Studies on strobilation in the Scyphozoa. J. Biol. Ed. 1, 79–81.

da Silveira, F.L., Morandini, A.C., 1998. Asexual reproduction of *Linuche unguiculata* (Swartz, 1788) (Scyphozoa: Coronate) by planuloid formation trough strobilation and segmentation. Proc. Biol. Soc. Wash. 111, 781–794.

Dawson, M.N., Martin, L.E., 2001. Geographic variation and ecological adaptation in *Aurelia* (Scyphozoa, Semaeostomeae): some implications from molecular phylogenetics. Hydrobiologia. 451, 259–273.

Dawson, M.N., Martin, L.E., Penland, L.K., 2001. Jellyfish swarms, tourists, and the Christ-child. Hydrobiologia. 451, 131–144.

Desor, E. (1848). No title. Meeting of the November 1, 1848. Proceedings of the Boston Society of Natural History. 3(1848 to 1851): 73–77 (Desor part pp. 75–76).

Ding, G., Chen, J., 1981. The life history of *Rhopilema esculenta* Kishinouye. J. Fish. China 5 (2), 93–104.

Dong, J., Sun, M., Purcell, J.E., Chai, Y., Zhao, Y., Wang, A., 2014. Effect of salinity and light intensity on somatic growth and podocyst production in polyps of the giant jellyfish *Nemopilema nomurai* (Scyphozoa: Rhizostomeae). Hydrobiologia 754, 75–83.

Dror, H., Angel, D., 2023. Rising seawater temperatures affect the fitness of *Rhopilema nomadica* polyps and podocysts and the expansion of this medusa into the western Mediterranean. Mar. Ecol. Progr. Ser. 728, 123–143. https://doi.org/10.3354/meps14224.

Eckelbarger, K.J., 1994. Oocyte nutrition in the lower Metazoa: the Scyphozoa. In: Wilson, W.H., Stricker, S.A., Shinn, G.L. (Eds.), Reproduction and Development of Marine Invertebrates. John Hopkins University Press, Baltimore, pp. 15–28.

Eckelbarger, K.J., Larson, R.J., 1988. Ovarian morphology and oogenesis in *Aurelia aurita* (Scyphozoa: Semaeostomeae): ultrastructural evidence of heterosynthetic yolk formation in a primitive metazoan. Mar. Biol. 100, 103–115.

Eckelbarger, K.J., Larson, R., 1992. Ultrastructure of the ovary and oogenesis in the jellyfish *Linuche unguiculata* and *Stomolophus meleagris*, with a review of the ovarian structure in the Scyphozoa. Mar. Biol. 114, 633–643.

Eschscholtz, F., 1829. System der Acalephen. Eine ausführliche Beschreibung aller medusenartigen Strahltiere. Ferdinand Dümmler, Berlin, pp. 1-190.

Fautin, D.G., 2002. Reproduction of Cnidaria. Can. J. Zool. 80, 1735–1754.

Feng, S., Lin, J., Sun, S., Zhang, F., Li, C., 2018. Hyposalinity and incremental micro-zooplankton supply in early-developed *Nemopilema nomurai* polyp survival, growth, and podocyst reproduction. Mar. Ecol. Progr. Ser. 591, 117–128.

Feng, S., Zhang, F., Sun, S., Wang, S., Li, C., 2015a. Effects of duration at low temperature on asexual reproduction in polyps of the scyphozoan *Nemopilema nomurai* (Scyphozoa: Rhizostomeae). Hydrobiologia 754, 97–111.

Feng, S., Zhang, G.-T., Sun, S., Zhang, F., Wang, S.-W., Liu, M.-T., 2015b. Effects of temperature regime and food supply on asexual reproduction in *Cyanea nozakii* and *Nemopilema nomurai*. Hydrobiologia 754, 201–214.

Fewkes, J.W., 1887. A new Rhizostomatous medusa from New England. Am. J. Sci. 33, series 3, 119–125.

Fitt, W.K., Costley, K., 1998. The role of temperature in survival of the polyp stage of the tropical rhizostome jellyfish *Cassiopea xamachana*. J. Exp. Mar. Biol. Ecol. 222, 79–91.
Fitt, W.K., Hofmann, D.K., 1985. Chemical induction of settlement and metamorphosis of the reef-dwelling coelenterate *Cassiopea andromeda*. Proc. 5th Int. Coral Reef. Symp. 5, 239–244.
Fitt, W.K., Hofmann, D.K., Kemp, D.W., Ohdera, A.H., 2021. Different physiology in the jellyfish *Cassiopea xamachana* and *C. frondosa* in Florida Bay. Oceans 2 (4), 811–821.
Fitt, W.K., Hofmann, D.K., Wolk, M., Rahat, M., 1987. Requirement of exogenous inducers for metamorphosis of axenic larvae and buds of *Cassiopea andromeda* (Cnidaria: Scyphozoa). Mar. Biol. 94, 415–422.
Fleck, J., Bischoff, A., 1993. Protein kinase C is possibly involved in chemical induction of metamorphosis in *Cassiopea* spp. (Cnidaria: Scyphozoa). Proceedings of the Seventh International Coral Reef Symposium. University of Guam Press, Guam, pp. 456–462.
Fleck, J., Fitt, W.K., 1999. Degrading mangrove leaves of *Rhizophora mangle* Linné provide a natural cue for settlement and metamorphosis of the upside-down jellyfish *Cassiopea xamachana* Bigelow. J. Agric. Food Chem. 234, 83–94.
Fleck, J., Hofmann, O.K., 1990. The efficiency of metamorphosis inducing oligopeptides in *Cassiopea* species (Cnidaria: Scyphozoa) depends on both primary structure and amino- and carboxiterminal substituents. Verh. DIsch. Zool. Ges. 83, 452–453.
Forskål, P. 1775. Descriptiones animalium avium, amphibiorum, piscium, insectorum, vermium; quae in intinere orientali observavit Petrus Forskål, Hauniae.
Fuentes, V., Straehler-Pohl, I., Atienza, D., Franco, I., Tilves, U., Gentile, M., et al., 2011. Life cycle of the jellyfish *Rhizostoma pulmo* (Scyphozoa: Rhizostomeae) and its distribution, seasonality and inter-annual variability along the Catalan coast and the Mar Menor (Spain, NW Mediterranean). Mar. Biol. 158, 2247–2266.
Galil, B.S., Spanier, E., Ferguson, W.W., 1990. The scyphomedusae of the Mediterranean coast of Israel, including two lessepsian migrants new to the Mediterranean. Zool. Med. 64, 95–105.
Gamero-Mora, E., Lawley, J.W., Marona, M.M., Stampar, S.N., Morandini, A.C., 2024. Morphological and molecular data in the study of the evolution, population genetics and taxonomy of Rhizostomeae. Adv. Mar. Biol. 98.
Gershwin, L., Zeidler, W., Davie, P.J.F., 2010. Medusae (Cnidaria) of Moreton Bay, Queensland, Australia. Mem. Queensland Mus. 54, 47–108.
Giangrande, A., Geraci, S., Belmonte, G., 1994. Life-cycle and life-history diversity in marine invertebrates and the implications in community dynamics. Oceanogr. Mar. Biol.: An Annu. Rev. 32, 305–333.
Gmelin, J.F., 1791. In: Tomus, I., Pars, V.I. (Eds.), Systema naturae per regna tria naturae: secundum classes, ordines, genera, species cum characteribus, differentiis, sinonimis, locis, 13th ed. G.E. Beer, Leipzig.
Goette, A, 1886. Verzeichnis der Medusen, welche von Dr. Sander, Stabsarzt auf S.M.S. "Prinz Adalbert" gesammelt wurden. Sitzber K Preuss Akad. wiss. Berlin 39, 831–837.
Goette, A., 1887. Entwickelungsgeschichte der *Aurelia aurita* und *Cotylorhiza tuberculata*. Abhandlungen zur Entwickelungsgeschichte der Tiere 4, 1–79.
Gohar, H.A.F., Eisawy, A.M., 1960. The development of *Cassiopea andromeda* (Scyphomedusae). Publications Marine Biological Station. Al-Ghardaqa 11, 148–190.
González-Valdovinos, M., 2010. Alimentación y desarrollo en la fase pólipo de la medusa bola de cañón *Stomolophus meleagris* L. Agassiz, 1860 (Scyphozoa, Rhizostomida). PhD Thesis, Departamento Académico de Biología Marina. Universidad Autónoma de Baja California Sur, La Paz, pp. 47. In Spanish.
Gröndahl, F., 1988. A comparative ecological study on the scyphozoans *Aurelia aurita*, *Cyanea capillata* and *C. lamarckii* in the Güllmar Fjord, western Sweden, 1982 to 1986. Mar. Biol. 97, 541–550.

Gueroun, S.K.M., Torres, T.M., dos Santos, A., Vasco-Rodríguez, N., Canning-Clode, J., Andrade, C., 2021. *Catostylus tagi* (Class: Scyphozoa, Order: Discomedusae, Suborder: Rhizostomida, Family: Catostylidae) life cycle and first insight into its ecology. PeerJ 9, e12056.

Guo, P., 1990. Effect of nutritional condition on the formation and germination of the podocyst of scyphistomae of *Rhopilema esculenta* Kishinouye. J. Fish. China 14, 206–211.

Haacke, W., 1887. Die Scyphomedusae des St. Vincent Golfes. Jenaischen Z. für Naturwissenschaft 20 (13), 588–638.

Haeckel, E., 1869. Ueber die Crambessiden, eine neue Meduse-Familie aus der Rhizostomeengruppe. Zeitschrift für wissenschaftliche Zoologie 19, 509–537.

Haeckel, E., 1880. Das System der Medusen. I, 2: System der Acraspeden. Gustav Fischer, Jena.

Hand, C., Uhlinger, K.R., 1995. Asexual reproduction by transverse fission and some anomalies in the sea anemone *Nematostella vectensis*. Invert. Biol. 114 (1), 9–18.

Hashimoto, S., Ohtsuka, S., Kondo, Y., Iida, K., Yonetani, M., Iwasaki, S., et al., 2016. Strobilation of *Netrostoma setouchianum* (Kishinouye, 1902) induced by 5-methoxy-2-methylindole. Bull. Setouchi Field Sci. Cent. Graduate Sch. Biosphere Sci. Hiroshima Univ. 14, 1–6.

He, J., Zheng, L., Zhang, W., Lin, Y., 2015. Life cycle reversal in *Aurelia* sp.1 (Cnidaria, Scyphozoa). PLoS One 10, e0145314.

Heins, A., Glatzel, T., Holst, S., 2015. Revised descriptions of the nematocysts and the asexual reproduction modes of the scyphozoan jellyfish *Cassiopea andromeda* (Forskål, 1775). Zoomorphology 134, 351–366.

Helm, R.R., 2018. Evolution and development of scyphozoan jellyfish. Biol. Rev. 93 (2), 1228–1250.

Helm, R.R., Dunn, C.W., 2017. Indoles induce metamorphosis in a broad diversity of jellyfish, but not in a crown jelly (Coronatae). PLoS One 12, e0188601.

Hernroth, L., Gröndahl, F., 1983. On the biology of *Aurelia aurita* (L.) 1. Release and growth of *Aurelia aurita* (L.) ephyrae in the Gullmar Fjord, western Sweden, 1982–83. Ophelia 22, 189–199.

Hernroth, L., Gröndahl, F., 1985. On the biology of *Aurelia aurita* (L.): 2. Major factors regulating the occurrence of ephyrae and young medusae in the Gullmar Fjord, western Sweden. Bull. Mar. Sci 37, 567–576.

Hofmann, D.K., Brand, U., 1987. Induction of metamorphosis in the symbiotic scyphozoan *Cassiopea andromeda*: role of marine bacteria and of biochemicals. Symbiosis 4, 99–116.

Hofmann, D.K., Crow, G., 2002. Induction of larval metamorphosis in the tropical scyphozoan *Mastigias papua*: striking similarity with upside down-jellyfish *Cassiopea* spp (notes Relat. species). Vie et. Milieu 52 (4), 141–147.

Hofmann, D.K., Fitt, W.K., Fleck, J., 1996. Checkpoints in the life-cycle of *Cassiopea* spp.: control of metagenesis and metamorphosis in a tropical jellyfish. J. Dev. Biol. 40, 331–338.

Hofmann, D., Gottlieb, M., 1991. Bud formation in the scyphozoan *Cassiopea andromeda*: epithelial dynamics and fate map. Hydrobiologia 216/217, 53–59.

Hofmann, D.K., Hadfield, M.G., 2002. Hermaphroditism, gonochorism, and asexual reproduction in *Cassiopea* sp.—an immigrant in the islands of Hawai'i. Invertebrate Reprod. Dev. 41 (1-3), 215–221.

Hofmann, D.K., Honegger, T.G., 1990. Bud formation and metamorphosis in *Cassiopea andromeda* (Cnidaria: Scyphozoa): a developmental and ultraestructural study. Mar. Biol. 105, 509–518.

Hofmann, D.K., Kremer, B.P., 1981. Carbon metabolism and strobilation in *Cassiopea andromeda* (Cnidaria: Scyphozoa): significance of endosymbiotic dinoflagellates. Mar. Biol. 65, 25–33.
Hofmann, D., Manitz, B., Reckenfelderbilumer, B., 1992. Metamorphosis in the scyphozoan *Cassiopea andromeda*: cellular and morphogenetic responses to exogenous inducers. Proc. 7th Int. Coral Reef. Symposium 463–470.
Hofmann, D.K., Neumann, R., Henne, K., 1978. Strobilation, budding and initiation of Scyphistoma morphogenesis in the Rhizostome *Cassiopea andromeda* (Cnidaria: Scyphozoa). Mar. Biol. 47, 161–176.
Holst, S., 2012a. Morphology and development of benthic and pelagic life stages of North Sea jellyfish (Scyphozoa, Cnidaria) with special emphasis on the identification of ephyra stages. Mar. Biol. 159, 2707–2722.
Holst, S., 2012b. Effects of climate warming on strobilation and ephyra production of North Sea scyphozoan jellyfish. Hydrobiologia 690 (1), 127–140.
Holst, S., Jarms, G., 2006. Responses of solitary and colonial coronate polyps (Cnidaria, Scyphozoa, Coronatae) to sedimentation and burial. J. Exp. Mar. Biol. Ecol. 329, 230–238.
Holst, S., Jarms, G., 2007. Substrate choice and settlement preferences of planula larvae of five Scyphozoa (Cnidaria) from German Bight, North Sea. Mar. Biol. 151, 863–871.
Holst, S., Kaiser, L.-R., Sötje, S., 2024b. Planula settlement and polyp morphogenesis in two bloom forming jellyfish species of the genus *Cyanea* Péron and Lesueur, 1810 and effects of abiotic factors on planulocysts. Mar. Biol. 171, 6.
Holst, S., Sötje, I., Tiemann, H., Jarms, G., 2007. Life cycle of the rhizostome jellyfish *Rhizostoma octopus* (L.) (Scyphozoa, Rhizostomeae), with studies on cnidocysts and statoliths. Mar. Biol. 151, 1695–1710.
Holst, S., Tiseo, G.R., Djeghri, N., Sötje, I., 2024a. Approaches and findings in histological and micromorphological research on Rhizostomeae. Adv. Mar. Biol. 98.
Hyman, L.H., 1940. The Invertebrates: Protozoa through Ctenophora. McGraw-Hil Book Company. Inc New York.
Ikeda, H., Ohtsu, K., Uye, S.-I., 2011. Structural changes of gonads during artificially induced gametogenesis and spawning in the giant jellyfish *Nemopilema nomurai* (Scyphozoa: Rhizostomeae). J. Mar. Biol. Assoc. U. K. 91, 215–227.
Ikegami, S., Honji, N., Yoshida, M., 1978. Light-controlled production of spawning-inducing substance in jellyfish ovary. Nature. 272, 611–612.
Jacobson, C.O., 2016. Efecto de temperaturas oscilantes en la fisiología de pólipos de medusa *Stomolophus meleagris* (Rhizostomeae: Stomolophidae). MSc Thesis. Centro de Investigaciones Biológicas del Noroeste, Baja California Sur, La Paz, pp.67.
Jarms, G., 2010. The early life history of Scyphozoa with emphasis on Coronatae. Verh. Naturwiss. Ver. Hambg. 45, 17–31.
Jarms, G., Morandini, A., 2019. World Atlas of Jellyfish. Dölling und Galitz Verlag.
Jiang, S., Lu, N., Chen, J., 1993. Effect of temperature, salinity and light on the germination of the podocyst of *Rhopilema esculenta*. Fish. Sci. (Tokyo) 12, 1–4.
Jufri, A.W., 2001. Description of the asexual reproduction and strobilation of the scyphistomae stage of *Phyllorhiza punctata* (Scyphozoa: Rhizostomae). Biota VI, 49–56.
Kakinuma, Y., 1975. An experimental study of the life cycle and organ differentiation of *Aurelia aurita* Lamarck. Bull. Mar. Biol. Stn. Asamushi 15, 101–113.
Karande, A., Nakauchi, M., 1981. Culturing of the ascidian, *Aplidium multiplicatum* & its dimorphic larvae. Indian. J. Mar. Sci. 10, 93–95.
Kawahara, M., Ohtsu, K., Uye, S.I., 2013. Bloom or non-bloom in the giant jellyfish *Nemopilema nomurai* (Scyphozoa: Rhizostomeae): roles of dormant podocysts. J. Plankton Res. 35, 213–217.

Kawahara, M., Uye, S., Ohtsu, K., Iizumi, H., 2006. Unusual population explosion of the giant jellyfish *Nemopilema nomurai* (Scyphozoa: Rhizostomeae) in East Asian waters. Mar. Ecol. Progr. Ser. 307, 161–173.

Kienberger, K., Riera-Buch, M., Schönemann, A.M., Bartsch, V., Halbauer, R., Prieto, L., 2018. First description of the life cycle of the jellyfish *Rhizostoma luteum* (Scyphozoa: Rhizostomeae). PLoS One 13 (8), e0202093.

Kikinger, R., 1986. Development cycle of *Cotylorhiza tubercolata* (Scyphozoa: Rhizostomae). Nova Thalassia 8 (2), 73–82.

Kikinger, R., 1992. *Cotylorhiza tuberculata* (Cnidaria: Scyphozoa)—life history of a stationary population. PSZN. I: Mar. Ecol. 13 (4), 333–362.

Kishinouye, K., 1891. Bizen kurage. Zool. Mag. (Dobutsugaku zasshi) 3, 53.

Kishinouye, K., 1902. Some new scyphomedusae of Japan. J. Coll. Sci. Imp. Univ. 17, 1–17.

Kishinouye, K., 1922. Echizen kurage. Zool. Mag. (Dobutsugaku zasshi) 34, 343–346.

Kroiher, M., Siefker, B., Berking, S., 2000. Induction of segmentation in polyps of *Aurelia aurita* (Scyphozoa, Cnidaria) into medusae and formation of mirror-image medusa anlagen. Int. J. Dev. Biol. 44, 485–490.

Lamarck, J.B., 1816. Histoire naturelle des animaux sans vertèbres (Vol. Tome Second). Verdiére, Paris.

Lambert, F.J., 1935. Jellyfish difficulties study their life history other problems. Essex Nat. 25, 70–86.

Lange, J., Kaiser, R., 1995. The maintenance of pelagic jellyfish in the Zoo-Aquarium Berlin. Int. Zoo. Yb 34, 59–64.

Lee, H.E., Han, C.H., Kim, B., Yoon, W.D., 2017. Effects of temperature and salinity on the asexual reproduction of *Nemopilema nomurai* (Scyphozoa: Rhizostomeae). Ocean. Sci. J. 72, 573–579. https://doi.org/10.1007/s12601-017-0040-5.

Leitz, T., 1997. Induction of settlement and metamorphosis of cnidarian larvae: signals and signal transduction. Invertebrate Reprod. Dev. 31, 109–122.

Lesson, R.P., 1830. Voyage autour du monde, exécut par ordre du Roi, sur la corvette de Sa Magest La Coquille, pendant les années 1822–25. Vol. 2. Arthus Bertrand, Paris, 1828 pp.

Light, S.F., 1914. Some Philippine Scyphomedusae, including two new genera, five new species, and one new variety. Philip. J. Sci. 9 (section D 3), 195–231.

Linnaeus, C.v., 1758. Systema Naturae per regna tria naturae secundum classes, ordines, genera, species, cum characteibus, differentiis, synonymis, locis (10th ed. Vol. Tomus I, Editio Decima, reformata). Stockholm: L. Salvii.

Liu, W.C., Lo, W.T., Purcell, J.E., Chang, H.H., 2009. Effects of temperature and light intensity on asexual reproduction of the scyphozoan, *Aurelia aurita* (L.) in Taiwan. Hydrobiologia 616, 247–258.

López-Martínez, J., Álvarez-Tello, F.J., Porchas-Cornejo, M.A., Nevárez-López, C.A., Muhlia-Almazán, A., Urías-Padilla, K.V., 2022. Multiple reproduction forms in the polyps of the cannonball jellyfish *Stomolophus* sp. 2: probable life-cycle reversal. J. Exp. Zool. Part. A: Ecol. Integr. Physiol. 339 (3), 239–252.

Lotan, A., Ben-Hillel, R., Loya, Y., 1992. Life cycle of *Rhopilema nomadica*: a new immigrant scyphomedusan in the Mediterranean. Mar. Biol. 112, 237–242.

Lotan, A., Fine, M., Ben-Hillel, R., 1994. Synchronization of the life cycle and dispersal pattern of the tropical invader scyphomedusan *Rhopilema nomadica* is temperature dependent. Mar. Ecol. Prog. Ser. 109, 59–65.

Lu, N., Jiang, L., Chen, J., 1997. Effect of temperature, salinity and light on the podocyst generation of *Rhopilema esculenta* Kishinouye. Fish. Sci. 16, 3–8.

Lucas, C.H., 2001. Reproduction and life history strategies of the common jellyfish, *Aurelia aurita*, in relation to its ambient environment. Hydrobiologia. 451, 229–246.

Lucas, C.H., Dawson, M.N., 2014. What are jellyfishes and thaliaceans and why do they bloom? Jellyfish Blooms. Springer, pp. 9–44.

Lucas, C.H., Graham, W.M., Widmer, C., 2012. Jellyfish life histories: role of polyps in forming and maintaining scyphomedusa populations. In: Lesser, M. (Ed.), Advances in Marine Biology. Academic Press, pp. 133–196.

Maas, O., 1903. Die Scyphomedusen der Siboga-Expedition. *Siboga-Expeditie,* 11, 1-91, Pls I-XII.

Mackie, G.O., 2002. What's new in cnidarian biology? Can. J. Zool. 80, 1649–1653.

Macri, S., 1778. Nuove osservazioni intorno la storia naturale del polmone marino degli antichi. Napoli, pp. 1-36.

Magnusen, J.E., 1980. Epidermal cell movement during podocyst formation in *Chrysaora quinquecirrha*. In: Tardent, O., Tardent, R. (Eds.), Developmental and Cellular Biology of Coelenterates. Elsevier/North Holland Biomedical Press, Amsterdam, pp. 435–440.

Mammone, M., Bosch-Belmar, M., Milisenda, G., Castriota, L., Sinopoli, M., Allegra, A., et al., 2023. Reproductive cycle and gonadal output of the Lessepsian jellyfish *Cassiopea andromeda* in NW Sicily (Central Mediterranean Sea). PLoS One 18, 1–14.

Marques, A.C., Collins, A.G., 2004. Cladistic analysis of Medusozoa and cnidarian evolution. Invertebrate Biology 123, 23–42.

Martin, V.J., Chia, F.-S., 1982. Fine structure of a scyphozoan planula, *Cassiopea xamachana*. Biol. Bull. 163 (2), 320–328.

Martin, J.W., Gershwin, L., Burnett, J.W., Cargo, D.G., Bloom, D.A., 1997. *Chrysaora achlyos*, a remarkable new species of scyphozoan from the eastern Pacific. Biological Bulletin 193, 8–13.

Mayer, A.G., 1915. Medusae of the Philippines and of Torres Strait. Pap. Tort. Lab. 8, 157–202.

Mendoza Islas, H.M., 2015. Efecto de la luz en la reproducción de la fase pólipo de la medusa bola de cañón (*Stomolophus meleagris*). MSc thesis, Departamento Académico de Biología Marina. Universidad Autónoma De Baja California Sur, La Paz, pp. 49. (In Spanish).

Mills, C., Larson, R.J., 2007. Scyphozoa: Scyphomedusae, Stauromedusae, and Cubomedusae. In: Carlton, J.T. (Ed.), The Light and Smith Manual. Intertidal Invertebrates from Central California to Oregon. University of California Press, Los Angeles, pp. 168–173.

Miyake, H., Adachi, A., Ikeda, S., Nishikawa, J., Ohtsuka, S., Srinui, K., et al., 2018. New exhibitions of beautiful Southeast Asian jellyfish at aquaria: collection and life cycle. Proceedings of the 10th International Aquarium Congress. Fukushima, Japan, pp. 7–10.

Miyake, H., Honda, S., Nishikawa, J., Yusoff, F.M.J.A., 2021. Life cycle of edible jellyfish *Acromitus hardenbergi* Stiasny, 1934 (Scyphozoa: Rhizostomeae) inhabiting a brackish-water environment. Animals 11, 2138.

Morandini, A.C., Schiariti, A., Stampar, S.N., Maronna, M.M., Straehler-Pohl, I., Marques, A.C., 2016. Succession of generations is still the general paradigm for scyphozoan life cycles. Bull. Mar. Sci. 92, 343–351.

Morandini, A.C., 2024. Morphology of Rhizostomeae jellyfishes: what is known and what we advanced since the 1970s. Adv. Mar. Biol. 98.

Müller, W.A., Leitz, T., 2002. Metamorphosis in the Cnidaria. Can. J. Zool. 80, 1755–1771.

Nagata, R.M., D'Ambra, I., Lauritano, C., von Montfort, G.M., Djeghri, N., Jordano, M.A., Colin, S.P., Costello, J.H., Leoni, V., 2024. Physiology and functional biology of Rhizostomeae jellyfish. Adv. Mar. Biol. 98.

Neumann, R., 1979. Bacterial induction of settlement and metamorphosis in the planula larvae of *Cassiopea andromeda* (Cnidaria: Scyphozoa: Rhizostomeae). Mar. Ecol. Prog. Ser. 1, 21–28.

Neumann, R., Schmahl, G., Hofmann, D.K., 1980. Bud formation and control of polyp morphogenesis in *Cassiopea andromeda* (Scyphozoa). Developmental and Cellular Biology of Coelenterates. Elsevier/North-Holland Biomedical Press, Amsterdam, pp. 217–223.

Ohtsu, K., Kawahara, M., Ikeda, H., Uye, S.-I., 2007. Experimental induction of gonadal maturation and spawning in the giant jellyfish *Nemopilema nomurai* (Scyphozoa: Rhizostomeae). Mar. Biol. 152, 667–676.

Owen, R., 1843. Lectures on the comparative anatomy and physiology of the invertebrate animals.

Pallas, P.S., 1774. Spicilegia Zoologica quibus novae imprimis et obscurae animalium species iconibus, descriptionibus atque commentariis illustrantur. Fasciculus 10. Berolini, Prostant Apud Gottl, pp. 1–41.

Papenfuss, E.J., 1936. The utility of the nematocysts in the classification of certain Scyphomedusae. I. *Cyanea capillata, Cyanea palmstruchii, Dactylometra quinquecirrha, Dactylometra quinquecirrha* var. *chesapeakei*, and *Chrysaora hysoscella*. Acta Universitatis Lundensis, Nova Series 31, 19–26.

Paspaleff, B.W., 1938. Über die Entwicklung von *Rhizostoma pulmo* Agass. Tr. chernomorsk Biol. Sta Varna 7, 1–25.

Pauly, D., Graham, W.M., Libralato, S., Morissette, L., Palomares, M.L.D., 2009. Jellyfish in ecosystems, online databases, and ecosystem models. Hydrobiologia 616, 67–85.

Percival, E., 1923. On the strobilation of *Aurelia*. Q. J. Micro. Sci. 67, 85–100.

Pereyra, D.Y., 2015. Efecto de las variaciones de temperatura y de la disponibilidad de alimento sobre las tasas de estrobilación de *Lychnorhiza lucerna* (Cnidaria: Scyphozoa). MSc Thesis, Universidad de Mar del Plata, Mar del Plata, Argentina, pp. 59. (In Spanish).

Péron, F, Lesueur, CA, 1810. Tableau des caracteres génériques et spécifiques de toutes les espéces de Méduses connues jusqu'à ce jour. Annales du Muséum National d'Histoire Naturelle, Paris 14, 325–366.

Phuangsanthia, W., Choosri, S., Luangoon, N., 2018. Life cycle of a rhizostome jellyfish (*Catostylus townsendi* Mayer, 1915) under laboratory condition. J. Agric. Res. Ext. 35 (Suppl. 2), 1021–1033.

Phuangsanthia, W., Choosri, S., Fakrajang, S., Muthuvan, V., 2020. Life cycle of the edible jellyfish *Rhopilema hispidum* (Vanhöffen, 1888). Khon Kaen Agr. J. 48 (Suppl.1), 907–916.

Pitt, K.A., 2000. Life history and settlement preferences of the edible jellyfish *Catostylus mosaicus* (Scyphozoa: Rhizostomeae). Mar. Biol. 136 (2), 269–279.

Pitt, K.A., Kingsford, M.J., 2000. Reproductive biology of the edible jellyfish *Catostylus mosaicus* (Rhizostomeae). Mar. Biol. 137, 791–799.

Pitt, K.A., Kingsford, M.J., 2003. Temporal variation in the virgin biomass of the edible jellyfish, *Catostylus mosaicus* (Scyphozoa, Rhizostomeae). Fish. Res. 63, 303–313.

Prieto, L., Astorga, D., Navarro, G., Ruiz, J., 2010. Environmental control of phase transition and polyp survival of a massive-outbreaker jellyfish. PLoS One 5, e13793. https://doi.org/10.1371/journal.pone.0013793.

Purcell, J.E., Atienza, D., Fuentes, V., Olariaga, A., Tilves, U., Colahan, C., et al., 2012. Temperature effects on asexual reproduction rates of scyphozoan species from the northwest Mediterranean Sea. Hydrobiologia 690, 169–180.

Purcell, J.E., Hoover, R.A., Schwarck, N.T., 2009. Interannual variation of strobilation by the scyphozoan *Aurelia labiata* in relation to polyp density, temperature, salinity, and light conditions in situ. Mar. Ecol. Prog. Ser. 375, 139–149.

Purcell, J.E., White, J.R., Nemaziel, D.A., Wright, D.A., 1999. Temperature, salinity and food effects on asexual reproduction and abundance of the scyphozoan *Chrysaora quinquecirrha*. Mar. Ecol. Prog. Ser. 180, 187–196.

Quoy, J. R. C., Gaimard, J. P., 1824. Voyage autour du monde, entrepis par ordre du Roi, sous les ministère et conformément aux instructions de S. Exc. M. Le Vicomte du Bouchage, secrétaire d`état au départment de la Marine, exécuté sur les corvettes de S. M. l'Uranie et la Physicienne, pendant les années 1817, 1818, 1819 et 1820 (Vol. Zoologie). Paris: Chez Pillet Aîné.

Quoy, J. R. C., Gaimard, J. P., 1827. Observations zoologiques faites à bord de l'*Astrolabe*, en mai 1826, dans le détroit de Gibraltar. Annales de Sciences Naturelles, *10*, 1-10, 172-193, pls 171-172, 174-179.

Rahat, M., Adar, O., 1980. Effect of symbiotic zooxanthellae and temperature on budding and strobilation in *Cassiopeia andromeda* (Eschscholtz). Biol. Bull. 159, 394–401.

Rato, L.D., Pinto, C., Duarte, I.M., Leandro, S.M., Marques, S.C., 2021. Euryhalinity and thermal tolerance of *Phyllorhiza punctata* (Scyphozoa) scyphostomae: life history and physiological trade-offs. Mar. Biol. 168, 1–12.

Reynaud, A.A.M., 1830. *Rhizostoma* (*Medusa*) *rosea*. In: Lesson, R.P. (Ed.), Centurie Zoologique, ou choix d'animaux rares, nouveaux ou imparfaitement connus. F.G. Levrault, Paris, pp. 97–98.

Richardson, A.J., Bakun, A., Hays, G.C., Gibbons, M.J., 2009. The jellyfish joyride: causes, consequences and management responses to a more gelatinous future. TREE 24, 312–322.

Rippingale, R.J., Kelly, S.J., 1995. Reproduction and survival of *Phyllorhiza punctata* (Cnidaria: Rhizostomeae) in a seasonally fluctuating salinity regime in western Australia. Mar. Fresh. Res. 46, 1145–1151.

Robinson, K.L., Ruzicka, J.J., Decker, M.B., Brodeur, R.D., Hernandez, F.J., Quiñones, J., et al., 2014. Jellyfish, forage fish, and the world's major fisheries. Oceanography 27, 104–115.

Rottini Sandrini, L., Avian, M., 1983. Biological cycle of *Pelagia noctiluca*: morphological aspects of the development from planula to ephyra. Mar. Biol. 74, 169–174.

Rouse, G.W., Pitt, K., 2000. Ultrastructure of the sperm of *Catostylus mosaicus* and *Phyllorhiza punctata* (Scyphozoa, Cnidaria): implications for sperm terminology and the inference of reproductive mechanisms. Invertebr. Reprod. Dev. 38, 23–34.

Russell, F.S., 1964. On scyphomedusae of the genus *Pelagia*. Journal of the Marine Biological Association of the United Kingdom 44, 133–136.

Schiariti, A., Christiansen, E., Morandini, A.C., da Silveira, F.L., Agustin, G., Walter, D., et al., 2012. Reproductive biology of *Lychnorhiza lucerna* (Cnidaria: Scyphozoa: Rhizostomeae): individual traits related to sexual reproduction. Mar. Biol. Res. 8, 255–264.

Schiariti, A., Dutto, M.S., Morandini, A.C., Nagata, R.M., Pereyra, D.Y., Puente Tapia, A., et al., 2018. An overview of the Medusozoa from the Southwestern Atlantic. In: Hoffmeyer, M. (Ed.), Plankton Ecology of the Southwestern Atlantic. Springer, pp. 413–449.

Schiariti, A., Kawahara, M., Uye, S., Mianzan, H., 2008. Life cycle of the jellyfish *Lychnorhiza lucerna* (Scyphozoa: Rhizostomeae). Mar. Biol. 156, 1–12.

Schiariti, A., Melica, V., Kogovšek, T., Malej, A., 2015. Density-dependent effects control the reproductive strategy and population growth of *Aurelia aurita* s.l. scyphistomae. Mar. Biol. 162, 1665–1672.

Schiariti, A., Morandini, A.C., Jarms, G., von Glehn Paes, R., Franke, S., 2014. Asexual reproduction strategies and blooming potential in Scyphozoa. Mar. Ecol. Prog. Ser. 510, 241–253.

Schmahl, G., 1985. Bacterially induced stolon settlement in the scyphopolyp of *Aurelia aurita* (Cnidaria, Scyphozoa). Helgolaender Meeresuntersuchungen 39, 33–42.

Sicarello, L., Minasian Jr., L.L., Bigger, C.H., Mariscal, R.N., 1979. Asexual reproduction in scyphistomae of *Cassiopea xamachana* (Scyphozoa: Rhizostoma): exogenous effects and the interaction of budding and strobilation, American Zoologist. Amer Soc. Zool. 66044, 985–985.

Smith, H.G., 1936. Contribution to the anatomy and physiology of *Cassiopea frondosa*. Publ.—Carnegie Instit. Wash. 31, 18–52.

Sötje, I., Jarms, G., 1999. Detailed description of *Thecoscyphus zibrowii* Werner, 1984 (Scyphozoa, Coronatae) with remarks on the life cycle. Mitt. Hamb. Zool. Mus. Inst. 96, 5–13.

Spangenberg, D.B., 1965. A study of strobilation in *Aurelia aurita* under controlled conditions. J. Exp. Zool. 160, 1–9.

Spangenberg, D.B., 1968. Recent studies of strobilation in jellyfish. Oceanogr. Mar. Biol. Ann. Rev. 6, 231–247.

Spangenberg, D.B., 1974. Thyroxine in early strobilation in *Aurelia aurita*. Amer. Zool. 14, 825–831.

Stearns, S.C., 1992. The evolution of life histories. Interprint Ltd., Malta.

Stiasny, G., 1920. Die Scyphomedusen-Sammlung des Naturhistorischen Reichsmuseums in Leiden. III. Rhizostomeae. Zool. Meded. 5, 213–230.

Stiasny, G., 1934. Acromitus hardenbergi nov. spec., eine neue rhizostome meduse aus dem Malayischen Archipel. Zool. Meded. 17, 1–7.

Straehler-Pohl, I., 2009. Die Phylogenie der Rhopaliophora (Scyphozoa und Cubozoa) und die Paraphylie der 'Rhizostomeae', Department Biologie der Fakultät für Mathematik, Informatik und Naturwissenschaften. Universität Hamburg, Hamburg, pp. 373.

Straehler-Pohl, I., Jarms, G., 2005. Life cycle of *Carybdea marsupialis* Linnaeus, 1758 (Cubozoa, Carybdeidae) reveals metamorphosis to be a modified strobilation. Mar. Biol. 147, 1271–1277.

Straehler-Pohl, I., Jarms, G., 2022. Back to the roots, Part 1. Early life cycle data of Rhopaliophora (Scyphozoa, Cubozoa and Staurozoa). Plank. Ben. Res. 17, 1–33.

Sugiura, Y., 1962. Strobilation and zooxanthellae of *Mastigias papua* and *Cephea cephea*. Zool. Mag. 71, 410–411.

Sugiura, Y., 1963. On the life-history of rhizostome medusae I. *Mastigias papua* L. Agassiz. Annot. Zool. Jpn. 36 (4), 194–202.

Sugiura, Y., 1964. On the life-history of rhizostome medusae II. Indispensability of Zooxanthellae for strobilation in *Mastigias papua*. Embryologia 8, 223–233.

Sugiura, Y., 1965. On the life-history of rhizostome medusae. III. On the effects of temperature on the strobilation of *Mastigias papua*. Biol. Bull. 128, 493–496.

Sugiura, Y., 1966. On the life-history of rhizostome medusae IV. *Cephea cephea*. Embryologia. 9 (2), 105–122.

Sugiura, Y., 1969. On the life-history of rhizostome medusae V. On the relation between zooxanthellae and the strobilation of *Cephea cephea*. Bull. Mar. Biol. Stat. Asamushi 13, 227–233.

Sun, M., Chen, F., Duan, Y., Sun, J., 2022. Effects of temperature and salinity on podocyst recycling of the edible jellyfish *Rhopilema esculentum* (Kishinouye, 1891). Sustainability 14, 5202.

Takao, M., Uye, S.-i., 2018. Effects of low salinity on the physiological ecology of planulae and polyps of scyphozoans in the East Asian marginal seas: potential impacts of monsoon rainfall on medusa population size. Hydrobiologia. 815, 165–176.

Takauchi, S., Miyake, H., Hirata, N., Nagai, M., Suzuki, N., Ogiso, S., et al., 2021. Morphological characteristics of ephyrae of *Aurelia coerulea* derived from planula strobilation. Fish. Sci. 87, 671–679.

Technau, U., Genikhovich, G., Kraus, J.E., 2015. Cnidaria. Evolutionary developmental biology of invertebrates 1: Introduction, Non-Bilateria, Acoelomorpha, Xenoturbellida, Chaetognatha. Springer, Vienna, pp. 115–163.

Thein, H., Ikeda, H., Uye, S.I., 2012. The potential role of podocysts in perpetuation of the common jellyfish *Aurelia aurita* s.l. (Cnidaria: Scyphozoa) in anthropogenically perturbed coastal waters. Hydrobiologia 690, 157–167.

Thibault, D., Kuplik, Z., Prieto, L., Enrique-Navarro, A., Brown, M.K., Uye, S.-I., Doyle, T.K., Pitt, K., Fitt, W.K., Gibbons, M.J., 2024. Ecology of Rhizostomeae. Adv. Mar. Biol. 98.

Thiel, M.E., 1938. Semaeostomeae [Ontogenie], Dr. H.G. Bronns. Klassen und Ordnungen des Tierreichs. Zweiter Band: Spongiaria, Coelenterata, Echinodermata, II. Abteilung: Coelenterata, 2 buch Scyphomedusae, 5. Lieferung. Akademische Verlagsgesellschaft m.b. H., Leipzig, pp. 677–856.

Thiel, M.E., Reinicke, G.B., Holst, S., Morandini, A.C., Sötje, I., Straehler-Pohl, I., et al., 2024. Rhizostomeae—a remaining manuscript to complete the Monograph on Scyphozoa by ME Thiel (published 1936–1977). Adv. Mar. Biol. 98 (electronic supplement), doi....

Thieme, C., Hofmann, D.K., 2003. Control of head morphogenesis in an invertebrate asexually produced larva-like bud (*Cassiopea andromeda*; Cnidaria: Scyphozoa). Dev. Genes Evol. 213, 127–133.

Tiseo, G.R., 2022. Sexual reproduction in jellyfish species (Scyphozoa, Cnidaria) from the Southeastern Brazilian Coast. PhD Thesis, Universidade de São Paulo, São Paulo, Brazil, pp. 257. (In Portuguese).

Tomaru, A., Sasaki, R., Miyahara, H., Okino, A., Ogawa, N., Hamasaki, K., 2014. Settlement of planulae of the moon jellyfish *Aurelia aurita* onto hydrophilic polycarbonate plates modified by atmospheric plasma treatment. PLoS One 9, e85569.

Uchida, T., 1926. The anatomy and development of a rhizostome medusa, *Mastigias papua* L. Agassiz, with observations on the phylogeny of Rhizostomeae. J. Fac. Sci. Uni. Tokyo Sect. 4, Zool. 45 (1), 95.

Uchida, T., 1927. Report of the Biological Survey of Mutzu Bay. 2. Medusae of Mutsu Bay. Sci. Rep. Tôhoku Univ. 42, 215–238.

Vanhöffen, E., 1888. Untersuchungen über semäostome und rhizostome Medusen. Bibl. Zool. 1, 5–52.

von Lendenfeld, R., 1884. The scyphomedusae of the southern hemisphere. Part III. Proc. Linn. Soc. NSW 9, 259–306.

Watanabe, T., Ishii, H., 2001. In situ estimation of ephyrae liberated from polyps of *Aurelia aurita* using settling plates in Tokyo Bay, Japan. Hydrobiologia. 451, 247–258.

Werner, B., 1973. New investigations on systematics and evolution of the class Scyphozoa and the phylum Cnidaria. Publ. Seto Mar. Biol. Lab. 20, 35–61.

Werner, B., 1984. Stamm Cnidaria, Nesseltiere. In: Lehrbuch der speziellen Zoologie. Erster Hrsg. Kaestner A. Stuttgart 1./2, Fischer. I/2.Teil: 11— 305.

Widersten, B., 1965. Genital organs and fertilization in some Scyphozoa. Zoologiska Bidr. Från Upps. 37, 45–58.

Willcox, S., Moltschaniwskyj, N.A., Crawford, B., 2007. Asexual reproduction in scyphistomae of *Aurelia* sp.: effects of temperature and salinity in an experimental study. J. Exp. Mar. Biol. Ecol. 353, 107–114.

Wolk, M., Rahat, M., Fitt, W.K., Hofmann, D.K., 1985. Cholera toxin and thyrotropine can replace natural inducers required for the metamorphosis of larvae and buds of the scyphozoan *Cassiopea andromeda*. Wilhelm. Roux's Arch. Develop. Biol. 194, 487–490.

Wu, Y., Li, S.Y., Yan, L.P., Jiang, Y.Z., Cheng, J.H., 2009. Effects of temperature and salinity on asexual reproduction of *Rhopilema esculenta*. J. Anhui Agric. Sci. 37, 11414–11418.

Yamamori, L., Okuizumi, K., Sato, C., Ikeda, S., Toyohara, H., 2017. Comparison of the inducing effect of indole compounds on medusa formation in different classes of Medusozoa. Zool. Sci. 34, 173–178.

Yasuda, T., 1975. Ecological studies on the jellyfish, *Aurelia aurita* (Linné), in Urazoko Bay, Fukui Prefecture-XI. An observation on ephyra formation. Publ. Seto Mar. Biol. Lab. 22, 75–80.

You, K., Ma, C., Gao, H., Li, F., Zhang, M., Qiu, Y., et al., 2007. Research on the jellyfish (*Rhopilema esculentum* Kishinouye) and associated aquaculture techniques in China: current status. Aquac. Int. 15, 479–488.

You, K., Ma, C.H., Wang, S.J., Gao, T.X., Yao, Z.H., Wang, X.W., 2010. The effect of short-term low temperature on strobilation of jellyfish, *Rhopilema esculentum* Kishinouye. Acta Hydrobiol. Sin. 34, 1223–1227.

CHAPTER FIVE

Physiology and functional biology of Rhizostomeae jellyfish

Renato M. Nagata[a,b,*], Isabella D'Ambra[c,d], Chiara Lauritano[e], Guilherme M. von Montfort[a,b], Nicolas Djeghri[f,g], Mayara A. Jordano[h], Sean P. Colin[i,j], John H. Costello[j,k], and Valentina Leoni[l]

[a]Laboratório de Zooplâncton, Instituto de Oceanografia, Universidade Federal do Rio Grande (FURG), Rio Grande, RS, Brazil
[b]Programa de Pós-graduação em Oceanografia Biológica (PPGOB), Instituto de Oceanografia, Universidade Federal do Rio Grande (FURG), Rio Grande, RS, Brazil
[c]Integrative Marine Ecology Department, Stazione Zoologica Anton Dohrn, Napoli, Italy
[d]National Center for Future Biodiversity (NFBC), Palermo, Italy
[e]Ecosustainable Marine Biotechnology Department, Stazione Zoologica Anton Dohrn, Napoli, Italy
[f]The Marine Biological Association, The Laboratory, Citadel Hill, Plymouth, United Kingdom
[g]University of Brest (UBO), Institut Universitaire Européen de la Mer (IUEM), Laboratoire des sciences de l'environnement marin (LEMAR, UMR 6539), Plouzané, France
[h]Departamento de Zoologia, Instituto de Biociências, Universidade de São Paulo, São Paulo, SP, Brazil
[i]Marine Biology and Environmental Science, Roger Williams University, Bristol, RI, United States
[j]Whitman Center, Marine Biological Laboratory, Woods Hole, MA, United States
[k]Biology Department, Providence College, Providence, RI, United States
[l]CoNISMa, Consorzio Nazionale Interuniversitario per le Scienze del Mare, Rome, Italy
*Corresponding author. e-mail address: renatonagata@gmail.com

Contents

Advances in Marine Biology, Volume 98
ISSN 0065-2881, https://doi.org/10.1016/bs.amb.2024.07.007

Abstract

Rhizostomeae species attract our attention because of their distinctive body shape, their large size and because of blooms of some species in coastal areas around the world. The impacts of these blooms on human activities, and the interest in consumable species and those of biotechnological value have led to a significant expansion of research into the physiology and functional biology of Rhizostomeae jellyfish over the last years. This review brings together information generated over these last decades on rhizostome body composition, locomotion, toxins, nutrition, respiration, growth, among other functional parameters. Rhizostomes have more than double the carbon content per unit of biomass than jellyfish of Semaeostomeae. They swim about twice as fast, and consume more oxygen than other scyphozoans of the same size. Rhizostomes also have faster initial growth in laboratory and the highest body growth rates measured in nature, when compared to other medusae groups. Parameters such as body composition, nutrition and excretion are highly influenced by the presence of symbiotic zooxanthellae in species of the Kolpophorae suborder. These physiological and functional characteristics may reveal a wide range of adaptive responses, but our conclusions are still based on studies of a limited number of species. Available data indicates that Rhizosotomeae jellyfish have a higher energy demand and higher body productivity when compared to other jellyfish groups. The information gathered here can help ecologists better understand and make more assertive predictions on the role of these jellyfish in their ecosystems.

1. Introduction

Thiel's section on the physiology of Rhizostomeae, in the review up to 1970 (hereafter referred to as Thiel et al., 2024), brings together beyond physiology, aspects of the anatomy, behavior, functional morphology and biochemistry of Rhizostomeae. Much of the information contained in Thiel et al. (2024), especially in the case of papers written in German, has been overlooked in the studies published over the last 50 years, even though they may contain relevant data or, in many cases, the original descriptions on processes in question. An aspect of Thiel's review on physiology that will be new to most readers is the historical perspective that describes how the understanding about physiological processes and on the functioning of jellyfish have changed over time (see Thiel et al., 2024).

One example refers to the evolution in the understanding of the respiration in rhizostome medusae. The pulsating movements of these medusae when swimming have historically been compared to the respiration of the human thorax. The name "poumons de mer" (sea lungs), used by French fishermen to refer to *Rhizostoma pulmo* (Macri, 1778), is an additional historical testament of the human perception on the rhizostome respiration (Thiel et al., 2024). The first anatomy studies attributed the respiratory function to the complex network of rhizostome umbrellar canals, by comparing these canals to blood vessels and arteries (Réaumur, 1710). Earlier authors, however, believed that gas exchange would take place in the subgenital ostia, as these regions are open to the surrounding water (Tilesius, 1829). This view was refused by Agassiz (1862), because of the absence of a connection between the gastrovascular cavity and the fluids that fill the subgenital ostia. The first experiments on respiration in rhizostomes were only carried out by Vernon (1895), who measured O_2 pressure in the body fluids of *R. pulmo in situ* and in aquaria. Vernon (1895) noted that the respiratory activity was significantly lower compared to animals like bony fish. In addition, respiration was comparatively more intense in smaller medusae than in larger ones. This was also found by McClendon (1917) for *Cassiopea xamachana* Bigelow, 1892 and these patterns were attributed to the jelyfish's growing and immaturity. Later, Backmann (1915) demonstrated that oxygen uptake occurred in the gastrovascular cavities, since the oxygen pressure was higher in the water surrounding the animals than in fluids of the gastrovascular cavity.

The force that generated the entrance of fluids into the gastrovascular cavity was another point of debate. Thiel criticized the repetition of the unproven idea of the presence of "suction organs" that would be the driving force behind the ingestion and internal circulation: "Even today, people speak of the "sucking mouths", "sucking ruffs" or even "sucking cups" of the rhizostomes, without considering that, given the structure of these organs, proper sucking is not possible at all and that the whole theory has only been developed on the basis of the anatomical conditions, while physiological experiments on rhizostomes have not been carried out" (Thiel et al., 2024). Today, we recognize the role of the ciliated epidermis in the transport of fluids of the gastrovascular cavities (Avian et al., 2022; Jordano et al., 2024), a process which was originally described by Hamann (1881). In this way, Thiel et al. (2024) is more than a compilation of the information that has existed to date, but a review of the development of ideas about the biology of the group, which gives us the opportunity to give credit to the pioneers of Rhizostomeae biology.

Thiel et al. (2024) gives us a glimpse of the immense technological advances that biological sciences has undergone since the first observations made by earlier naturalists who had little more than a clinical eye, their logic, and a good deal of improvisation at disposal. At a time when omics sciences can reveal structure and functional properties of toxins and enzymes (Li et al., 2014), it is still interesting to know that the first observations on the digestion of food by jellyfish were made by placing gastric cirri on the flesh of fish and verifying its degradation after time, in experiments performed by Müller (1858). Now that technologies such as high-speed cameras are used to describe fast behavioral processes (e.g. Costello et al., 2021), or endocast techniques can be employed to visualize the internal anatomical structure (Avian et al., 2022), we can still recognize the inventiveness of Cuvier (1799) who used milk as a dye to trace that allowed him to first describe internal circulation within the medusa body.

Some of the topics contained in the physiology section of Thiel's review (Thiel et al., 2024) were included in other chapters in this volume, such as sexual reproduction and strobilation, which were reviewed in the chapter about life cycles (Schiariti et al., 2024). In some areas, the focus of our review deviates considerably from the text in Thiel et al. (2024), when Thiel focuses on physiological studies at the most basic tissue level, which is not much covered in our text. Here, we focus on functional aspects of Rhizostomeae biology and prioritize higher levels of organization, at the systems, individual and sometimes populational level, which makes our text more connected to the ecology of the group. Instead of detailing the physiology of muscle function, and how nerve stimuli generate umbrella contractions (see Holst et al., 2024 for a review on rhizostome nervous system, sensing and effector structures), our text explains how umbrella pulsation generates locomotion and discusses how locomotion affects food capture, in order to integrate the different systems (*i.e.* locomotor and alimentary). This is partly due to our desire to understand how functional biology can explain the ecological role and, perhaps, bring answers about environmental drivers of blooms by these organisms.

The number of studies contained in the Thiel et al. (2024) review is not evenly distributed through time. During the first decades of the 20th century, there was a greater number of contributions on physiology, which was interrupted by the period of the World Wars, until 1945. Initial studies on the physiology of Rhizostomeae were carried out with species available around European and US research centers, so that practically all the knowledge on feeding, swimming, respiration, nervous system function and growth was generated for the species *C. xamachana*, *Cotylorhiza tuberculata* (Macri, 1778) and *R. pulmo* (Thiel et al., 2024). Studies with non-European species involved the reproduction of *Phyllorhiza punctata* von Lendenfeld, 1884 from Australia, studies on digestion, enzyme function, musculature and swimming of *Stomolophus meleagris* Agassiz, 1860 from North America. The amount of data on Rhizostomeae species, even among the most frequently studied taxa, is still only a fraction of what has been studied on Semaeostomeae jellyfish, particularly species of the genera *Aurelia, Chrysaora* and *Cyanea* (Arai, 1997; Purcell et al., 2010; von Montfort et al., 2023). There is still no representative of Rhizostomeae that can be considered a model species for the group, even if these species have the desirable characteristics of model species, as ecological relevance, wide distribution, and ease of experimental manipulation. One of the most studied species of Rhizostomeae may be *C. xamachana* and it has been considered as an emerging model for investigating cnidarian-algal symbiosis (Medina et al., 2021). The presence of symbiotic zooxanthellae in species of the genus *Cassiopea* (like in other Kolpophorae species), however, makes them totally different from non-symbiotic rhizostomes, concerning characteristics of development, nutrition, biochemical composition, swimming, excretion and respiration. In addition, jellyfish of *Cassiopea* are benthic, which is also a key ecological difference from the other pelagic jellyfish in the group. It is therefore necessary to have also a pelagic, non-symbiotic species of Rhizostomeae to be used as an experimental model. This should be feasible because characteristics such as swimming patterns (von Montfort et al., 2023) and respiration rates (Purcell et al., 2010) seem to be similar between different species of Rhizostomeae.

Morphological and functional features of rhizostome medusae stand out when compared with other larger medusae. Even so, rhizostome medusae are often grouped with other gelatinous organisms in ecological studies. Our review asks whether differences in the morphology, mechanics of locomotion and filter-feeding behavior between rhizostome and semaeostome medusae have connections with possible differences in functional parameters such as biochemical composition, growth rates, diet, feeding rates, toxins, and other functional parameters. We tested this concept, considering the much smaller amount of work and information is available on the Rhizostomeae. Besides this lack of information, it will be also important to consider what is the phylogenetic coverage of this information, in a group of 91 known species. Each of the following six sections of this review can be read independently. We hope that physiological and functional properties gathered here will allow other researchers to draw more precise models for trophodynamics and population dynamics of these organisms. Finally, we also discuss some information bottlenecks, perspectives for research and major findings that should drive future work on the group.

2. Body composition of Rhizostomeae

Between 95% and 98% of body volume of large medusae of Scyphozoa is composed of water, of which 40% is a hydrated extracellular matrix of their mesoglea (Larson, 1986; Arai, 1997; Kogovšek et al., 2014). Because jellyfish are osmoconformers, the mesogleal extracellular fluid contains salt in similar concentrations as its surrounding fluids. Hence salt is the second main component of body biomass. A medusa of 1 kg contains ~960 g of water and 40 g of dried material. If the surrounding salinity is 35, the 960 g of water should contain ~33.6 g of salts. Thus, about ~84% of the dried material should be salts (Hubot et al., 2022). Only the third major component of medusan biomass is organic content, represented by the sum of macromolecules of protein, lipids, and carbohydrates. The proportion of organic material in tissues of scyphozoan medusae bodies can be up to two orders of magnitude lower than in non-gelatinous organisms, such as crustaceans (Acuña et al., 2011; McConville et al., 2017). Thus, the amount of organic material in the bodies of large scyphozoan medusae can be equal to that of much smaller crustaceans or fishes.

2.1 Elemental composition

The relative proportions of elements can unravel functional characteristics of marine organisms. For example, nitrogen is primarily present in proteins (Arai, 1997; Ikeda, 2014), so low concentrations of this element indicate reduced musculature, hence low locomotion capacity (Ikeda, 2014).

Phosphorus is primarily present in lipids (*i.e.* energy reserves) and its accumulation in gelatinous organisms is negligible compared to crustaceans and fishes (Ikeda, 2014). Moreover, the proportion of C and N in tissues of medusae and ctenophores (as C% of dry mass, C%DM) apparently decrease with increase of habitat temperature (Ikeda, 2014). Jellyfish elemental composition may also have ecological consequences as their organic matter represents nitrogen-rich storage that can supply the environment through excretion of inorganic nutrients and secretion of mucus (Condon et al., 2011; Hubot et al., 2022).

Carbon content (C% of wet mass, C%WM) is a basic metric used to evaluate an organism's functional properties. Analysis of carbon content variations across species may provide insights into how gelatinous tissues evolved within six of the zooplankton phyla, how they provide adaptations and increase in fitness on pelagic environments. For example, respiration rates of marine zooplankton are different on wet-weight basis, but similar in terms of body carbon content (Schneider, 1992). Although respiration of jellyfish conforms to the rates of other animals, in terms of carbon-based allometric relationships, other functional characteristics deviate when body size is transformed to carbon content. Some gelatinous taxa, like tunicates, have higher carbon content-specific feeding rates than other non-gelatinous zooplankton (Acuña, 2001) and similar clearance rates to fish (Acuña et al., 2011). Moreover, pooled gelatinous zooplankton taxa, including rhizostomeae medusae, have growth rates twice as fast, excretion rates ten times slower, and, because of their larger body sizes, can operate in Reynolds number flows ~10 times greater than other pelagic organisms of equivalent carbon content (Hirst et al., 2003; Pitt et al., 2013).

Rhizostomeae species are often pooled together with medusae species of other groups (i.e. Semaeostomeae and Coronatae) in studies that evaluate the functional and ecological effects of carbon and water contents (e.g. Ikeda, 2014; McConville et al., 2017; Pitt et al., 2013). However, a compilation of carbon content of non-gelatinous (Copepoda, other Crustacea), semi-gelatinous taxa (Mollusca, Polychaeta), and "true" gelatinous zooplankton (Cnidaria, Ctenophora, Thaliacea) demonstrates that the greatest variability of carbon content within groups occurs in Scyphozoa, ranging from 0.02, in *Poralia rufescens* Vanhöffen, 1902, to 1.2, in *Catostylus mosaicus* (Quoy and Gaimard, 1824) (C%WM) (McConville et al., 2017). Furthermore, when we reanalyzed published data of gelatinous zooplankton taxa (Lucas et al., 2011; McConville et al., 2017; Pitt et al., 2013), variability in carbon content seems to be related to the three traditional scyphozoan groups (i.e. Rhizostomeae, Semaeostomeae and Coronatae). Carbon content (C%WM) of rhizostome medusae ranges from 0.34, for *R. pulmo,* to 1.21 for *Rhopilema nomadica* Galil, Spanier and Ferguson, 1990 (Table 1). Rhizostome carbon content (C%WM mean: 0.69 ± 0.3, *n* = 9 species) is more similar to those of coronate medusae (C%WM: 0.70 ± 0.09, *n* = 2 species), although the number of species quantified of the latter group is limited. Carbon content of rhizostome medusae is on average higher by a factor of 2.6, compared to semaeostome medusae (mean: 0.27 ± 0.17, *n* = 9 species) (McConville et al., 2017; Pitt et al., 2013; Purcell et al., 2010). Higher proportion of carbon in Rhizostomeae than in Semaeostomeae seems to be contradictory to the pattern that carbon content of jellyfish (pelagic cnidarians and ctenophores) decreased with the increase of habitat temperature (Ikeda, 2014). Because rhizostomes inhabit tropical and subtropical environments, whereas semaeostomes can inhabit colder temperate areas, variation in the proportion of body carbon of scyphozoan medusae may be related to a factor other than the habitat temperature.

Changes on carbon content may also occur through ontogeny, as demonstrated in the rhizostome *R. nomadica,* of which juvenile (<16 mm) had higher proportions (~8.7%) compared to larger (>23 mm) specimens (4.8%) (Kuplik et al., 2021). Despite the limitation that the evaluated size range includes only juvenile medusae, similar results were previously reported for the semaeostome *Aurelia aurita* (Linnaeus, 1758) (Lucas, 1994). An apparently inverse pattern was observed for *Rhizostoma luteum* (Quoy and Gaimard, 1827) in which dry mass (DM%WM) increased from 3.2%, in juvenile (35 days-old) medusae of 6.2 mm of bell diameter (BD), to 31%, in medusae (109 days-old) of 9.2 mm (Prieto et al., 2018). These are limited size ranges compared to the dramatic increase in size that some species may undergo through development. Thus, alterations in carbon content during development of rhizostome medusae requires further quantification. Importantly, the use of conversion factors (Table 1) should prioritize phylogenetically related species and further research should explore how the variability in biochemical composition affects functional and ecological traits between groups of Scyphozoa.

2.2 Proximate composition

The proximate composition (proteins, lipids and carbohydrates) of rhizostome medusae is similar to other medusae in the sense that proteins are the major component, comprising up to 40% of dried mass (Leone et al., 2015). Proximate composition data (as % of WM or DM) of rhizostome species in the

Table 1 Biometric conversions for Rhizostomeae medusae in terms of dry mass % of wet mass (DM%WM), carbon mass % dry mass (C% DM), carbon mass % wet mass (CW%WM) and environmental salinity. ND means "no data".

Species	DM%WM	C%DM	C%WM	Salinity	Source
Cassiopea xamachana	4.5	11.9	ND	ND	Lutcavage and Lutz (1986)
Catostylus mosaicus	ND	ND	1.17	ND	Kiørboe (2013)
Catostylus mosaicus	ND	14.1	ND	ND	Lucas et al. (2011)
Cotylorhiza tuberculata	4.2	ND	ND	34–37	Kikinger (1992)
	ND	ND	0.55	ND	Pitt et al. (2013)
Eupilema inexpectata	ND	2.2 ± 0.2	ND	ND	Hubot et al. (2022)
Mastigias papua	4.7	14.3	0.67	35	Lucas et al. (2011)
Nemopilema nomurai	4.9	11.2	0.55	35	Lucas et al. (2011)
	ND	ND	0.60	ND	Purcell et al. (2010)
	4.2	6.6	ND	ND	Iguchi et al. (2017)
Phyllorhiza punctata	3.8	12	0.46	29–32.5	Graham et al. (2003)
Rhizostoma luteum	3.5%–31%	5.9 ± 0.3	ND	ND	Prieto et al. (2018)
Rhizostoma octopus	ND	ND	0.35	ND	Pitt et al. (2013)

(continued)

Table 1 Biometric conversions for Rhizostomeae medusae in terms of dry mass % of wet mass (DM%WM), carbon mass % dry mass (C%DM), carbon mass % wet mass (CW%WM) and environmental salinity. ND means "no data". (*cont'd*)

Species	DM%WM	C%DM	C%WM	Salinity	Source
Rhizostoma pulmo	ND	ND	0.98	ND	Kiørboe (2013)
	ND	9.3 ± 0.4	ND	ND	Hubot et al. (2022)
	6.3	5.6	0.34	38	Purcell et al. (2010)
Rhopilema esculentum	5.5	22	1.21	ND	Lucas et al. (2011)
Rhopilema hispidum	5.2	14.6	0.76	ND	Pitt et al. (2013)
Rhopilema nomadica	ND	5–8	ND	39.5–40	Kuplik et al. (2021)

different body regions (e.g. oral arms, bell) were compiled in Table 2. The concentrations of carbohydrates and lipids are usually one order of magnitude lower than proteins (Arai, 1997; Khong et al., 2016). The main proteinaceous content is comprised of collagen, which can account for up to 35% of the lyophilized dry mass (Nagai et al., 2000). A detailed review of the collagen content of rhizostome species is available in Brotz et al. (2024). Several biochemical characterizations have been recently published for *R. luteum* (Prieto et al., 2018), *R. pulmo* (Leone et al., 2015), *C. tuberculata* (Leone et al., 2013), and *Cassiopea andromeda* (Forskål, 1775) (De Dominico et al., 2023). Higher lipid content of *C. tuberculata* has been related to photosynthetic membranes of zooxanthellae (Prieto et al., 2018). The relative proportions of polyunsaturated fatty acids were usually higher than those of saturated, monounsaturated fatty acids (De Dominico et al., 2023; Leone et al., 2013; Prieto et al., 2018).

The proportion of organic components differ among medusae body parts (umbrella, oral arms, gonads), with the umbrella usually containing a lower proportion of organic material. For *Rhizostoma octopus* (Gmelin, 1791), ash-free dry mass (AFDM) was not significantly different among body parts, however, protein content was twice as high in gonads (12.1%DM) and oral arms (13.4%DM) than in umbrella (6.6%DM). Lipid content was twice as high in the gonads as in oral arms and umbrella (Doyle et al., 2007). Thus, as expected, energy content per unit of biomass was significantly higher in gonads (mean: 5.59 kJ g DM^{-1}) than oral arms (mean:3.62 kJ g DM^{-1}) and bell (mean: 1.87 kJ g DM^{-1}) (Doyle et al., 2007). For the species *Acromitus hardenbergi* Stiasny, 1934, *Rhopilema hispidum* (Vanhöffen, 1888), and *Rhopilema esculentum* Kishinouye, 1891, the protein and energy content were approximately twice as high in oral arms than in umbrella (Khong et al., 2016). For *R. pulmo* about three times more protein was found in oral arms than in the umbrella (Carli et al., 1991). A comparison indicated similar biochemical composition of scyphozoan body and mucus in terms of macromolecules and elements, although mucus has much less energy content (Hubot et al., 2022). The composition of macromolecules of rhizostome medusae have been recently studied because of the potential industrial application as food products, including collagen extraction, and because of the present toxins and bioactivities in molecules (e.g. antioxidant, anticoagulant capacity), that are exemplified in Brotz et al. (2024).

2.3 The impact of zooxanthellae on biochemical composition

The Rhizostomeae medusae of the suborder Kolpophorae (representing roughly half of Rhizostomeae diversity) harbor symbiotic dinoflagellates

Table 2 Biochemical composition (mean ± SD) as a percentage of dry (DM) or wet mass (WM) in different body parts of Rhizostomeae.

Species	Body part	Carb.	Lip.	Prot.	Carb.	Lip.	Prot.	References
		(% medusa WM)			(% medusa DM)			
Acromitus hardenbergi	B			0.8 ± 1.2	17.7	0.4 ± 0.2	21.4 ± 0.3	Khong et al. (2016)
	OA			1.3 ± 1.0	6	1.1 ± 0.2	33.7 ± 1.1	
Catostylus tagi	W		0.4					Morais et al. (2009)
	OA		0.5	4.3				
	B		0.2	1.8				
Cotylorhiza tuberculata	W					12.3 ± 0.7	2.2	Leone et al. (2015)
	OA					6.4	20	Carli et al. (1991)
	B					0.7	12	
	BM					0.5	7.6	
	G					6	36.8	
Rhizostoma octopus	W				0.8	0.3	12.8 ± 2.3	Doyle et al. (2007)
	OA				0.7 ± 0.3	0.3 ± 0.1	13.4 ± 0.4	
	B				0.7 ± 0.01	0.3 ± 0.1	6.6 ± 2.3	
	G				0.9 ± 0.03	0.6 ± 0.4	12.1 ± 9.8	

Rhizostoma pulmo	W			2.3		Gubareva et al. (1983)
	W			4.0 ± 0.1	6	Leone et al. (2015)
	B			0.7	8.7	Carli et al. (1991)
	BM			1	13.7	
	G			1.2	18	
	OA			0.8	27	
Rhizostoma luteum	W				0.8 – 1.9	Prieto et al. (2018)
Rhopilema hispidum	OA	2.0 ± 1.6	10.7	1.4 ± 0.2	43.8 ± 0.2	Khong et al. (2016)
	B	0.5 ± 0.2	18.2	0.5 ± 0.3	19.9 ± 0.7	
Rhopilema esculentum	OA	2.7 ± 0.9	7.7	1.8 ± 0.3	53.9 ± 2.1	
	B	1.6 ± 0.8	8.9	0.6 ± 0.1	38.1 ± 1.1	
Stomolophus meleagris	B	1.1 ± 0.2				Huang (1988)
	M	1.0 ± 0.1				

B, bell; BM, bell margin; Carb., carbohydrates; G, gonads; Lip., lipids; M, manubrium; OA, oral arms; Prot., proteins; T, tentacles; W, whole body.

(Symbiodiniaceae), called zooxanthellae in their tissues (see Holst et al., 2024). The density of symbionts in medusae tissues is typically in the order of few millions of cells per gram of wet mass (see Holst et al., 2024). Due to exchanges of molecules, such densities affect the biochemical composition, nutrition and physiology of the holobiont and of the host itself. In terms of elemental stoichiometry, zooxanthellate Kolpophorae are generally enriched in carbon with C:N ratios often superior to 5 (Djeghri et al., 2020; 2021) as compared to generally less than 4 for other gelatinous zooplankton (Ikeda, 2014; Molina-Ramírez et al., 2015). These high C:N ratios are likely linked to the tendency of zooxanthellae to accumulate carbon rich lipid reserves (e.g. Jiang et al., 2014). Lipid composition is thus also affected by zooxanthellae. Zooxanthellate Kolpophorae generally present high proportions of neutral lipids (Djeghri et al., 2021) and of fatty acids typically found in dinoflagellates such as the docosahexaenoic acid (22:6$n-3$; in *Cassiopea* sp., Mortillaro et al., 2009; in *Mastigias papua* (Lesson, 1830), Djeghri et al., 2021; in *C. tuberculata*, Leone et al., 2015).

2.4 Body composition and marine pollution

Body composition is affected by environmental concentrations of trace elements and is particularly concerning because higher accumulation of multiple pollutants has been demonstrated in some rhizostome medusae. Concentrations of trace elements (Al, Ti, Cr, Mn, Fe, Ni, Cu, Zn, As, Cd, Sn, and Pb) in *R. pulmo* were, in most cases, two orders of magnitudes higher than in the seawater in a coastal lagoon in Southeastern Spain (Muñoz-Vera et al., 2016). Concentrations were up to 4- to 2-fold higher in the oral arms than in the umbrella (Muñoz-Vera et al., 2016). Concentrations of Al, As, Ba, Ca, Cd, Cr, Cu, Fe, Li, Mg, Mn, Pb, Sr in tissues were determined for *Mastigias* sp., *Cassiopea* sp., and *Netrostoma* sp. along the Great Barrier Reef, Australia (Templeman and Kingsford, 2012). *Cassiopea* sp. accumulated Ba, Cu, Mn, Pb and Zn (from 28 to 221 times seawater concentrations). *Mastigias* sp. accumulated Cu and Zn (37x and 193x seawater), while *Netrostoma* sp. accumulated only Cr (2x seawater).

Concentrations of trace elements in tissues reflect spatial heterogeneity and localized sources, potentially providing unique signatures among locations (Epstein et al., 2016). Populations of *R. pulmo* from different regions of Mar Menor lagoon in Spain have significantly different tissue compositions on south-north and east-west axes of the Lagoon (Muñoz-Vera et al., 2016). Concentrations of Cu and Zn in tissues of *Cassiopea maremetens* Gershwin, Zeidler & Davie, 2010 increased linearly along sites with increased urban gradient factors, in estuarine areas from northeast Australia (Epstein et al., 2016). In areas of the Great Barrier Reef, Australia, concentrations of Al, As, Cu, Zn and Fe in tissues of *Mastigias* sp. and *Netrostoma* sp. decreased with increase of the distance from the coast in (Templeman and Kingsford, 2012).

The sequestration of some elements may be related, as correlations in tissue concentrations have been detected for Al-Pb, Al-Fe, Ti-Zn, Ti-As, Cr-Fe, Mn-Fe, and Zn-As, in *C. tuberculata* (Muñoz-Vera et al., 2015). Accumulation of Cd were experimentally demonstrated for *C. andromeda,* as tissue concentrations increased linearly after 7 days of exposure (Béziat and Kunzmann, 2022). The species were apparently capable of regulating intake of Pb, and of taking up anthropogenic Gd complexes without bioaccumulation (Béziat and Kunzmann, 2022). Biokinetic models were applied to predict accumulation of Cu and Zn in *C. maremetens* based in a 28 days' experiment, that included 14 days of exposure and 14 days of clearance of these metals (Templeman and Kingsford, 2015). Concentrations of Cu rapidly increased (500% in the first 24 h) reaching up to 33.78 $\mu g\ g^{-1}$ of DM, and was also rapidly depurated (half-life of 1.68 days). The accumulation and depuration of Zn were slower and linear, reaching concentrations up to 125.1 $\mu g\ g^{-1}$ of DM, and decreasing to 63.32 $\mu g\ g^{-1}$ DM, which was approximately double the background concentration (Templeman and Kingsford, 2015). The presence of nutrients has the potential to increase rates of accumulation by binding with metals, as the uptake of Cu under high nutrient conditions significantly increased (~18%) compared to ambient nutrient concentrations (Epstein et al., 2016). Such results demonstrate the importance of multiple stressors (e.g. eutrophication, temperature) to bioaccumulation of trace metals.

Some features of rhizostome medusae are particularly worrying with regard to environmental pollution and bioaccumulation. Rhizostome medusae are capable of clearing enormous volumes of water (cf. 5.2 Feeding rates and feeding impacts) and have rapid turnover of biomass (Epstein et al., 2016). Human consumption of rhizostome jellyfish (reviewed in Brotz et al., 2024) is particularly concerning because of higher levels of Al found in food samples from China, that is probably related to production method (Raposo et al., 2022). The presence of Al was found in concentrations up to one order of magnitude higher than recommended (100 $mg\ kg^{-1}$) and there are innumerous potential connections of Al and diseases such as anemia, oxidative processes, neurodegenerative disorders, etc. (Raposo et al., 2022). Rhizostome medusae are widespread in tropical and subtropical shallow waters, estuaries and coastal lagoons that have been disproportionally affected by human-induced stressors. The number of known jellyfish predators (reviewed in Thibault et al., 2024) poses reason for concern about trophic web contamination. Particularly in the case of benthic medusae of *Cassiopea* spp. its behavior of continuously pumping interstitial waters may both concentrate or release contaminants towards the surface (Béziat and Kunzmann, 2022). Such studies demonstrate the potential use of rhizostome medusae, particularly of the benthic medusa genus *Cassiopea*, as bioindicators. These medusae are easy to maintain in laboratory conditions, have low mortality rates, can rapidly absorb trace metals and reflect fine spatial and temporal scales of continuous or pulse-contamination (Epstein et al., 2016; Templeman et al., 2021).

2.5 Methodological challenges and outlook

Environmental salinity may affect jellyfish dry mass (DM) quantification (Hirst and Lucas, 1998). Kogovšek et al. (2014) demonstrated that organic matter determined after combustion was overestimated by 51% (±13%) on average, when compared to material that was previously dialyzed. They concluded that traditional processing methods (without dialyzing) might affect the determination of C and N, with a possible increase in estimates of 1% per 10 salinities unities. Thus, environmental salinity where animals were collected should be reported, and corrections should be applied, if possible.

Organic content of tissues should be the same as the AFDM, which represents the burnt mass of a tissue sample that is ashed in a furnace. In gelatinous zooplankton, however, residual or bound water remains after tissues are dried (60 °C) to constant mass, and this residual water accounts for the difference between true organic content and AFDM (Arai, 1997; Kogovšek et al., 2014). The bound water of jellyfish tissues represents a methodological challenge for quantification of organic component and elemental measurements (carbon, nitrogen and phosphorus) (Arai, 1997). Therefore, measurements of organic proportions require corrections, for example, assuming that bound water represents a proportion (e.g. 11.7%) of AFDM (Doyle et al., 2007). Such estimates of bound water content, however, are not available for rhizostome medusae.

3. Nematocysts and toxins

All cnidarians possess cnidae, specialized organelles which are full of venomous compounds used to subdue prey and discourage predators. The presence of these highly specialized cells (see Holst et al., 2024) is unique in this group within the living and extinct organisms on Earth, and it has been recently suggested as one of the keys to the adaptive success of cnidarians across time (Guo et al., 2022). The structure of cnidae is similar across Cnidaria (Holst et al., 2024). The capsule has a bulb-like shape surrounded by collagen walls and is filled with venom. A coiled, hollow thread-like tubule is everted after a mechanical stimulation of the cnidocil. Many tubules are able to penetrate the skin and inject the venom contained in the capsule, particularly nematocysts (Mariscal, 1974). Some threads have an enlarged basal portion (the shaft) and bear numerous spines. The classification of cnidae is based on the presence and shape of the shaft and the type and distribution of spines (Holst et al., 2024; Mariscal, 1974; Östman, 2000). Cnidae are classified into nematocysts, spirocysts and ptychocytes (Mariscal, 1974; Mariscal et al., 1977; Östman, 2000). Scyphozoa, like Hydrozoa, Cubozoa and Staurozoa, possess only nematocysts, which have two walls surrounding the capsule and are sub-classified into about 30 different types (Mariscal, 1974; Östman, 2000).

3.1 Nematocysts

Although the cnidome of most rhizostomes is unknown at present, yet the few described cnidomes indicate that at least two types of different nematocysts can be found within the same medusa. More details on the composition of the rhizostome cnidome are contained in the chapter on the histology of Rhizostomeae in this volume (Holst et al., 2024). The same type of nematocyst can vary in size, as it was found in *Lobonemoides robustus* Stiasny, 1920 (Kondo et al., 2020), *Nemopilema nomurai* Kishinouye, 1922 (Kubota et al., 2006), and *P. punctata* (Peach and Pitt, 2005). The functions of the different sizes and types of nematocysts remain obscure. For example, the different types of nematocysts did not show any correlation with the number of captured zooplankton prey (Peach and Pitt, 2005). Possessing a portfolio of nematocyst types and sizes appears to be a common trait among Rhizostomeae as well as Semaeostomeae (Avian et al., 1991; Calder, 1977; Colin and Costello, 2007). These nematocyst portfolios can vary across different life stages within the life cycle (Ballesteros, et al. 2021) and defining the cnidomes of more species will help understanding ecological roles played by rhizostome medusae.

Recently, a new structure containing nematocysts has been described in *C. xamachana, M. papua, P. punctata* and *C. mosaicus*, which was named cassiosomes (Ames et al., 2020; Holst et al., 2024). The venom released by cassiosomes is likely made of a proteinaceous matrix, but further studies are required to better describe it (Ames et al., 2020). Cassiosomes of *C. xamachana*, *M. papua* and *P. punctata* are very similar and highly motile, while *C. mosaicus* discharges non-motile cassiosomes which do not contain microalgae of the genus *Symbiodinium*, but unidentified microalgae which are distributed homogenously throughout the cells (Ames et al., 2020). Considering that Rhizostomeae lack marginal tentacles (see Morandini, 2024), which are common in Semaeostomeae and are used for prey capture, the presence of cassiosomes within the appendages of rhizostomes appears to have a similar function to the fishing tentacles of semaeostomes. However, the identification of proteinaceous compounds within cassiosomes suggests that these specialized and complex structures may have unknown additional functions (Ames et al., 2020).

The mechanism of nematocyst discharge is activated by chemical and mechanical stimulation. When the cnidocil is stimulated, the thread, which is enveloped inside the capsule, is pushed by osmotic pressure outside the cell and eventually penetrates the object which stimulated the cnidocil (Arai, 1997). In most organisms, the injection induces paralysis, which facilitates the following capture by the scyphozoan. The prey capture mechanism described above is basically the same across life stages, from polyp to ephyra and medusa alike (Fautin, 2009).

3.2 Toxins

Most rhizostome scyphozoan medusae induce only mild stings in humans, while species like *C. tuberculata* and *L. lucerna* are known to be completely harmless to humans (Mariottini and Pane, 2010; 2016). This may explain why the venoms of this group have been comparatively less studied than other Cnidaria groups, such as Cubozoa and Anthozoa (D'Ambra and Lauritano, 2020). Nevertheless, some toxins have been identified within Rhizostomeae (Table 3). Although the composition of the venoms is often not known in detail, they appear to contain mainly a variety of proteinaceous compounds (peptides, proteins, enzymes and proteinase inhibitors), likely as a consequence of the fact that proteins account for most of the organic matter in all scyphozoan medusae (Merquiol et al., 2019). In addition to proteinaceous compounds, non-proteinaceous compounds (purines, quaternary ammonium compounds, biogenic amines and betaines) have been identified within toxins of Rhizostomeae (Table 3). The heaviest compound (55–95 kDa) is hyaluronidase and has been identified in *R. esculentum* (Lee et al., 2011).

Table 3 Toxins identified in Rhizostomeae, including their molecular mass (kDa) and biological activity. N/A means not available.

Species	Compound	Molecular mass (kDa)	Biological activity	References
Cassiopea andromeda	Phospholipase A_2	N/A	Hemolytic, dermonecrotic, local skin irritation	Radwan et al. (2001)
Cassiopea xamachana	Phospholipase A_2	N/A	Hemolytic, dermonecrotic, local skin irritation	Radwan et al. (2001)
	CassTX-A, CassTX-B, CassTX-C	N/A	N/A	Ames et al. (2020)
Cotylorhiza tuberculata	Raw extract	N/A	Cytotoxic	Riccio et al. (2022)
Nemopilema nomurai	Metalloproteinases	28–36	Gelatinolytic, caseinolytic, fibrinolytic	Lee et al. (2011)
		20–40/10–15	Cytotoxic, hemolytic	Kang et al. (2009)
Phyllorhiza punctata	Phospholipase A2	N/A	Neurotoxic	Carneiro et al. (2011)
Rhizostoma pulmo	Rhizoprotease	95	Proteolytic, hemolytic	Rastogi et al. (2017)
	Rhizolysin	260	Hemolytic	Cariello et al. (1988)
		N/A	Cytotoxic, hemolytic	Allavena et al. (1998)

(continued)

Table 3 Toxins identified in Rhizostomeae, including their molecular mass (kDa) and biological activity. N/A means not available. (*cont'd*)

Species	Compound	Molecular mass (kDa)	Biological activity	References
Rhopilema esculentum	Metalloproteinases	N/A	Gelatinolytic, caseinolytic, fibrinolytic	Lee et al. (2011)
	Hyaluronidase	55–95	Degradation of extracellular matrix components	Lee et al. (2011)
		N/A	Proteolytic, cytotoxic, hemolytic	Li et al. (2005); Yu et al. (2007)
Rhopilema nomadica	Phospholipase A2	N/A	Hemolytic	Gusmani et al. (1997)
	Serine protease	N/A	Local skin damage	Yoffe and Baruchin (2004)
Rhopilema sp.	Crude venom	N/A	Hemolytic	Jafari (2019)
Stomolophus meleagris	SmP90	90	Radical scavenging	Li et al. (2014)
	Phospholipase A_2	N/A	Cytotoxic, cytolytic, hemolytic, local tissue damage	Li et al. (2014)
	218 toxins including C-type lectin, phospholipase A2, potassium channel inhibitors, serine proteinase inhibitors, metalloproteinases and hemolysins	N/A	Voltage-gated potassium channel inhibitor	Li et al. (2014)

Phospholipase A_2 has been identified in six species out of the eight for which toxic compounds have been analysed (Table 3). This finding is in line with the fact that this compound is common in cnidarians (D'Ambra and Lauritano, 2020). This compound is found not only in marine animals such as cnidarians and cephalopods, but also in terrestrial ones, such as mammals, reptiles (snakes and anguimorph lizards), centipedes, insects, arachnids (scorpions, spiders, and ticks) (Fry et al., 2009; Nevalainen et al., 2004).

Phospholipase A_2 induces the breakdown of glycerophospholipids, which results in the production of lysophospholipids and fatty acids, such as arachidonic acid (Murakami and Kudo, 2002). Arachidonic acid produces metabolites (prostaglandins, thromboxanes and leukotrienes) which control several different cellular functions, including the catabolism of lipids derived from the diet, metabolic processes within the cell membrane and inflammatory processes (Romero et al., 2010). In rhizostomes, the biological activity of phospholipase A_2 is associated mainly with hemolytic and neurotoxic activities (Table 3), as already detected in other scyphozoans (D'Ambra and Lauritano, 2020) and it is consistent with the general effects determined for this compound (Romero et al., 2010). Given the similarity of effects, this compound may have different functions in rhizostomes, such as defense, immobilization and digestion of prey, as suggested for Cnidaria in general (Nevalainen et al., 2004).

Basically, metalloproteinases are proteinase enzymes, which bear a metal atom to perform their catalytic activity. They include a variety of different compounds. Metalloproteinases are found in the venom of terrestrial animals, such as centipedes, snakes and ticks (Fry et al., 2009; Undheim and King, 2011). They induce hemorrhagia and necrosis by degrading the extracellular matrix and preventing blood clot formation (Fox and Serrano, 2005; Fry et al., 2009). These effects are very similar to those recorded after a typical jellyfish sting (skin damage, edema, blister formation, myonecrosis and inflammation) (Fox and Serrano, 2005). Metalloproteinases were found in stinging rhizostomes such as *S. meleagris* (Li et al., 2014), *N. nomurai* and *R. esculentum*, in which gelatinolytic, caseinolytic and fibrinolytic effects (Lee et al., 2011). Purification of the extract from the oral arms of the Mediterranean rhizostome *R. pulmo* lead to the isolation of a zinc metalloproteinase which was named Rhizoproteinase, with a hemolytic activity (Rastogi et al., 2017). In addition to Rhizoproteinase, the hemolytic activity shown by extracts from *R. pulmo* was attributed to a lysin proteinase, which was named Rhizolysin after the species from which it was isolated (Cariello et al., 1988) and an unnamed compound which had cytolytic and clastogenic activities in addition to the hemolytic ones (Allavena et al., 1998).

To date, *S. meleagris* is the only rhizostome species in which ion channel inhibitors have been identified (Li et al., 2014). This group includes different types with specific physiological, pharmacological and structural characteristics (Messerli and Greenberg, 2006). Voltage-gated ion channels regulate the excitability of cells and neuromuscular transmission of signals. Voltage-gated ion channels activate non-selective pores within membranes through which ions move using the electrochemical gradient across the membrane itself (Messerli and Greenberg, 2006). The transmission of signals through the neurons and muscles is highly sensitive to alterations in the gradient and passage of ions, which may result in paralysis. Recently, Riccio et al. (2022) tested the effects of the methanolic extract and enriched fractions extracted from *C. tuberculata* on various human tumor cell lines and found that they had cytotoxic activity against melanoma cells, with potential applications for bioprospecting, as better described in the chapter anthropogenic use by Brotz et al. (2024).

3.3 Omics

Recently, different omics technologies have been applied to rhizostome species providing new resources to understand better their ecology and possible biotechnological applications. Searching in the public database GenBank, only four genomes are available for species belonging to Rhizostomeae (Table 4. All of them have been deposited very recently (https://www.ncbi.nlm.nih.gov/genome/?term=Rhizostomeae; accessed on February 23rd, 2024), showing that genomic approaches are recent for this group. Transcriptomes are available for *R. esculentum*, *N. nomurai*, *C. xamachana*, *C. andromeda*, and *S. meleagris* (deposited on the SRA database in Genbank; https://www.ncbi.nlm.nih.gov/sra/?term=Rhizostomeae, by using RNA as filter, accessed on February 23rd, 2024, Table 5). Examples

Table 4 Genomes of Rhizostomeae deposited in Genbank, their assembly accession numbers (accessed on February 23rd, 2024).

Species	Accession number	Date
Cassiopea andromeda	GCA_018155075.1	28/04/2021
C. xamachana	GCA_900291935.1	24/02/2018
Nemopilema nomurai	GCA_003864495.1	05/12/2018
Rhopilema esculentum	GCA_013076305.1	13/05/2020

Table 5 Transcriptomic data available for Rhizostomeae deposited on the SRA database in Genbank, including release data, accession numbers, library strategy, platform model, and species scientific name (accessed on February 23rd, 2024). RNA-seq stands for RNA-sequencing, while EST stands for expressed sequence tag.

Release date	Accession number	Library strategy	Platform model	Scientific name
02/06/17 17:30	ERX2051540	RNA-Seq	Illumina HiSeq 2500	*Cassiopea xamachana*
02/06/17 17:30	ERX2051541	RNA-Seq	Illumina HiSeq 2500	*Cassiopea xamachana*
02/06/17 17:30	ERX2051542	RNA-Seq	Illumina HiSeq 2500	*Cassiopea xamachana*
02/06/17 17:30	ERX2051543	RNA-Seq	Illumina HiSeq 2500	*Cassiopea xamachana*
02/06/17 17:30	ERX2051544	RNA-Seq	Illumina HiSeq 2500	*Cassiopea xamachana*
02/06/17 17:30	ERX2051545	RNA-Seq	Illumina HiSeq 2500	*Cassiopea xamachana*
02/06/17 17:30	ERX2051546	RNA-Seq	Illumina HiSeq 2500	*Cassiopea xamachana*
02/06/17 17:30	ERX2051547	RNA-Seq	Illumina HiSeq 2500	*Cassiopea xamachana*
02/06/17 17:30	ERX2051548	RNA-Seq	Illumina HiSeq 2500	*Cassiopea xamachana*
02/06/17 17:30	ERX2051549	RNA-Seq	Illumina HiSeq 2500	*Cassiopea xamachana*
02/06/17 17:30	ERX2051550	RNA-Seq	Illumina HiSeq 2500	*Cassiopea xamachana*
02/06/17 17:30	ERX2051551	RNA-Seq	Illumina HiSeq 2500	*Cassiopea xamachana*
12/05/14 00:00	SRX470083	RNA-Seq	Illumina HiSeq 2000	*Stomolophus meleagris*
14/04/16 22:23	SRX1692760	EST	Illumina HiSeq 2500	*Rhopilema esculentum*
14/04/16 22:23	SRX1692774	EST	Illumina HiSeq 2500	*Rhopilema esculentum*
14/04/16 22:22	SRX1692805	EST	Illumina HiSeq 2500	*Rhopilema esculentum*

(continued)

Table 5 Transcriptomic data available for Rhizostomeae deposited on the SRA database in Genbank, including release data, accession numbers, library strategy, platform model, and species scientific name (accessed on February 23rd, 2024). RNA-seq stands for RNA-sequencing, while EST stands for expressed sequence tag. (*cont'd*)

Release date	Accession number	Library strategy	Platform model	Scientific name
01/07/17 23:47	SRX1894802	RNA-Seq	Illumina HiSeq 4000	*Rhopilema esculentum*
29/06/17 00:03	SRX1883973	RNA-Seq	Illumina HiSeq 4000	*Rhopilema esculentum*
15/08/17 00:03	SRX3033117	RNA-Seq	Illumina HiSeq 2500	*Rhopilema esculentum*
28/07/19 01:21	SRX4329763	RNA-Seq	Illumina HiSeq 2000	*Rhopilema esculentum*
28/07/19 01:21	SRX4329760	RNA-Seq	Illumina HiSeq 2000	*Rhopilema esculentum*
28/07/19 01:21	SRX4329761	RNA-Seq	Illumina HiSeq 2000	*Rhopilema esculentum*
28/07/19 01:21	SRX4329762	RNA-Seq	Illumina HiSeq 2000	*Rhopilema esculentum*
28/07/19 01:21	SRX4329759	RNA-Seq	Illumina HiSeq 2000	*Rhopilema esculentum*
28/07/19 01:21	SRX4329758	RNA-Seq	Illumina HiSeq 2000	*Rhopilema esculentum*
07/12/18 09:50	SRX3399143	RNA-Seq	Illumina HiSeq 2500	*Nemopilema nomurai*
07/12/18 09:50	SRX3399140	RNA-Seq	Illumina HiSeq 2500	*Nemopilema nomurai*
20/09/19 00:22	SRX4610615	RNA-Seq	Illumina HiSeq 2500	*Nemopilema nomurai*
20/09/19 00:22	SRX4610616	RNA-Seq	Illumina HiSeq 2500	*Nemopilema nomurai*
20/09/19 00:22	SRX4610617	RNA-Seq	Illumina HiSeq 2500	*Nemopilema nomurai*
01/01/19 00:20	SRX4928468	RNA-Seq	Illumina HiSeq 2500	*Nemopilema nomurai*
01/01/19 00:20	SRX4928469	RNA-Seq	Illumina HiSeq 2500	*Nemopilema nomurai*
01/01/19 00:20	SRX4928470	RNA-Seq	Illumina HiSeq 2500	*Nemopilema nomurai*

12/12/19 05:58	SRX5013162	RNA-Seq	Illumina HiSeq 4000	*Rhopilema esculentum*
12/12/19 05:58	SRX5013161	RNA-Seq	Illumina HiSeq 4000	*Rhopilema esculentum*
12/12/19 05:58	SRX5013160	RNA-Seq	Illumina HiSeq 4000	*Rhopilema esculentum*
13/12/19 06:16	SRX5014510	miRNA-Seq	Illumina HiSeq 2500	*Rhopilema esculentum*
13/12/19 06:16	SRX5014509	miRNA-Seq	Illumina HiSeq 2500	*Rhopilema esculentum*
13/12/19 06:16	SRX5014508	miRNA-Seq	Illumina HiSeq 2500	*Rhopilema esculentum*
11/05/20 07:46	SRX6356608	miRNA-Seq	Illumina HiSeq 2500	*Rhopilema esculentum*
11/05/20 07:46	SRX6356607	miRNA-Seq	Illumina HiSeq 2500	*Rhopilema esculentum*
11/05/20 07:46	SRX6356606	miRNA-Seq	Illumina HiSeq 2500	*Rhopilema esculentum*
30/01/19 18:13	SRX5005072	RNA-Seq	BGISEQ-500	*Nemopilema nomurai*
06/02/20 01:36	SRX5211048	EST	Illumina HiSeq 2500	*Rhopilema esculentum*
06/02/20 01:36	SRX5211051	EST	Illumina HiSeq 2500	*Rhopilema esculentum*
06/02/20 01:36	SRX5211050	EST	Illumina HiSeq 2500	*Rhopilema esculentum*
06/02/20 01:36	SRX5211047	EST	Illumina HiSeq 2500	*Rhopilema esculentum*
06/02/20 01:36	SRX5211046	EST	Illumina HiSeq 2500	*Rhopilema esculentum*
06/02/20 01:36	SRX5211049	EST	Illumina HiSeq 2500	*Rhopilema esculentum*
06/02/20 01:36	SRX5211043	EST	Illumina HiSeq 2500	*Rhopilema esculentum*
06/02/20 01:36	SRX5211042	EST	Illumina HiSeq 2500	*Rhopilema esculentum*
06/02/20 01:36	SRX5211053	EST	Illumina HiSeq 2500	*Rhopilema esculentum*

(continued)

Table 5 Transcriptomic data available for Rhizostomeae deposited on the SRA database in Genbank, including release data, accession numbers, library strategy, platform model, and species scientific name (accessed on February 23rd, 2024). RNA-seq stands for RNA-sequencing, while EST stands for expressed sequence tag. (*cont'd*)

Release date	Accession number	Library strategy	Platform model	Scientific name
06/02/20 01:36	SRX5211052	EST	Illumina HiSeq 2500	*Rhopilema esculentum*
06/02/20 01:36	SRX5211045	EST	Illumina HiSeq 2500	*Rhopilema esculentum*
06/02/20 01:36	SRX5211044	EST	Illumina HiSeq 2500	*Rhopilema esculentum*
01/12/20 02:27	SRX7088781	RNA-Seq	Illumina HiSeq 2000	*Cassiopea andromeda*
01/12/20 02:27	SRX7088780	RNA-Seq	Illumina HiSeq 2000	*Cassiopea andromeda*
30/04/22 02:44	SRX14245676	RNA-Seq	HiSeq X Ten	*Rhopilema esculentum*
30/04/22 02:44	SRX14245677	RNA-Seq	HiSeq X Ten	*Rhopilema esculentum*
30/04/22 02:44	SRX14245678	RNA-Seq	HiSeq X Ten	*Rhopilema esculentum*
30/04/22 02:44	SRX14245679	RNA-Seq	HiSeq X Ten	*Rhopilema esculentum*
30/04/22 02:44	SRX14245680	RNA-Seq	HiSeq X Ten	*Rhopilema esculentum*
30/04/22 02:44	SRX14245681	RNA-Seq	HiSeq X Ten	*Rhopilema esculentum*
30/04/22 02:44	SRX14245673	RNA-Seq	HiSeq X Ten	*Rhopilema esculentum*
30/04/22 02:44	SRX14245674	RNA-Seq	HiSeq X Ten	*Rhopilema esculentum*
30/04/22 02:44	SRX14245675	RNA-Seq	HiSeq X Ten	*Rhopilema esculentum*
26/09/23 15:16	SRX20220662	RNA-Seq	Illumina HiSeq 2500	*Cassiopea xamachana*
26/09/23 15:16	SRX20220661	RNA-Seq	Illumina HiSeq 2500	*Cassiopea xamachana*
26/09/23 15:16	SRX20220660	RNA-Seq	Illumina HiSeq 2500	*Cassiopea xamachana*

are RNA-sequencing data for *R. esculentum* 72 h after exposure to microplastics/nanoplastics (Accession numbers from SRX14245673 to SRX14245681) or body parts, such as gonads, umbrella and oral tissue (SRX1692774, SRX1692805 and SRX1692760, respectively), RNA-seq data of *N. nomurai* tentacles and juveniles (Accession numbers: SRX5005072 and SRX4928470, respectively), RNA-seq data of *C. xamachana* cassiosomes, oral arms, gonads and bell (SRX20220662, SRX20220661, SRX20220660 and SRX20220659, respectively), young *C. andromeda* collected from shallow water (SRX7088780 and SRX7088781) and *S. meleagris* tentacle transcriptome data (SRX470083).

A combined transcriptomic-proteomic approach was used by Li et al. (2014) to study the rhizostome *S. meleagris*. In particular, tryptic digestion of the crude venom followed by RP-HPLC separation and MS/MS analysis was used for venom proteomics, while the Illumina sequencing platform HiSeq™ 2000 was used for the transcriptome sequencing. They identified 218 toxins, including the C-type lectin, phospholipase A_2, potassium channel inhibitors, serine proteinase inhibitors, metalloproteinases and hemolysins (Li et al., 2014).

4. Locomotion and feeding mechanics

The ecological role played by rhizostome medusae depends upon prey capture as they swim through the water column. For scyphozoan medusae, swimming and prey capture occur simultaneously (Costello and Colin, 1995) so biomechanical patterns and ecological roles are intricately linked together. The mechanics of locomotion and feeding of Rhizostomeae species are reviewed and some ecological aspects of these functions are discussed.

4.1 The biomechanics of locomotion and animal-fluid interactions

A typical medusa body has a hemispherical bell shape, termed umbrella. On the underside (subumbrella), the subumbrellar cavity, is filled with fluid from the external environment. Rhythmic bell pulsations are divided between a contraction phase, when fluid moves out of the subumbrellar cavity, and a relaxation phase, when fluid flows into the subumbrellar cavity. In large jellyfish, these movements create a series of interconnected vortices, due to a fine control of the flexible bell margin (Costello et al., 2008; Dabiri et al., 2005). Adult Rhizostomeae medusae are typically large in size (>5 cm) with more flattened bodies, or oblate bells, and their swimming mode is termed *rowing-propulsion,* and involves the creation of counter-rotating vortices in the wake in a similar manner to any other large oblate medusa of the three classes of pelagic medusae. During bell contraction the oblate bell movement generates a negative pressure region along the exumbrellar surface (Fig. 1) near the bell margin that entrains a large volume of fluid that is channeled downward into the wake termed the "starting vortex" (Colin et al., 2012, Costello et al., 2021, Dabiri et al., 2005, Gemmell et al., 2013). It is this negative pressure region that is the primary source of thrust during the contraction phase that acts to pull the medusae forward and is termed "suction thrust" (Colin et al., 2012, Gemmell et al., 2015). At the end of the contraction phase, the subumbrellar muscles relax and the bell returns to its initial relaxed shape as fluid refills the subumbrellar cavity (Megill et al., 2005). During the relaxation phase, a "stopping vortex" is formed inside the bell, with opposite rotation in relation to the starting vortex. As the starting vortex moves away from the bell margin, it meets the vortex that remains in the wake from the previous cycle. That wake vortex offers resistance to the new starting vortex motion, and this vortex–vortex interaction underneath the bell is known as virtual wall effect (Gemmell et al., 2021). Eventually, the starting vortex separates from the bell margin, merging with, and largely canceling, the previous stopping vortex (Costello et al., 2021). The interactions of the starting and stopping vortices are important for enhancing the thrust and efficiency of rowing propulsion.

The mesoglea of cnidarians is an extracellular matrix between the ectodermal and endodermal epithelia, and it represents a high percentage of the animal's volume. It consists of collagen and elastic fibers embedded in a hydrated matrix, which may also contain cells, such as epitheliomuscular cells and zooxanthellae hosts cells in symbiotic species (*e.g.* genera *Cassiopea* and *Cotylorhiza*). In medusae of Rhizostomeae, the subumbrellar layer below the gastrovascular cavity serves as the base for the swimming muscles (Arai, 1997). Above the gastrovascular cavity, a thick exumbrellar mesoglea is involved in elastic recoil following the muscular contraction. These layers are joined by radial anchoring ridges located among the channels of the gastrovascular system (Arai, 1997). The mesoglea may have a variety of further functions, including acting as a skeleton, maintenance of buoyancy, and transparency (Arai, 1997; Chapman, 1959; Gutmann, 1965, 1966). Collagen fibers embedded in the mesogleal tissue store potential energy from the bell contraction. Since the mesoglea acts antagonistically to the

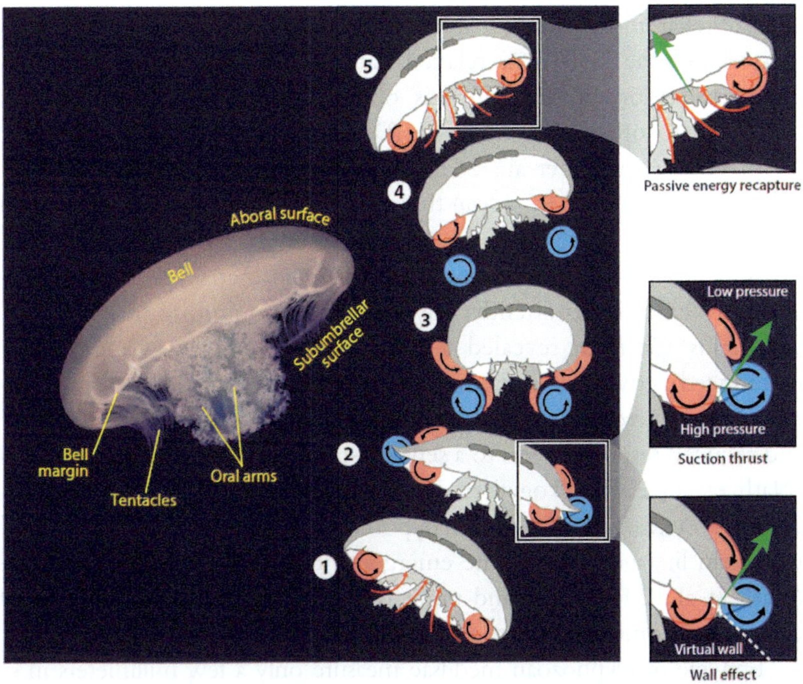

Fig. 1 Jellyfish (*Aurelia* spp.) anatomy and critical vortex interactions allowing efficient jellyfish swimming. Red areas represent stopping vortices, and blue areas represent starting vortices. Black arrows denote the direction of fluid circulation. Green arrows represent the direction of thrust contribution provided by the different vortex interactions. (1) During the relaxation phase of a pulsation cycle, a stopping vortex resides within the subumbrellar cavity and induces flow upward against the subumbrellar surface in a process known as passive energy recapture (Gemmell et al., 2013). (2) The bell margin bends, forming a vortex at the aboral surface. This vortex creates a pressure gradient across the bell margin that results in suction thrust along a high-to-low pressure gradient (Colin et al., 2012; Gemmell et al., 2015a). There is also vortex–vortex interaction underneath the bell that creates a virtual wall effect (Gemmell et al., 2021). (3) As the starting vortex is pushed away from the bell margin, it encounters the vortex remaining in the wake from the previous cycle. That wake vortex offers resistance to the new starting vortex motion and acts as a virtual wall in a process termed the wall effect. Ultimately, the starting vortex separates from the bell margin, merging with and largely canceling the previous stopping vortex (Dabiri et al., 2005). (4) The bell expands outward, and the newly formed stopping vortex gains circulation as it flows inside the subumbrellar cavity. That stopping vortex then generates a high-pressure region that pushes against the subumbrellar surface as passive energy recapture. (5) The jellyfish returns to the relaxation phase, and the entire cycle then repeats as it continues swimming. *Reproduced from Costello, J.H., Colin, S.P., Dabiri, J.O., Gemmell, B.J., Lucas, K.N., Sutherland, K.R., 2021. The hydrodynamics of jellyfish wwimming. Annu. Rev. Marine Sci. 13, 375–396. https://doi.org/10.1146/annurev-marine-031120, Fig. 5, with kind permission of Annual Reviews.*

subumbrellar musculature, harder and sturdier bells may generate faster recovery of the bell to its relaxed state, which shorten pulsation times. Faster relaxation of the bell was observed in Rhizostomeae medusae, compared to other Discomedusae (D'Ambra et al., 2001; Nagata et al., 2016; von Montfort et al., 2023) and this may be related to their high collagen content (*e.g. R. pulmo*) that can be up to ten times higher than for Semaeostomeae medusae of the genus *Aurelia* (Addad et al., 2011).

Although it has traditionally been accepted that scyphozoan medusae generally swim intermittently, day and night monitoring of specimens of the genus *Cassiopea* revealed a behavioral quiescence at night, a state of reduced activity or rest, that is rapidly reversible. When deprived of nighttime quiescence, medusae of *Cassiopea* exhibited decreased activity and reduced responsiveness to a sensory stimulus during the subsequent day (Nath et al., 2017). Together, these results indicate that *Cassiopea* has a sleeplike state, supporting the hypothesis that sleep arose early in the metazoan lineage, prior to the emergence of a centralized nervous system (Nath et al., 2017). Day and night observations of pulsation patterns for other rhizostome species are yet to be conducted (Arai, 1997).

Ephyrae of scyphozoan medusae measure only a few millimeters in size, while juveniles and adults can reach much larger sizes (Jordano et al., 2022; Russel, 1970). During early ontogeny, ephyrae undergo morphological and functional transitions (see Morandini, 2024) that accompany changes on propulsive strategies (Nagata et al., 2016; von Montfort et al., 2023). Whereas ephyrae experience viscous environments (Reynolds number < 10), that constrain their locomotion and cause rapid decelerations, adult medusa fluid flows are dominated by inertial forces (Reynolds numbers > 100; Feitl et al., 2009; Higgins et al., 2008). The transition in the fluid environment that occurs during development brings about morphological and kinematic changes such as changes in bell shape and pulsation rates (Nagata et al., 2016; von Montfort et al., 2023). Morphological and functional changes occur earlier in rhizostome species than in semaeostome species, when kept in similar laboratory conditions (Jordano et al., 2022; von Montfort et al., 2023). Besides higher pulsation rates (>3 Hz), ephyrae pulse with broader bell strokes that involve the entire bell, whereas adults pulsation movements became more localized to the bell margin (D'Ambra et al., 2001; Feitl et al., 2009; Higgins et al., 2008; Nagata et al., 2016). Furthermore, different magnesium concentrations can affect pulsation rates in *C. andromeda*, with low magnesium resulting in higher pulse rates, whereas high concentrations resulted in lower pulsation rates, as both decreased bell diameter and also produced oral arms malformation. Additionally, both low and high magnesium exposure alters body color from pale blue to brown, which indicates the loss of zooxanthellae (Evans et al., 2020).

4.2 Swimming performance and costs of swimming

Pelagic medusae exhibit lower respiration rates during swimming compared to other metazoans of the same size (on a wet weight basis), due to the large proportion of metabolically inactive tissue in the mesoglea (Acuña et al., 2011; Gemmell et al., 2013). In addition to the low carbon content in jellyfish bodies, they have ≤1% of the body mass represented by muscle (Larson, 1987). Lower energy expenditure to generate propulsive motion may be an adaptive advantage for gelatinous zooplankton (Acuña, 2001), but also imposes physiological constraints in which low body carbon and muscle mass limits propulsive strategies for large jellyfish, impeding higher-performance swimming modes (Costello et al., 2008). A recent review of the swimming kinematics of 17 scyphozoan species (eight rhizostomes, eight semaeostomes and one coronate) showed that rhizostome medusae swim about twice as fast, and pulse about twice as frequent (3.5 cms^{-1} and 1.5 Hz) than semaeostome counterparts (1.5 cms^{-1} and 0.6 Hz), when comparing individuals of the same size (~6 cm) (von Montfort et al., 2023) (Fig. 2A,B and C). Despite the overall higher swimming velocities of the rhizostomes, they seem to travel similar distances per pulsation cycle (Fig. 2D) as semaeostome medusae. This suggests that each pulsation is hydrodynamically similar and higher velocities are achieved through higher frequency pulsation rates. This higher frequency pulsation pattern is a conservative trait among rhizostome lineages (von Montfort et al., 2023) and brings other important ecological consequences. On the one hand, slower swimming velocities of semaeostomes (Cyaneidae, Pelagiidae, and *Aurelia* spp.) generate lower costs of transport, which indicates these lineages developed higher efficiency (measured as cost of transport) compared to rhizostomes (Gemmell et al., 2013; von Montfort et al., 2023). On the other hand, more rapid pulsation by rhizostome medusae allows the processing of large fluid volumes for suspension feeding per unit time. In addition, they produce comparatively higher velocities of the fluid entrained toward oral arms capture surfaces, which hypothetically allow the capture of more evasive prey (D'Ambra et al., 2001; Nagata et al., 2016; Wagner et al., 2020). Whether such biomechanical differences result in ecological niche differentiation remain uncertain, mostly due to the lack of ecological data on the rhizostome species (cf. below).

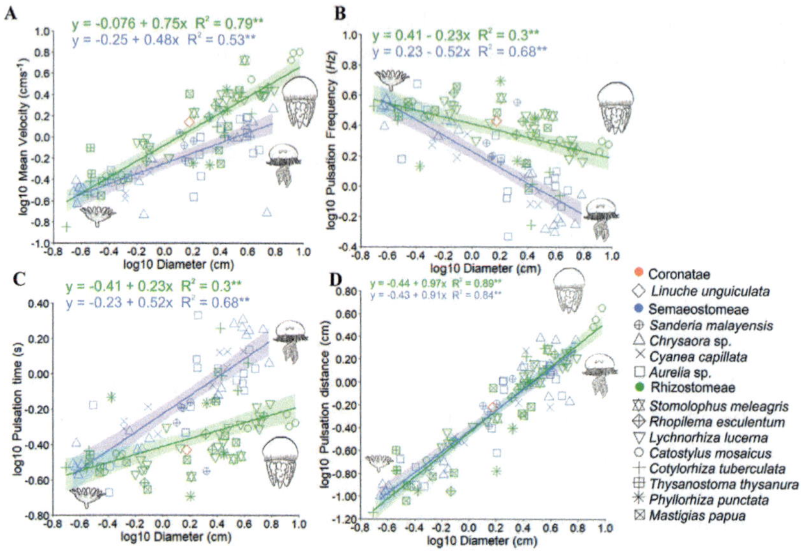

Fig. 2 Scyphozoa orders relationship between diameter (cm) and: (A) Mean velocities; (B) Pulsation frequency; (C) Pulsation time; and (D) Pulsation distance; Label displays the shape of each plotted species. In red is the Coronatae = *Linuche*; in blue are the "Semaeostomeae" = *Sanderia, Chrysaora, Cyanea and Aurelia*; in green are the Rhizostomeae = *Stomolophus, Rhopilema, Lychnorhiza, Catostylus, Cotylorhiza, Thysanostoma, Phyllorhiza, and Mastigias*. Images of representative morphologies were used, common generalized ephyrae, *Aurelia* for "Semaeostomeae" and *Catostylus* for Rhizostomeae. *Re-drawn from von Montfort, G.M., Costello, J.H., Colin, S.P., Morandini, A.C., Migotto, A.E., Maronna, M.M., Reginato, M., Miyake, H., Nagata, R.M., 2023. Ontogenetic transitions, biomechanical trade-offs and macroevolution of scyphozoan medusae swimming patterns. Sci. Rep. 13 (1), 9760. https://doi.org/10.1038/s41598-023-34927-w, with kind permission from Springer Nature.*

Fine-scale fluid dynamics related to these two pulsation patterns affect swimming performances differently. The stopping vortices produced by bell relaxation direct flows into the umbrella, increasing pressure inside the subumbrella, while it pushes the bell forward. This generates a passive thrust called passive energy recapture (PER), that occurs during elastic recoil of the bell and it is followed by a short pause in bell motion between the subsequent pulsation, resulting in a secondary velocity peak (Gemmell et al., 2013). High contribution of PER (~30%) to the total locomotion during pulsation cycle allows medusae of *Aurelia* to be one of the most energetically efficient swimmers known, in terms of energetic costs related to biomass and distance transported (Gemmell et al., 2018, 2013). By contrast, medusae of Rhizostomeae usually have much shorter pauses between pulsations, thus PER contributes less to total locomotion. von Montfort et al. (2023) demonstrated that the Semaeostomeae *Aurelia* spp. is actually an intermediate user of PER, whereas the rhizostome *C. tuberculata*, together with the semaeostomes *Chrysaora pacifica* (Goette, 1886) *Chrysaora lactea* Eschscholtz, 1829 may potentially have even lower energetic costs of swimming, due to lower pulsation frequencies and higher contribution of PER to total displacement. Such mechanisms, however, are still poorly understood and further investigations are needed to quantify the contributions of this effect in a wider range of medusae species.

Pelagic animals rarely swim near solid surfaces, however, evaluations of medusae starting and stopping vortices, suggest that the interface between the interaction of those, acts similarly as a solid boundary and results in greater thrust and reduced cost of transport (Fernández-Prats et al., 2015), due to the virtual wall effect (Costello et al., 2021; Gemmell et al., 2021). Since rhizostomes have higher pulsation frequencies (Fig. 2D) and, consequently, more closely spaced vortices (Neil and Askew, 2018), higher swimming performance may be aided through virtual wall effects. Such an effect, however, has never been described in a rhizostome species. Few rhizostome medusae species (<10% of known species) have been studied using flow quantification methods such as Digital Particle Image Velocimetry (DPIV), that are necessary to evaluate the details of fluid interactions, and experiments of oxygen consumption, in order to measure metabolic demands of swimming. Although observations on biomechanical phenomena still involve few species, the apparent uniformity in swimming patterns among rhizostomes (von Montfort et al., 2023) may be evidence that the species in the group share other common functional characteristics.

4.3 Oriented swimming in Rhizostomeae medusae?

Jellyfish, as well as other zooplankton, are generally described as organisms with low locomotion capacity, passively drifting along oceanic currents. This understanding may not be so precise, since active swimming and complex vertical movements are observed and may help jellyfish to better exploit their environments (Hays et al., 2012). This is particularly important for rhizostome medusae which exhibit more robust swimming patterns than other medusae groups (von Montfort et al., 2023). The most iconic case of oriented swimming is that of the golden jellyfish of the genus *Mastigias* from marine lakes in Palau, that migrate horizontally, avoiding shadows (see also Holst et al., 2024), in order to maintain their zooxanthellae under the sunlight (Dawson and Hamner, 2003; Hamner and Hauri, 1981). Individuals of *R. octopus* were equipped with acceleration data loggers that recorded their diving behavior, activity, and body orientation revealed the medusae actively swam in response to current drift, especially during the ebb tide and the flood tide (Fossette et al., 2015). Oriented swimming behavior may be achieved by direct detection of current shear through their body surface (Rakow & Graham, 2006), in which *R. octopus* passed most of the time swimming horizontally near the water surface (0–2 m) (Fossette et al., 2015). Near surface swimming was also observed in *S. meleagris* (Shanks and Graham, 1987) whose swimming direction was directly correlated with the direction of the wind. Together, those data indicate that rhizostome medusae are capable of actively orienting their swimming to environmental cues such as tidal currents and wind-generated waves. Oriented swimming can have adaptive value by affecting survival, population connectivity, and recruitment. By staying within favorable water currents for growth and survival, they may be capable to avoid drifting into areas with unfavorable thermal conditions that could threaten their survival (Fossette et al., 2015). Furthermore, oriented swimming may also be widespread among scyphozoan species that form blooms, as it can help keep them close rather than dispersing or washing ashore.

An experimental study examined six different species of rowing jellyfish (two Rhizostomeae and four Semaeostomeae) swimming under various background flow speeds (Yang et al., 2018). The study found that pulsation frequencies were independent of the medusae body orientation and the speed of the background flow, suggesting that physiological and phylogenetic traits constrain pulsation frequency patterns. It also implies that Rhizostomeae medusae do not adjust their kinematics in response to changing environmental flows, which partly contrasts to previous in situ observations (e.g. Fossette et al., 2015; Shanks and Graham, 1987). These issues underscore the need for additional research to understand the mechanisms behind oriented swimming in scyphozoan medusae and its impact on the fitness and survival of the species. This requires a more extensive field and experimental studies of a broader range of medusa species.

4.4 The mechanics of food capture and ingestion in Rhizostomeae medusae

One of the most conspicuous features of rhizostome medusae is the lack of a larger central mouth on the subumbrella and the presence of a complex canal system inside the oral arms, that opens to the external environment by means of innumerous millimeter-sized mouth openings (Fig. 3). The highly modified architecture of oral arms (see Morandini, 2024) provides a complex mechanism of food particle retention and determines an

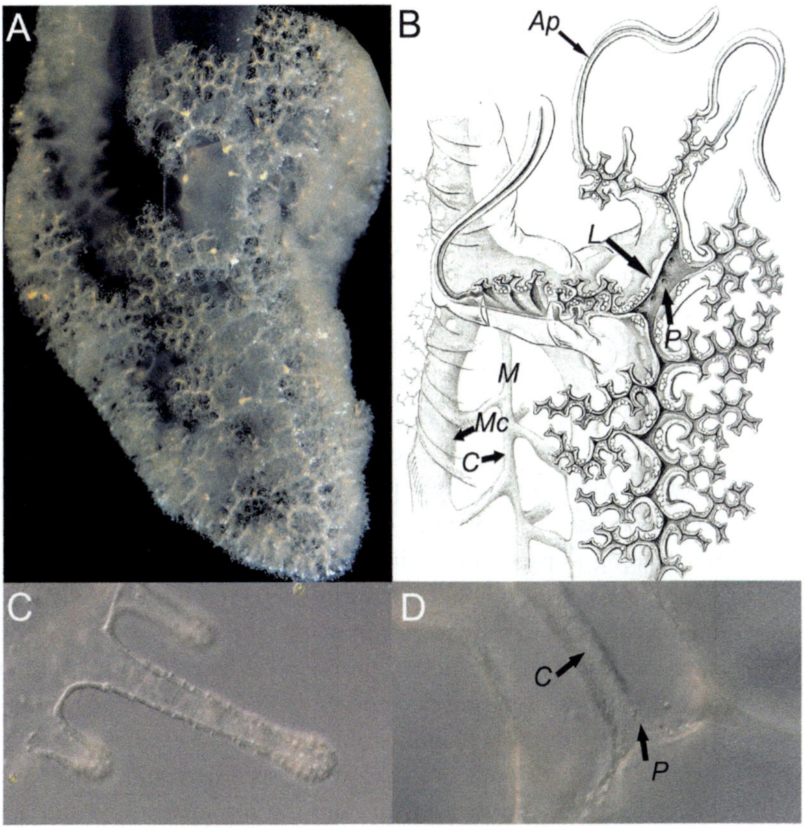

Fig. 3 Oral arms and feeding structures of *Lychnorhiza lucerna* (Scyphozoa, Rhizostomeae). (A) Conic-shaped oral arm of a 3-month old medusa (~3 cm bell diameter) (B) Schematic view of the oral arm canal system leading to the small mouth apertures, *M*; terminal appendages, *Ap*; small pore of the canal inlet, *P*; lateral canals leading to the apertures, *L*; main canal of the internal wing, *Mc*; central canal of the oral arm, *C*; the digitata are not shown. (C) Tree finger-like projections, armed with terminal nematocyst clusters, called digitata. (D) The small mouth aperture, showing the pore, *P*, that connects the external environment to the internal canal, *C*, with dense epidermal motile cilia that directs food particles into the oral arm. Scale bars: (A) 5 mm, (B, C) 0,1 mm. *Panels A and B were reproduced from Nagata, R.M., Morandini, A.C., Colin, S.P., Migotto, A.E., Costello, J.H. 2016. Transitions in morphologies, fluid regimes, and feeding mechanisms during development of the medusa Lychnorhiza lucerna. Marine Ecol. Progr. Ser. 557, 145–159. https://doi.org/10.3354/meps11855, Fig. 2, with kind permission from Inter-Research Science Center.*

important size-constraint to food ingestion. Even larger medusae (*e.g. N. nomurai*) bear mouth apertures of <3 mm in diameter (Lee et al., 2008) that prevents rhizostome medusae from feeding on prey, such as fish, larger crustaceans and gelatinous organisms consumed by coronate medusae, semaeostome and larger hydromedusae (Larson, 1987; Zamponi and Mianzan, 1985). Food capture mechanisms of rhizostome medusae have been studied under morphological and functional perspectives, applying anatomical description (Smith, 1936), direct behavioral observations (Bigelow, 1909; Larson, 1979), and more recently, by means of high-speed videography (D'Ambra et al., 2001; Nagata et al., 2016; Wagner et al., 2020) and microtomography techniques (Avian et al., 2022). Because many rhizostomes can occur in blooms over coastal areas worldwide (Thibault et al., 2024) it is necessary to increase the knowledge about their morphology, biomechanics, physiology and ecology in order to better understand the dynamics of their mass occurrences and their structuring effects to their ecosystems. In this regard, functional morphology and biomechanics studies are valuable tools to understand how these animals operate to feed and how potential prey interact with them (Bezio et al., 2018; Nagata et al., 2016; Wagner et al., 2020), which ultimately determine their feeding impacts and their ecological roles.

Early morphological studies of rhizostome medusae have considered the size limitation imposed to food ingestion due to the small month apertures, while other authors have questioned previous findings of gastric contents on rhizostome medusae, stating that the ingestion of solid food would be impossible for these animals (Thiel et al., 2024). The mechanical basis of food ingestion in rhizostome medusae were first investigated by Cuvier (1799), who described the suction through the innumerous mouth apertures and how food is delivered through the complex canal systems. The function of a ciliated epithelium as the driving force for internal food transport towards the cruciform stomach was first described by Hamann (1881), and it was recognized in many rhizostome species in studies based on direct observations with the aid of stain injections (Avian et al., 2022; Bigelow, 1900; Larson, 1997; Smith, 1936). Larson (1997) described that the axial canals of oral arms in *Cassiopea* sp. have bidirectional ciliary tracts, as in other scyphozoan medusae (Southward, 1955), and transport of prey items took about 5 min from the mouths on the tip of oral arms to the cruciform central stomach on the umbrella. Later, Avian et al. (2022) described centrifugal and centripetal flows, with the inward flow concentrated in the internal oral wing of the oral arms, while outward flow involves only the two outermost oral wings.

The role of the rhythmic bell pulsations of rhizostome medusa to draw food, toward their oral arms was first documented by Bigelow (1909), based on observations of individuals of *Cassiopea*. This feeding behavior is not exclusive to rhizostome medusae, but it is a feature of medusae with more oblate bells, like those of the orders Leptothecata (Hydrozoa) and Semaeostomeae, for instance (Costello and Colin, 1995; Costello et al., 2008). Although the oblate shape increases drag resistance to swimming, it facilitates the transport of large volumes of water, creating feeding currents that drive food particles toward oral arms, increasing contact rates of surrounding prey with the medusae (Costello and Colin, 1994; D'Ambra et al., 2001; Larson, 1991; Nagata et al., 2016). The entrainment of fluid that make up the feeding current occurs both when the medusa contracts and expands its bell, creating respectively the starting and the stopping vortices (Fig. 4). The highest fluid velocities in the feeding current occur when the medusa contracts its bell. While the medusa pulses and advances, these vortices continuously pull fluids that interact with feeding structures on oral arms (Dabiri et al., 2005).

Although the shape and kinematic properties of vortices produced by bell pulsations have been well described among rhizostome medusae (D'Ambra et al., 2001; Nagata et al., 2016; Santhanakrishnan et al., 2012) the fluid interactions with oral arms that initiate prey capture remain poorly studied. This feeding mechanism has important functional and ecological implications for prey selectivity, since the strength of the feeding currents may affect the

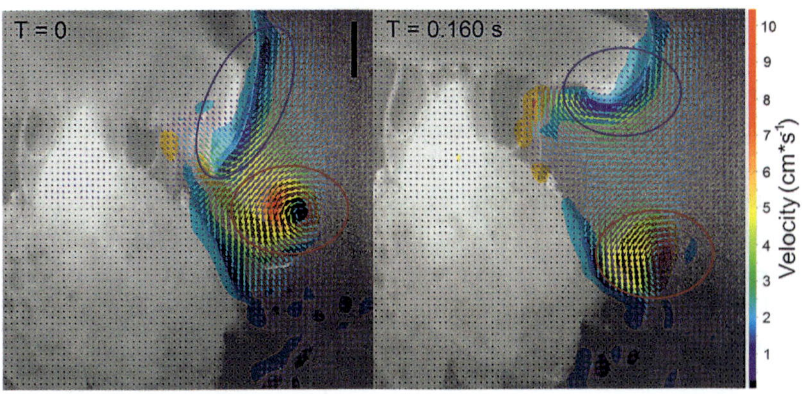

Fig. 4 Sequence of particle image velocimetry showing velocity of fluids ($cm \times s^{-1}$) driven by the pulsation of the rhizostome medusa *Lychnorhiza lucerna*. The left panel show the end of the contraction movement, when the stopping vortex (blue) starts to be formed below the bell margin and the starting vortex (red) was formed. During this phase, the highest flow velocities (~8 $cm \times s^{-1}$) of the pulsation cycle are found. The right panel show the end of the expansion movement, when the stopping vortex (blue) is well formed filling the subumbrellar cavity. The starting vortex (red) continuously pulls fluids and prey toward the oral arms external surface.

type of entrained prey (Costello and Colin, 1994; Nagata et al., 2016; Wagner et al., 2020). Additionally, feeding performances may be affected, because the volume of cleared water is associated with the amount of entrained water through bell pulsations (Acuña et al. 2011; Katija, 2015; Olesen, 1995). Faster bell kinematics result in faster water velocities around the bell with fluid hydrodynamics dominated by inertial forces (Nagata et al., 2016; Wagner et al., 2020). Faster fluxes and more inertial regimes may facilitate the penetration of fluid through their capture sieve and set up the hydrodynamic requirements of sieving and direct interception of small prey by this capture surface (Rubenstein and Koehlt, 1977).

Prey retention by the oral arms in medusae of Rhizostomeae is carried out by small (200–300 μm) finger-like capitate tentacles called "digitata". The digitata bear clusters of nematocysts on their tips and bases that discharge when contacted by prey items, rapidly immobilizing them (see also Section 5.6 in Holst et al., 2024). The dimensions and spacing of digitata are regular among species of Rhizostomeae (Nagata et al., 2016). However, it has been suggested that changes in fluid viscosity brought about by changes in temperature can alter the functional dimensions of these feeding structures (Jordano et al. 2020). The similar design features of these structures among different species suggest natural selection for efficient filtration under shared fluid dynamic conditions.

Faster growth and initial development in rhizostome medusae suggest they undergo earlier transitions on diet, since the ability to entrain more evasive prey can increase with the strength of bell pulsations (among other variables), that scale with increasing size and swimming velocities (D'Ambra et al., 2001; Nagata et al., 2016). This assumption is supported by conflicting evidence of gut content data, as discussed below. There is much to be understood about small-scale animal-fluid interactions, for example, on investigating the role of flows caused by epidermal cilia on particle removal and exchanges with the environment, as it has been demonstrated in corals (Shapiro et al., 2014) and medusae (Jordano et al., 2024).

Direct observations of rhizostome capture mechanisms suggest that feeding particles are removed both by sieving the feeding currents that enter the center of the oral arms and by inertial impact of particles that are transported by bell pulsation vortices toward external oral arm surface (Larson, 1991; Nagata et al., 2016). The polygonal spaces created by the lateral branches of oral arms produce a sieving surface, with millimeter sized pores on the external surface of oral arms (Fig. 3). These spaces also can be seen in oral arms of medusae of Ulmaridae family (Jarms and Morandini, 2019).

Fluorescein dye visualization revealed that fluid entrained by bell pulsation, on the upper part of the oral arms, flows distally to the outside of the oral arms (Fig. 5), thought these polygonal gaps (Nagata et al., 2016). Fluxes over the internal surface of the oral arms are probably enhanced by epidermal ciliary

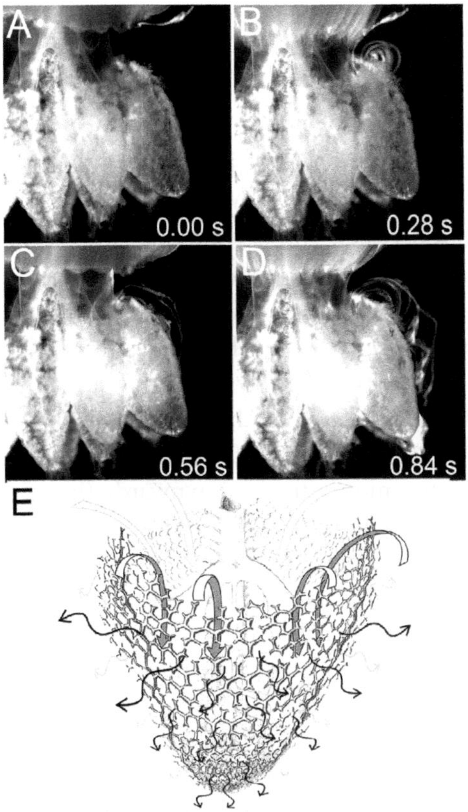

Fig. 5 Sequence of vortice formation during 2 bell pulsation cycles (A–D, $T = 0.84$ s) and interaction of vortex flows with oral arms of *Lychnorhiza lucerna.* (A) Fluorescein dye entrained within subumbrellar cavity, below bell margin at the first expansion. (B) Subsequent bell contraction and formation of a starting vortex, near the beginning of prey capture surfaces on the oral arms. (C) The next bell expansion and the previous starting vortex translating to inside the oral arms. (D) The next bell contraction and formation of the second starting vortex, while fluid entrained during the first pulsation cycle reaches the tip of the oral arm. (E) A schematic of the filtration mechanism of an oral arm, in which fluid passes through polygonal gaps formed by oral arm branches. Scale bar = 1 cm. *(a) Reproduced from Nagata, R.M., Morandini, A.C., Colin, S.P., Migotto, A.E., Costello, J.H. 2016. Transitions in morphologies, fluid regimes, and feeding mechanisms during development of the medusa Lychnorhiza lucerna. Marine Ecol. Progr. Ser. 557, 145–159. https://doi.org/10.3354/meps11855, Fig. 8, with kind permission from Inter-Research Science Center.*

beating that promotes a fluid transport towards the tips of the digitata (Avian et al., 2022; Jordano et al., 2024; Nagata et al., 2016). Despite the large volumes of fluid that are transported by bell pulsation, a considerable proportion of the entrained prey (~50%) does not come into contact with their oral arms (Bezio et al., 2018). If prey touches the oral arm surface, however, the likelihood of capture is high, because of the high density of the nematocysts and their effectiveness in retaining prey (Bezio et al., 2018).

The spatial arrangement of oral arms is highly diverse among rhizostome medusae, particularly the relative size and oral arm posture related to the bell (Jarms & Morandini, 2019). These diverse plans may lead to different interactions between vortices and oral arms. In species in which the oral arms extend distally beyond the bell margin (*e.g. Cassiopea* and *Cotylorhiza*) entire vortices translate through the oral arms (Hamlet et al., 2011) and may be rapidly dissipated by the interception of these structures. In contrast, for species with relatively short oral arms (*Stomolophus* spp.) or those with oral arms confined to the bell center, (e.g. *Lychnorhiza lucerna* Haeckel, 1880), the outer edge of vortex contacts oral arms while the center of the vortex moves freely downwards (Fig. 5). It is not known whether this biomechanical diversity leads to prey selection variations or further ecological consequences.

In addition to the digitata, the oral arms of some rhizostome medusae bear innumerous small vesicles that may play an important role in the feeding process. It has been suggested that these vesicles would bend to trap small prey within the funnel-shaped space of the mouthlets (Bigelow, 1909) or that these vesicles would bend to transfer immobilized prey on external mucus to oral arms, in medusae of *Cassipea* (Smith, 1936). Such active behaviors of vesicles, however, have never been observed by later authors who studied medusae of the genus. On the other hand, the role of oral arms vesicles in releasing bodies of nematocysts and mucus has been described by several authors in species of *C. xamachana*, when the medusae is stimulated by the presence of food, like mesozooplankton crustaceans (Smith, 1936). This released mucus forms stinging strings that entrap and rapidly immobilize prey. Afterwards, the prey is transferred to the digitata of oral arms to be ingested (Durieux and Gemmell, 2022; Larson, 1997; Smith, 1936). A more recent description of ultrastructure, development and function of these bodies of nematocysts has named them cassiosomes (Ames et al., 2020; Holst et al., 2024). Although some rhizostome species are usually found without mucus in the field (Larson,

1991; Nagata and Morandini, 2018), increasing evidence suggests that the release of mucus (and cassiosomes) is a response to the presence of food or to mechanical disturbance, which probably represents a defense mechanism in these medusae (Ames et al., 2020; Shanks and Graham, 1988). Furthermore, the mucous with cassiosomes was also associated to the stinging water reported by divers from mangroves inhabited by medusae of *Cassiopea* (Ames et al., 2020).

It is important to note that the feeding mechanism based on fluid and prey transport, through bell pulsations develops throughout the initial growth of the jellyfish. The early development of rhizostome ephyrae involves transitions in body plans (see Morandini, 2024) that affects feeding strategies. The fluid environment of millimeter sized ephyrae is dominated by viscosity (Reynolds number < 100) and fluid transport by ephyra bell pulsations is inefficient to bring prey to distal capturing surfaces of oral arms. In contrast, adult medusan fluid flows are dominated by inertial forces, and bell pulsations effectively transports fluids and entrained prey toward the oral arms. These properties also contribute to higher downstream fluxes of fluid, thus higher probabilities of transportation of entrained prey to their capture surfaces (D'Ambra et al., 2001; Nagata et al., 2016; Wagner et al., 2020). Initially it was predicted that medusae foraging mode would exert trophic impacts primarily upon prey that are either slow swimmers or stationary relative to the medusan motions (Costello and Colin, 1994, 1995; Hansson et al., 2005; Sullivan et al., 1994). Feeding mechanisms of rhizostome medusae have been evaluated in detail only for a few species (Costello and Colin, 1995; D'Ambra et al., 2001; Nagata et al., 2016; Wagner et al., 2020) and the exact feeding mode of the group remain poorly understood. Wagner et al. (2020) analysed the predation of four scyphozoan medusae (two Rhizostomeae and two "Semaeostomeae" species) on the copepod *Acartia tonsa* Dana, 1849 observed that captured prey does not originate from a particular part of the umbrellar flow. The proportion of prey that reacted to the medusan generated flow varied between 38% and 78% and the overall capture was about 18.0% (±6.8). Similar observations on *L. lucerna* feeding on *Temora turbinata* (Dana, 1849) (Fig. 6) found that predator efficiency in transporting the prey was about 17% (Nagata et al., 2016). Although the feeding currents produced by the medusa species are much slower than copepod velocities and generated fluid shear deformation rates that are well above the detection limits of copepods (Wagner et al., 2020), the copepods often fail to elicit escape responses. Those studies

Fig. 6 Interaction of a juvenile (~1-month old) of the medusa *Lychnorhiza lucerna* and a calanoid copepod *Temora turbinata* (from laboratory high-speed video recordings). The medusa moves its bell from a fully expanded (translucent) position, initiating a contraction movement. As soon as the copepod senses the flows generated by the umbrella, it quickly closes its antennae and triggers an escape response away from the medusa. The translucent figures of the copepod illustrate its position every 15 ms, until the final position after 45 ms. Scale bar: 0.5 mm.

suggest that the behavior of the copepods have only a moderate impact against medusa capture mechanism and demonstrates the complexity of this predator-prey interaction. Further investigations are necessary for understanding how Rhizostomeae effectively capture potentially evasive prey and how these react to fluid motions. Furthermore, ontogenetic transitions in scyphozoan medusae may have implications for foraging modes and mechanisms used for encountering and capturing evasive prey. These factors may be determinants for trophic diversification of scyphozoan medusae (Costello et al., 2008). The following section will discuss results of studies about feeding habitats of rhizostome jellyfish.

5. Nutrition

Nutrition in rhizostomes is diverse, ranging from predation on micro and mesozooplankton in azooxanthellate species (Fancett, 1988; Larson, 1991; Nagata and Morandini, 2018) to zooplanktivory along with autotrophy (mixotrophy) in zooxanthellate species (Smith, 1936; García and Durbin, 1993). Here, we review the knowledge about rhizostome nutrition from a historical, methodological and physiological point of view. Ecological implications are discussed briefly, as more details can be found in Thibault et al., (2024).

5.1 Diet and prey selectivity

A survey of studies ($N = 38$) on the diet of scyphozoan jellyfish published after 1961 has shown that Rhizostomeae species were much less studied (16% of the studies) than Semaeostomeae species (82% of the studies) (Nagata et al., 2015). So far, of the 91 known rhizostome species, diet studies were carried out in 20 species (~22% of known species), of which only 7 species have been extensively quantified by counting all prey items in the gastric contents. The first description of the diet of a Rhizostomeae species was of *Cassiopea frondosa* (Pallas, 1774) by Bigelow (1892), in which copepods and other small crustaceans where found, as well as diatoms, crab zoea, nematodes, foraminifera and a compact greenish mass. Early qualitative diet observations (that did not quantify prey numbers) in medusae of Rhizostomeae often reported copepods, crab zoea, mollusc larvae and detritus in gastric contents of the species *Cassiopea andromeda* (Forskål, 1775) investigated by Gohar and Eisawy (1961), *C. frondosa* (Larson, 1997; Smith, 1936, 1934), and *C. xamachana* (Larson, 1997). The absence of food or the occurrence of very few items has also been found in *C. xamachana* (Bigelow, 1892) and *C. tuberculata* (Pérez-Ruzafa et al., 2002). A study of the diets of 6 Rhizostomeae species unexpectedly found all individuals of *Acromitus flagellatus* (Maas, 1903) with empty stomachs, while the other 5 species had between 69% and 100% of individuals with prey in their stomach content (Syazwan et al., 2021). The absence or scarcity of food is more frequent in juveniles (Leoni et al., 2022) and in senile individuals that stop feeding, as this is a post-reproductive condition in which medusae shrink and disintegrate feeding structures of oral arms before perishing (Kuplik and Angel, 2020; Leoni et al., 2022).

Studies that presented proportions of prey items quantified found that micro and mesozooplankton are the major food source (Fancett, 1988;

Larson, 1991; Nagata and Morandini, 2018), but larger zooplankton (>700 μm) has also been found in minor proportion (Leoni et al., 2022). The diversity of zooplankton prey described in diet studies is wide and proportion of ingestion differs greatly among species (Table 6). Diets of rhizostome medusae are highly influenced by zooplankton prey availability and may follow seasonal succession of planktonic prey items (Leoni et al., 2022). This does not mean that Rhizostomeae medusae are non-selective consumers as prey selectivity patterns have often been demonstrated (Larson, 1991; Leoni et al., 2022; Nagata and Morandini, 2018).

Predation on fish eggs and larvae is common when they are available in the field (Álvarez-Tello et al. 2016; Graham et al. 2003; Leoni et al. 2022; Padilla-Serrato et al. 2013), or in laboratory experiments (Cruz et al., 2021). These prey items, however, did not comprise the diet of *L. lucerna, N. nomurai,* and *R. nomadica,* probably because of their absence in the field during jellyfish samplings (Kuplik and Angel, 2020, Nagata and Morandini, 2018; Wang et al., 2023).

The jellyfish *L. lucerna* and *R. nomadica* ingested calanoid copepods in similar proportions to their availability in the field (Kuplik and Angel, 2020; Nagata and Morandini, 2018) while *Pseudorhiza haeckeli* Haacke, 1884 and juveniles of *R. pulmo* showed positive selection, consuming copepods in a greater proportion than their availability (Fancett, 1988; Leoni et al., 2022). In medusae of *N. nomurai,* copepods represented 99% of prey found in gut contents (Wang et al. 2023). The species *P. punctata, R. esculentum, L. robustus,* and *Lychnorhiza malayensis* Stiasny, 1920 were also considered as major copepod feeders (Syazwan et al., 2021). Thus, copepods represent the main prey items for most studied rhizostome medusae. On the other hand, an apparent low efficiency to feed on copepods has been shown in studies with *S. meleagris,* when diet was compared to environmental variability. The species feeds mostly on non-evasive prey, such as mollusc veligers, cirriped larvae, and fish eggs (Larson, 1991; Padilla-Serrato et al., 2013; Álvarez-Tello et al., 2016). Studies with *P. punctata* showed contrasting results, with the species diet dominated by the copepod *A. tonsa* in the Laguna Joyuda in Puerto Rico (García and Durbin, 1993), whereas low proportions of copepods in stomach contents was also find, with feeding dominated by microzooplankton (tintinnids) and non-motile prey (fish eggs) in the Gulf of Mexico (Graham et al., 2003). Other rhizostomes, like *C. tuberculata, Crambione mastigophora* Maas, 1903, and adults of *R. pulmo,* also feed mainly on non-evasive prey, such as mollusc veligers, diatoms, tintinnids, and fish and invertebrate eggs (Table 6). This

Table 6 Percentage (%) of prey items of the gut contents of rhizostome medusae; (–) indicates significant negative selectivity and (+) indicates significant positive selectivity for this prey item.

	Molluscs	Diatoms	Ciliates	Fish eggs	Cladocerans	Cirriped larvae	Copepods	Decapod larvae	Other
Pseudorhiza haeckeli[1]	0	0	0	40.8+	4	0	33+	4.9+	17.3
Crambione mastigophora[2]	31	8	1	0	0	0	5	0.9	55
Lychnorhiza. lucerna[3]	9	4.3	1.7	0.1	0.6−	0	80.8	0.4	3
Stomolophus meleagris[4]	71.2+	0	9.3	< 1	< 1	< 1	7.3−	1	< 1
Stomolophus meleagris[5]	26.4	0	0	65.3	0.1	4.1	1.5	2	3
Stomolophus meleagris[6]	29.9	< 0.1	0.1	49.8	< 0.1	5.7	8.3	1.6	4.6
Cotylorhiza. tuberculata[7]	6	85	5	0	0	0	3	0	0
Rhizostoma pulmo[7]	3	65	30	0	0	0	1	0	0
Rhizostoma pulmo[8]	4	0	31	5	0	0	27#	0	33
Phillorhiza punctata[9]	35	0	23	15	0	0	23	0	4
Rhopilema nomadica[10]	31	2	1	0	3	0	38	0	25

1—Fancett (1988), 2—Keesing et al. (2016) 3—Nagata and Morandini, (2018), 4—Larson, (1991), 5—Padilla-Serrato et al. (2013), 6—Álvarez-Tello et al. (2016), 7—Pérez-Ruzafa et al. (2002), 8—Dönmez and Levent (2019) 9—Graham et al. (2003) and 10—Kuplik and Angel, (2020). *The authors found positive selectivity for calanoid and harpacticoid copepods and negative selectivity for cyclopoid copepods. # The authors reported 27% of Crustacea, of which copepods were the most abundant taxa.

trophic variability and these specific prey selection patterns reinforce the perception that broad generalizations on the feeding habits of Rhizostomeae species may oversimplify their trophic roles.

Rhizostomeae medusae change their diet in different ways throughout their ontogeny. This is not surprising since many species undergo dramatic size changes, from a few millimeter large ephyrae to adults with more than 1 m in bell diameter. *L. lucerna* decreased the diversity of its diet during growth (Nagata and Morandini, 2018), whereas *S. meleagris* increased the diversity of its diet (Álvarez-Tello et al., 2016). *L. lucerna* and *R. nomadica* exhibited an apparent increase in the consumption of evasive prey (copepods) with increase in their size, whereas the consumption of smaller sized and non-motile prey was higher in smaller sized individuals (Kuplik and Angel, 2020; Nagata and Morandini, 2018). These changes follow the model proposed by Costello and Colin (1995) for the Semaeostomeae *A. aurita,* in which the ontogenetic increase in the strength of the feeding currents produced by bell pulsations would allow larger medusae to access fast escaping prey. On the other hand, a different type of ontogenetic diet shift was described in *R. pulmo,* in which smaller jellyfish (0–5 cm BD) showed positive selectivity for fish eggs and calanoid copepods and increased the ingestion of ciliates throughout development. A possible explanation for this last pattern is that the greater fluid disturbances around the bell margin of adult medusae would be more easily detectable, thus allowing copepods to avoid predation (Leoni et al., 2022). Observations of medusa-copepod interactions support the importance of pre-encounter factors such as prey detection capability that affect the likelihood of prey to elicit fast escape responses (Nagata et al., 2016; Suchman and Sullivan, 1998; Wagner et al., 2020). Post-encounter features such as the morphology of the jellyfish capture surfaces (Bezio et al., 2018) and selective discharge of nematocysts (Nagata et al., 2016) that affect prey encounter and prey retention, respectively, may also contribute to determine prey-selective patterns in Rhizostomeae medusae. The scarcity of studies about gut contents and prey selection, as well as of direct observations of this prey-predator interaction (discussed in the previous section) limits general conclusions about possible ontogenetic changes on their diets.

The diet of Rhizostomeae polyps in the wild has not been studied, because most polyps are known only from laboratory cultures (see Schiariti et al., 2024). In the laboratory, polyps are maintained with nauplii of *Artemia* sp. or with rotifers. Moreover, polyps are known to feed on bacteria, organic particulate matter, small crustaceans, microalgae and tintinnids (Östman, 1997). It has been reported that the polyps are able to absorb free amino acids, such as glycine and alanine, which are a supplemental nutritional source necessary for normal strobilation and development (Shick, 1975). Their ability to feed on relatively large prey of almost the same size as themselves, is an asset of the retention efficiency of their tentacles and of the expansion capability of their mouths (Pitt, 2000). Experiments with scyphistomae of *Catostylus tagi* (Haeckel, 1869) have shown that ephyrae production is not affected by feeding regimes with *Artemia* sp. nauplii or rotifers (Gueroun et al., 2021). The quantity of food, however, affected strobilation in *C. mosaicus, C. tagi* and *R. esculenta* as higher feeding rates induce strobilation, whereas less frequent feeding regimes decrease or cease strobilation, or even induce podocyst formation (Chen et al., 1985; Pitt, 2000; Gueroun et al., 2021). More studies are necessary to evaluate how much of the dietary diversity in Rhizostomeae medusae and polyps is influenced by prey availability, ontogenetic changes, among others, and how diet affects survival, and reproductive strategies (see Schiariti et al., 2024).

5.2 Feeding rates and predatory impacts

Trying to quantify how much zooplankton is consumed by jellyfish has always been a major concern in understanding marine carbon cycling, as their blooms may compromise a great proportion of pelagic production. The effectiveness of the feeding behavior of Rhizostomeae medusae is a key factor to efficiently exploit smaller-sized zooplankton allowing rhizostomes to thrive in the most diverse environments from hyperoligotrophic coastal regions of the Mediterranean Sea, to eutrophic estuaries and coastal lagoons worldwide. Individual feeding rates can be determined by analysing gut contents of field-captured medusae along with measuring the digestion rates at a given temperature. Alternatively, feeding rates can be estimated as prey removal in laboratory containers or field enclosures (reviewed in Hansson, 2006; Purcell, 2009). Here, we examine the few results obtained by both approaches with Rhizostomeae medusae and discuss limitations, difficulties, and recommendations for future studies.

Because rhizostome medusae are stronger and faster swimmers than medusae of other groups (D'Ambra et al., 2001; Nagata et al., 2016) higher metabolic demand and, consequently, higher feeding rates are expected, at least on a wet mass basis. Gut content based ingestion rates (in situ method), however, are still rare in Rhizostomeae because most studies evaluated diets qualitatively (e.g. Keesing et al., 2016; Graham et al., 2003). Only a few works carried out complete dissections (within oral arms, gut content and umbrella canals) that resulted in estimates of total ingestion (G) (e.g. Larson, 1991; Leoni et al., 2022; Nagata and Morandini, 2018; Wang et al., 2023). The number of copepods found in gut contents of medusae of *N. nomurai* increased from 74 in a 5 cm medusa individual, to 1660 in a 47 cm individual (Wang et al., 2023). In *L. lucerna,* total prey ingestion ranged from 17 to 12,000 organisms in individuals from 5 to 30 cm (Nagata and Morandini, 2018). Even for the smaller jellyfish *S. meleagris* total prey ingestion ranged from 700 to 6000 organisms in individuals between 9 and 325 g (Larson, 1991). Beyond possible ecological impacts, these numbers show how laborious in situ diet quantification can be. It is a task that becomes impractical when dealing with large individuals (>30 cm), as is the case with many rhizostomes. This method, however, is simple, easily replicable and requires minimum laboratory facilities, therefore, suggestions to overcome the tediousness of these countings will be discussed below.

In situ ingestion rates (I, prey consumed day^{-1}) can be estimated when total ingestion (G) is corrected by the time (D) that it takes to completely digest a given prey, as: $I = G/D \times 24$. Further details on the digestion process and on experimentally estimated digestion times will be discussed in the Section 6 Digestion and Metabolism. The approach of *in situ* ingestion times was applied to medusae of *L. lucerna* (Nagata and Morandini, 2018), *R. pulmo* (Leoni et al., 2022), *S. meleagris* (Larson, 1991), and *N. nomurai* (Wang et al., 2023), of which only the last two species had their digestion times (D) measured experimentally. Ingestion rates can also be estimated by means of experimental incubations in the laboratory or in the field. Though carrying out such experiments to estimate feeding rates can be problematic since rhizostome medusae are usually large and containers of limited volume may reduce ingestion and affect natural behavior of both predators and prey (Hansson, 2006; Purcell, 2009). Container wall effects would be avoided only when the volume of the tank exceeds 15,000 times the volume of the predator, which means that for a 15 cm medusa, a container of 2 metric tons of water would be necessary (Hansson, 2006). Such devices are impractical for most laboratories and they also may introduce heterogeneous distribution of prey during incubations. Thus, experiments with small medusae and ephyrae are more feasible with ordinary chambers. Ingestion rates as clearance rates C can be estimated from incubations as: $C = \{(\ln\ (n_i\ -\ \text{In}\ (n_f))\ V\}\ N\ T^{-1},$ where C = clearance rate (litres × medusa^{-1} h^{-1}), n_i = initial number of prey, n_f = number of prey at the end of the experiment, V = volume of experimental container (in litres), N = number of medusae and T = duration of the experiment (h). Few studies, however, have applied this approach in Rhizostomeae species, as follows.

Ingestion rates of juveniles (3–5 cm of BD) of *C. tagi* demonstrated a nearly isometric increase of ingestion with increasing prey concentrations, reaching rates up to 270.8 mesozooplankton ind.$^{-1}$ h^{-1} at a prey density of 389.2 ind. L^{-1} (Cruz et al. 2021). In *P. punctata,* the effects of body size and prey densities were tested for the ingestion of copepods. Only body size was significant, since clearance rates scaled from 16 L medusa^{-1} h^{-1} in a 4 cm medusa to 177 L medusa^{-1} h^{-1} in a 24 cm medusa (García and Durbin, 1993). In *P. haeckeli,* ingestion rates for calanoid copepods, decapods and fish eggs and yolk-sac larvae increased with prey density and body size (Fancett and Jenkins, 1988). The effect of body size was shown for clearance rates of *R. nomadica* feeding on *Artemia* sp. nauplii that scaled from ~80 nauplii medusa^{-1} h^{-1} in a 1 cm medusa up to 900 nauplii medusa^{-1} h^{-1} in a 5 cm medusa at a density of 100 *Artemia* nauplii L^{-1} (Kuplik et al., 2021). Predation behavior on *Artemia* nauplii did not necessarily represent functional responses in nature, but the linear response of *C* to extreme prey densities demonstrates the enormous feeding potential. Feeding experiments suggest that rhizostome medusae, as well as other gelatinous zooplankton predators, can feed at maximum rate when exposed to extremely high prey concentrations that are not usually found in the wild. This is not surprising to cultivaters of these jellyfish who have witnessed the speed in which provided food is cleared (Fig. 7).

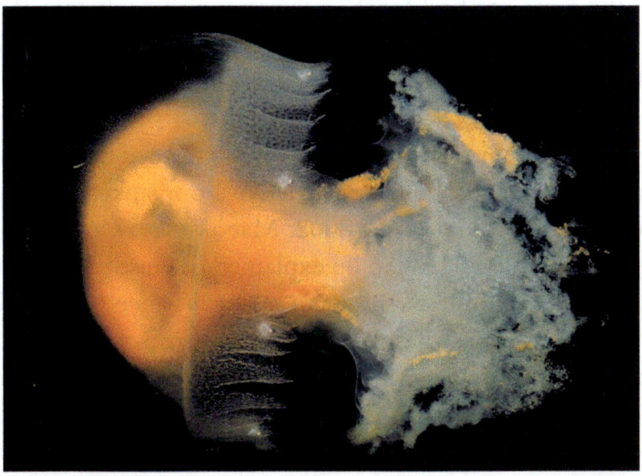

Fig. 7 Medusa of *Lycnorhiza lucerna* after feeding on *Artemia* nauplii. The entire volume of the gastrovascular cavity, including the canals of the oral arms, is filled with *Artemia* sp. nauplii, which illustrate the feeding potential of *Rhizostomeae medusae.* Photo: Alvaro E. Migotto.

Temperature effects on feeding (e.g. Purcell et al., 2010) are still to be explored in rhizostomes. Their predatory effects over natural populations have been much less studied than temperate semaeostome medusae or the ctenophore *Mnemiopsis leidyi* (Purcell, 2009).

Individual feeding rates (from gut content analyses or incubation experiments) can be applied to quantify how much of a given prey population is compromised and which prey populations within a community are positively selected for. Prey consumption (*PC*) is the number of prey consumed per unit of water per day and can be calculated as: $PC = M_{Ab} \times I$, where M_{Ab} is the medusa density (ind.km^2 or ind.m^3) and I is the individual feeding rate (e.g. Wang et al., 2023). In order to estimate the percentage of prey population (*i.e.* standing stock) that is removed by a medusa population, predatory impact (P_I) can be calculated through the inclusion of data on predator and prey densities as: $P_I = I\ (D_{\text{Pred}}\ D_{\text{Prey}}{}^{-1}) \times 100$, P_I represents the percentage of prey standing stock consumed by the predator day^{-1}, I represents ingestion rates, D_{Pred} = predator density (ind.m^{-3}), and D_{Prey} = prey density (ind.m^{-3}) (e.g. Purcell, 2003).

Predatory impacts of *P. haeckeli* were up to 4.8% of copepods and 3.8% of fish eggs and larvae standing stock consumed day^{-1} in Port Phillip Bay, Australia (Fancett and Jenkins, 1988). A population of *R. pulmo* consumed about 5% of copepods, 5% of fish eggs and 8% of ciliates' standing stock day^{-1} at the Bages Sigean lagoon in France (Leoni et al., 2022). Estimates based on the same method for *N. nomurai* found negligible P_I of only up to 0.1% of copepods' standing stock day^{-1} in East China Sea (Wang et al., 2023). The P_I estimated for rhizostomes so far were much lower than those calculated for semaeostome counterparts, such as *Aurelia* spp. and *Chrysaora* spp., of which some populations exhibited a daily consumption between 20% and 60% of the available zooplankton (Behrends and Schneider, 1995; Brodeur et al., 2002; Matsakis and Conover, 1991; Uye and Shimauchi, 2005). This can be explained because populations of small individuals of rhizostome species were studied (Fancett and Jenkins, 1988; Wang et al., 2023) or because the densities of rhizostome medusae were lower than those previously found in studies with other gelatinous zooplankton. The only feeding rate estimate for rhizostome medusae through incubation experiments that indicated higher P_I was carried out with the species *P. puncatata* and it found that large seasonal population of the medusa (>20 cm diameter) would have the capacity of filtering up to ~22% and 35% of the volume of the Laguna Joyuda day^{-1} (García and Durbin, 1993).

5.3 Nutrition in azooxanthellate rhizostome medusae

It has been a major question whether ingestion rates (*I*) of jellyfish provide enough carbon to sustain their metabolic demands and somatic growth. In addition, Rhizostomeae jellyfish may also produce a considerable amount of mucous either spontaneously (Camacho-Pacheco et al., 2022) or as a response to stress (Shanks and Graham, 1987), that demand the oxidation of ingested carbon. The previous section dealt with the total intake of food, while here we will analyse how much intake is necessary for optimal functioning of biological processes (metabolism, growth and reproduction).

Because respiration rates (*RR*) of medusae and of other gelatinous zooplankton are easily measured in laboratory experiments these have been quantified in order to evaluate minimum food requirements (reviewed by Purcell et al., 2010). Although, the extent of anaerobic metabolism is unknown for scyphomedusae (Arai, 1997), *RR* have been used to calculate the minimum carbon requirement (MCR). The MCR represents the minimum carbon or energy required to maintain an animal at constant weight. It can be estimated by multiplying the respiratory quotient (RQ = 0.8) that estimate carbon budgets (CO_2 produced/O_2 consumed) for zooplanktonic carnivores (Larson, 1987; Purcell et al., 2010) with the *RR*. Then, MCR can be converted in terms of the minimum numbers of prey to be ingested daily, when values of prey carbon content (e.g. Ara, 2001, Harris et al., 2000) are available.

Early diet studies on rhizostome medusae have not found substantial numbers of prey in gastric contents (Eysenhardt, 1821) and the nutritional sources for these animals became an intriguing research question (see Thiel et al., 2024 Physiology section). The capacity of the rhizostome *R. pulmo* to feed on solid food was even questioned, because of the lack of a central mouth and because of their supposed inability to clear large volumes to obtain enough plankton, which was based on an estimate of the MCR by Pütter (1909). According to the author, a medusa of *R. pulmo* of 232 cm^3 of volume that consumed 5.1 mg O_2 (122 mg/day) at 26 °C, would require about 24,400 adult copepods of *Eurytemora* or 490,000 of its nauplii daily (Pütter, 1909). Thus, a specimen of 50 cm of bell diameter would require about 39,200 copepods of *Eurytemora* or 263,0000 of *Oithona* to cover its MCR (Pütter, 1909). These numbers were far beyond what was initially found by early studies in the rhizostome gut contents. Further, gut content analysis of rhizostome species that have found the highest ingestion rates of micro- and mesozooplankton, such as 50,000 mollusc veligers ingested day^{-1} (Larson, 1991) or ~100,000 copepods ingested day^{-1} (Nagata and Morandini, 2018). Even though these ingestion rates stand out as the highest described for scyphozoan medusae, individual metabolic demands, in terms of MCR, were not met by these estimates (Larson, 1991; Nagata and Morandini, 2018). For *L. lucerna,* MCR for medusae between 5 and 30 cm (14–2900 g) would require approximately between 5.7 and 818.4 mg C day^{-1}, and carbon ingestion estimated from gut content on average only explained 17% of MCR (Nagata and Morandini, 2018). Thus, other items that could not be quantified in the gut content as rapidly digested microzooplankton or dissolved and particulate organic matter (e.g. Skikne et al. 2009) should be considered as potential nutritional sources.

The MCR can be expressed as a carbon turnover rate or the % of body carbon that is changed day^{-1}. Larson (1987) compiled such values for medusae groups under temperatures of 3 °C to 22 °C and found that for hydromedusae between 2% and 5% (average = 3.8%) of carbon can be changed day^{-1}. For semaeostomes, values ranged between 3% and 10% (average = 6.1%) and for rhizostomes between 4% and 7%. The MCR for medusae of *R. nomadica,* based on measured respiration rates (RQ = 0.8), were estimated to be about 10% at temperatures between 21 °C and 29 °C (Kuplik et al., 2021). The results were higher than expected and were explained by the higher experimental temperatures. Similar results were also achieved by studies on *C. andromeda* that showed a respiration rate of 2 ml of O_2 per g of wet weight^{-1}day^{-1} (Cates, 1975; Mergner and Svoboda, 1977). Assuming a similar rate and using a respiratory quotient of 0.8 (Ishii and Tanaka, 2006; Purcell et al., 2010), a 150 g *Cassiopea* would have a daily respirational carbon loss of about 0.1 g. From this, the quantified prey ingestion equaling only to 0.01 g carbon, would be insufficient to meet metabolic needs by about one order of magnitude. Thus, for a zooxanthellate organism, like the species of the family Mastigiidae, nourishment from the zooxanthellae will account for most of its carbon demands, if the above assumptions are realistic. Further discussion on the nutrition of zooxanthellate Rhizostomeae will follow in the next section. Evaluating minimum carbon intake required for sustenance through experiments of ingestion rates or through estimates derived from respiration rates provide insights into the nutrient cycle and energy transfer within an ecosystem. Thus, the effects of large populations (blooms) of jellyfish on marine ecosystems and their role in shaping ecosystem dynamics and structure can be assessed by estimating these individual metabolic rates.

5.4 Nutrition of zooxanthellate Rhizostomeae

The Kolpophorae-zooxanthellae symbiosis (see Holst et al., 2024) is mutualistic and similar to the more widely known symbiosis in corals (e.g. Muller-Parker et al., 2015): zooxanthellae are hosted in the jellyfish and they access a unique environment, and escape predators and competitors. Transfers of nutrients take place between the host and the symbiont: zooxanthellae provide autotrophically produced organic matter while using organic matter and excretion products obtained from the jellyfish heterotrophy (Freeman et al., 2017; Lyndby et al., 2020; Mortillaro et al., 2009; Röthig et al., 2021). As a result, the holobiont is a mixotroph.

The amount of carbon fixed through photosynthesis is generally important, often exceeding respiration and even growth demands (see for *Cassiopea* spp. Cates, 1975; Drew, 1972; Jantzen et al., 2010; Mammone et al., 2021; Mergner and Svoboda, 1977; Verde and McCloskey, 1998; Welsh et al., 2009; for *M. papua*, McCloskey et al. 1994; for *C. tuberculata*, Kikinger, 1992). The autotrophy of the zooxanthellae requires inorganic nutrients not used by azooxanthellate medusae. Though the excretion products from the medusae contribute to the demand of the zooxanthellae, they do not meet the sufficient amount. The holobiont, therefore, often takes up nutrients from its surrounding water (reviewed in Pitt et al., 2009): the uptake of nitrogen is preferred in the form of ammonium over nitrate/nitrite, while phosphorus is acquired as phosphates (in *Cassiopea* spp., Cates and McLaughlin, 1976; Freeman et al., 2016; Jantzen et al., 2010; Lyndby et al., 2020; Röthig et al., 2021; Todd et al., 2006; Welsh et al., 2009; in *M. papua,* Muscatine and Marian, 1982; in *P. punctata*, Pitt et al., 2005). Environmental factors such as nutrient concentrations and light can affect uptake rates. For instance, ammonium uptake may be particularly dependent on light (Cates and McLaughlin, 1976; Pitt et al., 2005; Welsh et al., 2009), while the uptake of nitrate and nitrite might be controlled by bacterial communities associated with the jellyfish (Freeman et al., 2016; 2017; Röthig et al., 2021; Welsh et al., 2009). However, these are not strict rules and exceptions might exist for instance due to peculiar environmental availability or behavioral adaptations that help control the access to nutrients. For example, night time accession to a nitrogen rich chemocline in the marine lake Ongeim'l Tketau, Palau, by *M. papua* (Muscatine and Marian, 1982) and pumping of nutrient rich interstitial porewater by *Cassiopea* spp. (Jantzen et al., 2010; Durieux et al., 2023).

This general overview is nuanced by the variability in the relative importance of zooxanthellae for total nutrition in zooxanthellate Kolpophorae. This variability emerges during the life cycle (i.e. contrasting polyp, strobila, and medusa) but also intra- and inter-specifically. At the polyp stage, zooxanthellae appear to have little nutritive relevance (reviewed in Djeghri et al., 2019). Several lines of evidence show that absence or inactivity of zooxanthellae generally do not affect budding rates or survival of Kolpophorae polyps (e.g. in *Cassiopea* spp., Djeghri, 2019; Hofmann et al., 1978; Mellas et al., 2014; Rahat and Adar, 1980; in *Mastigias papua*, Sugiura, 1963; in *C. tuberculata*, Prieto et al., 2010). Zooxanthellae however become more important in the latter stage of life, most notably during strobilation where their presence facilitates or is even necessary to induce strobilation (reviewed in Djeghri et al., 2019, in *Cassiopea* spp., Hofmann et al., 1978; Rahat and Adar, 1980; in *M. papua*, Sugiura, 1964; in *Cephea cephea* (Forskål, 1775), Sugiura, 1969). Their role, however, is most likely not limited to nutritional supply, since the presence of inactive zooxanthellae also initiate strobilation (Hofmann and Kremer, 1981; Sugiura, 1969). It is known that strobilation can be induced by certain molecules (e.g. indoles, Helm and Dunn, 2017) and zooxanthellae might induce strobilation in a similar way. At the medusa stage, intra- and inter-specific differences in the importance of zooxanthellae become evident.

Intra-specific variability in the importance of zooxanthellae in nutrition are exemplified by cases where zooxanthellate and azooxanthellate populations exist within the same jellyfish species. This has been reported in *M. papua* (Dawson et al., 2001; Djeghri et al., 2021) and *P. punctata* (Bolton and Graham, 2004), but may also occur in other taxa (e.g. *Cotylorhiza erythraea* Stiasny, 1920 and *Versuriga* spp., Djeghri et al., 2019). Subtler variations have been shown: for instance, although generally associated with very shallow environments, *Cassiopea* sp. jellyfish have also been observed at nearly 100 m depth in the Red Sea where less than 1% of surface light is available (Fricke, 1996). The nutrition of *M. papua* has been shown to be highly plastic, spanning a spectrum from pure heterotrophy and dominant autotrophy (Djeghri et al., 2021). Likewise, the photosynthesis of *Cassiopea* is plastic to light conditions (Mammone et al. 2021). All this indicates that zooxanthellate Kolpophorae can display important intra-specific variability and plasticity in the importance of zooxanthellae for their nutrition. This plasticity likely helps zooxanthellate Kolpophorae to thrive in a wide variety of environmental conditions explaining their success in disturbed, human impacted sites (e.g. Stoner et al. 2022; Thé et al. 2020; Tilstra et al. 2022). Other examples of the phenotypic plasticity of these organisms are discussed in the Section 7 Growth and Development.

Inter-specific differences are also found. *Cassiopea* spp. are generally considered very dependent on its zooxanthellae with low levels of light or bleaching inducing shrinking or even death (Klein et al., 2019; Mortillaro et al., 2009; Röthig et al., 2021). This contrasts with the Mastigiidae *M. papua* and *P. punctata* where some populations are azooxanthellate (Bolton and Graham, 2004; Dawson et al., 2001; Djeghri et al., 2021). The contrast is even greater in *C. cephea*, and possibly other Cepheidae (*Netrostoma*), who lose their symbionts when reaching full development (Djeghri et al., 2019; Sugiura, 1969). This has prompted Djeghri et al. (2019) to provide the hypothetical order of the main Kolpophorae associated families from the most to the less reliant on zooxanthellae for nutrition at the medusa stage: Cassiopeidae > Mastigiidae > Cepheidae.

5.5 Recommendations for studies on feeding habits

Most studies about Rhizostomeae medusae diet were carried out without standardization or the effort to improve the sampling/quantification protocols. Both of these aspects, however, are paramount to improve the quality of data (Gibbons and Richardson, 2013; Nagata and Morandini, 2018). Recommendations for future gut content studies with Rhizostomeae medusae include a careful sampling with a solid or fine mesh sampler (bucket, plastic bag or dip net), which may avoid the loss of gut content or cod-end feeding inherent in net sampling (Barz and Hirche, 2005; Matsakis and Conover, 1991). Careful handling and immediate preservation of medusae specimens on formalin at 4% (or higher concentration) for further gut content analysis in the laboratory are essential for precise quantification. Thereby, extracting gut content of the jellyfish in situ becomes inappropriate because of the possibility of loss of material and because such manipulation, even if careful, induces mucous production. Mucous is produced in response to minor physical stress (Camacho-Pacheco et al., 2022) and makes gut content quantification much more laborious by clumping prey items. By preventing this clumping, subsampling of gut-content samples becomes suitable, since a biased distribution is avoided. The huge effort invested on the quantification of total ingestion limits sample-size and thus the spatial scale of in situ feeding studies that can be managed (Purcell, 2009). Therefore, strategies to subsample stomach content should be considered in future work (Nagata and Morandini, 2018). Other methods like metabarcoding of gut contents (e.g. Sun et al., 2021) may also be applicable by allowing the evaluation of the role of other sources than micro and mesozooplankton, such as non-quantifiable microzooplankton and phytoplankton (e.g., Arai, 1997), but this has not been tested for Rhizostomeae species yet. The diet reconstruction technique based on the analysis of stable isotopes of C and N, which has been extensively used in studies of the trophic ecology of jellyfish (revised in Pitt et al., 2009), still needs support in experiments that quantify the specific fractionation rates for Rhizostomeae species (e.g. D'Ambra et al., 2014, for a semaeostome).

Experimental studies on growth in different feeding regimes could generate additional evidence on the importance of alternative nutrition sources. However, such studies should focus on realistic concentrations. The development of feeding rate models also depends on basic parameters, mainly the influence of temperature on ingestion rates, digestion rates (i.e. Q_{10}) and the effect of densities, which are all still largely unknown for species in the group (e.g. Larson, 1987; Purcell, 2009). Despite important advances, the idea that jellyfish research is in its "infancy" is still valid, thus improvement of sampling, quantification, experimental protocols to obtain better quality data is necessary (Gibbons and Richardson, 2013).

6. Digestion and metabolism

Once the food is captured and ingested, the digestion process begins. The enzymatic composition is an indication of which substrates can or cannot be used as food. The rate or time of digestion can be easily measured in the laboratory and the associated total amount of food ingested (G) can be used to estimate daily food intake. The organic material is oxidized and adenosine triphosphate (ATP) molecules are produced, which store energy and are used for cellular metabolism. The total metabolic rate represents the sum of ATP production by oxygen-independent processes (i.e. anaerobic metabolism) and oxygen-dependent processes (i.e. aerobic metabolism). The contribution of anaerobic metabolism is still unknown in scyphozoan medusae. Therefore, because it is easily measurable in the laboratory, the rate of oxygen consumption is often used as an index for the total metabolic rate. The catabolism of proteins results in nitrogenous compounds that are excreted as waste. This section is a summary of the information available on these subjects for species of Rhizostomeae.

6.1 Enzymology

Estimates of the digestive enzyme activity have proven valuable in comprehending the digestive physiology and nutritional needs of marine crustaceans (Lemos et al., 2002; Lovett and Felder, 1990) and fishes (Alvarez-González et al., 2005), but have rarely been applied to jellyfish. Rhizostome medusae are carnivorous organisms (cf. Section 5 Nutrition) and digestion involves an extracellular phase followed by an intracellular phase (Chuin, 1929; Yonge, 1931). The gastrovascular system of scyphozoan medusae consists of a central stomach surrounded by stomach pouches, umbrella and oral arm canal systems (compare Morandini, 2024). In rhizostome medusae, gastric cirri cover the surface of the central stomach and internal canals of the four gastric pillars. The gastrodermis of the gastric cirri contains serous cells concentrated in the basal region that produce digestive enzymes that are utilized to digest prey (Arai, 1997). Scyphozoan polyps lack gastric cirri, but they can release enzymes concentrated on the longitudinal septa (Arai, 1997). Extracellular digestion occurs in the gastrodermal cells following endocytosis of particles derived from food. The enzymes carboxylesterase, N-acetylglucosaminidase, aminopeptidase and phosphatases are related to intracellular digestion (Hoeger and Mommsen, 1984). These enzymes break down food into small particles, and then ingest the particles through phagocytosis for intracellular digestion. After digestion, nutrients reach the ectoderm by diffusing through the mesoglea (Chapman and Pardy, 1972).

The enzymatic activity involved in the transformation of organic matter into usable metabolic fuels provides insights into the connection of enzymatic activity with the environmental available substrates. Enzyme synthesis is under genetic control in invertebrates, such as lobster and shrimp larvae and can be modulated through the diet, metabolic alterations, or via nutritional requirements in order to optimize digestive efficiency (Biesiot, 1986; Boucher et al., 1975; Lovett and Felder, 1990; Samain et al., 1980). In general, greater enzymatic activity implies greater food intake (Boucher et al., 1975; Hoeger and Mommsen, 1984).

A recent study investigated the digestive capabilities throughout the development of *S. meleagris* and changes at the different developmental stages (González-Valdovinos et al., 2019). All stages were capable of hydrolyze carbohydrates, protein, and lipid substrates. Nevertheless, higher amylase activity was found in polyps, which suggested that phytoplankton can be an important source, since it provides major digestive capacities to assimilate food items, such as dissolved and particulate organic material

(González-Valdovinos et al., 2019). The higher activities of amylase and aminopeptidase in polyps were associated with the limited opportunities to encounter zooplankton prey, thus, these enzymes allow the digestion of small particles and organisms with little mobility, such as phytoplankton and organic particulate matter (González-Valdovinos et al., 2019; Shick, 1975). In ephyrae and medusae, amylase and aminopeptidase activities decreased, which could be related to a diet transition inherent to the pelagic life (González-Valdovinos et al., 2019).

Phosphatases are responsible for catalyzing the separation of inorganic phosphorus from organic phosphate to produce energy and transport nutrients through the cells, such as glucose, proteins, and lipids and even water and ions in the case of alkaline phosphatase (Dupuis et al., 1991; Alvarez-González et al., 2005). In *S. meleagris*, the highest phosphatase activity, both acidic and alkaline, was found in late ephyrae. The highest activity of phosphatase and lipase in late ephyrae is related to exponential growth during the transition between the ephyra and young medusa stages. Lipids have about twice the energy content of carbohydrates and proteins per unit of biomass and are essential for egg production. Triglycerides are an important energy storage in *S. meleagris* and probably for all Rhizostomeae medusae during vitellogenesis (Carvalho-Saucedo et al., 2011). The different relative proportions of enzymes (proteases, lipases, phosphatases, and amylases) throughout development are probably related to the genetic regulation of enzyme synthesis and, consequently, it follows shifts in feeding habits (cf. Section 5 Nutrition). Polyps hydrolyze carbohydrates best, ephyrae hydrolyze proteins best and, in the juveline medusa stage, this shifts to lipids. Such changes in enzyme activity and expression may have adaptive value considering the variability in planktonic prey composition which might be reflected in the seasonal jellyfish mass occurrences (Boucher et al., 1975).

More recently, the lipolytic activity at the gastric pouches of *S. meleagris* adult medusae, in which lipolytic digestive enzymes were able to hydrolyze short-chain triacylglycerides better than long ones was observed for the first time. Enzymatic activity was halotolerant (0–4 M NaCl), occurring under strong alkaline pH (~11) with an optimal temperature between 50 °C and 60 °C. Enzymatic activity was independent of the presence of Ca^{2+} and Mn^{2+} ions, but reduction in activity was observed with Mg^{2+} presence (Martínez-Pérez et al., 2020). Cationic cofactors are necessary to activate lipase in vertebrates, but this activation has not been observed in invertebrates like the white leg shrimp *Litopenaeus vannamei* (Boone, 1931)

(Rivera-Pérez et al., 2011), and the Pacific flying squid *Todarodes pacificus* (Steenstrup, 1880) (Park et al., 2008). Yue et al. (2017) reported catalytic metalloproteinases and phospholipases activity in *N. nomurai* nematocysts venom. In general, enzymatic toxins are considered to be responsible for immobilization of prey, but pre-digestion properties are widely unknown in Rhizostomeae species (Kang et al., 2011). Further work with other species of Rhizostomeae should be conducted in order to study the ability (i.e. kinetics) to absorb and assimilate the main dietary nutrients and their response to environmental factors as an essential step toward understanding the mechanism of digestion of scyphozoan jellyfish. This is of particular interest with regard to blooming species that can rapidly consume available prey and generate structuring effects to marine food webs.

6.2 Digestion

Once prey have been captured, they need to be brought to the gastric pouches located in the center of the bell for digestion. While Semaeostomeae possess long and retractile tentacles which are able to bring prey directly to the central mouth in the sub umbrella, Rhizostomeae use ciliary movements within the complex network of channels running throughout their body within the mesoglea (Avian et al., 2022; Jordano et al., 2024; Morandini, 2024). Although extracellular digestion is known to occur on the oral arms of some semaeostomes, likely to facilitate the digestion of large prey (Arai, 1997, and references therein), it is unknown whether the digestion also starts after capture on the oral arms of rhizostomes. Within the gastric pouches, extracellular digestion is carried out by enzymes secreted by serous cells at the base of gastric cirri (Arai, 1997).

Digestion times have been measured mostly in semaeostomes, while *P. haeckeli, N. nomurai* and *S. meleagris* are the only rhizostomes for which published data of this metabolic process are available (Table 7). Digestion times of medusae are influenced by temperature, stomach fullness, predator size and whether the type of prey is hard or soft bodied (Martin and Båmstedt, 2001; Purcell, 2009). Such variables were never evaluated in species of Rhizostomeae. For semaeostomes digestion rates are faster with increasing temperature and predator size. As for prey type, the variability of digestion times confuses potential patterns related to this factor. However, digestion rates of *P. haeckeli* and *S. meleagris* are similar for the same prey type, with fish eggs being digested slower than copepods. At 28 °C–30 °C *S. meleagris* digests tintinnids, appendicularians, and copepods within 1.5 to 2 h, fish eggs within 3 h and mollusc velligers within 4 h (Larson, 1991).

Table 7 Digestion times (DT, h) of rhizostome medusae preying upon different plankton.

Species	Prey	DT (h)	References
Nemopilema nomurai	Copepods	3–4	Wang et al. (2023)
Pseudoehiza haeckeli		2	Fancett (1988)
Stomolophus meleagris		0.5–23.3	Larson (1991)
Stomolophus meleagris	Cirriped nauplii	2	Larson (1991)
Stomolophus meleagris	Mollusc veligers	2–4	Larson (1991)
Stomolophus meleagris	Tintinnids	1.5	Larson (1991)
Pseudoehiza haeckeli	Fish eggs	3	Fancett (1988)
Stomolophus meleagris		3	Larson (1991)
Stomolophus meleagris	Larvaceans	1.5	Larson (1991)

Digestion rates of *P. haeckeli* for copepods were about 2 h and for fish eggs about 3.5 h. Unfortunately, temperatures of these experiments were not provided (Fancett, 1988). At 18 °C the digestion of larger copepods (>1 mm) took about 4 h in *N. nomurai*, while small copepods were digested within 3 h (Wang et al., 2023). Such differences in digestion times, considering the experimental temperatures, indicated that the $Q_{10} = 2.08$, calculated by Purcell (2009) for the effect of temperature on the digestion times of scyphozoan species, is realistic for rhizostome species. Therefore, the application of published digestion rates, corrected by temperature, seems to be an alternative for estimating daily consumption rates in different rhizostome species (e.g. Leoni et al., 2022; Nagata and Morandini, 2018). Further experiments, however, are still necessary in order to evaluate the above-mentioned effects on digestion rates of rhizostome species (e.g. Martinussen and Båmstedt, 1999).

6.3 Oxygen consumption

Oxygen played a fundamental role in the evolution of complex life on Earth. Its increased presence in the atmosphere has enabled organisms to utilize the energy from a food source. Food intake supports growth and reproduction, while a part of it is allocated to respiration (Schneider, 1989;

Uye and Shimauchi, 2005). In this process, energy is released through the breakdown of organic material from digested food into adenosine triphosphate (ATP) and heat. This enzyme-mediated process is oxygen-dependent and involves the oxidation of organic substances, resulting in the production of water (H_2O) and carbon dioxide (CO_2) (Lane, 2015). While some organisms have evolved specialized respiratory organs to extract oxygen from the environment such as gills or lungs, cnidarians lack such structures. They rely on diffusion through the body epithelia to obtain oxygen. The outer cellular layer exposed to the external environment facilitates oxygen exchange, thereby satisfying the organism's oxygen requirements. Remarkably, despite their simple respiratory systems, some jellyfish (mainly of Scyphozoa and Cubozoa) can achieve significantly large sizes due to their relatively low cell numbers and high amount (~95%) of water content (Acuña et al., 2011).

Rates of oxygen consumption may be reported in volumetric, gravimetric or molar terms. Molar terms are preferred for calculations of molecular relationships (Gnaiger, 1983), but the majority of data on scyphozoans has been expressed in volumetric terms. Also, oxygen consumption rates have been expressed with reference to bell diameter (BD), wet mass (WM), dry mass (DM), and protein mass. Respiration rates (*RR*) of medusae increase with size and the relationship between *RR* and size may be expressed by the allometric equation $RR = a \times M^b$, where $RR = O_2$ consumption, $a =$ constant for the species and temperature, $M =$ individual mass, and $b =$ exponent for size and also the slope of the regression line in a log-log plot of respiration against body mass (Arai, 1997). If weight-specific respiration remains constant as size increases, then $b \approx 1$. The work by Purcell et al. (2010) was the latest extensive review on the respiration rates of scyphomedusae. To date, the relationship between *RR*s and mass have only been measured for seven rhizostome medusae (Table 8).

The *RR*s of various species of scyphozoan estimated under controlled laboratory conditions increased predictably with body mass (wet weight), but not with temperature (Purcell et al., 2010). This suggests that as the organisms grow, there is an expected increase of metabolic demand and consequently a demand for higher feeding rates. However the effects of temperature could not be modeled with available data of scyphozoan medusae. Purcell et al. (2010) reported that *RRs* of both scyphozoan groups increased in a proportional manner (1:1) to body size, rather than following the ¾ power rule known as "Kleiber's law", which is commonly observed in vertebrates and benthic animals (Glazier, 2006; Nagy, 2005; Nagy et al., 1999). Moreover,

Table 8 Respiration rates (RRs) of Rhizostomeae medusae measured at near-ambient temperatures (given in °C). Purcell et al. (2010) Lilley and Lombard (2015) Kuplik et al. (2021).

Species	T (°C)	N	R^2	Respiration rate equation	References
Cassiopena xamanchana[a]	24	41	0.74	RR (μg O_2 mg $prot^{-1}$ d^{-1}) = 607.91 × mg prot−0.26	Verde and McCloskey (1998)
	24	41	0.74	RR (ml O_2 d^{-1}) = 0.4365 × g WW0.740	
	30	27	0.38	RR (μg O_2 mg $prot^{-1}$ d^{-1}) = 409.75 × mg prot−0.15	
	30	27	0.38	RR (ml O_2 d^{-1}) = 0.1658 × g WW0.850	
Catostylus mosaicus	25	16	–	RR (μl O_2 kg WW^{-1} d^{-1}) = (11.6) × 1.54 × kg WW	Pitt et al. (2005)[b]
Mastigias sp.[a]	29	22	0.98	RR (mg O_2 d^{-1}) = 19.07 × g DW0.685	McCloskey et al. (1994)
Nemopilema nomurai	24	36	0.58	RR (ml O_2 d^{-1}) = 0.781 × g WW0.675	Purcell et al. (2010)
	24	12	0.77	RR (ml O_2 d^{-1}) = 1.491 × g WW0.880	
	24	48	0.99	RR (ml O_2 d^{-1}) = 0.549 × g WW1.0 combined	
Rhizostoma pulmo	24.5–27	14	0.66	RR (ml O_2 h^{-1}) = 0.025 × g WW0.993	Purcell et al. (2010)
Rhizostoma pulmo	20	48	0.72	RR (ml O_2 d^{-1}) = 0.287 × g WW0.8661	Purcell et al. (2010)
Rhizostoma pulmo	16.1	7	0.98	RR (ml O_2 d^{-1}) = 0,86 × g WW1,38	Lilley and Lombard (2015)
Stomolophus meleagris	30	71	0.97	RR (ml O_2 g $WW^{-1}h^{-1}$) = 0.05 × g WW0.99	Larson (1991)
Rhopilema nomadica	21–29	21	0.97	RR (ml O_2 $mgDW^{-1}$ d^{-1}) = 0.0241 × DW0.8575	Kuplik et al. (2021)

[a]With symbiotic zooxanthellae (algal/total RR *3%; McCloskey et al., 1994).
[b]Excretion rates were converted to respiration by the O:N atomic ratios of 11.6. Extracted from Purcell et al. (2010).
Compiled from Purcell et al. (2010), Lilley and Lombard (2015) and Kuplik et al. (2021).

rhizostomes generally have higher slopes (1.09; $P < 0.001$) compared to semaeostomes (~0.91; $P < 0.001$). Rhizostomeae medusae consistently have higher individual oxygen consumption rates based on wet mass than the semaeostome counterparts (Purcell et al., 2010). Consequently, rhizostomes require higher individual feeding rates to sustain their higher metabolic demands for the same size medusae of semaeostomes. This can partly be explained by the higher carbon content in their tissues (cf. Section 1 Body composition) compared to semaeostome medusae and because they have a more robust swimming pattern (cf., Section 4 Locomotion and feeding mechanisms), which makes rhizostome medusae energetically costlier animals than semaeostomes.

The *RR* of most marine invertebrates tends to increase about 2.5-fold with every 10 °C temperature increase ($Q_{10} = 2.5$) within their survival limits (Arai, 1997). The baseline, or inactive metabolism steadily rises with temperature until reaching lethal levels. The active metabolism can either increase until levelling off or reach a peak beyond which the organisms cannot maintain higher metabolism levels (Burggren and Roberts, 1991). Within their lethal range, animals acclimatize to their environment's temperature through metabolic and behavioral adjustments. The reported Q_{10} values for the rhizostome *R. pulmo* was 2.6 (Larson, 1987), while it was 2.9 and 3.4 for the semaeostome medusae *A. aurita* and *C. capillata*, respectively (Arai, 1997). In order to make more comprehensive comparisons among scyphomedusae additional experimental studies are needed.

Zooxanthellate rhizostome medusae have an additional energy source by utilizing the photosynthesis of their symbiotic zooxanthellae (reviewed by Djeghri et al. 2019, Section 5.4). This symbiosis partially explains the lower *RR* slopes for *Cassiopea* sp. (a = 0,74) and *Mastigias* sp. (a = 0,68), whereas azooxanthellate species, such as *Catostylus* sp. and *Rhizostoma* sp. have higher *RR* slopes of 1.54 and 1.38, respectively (Table 8). The relationship between the oxygen produced/gWW and the consumed oxygen/gWW (P/R ratio) mean ratio values for *Cassiopea* are 1.67 ($n = 9$) (Cates, 1975), 1.5 (n = NA) (Mergner and Svoboda, 1977), 2.04 (Welsh et al., 2009) and 2.6 (Jantzen et al., 2010), in which the contribution of zooxanthellae to animal respiration (CZAR) satisfies up to 169% of the medusae metabolic carbon demand (Verde and McCloskey, 1998). For *C. tuberculata* P/R ratio ranged from 0.3 to 1.17 with higher production rates in larger (8–12 cm) specimens (Kikinger, 1992); for *Mastigias* sp. (bell diameter range from 2.2–16 cm) the CZAR satisfies a mean of 97.3% of the medusae's metabolic demands ($n = 22$) (McCloskey et al., 1994). In conclusion, while some studies suggest that photosynthesis can fulfill the carbon needs of the organism (e.g. Schmid et al., 1999), it appears that predation is essential to satisfy the organism's nitrogen and phosphorus requirements (Welsh et al., 2009).

6.4 Excretion rates

Excretion is the process by which organisms remove the unwanted compounds produced during catabolism from their bodies. In general, marine invertebrates excrete nitrogen compounds as a result of catabolism of proteins and nucleic acids. Because the organic matter of scyphomedusae is mainly made up of proteins (Table 2), the majority of excreted waste consists of nitrogen compounds, but the fact that measurements of excretion rates were made based on ammonium alone may bias the current understanding of excretion in scyphomedusae and lead to underestimating the importance of other nitrogenous and non-nitrogenous compounds such as phosphorus. This lack of data also limits the ability to follow nutrient pathways in the water column, particularly nutrient recycling (Biggs, 1977). Schneider (1990) calculated that the excretion rate of the Semaeostomeae *A. aurita* was comparable to the rate of other marine invertebrates when converted into carbon basis. This suggests that excretion by scyphozoan medusae, particularly during outbreaks or in oligotrophic waters, may be an important source of recycled nutrients for other organisms.

As highlighted above for other processes, current knowledge of metabolic processes relies mostly on studies carried out on species of Semaeostomeae, while very few data are available for Rhizostomeae (Table 9), with the exception of *C. mosaicus* and *P. punctata* (Pitt et al., 2005). Conversely, rates in semaeostome species have been determined under different conditions and modeled into equations which are based either on wet mass for *A. aurita* (Shimauchi and Uye, 2007) and *P. noctiluca* (Forsskål, 1775) (Lilley et al., 2014), or dry mass for *Chrysaora quinquecirrha* (Desor, 1848) (Nemazie et al., 1993).

Even within Semaeostomeae comparisons are difficult due to the different conditions and units of measurements used. The mass of scyphozoan medusae is the most frequently factor used to model excretion rates, as most of equations mentioned above rely on this parameter. However, the study by Pitt et al. (2005) highlighted the importance of symbiotic algae in the excretion process. Medusae of *P. punctata*, which hosts symbiotic algae, excreted less NH_3 but more NO_x than the non or little symbiotic medusae

Table 9 Excretion rates of nitrogen (as NH_3, NH_4^+ or NO_x) and phosphorus compounds (as PO_4) of Semaeostomeae and Rhizostomeae medusae.

Group	Species	Excretion rate	Unit	T (°C)	References
Nitrogen compounds—NH_3					
Rhizostomeae	*Catostylus mosaicus*	1555–1004	μg NH_3 kg $WM^{-1}h^{-h}$	22–28	Pitt et al. (2005)
Rhizostomeae	*Phyllorhiza punctata*	49	μg NH_3 kg $WM^{-1}h^{-h}$	22–28	Pitt et al. (2005)
Nitrogen compounds—NH_4					
Semaeostomeae	*Aurelia aurita*	0.2	μg NH_4 + mg $protein^{-1}h^{-1}$		Biggs (1977)
		117.25 – 649.38	μg NH_4 + h^{-1}	15	Schneider (1989)
		46,800	nM NH_4 + g DM^{-1} d^{-1}		Schneider and Weisse (1985)
Semaeostomeae	*Chrysaora quinquecirrha*	84.0 – 120.0	μg NH_4 + g DM^{-1} d^{-1}		Nemazie et al. (1993)
		0.456 – 213.36	μg NH_4 + d^{-1}		
Semaeostomeae	*Pelagia noctiluca*	ln(NR) = 78.24	μg NH_4 + d^{-1}	21	Morand et al. (1987)
		535.74 ± 185.80	μg NH_4 + g DM^{-1} d^{-1}	12	Malej (1989)
		250.73 ± 72.15	μg NH_4 + g DM^{-1} d^{-1}	16	
		519.51 ± 119.05	μg NH_4 + g DM^{-1} d^{-1}	19	
		584.45 ± 245.32	μg NH_4 + g DM^{-1} d^{-1}	23	
		579.03 ± 113.64 – 873.0-6 ± 120.86	μg NH_4 + g DM^{-1} d^{-1}	25	
		2.4	NH_4 + %N d^{-1}	16.5	Morand et al. (1987)

(continued)

Table 9 Excretion rates of nitrogen (as NH_3, NH_4^+ or NO_x) and phosphorus compounds (as PO_4) of Semaeostomeae and Rhizostomeae medusae. (*cont'd*)

Group	Species	Excretion rate	Unit	T (°C)	References
		3.3	NH_4 + %N d^{-1}	21	
		6.7	NH_4 + %N d^{-1}	24	
		0.8 – 1.2 ± 0.6	μg NH_4 + mg $protein^{-1}h^{-1}$		Biggs (1977)
		28.68 – 619.08	μg NH_4 + d^{-1}	20	Lilley et al. (2014)
Nitrogen compounds—NOx					
Rhizostomeae	*Catostylus mosaicus*	26	μg NOx kg WM^{-1} h^{-h}	22–28	Pitt et al. (2005)
Rhizostomeae	*Phyllorhiza punctata*	52 – 80	μg NOx kg $WM^{-1}h^{-h}$	22–28	Pitt et al. (2005)
Phosphorus—PO_4					
Semaeostomeae	*Aurelia aurita*	5.28	nM PO4 mg DM^{-1} d^{-1}		Schneider and Weisse (1985)
		1.4 – 5.7	μmol $PO4h^{-1}$	15	Schneider (1989)
Semaeostomeae	*Pelagia noctiluca*	3.6 ± 2.3	μmol PO4 g DM^{-1} d^{-1}	12	Malej (1989)
		1.8 ± .07	μmol PO4 g DM^{-1} d^{-1}	16	
		2.7 ± 0.7	μmol PO4 g DM^{-1} d^{-1}	19	
		3.6 ± 0.6	μmol PO4 g DM^{-1} d^{-1}	23	
		3.0 ± 0.4 – 6.4 ± 0.9	μmol PO4 g DM^{-1} d^{-1}	25	
Rhizostomeae	*Catostylus mosaicus*	173 – 104	μg PO4 kg WM^{-1} h^{-1}	22–28	Pitt et al. (2005)
Rhizostomeae	*Phyllorhiza punctata*	0	μg PO4 kg WM^{-1} h^{-1}	22–28	Pitt et al. (2005)

of *C. mosaicus*. This is because in *P. punctata*, NH_3 may be recycled directly by endosymbiotic algae within the context of the mutual exchange with their host. Therefore, the presence and quantity of holosymbionts in Rhizostomeae jellyfish affects excretion rates, potentially generating ecological implications related to the supply of recycled nutrients.

7. Growth and development

Over the past decades, increasing attention has been devoted to the growth of rhizostome medusae to improve our understanding on their blooms and on modeling population dynamics of commercially exploited species. Growth rates represent the change of individual size, relative to its age. For rhizostomes such estimates have been developed in the laboratory, or through cohort analysis (i.e. through the increase in bell diameter of a given population during a season). Laboratory growth estimates have been carried out as part of life cycle (e.g. Astorga et al., 2012) and early developmental studies (e.g. Jordano et al., 2022; Straehler-Pohl and Jarms, 2011). As medusae develop, however, the volume of laboratory containers and food offered may limit growth (Holst et al., 2007). For larger medusae, the monitoring of cohorts becomes the only feasible measure of growth (e.g. Kikinger, 1992; Leoni et al. 2021; Pitt and Kingsford, 2003). More recently, another indirect method has been proposed to estimate age by analysing growth of statoliths in medusae bell margins, as they increase in number and size with age (Holst et al., 2024; Holst et al., 2007; Holst et al., 2016; Sötje et al., 2017). The accuracy and limitations of this method have been experimentally tested and discussed (Heins et al., 2018; Hopf and Kingsford, 2013; López-Martínez et al., 2020). These estimates allow us to make inferences on parameters (e.g. mortality rates, population biomass) necessary to include jellyfish data in ecosystem models (Palomares and Pauly, 2009). Aspects related to growth, morphological development, or any related to population dynamics of rhizostome medusae are reviewed here.

7.1 Early development and morphogenesis in ephyrae of Rhizostomeae

Studies on the life cycles of rhizostome species traditionally devote more attention to issues such as the development and reproduction of adult jellyfish, settlement and reproduction modes of polyps (Schiariti et al., 2024) than to the initial morphogenesis of ephyrae. More recently, comparative analyses of morphologies of scyphozoan ephyrae were carried out by Straehler-Pohl and Jarms (2010) and Holst (2012) and patterns of development of the ephyrae were studied by Jordano et al. (2022). The initial morphogenesis of Rhizostomeae ephyrae is marked by key events in almost all species: (i) the filling of the inter-lobe gaps; (ii) the thickening of the mesoglea; (iii) the development of the 8 oral arms, often with terminal appendages; and (iv) the closure of the centrial mouth of the ephyrae, which comes with a great increase in size, e.g. in *L. lucerna* (Schiariti et al., 2008) and in *R. octopus* (Holst et al., 2007). The only exception to this developmental pattern occurs in species of the genus *Cassiopea,* in which the ephyrae are released with four oral arms already formed and without inter-lobe gaps (Jordano et al., 2022). Therefore, ephyrae of this genus only undergo an increase in size and in the number of oral arms as well as in the appearance of the vesicles on the oral arms.

The time taken for structures to appear and develop in ephyrae of the same species varies greatly when considering individuals of the same age, but it is more constant in relation to individuals of the same bell diameter. Therefore, the increase in bell diameter, rather than in age (days), is in general better to identify the stage of development and to predict events of morphogenesis in scyphozoan ephyrae (Jordano et al., 2022). This is probably because jellyfish can shrink and grow due to starvation and other physiological conditions (see below), thus morphological responses to environmental changes (i.e. phenotypic plasticity) affect the timing of emergence of structures and developmental patterns (Jordano et al., 2020, 2022; Lilley et al., 2014).

Moreover, observations from laboratory cultures suggested that morphogenesis events occur earlier in rhizostome than in semaeostome medusae (Jordano et al., 2022). The bell diameter in which it is possible to observe the emergence of the digitata on oral arms of the Semaeostomeae species *Aurelia coerulea* von Lendenfeld, 1884 and *Aurelia malayensis* Lawley, Gamero-Mora, Maronna, Chiaverano, Stampar, Hopcroft, Collins Morandini, 2021 was on average 7.23 and 6.65 mm, while the digitata appeared in ephyrae of about half of this size in the following Rhizostomeae species: *P. punctata* (2.89 mm), *R. esculentum* (3.38 mm), *C. mosaicus* (3.74 mm), *C. towsendi* (3.84 mm), *M. papua* (5.23 mm), *L. lucerna* (1.90 mm) and *C. tuberculata* (2.62 mm) (Jordano et al., 2022). The emergence of the velar lappets in *A. coerulea* (5.52 mm) and *A. malayensis* (4.62 mm) also occurred later and in the more developed ephyrae compared to ephyrae of the rhizostomes *P. punctata* (3.30 mm), *R. esculentum* (4.12 mm), *C. mosaicus*

(3.95 mm), *C. towsendi* (4.17 mm), *M. papua* (4.42 mm), and *L. lucerna* (4.3 mm). Likewise, the filling of the inter-lobe gaps of ephyrae to form a continuous bell margin occurs in smaller sized individuals of rhizostomes compared to semaeostomes species (Jordano et al., 2022). This can be associated with the fluid dynamics around the bell; with a larger influence of the inertial regimes around the rhizostome ephyrae at the beginning of development, when compared to the slower swimming and less inertial regime influence over the ephyrae of semaeostome species (von Montfort et al., 2023).

The pattern of development of the ephyrae may also demonstrate phylogenetic signals in scyphozoan medusae. The pattern of filling of the inter-lobe gaps of the Rhizostomeae ephyrae is similar to that found in species of Ulmaridae (Semaeostomeae), which are considered sister groups (Daglio and Dawson, 2017). The filling of the inter-lobe gaps in these two clades occurs through the emergence of "differentiated" velar lappets, which develop distally until they form a continuous bell margin (Jordano et al., 2022). Conversely, for ephyrae of other semaeostome families (Pelagiidae, Cyaneidae, and Phacellophoridae) the velar lappets develop in an undifferentiated way, i.e. inter-lobe gaps are gradually filled without the distal development of discernible velar lappets. Another characteristic that reflects the evolutionary history of scyphozoans is the presence of digitata in the oral arms of medusae of Rhizostomeae and Ulmaridae. In addition, the marginal lappets of the ephyrae of Kolpophorae (Rhizostomeae) and Ulmaridae (Semaeostomeae) have a smooth contour, unlike the ephyrae of Dactyliophorae (Rhizostomeae). The latter have serrated marginal lappets (Jordano et al., 2022). Whether these differences on morphology and development represent any functional or performance differences remain unknown. However, the tendency for Rhizostomeae jellyfish to have faster growth and initial morphological development compared to Semaeostomeae jellyfish implies that the former must achieve higher swimming performance and operate at higher Reynolds numbers (Re) earlier during development. Ecological implications of these differences should be studied in the future.

7.2 Phenotypic plasticity

When environmental variability drives distinct morphological development that cannot be explained by random phenotypic variation, morphological variability is associated with phenotypic plasticity (Travis, 1994). In scyphomedusae, evidences indicates that phenotypic plasticity may be triggered by changes in temperature, water salinity, food, or in demographics. Changes in temperature are also associated with changes in water viscosity and changes in the rate of solar incidence (Anthony et al., 2022; Båmstedt et al., 1999; Chiaverano and Graham, 2017; Feitl et al., 2009; Jordano et al., 2020; Nawroth et al., 2010; Swift and Dawson, 2020; Widmer, 2005).

Morphogenesis of ephyrae bell to young medusae shape can be triggered by their surrounding fluid regime (Nawroth et al., 2010) resulting in different morphotypes. Ephyrae of Rhizostomeae and Semaeostomeae show differences in bell development under different temperatures (Jordano et al., 2020; Nawroth et al., 2010; Widmer, 2005). Ephyrae of the rhizostomes *L. lucerna* and *C. Andromeda,* and of the semaeostome *Aurelia* sp. that developed under lower temperatures reached smaller bell diameters than those, of the same age that developed under higher temperatures. Thus, higher temperatures accelerated growth (Jordano et al., 2020; Nawroth et al., 2010; Widmer, 2005). In addition, under lower temperatures, the inter-lobe gaps were filled later. This has been explained by lower temperatures and consequently lower Reynolds numbers around those species lead to the extent of the boundary layer (region of fluid deceleration near solid surfaces) around the lobes being greater. Thus, the boundary layer around one lobe overlaps with the next lobe, forming a virtually closed structure. Due to the overlap of consecutive boundary layers, the clefted ephyrae bell margin operates as if it were completely filled with tissue (Feitl et al. 2009; Jordano et al., 2020; Nawroth et al., 2010). By contrast, at higher temperatures, boundary layers around the lobes are reduced because of the less viscous environment (i.e. higher Reynolds numbers). Under such conditions, it was observed that the inter-lobe spaces close earlier and the medusae fill in these gaps earlier.

Environmental temperatures also affected the development of the capturing structures—i.e. digitata—of rhizostome ephyrae (Jordano et al., 2020). Ephyrae of *L. lucerna* cultured under lower temperatures developed thicker and, especially, more widely spaced digitata than ephyrae cultured under higher temperatures. Ephyrae of *C. andromeda,* which already bear the digitata when released from the strobila, developed longer digitata when cultured under lower temperatures. Both of these patterns can be explained by the fact that under lower temperatures, increased boundary layers around the digitata would potentially compromise filtration volumes through spaces between these structures. Thus, the maintenance of functional filtering structures at lower temperatures occurred through the development of digitata with larger spaces between one another (*L. lucerna*) or longer enough to enable filtration fluxes (*Cassiopea*). This prevented the overlapping of boundary layers of adjacent digitata from reducing flow rates and therefore filtration performance (Jordano et al., 2020).

The rate of sunlight reaching the jellyfish can also affect their development (Anthony et al., 2022). It was observed that medusae of *Cassiopea ornata* Haeckel, 1880 that developed in clearer water (collected in Piti Channel, Guam) and, therefore, with a higher incidence of sunlight, showed the pigment Cassio Blue and larger vesicles. Medusae from turbid waters (collected in Cocos Lagoon, Guam) and less sunlight lacked the pigment and had smaller vesicles. These differences could be an acclimatization response to light that increases protection of the photosystems and regulates the photophysiology of the endosymbiont (Symbiodinaceae) present in these medusae. The absence of large vesicles in individuals from the Cocos Lagoon population suggests that the development of these structures would be an unnecessary investment in a turbid environment (Anthony et al. 2022).

In addition, differences were also found between the cassiosomes (see Section 3.1 and Holst et al. 2024) of the Cocos Lagoon and Piti Channel populations (Anthony et al., 2022). The cassiosomes are released into the water column and swim with the aid of flagella, what means that prey is captured without encountering the medusa (Ames et al., 2020; Muffett et al., 2021). The differences between the cassiosomes of the *C. ornata* populations were related to: (1) the presence of Symbiodiniaceae, (2) the size of the O-isorhiza and (3) the abundance of nematocysts. Cassiosomes of the population from the turbid waters of Cocos Lagoon had endosymbionts and larger and more abundant nematocysts than those of the Piti Channel. These differences could be a by-product of trophic strategies. The Symbiodiniaceae present in the cassiosome probably provide nutrients, feeding the flagella and increasing the longevity of the cassiosome in the water column (Ames et al., 2020). This suggests that individuals from the Cocos Lagoon have a greater investment in the function of the cassiosome and a lower investment in photoprotective structures (i.e. large vesicles and Cassio Blue) for the endosymbionts, because of their development in turbid waters. In addition, this region has a greater supply of nutrients, which increases the availability of plankton and symbiont productivity. The increase in symbiont productivity in cassiosomes increases their longevity and the greater number of nematocysts promotes greater success in capturing prey. These adaptations are the result of a selective pressure that favors the heterotrophic strategy. Individuals from the clearer waters of the Piti Channel population, on the other hand, developed under higher sunlight incidence, so they were selected to invest more in photoprotective structures for their endosymbionts to the detriment of the cassiosomes, thus, relying more on the photoautotrophy than individuals from turbid waters.

Demographic and environmental disturbances can also result in phenotypic plasticity of morphological and behavioral characteristics (Swift and Dawson, 2020). Populations of *M. papua* from Ngermeuangel Lake (Palau) were compared before and after a demographic decline. This population declined from ~4–5 million jellyfish to 0, after an increase in temperature and salinity of the upper habitable layer of the lake. Years later, the population recovered, but with fewer individuals (from 5–7 jellyfish m^{-3} to <0.05 jellyfish m^{-3}) and with different morphological characteristics and swimming behaviors than the previous population. Jellyfish from the low density population had larger bell diameters and pigmentations, longer and tricorn-shaped terminal clubs, a greater number of intermediate filaments, thinner mesoglea and a greater internal canal complexity. They also had a lower pulsation rate and higher swimming speeds. These responses to changes in the environment were interpreted as phenotypic plasticity, since the two populations were genetically very similar (Swift and Dawson, 2020). The origin of this variation is unknown, but these results suggest consequences for the ecosystem, because higher pulsation rates, as well as the larger bell size of the jellyfish implies a higher clearance rate. All of these morphological and behavioral changes in response to changes in the environment showed the resilience of these organisms and their acclimatisation capacity, both of which allow them to maintain their populations in environments under continuous changes.

7.3 Regeneration

Regeneration capacities of cnidarians have been recognized for a long time. The polyp of *Hydra* spp. (Cnidaria, Hydrozoa) is the classical example, where a small piece of tissue can originate a whole new individual (Gold and Jacobs, 2013). Studies with *Hydra* spp. have provided the foundation for understanding stem cells, tissue organization and differentiation during regeneration in metazoans (Gold and Jacobs, 2013). In Scyphozoa, regeneration of new individuals from small pieces of tissue have been observed in scyphistomae of the genera *Aurelia, Cassiopea* and *Chrysaora* (Arai, 1997; Gamero-Mora et al., 2019). Scyphistomae hold a higher potential of regeneration than ephyrae since new scyphistoma individuals regenerate from small pieces of cut tentacles (Laurie-Lesh and Corriel, 1973), yet this capacity has rarely been observed in ephyrae. However, a peculiar mode of regeneration has been described for ephyrae of *Aurelia*, *C. pacifica*, *Mastigias* sp., and *C. tuberculata*, in which the cut marginal lobes or oral arms do not regenerate, but the animal spatially reorganizes remaining body parts to regain radial symmetry in a process named symmetrisation. This process occurrs within 1 to 4 days of the initial removal of marginal lobes or oral arms (Abrams et al., 2015). This fast recovery resulted in animals with a smaller number of structures, but recovering the initial body symmetry still allowed effective swimming. The mechanics of muscle cell proliferation and mesoglea reorganization are still to be understood and questions have been raised with regard to how the animals sense their lost parts and drive the morphogenesis in order to regain symmetry and locomotory function (Abrams et al., 2015).

The regeneration of lost structures in medusae has been observed in *Cassiopea* (Gamero-Mora et al., 2019). Adult medusae of *C. xamachana* were observed regenerating new sets of body structures after an injury of the umbrella (Gamero-Mora et al., 2019). In the place of the injury started a tissue regeneration, with closure of the wound, but subsequently new tissues were generated, with gastrovascular cavities and radial canals. Reorganization followed by the development of two new sets of body parts from the original individual. Both sets possessed oral arms, oral appendages, radial canals, subgenital ostia and wandering cells. Gastrovascular cavities remained connected and fission was not observed, so that asexual reproduction (e.g. budding) was not characterized therein. However, a more recent experiment demonstrated that *Cassiopea* medusae cut into two (one piece holding the oral arms, while the other was without could regenerate into two new individuals (Ostendarp et al., 2022)). This demonstrated the outstanding regenerative capacity of *Cassiopea,* in which autotrophy plays a central role sustaining nutritional demands of the umbrella pieces that have lost their oral arms. Such experiment also supported the hypothesis of trans-differentiation potential in umbrella striated muscle cells and included regenerative capacities of medusae of *Cassiopea* at similar levels to those previously described for hydromedusae (Sinigaglia et al., 2020). The fragmentation of individuals represents asexual reproduction and implies in great resilience of populations may contribute to the great invasive potential of *Cassiopea* (Ostendarp et al., 2022). Such findings may lead to new investigations on gene expression, cellular cues, mechanisms of cells recruitment, and molecular inductors of tissue proliferation, differentiation or transdifferentiation (Gamero-Mora et al., 2019; Ostendarp et al., 2022; Sinigaglia et al., 2020).

7.4 Physiological effects of starvation on growth and reproduction

Periods of starvation may cause physiological disturbances in azooxanthellate medusae, affecting their development. It can cause a reduction in size (i.e. degrowth), that can be followed by a period of recovery to its normal size (Jordano et al., 2022; Lilley et al., 2014). This ability is surprising given the energetic storage in medusae tissue. This storage is comparatively limited because of the low lipid content and the absence of storage lipids as wax esters (see Section 2. Body Composition). Energetic reserves of scyphozoans consist of the protein content, but also of glucose, glycogen, and lipid droplets inside amoebocytes within the extracellular matrix of the mesoglea (Schmid et al., 1999; Nevárez-Lópes et al., 2020). The effects of starvation in scyphozoans are likely to be diverse, depending on the stage of the life cycle and the species in question. Scyphistomae and ephyrae hold a higher capacity to delay death for months when starved, whereas mature medusae rapidly decline under starvation. In polyps of *R. octopus,* growth was negatively affected by starvation (Holst et al., 2007). Starved polyps of *Acromitus hardenbergi* Stiasny, 1934 developed abnormally longer stalks, compared to those that were fed. Fed individuals developed bowl or goblet-shaped stalks (Miyake et al., 2021). These longer stalks create larger encounter areas for polyps (Miyake et al., 2021), but further functional consequences of this possible phenotypic plasticity are yet to be investigated.

Effects of starvation on strobilation and on other asexual reproductive pathways of rhizostome polyps (i.e. production of planuloids and podocytsts) are, in general, negative. Starvation inhibited strobilation in polyps of *C. mosaicus* (Pitt, 2000) and probably in *R. esculenta,* as the species started strobilating only after being fed daily (Chen et al., 1985). The production of free-swimming buds and budding by polyps of *Cassiopea* sp., *C. cephea, M. papua, and P. punctata* was decreased or ceased under starvation (Schiariti et al., 2014). This contrasted with the observation that short periods of starvation would trigger strobilation in low-light treatments with scyphistomae of the semaeostome *Aurelia labiata* Chamisso and Eysenhardt, 1821 (Purcell et al., 2007). Effects of starvation on asexual reproduction in Semaeostomeae species like *Aurelia* spp. are well known (Lucas et al., 2012; Wang et al., 2023), for rhizostome species, experimental evidence is much more limited, but existing data indicates that starvation negatively affects asexual reproduction (Holst et al., 2007; Schiariti et al., 2014).

It was previously speculated that podocysts represented a resistance/dormancy mechanism which allowed populations to persist during adverse periods, and their production was triggered by stressors, such as reduced food supply or the presence of predators (Arai, 2009). Some experimental results supported this idea, as starvation increased the production of podocysts in polyps of *C. mosaicus* (Pitt, 2000) and *A. aurita* (Thein et al., 2012). However, for other rhizostome species like *L. lucerna*, *R. pulmo*, *R. esculentum*, *N. nomurai* and *S. meleagris,* the production of podocysts was higher under increased food supply (Schiariti et al., 2014). In *R. esculentum,* podocyst production decreased during longer periods of starvation (Guo, 1990). Environmental stressors, such as abnormally high temperatures, low salinities and hypoxia also increased the production of podocysts in polyps of *N. nomurai* (Kawahara et al., 2013). These mixed results on the effect of starvation and other stressors on podocyst formation reflect species or population-specific differences (Lucas et al., 2012). In any case, the formation of podocysts represent a normal reproductive pathway (Schiariti et al., 2024) and the amount of podocysts formed is a key-factor that may increase the size of the jellyfish population of the next reproductive season. Hence they may trigger the formation of blooms (Arai, 2009; Kawahara et al., 2013). Further discussion about podocyst formation and on the consequences to population dynamics are found in Schiariti et al. (2024).

Ephyrae of the semaeostome *Aurelia* spp. can usually survive for months, and up to 100 days without feeding. Their point of no return from starvation (for about 50% of individuals) was estimated between 33 to 58 days (Fu et al., 2014). These values are one order of magnitude longer than the capacity of crustaceans, molluscs and fish larvae (Fu et al., 2014). Experiments with ephyrae of the rhizostome *C. xamachana* demonstrated higher survivability (~60%) of individuals under starvation and low light conditions for 28 days, which was followed by a resumption of normal growth (Muffett et al., 2022). In addition, these ephyrae could keep their zooxanthellae during this unfavorable period and this survival capacity demonstrated the potential of ephyrae to withstand long journeys in ballast water, or in hiperoligotrophic waters, traveling on currents that can eventually transport them to new locations (Muffett et al., 2022). Because experiments evaluating starvation for scyphistomae and ephyrae usually employed filtered seawater, dissolved organic matter and bacteria that can go through the filtering process could have comprised non-quantified nutritional sources (Skikne et al., 2009). Although previous experiments did not demonstrate the capacity of ephyrae to develop and grow relying solely on dissolved organic matter, they exhibited a better physiological condition than starved ephyrae (Skikne et al., 2009). Therefore, artificial seawater should be used in experiments evaluating the role of starvation (e.g. Stampar et al., 2008).

For the zooxanthellate *Cassiopea,* suppression of heterotrophy can be compensated by the production of their zooxanthellae, as some medusae were kept alive for years in the laboratory under constant light and without feeding (Schmid et al., 1999). For azooxanthellate medusae, however, starvation rapidly decreases respiration rates (Arai, 1997; Purcell et al., 2010) and initiates shrinkage of body parts as quickly as 24 h after the animal has been deprived of food. Size reduction might be an adaptive behavior that triggers catabolic consumption of reproductive and somatic tissues, allowing individuals to survive for a set period of time when food is scarce or absent (Toullec et al., 2024). The capacity of scyphozoan medusae to reduce or increase the bell size depending on feeding regimes indicated that the extracellular matrix (i.e. mesoglea) could be degraded and used as a nutritional source (Schmid et al., 1999). Experiments with *Cassiopea* jellyfish demonstrated that bell volume decrease under starvation and heat stress. This decrease is most likely caused by the catabolic degradation of the mucopolysaccharide and collagen matrix of the mesoglea (Toullec et al., 2024). The oxidation of structural carbohydrates and proteins is followed by the loss of water content resulting in shrinkage (Aljbour et al., 2017; Chapman, 1953; Toullec et al., 2024). Degradation of feeding structures of the digitata and reduction of oral arm length has been described after 48 h of starvation in *S. meleagris* (Camacho-Pacheco et al., 2022). This can cause a loss of feeding capacity, due to the inability to capture prey. In adult medusae gonads are resorbed and show dedifferentiation, but the spermatogonia on more advanced stages of maturation continue to mature, but the remaining gonads decrease in size (Arai, 1997). Even under starvation, medusae of *P. noctiluca* demonstrated a remarkable capacity to release eggs continuously over a 28-day period, in which between 76% and 86% of wet mass was lost (Lilley et al., 2014). This is similar to the shrinkage inherent of the aging process in hydromedusae, in which the bell is completely reabsorbed, while gonads on the manubrium continuously produce gametes until the animal completely disintegrates

(Schmid and Tardent, 1969). These observations reinforce that medusae under food scarcity not only reduce growth rates but also prioritize energy allocation toward sexual reproduction (Goldstein and Riisgård, 2016).

7.5 Field estimations of growth rates

Growth rate estimations are crucial for biology and ecology, but also for stock assessments of harvested populations (Sparre and Venema, 1998). Despite the increasing commercial interest in rhizostome jellyfish, studies on their growth are still scarce compared to crustacean or fish populations (Palomares and Pauly, 2009). With the current alleged increase in their blooms, an understanding of rhizostome jellyfish life-history traits is key to comprehending these events and to provide early warnings. Technical constrains on their collection and manipulation (Purcell, 2009) result in a predominance of laboratory studies (e.g., Acevedo et al., 2013; Astorga et al., 2012) in contrast to field monitoring (López-Martínez et al., 2020). Laboratory growth estimations are useful for evaluating the effects of diverse factors on growth, but they do not necessarily reflect the complexity of multiple factor interactions, as environmental conditions and food supply shape growth patterns of marine populations (Bhaud et al., 1995). Moreover, laboratory analysis can only be used up to a certain body size, thus, growth rates of larger (>15 cm) Rhizostomeae species can only be estimated in the field.

The most implemented method to assess jellyfish growth rates in the field is the cohort identification and then tracking their size increase through time. This method possesses the advantage of not requiring jellyfish death (e.g., Pitt and Kingsford, 2003), as measurements can be done on the water or on board. The followed populations can present one (Brewer, 1989; Fernández-Alías et al., 2020; Kikinger, 1992; Marques et al., 2015; Rosa et al., 2013) or more (2, 3 or 4 in some cases) cohorts during the year. This multi-cohort pattern has been observed in species of Rhizostomeae (Fernández-Alías et al., 2020; García, 1990; Kikinger, 1992; Leoni et al., 2021; López-Martínez et al., 2020), Semaeostomeae (Malej and Malej, 1992) and Cubozoa (Gordon et al., 2004). The presence of multiple cohorts occurs when the population undergoes consecutive strobilations after the first strobilation event (Calder, 1982; Purcell et al., 2012), which result in multiple recruitments. This pattern has been observed in species with both monodisc (i.e. one ephyra is produced) and polydisc (i.e. two or more ephyrae are produced) strobilation (Table 10).

Table 10 Growth rates (mm day^{-1}) of Rhizostomeae jellyfish species in the wild. Reference of strobilation type, when different from the reference of growth rate assessment: [1]Pitt (2000) [2]Jarms and Morandini (2023) [3]Calder (1982) [4]Fuentes et al. (2011) [5]Holst et al. (2007) [6]Ding and Chen (1981).

Species	Strobilation	Growth rate (mm day^{-1})	Method	N cohorts per year	Measure	Period of time followed	Study area	Recruitment	Medusae size (mm)	Lifespan (months)	References
Catostylus mosaicus	Monodisc/ polydisc[1]	0.39–4.81	Cohort	3 (6–7 smaller)	Bell diameter	96 months	Botany Bay (Australia)	December, March–July	20–280	13	Pitt and Kingsford (2003)
		0.10–2.90	Cohort	2 or 3 (6–7 smaller)	Bell diameter	30 months	Lake Illawarra (Australia)	February–July	20–300	13	Pitt and Kingsford (2003)
Cotylorhiza tuberculata	Monodisc	5.43	Cohort	1	Bell diameter	6 months	Bay of Vlyho (Greece)	August	28–304	6	Kikinger (1992)
		5.00–5.14	Cohort	2	Bell diameter	4 months	Bay of Vlyho (Greece)	July and September	28–304 (maximum size 400)	6	Kikinger (1992)
Phyllorhiza punctata	Monodisc[2]	2.00	Cohort	1 or 2	Bell diameter	12 months	Laguna Joyuda (Puerto Rico)	March to July, max. in June	5500	12	(García, 1990)
Stomolophus meleagris	Polydisc[3]	3.10	Cohort	3	Total length	15 months	Gulf of California (Mexico)	December, January, February	4.2–170	8	López-Martínez et al. (2020)

Rhizostoma pulmo	Polydisc[4]	1.50–2.30	Cohort	3	Bell diameter	10 months	Bages Sigean lagoon (France)	April to June	6.7–340	7	Leoni et al. (2021)
Rhizostoma octopus (as *R. pulmo*)	Polydisc[5]	8.25	Cohort	2	Bell diameter	6 months	France (La Rochelle)	July and December	5–850	8	Husson and Fay (1984)
Rhopilema esculentum	Polydisc[6]	7.8	Jellyfish release and recapture	NA	Bell diameter	3 months	Liaodong Bay (China)	Release between May and June	20–700	NA	Dong et al. (2009)
		4.98	NA	NA	NA	NA	Liaodong Bay (China)	NA	1.5–450.5		Ding and Chen (1981)[a]
		5.69	NA	NA	NA	NA	Ariake Sea (Japan)	NA	17–700	12	Nakano (1980)[a]

[a]References from Omori and Nakano (2001)

7.6 The cohort method

Bell diameter (BD) is the measure considered in most field growth rate estimations of Rhizostomeae. Yet, the total length (TL) is also an appropriate feature, as oral arms are an important part of the jellyfish biomass and organic content (see Section 1 Body composition). In the cannonball jellyfish *S. meleagris,* TL has been considered as the most appropriate feature due to the medusae bell rigidity that prevents precise BD determination (López-Martínez et al., 2020). Moreover, because the swimming orientation of some rhizostome medusae such as *C. mosaicus* or *R. pulmo* is usually parallel to the water surface (Raoult and Gaston, 2018), TL can easily be obtained from remote images (Cimino et al., 2018; Raoult and Gaston, 2018; Schaub et al., 2018).

The frequency distribution of BD or TL has been used as a proxy to estimate the population growth (Houghton et al., 2006; Palomares and Pauly, 2009). The cohort number can be determined by multi-modal analyses by fitting a series of normal distributions to size-frequency histograms (Fernández-Alías et al., 2020; Pitt and Kingsford, 2003). Commonly, specimens are grouped in classes with a span of 5 mm (Leoni et al., 2021; López-Martínez et al., 2020). Finally, the number of cohorts by month or sampling date, is equal to the number of normal distribution curves observed. Once cohorts were defined, diverse authors supported the use of the von Bertalanffy Growth Model (VBGM) to assess jellyfish growth (Lee et al., 2020; Palomares and Pauly, 2009) including periods when multiple cohorts coexist. This model has not been widely applied to Rhizostomeae species yet, but has been considered as a good approach for rhizostome medusae (Fernández-Alías et al., 2020; Leoni et al., 2021; López-Martínez et al., 2020; Palomares and Pauly, 2009). Palomares and Pauly (2009) estimated the growth of jellyfish based on this method collecting published data in BD of wild jellyfish populations through time that included five species of Rhizostomeae. The results allowed multi-taxa comparisons of growth parameters such as the growth coefficient k (year^{-1}). Growth coefficients estimated for Rhizostomeae varied from 0.60 to 5.79 ($n_{\text{species}} = 7$; Fernández-Alías et al., 2020; Leoni et al., 2021; López-Martínez et al., 2020; Palomares and Pauly, 2009), which were a little higher than those of Semaeostomeae species (from 0.54 to 4.30, $n_{\text{species}} = 5$) (Palomares and Pauly, 2009). According to these findings, jellyfish growth rates are comparable to those of small fishes, such as guppies and anchovies (Palomares and Pauly, 2009; Uye and Brodeur, 2017).

Commonly, the growth rate of jellyfish varies over time, starting with higher rates in the first medusae stages and first cohorts (Kikinger, 1992; Leoni et al., 2021; Pitt and Kingsford, 2003), with a reduction in larger sizes after sexual maturation. Which factors are responsible for this variability are not yet clear, but food availability, temperature and resources reallocation to reproductive tissue during adult stages could explain this development (Lucas and Lawes, 1998). Moreover, the differences in growth rates among cohorts and the diminishing size that adults of the last cohorts usually achieve could be explained by the natural variability on zooplankton composition and food competition.

One particular physio-ecological feature of medusae is the degrowth potential under starvation and regrowth with subsequent feeding. This shrink potential has been considered a reason for the implementation of different growth models (e.g., sigmoid curves, Dong et al., 2009), which is able to capture size shrink periods in adult stages. Some authors, however, propose incorporating uncertainties through standard deviations of each cohort in the VBGM method to account for the degrowth and regeneration processes (López-Martínez et al., 2020). Another controversial point is the comparatively higher amount of water in medusae tissues. Palomares and Pauly (2009) suggested a correction in order to compare growth rates among diverse groups. A new asymptotic weight (W_{∞}) can be estimated to scale the same water content as in jellyfishes and fishes. To re-scale W_{∞}, the species-specific water content should ideally be known, otherwise, the assumption of 96% as a proxy for gelatinous zooplankton (Larson, 1986), or proportions obtained from phylogenetically close taxa can be considered.

A literature review on the growth rate of jellyfish species (as mm day^{-1}) revealed that rhizostome species stood out with the highest rates and bigger body sizes among groups (Fig. 8, Table 10). Most studied species belong to Semaeostomeae, mainly to the genus *Aurelia*. The studies on Rhizostomeae growth rate in the wild include seven species (of a total of 91 described species, Table 10).

7.7 Rhizostomeae lifespan in the wild

Lifespan of jellyfish has not been elucidated yet, but in the wild, rhizostome medusae have been observed to live between 6 and 13 months (Table 10). These observations are similar to the estimated lifespan of semaeostome medusae. Medusae of the genus *Aurelia* usually live from 6 to 8 months (*A. coerulea:* 7–8 months; Bonnet et al., 2012; Marques et al., 2015; *A. solida:* 6–7 months; Gueroun et al., 2020), but a lifespan of up to 14 months has been recorded (*A. aurita*; Arai, 1997). Senescence and death usually occur after spawning, but it can also be accelerated by low temperatures or low food availability, predation, parasitism or diseases. However, some exceptions exist as

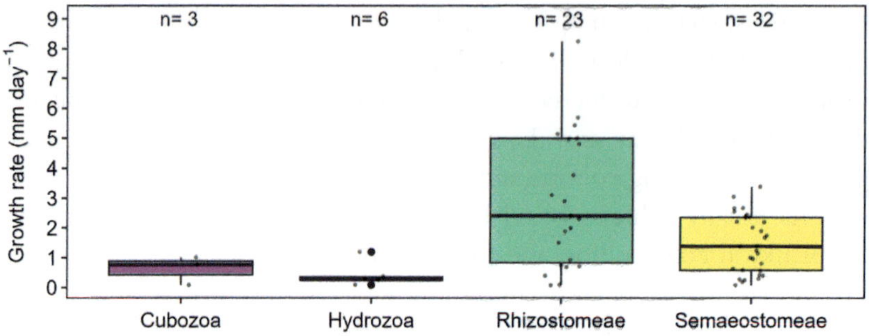

Fig. 8 Growth rate (mm day^{-1}) for different Medusozoa taxa, including laboratory and field estimations. Dots represent each data (n). *Modified from Leoni, V., Molinero, J.C., Meffre, M., Bonnet, D., 2021. Variability of growth rates and thermohaline niches of Rhizostoma pulmo's pelagic stages (Cnidaria: Scyphozoa). Mar. Biol. 168, 107. https://doi.org/10.1007/s00227-021-03914-y.*

some species can, apparently overwinter, e.g. *R. pulmo* (Houghton et al., 2007; Fuentes et al., 2011), *P. punctata* (Haddad and Nogueira, 2006) and species of the genera *Aurelia* (Albert, 2005) and *Chrysaora* (Brodeur et al., 2008). Indeed, some species kept in captivity under optimal conditions may live for multiple years (e.g., *Cassiopea* sp. up to 4 years; *A. aurita* up to 2 years; Zahn, 1981).

Most populations of scyphozoan polyps have not been found in the wild yet. From the 91 Rhizostomeae species, the life cycle of 62 species is still unknown and the 28 species with known life cycle they all present a polyp stage, with no record of holoplanktonic species (estimated from Jarms and Morandini, 2019, 2023). As far as we know, only polyps of *C. xamachana* (Bigelow, 1900), *C. tuberculata* (Kikinger, 1992), *M. papua* (Dawson et al., 2001), and *C andromeda* (Morandini et al. 2017) have been found in the wild. This lack of knowledge hampers the possibility to define the lifespan of polyps and whether they are perennial populations with seasonal abundance variability (as *A. coerulea*, Marques et al., 2019) or survive as podocysts during unfavorable periods (e.g. *Cyanea*; Brewer and Feingold, 1991). The lack of understanding of the parameters of the population dynamics of polyps (e.g. lifespan, asexual reproduction, mortality) still represents a huge gap in the understanding of the occurrence patterns of jellyfish and their bloom events.

8. Concluding remarks

We analyzed functional characteristics of Rhizostomeae species and discussed whether these features distinguish them from other scyphozoans or from other gelatinous zooplankton. Body composition of rhizostome medusae have on average more than double the carbon content per unit of biomass than semaeostome counterparts. Rhizostome species of Kolpophorae typically harbor a few millions of symbiotic zooxanthellae per gram of wet mass in their tissues, which also affect their body composition. Because of the tendency of zooxanthellae to accumulate carbon rich lipid reserves, zooxanthellate Kolpophorae are generally enriched in carbon with C:N ratios often superior to 5 as compared to generally less than 4 for other gelatinous zooplankton.

Body composition of rhizostome medusae can be an indicator of marine pollution. Particularly, tissues of the benthic *Cassiopea* spp. jellyfish reflect environmental gradients and local sources of pollution. These medusae are easy to keep in laboratory, have low mortality rates, and rapidly absorb trace metals, which can be useful biondicator of continuous or pulse-contamination. Body composition affects physiology, nutrition and other functional traits, therefore, in order to include rhizostome species in biogeochemical or trophic models, species-specific data of the biochemical composition, or from phylogenetically related species, should be used to make these models more accurate.

Just like other scyphozoan medusae, swimming and prey capture of rhizostome species occur simultaneously. As the medusae pulse, they transport large volumes of water, creating feeding currents that drive food particles toward oral arms. Rhizostome medusae swim about twice as fast as semaeostome species and such higher velocities are achieved because they pulse about twice as frequent. Rhizostomeae medusae perform faster relaxation of the bell, compared to other Discomedusae. This may be related to their high collagen content that can be up to ten times higher in rhizostomes than for semaeostomes. Therefore, harder and sturdier bells in rhizostomes generate faster recovery of the bell, which shorten pulsation times. In addition, they produce comparatively higher velocities of the fluid entrained toward oral arms capture surfaces, which hypothetically allow the capture of more evasive prey. The more frequent pulsation performed by rhizostomes allows the processing of larger fluid volumes for suspension feeding per unit of time. Future research should evaluate whether such biomechanical differences in swimming patters result in niche differentiation.

The number of studies on the feeding habits of rhizostomes is only a fraction of what has been studied about species of semaostome genera *Chrysaora, Aurelia* and *Cyanea*. The diets of rhizostome medusae are highly influenced by zooplankton prey availability and may follow seasonal succession of planktonic prey items, but some evidence of prey selectivity patterns has been discussed. Very few studies of rhizostomes estimated individual ingestion rates corrected by digestion times and provided predatory impacts estimations. Factors that affect feeding habits such as ontogenetic development, digestion times, temperature, prey density are very little known among rhizostomes.

Enzymatic composition of rhizostomes revealed that all stages were capable of hydrolyze carbohydrates, protein, and lipids. Nevertheless, higher amylase and aminopeptidase activity in polyps suggested that phytoplankton, particulate or dissolved organic matter can be important sources during this stage. Conversely, in ephyrae and medusae, amylase and aminopeptidase activities decreased, which could be related to a diet shift in which carnivory becomes more important during the pelagic life. Estimates of digestion times of rhizostome species are different for each type of prey. Experiments are still needed to assess whether digestion times are similar among rhizostome species and to model the effect of temperature, prey and predator size, and stomach fullness, etc.

Rhizostomeae jellyfish consistently have higher oxygen consumption rates compared to those of Semaeostomeae of the same size. These rates increase more rapidly with individual size in rhizostomes than among semaeostomes. Consequently, rhizostomes must have comparatively higher individual feeding rates to sustain their metabolic demands. The higher oxygen demand may be related to the higher carbon content in their tissues and to the more robust swimming, which makes rhizostome medusae energetically costlier animals than semaeostomes. In zooxanthelllate rhizostomes the energy provided by the photosynthesis of their symbionts can partially or totally sustain metabolic carbon demands, depending on the species. Even so, predation is still essential to satisfy nitrogen and phosphorus requirements.

Rhizostomeae medusae tend to have faster growth and initial morphological development compared to Semaeostomeae medusae. Laboratory and field observations demonstrate that rhizostomes exhibit plasticity in the development in response to factors such as temperature and rate of sunlight. Rhizostome species stood out with the highest growth rates and larger body sizes among medusozoan groups. Faster body growth of rhizostome species means that they achieve higher swimming performance and operate at higher Reynolds numbers (Re) earlier during development, which can provide adaptive responses to quickly explore their environments. Ecological implications of these differences should be studied in the future.

Our review demonstrated that Rhizostomeae species combine a set of physiological and functional features relative to the body composition, swimming and feeding behaviors that apparently makes them more energetically demanding, when compared to other Discomedusae species. The presence of zooxanthellae in Kolpophorae changes various functional parameters in relation to non-symbiont species which enhances trophic and ecologic diversity of the group. It is necessary to evaluate whether the apparent higher body productivity and asexual reproduction rates affect population growth and whether these factors are related to the numerous rhizostomeae species that exhibit blooms in coastal areas. Broader conclusions about the functional biology of rhizostomes are still preliminary, as a large proportion of the species have not been studied. There are large gaps in knowledge about the biology of polyps (e.g. lifespan, asexual reproduction, mortality) and medusae (e.g. nutrition, feeding rates) that still limits our understanding about the ecology and functional biology of the Rhizostomeae.

Acknowledgment

The authors are grateful for the comments and corrections made by the Editors, André C. Morandini, Sabine Holst, and Amanda A. Wiesenthal, which greatly improved the quality of this review. RMN received a grant from the National Council of Scientific and Technological Development (CNPq 438805/2018-6). MAJ Received a scholarchip from the Coordenação de Aperfeiçoamento de Pessoal de Nível Superior – Brazil (CAPES). GMM received a scholarchip from the National Council of Scientific and Technological Development - Brazil (CNPq). ND is supported by ISblue project, Interdisciplinary graduate school for the blue planet (ANR-17-EURE-0015) and co-funded by a grant from the French government under the programas "Investissements d'Avenir" embedded in France 2030. SPF received grants from the National Science Foundation (NSF-IOS 2114169 and NSF-CBET 2100156). JHC received grants from the National Science Foundation (NSF-2114171 and NSF-2100705).This is also a contribution from the Brazilian National Institute of Science and Technology - INCT-Mar COI (CNPq 400551/2014-4). VL was supported by EU Horizon Euro Research and innovation program ACTNOW ("Advancing understanding of cumulative impacts on European marine biodiversity, ecosystem functions and services for human wellbeing", grant agreement No. 101060072).

References

Abrams, M.J., Basinger, T., Yuan, W., Guo, C.L., Goentoro, L., 2015. Self-repairing symmetry in jellyfish through mechanically driven reorganization. Proc. Natl. Acad. Sci. 112 (26), E3365–E3373. https://doi.org/10.1073/pnas.1502497112.

Acevedo, M.J., Fuentes, V.L., Olariaga, A., Canepa, A., Belmar, M.B., Bordehore, C., et al., 2013. Maintenance, feeding and growth of *Carybdea marsupialis* (Cnidaria: Cubozoa) in the laboratory. J. Exp. Mar. Biol. Ecol. 439, 84–91. 10.1016/j.jembe.

Acuña, J.L., 2001. Pelagic tunicates: why gelatinous? Am. Nat. 158 (1), 100–107. https://doi.org/10.1086/320864.

Acuña, J.L., López-Urrutia, Á., Colin, S., 2011. Faking giants: The evolution of high prey clearance rates in jellyfishes. Science 333 (6049), 1627–1629. https://doi.org/10.1126/science.1205134.

Addad, S., Exposito, J.Y., Faye, C., Ricard-Blum, S., Lethias, C., 2011. Isolation, characterization and biological evaluation of jellyfish collagen for use in biomedical applications. Mar. Drugs 9 (6), 967–983. https://doi.org/10.3390/md9060967.

Agassiz, A., 1862. The Acalephan fauna of the southern coast of Massachusetts (Buzzard's Bay). P. Boston Soc. Nat. Hist. 8, 224–226.

Albert, D., 2005. Reproduction and longevity of *Aurelia labiata* in Roscoe Bay, a small bay on the Pacific coast of Canada. J. Mar. Biol. Assoc. 85 (3), 575–581. https://doi.org/10.1017/S0025315405011501.

Aljbour, S.M., Zimmer, M., Kunzmann, A., 2017. Cellular respiration, oxygen consumption, and trade-offs of the jellyfish *Cassiopea* sp. response Temp. change. J. Sea Res. 128, 92–97. https://doi.org/10.1016/%20j.seares.2017.08.006.

Allavena, A., Mariottini, G.L., Carli, A.M., Contini, S., Martelli, A., 1998. In vitro evaluation of the cytotoxic, hemolytic and clastogenic activities of *Rhizostoma pulmo* toxin (s). Toxicon 36, 933–936. https://doi.org/10.1016/s0041-0101(97)00171-2.

Alvarez-González, C.A., Cervantes-Trujano, M., Tovar-Ramírez, D., Conklin, D.E., Nolasco, H., Gisbert, E., et al., 2005. Development of digestive enzymes in California halibut *Paralichthys californicus* larvae. Fish. Physiol. Biochem. 31, 83–93.

Álvarez-Tello, F.J., López-Martínez, J., Lluch-Cota, D.B., 2016. Trophic spectrum and feeding pattern of cannonball jellyfish *Stomolophus meleagris* (Agassiz, 1860) from central Gulf of California. J. Mar. Biol. Assoc. UK 96 (6), 1217–1227. https://doi.org/10.1017/S0025315415001605.

Ames, C.L., Klompen, A.M.L., Badhiwala, K., Muffett, K., Reft, A.J., Kumar, M., 2020. Cassiosomes are stinging-cell structures in the mucus of the upside-down jellyfish *Cassiopea xamachana*. Comm. Biol. 3, 67. https://doi.org/10.1038/s42003-020-0777-8.

Anthony, C.J., Heagy, M., Bentlage, B., 2022. Phenotypic plasticity in *Cassiopea ornata* (Cnidaria: Scyphozoa: Rhizostomeae) suggests environmentally driven morphology. Zoomorphology 141 (2), 115–131. https://doi.org/10.1007/s00435-022-00558-4.

Ara, K., 2001. Length-weight relationships and chemical content of the planktonic copepods in the Cananéia Lagoon estuarine system, São Paulo, Brazil. Plankton. Biol. Ecol. 48, 121–127.

Arai, M.N., 1997. A Functional Biology of Scyphozoa. Springer Netherlands.

Arai, M.N. 2009. The potential importance of podocysts to the formation of scyphozoan blooms: a review. In: Jellyfish Blooms: Causes, Consequences, and Recent Advances: Proceedings of the Second International Jellyfish Blooms Symposium, held at the Gold Coast, Queensland, Australia, 24–27 June, 2007. Springer Netherlands, pp. 241–246. https://doi.org/10.1007/978–1-4020–9749-2_16.

Astorga, D., Ruiz, J., Prieto, L., 2012. Ecological aspects of early life stages of *Cotylorhiza tuberculata* (Scyphozoa: Rhizostomae) affecting its pelagic population success. Jellyfish Bloom. IV: Interact. Hum. Fish. 141–155. https://doi.org/10.1007/978-94-007-5316-7_11.

Avian, M., Del Negro, P., Rottini-Sandrini, L., 1991. A comparative analysis of nematocysts in *Pelagia noctiluca* and *Rhizostoma pulmo* from the north adriatic sea. Hydrobiologia 216-217, 615–621.

Avian, M., Mancini, L., Voltolini, M., Bonnet, D., Dreossi, D., Macaluso, V., et al., 2022. A novel endocast technique providing a 3D quantitative analysis of the gastrovascular system in *Rhizostoma pulmo*: an unexpected through-gut in cnidaria. PLoS One 17 (8). https://doi.org/10.1371/journal.pone.0272023.

Backmann, E.L., 1915. Über die Tension des Sauerstoffes in der Körperflüssigkeit der Coelenteraten im freien Zustande, im Meere und nach Haltung in Aquarien.—Upps. Läkareförenings Förh, (NF) Tjug. Bon 287–300 (Uppsala).

Ballesteros, A., Östman, C., Santín, A., Marambio, M., Narda, M., Gili, J.M., 2021. Cnidome and morphological features of *Pelagia noctiluca* (cnidaria: Scyphozoa) throughout the different life cycle stages. Front. Mar. Sci. 8. https://doi.org/10.3389/fmars.2021.714503.

Båmstedt, U., Lane, J., Martinussen, M.B., 1999. Bioenergetics of ephyra larvae of the scyphozoan jellyfish *Aurelia aurita* in relation to temperature and salinity. Mar. Biol. 135, 89–98. https://doi.org/10.1007/s002270050605.

Barz, K., Hirche, H.J., 2005. Seasonal development of scyphozoan medusae and the predatory impact of *Aurelia aurita* on the zooplankton community in the Bornholm Basin (central Baltic Sea). Mar. Biol. 147, 465–476. https://doi.org/10.1007/s00227-005-1572-2.

Behrends, G., Schneider, G., 1995. Impact of *Aurelia aurita* medusae (Cnidaria, Scyphozoa) on the standing stock and community composition of mesozooplankton in the Kiel Bight (western Baltic Sea). Mar. Ecol. Prog. Ser. 127, 39–45. htps://doi.org/10.3354/meps127039.

Béziat, P., Kunzmann, A., 2022. Under pressure: *Cassiopea andromeda* jellyfish exposed to increasing water temperature or lead, cadmium and anthropogenic gadolinium contamination. Mar. Biol. Res. 18 (1-2), 48–63. htps://doi.org/10.1080/17451000.2022.2066132.

Bezio, N., Costello, J.H., Perry, E., Colin, S.P., 2018. Effects of capture surface morphology on feeding success of scyphomedusae: a comparative study. Mar. Ecol. Prog. Ser. 596, 83–93. https://doi.org/10.3354/meps12549.

Bhaud, M., Cha, J.H., Duchêne, J.C., Nozais, C., 1995. Influence of temperature on the marine fauna: what can be expected from a climatic change. J. Therm. Biol. 20, 91–104. https://doi.org/10.1016/0306-4565(94)00031-D.

Biesiot, P.M., 1986. Changes in midgut gland morphology and digestive enzyme activities associated with development in early stages of the American lobster *Homarus americanus*. Doctoral thesis, Massachusetts Institute of Technology and Woods Hole Oceanographic Institution. Woods Hole Open Access Server. https://doi.org/10.1575/1912/3478.

Bigelow, R.P., 1892. On a new species of *Cassiopea* from Jamaica. Zool. Anz. 393, 212–214.

Bigelow, R.P., 1900. The anatomy and development of *Cassiopea xamachana*.—Mem. Boston Soc. Nat. Hist. 5 (6), 191–236 (Boston).

Bigelow, H.B., 1909. The Medusae, Rep. Sci. Res. Eastern Trop. Pac. Exped. US fish Comm. Steamer "Albatross" 1904–1905. Mem. Mus. Cornp. Zool. Harv. Coll. 37: 1–243, Cambridge.

Biggs, D.C., 1977. Respiration and ammonium excretion by open ocean gelatinous zooplankton 1. Limno. Oceanogr. 22 (1), 108–117. https://doi.org/10.4319/lo.1977.22.1.0108.

Bolton, T.F., Graham, W.M., 2004. Morphological variation among populations of an invasive jellyfish. Mar. Ecol. Prog. Ser. 278, 125–139. https://doi.org/10.3354/meps278125.

Bonnet, D., Molinero, J.C., Schohn, T., Daly, Y.M.N., 2012. Seasonal changes in the population dynamics of *Aurelia aurita* in Thau lagoon. Cah. Biol. Mar. 53, 343–347.

Boucher, J., Laurec, A., Samain, J.F., Smith, S.L., 1975. Etude de la nutrition, du régime et du rythme alimentaire du zooplancton dans les conditions naturelles, par la mesure des activités enzymatiques digestives. In: Proceedings of the 10th European Symposium on Marine Biology (Vol. 2, pp. 85–110).

Brewer, R.H., 1989. The annual pattern of feeding, growth, and sexual reproduction in Cyanea (Cnidaria: Scyphozoa) in the Niantic River Estuary, Connecticut. Biol. Bull. 176, 272–281. https://doi.org/10.2307/1541985.

Brewer, R.H., Feingold, J.S., 1991. The effect of temperature on the benthic stages of *Cyanea* (Cnidaria: Scyphozoa), and their seasonal distribution in the Niantic River estuary, Connecticut. J. Exp. Mar. Biol. Ecol. 152 (1), 49–60. https://doi.org/10.1016/0022-0981(91)90134-I.

Brodeur, R.D., Decker, M.B., Ciannelli, L., Purcell, J.E., Bond, N.A., Stabeno, P.J., et al., 2008. Rise and fall of jellyfish in the eastern Bering Sea in relation to climate regime shifts. Prog. Oceanogr. 77, 103–111.

Brodeur, R.D., Sugisaki, H., Hunt, 2002. Increases in jellyfish biomass in the Bering Sea: implications for the ecosystem. Mar. Ecol. Prog. Ser. 233, 89–103. https://doi.org/10.3354/meps233089.

Brotz, L., Angel, D.L., D'Ambra, I., Enrique-Navarro, A., Lauritano, C., Thibault, D., et al., 2024. Rhizostomes as a resource: The expanding exploitation of jellyfish by humans. Adv. Mar. Biol. 98.

Burggren, W., Roberts, J., 1991. Respiration and metabolism. Env. Met. Anim. Physiol. 353–435.

Calder, D.R., 1977. Nematocysts of the ephyra stages of *Aurelia*, *Chrysaora*, *Cyanea*, and *Rhopilema* (cnidaria, scyphozoa). T Am. Microsc. Soc. 96, 13–19.

Calder, D.R., 1982. Life history of the cannonball jellyfish, *Stomolophus meleagris* L. Agassiz, 1860 (Scyphozoa, Rhizostomida). Biol. Bull. 162, 149–162. https://doi.org/10.2307/1540810.

Camacho-Pacheco, A.V., Gómez-Salinas, L.C., Cisneros-Mata, M.Á., Rodríguez-Félix, D., Díaz-Tenorio, L.M., Unzueta-Bustamante, M.L., 2022. Feeding behavior, shrinking, and the role of mucus in the cannonball jellyfish *Stomolophus* spp. 2 in captivity. Diversity 14 (2), 103. https://doi.org/10.3390/d14020103.

Cariello, L., Romano, G., Spagnuolo, A., Zanetti, L., 1988. Isolation and partial characterization of rhizolysin, a high molecular weight protein with hemolytic activity, from the jellyfish *Rhizostoma pulmo*. Toxicon 26, 1057–1065.

Carli, A., Pane, L., Valente, T., Cotta, S., 1991. Lipid and protein content of jellyfish from the Ligurian Sea. First results, in UNEP: Jellyfish Blooms in the Mediterranean. Proceedings of the II workshop on Jellyfish in the Mediterranean Sea, MAP Technical Reports Series No. 47, UNEP, Athens, pp. 236–240.

Carneiro, R.F.V., Nascimento, N.R.F., Costa, P.P.C., Gomes, V.M., de Souza, A.J.F., de Oliveira, S.C.B., 2011. The extract of the jellyfish *Phyllorhiza punctata* promotes neurotoxic effects. J. Appl. Toxicol. 31, 720–729.

Carvalho-Saucedo, L.J., López-Martínez, F., García-Domínguez, C., Rodríguez-Jaramillo, J., Padilla-Serrato, 2011. Biología reproductiva de la medusa bola de cañón *Stomolophus meleagris* en la laguna Las Guasimas, Sonora, Mexico. Hidrobiológica 21, 77–88.

Cates, N., 1975. Productivity and organic consumption in *Cassiopea* and *Condylactus*. J. Exp. Mar. Biol. Ecol. 18, 55–59.

Cates, N., McLaughlin, J.J.A., 1976. Differences of ammonia metabolism in symbiotic and aposymbiotic *Condylactus* and *Cassiopea* spp. J. Exp. Mar. Biol. Ecol. 21, 1–5.

Chapman, G., 1953. Studies of the mesogloea of colenterates. I. Histology and chemical properties.—Quart. J. Microsc. Sci. 94 (3. Ser.), 155–176 (London).

Chapman, G., 1959. The mesogloea of *Pelagia noctiluca*.—Quart. J. Microsc. Sci. 100 (4), 599–610 (London).

Chapman, G., Pardy, R.L., 1972. The movement of glucose and glycine through the tissues of *Corymorpha palma* torrey (Coelenterata, Hydrozoa). J. Exp. Biol. 56, 639–645.

Chen, J., Ding, G., Liu, C., 1985. Effect of nutritional conditions on the strobilation of edible medusa, *Rhopilema esculenta* Kishinouye. J. Fish. China (Sui Chan Xue Bao) 9, 321–329.

Chiaverano, L.M., Graham, W.M., 2017. Morphological plasticity in *Aurelia* polyps, with subsequent effects on asexual fecundity and morphology of young medusae. Mar. Ecol. Prog. Ser. 582, 79–92. https://doi.org/10.3354/meps12314.

Chuin, T., 1929. La digestion chez le scyphistome de *Chrysora*. Cr. Soc. Biol. 102, 557–558.

Cimino, M.A., Patris, S., Ucharm, G., Bell, L.J., Terrill, E., 2018. Jellyfish distribution and abundance in relation to the physical habitat of Jellyfish Lake, Palau. J. Trop. Ecol. 34 (1), 17–31. https://doi.org/10.1017/S0266467418000044.

Colin, S.P., Costello, J.H., 2007. Functional characteristics of nematocysts found on the scyphomedusa *Cyanea capillata*. J. Exp. Mar. Biol. Ecol. 351, 114–120.

Colin, S.P., Costello, J.H., Dabiri, J.O., Villanueva, A., Blottman, J.B., Gemmell, B.J., et al., 2012. Biomimetic and live medusae reveal the mechanistic advantages of a flexible bell margin. PLoS One 7 (11), e48909. https://doi.org/10.1371/journal.pone.0048909.

Condon, R.H., Steinberg, D.K., Del Giorgio, P.A., Bouvier, T.C., Bronk, D.A., Graham, W.M., et al., 2011. Jellyfish blooms result in a major microbial respiratory sink of carbon in marine systems. Proc. Natl. Acad. Sci. 108 (25), 10225–10230. https://doi.org/10.1073/pnas.101578210.

Costello, J.H., Colin, S.P., Dabiri, J.O., 2008. Medusan morphospace: phylogenetic constraints, biomechanical solutions, and ecological consequences. Invertebr. Biol. 127 (3), 265–290. https://doi.org/10.1111/j.1744-7410.2008.00126.x.

Costello, J.H., Colin, S.P., Dabiri, J.O., Gemmell, B.J., Lucas, K.N., Sutherland, K.R., 2021. The hydrodynamics of jellyfish swimming. Annu. Rev. Mar. Sci. 13, 375–396. https://doi.org/10.1146/annurev-marine-031120.

Costello, J.H., Colin, S.R., 1994. Morphology, fluid motion and predation by the scyphomedusa *Aurelia aurita*. Mar. Biol. 121, 327–334.

Costello, J.H., Colin, S.P., 1995. Flow and feeding by swimming scyphomedusae. Mar. Biol. 124, 399–406.

Cruz, J., Cerveira, I., Andrade, I., Baptista, V., Teodósio, M.A., 2021. Prey selectivity and feeding rates of the scyphozoan *Catostylus tagi* (Haeckel, 1869). J. Plankton Res. 43 (6), 986–990. https://doi.org/10.1093/plankt/fbab065.

Cuvier, G.D., 1799. Mémoire sur l'organisation de la méduse. Bull. Sci. Soc. Philom. Paris 2, 69.

D'Ambra, I., Carmichael, R.H., Graham, W.M., 2014. Determination of δ13C and δ15N and trophic fractionation in jellyfish: implications for food web ecology. Mar. Biol. 161, 473–480. https://doi.org/10.1023/A:1011832222174.

D'Ambra, I., Costello, J.H., Bentivegna, F., 2001. Flow and prey capture by the scyphomedusa *Phyllorhiza punctata* von Lendenfeld, 1884. Hydrobiologia 451, 223–227. https://doi.org/10.1023/A:1011832222174.

D'Ambra, I., Lauritano, C., 2020. A review of toxins from cnidaria. Mar. Drugs. 18, 507.

Dabiri, J.O., Colin, S.P., Costello, J.H., Gharib, M., 2005. Flow patterns generated by oblate medusan jellyfish: Field measurements and laboratory analyses. J. Exp. Biol. 208 (7), 1257–1265. https://doi.org/10.1242/jeb.01519.

Daglio, L.G., Dawson, M.N., 2017. Species richness of jellyfishes (Scyphozoa: Discomedusae) in the Tropical Eastern Pacific: missed taxa, molecules, and morphology match in a biodiversity hotspot. Invertebr. Syst. 31 (5), 635–663. https://doi.org/10.3390/d14020103.

Dawson, N.M., Hamner, W.M., 2003. Geographic variation and behavioral evolution in marine plankton: the case of *Mastigias* (Scyphozoa, Rhizostomeae). Mar. Biol. 143, 1161–1174. https://doi.org/10.1007/s00227-003-1155-z.

Dawson, M.N., Martin, L.E., Penland, L.K., 2001. Jellyfish swarms, tourists, and the Christ-child. In: Jellyfish Blooms: Ecological and Societal Importance: Proceedings of the International Conference on Jellyfish Blooms, held in Gulf Shores, Alabama, 12–14 January 2000 (pp. 131–144). Springer Netherlands. https://doi.org/10.1007/978–94-010–0722-1_12.

De Dominico, S., De Rinaldis, G., Mammone, M., Bosch-Belmar, M., Piraino, S., Leone, A., 2023. The zooxanthellate jellyfish holobiont *Cassiopea andromeda*, a source of soluble bioactive compounds. Mar. Drugs. 21, 272.

Ding, G.W., Chen, J.K., 1981. The life history of *Rhopilema escuenta* Kishinouye. J Fish China. 5: 93–102 (in Chinese) in Omori M, Nakano E (2001) Jellyfish fisheries in southeast Asia. Hydrobiologia 451, 19–26. https://doi.org/10.1007/978-94-010-0722-1_3.

Djeghri, N., 2019. Variability and plasticity of the nutrition of zooxanthellate jellyfishes: insights from experimental and field studies. PhD Thesis, Université de Bretagne Occidentale, Brest, France.

Djeghri, N., Pondaven, P., Le Grand, F., Bideau, A., Duquesne, N., Stockenreiter, M., et al., 2021. High trophic plasticity in the mixotrophic *Mastigias papua*-Symbiodiniaceae holobiont: implications for the ecology of zooxanthellate jellyfishes. Mar. Ecol. Prog. Ser. 666, 73–88.

Djeghri, N., Pondaven, P., Stibor, H., Dawson, M.N., 2019. Review of the diversity, traits, and ecology of zooxanthellate jellyfishes. Mar. Biol. 166, 147.

Djeghri, N., Stibor, H., Lebeau, O., Pondaven, P., 2020. $\delta^{13}C$, $\delta^{15}N$, and C:N ratios as nutrition indicators of zooxanthellate jellyfishes: insights from an experimental approach. J. Exp. Mar. Biol. Ecol. 522, 151257.

Dong, J., Jiang, L., Tan, K., Liu, H., Purcell, J.E., Li, P., et al., 2009. Stock enhancement of the edible jellyfish (*Rhopilema esculentum* Kishinouye) in Liaodong Bay, China: a review. Hydrobiologia 616, 113–118. https://doi.org/10.1007/s10750-008-9592-9.

Dönmez, M.A., Levent, B.A.T., 2019. Detection of feeding dietary *Rhizostoma pulmo* (Macri, 1778) in Samsun coasts of the Black Sea, Turkey. Ege J. Fish. Aqua. Sci. 36 (2), 135–144. https://doi.org/10.12714/egejfas.2019.36.2.05.

Doyle, T.K., Houghton, J.D., McDevitt, R., Davenport, J., Hays, G.C., 2007. The energy density of jellyfish: estimates from bomb-calorimetry and proximate-composition. J. Exp. Mar. Biol. Ecol. 343 (2), 239–252. https://doi.org/10.1016/j.jembe.2006.12.010.

Drew, E.A., 1972. The biology and physiology of alga-invertebrate symbioses I Carbon fixation in *Cassiopea* sp. at Aidabra Atoll. J. Exp. Mar. Biol. Ecol. 9, 65–69.

Dupuis, Y., Tardivel, S., Porembska, Z., Fournier, P., 1991. Effect of some alkaline phosphatase inhibitors on intestinal calcium transfer. Int. J. Biochem. 23 (2), 175–180. https://doi.org/10.1016/0020-711x(91)90186-q.

Durieux, D.M., Gemmell, B.J., 2022. Aggregation and defence. bioRxiv, 2022–10. https://doi.org/10.1101/2022.10.25.513673.

Durieux, D.M., Scrogham, G.D., Fender, C., Lewis, D.B., Deban, S.M., Gemmel, B.J., 2023. Benthic jellyfish acts as suction pumps to facilitate release of interstitial porewater. Sci. Rep. 13, 3770.

Epstein, H.E., Templeman, M.A., Kingsford, M.J., 2016. Fine-scale detection of pollutants by a benthic marine jellyfish. Mar. Pollut. Bull. 107, 340–346.

Evans, L.E., Hirst, A.G., Kratina, P., Beaugrand, G., 2020. Temperature-mediated changes in zooplankton body size: large scale temporal and spatial analysis. Ecography 43 (4), 581–590. https://doi.org/10.1111/ecog.04631.

Eysenhardt, C.G., 1821. Zur Anatomie und Naturgeschichte der Quallen. I. Von dem *Rhizostoma cuvieri* LAM.—Nova Acta Acad. Leopold. Carol 10. pp. 377–410 (Bonn).

Fancett, M.S., 1988. Diet and prey selectivity of scyphomedusae from Port Phillip Bay, Australia. Mar. Biol. 98, 503–509. https://doi.org/10.1007/BF00391541.

Fancett, M.S., Jenkins, G.P., 1988. Predatory impact of scyphomedusae on ichthyoplankton and other zooplankton in Port Phillip Bay. J. Exp. Mar. Biol. Ecol. 116 (1), 63–77. https://doi.org/10.1016/0022-0981(88)90246-8.

Fautin, D., 2009. Structural diversity, systematics, and evolution of cnidae. Toxicon 54, 1054–1064.

Feitl, K.E., Millett, A.F., Colin, S.P., Dabiri, J.O., Costello, J.H., 2009. Functional morphology and fluid interactions during early development of the scyphomedusa *Aurelia aurita*. Biol. Bull. 217 (3), 283–291. https://doi.org/10.1086/BBLv217n3p283.

Fernández-Alías, A., Marcos, C., Quispe, J.I., Sabah, S., Pérez-Ruzafa, A., 2020. Population dynamics and growth in three scyphozoan jellyfishes, and their relationship with environmental conditions in a coastal lagoon. Estuar. Coast. Shelf. Sci. 243, 106901. https://doi.org/10.1016/j.ecss.2020.106901.

Fernández-Prats, R., Raspa, V., Thiria, B., Huera-Huarte, F., Godoy-Diana, R., 2015. Large-amplitude undulatory swimming near a wall. Bioinspir Biomim. 10 (1), 016003. https://doi.org/10.1088/1748-3190/10/1/016003.

Fossette, S., Gleiss, A., Chalumeau, J., Bastian, T., Armstrong, C.D., Vandenabeele, S., et al., 2015. Current-oriented swimming by jellyfish and its role in bloom maintenance. Curr. Biol. 25 (3), 342–347. https://doi.org/10.1016/j.cub.2014.11.050.

Fox, J.W., Serrano, S.M.T., 2005. Structural considerations of the snake venom metalloproteinases, key members of the m12 reprolysin family of metalloproteinases. Toxicon 45, 969–985. https://doi.org/10.1016/j.toxicon.2005.02.012.

Freeman, C.J., Stoner, E.W., Easson, C.G., Matterson, K.O., Baker, D.M., 2016. Symbiont carbon and nitrogen assimilation in the *Cassiopea-Symbiodinium* mutualism. Mar. Ecol. Prog. Ser. 544, 281–286. https://doi.org/10.3354/meps11605.

Freeman, C.J., Stoner, E.W., Easson, C.G., Matterson, K.O., Baker, D.M., 2017. Variation in $\delta^{13}C$ and $\delta^{15}N$ values suggests a coupling of host and symbiont metabolism in the Symbiodinium-*Cassiopea* mutualism. Mar. Ecol. Prog. Ser. 571, 245–251. https://doi.org/10.3354/meps12138.

Fricke, H., 1996. Deep-water exploration of the Red Sea by submersible. In: Uiblein, F., Ott, J., Stachowitsch, M. (Eds.), Deep Sea and Extreme Shallow-water Habitats: Affinities and Adaptations. Biosystematics and Ecology Series, 11. Österreichische Akad der Wiss, Wien, pp. 67–89.

Fry, B.G., Roelants, K., Champagne, D.E., Scheib, H., Tyndall, J.D.A., King, G.F., 2009. The toxicogenomic multiverse: convergent recruitment of proteins into animal venoms. Annu. Rev. Genom. Hum. G. 10, 483–511. https://doi.org/10.1146/annurev.genom.9.081307.164356.

Fu, Z., Shibata, M., Makabe, R., Ikeda, H., Uye, S.I., 2014. Body size reduction under starvation, and the point of no return, in ephyrae of the moon jellyfish *Aurelia aurita*. Mar. Ecol. Prog. Ser. 510, 255–263. https://doi.org/10.3354/meps10799.

Fuentes, V., Straehler-Pohl, I., Atienza, D., Franco, I., Tilves, U., Gentile, M., et al., 2011. Life cycle of the jellyfish *Rhizostoma pulmo* (Scyphozoa: Rhizostomeae) and its distribution, seasonality and inter-annual variability along the Catalan coast and the Mar Menor (Spain, NW Mediterranean). Mar. Biol. 158, 2247–2266. https://doi.org/10.1007/s00227-011-1730-7.

Gamero-Mora, E., Halbauer, R., Bartsch, V., Stampar, S.N., Morandini, A.C., 2019. Regenerative capacity of the upside-down jellyfish *Cassiopea xamachana*. Zool. Stud. 58, 37. https://doi.org/10.6620/ZS.2019.58-37.

García, J.R., 1990. Population dynamics and production of *Phyllorhiza punctata* (Cnidaria: Scyphozoa) in Laguna Joyuda, Puerto Rico. Mar. Ecol. Prog. Ser. 64, 243–251. https://doi.org/10.3354/meps0 64243.

García, J.R., Durbin, E., 1993. Zooplanktivorous predation by large scyphomedusae *Phyllorhiza punctata* (Cnidaria: Scyphozoa) in Laguna Joyuda. J. Exp. Mar. Biol. Ecol. 173 (1), 71–93. https://doi.org/10.1016/0022-0981(93)90208-6.

Gemmell, B.J., Colin, S.P., Costello, J.H., 2018. Widespread utilization of passive energy recapture in swimming medusae. J. Exp. Biol. 221 (1). https://doi.org/10.1242/jeb.168575.

Gemmell, B.J., Costello, J.H., Colin, S.P., Stewart, C.J., Dabiri, J.O., Tafti, D., et al., 2013. Passive energy recapture in jellyfish contributes to propulsive advantage over other metazoans. Proc. Natl. Acad. Sci. USA 110 (44), 17904–17909. https://doi.org/10.1073/pnas.1306983110.

Gemmell, B.J., Costello, J.H., Colin, S.P., Stewart, C.J., Dabiri, J.O., Tafti, D., Priya, S., 2013. Passive energy recapture in jellyfish contributes to propulsive advantage over other metazoans. P. Natl. Acad. Sci. USA 110 (44), 17904–17909. https://doi.org/10.1073/pnas.1306983110.

Gemmell, B.J., Du Clos, K.T., Colin, S.P., Sutherland, K.R., Costello, J.H., 2021. The most efficient metazoan swimmer creates a "virtual wall" to enhance performance. P. Roy. Soc. B-Biol. Sci. 288, 20202492. https://doi.org/10.1098/rspb.2020.2494.

Gemmell, B.J., Troolin, D.R., Costello, J.H., Colin, S.P., Satterlie, R.A., 2015. Control of vortex rings for maneuverability. J. R. Soc. Interface. 12 (108), 20150389. https://doi.org/10.1098/rsif.2015.0389.

Gibbons, M.J., Richardson, A.J., 2013. Beyond the jellyfish joyride and global oscillations: advancing jellyfish research. J. Plankton Res. 35 (5), 929–938. https://doi.org/10.1093/plankt/fbt063.

Glazier, D.S., 2006. The-power law is not universal: evolution of isometric, ontogenetic metabolic scaling in pelagic animals. BioScience 56, 325–332.

Gnaiger, E., 1983. Calculation of energetic and biochemical equivalents of respiratory oxygen consumption. In: Polarographic Oxygen Sensors: Aquatic and Physiological Applications (pp. 337–345). Berlin, Heidelberg: Springer Berlin Heidelberg. https://doi.org/10.1007/978-3-642-81863-9_30.

Gohar, H.A.F., Eisawy, A.M., 1961. The development of *Cassiopea andromeda* (Scyphomedusae). Publ. Mar. Biolo. Sta. Al-Ghardaqa 11, 148–190.

Gold, D.A., Jacobs, D.K., 2013. Stem cell dynamics in Cnidaria: are there unifying principles? Dev. Genes. Evol. 223, 53–66. https://doi.org/10.1007/s00427-012-0429-1.

Goldstein, J., Riisgård, H.U., 2016. Population dynamics and factors controlling somatic degrowth of the common jellyfish, *Aurelia aurita*, in a temperate semi-enclosed cove (Kertinge Nor, Denmark). Mar. Biol. 163, 1–12. https://doi.org/10.1007/s00227-015-2802-x.

González-Valdovinos, M., Ocampo, L., Tovar-Ramírez, D., 2019. Evaluation of digestive capacity in the polyp, ephyrae, and medusae stages of the cannonball jellyfish *Stomolophus meleagris*. Hydrobiologia 828, 259–269. https://doi.org/10.1007/s10750-018-3817-3.

Gordon, M., Hatcher, C., Seymour, J., 2004. Growth and age determination of the tropical Australian cubozoan *Chiropsalmus* sp. Hydrobiologia 530 (531), 339–345. https://doi.org/10.1007/s10750-004-2655-7.

Graham, W.M., Martin, D.L., Felder, D.L., Asper, V.L., Perry, H.M., 2003. Ecological and economic implications of a tropical jellyfish invader in the Gulf of Mexico. Mar. Bioinvasions: Patterns, Process. Perspect. 53–69. https://doi.org/10.1007/978-94-010-0169-4_6.

Gubareva, A.E., Pustovalova, L.M., Ryzhkov, Y.D., et al., 1983. The features of the chemical composition of medusae in the Sea of Azov. Izvestiya Severo-Kavkazskogo Nauchnogo Tsentra Vysshei Shkoly Estestvennye Nauki 0, 22–25 (In Russian).

Gueroun, S.K.M., Molinero, J.C., Piraino, S., Yahia, M.N.D., 2020. Population dynamics and predatory impact of the alien jellyfish *Aurelia solida* (Cnidaria, Scyphozoa) in the Bizerte Lagoon (southwestern Mediterranean Sea). Mediterr. Mar. Sci. 21, 22–35. https://doi.org/10.12681/mms.17358.

Gueroun, S.K., Torres, T.M., Dos Santos, A., Vasco-Rodrigues, N., Canning-Clode, J., Andrade, C., 2021. *Catostylus tagi* (Class: Scyphozoa, Order: Discomedusae, Suborder: Rhizostomida, Family: Catostylidae) life cycle and first insight into its ecology. PeerJ 9, e12056. https://doi.org/10.7717/peerj.12056.

Guo, P., 1990. Effect of nutritional condition on the formation and germination of the podocyst of scyphistomae of *Rhopilema esculenta* Kishinouye. J. Fish. China 14, 206–211 (In Chinese; English summary).

Guo, Q., Whipps, C.M., Zhai, Y., Li, D., Gu, Z., 2022. Quantitative insights into the contribution of nematocysts to the adaptive success of cnidarians based on proteomic analysis. Biology 11, 91. https://doi.org/10.3390/biology11010091.

Gusmani, L., Avian, M., Galil, B., Patriarca, P., Rottini, G., 1997. Biologically active polypeptides in the venom of the jellyfish *Rhopilema nomadica*. Toxicon 35, 637–648.

Gutmann, W.F.R., 1965. Rückstoss und Ruderschlag der Quallen. Senckenberg am Meer 232: 455–464.

Gutmann, W.F.R., 1966. Zu Bau und Leistung von Tierkonstruktionen. Senckenberg am Meer 234, 103–108.

Haddad, M.A., Nogueira Jr, M., 2006. Reappearance and seasonality of *Phyllorhiza punctata* von Lendenfeld (Cnidaria, Scyphozoa, Rhizostomeae) medusae in southern Brazil. Rev. Bras. Zool. 23 (3). https://doi.org/10.1590/S0101-81752006000300030.

Haddad Jr. V., 2016. Medical emergencies caused by aquatic animals. A zoological and clinical guide. Springer International Publishing Switzerland https://doi.org/10.1007/978-3-319-20288-4_1.

Hamann, O., 1881. Die Mundarme der Rhizostomeen und ihre Anhangsorgane. Jen. Zs. Naturw. 15, 243–285.

Hamlet, C., Santhanakrishnan, A., Miller, L.A., 2011. A numerical study of the effects of bell pulsation dynamics and oral arms on the exchange currents generated by the upside-down jellyfish *Cassiopea xamachana*. J. Exp. Biol. 214 (11), 1911–1921. https://doi.org/10.1242/jeb.052506.

Hamner, W.M., Hauri, I.R., 1981. Long-distance horizontal migrations of zooplankton (Scyphomedusae: *Mastigias*). Limnol. Oceanogr. 26, 414–423. https://doi.org/10.4319/lo.1981.26.3.0414.

Hansson, L.J., 2006. A method for in situ estimation of prey selectivity and predation rate in large plankton, exemplified with the jellyfish *Aurelia aurita* (L.). J. Exp. Mar. Biol. Ecol. 328 (1), 113–126. https://doi.org/10.1016/j.jembe.2005.07.002.

Hansson, L.J., Moeslund, O., Kiørboe, T., Riisgård, H.U., 2005. Clearance rates of jellyfish and their potential predation impact on zooplankton and fish larvae in a neritic ecosystem (Limfjorden, Denmark). Mar. Ecol. Prog. Ser. 304, 117–131. https://doi.org/10.3354/meps304117.

Harris, R., Wiebe, P., Lenz, J., Skjoldal, H.R., Huntley, M., 2000. ICES Zooplankton Methodology Manual. Academic Press, New York.

Hays, G.C., Bastian, T., Doyle, T.K., Fossette, S., Gleiss, A.C., Gravenor, M.B., et al., 2012. High activity and Lévy searches: jellyfish can search the water column like fish. P. Roy. Soc. B-Biol Sci. 279 (1728), 465–473. https://doi.org/10.1098/rspb.2011.0978.

Heins, A., Sötje, I., Holst, S., 2018. Assessment of investigation techniques for scyphozoan statoliths, with focus on early development of the jellyfish *Sanderia malayensis*. Mar. Ecol. Progr. Ser. 591, 37–56. https://doi.org/10.3354/meps12292.

Helm, R.R., Dunn, C.W., 2017. Indoles induce metamorphosis in a broad diversity of jellyfish, but not in a crown jelly (Coronatae). PlosOne 12, e0188601. https://doi.org/10.1371/journal.pone.0188601.

Higgins III, J.E., Ford, M.D., Costello, J.H., 2008. Transitions in morphology, nematocyst distribution, fluid motions, and prey capture during development of the scyphomedusa *Cyanea capillata*. Biol. Bull. 214 (1), 29–41. https://doi.org/10.2307/25066657.

Hirst, A.G., Lucas, C.H., 1998. Salinity influences body weight quantification in the scyphomedusa *Aurelia aurita*: Important implications for body weight determination in gelatinous zooplankton. Mar. Ecol. Prog. 1998 (165), 259–269.

Hirst, A.G., Roff, J.C., Lampitt, R.S., 2003. A synthesis of growth rates in marine epipelagic invertebrate zooplankton. Adv. Mar. Biol. 44, 1–142. https://doi.org/10.1016/S0065-2881(03)44002-9.

Hoeger, U., Mommsen, T.P., 1984. Hydrolytic enzymes in the two North Sea ctenophores *Pleurobrachia pileus* and *Beroe gracilis*. Mar. Biol. 81, 123–130. https://doi.org/10.1007/BF0039311.

Hofmann, D.K., Kremer, B.P., 1981. Carbon metabolism and strobilation in *Cassiopea andromedea* (Cnidaria: Scyphozoa): significance of endosymbiotic dinoflagellates. Mar. Biol. 65, 25–33. https://doi.org/10.1007/BF00397064.

Hofmann, D.K., Neumann, R., Henne, K., 1978. Strobilation budding and initiation of scyphistome morphogenesis in the rhizostome *Cassiopea andromeda* (Cnidaria: Scyphozoa). Mar. Biol. 47, 161–176.

Holst, S., 2012. Morphology and development of benthic and pelagic life stages of North Sea jellyfish (Scyphozoa, Cnidaria) with special emphasis on the identification of ephyra stages. Mar. Biol. 159, 2707–2722. https://doi.org/10.1007/s00227-012-2028-0.

Holst, S., Michalik, P., Noske, M., Krieger, J., Sötje, I., 2016. Potential of x-ray micro-computed tomography for soft-bodied and gelatinous cnidarians with special emphasis on scyphozoan and cubozoan statoliths. J. Plankton Res. 38, 1225–1242. https://doi.org/10.1093/plankt/fbw054.

Holst, S., Sötje, I., Tiemann, H., Jarms, G., 2007. Life cycle of the rhizostome jellyfish *Rhizostoma octopus* (L.) (Scyphozoa, Rhizostomeae), with studies on cnidocysts and statoliths. Mar. Biol. 151, 1695–1710. https://doi.org/10.1007/s00227-006-0594-8.

Holst, S., Tiseo, G.R., Djeghri, N., Sötje, I., 2024. Approaches and findings in histological and micromorphological research on Rhizostomeae since 1970. Adv. Mar. Biol. 98.

Hopf, J.K., Kingsford, M.J., 2013. The utility of statoliths and bell size to elucidate age and condition of a scyphomedusa (*Cassiopea* sp.). Mar. Biol. 160, 951–960. https://doi.org/10.1007/s00227-012-2146-8.

Houghton, J., Doyle, T., Davenport, J., Hays, G., 2006. Developing a simple, rapid method for identifying and monitoring jellyfish aggregations from the air. Mar. Ecol. Prog. Ser. 314, 159–170. https://doi.org/10.3354/meps314159.

Houghton, J.D.R., Doyle, T.K., Davenport, J., Lilley, M.K.S., Wilson, R.P., Hays, G.C., 2007. Stranding events provide indirect insights into the seasonality and persistence of jellyfish medusae (Cnidaria: Scyphozoa). Hydrobiologia 589, 1–13. https://doi.org/10.1007/s10750-007-0572-2.

Huang, Y. A-W., 1988. Cannonball jellyfish (*Stomolophus meleagris*) as a food resource. J. Food. Sci. 53 (2), 341–343. https://doi.org/10.1111/j.1365-2621.1988.tb07701.x.

Hubot, N., Giering, S.L., Lucas, C.H., 2022. Similarities between the biochemical composition of jellyfish body and mucus. J. Plankton Res. 44 (2), 337–344. https://doi.org/10.1093/plankt/fbab091.

Husson, B., Fay, S., 1984. Répartition, fréquence et régimen alimentaire de la méduse Rhizostoma pulmo. Annales 7, 231–252.

Iguchi, N., Iwatani, H., Sugimoto, K., Kitajima, S., Honda, N., Katoh, O., 2017. Biomass, body elemental composition, and carbon requirement of *Nemopilema nomurai* (Scyphozoa: Rhizostomeae) in the southwestern Japan Sea. Plankton and Benthos Res. 12 (2), 104–114. https://doi.org/10.3800/pbr.12.104.

Ikeda, T., 2014. Synthesis toward a global model of metabolism and chemical composition of medusae and ctenophores. J. Exp. Mar. Biol. Ecol. 456, 50–64.

Ishii, H., Tanaka, F., 2006. Respiration rates and metabolic demands of *Aurelia aurita* in Tokyo Bay with special reference to large medusae. Plankton Benthos Res. 1 (1), 64–67. https://doi.org/10.3800/pbr.1.64.

Jafari, H., 2019. Identification and hemolytic activity of jellyfish (*Rhopilema* sp., Scyphozoa: Rhizostomeae) venom from the Persian Gulf and Oman Sea. Biodiversitas 20, 1228–1232. https://doi.org/10.13057/BIODIV/D200440.

Jantzen, C., Wild, C., Rasheed, M., El-Zibdah, M., Richter, C., 2010. Enhanced pore-water nutrient fluxes by the upside-down jellyfish *Cassiopea* sp. in a Red Sea coral reef. Mar. Ecol. Prog. Ser. 411, 117–125.

Jarms, G., Morandini, A.C., 2023. World atlas of jellyfish—Scyphozoa except Stauromedusae. Abhandlungen des Naturwissenschaftlichen Vereins in Hamburg, Special Volume. Dölling und Galitz Verlag, Hamburg, 827 pp.

Jarms, G.,Morandini, A.C. (Eds.), 2019. World atlas of jellyfish: Scyphozoa except Stauromedusae. Abhandlungen des Naturwissenschaftlichen Vereins in Hamburg (N.F.). Dölling und Galitz Verlag, Hamburg, 816 pp. ISBN 9783862180820.

Jiang, P.L., Pasaribu, B., Chen, C.S., 2014. Nitrogen-deprivation elevates lipid levels in *symbiodinium* spp. by lipid droplet accumulation: morphological and compositional analyses. PLoS ONE 9 (1), e87416. https://doi.org/10.1371/journal.pone.0087416.

Jordano, M. de A., Miyake, H., Nagata, R.M., Morandini, A.C., 2022. Patterns of morphological development in Scyphozoa ephyrae (Cnidaria, Medusozoa). Mar. Biodivers. 52 (3). https://doi.org/10.1007/s12526-022-01269-1.

Jordano, M. de A., Morandini, A.C., Nagata, R.M., 2020. Is phenotypic plasticity determined by temperature and fluid regime in filter-feeding gelatinous organisms? J. Exp. Mar. Biol. Ecol. 522. https://doi.org/10.1016/j.jembe.2019.151238.

Jordano, M.d.A., Nagata, R.M., Morandini, A.C., 2024. A review of the role played by cilia in medusozoan feeding mechanics. Biol. Rev. 99, 950–964. https://doi.org/10.1111/brv.13052.

Kang, C., Kim, Y.K., Lee, H., Cha, M., Sohn, E.T., Jung, E.S., et al., 2011. Target organ identification of jellyfish envenomation using systemic and integrative analyses in anesthetized dogs. J. Pharmacol. Tox. Met. 64 (2), 173–179. https://doi.org/10.1016/j.vascn.2011.07.004.

Kang, C., Munawir, A., Cha, M., Sohn, E.-T., Lee, H., Kim, J.-S., et al., 2009. Cytotoxicity and hemolytic activity of jellyfish *Nemopilema nomurai* (Scyphozoa: Rhizostomeae) venom. Comp. Biochem. Physiol. Part. C: Toxicol. Pharm. 150, 85–90. https://doi.org/10.1016/j.cbpc.2009.03.003.

Katija, K., 2015. Morphology alters fluid transport and the ability of organisms to mix oceanic waters. Integr. Comp. Biol. 55 (4), 698–705. https://doi.org/10.1093/icb/icv075.

Kawahara, M., Ohtsu, K., Uye, S.I., 2013. Bloom or non-bloom in the giant jellyfish *Nemopilema nomurai* (Scyphozoa: Rhizostomeae): roles of dormant podocysts. J. Plankton Res. 35 (1), 213–217. https://doi.org/10.1093/plankt/fbs074.

Keesing, J.K., Gershwin, L.A., Trew, T., Strzelecki, J., Bearham, D., Liu, D., et al., 2016. Role of winds and tides in timing of beach strandings, occurrence, and significance of swarms of the jellyfish *Crambione mastigophora* Mass 1903 (Scyphozoa: Rhizostomeae: Catostylidae) in north-western Australia. Hydrobiologia 768, 19–36. https://doi.org/10.1007/s10750-015-2525-5.

Khong, N.M., Yusoff, F.M., Jamilah, B., Basri, M., Maznah, I., Chan, K.W., et al., 2016. Nutritional composition and total collagen content of three commercially important edible jellyfish. Food Chem. 196, 953–960. https://doi.org/10.1016/j.foodchem.2015.09.094.

Kikinger, R., 1992. *Cotylorhiza tuberculata* (Cnidaria: Scyphozoa) life history of a stationary population. Mar. Ecol. 13, 333–362. https://doi.org/10.1111/j.1439-0485.1992.tb00359.x.

Kiørboe, T., 2013. Zooplankton body composition. Limnol. Oceanogr. 58 (5), 1843–1850. https://doi.org/10.4319/lo.2013.58.5.1843.

Klein, S.G., Pitt, K.A., Lucas, C.H., Hung, S.H., Schmidt-Roach, S., Aranda, M., Duarte, C.M., 2019. Night-time temperature reprieves enhance the thermal tolerance of a symbiotic cnidarian. Front. Mar. Sci. 6, 460582. https://doi.org/10.3389/fmars.2019.00453.

Kogovšek, T., Tinta, T., Klun, K., Malej, A., 2014. Jellyfish biochemical composition: Importance of standardised sample processing. Mar. Ecol. Prog. Ser. 510, 275–288. https://doi.org/10.3354/meps10959.

Kondo, Y., Suzuki, Y., Ohtsuka, S., Nagai, H., Tanaka, H., Srinui, K., 2020. Differences in the cnidomes and toxicities of the oral arms of two commercially harvested rhizostome jellyfish species in Thailand. J. Mar. Biol. Assoc. UK 100, 701–711. https://doi.org/10.1017/S002531542000065X.

Kubota, S., Kawamura, M., Ueno, S., 2006. First occurrence of a mature medusa of *Nemopilema nomurai* (Cnidaria, Scyphozoa, Rhizostomae) in tanabe bay, wakayama prefecture, Japan. 南紀生物 48:57–59.

Kuplik, Z., Angel, D.L., 2020. Diet composition and some observations on the feeding ecology of the rhizostome *Rhopilema nomadica* in Israeli coastal waters. J. Mar. Biol. Assoc. UK 100 (5), 681–689. https://doi.org/10.1017/S0025315420000697.

Kuplik, Z., Kerem, D., Angel, D.L., 2021. Respiration rates, metabolic demands and feeding of ephyrae and young medusae of the Rhizostome *Rhopilema nomadica*. Diversity 13 (7), 320. https://doi.org/10.3390/d13070320.

Lane, N., 2015. Vital question: energy, evolution, and the origins of complex life. WW Norton Company.

Larson, R.J., 1979. Feeding in coronate medusae (Class Scyphoza, Order Coronatae). Mar. Freshw. Behav. Phy. 6 (2), 123–129. https://doi.org/10.1080/10236247909378559.

Larson, R.J., 1986. Water content, organic content, and carbon and nitrogen composition of medusae from the northeast Pacific. J. Exp. Mar. Biol. Ecol. 99 (2), 107–120. https://doi.org/10.1016/0022-0981(86)90231-5.

Larson, R.J., 1987. Costs of transport for the scyphomedusa *Stomolophus meleagris* L. Agassiz. Can. J. Zool. 65, 2690–2695. https://doi.org/10.1139/z87-408.

Larson, R.J., 1991. Diet, prey selection and daily ration of *Stomolophus meleagris*, a filter-feeding scyphomedusa from the NE Gulf of Mexico. Estuar. Coast. Shelf S. 32 (5), 511–525. https://doi.org/10.1016/0272-7714(91)90038-D.

Larson, R.J., 1997. Feeding behaviour of Caribbean scyphomedusae: *Cassiopea frondosa* (Pallas) and *Cassiopea xamachana* Bigelow. Stud. Nat. History Caribb. Reg. 73 (1), 43–54.

Laurie-Lesh, G.E., Corriel, R., 1973. Scyphistoma regeneration from isolated tentacles in *Aurelia aurita*. J. Mar. Biol. Assoc. UK 53 (4), 885–894. https://doi.org/10.1017/S0025315400022542.

Lee, L., Atkinson, D., Hirst, A.G., Cornell, S.J., 2020. A new framework for growth curve fitting based on the von Bertalanffy growth function. Sci. Rep. 10, 7953. https://doi.org/10.1038/s41598-020-64839-y.

Lee, H., Jung, E., Kang, C., Yoon, W.D., Kim, J.S., Kim, E., 2011. Scyphozoan jellyfish venom metalloproteinases and their role in the cytotoxicity. Toxicon 58, 277–284. https://doi.org/10.1016/j.toxicon.2011.06.007.

Lee, H.E., Yoon, W.D., Lim, D., 2008. Description of feeding apparatus and mechanism in *Nemopilema nomurai* Kishinouye (Scyphozoa: Rhizostomeae). Ocean Sci. J. 43, 61–65. https://doi.org/10.1007/BF03022432.

Lemos, D., García-Carreño, F.L., Hernández, P., Del Toro, A.N., 2002. Ontogenetic variation in digestive proteinase activity, RNA and DNA content of larval and post-larval white shrimp *Litopenaeus schmitti*. Aquaculture 214, 363–380. https://doi.org/10.1016/S0044-8486(02)00253-3.

Leone, A., Lecci, R.M., Durante, M., Piraino, S., 2013. Extract from the zooxanthellate jellyfish *Cotylorhiza tuberculata* modulates gap junction intercellular communication in human cell cultures. Mar. Drugs 11 (5), 1728–1762. https://doi.org/10.3390/md11051728.

Leone, A., Lecci, R.M., Durante, M., Meli, F., Piraino, S., 2015. The bright side of gelatinous blooms: nutraceutical value and antioxydant properties of three Mediterranean jellyfish (Scyphozoa). Mar. Drugs. 13, 4654–4681. https://doi.org/10.3390/md13084654.

Leoni, V., Molinero, J.C., Crochemore, S., Meffre, M., Bonnet, D., 2022. Ontogenetic dietary shifts of the medusa *Rhizostoma pulmo* (Cnidaria: Scyphozoa). Hydrobiologia 849 (13), 2933–2948. https://doi.org/10.1007/s10750-022-04903-y.

Leoni, V., Molinero, J.C., Meffre, M., Bonnet, D., 2021. Variability of growth rates and thermohaline niches of *Rhizostoma pulmo*'s pelagic stages (Cnidaria: Scyphozoa). Mar. Biol. 168, 107. https://doi.org/10.1007/s00227-021-03914-y.

Li, C., Yu, H., Liu, S., Xing, R., Guo, Z., Li, P., 2005. Factors affecting the protease activity of venom from jellyfish *Rhopilema esculentum* kishinouye. Bioorg. Med. Chem. Lett. 15, 5370–5374. https://doi.org/10.1016/j.bmcl.2005.09.010.

Li, R., Yu, H., Xue, W., Yue, Y., Liu, S., Xing, R., 2014. Jellyfish venomics and venom gland transcriptomics analysis of *Stomolophus meleagris* to reveal the toxins associated with sting. J. Proteom. 106, 17–29. https://doi.org/10.1016/j.jprot.2014.04.011.

Li, R., Yu, H., Yue, Y., Liu, S., Xing, R., Chen, X., et al., 2014. In depth analysis of the in vivo toxicity of venom from the jellyfish *Stomolophus meleagris*. Toxicon 92, 60–65. https://doi.org/10.1016/j.toxicon.2014.10.002.

Lilley, M.K., Elineau, A., Ferraris, M., Thiéry, A., Stemmann, L., Gorsky, G., et al., 2014. Individual shrinking to enhance population survival: quantifying the reproductive and metabolic expenditures of a starving jellyfish, *Pelagia noctiluca*. J. Plankton Res. 36 (6), 1585–1597. https://doi.org/10.3354/meps10854.

Lilley, M.K., Ferraris, M., Elineau, A., Berline, L., Cuvilliers, P., Gilletta, L., et al., 2014. Culture and growth of the jellyfish *Pelagia noctiluca* in the laboratory. Mar. Ecol. Prog. Ser. 510, 265–273. https://doi.org/10.3354/meps10854.

Lilley, M.K.S., Lombard, F., 2015. Respiration of fragile planktonic zooplankton: extending the possibilities with a single method. J. Exp. Mar. Biol. Ecol. 471, 226–231. https://doi.org/10.1016/j.jembe.2015.06.013.

López-Martínez, J., Arzola-Sotelo, E.A., Nevárez-Martínez, M.O., Álvarez-Tello, F.J., Morales-Bojórquez, E., 2020. Modeling growth on the cannonball jellyfish *Stomolophus meleagris* based on a multi-model inference approach. Hydrobiologia 847, 13991422. https://doi.org/10.1007/s10750-020-04182-5.

Lovett, D.L., Felder, D.L., 1990. Ontogeny of kinematics in the gut of the white shrimp *Penaeus setiferus* (Decapoda: Penaeidae). J. Crustacean Biol. 10 (1), 53–68. https://doi.org/10.1163/193724090X00249.

Lucas, C.H., 1994. Biochemical composition of *Aurelia aurita* in relation to age and sexual maturity. J. Exp. Mar. Biol. Ecol. 183 (2), 179–192.

Lucas, C.H., Graham, W.M., Widmer, C., 2012. Jellyfish life histories: role of polyps in forming and maintaining scyphomedusa populations. Adv. Mar. Biol. 63, 133–196. https://doi.org/10.1016/b978-0-12-394282-1.00003-x.

Lucas, C.H., Lawes, S., 1998. Sexual reproduction of the scyphomedusa *Aurelia aurita* in relation to temperature and variable food supply. Mar. Biol. 131, 629–638. https://doi.org/10.1007/s002270050355.

Lucas, C.H., Pitt, K.A., Purcell, J.E., Lebrato, M., Condon, R.H., 2011. What's in a jellyfish? Proximate and elemental composition and biometric relationships for use in biogeochemical studies: Ecological Archives E092-144. Ecology 92 (8), 1704. https://doi.org/10.1890/11-0302.1.

Lutcavage, M., Lutz, P.L., 1986. Metabolic rate and food energy requirements of the leatherback sea turtle, *Dermochelys coriacea*. Copeia 1986 (3), 796–798. https://doi.org/10.2307/1444962.

Lyndby, N.H., Rädecker, N., Bessette, S., Søgaard Jensen, L.H., Escrig, S., Trampe, E., et al., 2020. Amoebocytes facilitate efficient carbon and nitrogen assimilation in the *Cassiopea*-Symbiodiniaceae symbiosis. Proc. R. Soc. B. 287, 20202393. https://doi.org/10.1098/rspb.2020.2393.

Malej, A., 1989. Respiration and excretion rates of *Pelagia noctiluca* (Semaeostomeae, Scyphozoa). In: Proceedings of the Twenty First European Marine Biology Symposium: Gdańsk, 14–19 September 1986, Poland (Vol. 21, p. 107). Polish Academy of Sciences, Institute of Oceanology.

Malej, A., 1989. Shifts in size-classes and structure of net zooplankton in the presence of the jellyfish *Pelagia noctiluca* (Scyphozoa).

Malej, A., Malej, M., 1992. Population dynamics of the jellyfish *Pelagia noctiluca* (Forsskål, 1775). Marine eutrophication and population dynamics. G. Colomboet al., 215–219, Olsen Olsen, Fredensborg.

Mammone, M., Ferrier-Pagés, C., Lavorano, S., Rizzo, L., Piraino, S., Rossi, S., 2021. High photosynthetic plasticity lay reinforce invasiveness of upside-down zooxanthellate jellyfish in Mediterranean coastal waters. PloSONE 16, e0248814. https://doi.org/10.1371/journal.pone.0248814.

Mariottini, G.L., Pane, L., 2010. Mediterranean jellyfish venoms: a review on scyphomedusae. Mar. Drugs. 8, 1122–1152. https://doi.org/10.3390/md8041122.

Mariscal, R.N., 1974. Nematocysts. In: L. Muscatine and H.M. Lenhoff (Eds.), Coelenterate Biology: Rev. N. Perspect. Academic Press, New York, 129–178.

Mariscal, R.N., Conklin, E.J., Bigger, C.H., 1977. The ptychocyst, a major new category of cnida used in tube construction by a cerianthid anemone. Biol. Bull. 152, 392–405. https://doi.org/10.2307/1540427.

Marques, R., Albouy-Boyer, S., Delpy, F., Carré, C., Le Floc'h, E., Roques, C., et al., 2015. Pelagic population dynamics of *Aurelia* sp. in French Mediterranean lagoons. J. Plankton Res. 37, 1019–1035. https://doi.org/10.1093/plankt/fbv059.

Marques, R., Darnaude, A.M., Schiariti, A., Tremblay, Y., Molinero, J.C., Soriano, S., et al., 2019. Dynamics and asexual reproduction of the jellyfish *Aurelia coerulea* benthic life stage in the Thau lagoon (northwestern Mediterranean). Mar. Biol. 166, 74. https://doi.org/10.1007/s00227-019-3522-4.

Martin, M.B., Båmstedt, U., 2001. Digestion rate in relation to temperature of two gelatinous planktonic predators. Sarsia 86 (1), 21–35. https://doi.org/10.1080/00364827.2001.10420458.

Martínez-Pérez, R.B., Rodríguez, J.A., Soto, L.A.L., Gortáres-Moroyoqui, P., Diaz-Tenorio, L.M., 2020. Cannonball jellyfish digestion: an insight into the lipolytic enzymes of the digestive system. PeerJ 8, e9794. https://doi.org/10.7717/peerj.9794.

Martinussen, M.B., Båmstedt, U., 1999. Nutritional ecology of gelatinous planktonic predators. Digestion rate in relation to type and amount of prey. J. Exp. Mar. Biol. Ecol. 232 (1), 61–84. https://doi.org/10.1016/S0022-0981(98)00101-4.

Matsakis, S., Conover, R.J., 1991. Abundance and feeding of medusae and their potential impact as predators on other zooplankton in Bedford Basin (Nova Scotia, Canada) during spring. Can. J. Fish. Aquat. Sci. 48 (8), 1419–1430. https://doi.org/10.1139/f91-169.

McClendon, J.F., 1917. The direct and indirect calorimetry of *Cassiopea xamachana*. The effect of stretching on the rate of the nerve impulse. Jour. Biol. Chem. 32, 275.

McCloskey, L.R., Muscatine, L., Wilkerson, F.P., 1994. Daily photosynthesis, respiration, and carbon budgets in a tropical marine jellyfish (*Mastigias* sp.). Mar. Biol. 119, 13–22. https://doi.org/10.1007/BF00350101.

McConville, K., Atkinson, A., Fileman, E.S., Spicer, J.I., Hirst, A.G., 2017. Disentangling the counteracting effects of water content and carbon mass on zooplankton growth. J. Plankton R. 39 (2), 246–256. https://doi.org/10.1093/plankt/fbw094.

Medina, M., Sharp, V., Ohdera, A., Bellantuono, A., Dalrymple, J., Gamero-Mora, E., et al., 2021. The upside-down jellyfish *Cassiopea xamachana* as an emerging model system to study cnidarian–algal symbiosis. In: Handbook of Marine Model Organisms in Experimental Biology (pp. 149–171). CRC Press.

Megill, W.M., Gosline, J.M., Blake, R.W., 2005. The modulus of elasticity of fibrillin-containing elastic fibres in the mesoglea of the hydromedusa *Polyorchis penicillatus*. J. Exp. Biol. 208 (20), 3819–3834. https://doi.org/10.1242/jeb.01765.

Mellas, R.E., McIlroy, S.E., Fitt, W.K., Coffroth, M.A., 2014. Variation in symbiont uptake in the early ontogeny of the upside-down jellyfish, *Cassiopea* spp. J. Exp. Mar. Biol. Ecol. 459, 38–44. https://doi.org/10.1016/j.jembe.2014.04.026.

Mergner, H., Svoboda, A., 1977. Productivity and seasonal changes in selected reef areas in the Gulf of Aquaba (Red Sea). Helgol. Meeresunters 30, 383–399. https://doi.org/10.1007/BF02207849.

Merquiol, L., Romano, G., Ianora, A., D'Ambra, I., 2019. Biotechnological applications of scyphomedusae. Mar. Drugs. 17, 604. https://doi.org/10.3390/md17110604.

Messerli, S.M., Greenberg, R.M., 2006. Cnidarian toxins acting on voltage-gate ion channels. Mar. Drugs. 4, 70–81. https://doi.org/10.3390/md403070.

Miyake, H., Honda, S., Nishikawa, J., Yusoff, F.M., 2021. Life cycle of edible jellyfish *Acromitus hardenbergi* Stiasny, 1934 (Scyphozoa: Rhizostomeae) inhabiting a brackish-water environment. Animals 11 (7), 2138. https://doi.org/10.3390/ani11072138.

Molina-Ramírez, A., Cáceres, C., Romero-Romero, S., Bueno, J., González-Gordillo, J.I., Irigoien, X., et al., 2015. Functional differences in the allometry of the water, carbon and nitrogen content of gelatinous organisms. J. Plankton Res. 37, 989–1000. https://doi.org/10.1093/plankt/fbv037.

Morais, Z.B., Pintao, A.M., Costa, I.M., Calejo, M.T., Bandarra, N.M., Abreu, P., 2009. Composition and in vitro antioxidant effects of jellyfish *Catostylus tagi* from Sado Estuary (SW Portugal). J. Aquat. Food Prod. T. 18 (1-2), 90–107. https://doi.org/10.1080/10498850802581799.

Morand, P., Carré, C., Biggs, D.C., 1987. Feeding and metabolism of the jellyfish *Pelagia noctiluca* (Scyphomedusae, Semaeostomae). J. Plankton R. 9 (4), 651–665. https://doi.org/10.1093/plankt/9.4.651.

Morandini, A.C., 2024. Morphology of Rhizostomeae jellyfishes: what is known and what we advanced since the 1970s. Adv. Mar. Biol. 98, 61–98.

Morandini, A.C., Stampar, S.N., Maronna, M.M., Silveira, F.L.D., 2017. All non-indigenous species were introduced recently? The case study of *Cassiopea* (Cnidaria: Scyphozoa) in Brazilian waters. J. Mar. Biol. Assoc. UK 97 (2), 321–328. https://doi.org/10.1017/S0025315416000400.

Mortillaro, J.M., Pitt, K.A., Lee, S.Y., Meziane, T., 2009. Light intensity influences the production and translocation of fatty acids by zooxanthellae in the jellyfish *Cassiopea* sp. J. Exp. Mar. Biol. Ecol. 378, 22–30. https://doi.org/10.1016/j.jembe.2009.07.003.

Muffett, K.M., Aulgur, J., Miglietta, M.P., 2022. Impacts of light and food availability on early development of *Cassiopea* medusae. Front. Mar. Sci. 8, 783876. https://doi.org/10.3389/fmars.2021.783876.

Muffett, K.M., Klompen, A.M., Collins, A.G., Lewis Ames, C., 2021. Raising awareness of the severity of "Contactless Stings" by *Cassiopea* jellyfish and kin. Animals 11 (12), 3357. https://doi.org/10.3390/ani11123357.

Müller, F.R., 1858. Die Magenfäden der Quallen. Z. Wiss. Zool. 9. pp. 542–543 (Leipzig).

Muller-Parker, G., D'Elia, C.F., Cook, C.B., 2015. Interactions between corals and their symbiotic algae. In: Birkeland, C. (Ed.), Coral Reefs in the Anthropocene 543. Springer Netherlands, Dordrecht, pp. 99–116. https://doi.org/10.1007/978-94-017-7249-5_5.

Muñoz-Vera, A., Castejón, J.M.P., García, G., 2016. Patterns of trace element bioaccumulation in jellyfish *Rhizostoma pulmo* (Cnidaria, Scyphozoa) in a Mediterranean coastal lagoon from SE Spain. Mar. Poll. Bull. https://doi.org/10.1016/j.marpolbul.2016.06.069.

Muñoz-Vera, A., García, G., García-Sánchez, A., 2015. Metal bioaccumulation pattern by *Cotylorhiza tuberculata* (Cnidaria, Scyphozoa) in the Mar Menor coastal lagoon (SE Spain). Environ. Sci. Pollut. Res. 22, 19157–19169. https://doi.org/10.1007/s11356-015-5119-x.

Murakami, M., Kudo, I., 2002. Phospholipase a2. J. Biochem. Vol. 131, 285–292. https://doi.org/10.1093/oxfordjournals.jbchem.a003101.

Muscatine, L., Marian, R.E., 1982. Dissolved inorganic nitrogen flux in symbiotic and nonsymbiotic medusae. Limnol. Oceanogr. 27, 910–917. https://doi.org/10.4319/lo.1982.27.5.0910.

Nagai, T., Worawattanamateekul, W., Suzuki, N., Nakamura, T., Ito, T., Fujiki, K., et al., 2000. Isolation and characterization of collagen from rhizostomous jellyfish (*Rhopilema asamushi*). Food Chem. 70, 205–208. https://doi.org/10.1016/S0308-8146(00)00081-9.

Nagata, R.M., Morandini, A.C., 2018. Diet, prey selection, and individual feeding rates of the jellyfish *Lychnorhiza lucerna* (Scyphozoa, Rhizostomeae). Mar. Biol. 165 (12), 187. https://doi.org/10.1007/s00227-018-3445-5.

Nagata, R.M., Morandini, A.C., Colin, S.P., Migotto, A.E., Costello, J.H., 2016. Transitions in morphologies, fluid regimes, and feeding mechanisms during development of the medusa *Lychnorhiza lucerna*. Mar. Ecol. Prog. Ser. 557, 145–159. https://doi.org/10.3354/meps11855.

Nagata, R.M., Moreira, M.Z., Pimentel, C.R., Morandini, A.C., 2015. Food web characterization based on δ15N and δ13C reveals isotopic niche partitioning between fish and jellyfish in a relatively pristine ecosystem. Mar. Ecol. Prog. Ser. 519, 13–27. https://doi.org/10.3354/meps11071.

Nagy, K.A., 2005. Field metabolic rate and body size. J. Exp. Biol. 208 (9), 1621–1625. https://doi.org/10.1242/jeb.01553.

Nagy, K.A., Girard, I.A., Brown, T.K., 1999. Energetics of free-ranging mammals, reptiles, and birds. Annu. Rev. Nutr. 19 (1), 247–277. https://doi.org/10.1146/annurev.nutr.19.1.247.

Nakano, 1980. In Omori, M., 1981. Edible jellyfish (Scyphomedusae: Rhizostomeae) in the Far East waters: a brief review of the biology and fishery. Bull. Plankton Soc. Japan 28: 1–11 (in Japanese). https://doi.org/10.1007/978-94-010-0722-1_3.

Nath, R.D., Bedbrook, C.N., Abrams, M.J., Basinger, T., Bois, J.S., Prober, D.A., et al., 2017. The jellyfish *Cassiopea* exhibits a sleep-like state. Curr. Biol. 27 (19), 2984–2990. https://doi.org/10.1016/j.cub.2017.08.014.

Nawroth, J.C., Feitl, K.E., Colin, S.P., Costello, J.H., Dabiri, J.O., 2010. Phenotypic plasticity in juvenile jellyfish medusae facilitates effective animal-fluid interaction. Biol. Lett. 6 (3), 389–393. https://doi.org/10.1098/rsbl.2010.0068.

Neil, T.R., Askew, G.N., 2018. Jet-paddling jellies: Swimming performance in the Rhizostomeae jellyfish *Catostylus mosaicus*. J. Exp. Biol. 221 (24). https://doi.org/10.1242/jeb.191148.

Nemazie, D.A., Purcell, J.E., Glibert, P.M., 1993. Ammonium excretion by gelationous zooplankton and their contribution to the ammonium requirements of microplankton in Chesapeake Bay. Mar. Biol. 116, 451–458. https://doi.org/10.1007/BF00350062.

Nevalainen, T.J., Peuravuori, H.J., Quinn, R.J., Llewellyn, L.E., Benzie, J.A.H., Fenner, P.J., 2004. Phospholipase a_2 in cnidaria. Comp. Biochem. Phys. B. 139, 731–735. https://doi.org/10.1016/j.cbpc.2004.09.006.

Olesen, N.J., 1995. Clearance potential of jellyfish *Aurelia aurita*, and predation impact on zooplankton in a shallow cove. Mar. Ecol. Prog. Ser. 124, 63–72. https://doi.org/10.3354/meps124063.

Omori, M., Nakano, E., 2001. Jellyfish fisheries in southeast Asia. Hydrobiologia 451, 19–26. https://doi.org/10.1023/A:1011879821323.

Ostendarp, M., Plewka, J., Flathmann, J., Tilstra, A., El-Khaled, Y.C., Wild, C., 2022. Complete and rapid regeneration of fragments from the upside-down jellyfish *Cassiopea*. Front. Mar. Sci. 9, 949233. https://doi.org/10.1016/j.cbpc.2004.09.006.

Östman, C., 1997. Abundance, feeding behaviour and nematocysts of scyphopolyps (Cnidaria) and nematocysts in their predator, the nudibranch *Coryphella verrucosa* (Mollusca). Hydrobiologia 355, 21–22.

Östman, C., 2000. A guideline to nematocyst nomenclature and classification, and some notes on the systematic value of nematocysts. Sci. Mar. 64, 31–46.

Padilla-Serrato, J.G., López-Martínez, J., Acevedo-Cervantes, A., Alcántara-Razo, E., Rábago-Quiroz, C.H., 2013. Feeding of the scyphomedusa *Stomolophus meleagris* in the coastal lagoon Las Guásimas, northwest Mexico. Hidrobiológica 23 (2), 218–226.

Palomares, M.L.D., Pauly, D., 2009. The growth of jellyfishes. Hydrobiologia 616, 11–21. https://doi.org/10.1007/s10750-008-9582-y.

Park, J.W., Cho, S.Y., Choi, S.J., 2008. Purification and characterization of hepatic lipase from *Todarodes pacificus*. BMB Rep. 41 (3), 254–258.

Peach, M.B., Pitt, K.A., 2005. Morphology of the nematocysts of the medusae of two scyphozoans, *Catostylus mosaicus* and *Phyllorhiza punctata* (Rhizostomeae): Implications for capture of prey. Invertebr. Biol. 124, 98–108.

Pérez-Ruzafa, A., Gilabert, J., Gutiérrez, J.M., Fernández, A.I., Marcos, C., Sabah, S., 2002. Evidence of a planktonic food web response to changes in nutrient input dynamics in the Mar Menor coastal lagoon, Spain. In: Nutrients and Eutrophication in Estuaries and Coastal Waters: Proceedings of the 31st Symposium of the Estuarine and Coastal Sciences Association (ECSA), held in Bilbao, Spain, 3–7 July 2000 (pp. 359–369). Springer Netherlands. https://doi.org/10.1007/978-94-017-2464-7_26.

Pitt, K., 2000. Life history and settlement preferences of the edible jellyfish *Catostylus mosaicus* (Scyphozoa: Rhizostomeae). Mar. Biol. 136, 269–279. https://doi.org/10.1007/s002270050685.

Pitt, K.A., Duarte, C.M., Lucas, C.H., Sutherland, K.R., Condon, R.H., Mianzan, H., et al., 2013. Jellyfish body plans provide allometric advantages beyond low carbon content. PloSOne 8 (8), e72683. https://doi.org/10.1371/journal.pone.0072683.

Pitt, K., Kingsford, M.J., 2003. Temporal and spatial variation in recruitment and growth of medusae of the jellyfish, *Catostylus mosaicus* (Scyphozoa, Rhizostomeae). Mar. Freshw. Res. 54, 117. https://doi.org/10.1071/MF02110.

Pitt, K.A., Koop, K., Rissik, D., 2005. Contrasting contributions to inorganic nutrient recycling by the co-occuring jellyfishes, *Catostylus mosaicus* and *Phyllorhiza punctata* (Scyphozoa, Rhizostomeae). J. Exp. Mar. Biol. Ecol. 315, 71–86. https://doi.org/10.1016/j.jembe.2004.09.007.

Pitt, K.A., Welsh, D.T., Condon, R.H., 2009. Influence of jellyfish blooms on carbon, nitrogen and phosphorus cycling and plankton production. Hydrobiologia 616, 133–149.

Prieto, L., Astorga, D., Navarro, G., Ruiz, J., 2010. Environmental control of phase transition and polyp survival of a massive-outbreaker jellyfish. PLoSOne 5, e13793.

Prieto, L., Enrique-Navarro, A., Volsi, R.L., Ortega, M.J., 2018. The large jellyfish *Rhizostoma luteum* as sustainable a resource for antioxidant properties, nutraceutical value and biomedical applications. Mar. Drugs 16, 396. https://doi.org/10.3390/md16100396.

Purcell, J.E., 2003. Predation on zooplankton by large jellyfish, *Aurelia labiata*, *Cyanea capillata* and *Aequorea aequorea*, in Prince William Sound, Alaska. Mar. Ecol. Prog. Ser. 246, 137–152. https://doi.org/10.3354/meps246137.

Purcell, J.E., 2009. Extension of methods for jellyfish and ctenophore trophic ecology to large-scale research. Hydrobiologia 616, 2350. https://doi.org/10.1007/s10750-008-9585-8.

Purcell, J.E., Atienza, D., Fuentes, V., Olariaga, A., Tilves, U., Colahan, C., et al., 2012. Temperature effects on asexual reproduction rates of scyphozoan species from the northwest Mediterranean Sea. Hydrobiologia 690, 169–180. https://doi.org/10.1007/s10750-012-1047-7.

Purcell, J.E., Fuentes, V., Atienza, D., Tilves, U., Astorga, D., Kawahara, M., et al., 2010. Use of respiration rates of scyphozoan jellyfish to estimate their effects on the food web. Hydrobiologia 645, 135–152. https://doi.org/10.1007/s10750-010-0240-9.

Purcell, J.E., Uye, S.I., Lo, W.T., 2007. Anthropogenic causes of jellyfish blooms and their direct consequences for humans: a review. Mar. Ecol. Prog. Ser. 350, 153–174. https://doi.org/10.3354/meps07093.

Pütter, A., 1909. Die Ernährung der Wassertiere und der Stoffhaushalt. U. G. Fischer, Jena, pp. 166.

Radwan, F.F.Y., Burnett, J.W., Bloom, D.A., Coliano, T., Eldefrawi, M.E., Erderly, H., 2001. A comparison of the toxinological characteristics of two *Cassiopea* and *Aurelia* species. Toxicon 39, 245–257. https://doi.org/10.1016/S0041-0101(00)00121-5.

Rahat, M., Adar, O., 1980. Effect of symbiotic zooxanthellae and temperature on budding and strobilation in *Cassiopea andromeda* (Eschscholz). Biol. Bull. 159, 394–401. https://doi.org/10.2307/1541102.

Rakow, K.C., Graham, W.M., 2006. Orientation and swimming mechanics by the scyphomedusa *Aurelia* sp. in shear flow. Limnol. Oceanogr. 51 (2), 1097–1106. https://doi.org/10.4319/lo.2006.51.2.1097.

Raoult, V., Gaston, T.F., 2018. Rapid biomass and size-frequency estimates of edible jellyfish populations using drones. Fish. Res. 207, 160–164. https://doi.org/10.1016/j.fishres.2018.06.010.

Raposo, A., Alasqah, I., Alfheeaid, H.A., Alsharari, Z.D., Alturki, H.A., Raheem, D., 2022. Jellyfish as food: A narrative review. Foods 11 (18), 2773. https://doi.org/10.3390/foods11182773.

Rastogi, A., Sarkar, A., Chakrabarty, D., 2017. Partial purification and identification of a metalloproteinase with anticoagulant activity from *Rhizostoma pulmo* (barrel jellyfish). Toxicon 132, 29–39. https://doi.org/10.1016/j.toxicon.2017.04.006.

Réaumur, M.D.E., 1710. Sur les mouvements progressivs des Orties de Mer (Gelée de Mer.)—Mém. de l'Acad. Sci. de. Paris, Année 1710, 478–485 Paris 1712.

Riccio, G., Martinez, K.A., Martín, J., Reyes, F., D'Ambra, I., Lauritano, C., 2022. Jellyfish as an alternative source of bioactive antiproliferative compounds. Mar. Drugs. 20 (6), 350. https://doi.org/10.3390/md20060350.

Rivera-Pérez, C., García-Carreño, F.L., Saborowski, R., 2011. Purification and biochemical characterization of digestive lipase in whiteleg shrimp. Mar. Biotechnol. 13, 284–295. https://doi.org/10.1007/s10126-010-9298-7.

Romero, L., Marcussi, S., Marchi-Salvador, D., Silva, F., Fuly, A., Stabeli, R., 2010. Enzymatic and structural characterization of a basic phospholipase a2 from the sea anemone *Condylactis gigantea*. Biochimie 92, 1063–1071. https://doi.org/10.1016/j.biochi.2010.05.007.

Rosa, S., Pansera, M., Granata, A., Guglielmo, L., 2013. Interannual variability, growth, reproduction and feeding of *Pelagia noctiluca* (Cnidaria: Scyphozoa) in the Straits of Messina (Central Mediterranean Sea): linkages with temperature and diet. J. Mar. Syst. 111–112, 97–102. https://doi.org/10.1016/j.jmarsys.2012.10.001.

Röthig, T., Puntin, G., Wong, J.C.Y., Burian, A., McLeod, W., Baker, D., 2021. Holobiont nitrogen control and its potential for eutrophication resistance in an obligate photosymbiotic jellyfish. Microbiome 9, 127. https://doi.org/10.1186/s40168-021-01075-0.

Rubenstein, D.I., Koehlt, M.A.R., 1977. The mechanisms of filter feeding: some theoretical considerations. Am. Nat. 111 (981), 981–994. https://doi.org/10.1086/283227.

Russel, F.S., 1970. The medusae of the British Isles. Cambridge University Press, Cambridge.

Samain, J.F., Moal, J., Daniel, J.Y., Le Coz, J.R., Jezequel, M., 1980. The digestive enzymes amylase and trypsin during the development of *Artemia*: effect of food conditions. brine shrimp *Artemia* 2, 427–443.

Santhanakrishnan, A., Dollinger, M., Hamlet, C.L., Colin, S.P., Miller, L.A., 2012. Flow structure and transport characteristics of feeding and exchange currents generated by upside-down *Cassiopea* jellyfish. J. Exp. Biol. 215 (14), 2369–2381. https://doi.org/10.1242/jeb.053744.

Schaub, J., Hunt, B., Pakhomov, E., Holmes, K., Lu, Y., Quayle, L., 2018. Using unmanned aerial vehicles (UAVs) to measure jellyfish aggregations. Mar. Ecol. Prog. Ser. 591, 29–36. https://doi.org/10.3354/meps12414.

Schiariti, A., Holst, S., Straehler-Pohl, I., Tiseo G.R., Miyake, H., Morandini, A.C., 2024. Life cycles and reproduction of Rhizostomeae. Adv. Mar. Biol. 98.

Schiariti, A., Kawahara, M., Uye, S., Mianzan, H.W., 2008. Life cycle of the jellyfish *Lychnorhiza lucerna* (Scyphozoa: Rhizostomeae). Mar. Biol. 156, 1–12. https://doi.org/10.1007/s00227-008-1050-8.

Schiariti, A., Morandini, A.C., Jarms, G., von Glehn Paes, R., Franke, S., Mianzan, H., 2014. Asexual reproduction strategies and blooming potential in Scyphozoa. Mar. Ecol. Prog. Ser. 510, 241–253. https://doi.org/10.3354/meps10798.

Schmid, V., Ono, S.I., Reber-Müller, S., 1999. Cell-substrate interactions in Cnidaria. Microsc. Res. Tech. 44, 254–268.

Schmid, V., Tardent, P., 1969. Zur Gametogenese von *Podocoryne carnea* M. Sars. Rev. Suisse Zoo. l 76, 1071–1078.

Schneider, G., 1989. The common jellyfish *Aurelia aurita*: standing stock, excretion and nutrient regeneration in the Kiel Bight, Western Baltic. Mar. Biol. 100, 507–514. https://doi.org/10.1007/BF00394827.

Schneider, G., 1990. A comparison of carbon based ammonia excretion rates between gelatinous and non-gelatinous zooplankton: implications and consequences. Mar. Biol. 106, 219–225. https://doi.org/10.1007/BF01314803.

Schneider, G., 1992. A comparison of carbon specific respiration rates in gelatinous and non gelatinous zooplankton: a search for general rules in zooplankton metabolism. Helgoländer Meeresun 46, 377–388. https://doi.org/10.1007/BF02367205.

Schneider, G., Weisse, T., 1985. Metabolism measurements of *Aurelia aurita* planulae larvae, and calculation of maximal survival period of the free swimming stage. Helgoländer Meeresun 39, 43–47. https://doi.org/10.1007/BF01997518.

Shanks, A.L., Graham, W.M., 1987. Orientated swimming in the jellyfish *Stomolopus meleagris* L. Agassiz (Scyphozoan: Rhizostomida). J. Exp. Mar. Biol. Ecol. 108 (2), 159–169. https://doi.org/10.1016/S0022-0981(87)80020-5.

Shanks, A.L., Graham, W.M., 1988. Chemical defense in a scyphomedusa. Mar. Ecol. Prog. Ser. 45, 81–86. https://doi.org/10.3354/meps045081.

Shapiro, O.H., Fernandez, V.I., Garren, M., Guasto, J.S., Debaillon-Vesque, F.P., Kramarsky-Winter, E., et al., 2014. Vortical ciliary flows actively enhance mass transport in reef corals. Proc. Natl. Acad. Sci. 111 (37), 13391–13396. https://doi.org/10.1073/pnas.1323094111.

Shick, J.M., 1975. Uptake and utilization of dissolved glycine by *Aurelia aurita* scyphistomae: temperature effects on the uptake process; nutritional role of dissolved amino acids. Biol. Bull. 148, 117–140. https://doi.org/10.2307/1540654.

Shimauchi, H., Uye, S.I., 2007. Excretion and respiration rates of the scyphomedusa *Aurelia aurita* from the Inland Sea of Japan. J. Oceanogr. 63, 27–34. https://doi.org/10.1007/s10872-007-0003-z.

Sinigaglia, C., Peron, S., Eichelbrenner, J., Chevalier, S., Steger, J., Barreau, C., et al., 2020. Pattern regulation in a regenerating jellyfish. Elife 9, e54868. https://doi.org/10.7554/eLife.54868.

Skikne, S.A., Sherlock, R.E., Robison, B.H., 2009. Uptake of dissolved organic matter by ephyrae of two species of scyphomedusae. J. Plankton Res. 31 (12), 1563–1570. https://doi.org/10.1093/plankt/fbp088.

Smith, H.G., 1934. The feeding mechanism and digestion in *Cassiopea frondosa*. Yearb. Carneg. Inst. Wash 33. pp. 261–262 (Washington).

Smith, H.G., 1936. Contribution to the anatomy and physiology of *Cassiopea frondosa*. Pap. Tortugas Lab. Carneg. Inst. Wash 31. pp. 17–52 (Washington).

Sötje, I., Dishon, T., Hoffmann, F., Holst, S., 2017. New methods of morphometric analyses on scyphozoan jellyfish statoliths including the first direct evidence for statolith growth using calcein as a fluorescent marker. Microsc. Microanal. 23 (3), 553–568. https://doi.org/10.1017/S1431927617000344.

Southward, A.J., 1955. Observations on the ciliary currents of the jelly-fish *Aurelia aurita* L. J. Mar. Biol. Assoc. U. K. 34 (2), 201–216. https://doi.org/10.1017/S0025315400027570.

Sparre, P., Venema, S.C., 1998. Introduction to tropical fish stock assessment-Part 1: Manual. FAO, Rome, Italy.

Stampar, S.N., Silveira, F.L.D., Morandini, A.C., 2008. Food resources influencing the asexual reproductive cycle of coronate Scyphozoa. Cah. Biol. Mar. 49 (3), 247.

Stoner, E.W., Archer, S.K., Layman, C.A., 2022. Increased nutrient availability correlates with increased growth of the benthic jellyfish *Cassiopea* spp. Food Webs 31, e00231. https://doi.org/10.1016/j.fooweb.2022.e00231.

Straehler-Pohl, I., Jarms, G., 2010. Identification key for young ephyrae: a first step for early detection of jellyfish blooms. Jellyfish Blooms: N. Probl. Solut. 3–21. https://doi.org/10.1007/978-90-481-9541-1_2.

Straehler-Pohl, I., Jarms, G., 2011. Morphology and life cycle of *Carybdea morandinii*, sp. nov. (Cnidaria), a cubozoan with zooxanthellae and peculiar polyp anatomy. Zootaxa 2755 (2), 36–56. https://doi.org/10.11646/ZOOTAXA.2755.1.2.

Suchman, C.L., Sullivan, B.K., 1998. Vulnerability of the copepod *Acartia tonsa* to predation by the scyphomedusa *Chrysaora quinquecirrha*: effect of prey size and behavior. Mar. Biol. 132, 237–245. https://doi.org/10.1007/s002270050389.

Sugiura, Y., 1963. On the life-history of rhizostome medusae I. *Mastigias papua* L. Agassiz Annot. Zool. Jpn. 36, 194–202.

Sugiura, Y., 1964. On the life-history of rhizostome medusae II. Indispensability of zooxanthellae for strobilation in *Mastigias papua*. Embryologia 8, 223–233.

Sugiura, Y., 1969. On the life-history of rhizostome medusae V. On the relation between zooxanthellae and the strobilation of *Cephea cephea*. Bull. Mar. Biol. Stn. Asamushi 8, 227–233.

Sullivan, B.K., Garcia, J.R., Klein-MacPhee, G., 1994. Prey selection by the scyphomedusan predator *Aurelia aurita*. Mar. Biol. 121, 335–341. https://doi.org/10.1007/BF00346742.

Sun, T., Wang, L., Zhao, J., Dong, Z., 2021. Application of DNA metabarcoding to characterize the diet of the moon jellyfish *Aurelia coerulea* polyps and ephyrae. Acta Oceanol. Sin. 40, 160–167. https://doi.org/10.1007/s13131-021-1800-8.

Swift, H.F., Dawson, M.N., 2020. Demographic, environmental, and phenotypic change but genetic consistency in the jellyfish *Mastigias papua*. Biol. Bull. 239 (2), 80–94. https://doi.org/10.1086/710663.

Syazwan, W.M., Then, A.Y.H., Chong, V.C., Rizman-Idid, M., 2021. Trophic ecology of a tropical scyphozoan community in coastal waters: Insights from stomach content and stable isotope analyses. Cont. Shelf Res. 225, 104481. https://doi.org/10.1016/j.csr.2021.104481.

Templeman, M.A., Kingsford, M.J., 2012. Variation in soft tissue chemistry among scyphozoan and cubozoan jellyfishes from the Great Barrier Reef, Australia. Hydrobiologia 690, 279–290. https://doi.org/10.1007/978-94-007-5316-7_22.

Templeman, M.A., Kingsford, M.J, 2015. Predicting aqueous copper and zinc accumulation in the upside-down jellyfish *Cassiopea maremetens* through the use of biokinetic models. Environ. Monit. Assess. 187, 416. https://doi.org/10.1007/s10661-015-4657-5.

Templeman, M.A., McKenzie, M.R., Kingsford, M.J., 2021. The utility of jellyfish as marine biomonitors. Mar. Pollut. Bull. 173, 113056. https://doi.org/10.1016/j.marpolbul.2021.113056.

Thé, J., De Sousa Barroso, H., Mammone, M., Viana, M., Melo, C.S.B., Mies, M., et al., 2020. Aquaculture facilities promote populational stability throughout seasons and increase medusae size for the invasive jellyfish *Cassiopea andromeda*. Mar. Environ. Res. 162, 105161. https://doi.org/10.1016/j.marenvres.2020.105161.

Thein, H., Ikeda, H., Uye, S.I., 2012. The potential role of podocysts in perpetuation of the common jellyfish *Aurelia aurita* sl (Cnidaria: Scyphozoa) in anthropogenically perturbed coastal waters. Jellyfish Bloom. IV: Interact. Hum. Fish. 157–167. https://doi.org/10.1007/978-94-007-5316-7_12.

Thibault, D., Kuplik, Z., Priet, L., Enrique-Navarro, A., Brown, M.K., Uye, S., et al., 2024. Ecology of Rhizostomeae. Adv. Mar. Biol. 98.

Thiel, M.E., Reinicke, G.B., Holst, S., Morandini, A.C., Sötje, I., Straehler-Pohl, I., et al. 2024. Rhizostomeae—a remaining manuscript to complete the monograph on Scyphozoa by M.E. Thiel (published 1936–1977). Adv. Mar. Biol. 98. (Electronic supplement).

Tilesius, W.G., 1829. Beiträge zur Naturgeschichte der Medusen. Nova Acta physico-med 25, 249–288.

Tilstra, A., El-Khaled, Y.C., Meier, S., Wild, C., 2022. Invasive upside-down jellyfish tolerate organic eutrophication and warming. Bull. Mar. Sci. 98 (3), 381–392. https://doi.org/10.5343/bms.2021.0051.

Todd, B.D., Thornhill, D.J., Fitt, W.K., 2006. Patterns of inorganic phosphate uptake in *Cassiopea xamachana*: a bioindicator species. Mar. Pollut. Bull. 52, 515–521. https://doi.org/10.1016/j.marpolbul.2005.09.044.

Toullec, G., Lyndby, N.H., Banc-Prandi, G., Pogoreutz, C., Martin Olmos, C., Meibom, A., et al., 2024. Symbiotic nutrient exchange enhances the long-term survival of cassiosomes, the autonomous stinging-cell structures of *Cassiopea*. MSphere 9 (1), e00322–23. https://doi.org/10.1128/msphere.00322-23.

Travis, J., 1994. Evaluating the adaptive role of morphological plasticity. Ecol. Morphol. 99–122.

Undheim, E.A.B., King, G.F., 2011. On the venom system of centipedes (chilopoda), a neglected group of venomous animals. Toxicon 57, 512–524. https://doi.org/10.1016/j.toxicon.2011.01.004.

Uye, S., Brodeur, R.D., 2017. Report of Working Group 26 on jellyfish blooms around the North Pacific rim: causes and consequences. PICES, Sidney, British Columbia Canada.

Uye, S.I., Shimauchi, H., 2005. Population biomass, feeding, respiration and growth rates, and carbon budget of the scyphomedusa *Aurelia aurita* in the Inland Sea of Japan. J. Plankton Res. 27 (3), 237–248. https://doi.org/10.1093/plankt/fbh172.

Verde, E.A., McCloskey, L.R., 1998. Production, respiration, and photophysiology of the mangrove jellyfish *Cassiopea xamachana* symbiotic with zooxanthellae: effect of jellyfish size and season. Mar. Ecol. Prog. Ser. 168, 147–162. https://doi.org/10.3354/meps168147.

Vernon, H.M., 1895. The respiratory exchange of the lower marine invertebrates. J. Physiol. 19, 18–70.

von Montfort, G.M., Costello, J.H., Colin, S.P., Morandini, A.C., Migotto, A.E., Maronna, M.M., Reginato, M., Miyake, H., Nagata, R.M., 2023. Ontogenetic transitions, biomechanical trade-offs and macroevolution of scyphozoan medusae swimming patterns. Sci. Rep. 13 (1), 9760. https://doi.org/10.1038/s41598-023-34927-w.

Wagner, Z., Costello, J.H., Colin, S.P., 2020. Fluid and predator-prey interactions of Scyphomedusae fed calanoid copepods. Fluids 5 (2). https://doi.org/10.3390/fluids5020060.

Wang, P.P., Zhang, F., Sun, S., 2023. Predation effect on copepods by the giant jellyfish *Nemopilema nomurai* during the early occurrence stage in May in the northern East China Sea and southern Yellow Sea, China. Mar. Pollut. Bull. 186, 114462. https://doi.org/10.1016/j.marpolbul.2022.114462.

Welsh, D.T., Dunn, R.J., Meziane, T., 2009. Oxygen and nutrient dynamics of the upside-down jellyfish (*Cassiopea* sp.) and its influence on benthic nutrient exchanges and primary production. Hydrobiologia 635, 351–362. https://doi.org/10.1007/s10750-009-9928-0.

Widmer, C.L., 2005. Effects of temperature on growth of north-east Pacific moon jellyfish ephyrae, *Aurelia labiata* (Cnidaria: Scyphozoa). J. Mar. Biol. Assoc. U. K. 85 (3), 569–573. https://doi.org/10.1017/S0025315405011495.

Yang, P.J., Lemons, M., Hu, D.L., 2018. Rowing jellyfish contract to maintain neutral buoyancy. Theor. Appl. 8 (3), 147–152. https://doi.org/10.1016/j.taml.2018.03.001.

Yoffe, B., Baruchin, A.M., 2004. Mediterranean jellyfish (*Rhopilema nomadica*) sting. Burns 5, 503–504. https://doi.org/10.1016/j.burns.2004.01.013.

Yonge, C.M., 1931. Digestive processes in marine invertebrates and fishes. J. du. Cons. Internat. pour l'Explor. de. la. Mer. 6, 175–212.

Yu, H., Li, C., Li, R., Xing, R., Liu, S., Li, P., 2007. Factors influencing hemolytic activity of venom from the jellyfish *Rhopilema esculentum* kishinouye. Food Chem. Toxicol. 45, 1173–1178. https://doi.org/10.1016/j.fct.2006.12.025.

Yue, Y., Yu, H., Li, R., Xing, R., Liu, S., Li, K., et al., 2017. Biochemical and kinetic evaluation of the enzymatic toxins from two stinging scyphozoans *Nemopilema nomurai* and *Cyanea nozakii*. Toxicon 125, 1–12. https://doi.org/10.1016/j.toxicon.2016.11.005.

Zahn, V.M., 1981. Wie alt koennen Scyphomedusen werden? Zoolo. Beit. 27, 491–495.

Zamponi, M.O., Mianzan, H.W., 1985. La mecánica de captura y alimentación de *Olindias sambaquiensis* Müller, 1861 (Limnomedusae) en el medio natural y en condiciones experimentales. Hist. Nat. 5, 269–278.

CHAPTER SIX

Morphological and molecular data in the study of the evolution, population genetics and taxonomy of Rhizostomeae

Edgar Gamero-Mora[a,*], Jonathan W. Lawley[b], Maximiliano M. Maronna[c], Sérgio N. Stampar[d], Adriana Muhlia-Almazan[a], and André C. Morandini[e,f]

[a]Laboratorio de Bioenergética y Genética Molecular, Centro de Investigación en Alimentación y Desarrollo, Hermosillo, Sonora, Mexico
[b]School of Environment and Science, Centre for Planetary Health and Food Security, Griffith University, Gold Coast, QLD, Australia
[c]Instituto de Oceanografia, Universidade Federal do Rio Grande, Rio Grande, RS, Brazil
[d]Departamento de Ciências Biológicas, Universidade Estadual Paulista, Bauru, SP, Brazil
[e]Departamento de Zoologia, Instituto de Biociências, Universidade de São Paulo, São Paulo, SP, Brazil
[f]Centro de Biologia Marinha, Universidade de São Paulo, São Sebastião, SP, Brazil
*Corresponding author. e-mail address: egamero.mora@gmail.com

Contents

Abstract

Rhizostomeae research based on morphological approaches was reinforced and diversified by new techniques after the 1970s, including developing methodologies for phylogenetic analysis, the rise of the polymerase chain reaction, and the emergence of different sequencing technologies. Here, we summarize the contribution of morphological and molecular data to the study of the classification and phylogenetic relationships of Rhizostomeae in addition to the use of molecular data in studies at the population, species, and supraspecific levels. Throughout the history of the study

Advances in Marine Biology, Volume 98
ISSN 0065-2881, https://doi.org/10.1016/bs.amb.2024.09.001

of the Rhizostomeae systematics, morphological data have been neglected when it comes to phylogenetic inferences, which is reflected in the lack of a phylogenetic analysis of the taxa within Rhizostomeae based on phenotypic characters of the adult medusa. Concerning molecular data, *ca.* 3,200 nucleotide sequences are available in GenBank and are mainly used for discovering, delimiting, describing, and identifying species. Molecular approaches have also allowed species monitoring by qPCR and metabarcoding of environmental DNA, as well as unveiling the distribution and genetic diversity of jellyfish populations, shedding light on introduction events, conservation, and health of edible jellyfish stocks. Nucleotide sequences have also been key for the development of phylogenetic hypotheses that serve as basis for investigations on the origin and diversification of morphological, ecological, and behavioral traits within Cnidaria; however, despite the progress achieved, phylogenetic uncertainty still exists, especially within the formerly known superfamily Inscapulatae. Future directions in Rhizostomeae research involve generating molecular and morphological data of neglected taxa, which represents a golden opportunity to understand the evolution of Rhizostomeae.

1. Introduction

Rhizostomeae Cuvier, 1800 is an order within Scyphozoa Goette, 1887 that groups medusozoans whose medusa stage has numerous mouth openings and lacks marginal tentacles, while its polyp phase produces one ephyra per strobilation event (Avian and Ramšak, 2021) (but compare Schiariti et al., 2024). The first species of this order to be described was *Cassiopea frondosa* (Pallas, 1774), whose description represents the starting point for the study of the Linnaean taxonomy of this taxon (Pallas, 1774). Throughout the history of the Rhizostomeae taxonomic research, the period from 1774 to 1970 was a golden age since almost all of the supraspecific categories grouping the 91 valid species of Rhizostomeae were erected during this period (Collins and Morandini, 2024), except for Ptychophorae Gershwin and Davie, 2013, Bazingidae Gershwin and Davie, 2013, *Bazinga* Gershwin and Davie, 2013 and *Marivagia* Galil and Gershwin, 2010.

Since 1970, studies on the classification of biodiversity have benefited from the advent and popularization of phylogenetic systematics (Hennig, 1966), which made it possible to test if taxonomic groups created based on morphological affinities reflect evolutionary history. Concerning Rhizostomeae, there is scarce research involving phenotypic traits and phylogenetic analysis (Marques and Collins, 2004; Straehler-Pohl and Jarms, 2022). Multiple reasons could explain the limited implementation of morphological data to reconstruct phylogenies of Rhizostomeae. For example, the putative lack of phylogenetic signal of phenotypic characters due to the morphological crypsis and phenotypic plasticity shown by some genera of this taxon (Swift et al., 2016; Anthony et al., 2022) and the preference for other types of evidence for phylogenetic reconstructions, such as nucleic acid sequences (Lee and Palci, 2015).

Molecular data have revolutionized different fields of study in biology, including jellyfish taxonomy, by allowing the testing of hypotheses of classification in light of molecular evidence (Gómez Daglio and Dawson, 2019). The analysis of nucleotide sequences of single or multiple genes generated by Sanger sequencing or high throughput sequencing (HTS) has also allowed (1) a better understanding of jellyfish population genetics (Getino Mamet et al., 2019; Douek et al., 2024), (2) advances on discovering, delimiting, and identifying rhizostomes (Gómez Daglio and Dawson, 2017), and (3) the raising of phylogenetic inferences and hypotheses of trait evolution (Kayal et al., 2018; Djeghri et al., 2019).

This chapter aims to summarize research performed, from 1970 to date, updating Thiel's view (Thiel et al., 2024) on the population genetics, phylogenetic systematics, evolution, and classification of Rhizostomeae, based on morphological and molecular data, and to provide insights for future research on this group of jellyfish.

2. Morphological phylogenetics

It is part of human nature to categorize and classify objects of the natural world. Interrelationships of rhizostome jellyfish were always connected with morphological features, as summarized by Thiel in his unpublished manuscript (Thiel et al., 2024), so early jellyfish researchers provided classification systems and relationship diagrams based on their observations of morphological traits (see Section 4). However, it is uncertain whether these early classifications reflect the evolutionary relationships of jellyfish. Since phylogenetic systematics provides the theory and methods for grouping species based on common ancestry (Hennig, 1965, 1966), improving these classifications requires testing them under a phylogenetic framework.

Most of the work done from 1970 to date on the classification of jellyfish has focused on determining the phenotypic affinities between the major lineages of jellyfish (Werner, 1973; Salvini-Plawen, 1978; Brusca and Brusca, 1990; Schuchert, 1993; Ax, 1996; Fig. 1A–D); *i.e.*, to elucidate the

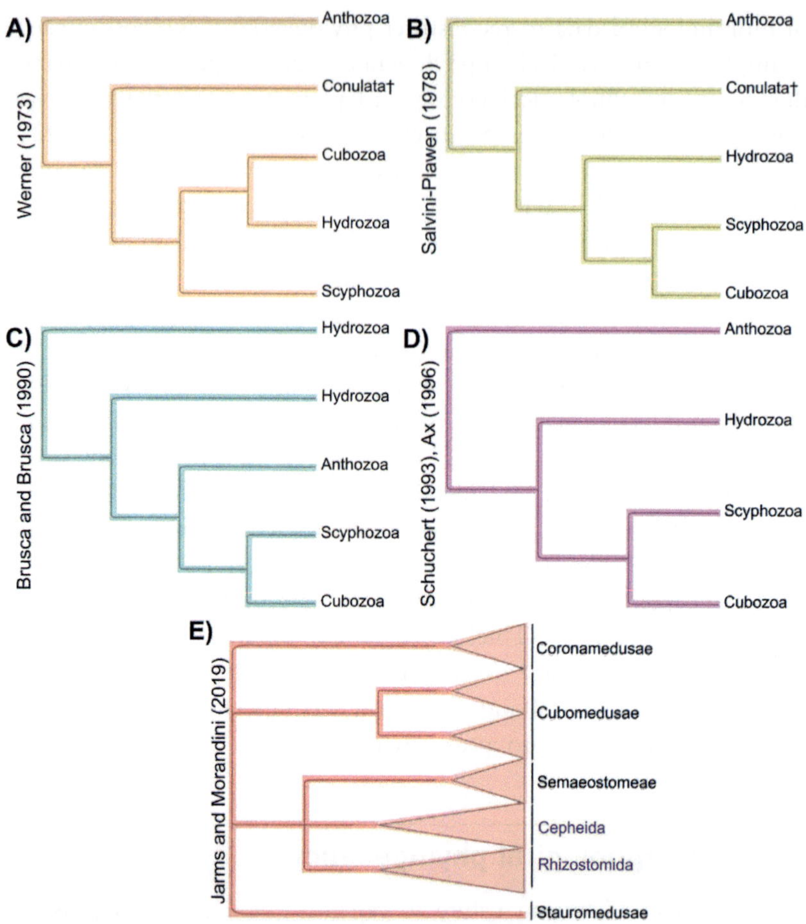

Fig. 1 Alternative morphological hypotheses of relationships among medusozoans according to different authors. (A) Werner (1973). (B) Salvini-Plawen (1978). (C) Brusca and Brusca (1990). (D) Schuchert (1993) and Ax (1996). (E) Jarms and Morandini (2019). In (E) the Rhizostomeae subgroups (Cepheida and Rhizostomida) were used as terminals in the tree and are highlighted in blue. † Extinct taxon.

relationships between Hydrozoa Owen, 1843, Staurozoa Marques and Collins, 2004, Cubozoa Werner, 1973 and Scyphozoa without considering a phylogenetic approach. Similarly, in their world atlas of jellyfish, but focused on Scyphozoa, Jarms and Morandini (2019) classified all rhizostome species as either cepheids (Cepheida Thiel, 1970) or rhizostomids (Rhizostomida Thiel, 1970) (Fig. 1E) based on a putative "biphyletic origin" of Rhizostomeae (Thiel, 1970).

Morphological research of Rhizostomeae jellyfish has a long history (summarized and reviewed in Morandini, 2024; Thiel et al., 2024), yet phylogenetic inferences based on phenotypic data within this group are scarce and their emergence date back to the 21st century (Marques and Collins, 2004). Morphological phylogenetic studies of Rhizostomeae can be divided into two types: (1) those that include "Rhizostomeae" as a terminal taxon in phylogenies of Medusozoa Petersen, 1979 and (2) those that show inferences about relationships within Rhizostomeae from research focused on Rhopaliophora Ax, 1995 (Scyphozoa, Cubozoa, and Staurozoa). Regarding the first type, it was not until the early 2000s that Marques and Collins (2004) included Rhizostomeae as a terminal taxon while investigating phylogenetic relationships among medusozoans using 87 morphological characters and life history features. These authors identified Rhizostomeae as monophyletic and the sister group of Semaeostomeae L. Agassiz, 1862 while noting that rhizostomes had been hypothesized as derived from within semaeostomes (citing Mayer, 1910; Uchida, 1926; Thiel, 1966; Collins, 2002) (Fig. 2A).

Following Marques and Collins (2004), Rhizostomeae was not used in any further morphological phylogenetic analysis of Medusozoa, perhaps because of the advent of molecular systematics in the early 2000s and because it was considered a monophyletic group (but no morphology-based approach formally tested the monophyly of the clade). On the other hand, a new hypothesis on phylogenetic relationships within Rhopaliophora using morphological data from early life cycle stages was published by Straehler-Pohl and Jarms (2022), which gave light to the phylogenetic relationships within Rhizostomeae. They analyzed 76 characters from 28 species using parsimony as the optimality criterion, which resulted in the reorganization of Rhizostomeae into a newly erected suborder Anterhizostomatida Straehler-Pohl and Jarms, 2022 along with members of the family Ulmaridae Haeckel, 1880 (*e.g.*, *Aurelia* Lamarck, 1816) from the polyphyletic order Semaeostomeae (Fig. 2B).

The scarcity of phylogenies based on phenotypic characters of Rhizostomeae reflects the limited use of morphological data to study this group's systematics. Likewise, it reflects the shift of interest towards molecular data for studying jellyfish relationships (Gómez Daglio and Dawson, 2019). Prospects for further studies in the phylogenetic systematics of Rhizostomeae include generating new morphological data (involving natural history collections and fieldwork) and coding phenotypic characters from already published morphological studies (such as taxonomic descriptions and morphological variation studies). While

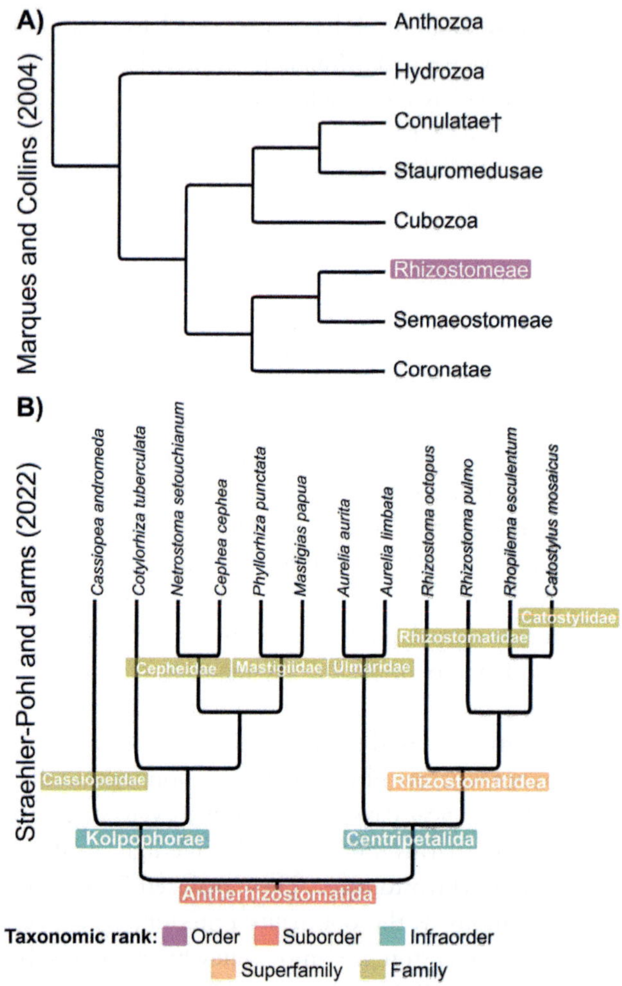

Fig. 2 Phylogenetic hypotheses of relationships among medusozoans and within Rhizostomeae. (A) Marques and Collins (2004)—Cladogram representation of the consensus tree showing the phylogenetic position of Rhizostomeae within Medusozoa. (B) Straehler-Pohl and Jarms (2022)—Phylogenetic hypothesis about the relationships within "Rhizostomeae" based on early life cycle data; Rhizostomeae here is indicated with quotation marks because the study does not recognize the taxon as monophyletic. † Extinct taxon.

there is evidence that morphological variability and phenotypic plasticity may hinder the study of jellyfish relationships at the species level (Lawley et al., 2021a; Anthony et al., 2022), the utility of morphological data for inferring relationships at higher taxonomic levels (like genera and families) remains unclear. Combining morphological and molecular data is essential to provide more robust phylogenetic hypotheses. For example, the monophyly of specific taxa has been recovered only after integrating morphological characters into molecular datasets in other animal groups (Cabra-García and Hormiga, 2020).

3. Molecular data

The advent and popularization of Sanger sequencing (Sanger et al., 1977) and the polymerase chain reaction (PCR) (Mullis and Faloona, 1987) provided solid grounds to promote the study of Rhizostomeae in different areas of the biological sciences. A total of 3,258 nucleotide sequences were available in NCBI GenBank for 20 genera of rhizostomes (out of 29 genera in total) as of August 2023 (Fig. 3; see supplementary material), which have revitalized the taxonomy, phylogenetic systematics, population genetics, and

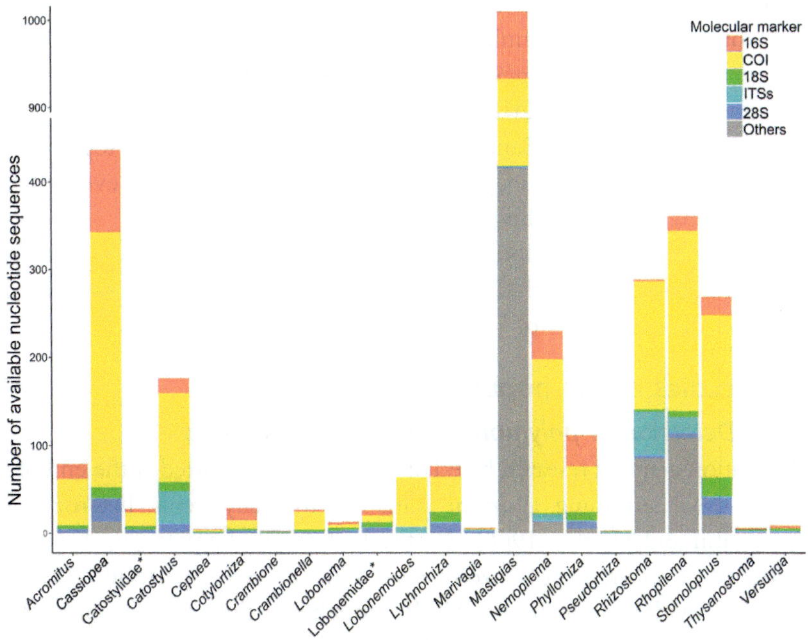

Fig. 3 Number of available sequences in NCBI's GenBank for each of the Rhizostomeae genera, where each color indicates a different molecular marker. The asterisks (*) indicate undescribed taxa for the families Catostylidae and Lobonemidae (according to Gómez Daglio and Dawson, 2017). Details on the methodology used to construct this graph can be found in the supplementary material.

evolutionary research of rhizostomes (*cf.* Maronna and Lawley, 2022). Molecular data provided a new line of evidence to solve different types of questions that had not been explored by the use of morphological data alone, such as (1) investigating jellyfish population structure (Dawson and Hamner, 2005), (2) recognizing complexes of cryptic species, unveiling geographic distribution patterns, studying morphological variation (Holland et al., 2004; Dawson 2005a), (3) inferring phylogenetic relationships at different taxonomic levels (Bridge et al., 1992), and (4) reinforcing hypotheses on the evolution of traits within Cnidaria (Collins, 2002; Collins et al., 2006; Dawson and Hamner, 2009).

HTS supperpassed Sanger sequencing concerning throughput and sequencing efficiency, showing a lower cost per sequenced base and a lower investment of time to generate them (Maronna and Lawley, 2022; Cheng et al., 2023). The rise of HTS platforms along with the development of bioinformatics tools, led to the sequencing and analysis of microsatellite markers, mitogenomes, transcriptomes, and complete genomes. These advances in sequencing enabled the detection of multiple nucleotide variations simultaneously (Getino Mamet et al., 2017), the reconstruction of phylogenetic inferences based on hundreds of gene sequences (Kayal et al., 2018), and the development of comparative analyses of jellyfish traits (Santander et al., 2022). All of this together has provided a broader and deeper picture of how jellyfish have evolved and, thus, a better understanding of the evolution of Metazoa.

Below we present a look at advances in Rhizostomeae research made with molecular data, including those generated by Sanger sequencing and HTS, in different areas of biology.

3.1 Studies at the population level

3.1.1 Detection of polymorphism in DNA fragments

Population genetics research based on molecular data started at the end of the 1960s by determining allozyme variation to study genic heterozygosity (Lewontin and Hubby, 1966). To the best of our knowledge, there is no evidence of research on Rhizostomeae using allozyme variation. However, there is evidence of using AFLP (Amplified Fragment Length Polymorphism) to study genetic variability in rhizostomes. AFLP is a PCR-based technique for analyzing DNA fragment polymorphisms that does not require sequencing (Sheeja et al., 2021). AFLP has helped to assess the genetic variation of populations of *Rhopilema esculentum* Kishinouye, 1891 in China, revealing more genetic differentiation within populations than among them, which can be explained because of the release of laboratory-grown young medusae in the sea to sustain the fishery of this edible jellyfish in China (Qiao et al., 2013). The selection of germplasm sources for artificial seeding is essential for the conservation and sustainable use of jellyfish that are a target for fisheries because it can determine the increase or decrease of genetic differentiation between and within populations and, thus, the adaptive capacity and survival of populations to environmental shifts.

3.1.2 Studies based on traditional markers

DNA sequences eventually replaced AFLP markers and allozyme variation, allowing the recognition of nucleotide variations with a higher resolution (Kreitman, 1983; Casillas and Barbadilla, 2017). The study of population genetics of Rhizostomeae species using Sanger sequencing for generating nucleotide sequences started in the first half of the 2000s (Dawson 2005b; Dawson and Hamner, 2005) and has helped to understand features of the distribution and genetic diversity of species in seven genera (Table 1).

Table 1 Table showing the different species of Rhizostomeae in which population genetic analyses have been published and each molecular marker used in the studies.

	12S	COI	COIII	AFLP	Calmodulin	ITSs	Microsatellite	Reference
Catostylus mosaicus		■				■		Dawson (2005b)
Lobonemoides robustus		■				■		Iida et al. (2021)
Mastigias sp.		■						Dawson and Hamner (2005)
Mastigias spp.		■	■					Swift et al. (2016)
Mastigias papua		■	■					Swift and Dawson (2020)
Nemopilema nomurai		■						Dong et al. (2016)
Rhizostoma octopus		■			■			Lee et al. (2013)
		■					■	Glynn et al. (2015)
Rhizostoma pulmo		■				■		Ramšak et al. (2012)
		■						Faleh et al. (2017)
Rhopilema esculentum				■				Qiao et al. (2013)
		■						Li et al. (2016)
							■	Zhu et al. (2015)
Rhopilema hispidum		■				■		Iida et al. (2021)
Rhopilema nomadica		■						Giallongo et al. (2021)
Stomolophus sp. 2 (Published as *S. meleagris*)	■	■						Nevárez-López et al. (2020)
Stomolophus spp.		■					■	Getino Mamet et al. (2019)

In early studies, the diversity and genetic structure of *Mastigias* sp. and *Rhizostoma pulmo* (Macri, 1778) were assessed using cytochrome c oxidase subunit I (COI) and internal transcribed spacer one (ITS1) sequences, which showed a lack of population structure in *R. pulmo* over the Mediterranean Sea (Ramšak et al., 2012) and a positive correlation between the genetic diversity of *Mastigias* sp. with habitat area in Palau (Dawson and Hamner, 2005). Other studies detected either high genetic differentiation in populations of *R. esculentum* (Liaodong Bay, the Ganghwado Island, Bohai and Yellow Seas and Sea of Japan) (Li et al., 2016), *R. pulmo* (Mediterranean Sea) (Faleh et al., 2017), and *Rhizostoma octopus* (Gmelin, 1791) (Irish Sea) (Lee et al., 2013) or low to moderate genetic differentiation for *Mastigias papua* (Lesson, 1830) (Indo-West Pacific) (Swift et al., 2016; Swift and Dawson, 2020), *Nemopilema nomurai* Kishinouye, 1922 (China) (Dong et al., 2016), and *Stomolophus* sp. 2 *sensu* Gómez Daglio and Dawson (2017) (published as *Stomolophus meleagris* Agassiz, 1860) (Gulf of California) (Nevárez-López et al., 2020). Furthermore, COI and ITS1 nucleotide sequences were also helpful in assessing genetic variance among populations of *Catostylus mosaicus* (Quoy and Gaimard, 1824) from southeast Australia and for testing different phylogeographic hypotheses (Dawson 2005b).

COI and ITS1 nucleotide sequences have also been employed to study the genetic diversity of other commercially important rhizostome jellyfish such as *Rhopilema hispidum* (Vanhöffen, 1888) and *Lobonemoides robustus* Stiasny, 1920. Contrary to *R. hispidum*, *L. robustus* has significant differentiation among populations from different locations in southeast Asia (Iida et al., 2021); this difference could be the result of the geological history, oceanographic patterns, and differences in biological characteristics of each species (*e.g.*, dispersal ability and reproductive strategies). Similarly, the study of the genetic variation of *Stomolophus* sp. 2 from the Gulf of California (Getino Mamet et al., 2019), using COI sequences, revealed significant genetic differentiation among four locations, which could reflect this species' life history traits.

Studying rhizostomes through population genetics has also been crucial in understanding aspects of invasive species. For example, the study of the population structure of the invasive species *Rhopilema nomadica* Galil, Spanier and Ferguson, 1990 on the Mediterranean coast of Israel, based on COI sequences, showed that all analyzed specimens belonged to a single unstructured population with high within-population and haplotype diversities, which suggested multiple introduction events or the existence of an open corridor connecting populations (Giallongo et al., 2021).

Populations of Rhizostomeae with different degrees of diversity, genetic structure, and gene flow reflect distinct demographic histories (*e.g.*, variable connectivity, migration and dispersal patterns, and population expansion and contraction events) and capacities to adapt to environmental changes. Therefore, population genetic studies are essential for detecting vulnerable species and for their conservation. Moreover, research on population genetics of commercially important species is vital for using edible jellyfish without compromising their genetic diversity and avoiding the local extinction of their populations, thus guaranteeing the food security of different nations through sustainable fisheries.

3.1.3 Studies based on microsatellite markers

Microsatellite markers have been employed to address questions about the population genetics of Rhizostomeae, because they are exceptionally variable. Using microsatellite markers enables the detection of variability that DNA barcodes such as COI and ITS1 do not detect (Terrones et al., 2022). Some studies have focused on the development and characterization of microsatellite markers for Rhizostomeae species (Zhu et al., 2015; Getino Mamet et al., 2017; Douek et al., 2020), while others on their applications to assess inbreeding, population genetic structure, and genetic diversity (Tian et al., 2017; Getino Mamet et al., 2019). For example, microsatellite markers have been used to study the inbreeding levels in a hatchery population and the estimation of genetic diversity in clones of *R. esculentum*, demonstrating that these markers can help analyze the identity of individuals and identification of kinship. This can be applied to enhance the stock of this edible jellyfish and conserve germplasm resources (Tian et al., 2017).

Another strategy used in genetic analyses of rhizostome populations is to combine microsatellite and traditional DNA barcodes to better understand population demography and structure. This strategy has been used to study populations of *R. octopus* and *Stomolophus* spp. As a result, in the case of *R. octopus*, it was concluded that combining mitochondrial and microsatellite markers provides a better understanding of the population genetics of jellyfish with large effective population sizes (Glynn et al., 2015). Considering *Stomolophus* spp., both COI and microsatellites revealed divergence between *Stomolophus* sp. 1 sensu Gómez Daglio and Dawson (2017) and *Stomolophus* sp. 2 and also that high genetic differentiation between different populations of *Stomolophus* sp. 2 exist in the Gulf of California (Getino Mamet et al., 2019). In both cases, using microsatellites coupled with traditional markers provided more robust inferences about these species' genetic diversity and population structure, which serves as basis for developing conservation programs and effective fisheries management.

Population genomic approaches are focused on the detection of large-scale polymorphism at multiple loci and on analysis of -omic datasets within or among populations, which in recent years have been used for studying various animal groups, including cnidarians (Aurelle et al., 2022). However, these approaches have yet to be applied to study Rhizostomeae populations. Therefore, whether genomic datasets support findings based on traditional and microsatellite markers is unknown (see Dawson et al., 2015). Consequently, developing a research agenda that considers generating genomic resources to validate previous findings and address other issues, such as studying genetic variability with higher resolution and developing genomic studies of local adaptation, is essential.

3.2 Studies at the species level

3.2.1 Species discovery, delimitation, and description

From 2010 to date, new species of *Marivagia*, *Crambionella* Stiasny, 1921 and *Cassiopea* Péron and Lesueur, 1810 have been described based on morphology but discovered and delimited by using DNA data (COI, 16S ribosomal RNA (16S) and nuclear 28S ribosomal RNA (28S)) (Galil et al., 2010; Nishikawa et al., 2015; Gamero-Mora et al., 2022). This framework for the discovery and delimitation of new species is based on the fact that newly generated nucleotide sequences are not assigned into well-supported monophyletic clades formed by already known species in phylogenetic analyses.

DNA sequences are also helpful for molecular taxonomy and have provided diagnostic characters in descriptions of some scyphozoans (Lawley et al., 2021a; Lawley et al., 2022). One of the earliest uses of molecular data for studying rhizostome jellyfish was the recognition and diagnosis of subspecies of *M. papua* (Dawson 2005c), as well as the identification and redescription of two subspecies of *C. mosaicus* based on morphological data, nucleotide sequences of ITS1 and diagnostic nucleotides of the mitochondrial gene COI (Dawson 2005d). Other taxonomical investigations that used molecular data were those conducted by De Souza and Dawson (2018) and Neethling et al. (2011), who included diagnostic nucleotides in the redescription of *M. papua* and determined the placement of *Crambionella stuhlmanni* (Chun, 1896) within the genus *Crambionella* for its redescription, respectively.

The use of diagnostic characters in descriptions of Rhizostomeae is scarce. However, such characters have proven helpful for species descriptions when morphological variability hinders diagnosability in other groups of jellyfish, *e.g.*, in the semaeostome *Aurelia* (Lawley et al., 2021a). Nowadays, multiple programs allow for obtaining molecular diagnostic characters (*e.g.*, Kühn and Haase, 2020; Fedosov et al., 2022), representing a way to describe cryptic species (Lawley et al., 2021a).

3.2.2 Species identification and monitoring

3.2.2.1 Sequencing-dependent species identification and monitoring

When discussing molecular data and species identification of Rhizostomeae, the primary approach used is identification *via* phylogenetic analysis. This approach involves assigning an identification to newly generated sequences based on their placement to clades already linked with rhizostomes species described in the literature. This framework has reduced the generation of unreproducible or less comparable findings (Muffett and Miglietta, 2023) by allowing to work with jellyfish identified at the species level of the genera *Cassiopea* and *Catostylus* Agassiz, 1862 *via* phylogenetic analyses and 18S ribosomal RNA (18S), 16S or COI nucleotide sequences (Parracho and Morais, 2015; Gamero-Mora et al., 2019; Anthony et al., 2022). For example, Mellas et al. (2014) used COI sequences to identify *Cassiopea* polyps; species identification would not be possible in this study without molecular data because morphological differentiation is not achievable for polyps of all *Cassiopea* species.

The sequencing and phenetic or phylogenetic analysis of fragments of the mitochondrial genes 16S and COI and even of the nuclear markers 18S, ITS1, and 28S (whose genetic variability is not always adequate to identify species; Maronna and Lawley, 2022) have helped to record species from various genera in new localities (Ullah et al., 2015; Rizman-Idid et al., 2016; Galil et al., 2017; Pourjomeh et al., 2018; Sun et al., 2019; Dagha et al., 2022). Molecular data has also been used to corroborate the presence of non-native species, including *Mastigias* sp. 1 *sensu* Bayha and Graham (2011) and the invasive *Cassiopea andromeda* (Forskål, 1775) (Holland et al., 2004; Morandini et al., 2017; Maggio et al., 2019; Stampar et al., 2020; Thé et al., 2021) in the Atlantic basin and the genus *Phyllorhiza* Agassiz, 1862 in multiple localities (Dong et al., 2019; Rosales-Catalan et al., 2021). Furthermore, molecular data and phylogenetic approaches have proven indispensable for discovering, identifying, and determining the phylogenetic relationships of cryptic species (Holland et al., 2004; Swift et al., 2016).

Thirty-eight species of jellyfish are considered edible, of which at least 29 are classified within the order Rhizostomeae (Brotz et al., 2017, 2024). The steps involved in traditional medusae processing include dehydration with salt and alum (Raposo et al., 2022), which causes a change in the organism's structure that makes subsequent morphological identification nearly impossible. Thus, most of the harvested species are identified by dealers and merchants based on the texture, size, form, and color of the commercialized products (Omori and Nakano, 2001). In this sense, molecular information has been used to identify jellyfish species in raw and processed food (Galimberti et al., 2013). For example, amplification and sequencing of the genes COI and 28S helped confirm the identification of *N. nomurai, R. esculentum*, and *R. pulmo* in dry-salted and ready-to-eat products (Armani et al., 2013, 2014).

DNA barcoding has been used for species-level identification as an alternative to phylogenetic analyses. Thus, by comparing generated COI sequences against the NCBI nucleotide or BOLD Systems databases, the BLAST algorithm allowed the identification of *N. nomurai* (in processed seafood products) (Chin et al., 2016), *Cassiopea* sp., and *Phyllorhiza punctata* von Lendenfeld, 1884 (Ortman et al., 2010).

In addition to Sanger sequencing, HTS metabarcoding has recently been used to detect jellyfish from DNA extractions of water samples, *i.e.*, without capturing and subsampling target organisms. Thus, environmental DNA (eDNA) metabarcoding (targeting 16S, COI, and/or 18S) was employed to assess the diversity of medusozoans from near-shore waters of the Florida Keys, USA and from a marine aquaculture pond in Liaodong Bay, China. As a result, the rhizostomes *C. andromeda, C. frondosa*, and *Cassiopea xamachana* Bigelow, 1892 were detected in Florida (Ames et al., 2021), while *N. nomurai* and *R. esculentum* in Liaodong Bay (Li et al., 2022). eDNA methods are usually employed for ecological studies (*e.g.*, monitoring species), but in some instances, studying eDNA could be helpful in the early detection of non-native rhizostome species before the medusa stage is observable by detecting recently introduced larvae, polyps, or ephyrae into new localities (transported in ballast water or ship hulls).

3.2.2.2 Non-sequencing-dependent species identification and monitoring

Molecular identification of rhizostomes is mainly based on PCR amplification and Sanger sequencing. However, other non-sequencing-dependent strategies have been used for species identification, including the polymerase chain reaction-restriction fragment length polymorphism (PCR-RFLP) for the identification of *R. esculentum* and *S. meleagris* (possibly *N. nomurai*) in dry products (Jiang et al., 2019).

Nowadays, real-time PCR has a wide range of applications, including gene expression analyses, molecular diagnostics, and environmental monitoring (Wang et al., 2017; Hariharan and Prasannath, 2021; Langlois et al., 2021). In this sense, there is evidence that real-time PCR, in conjunction with taxon-specific Taqman© probes targeting the 16S gene, allows the identification of polyps of *Phyllorhiza* sp. (and other species) from DNA extracted from benthic surfaces. Since jellyfish polyps are difficult to find in the field due to their small size and cryptic nature, this approach offers a possibility for detecting polyps of invasive species in marine infrastructure such as cargo ship hulls (Bayha and Graham, 2009). Similarly, the use of real-time PCR, in conjunction with conventional 16S primers and intercalating DNA dye, has been important for the identification and study of the spatiotemporal distribution and abundance of *N. nomurai* and *R. esculentum* from plankton samples. The latter study offers a complementary or alternative approach to the conventional monitoring of jellyfish populations and, thus, to the understanding of jellyfish bloom dynamics (Wang et al., 2021).

3.3 Studies at supraspecific levels

3.3.1 Phylogenies based on Sanger sequencing data

The available knowledge of the relationships within Rhizostomeae comes from phylogenetic analyses performed at least at the subclass level, so no phylogenetic studies are focused on elucidating the relationships within Rhizostomeae or any of its taxa. The earliest phylogenetic hypothesis that used nucleotide sequences of a rhizostome had the aim to unveil class-level relationships within Cnidaria; evidence from the 18S gene together with morphological and mitochondrial genome structure from *Cassiopea* sp. and other cnidarians (such as anemones, octocorals, hydrozoans, cubozoans), reinforced the reciprocal monophyly of Anthozoa Ehrenberg, 1834 and Medusozoa (Bridge et al., 1995). Another class-level phylogenetic study that involved molecular data from rhizostomes was focused on inferring the phylogenetic relationships within Medusozoa and the evolution of life cycles using 18S sequences. As a result, the monophyly of Rhizostomeae was recovered and the order Semaeostomeae is paraphyletic with respect to Rhizostomeae (Collins, 2002).

Further research at the subphylum- and class-level provided results based on the combination of two nuclear markers, reinforcing the hypothesis of the monophyly of Rhizostomeae, as well as its sister-group relationship with the paraphyletic Semaeostomeae (Dawson, 2004 (5S+28S); Collins et al., 2006 (18S+28S); Bayha et al., 2010 (18S+28S)). However, as the number of different rhizostome sequences and molecular markers increased, some internal nodes of Rhizostomeae remained unresolved (Bayha et al., 2010; Gómez Daglio and Dawson, 2017; Picciani et al., 2018). For example, the paraphyly of superfamily Inscapulatae Stiasny, 1920 became evident due to families Catostylidae Claus, 1883 and Lychnorhizidae Haeckel, 1880, which were not recovered as monophyletic (Fig. 4).

Ortman et al. (2010) published the first phenetic analysis based on mitochondrial data (COI) that included species from all major lineages of Medusozoa. The dataset included two species of Rhizostomeae, which were grouped in a single cluster. However, according to the authors, COI is not helpful for the reconstruction of phylogenies within Medusozoa (due to the limited resolution that it provides at deeper nodes) and, therefore, they show the results as a phenogram (*i.e.*, they display the similarity between nucleotide sequences and not evolutionary relationships) and discuss their findings in light of the usefulness of COI as a DNA barcode for Medusozoa species identification (see Section 3.2.2.1). Similarly, Armani et al. (2013) used a phenetic method to display sequence similarity among scyphozoan COI sequences (including rhizostomes), in which various families of Rhizostomeae were not recovered in a single cluster.

The most comprehensive phylogenetic analysis of Scyphozoa in terms of taxa coverage was presented by Gómez Daglio and Dawson (2017), based on concatenated nuclear (18S and 28S) and mitochondrial (16S) molecular markers, resulting in a better-resolved phylogeny that included a high number of rhizostome samples for unsampled areas. Regarding Rhizostomeae, the authors found no resolution in taxa within Inscapulatae, recovering the families Catostylidae, Lychnorhizidae, and Lobonemidae Stiasny, 1920 as non-monophyletic (Fig. 4B)—as also found by Bayha et al. (2010) (Fig. 4A). Compared to the concatenated 18S and 28S maximum likelihood (ML) phylogeny for Scyphozoa by Bayha et al. (2010), in their ML tree Gómez Daglio and Dawson (2017) found a different topology for the clade composed of the families Cassiopeidae Tilesius, 1831, Cepheidae L. Agassiz, 1862 and Mastigiidae Stiasny, 1920 (Fig. 4B).

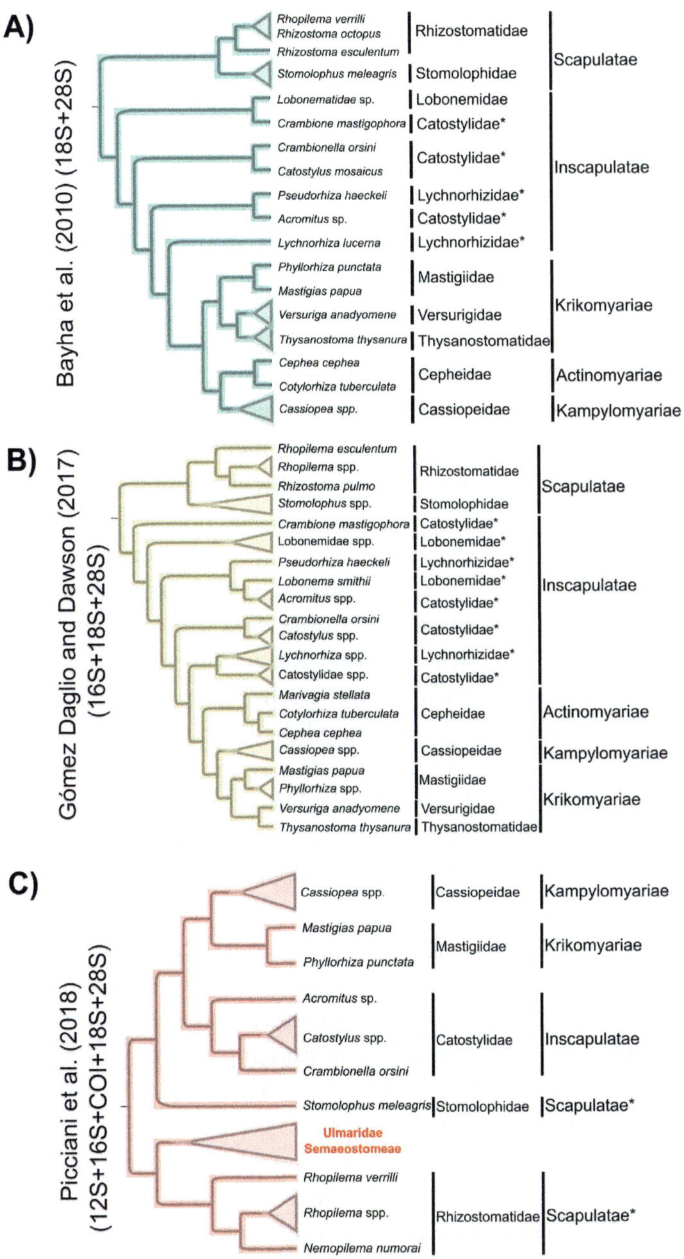

Fig. 4 Comparison of cladogram representations of molecular phylogenetic hypotheses for the relationships among Rhizostomeae based on concatenated datasets. (A) Bayha et al. (2010): 18S+28S. (B) Gómez Daglio and Dawson (2017): 16S+18S+28S—maximum likelihood tree. (C) Picciani et al. (2018): 12S+16S+COI+18S+28S. The asterisk (*) next to a taxon name indicates that it is not monophyletic. In (C), the family Ulmaridae (Semaeostomeae) is marked in red color because it appears in the middle of the rhizostome taxa.

A phylum-level multigene phylogeny (using previously published 18S, 28S, COI, 16S, and 12S ribosomal RNA (12S)) that included rhizostome sequences was published by Picciani et al. (2018), the largest phylogeny of cnidarians published to date used to understand eye origins in Cnidaria Verrill, 1865. In their phylogenetic tree, the semaeostome family Ulmaridae appears in the middle of the rhizostome taxa (Fig. 4C). Moreover, molecular divergence times were estimated for almost all major lineages of Cnidaria, including Rhizostomeae, whose estimated age was 184 million years.

3.3.2 HTS and phylogenetic reconstructions

In 2012, a study was published on the evolution of mitochondrial genomes in Medusozoa (Kayal et al., 2012) that included four mitogenomes of rhizostomes; these mitogenomes were sequenced using HTS platforms and eventually allowed further phylogenetic inferences based on a larger number of mitochondria protein amino acid sequences. Their analysis made it possible to infer the phylogenetic relationships of cnidarians, displaying the clustering of the semaeostome *Aurelia aurita* (Linnaeus, 1758) in one of the two clades that compose Rhizostomeae in the analysis performed by Kayal et al. (2013). From March 2017 to November 2023, nine mitogenomes of rhizostomes were published, which were accompanied by phenograms or phylogenetic trees to display the species placement within Medusozoa (Wang and Sun, 2017a,b; Lin et al., 2022; Ling et al., 2023; Xia et al., 2023). In total, 16 assembled rhizostome mitogenomes are deposited in NCBI's GenBank, and 13 have already been published as of November 2023 (Table 2).

Only one class-level study addresses -omics scale data of rhizostomes, specifically the transcriptomes of *S. meleagris* (possibly *N. nomurai*) and *C. xamachana* (Kayal et al., 2018). In that study, Kayal et al. (2018) showed a robust phylogeny of cnidarians (for studying the origins of organismal traits, such as endosymbiosis and coloniality), where Rhizostomeae was recovered as a monophyletic group.

As of August 2023, the number of sequenced transcriptomes and whole genomes of rhizostomes is scarce but increasing (Table 3). Its sequencing is propelling the development of more robust phylogenetic (phylogenomic) hypotheses constructed from hundreds of nucleotide characters that serve as basis for understanding the evolution of cnidarians and other early-branching metazoans.

Table 2 Assembled mitogenomes from Rhizostomeae species and different data (size of mitogenomes, integrity, genes sequenced, accession numbers in GenBank and reference) available in NCBI's GenBank.

Species (GenBank ID)	Size	Integrity	Protein-coding genes													rRNAs		tRNAs		Others				Accession	Reference
			COX1	COX2	ATP8	ATP6	COX3	NAD2	NAD5	NAD6	NAD3	NAD4L	NAD1	NAD4	Cytb	s-rRNA	l-rRNA	trna-Trp	trna-Met	Other tRNAs	Control region	ORFs	Polymerase		
Acromitus flagellatus	16,779	Complete																						OM457248	Lin et al. (2022)
Acromitus sp.	16,945	Partial																				2		OR400203	Ling et al. (2023)
Cassiopea andromeda	15,800	Partial																				2		JN700934	Kayal et al. (2012)
Cassiopea sp.	17,095	Complete																				2		OR400206	Ling et al. (2023)
Cassiopea xamachana	15,949	Complete																				2		JN700936	Kayal et al. (2012)
Catostylus mosaicus	14,847	Partial																				1		JN700940	Kayal et al. (2012)
Catostylus townsendi	16,308	Complete																				1		OK299144	Unpublished
Cotylorhiza tuberculata	16,590	Complete?	Non annotated																					OQ850984	Unpublished
Mastigias papua	16,560	Complete																				1		OQ695499	Xia et al. (2023)
Mastigias sp.	16,473	Complete																				2		OR400204	Ling et al. (2023)
Nemopilema nomurai	17,024	Complete																		4				KY454767	Wang and Sun (2017a)
Phyllorhiza punctata	16,967	Complete																				2		OR400201	Ling et al. (2023)
Phyllorhiza punctata	16,981	Complete																				2		OR400205	Ling et al. (2023)
Rhizostoma pulmo	5,648	Partial																						JN700988	Kayal et al. (2012)
Rhopilema esculentum	15,855	Complete																		3	1			KY454768	Wang and Sun (2017b)
Stomolophus sp.	17,006	Complete																						MK157198	Unpublished

Genes marked with light gray color represent annotated sequences. The numbers in the columns "Other tRNAs", "Control region" and "ORFs" indicate the number of additional tRNAs, putative control regions, and open reading frames.

3.3.3 Phylogenies and evolutionary research

Well-resolved molecular phylogenies have allowed disciplines such as comparative biology to bloom. Molecular phylogenies helped trace the phylogenetic distribution of life history features like the ability to aggregate, bloom, or swarm of medusozoans (Dawson and Hamner, 2009; Hamner and Dawson, 2009) and the origin and diversification of morphological, behavioral, and ecological traits in cnidarians (*e.g.*, the acquisition of endosymbionts and colonial forms, swimming patterns and canal system architectures) (Collins, 2002; Collins et al., 2006; Costello et al., 2008; Kayal et al., 2013, 2018; Djeghri et al., 2019; Von Montfort et al., 2023).

As medusozoans are early-diverging organisms in the animal phylogeny (Moon et al., 2023), comparative approaches can help understand homologies within cnidarians and deeper homologies within the animal tree of life. For example, the mining of genomes and transcriptomes has helped study pigment evolution among cnidarians (Lawley et al., 2021b), while new genomes, transcriptomes, and proteomes are shedding light on envenomation mechanisms and venom composition and evolution (Li et al., 2014; Choudhary et al., 2019; Wang et al., 2019; Li, R. et al., 2020; Klompen et al., 2021; Riyas et al., 2021). Additionally, the use of omic data from *C. xamachana*, *N. nomurai*,

Table 3 Rhizostomeae species for which whole genome, transcriptome and proteome data are available in NCBI's GenBank and PRIDE repositories.

Species	Data type	Accession	References
Cassiopea andromeda	Genome	PRJNA576371	Gacesa et al. (unpublished)
Cassiopea xamachana	Genome	PRJEB23739	Ohdera et al. (2019)
Nemopilema nomurai	Genome	PRJNA415234	Kim et al. (2019)
Rhopilema esculentum	Genome	PRJNA523480	Li, Y. et al. (2020)
	Genome	PRJNA505074	Nong et al. (2020)
Cassiopea andromeda	Transcriptome	PRJNA576371	Gacesa et al. (unpublished)
Cassiopea xamachana	Transcriptome	PRJEB21012	Kayal et al. (2018)
*Nemopilema nomurai**	Transcriptome	PRJNA498198	Khalturin et al. (2019)
	Transcriptome	PRJNA415234	Kim et al. (2019)
*Rhopilema esculentum**	Transcriptome	PRJNA318143	Li et al. (2017)
	Transcriptome	PRJNA394637	Ge et al. (2018)
	Transcriptome	PRJNA505074	Nong et al. (2020)
	Transcriptome	PRJNA512552	Li, Y. et al. (2020)
Stomolophus meleagris (possibly *N. nomurai*)	Transcriptome	PRJNA238144	Li et al. (2014)
Cassiopea xamachana	Proteome	PXD012177	Ames et al. (2020)
Lychnorhiza malayensis	Proteome	PXD022047	Riyas et al. (2021)
Nemopilema nomurai	Proteome	PXD011684	Wang et al. (2019)
	Proteome	PXD017680	Li, R. et al. (2020)

The asterisk (*) indicates that there are more projects associated with this species.

and *R. esculentum* provided valuable resources to gain insight into genome evolution (*e.g.*, gene loss and gain, synteny, small RNAs) and also for identifying features underlying body pattern formation (Khalturin et al., 2019; Kim et al., 2019; Ohdera et al., 2019; Li, Y. et al., 2020; Nong et al., 2020). Furthermore, transcriptomics and metabolomics helped study the metamorphosis of *R. esculentum*, which can be vital to understanding the evolution of metamorphosis in Metazoa (Ge et al., 2018, 2022; Li et al., 2017).

The classification of jellyfish has historically been challenging due to their morphological diversity and often uneven sampling effort of phenetic and molecular data (Gómez Daglio and Dawson, 2017). In this sense, robust inferences of phylogenetic relationships within Rhizostomeae are of paramount importance in the classification of these organisms (as for any other organisms), especially for those that are morphologically similar to each other but represent different evolutionary lineages (for example, cryptic taxa).

4. Classification of Rhizostomeae

The hierarchical classification of Rhizostomeae has been a matter of dispute since these animals were recognized as a distinct group of jellyfish. Rhizostome medusae were part of different classification schemes over time, and it is hard to track all those proposals, even if we check the original studies, due to inconsistencies in the nomenclature of certain taxa. However, Thiel provided a history of the classification and a summary of what was known until the 1970s (Thiel et al., 2024). Higher rank taxon names such as Medusae (class), Acalephae (phylum/class), Discophorae (order), Phanerocarpae (suborder), Umbrelliferae (order), Polystomae (suborder), and Acraspedae (subclass) they all could be considered like synonyms but with slight differences (and all embraced some rhizostome jellyfish as we know today).[1] Researchers like Cuvier (1800), Péron and Lesueur (1810), Eschscholtz (1829), Agassiz (1862), Haeckel (1880a, b), Vanhöffen (1888), Maas (1903), Mayer (1910), Stiasny (1920, 1921), Uchida (1926), and Kramp (1961) contributed with important information (organizing the diversity of jellyfish species) that were relevant to build the foundations of the classification system of Rhizostomeae taxa that we still

[1] The higher ranks indicated in brackets were considered as recognized nowadays.

use in present days. It is important to remember that Haeckel (1880b: XX; 1882: xii) presented the first hypothetical ancestral tree of the Acraspedae (comprising Scyphozoa, Cubozoa, and Staurozoa) and that the relationships between rhizostomes were depicted as "monophyletic". Stiasny (1921: 52) and Thiel (1970: 165, Fig. 21; 1976: 12, Fig. 9) also presented diagrams of relationships among rhizostomes, but they were not further explored and discussed in the literature.

As mentioned in Section 2 of this chapter, few morphological studies have tested the monophyletic nature of taxonomic categories erected without a phylogenetic basis. Thus, the classification of Rhizostomeae as we know today is derived mainly from the ideas of Vanhöffen (1888) and expanded by Stiasny (1921), Uchida (1926), and summarized by Kramp (1961)—the main idea was to correlate the evolution of the rhizostomes with the different patterns of the gastrovascular canal system within the different subgroups; evidently, rearrangements were proposed from one author to the other due to increasing knowledge about the development of rhizostome species. Despite the monophyletic origin of rhizostomes (also supported and defended in classical taxonomic publications, *e.g.*, Stiasny, 1921) and Thiel (1970, 1976) had distinct views and proposed that different groups of rhizostomes originated from different groups of semaeostomes. Furthermore, although morphological and molecular data support the monophyly of rhizostomes, Straehler-Pohl and Jarms (2022) recently resurrected the idea of a non-monophyletic Rhizostomeae.

An updated classification of Rhizostomeae is presented herein, with some nomenclatural corrections made by Calder (2009), also available in the World Register of Marine Species (WoRMS) (Collins and Morandini, 2024). This classification reflects findings from molecular phylogenies. However, the molecular data do not support the monophyly of the suborder Dactyliophorae Stiasny, 1920 and the families Catostylidae, Lobonemidae, and Lychnorhizidae, so the validity of those taxonomic categories needs to be assessed. Table 4 presents a comparative view of the classification presented here with Kramp (1961), Thiel (1970, 1976), and Straehler-Pohl and Jarms (2022). Thiel, in his unpublished manuscript (Thiel et al., 2024; electronic supplement), considered all rhizostomes under the Superorder Rhizostomoidea Thiel, 1976 with three orders (Cassiopeida, Cepheida, and Rhizostomida), but the lower ranks were similar (Table 4).

Table 4 Comparative classification of the order Rhizostomeae Cuvier, 1800. Names written in gray color are different from the present classification used in WoRMS (either hierarchical rank or spelling).

Collins and Morandini (2024) (WoRMS)	Kramp (1961)	Thiel (1970, 1976)	Straehler-Pohl and Jarms (2022)
			Infraclass Centrifugalia*
			Superorder Oppositida*
			Order Circulumcanalida*
Order Rhizostomeae	Order Rhizostomeae		Suborder Anterhizostomatida*
Suborder Kolpophorae	Suborder Kolpophorae	Order Cepheida	Infraorder Kolpophorae
	Infraorder Actinomyariae		Superfamily Cepheidea
Family Cepheidae	Family Cepheidae	Family Cepheidae	Family Cepheidae
	Infraorder Krykomyariae		
Family Leptobrachidae	Family Thysanostomatidae	Family Thysanostomatidae	Family Leptobrachidae
Family Mastigiidae	Family Mastigiidae	Family Mastigiadidae	Family Mastigiidae
Family Versurigidae	Family Versurigidae	Family Versurigidae	Family Versurigidae
	Infraorder Kampylomyariae	Order Cassiopeida	Superfamily Cassiopeidea
Family Cassiopeidae	Family Cassiopeidae	Family Cassiopeidae	Family Cassiopeidae

(continued)

Table 4 Comparative classification of the order Rhizostomeae Cuvier, 1800. Names written in gray color are different from the present classification used in WoRMS (either hierarchical rank or spelling). *(cont'd)*

Collins and Morandini (2024) (WoRMS)	Kramp (1961)	Thiel (1970, 1976)	Straehler-Pohl and Jarms (2022)
			Infraorder Centripedalida*
Suborder Dactyliophorae	Suborder Daktyliophorae	Order Rhizostomida	Superfamily Rhizostomatidea
	Superfamily Inscapulatae		
Family Catostylidae	Family Catostylidae	Family Catostylidae	Family Catostylidae
Family Lobonemidae	Family Lobonematidae	Family Lobonemidae	Family Lobonemidae
Family Lychnorhizidae	Family Lychnorhizidae	Family Lychnorhizidae	Family Lychnorhizidae
Superfamily Rhizostomatoidea	Superfamily Scapulatae		
Family Rhizostomatidae	Family Rhizostomatidae	Family Rhizostomatidae	Family Rhizostomatidae
Family Stomolophidae	Family Stomolophidae	Family Stomolophidae	Family Stomolophidae
Suborder Ptycophorae			
Family Bazingidae			

Check references and Collins and Morandini (2024) for authorities of the ranks. The symbol "*" means that only part of the group is represented (in the original paper, each of the groups marked includes other taxa, such as members of Semaeostomeae or Cubozoa); empty spaces mean that the group has no equivalent in the classification, or it was not known at the time of the classification proposal, or disregarded by the authors.

Order Rhizostomeae Cuvier, 1800
Suborder Kolpophorae Stiasny, 1920
Family Cepheidae L. Agassiz, 1862
Family Cassiopeidae L. Agassiz, 1862
Family Mastigiidae Stiasny, 1920
Family Leptobrachidae L. Agassiz, 1862
Family Versurigidae Kramp, 1961
Suborder Dactyliophorae Stiasny, 1920
Family Catostylidae Vanhöffen, 1888
Family Lobonemidae Stiasny, 1920
Family Lychnorhizidae Haeckel 1880a
Superfamily Rhizostomatoidea Cuvier, 1800
Family Rhizostomatidae Cuvier, 1800
Family Stomolophidae Haeckel, 1880
Suborder Ptycophorae Gershwin and Davie, 2013
Family Bazingidae Gershwin and Davie, 2013

Besides the classification above, a multigene phylogenetic hypothesis for Rhizostomeae, based on a concatenated molecular dataset (16S+COI+18S +28S), is presented herein (Fig. 5). This phylogeny is based primarily on

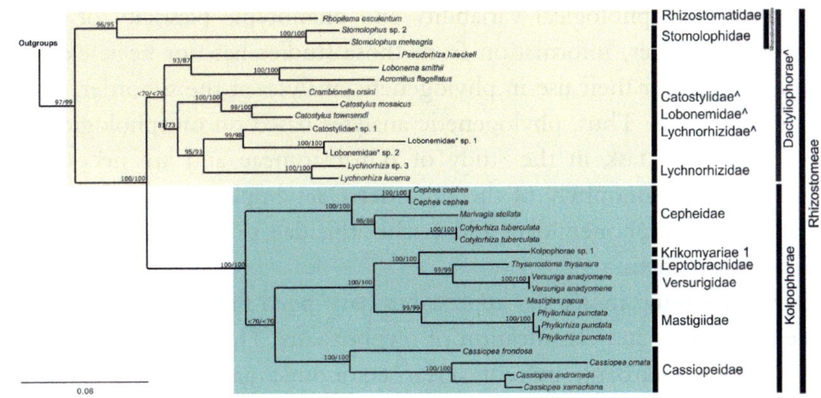

Fig. 5 Maximum likelihood tree (log likelihood = −33,332.26) based on a combined dataset (16S+COI+18S+28S) showing the phylogenetic relationships within the Rhizostomeae. The character caret (^) next to a taxon name indicates that it is not monophyletic. The asterisks (*) indicate undescribed taxa (according to Gómez Daglio and Dawson, 2017). Numbers at nodes are SH-aLRT support (%)/ultrafast bootstrap support (%). Details on the methodology used to construct this phylogenetic tree can be found in the supplementary material.

sequences available in GenBank (especially from Gómez Daglio and Dawson, 2017) but also includes new sequences generated for the suborder Kolpophorae (see supplementary material). Unlike Gómez Daglio and Dawson (2017), this phylogeny included data from the molecular marker COI. Although the mitochondrial COI marker may have limited resolution at deep nodes and its use as a single marker for phylogenetic inferences is questionable (Ortman et al., 2010; Bronstein et al., 2018), it provides robust and better resolved phylogenies when analyzed in conjunction with genes that have higher resolution power at deep nodes (*e.g.*, 18S and 28S). Similar to previous findings (*e.g.*, Bayha et al., 2010; Gómez Daglio and Dawson, 2017), the phylogenetic hypothesis presented here (based on a larger number of genes) displayed taxa within Dactyliophorae as non-monophyletic groups. In addition, relationships within the suborder Kolpophorae need to be confirmed, considering more molecular markers and increased species sampling.

5. Perspectives

Morphological studies have been essential for discovering, delimiting, describing, identifying, and classifying taxa and answering questions related to morphological variability and phenotypic plasticity of rhizostomes. However, information from those studies has not been extracted and codified for their use in phylogenetic analyses at the suborder, family, or genus levels. Thus, phylogenetic analyses based on morphological data are a pending task in the study of Rhizostomeae and are necessary to evaluate the monophyly of the suborder Dactyliophorae and the families Catostylidae, Lobonemidae, and Lychnorhizidae since molecular data do not support them.

Studies using molecular approaches have been pivotal to help advance the systematics and classification of scyphozoans. The genetic information in NCBI's GenBank presents clear taxonomic gaps in Rhizostomeae, highlighting the scarcity of molecular data for taxa such as Leptobrachidae, Versurigidae, and Bazingidae. Furthermore, additional molecular markers can help resolve some nodes that commonly sequenced markers cannot. Nowadays, sequencing of molecular data is more affordable; therefore, we suggest, instead of generating more sequences of the same species, defining candidate species for sequencing (considering the taxonomic gaps).

Having stablished the importance of morphological and molecular data to better comprehend the evolution, population genetics, and taxonomy of Rhizostomeae, the need to increase knowledge becomes evident. In this context, future advances will first depend on developing new tools and efficient strategies to obtain and analyze large amounts of integrated, new and existing, data to fill the gaps in the study of Rhizostomeae. However, the most significant contribution of a better understanding of Rhizostomeae in these disciplines will be to take advantage of our ability to describe and identify jellyfish species, understand their origin and the history of their populations, and protect them so that they continue to fulfill their ecological roles and allow the development of sustainable fisheries for commercially important species for human consumption.

Acknowledgments

The authors are grateful for the helpful comments made by the Editors, Allen G. Collins and Liza Gómez Daglio, which improved the quality of this chapter significantly. We also thank *Consejo Nacional de Humanidades, Ciencias y Tecnologías* (Conahcyt) for the post-doctoral fellowship granted to EGM (CVU 491802). SNS was supported by São Paulo Research Foundation (FAPESP) grant number 2022/16193-1 and National Council of Scientific and Technological Development (CNPq Research Productivity Scholarship) grant number 304267/2022-8. ACM received grants from FAPESP (2019/20042-6, 2023/12416-9) and CNPq (307832/2022-8); this is a NP-BioMar USP contribution.

Appendix A. Supporting information

Supplementary data associated with this article can be found in the online version at https://doi.org/10.1016/bs.amb.2024.09.001.

References

Agassiz, L., 1860. Contributions to the Natural History of the United States of America. Brown and Company, Boston.

Agassiz, L., 1862. Contributions to the natural history of the United States of America. IV. Second Monograph, In Five Parts, Acalephs in General, Ctenophorae, Discophorae, Hydroidae, Homologies of the Radiata. Little, Brown and Company, Boston.

Ames, C.L., Klompen, A.M., Badhiwala, K., Muffett, K., Reft, A.J., Kumar, M., et al., 2020. Cassiosomes are stinging-cell structures in the mucus of the upside-down jellyfish *Cassiopea xamachana*. Commun. Biol. 3, 67.

Ames, C.L., Ohdera, A.H., Colston, S.M., Collins, A.G., Fitt, W.K., Morandini, A.C., et al., 2021. Fieldable environmental DNA sequencing to assess jellyfish biodiversity in nearshore waters of the Florida Keys, United States. Front. Mar. Sci. 8, 640527.

Anthony, C.J., Heagy, M., Bentlage, B., 2022. Phenotypic plasticity in *Cassiopea ornata* (Cnidaria: Scyphozoa: Rhizostomeae) suggests environmentally driven morphology. Zoomorphology 141, 115–131.

Armani, A., Tinacci, L., Giusti, A., Castigliego, L., Gianfaldoni, D., Guidi, A., 2013. What is inside the jar? Forensically informative nucleotide sequencing (FINS) of a short mitochondrial COI gene fragment reveals a high percentage of mislabeling in jellyfish food products. Food Res. Int. 54, 1383–1393.

Armani, A., Giusti, A., Castigliego, L., Rossi, A., Tinacci, L., Gianfaldoni, D., et al., 2014. Pentaplex PCR as screening assay for jellyfish species identification in food products. J. Agric. Food Chem. 62, 12134–12143.

Aurelle, D., Pratlong, M., Oury, N., Haguenauer, A., Gélin, P., Magalon, H., et al., 2022. Species and population genomic differentiation in *Pocillopora corals* (Cnidaria, Hexacorallia). Genetica 150, 247–262.

Avian, M., Ramšak, A., 2021. Phylum Cnidaria: Classes Scyphozoa, Cubozoa, and Staurozoa. In: Schierwater, B., DeSalle, R. (Eds.), Invertebrate Zoology. CRC Press Taylor & Francis Group, New York, pp. 149–172.

Ax, P., 1995. Das System der Metazoa I: Ein Lehrbuch der phylogenetischen Systematik. Gustav Fischer Verlag, Stuttgart, Jena, New York.

Ax, P., 1996. Multicellular Animals: A New Approach to the Phylogenetic Order in Nature, first ed. Springer, Berlin, pp. 1–213.

Bayha, K.M., Graham, W.M., 2009. A new Taqman© PCR-based method for the detection and identification of scyphozoan jellyfish polyps. Hydrobiologia 616, 217–228.

Bayha, K.M., Dawson, M.N., Collins, A.G., Barbeitos, M.S., Haddock, S.H., 2010. Evolutionary relationships among scyphozoan jellyfish families based on complete taxon sampling and phylogenetic analyses of 18S and 28S ribosomal DNA. ICB 50, 436–455.

Bayha, K.M., Graham, W.M., 2011. First confirmed reports of the rhizostome jellyfish *Mastigias* (Cnidaria: Rhizostomeae) in the Atlantic basin. Aquat. Invasions 6, 361–366.

Bigelow, R.P., 1892. On a new species of Cassiopea from Jamaica. Zool. Anz. 15, 212–214.

Bridge, D., Cunningham, C.W., Schierwater, B., DeSalle, R., Buss, L.W., 1992. Class-level relationships in the phylum Cnidaria: evidence from mitochondrial genome structure. PNAS 89, 8750–8753.

Bridge, D., Cunningham, C.W., DeSalle, R., Buss, L.W., 1995. Class-level relationships in the phylum Cnidaria: molecular and morphological evidence. Mol. Biol. Evol. 12, 679–689.

Bronstein, O., Kroh, A., Haring, E., 2018. Mind the gap! The mitochondrial control region and its power as a phylogenetic marker in echinoids. BMC Evol. Biol. 18, 1–15.

Brotz, L., Angel, D.L., D'Ambra, I., Enrique-Navarro, A., Lauritano, C., Thibault, D., Prieto, L., 2024. Rhizostomes as a resource: The expanding exploitation of jellyfish by humans. Adv. Mar. Biol. 98.

Brotz, L., Schiariti, A., López-Martínez, J., Álvarez-Tello, J., Peggy Hsieh, Y.H., Jones, R.P., et al., 2017. Jellyfish fisheries in the Americas: origin, state of the art, and perspectives on new fishing grounds. Rev. Fish. Biol. Fish. 27, 1–29.

Brusca, R.C., Brusca, G.J., 1990. Invertebrates, first ed. Sinauer Associates, Massachusetts.

Cabra-García, J., Hormiga, G., 2020. Exploring the impact of morphology, multiple sequence alignment and choice of optimality criteria in phylogenetic inference: a case study with the Neotropical orb-weaving spider genus *Wagneriana* (Araneae: Araneidae). Zool. J. Linn. Soc. 188, 976–1151.

Calder, D.R., 2009. Cubozoan and Scyphozoan jellyfishes of the Carolinian biogeographic province, southeastern USA. ROM. Contributions Sci. 3, 1–58.

Casillas, S., Barbadilla, A., 2017. Molecular population genetics. Genetics 205, 1003–1035.

Cheng, C., Fei, Z., Xiao, P., 2023. Methods to improve the accuracy of next-generation sequencing. Front. Bioeng. Biotechnol. 11, 982111.

Chin, T.C., Adibah, A.B., Hariz, Z.D., Azizah, M.S., 2016. Detection of mislabelled seafood products in Malaysia by DNA barcoding: improving transparency in food market. Food Control. 64, 247–256.

Choudhary, I., Hwang, D.H., Lee, H., Yoon, W.D., Chae, J., Han, C.H., et al., 2019. Proteomic analysis of novel components of *Nemopilema nomurai* jellyfish venom: deciphering the mode of action. Toxins 11, 153.

Chun, C., 1896. Beiträge zur Kenntniss Ost-Afrikanischer Medusen und Siphonophoren. Mitt. Naturh. Mus. Hamb 13, 1–19.

Claus, C., 1883. Untersuchungen über die Organisation und Entwicklung der Medusen. Leipzig 1–96.

Collins, A.G., 2002. Phylogeny of Medusozoa and the evolution of cnidarian life cycles. J. Evol. Biol. 15, 418–432.

Collins, A.G., Morandini, A.C., 2024. World List of Scyphozoa. Rhizostomeae. Accessed through: World Register of Marine Species at: https://www.marinespecies.org/aphia.php?p=taxdetails&id=135224 on 2024-01-26.

Collins, A.G., Schuchert, P., Marques, A.C., Jankowski, T., Medina, M., Schierwater, B., 2006. Medusozoan phylogeny and character evolution clarified by new large and small subunit rDNA data and an assessment of the utility of phylogenetic mixture models. Syst. Biol. 55, 97–115.

Costello, J.H., Colin, S.P., Dabiri, J.O., 2008. Medusan morphospace: phylogenetic constraints, biomechanical solutions, and ecological consequences. Inv. Biol. 127, 265–290.

Cuvier, G., 1800. Mémoire sur l'organisation de quelques méduses. Bulletin des Sciences, Société Philomathique. Paris. Ser. I, Tome 2, 69.

Dagha, R.C., Kumari, K.A., Pawan-Kumar, A., Shah, N., 2022. Morphological and molecular documentation of Jellyfish *Rhopilema hispidum* (Vanhöffen, 1888) (Scyphozoa: Rhizostomatidae) from Mumbai coast, India. IJMS 50, 246–249.

Dawson, M.N., 2004. Some implications of molecular phylogenetics for understanding biodiversity in jellyfishes, with emphasis on Scyphozoa. Hydrobiologia 530, 249–260.

Dawson, M.N., 2005a. Morphological variation and systematics in the Scyphozoa: *Mastigias* (Rhizostomeae, Mastigiidae)–a golden unstandard? Hydrobiologia 537, 185–206.

Dawson, M.N., 2005b. Incipient speciation of *Catostylus mosaicus* (Scyphozoa, Rhizostomeae, Catostylidae), comparative phylogeography and biogeography in south-east Australia. J. Biogeogr. 32, 515–533.

Dawson, M.N., 2005c. Five new subspecies of *Mastigias* (Scyphozoa: Rhizostomeae: Mastigiidae) from marine lakes, Palau, Micronesia. J. Mar. Biol. Assoc. 85, 679–694.

Dawson, M.N., 2005d. Morphologic and molecular redescription of *Catostylus mosaicus conservativus* (Scyphozoa: Rhizostomeae: Catostylidae) from south-east Australia. J. Mar. Biol. Assoc. 85, 723–731.

Dawson, M.N., Hamner, W.M., 2005. Rapid evolutionary radiation of marine zooplankton in peripheral environments. PNAS 102, 9235–9240.

Dawson, M.N., Hamner, W.M., 2009. A character-based analysis of the evolution of jellyfish blooms: adaptation and exaptation. Hydrobiologia 616, 193–215.

Dawson, M.N., Cieciel, K., Decker, M.B., Hays, G.C., Lucas, C.H., Pitt, K.A., 2015. Population-level perspectives on global change: genetic and demographic analyses indicate various scales, timing, and causes of scyphozoan jellyfish blooms. Biol. Inv. 17, 851–867.

De Souza, M.R., Dawson, M.N., 2018. Redescription of *Mastigias papua* (Scyphozoa, Rhizostomeae) with designation of a neotype and recognition of two additional species. Zootaxa 4457, 520–536.

Djeghri, N., Pondaven, P., Stibor, H., Dawson, M.N., 2019. Review of the diversity, traits, and ecology of zooxanthellate jellyfishes. Mar. Biol. 166, 1–19.

Dong, Z., Liu, Z., Liu, D., Liu, Q., Sun, T., 2016. Low genetic diversity and lack of genetic structure in the giant jellyfish *Nemopilema nomurai* in Chinese coastal waters. Mar. Biol. Res. 12, 769–775.

Dong, Z., Morandini, A.C., Schiariti, A., Wang, L., Sun, T., 2019. First record of *Phyllorhiza* sp. (Cnidaria: Scyphozoa) in a Chinese coastal aquaculture pond. PeerJ 7, e6191.

Douek, J., Harbuzov, Z., Galil, B.S., Rinkevich, B., 2020. Developing novel microsatellite markers by NGS technology for *Rhopilema nomadica*, an invasive jellyfish. Mol. Biol. Rep. 47, 4821–4825.

Douek, J., Giallongo, G., Harbuzov, Z., Galil, B.S., Rinkevich, B., 2024. Long-term population genetic features of the *Rhopilema nomadica* jellyfish from the Israeli Mediterranean Coasts. J. Mar. Sci. Eng. 12, 171.

Ehrenberg, C.G., 1834. Beiträge zur physiologischen Kenntniss der Corallenthiere im allgemeinen, und besonders des rothen Meeres, nebst einem Versuche zur physiologischen Systematik derselben. Abh. König. Akad. Wiss. Berl. 1, 225–380.

Eschscholtz, F., 1829. System der Acalephen. Eine ausführliche Beschreibung aller Medusenartigen Strahlthiere. Ferdinand Dümmler, Berlin.

Faleh, A.R.B., Allaya, H., Armani, A., Shahin, A.A.B., 2017. Significant genetic differentiation among meroplanktonic barrel jellyfish *Rhizostoma pulmo* (Cnidaria: Scyphozoa) in the Mediterranean Sea. Afr. J. Mar. Sci. 39, 1–8.

Fedosov, A., Achaz, G., Gontchar, A., Puillandre, N., 2022. MOLD, a novel software to compile accurate and reliable DNA diagnoses for taxonomic descriptions. Mol. Ecol. Resour. 22, 2038–2053.

Forskål, P. 1775. Descriptiones animalium avium, amphibiorum, piscium, insectorum, vermium; quae in intinere orientali observavit Petrus Forskål, Hauniae.

Galil, B.S., Gershwin, L.A., Douek, J., Rinkevich, B., 2010. *Marivagia stellata* gen. et sp. nov. (Scyphozoa: Rhizostomeae: Cepheidae), another alien jellyfish from the Mediterranean coast of Israel. Aquat. Invasions 5, 331–340.

Galil, B.S., Gershwin, L.A., Zorea, M., Rahav, A., Rothman, S.B., Fine, M., et al., 2017. *Cotylorhiza erythraea* Stiasny, 1920 (Scyphozoa: Rhizostomeae: Cepheidae), yet another erythraean jellyfish from the Mediterranean coast of Israel. Mar. Biodivers. 47, 229–235.

Galil, B.S., Spanier, E., Ferguson, W.W., 1990. The Scyphomedusae of the Mediterranean coast of Israel, including two Lessepsian migrants new to the Mediterranean. Zool. Meded. 64, 95–105.

Galimberti, A., De Mattia, F., Losa, A., Bruni, I., Federici, S., Casiraghi, M., et al., 2013. DNA barcoding as a new tool for food traceability. Food Res. Int. 50, 55–63.

Gamero-Mora, E., Halbauer, R., Bartsch, V., Stampar, S.N., Morandini, A.C., 2019. Regenerative capacity of the upside-down jellyfish *Cassiopea xamachana*. Zool. Stud. 58, e37.

Gamero-Mora, E., Collins, A.G., Boco, S.R., Geson, S.M., Morandini, A.C., 2022. Revealing hidden diversity among upside-down jellyfishes (Cnidaria: Scyphozoa: Rhizostomeae: *Cassiopea*): distinct evidence allows the change of status of a neglected variety and the description of a new species. Invert. Syst. 36, 63–89.

Ge, J., Liu, C., Tan, J., Bian, L., Chen, S., 2018. Transcriptome analysis of scyphozoan jellyfish *Rhopilema esculentum* from polyp to medusa identifies potential genes regulating strobilation. Dev. Genes. Evol. 228, 243–254.

Ge, J., Chen, X., Liu, C., Tan, J., Bian, L., Chen, L., et al., 2022. Metabolomics provide insights into the endogenous mechanism of strobilation in the scyphozoan jellyfish *Rhopilema esculentum*. J. Oceanol. Limnol. 40, 226–234.

Gershwin, L., Davie, P.J.F., 2013. A remarkable new jellyfish (Cnidaria: Scyphozoa) from coastal Australia, representing a new suborder within the Rhizostomeae. Mem. Queensl. Mus. 56, 625–630.

Getino Mamet, L.N., Valdivia-Carrillo, T., Gómez Daglio, L., García-De León, F.J., 2017. Isolation and characterization of 14 tetranucleotide microsatellite loci for the cannonball jellyfish (*Stomolophus* sp.) by next generation sequencing. Mol. Biol. Rep. 44, 257–260.

Getino Mamet, L.N., Gómez Daglio, L., García-De León, F.J., 2019. High genetic differentiation in the edible cannonball jellyfish (cnidaria: Scyphozoa: *Stomolophus* spp.) from the Gulf of California, Mexico. Fish. Res. 219, 105328.

Giallongo, G., Douek, J., Harbuzov, Z., Galil, B.S., Rinkevich, B., 2021. Long-term changes in population genetic features of a rapidly expanding marine invader: implication for invasion success. Biol. Invasions 23, 2541–2552.

Glynn, F., Houghton, J.D., Provan, J., 2015. Population genetic analyses reveal distinct geographical blooms of the jellyfish *Rhizostoma octopus* (Scyphozoa). Biol. J. Linn. Soc. 116, 582–592.

Gmelin, J.F., 1791. Vermes. In Gmelin, J.F. (Ed.), Caroli a Linnaei Systema Naturae per Regna Tria Naturae, Ed. 13. Tome 1(6). G.E. Beer, Lipsiae [Leipzig], pp. 3021–3910.

Goette, A., 1887. Entwickelungsgeschichte der *Aurelia aurita* und *Cotylorhiza tuberculata*. Abh. Entwick. Tiere 4, 1–79.

Gómez Daglio, L., Dawson, M.N., 2017. Species richness of jellyfishes (Scyphozoa: Discomedusae) in the Tropical Eastern Pacific: missed taxa, molecules, and morphology match in a biodiversity hotspot. Invert. Syst. 31, 635–663.

Gómez Daglio, L.G., Dawson, M.N., 2019. Integrative taxonomy: ghosts of past, present and future. J. Mar. Biol. Assoc. 99, 1237–1246.

Haeckel, E., 1880a. Das System der Medusen. I, 2: System der Acraspeden. Gustav Fischer, Jena. 361–672.

Haeckel, E., 1880b. Ueber die Organisation und Classification der Discomedusen. Sitzber. Jenaisch. Gesellsch. Med. Naturwiss. Jena. 1880, 1–4.

Haeckel, E. 1882. Report on the deep-sea medusae dredged by H.M.S. Challenger during the years of 1873–1876. Report on the scientific results of the voyage of H.M.S. Challenger during the years 1873–1876. Zoology, IV, i-cv, 1–154.

Hamner, W.M., Dawson, M.N., 2009. A review and synthesis on the systematics and evolution of jellyfish blooms: advantageous aggregations and adaptive assemblages. Hydrobiologia 616, 161–191.

Hariharan, G., Prasannath, K., 2021. Recent advances in molecular diagnostics of fungal plant pathogens: a mini review. Front. Cell. Infect. Microbiol. 10, 600234.

Hennig, W., 1965. Phylogenetic systematics. Annu. Rev. Entomol. 10 (1), 97–116.

Hennig, W., 1966. Phylogenetic Systematics. University of Illinois Press, Chicago.

Holland, B.S., Dawson, M.N., Crow, G.L., Hofmann, D.K., 2004. Global phylogeography of *Cassiopea* (Scyphozoa: Rhizostomeae): molecular evidence for cryptic species and multiple invasions of the Hawaiian Islands. Mar. Biol. 145, 1119–1128.

Iida, A., Nohara, K., Yusoff, F.M., Srinui, K., Ha, T.M., Ohtsuka, S.E.B., et al., 2021. Genetic diversities of commercially harvested jellyfish, *Rhopilema hispidum* and *Lobonemoides robustus* in Southeast Asia. Plankton Benthos Res. 16 (4), 308–317.

Jarms, G., Morandini, A.C., 2019. World Atlas of Jellyfish. Dölling und Galitz Verlag, Hamburg.

Jiang, X.Y., Wang, F.L., Zhao, C.C., Yan, J.N., Han, J.R., Shang, W.H., et al., 2019. Identification of two jellyfish species (*Rhopilema esculentum* kishinouye and *Stomolophus meleagris*) in Liaoning Province of China by a rapid, simple PCR-RFLP method. Food Control. 105, 52–57.

Kayal, E., Bentlage, B., Collins, A.G., Kayal, M., Pirro, S., Lavrov, D.V., 2012. Evolution of linear mitochondrial genomes in medusozoan cnidarians. GBE 4, 1–12.

Kayal, E., Roure, B., Philippe, H., Collins, A.G., Lavrov, D.V., 2013. Cnidarian phylogenetic relationships as revealed by mitogenomics. BMC Evol. Biol. 13, 5.

Kayal, E., Bentlage, B., Sabrina Pankey, M., Ohdera, A.H., Medina, M., Plachetzki, D.C., et al., 2018. Phylogenomics provides a robust topology of the major cnidarian lineages and insights on the origins of key organismal traits. BMC Evol. Biol. 18, 68.

Khalturin, K., Shinzato, C., Khalturina, M., Hamada, M., Fujie, M., Koyanagi, R., et al., 2019. Medusozoan genomes inform the evolution of the jellyfish body plan. Nat. Ecol. Evol. 3, 811–822.

Kim, H.M., Weber, J.A., Lee, N., Park, S.G., Cho, Y.S., Bhak, Y., et al., 2019. The genome of the giant Nomura's jellyfish sheds light on the early evolution of active predation. BMC Biol. 17, 1–12.

Kishinouye, K., 1891. Bizen kurage. Zool. Mag. (Dobutsugaku zasshi) 3, 53.

Kishinouye, K., 1922. Echizen kurage. Zool. Mag. (Dobutsugaku zasshi) 34, 343–346.

Klompen, A.M., Kayal, E., Collins, A.G., Cartwright, P., 2021. Phylogenetic and selection analysis of an expanded family of putatively pore-forming jellyfish toxins (Cnidaria: Medusozoa). GBE 13, evab081.

Kramp, P.L., 1961. Synopsis of the medusae of the world. J. Mar. Biol. Assoc. 40, 7–382.

Kreitman, M., 1983. Nucleotide polymorphism at the alcohol dehydrogenase locus of *Drosophila melanogaster*. Nature 304, 412–417.

Kühn, A.L., Haase, M., 2020. QUIDDICH: quick identification of diagnostic characters. J. Zool. Syst. Evol. Res. 58, 22–26.

Lamarck, J.B., 1816. Histoire naturelle des animaux sans vertèbres. Tome Second. Verdiére, Paris. 1–568.

Langlois, V.S., Allison, M.J., Bergman, L.C., To, T.A., Helbing, C.C., 2021. The need for robust qPCR-based eDNA detection assays in environmental monitoring and species inventories. Environ. DNA 3, 519–527.

Lawley, J.W., Gamero-Mora, E., Maronna, M.M., Chiaverano, L.M., Stampar, S.N., Hopcroft, R.R., et al., 2021a. The importance of molecular characters when morphological variability hinders diagnosability: systematics of the moon jellyfish genus *Aurelia* (Cnidaria: Scyphozoa). PeerJ 9, e11954.

Lawley, J.W., Carroll, A.R., McDougall, C., 2021b. Rhizostomins: a novel pigment family from Rhizostome jellyfish (Cnidaria, Scyphozoa). Front. Mar. Sci. 8, 752949.

Lawley, J.W., Gamero-Mora, E., Maronna, M.M., Chiaverano, L.M., Stampar, S.N., Hopcroft, R.R., et al., 2022. Morphology is not always useful for diagnosis, and that's ok: Species hypotheses should not be bound to a class of data. Reply to Brown and Gibbons (S Afr J Sci. 2022; 118 (9/10), Art.# 12590). S. Afr. J. Anim. Sci. 118, 1–3.

Lee, P.L., Dawson, M.N., Neill, S.P., Robins, P.E., Houghton, J.D., Doyle, T.K., et al., 2013. Identification of genetically and oceanographically distinct blooms of jellyfish. J. R. Soc. Interface 10, 20120920.

Lee, M.S., Palci, A., 2015. Morphological phylogenetics in the genomic age. Curr. Biol. 25, R922–R929.

Lendenfeld, R.v., 1884. The scyphomedusae of the southern hemisphere. Part III. Proc. Linn. Soc. 9, 259–306.

Lesson, R.P., 1830. Voyage autour du monde, exécut par ordre du Roi, sur la corvette de Sa Magest La Coquille, pendant les années 1822–25. Zoologie, 2. Arthus Bertrand, Paris.

Lewontin, R.C., Hubby, J.L., 1966. A molecular approach to the study of genic heterozygosity in natural populations. II. Amount of variation and degree of heterozygosity in natural populations of *Drosophila pseudoobscura*. Genetics 54, 595.

Li, R., Yu, H., Xue, W., Yue, Y., Liu, S., Xing, R., et al., 2014. Jellyfish venomics and venom gland transcriptomics analysis of *Stomolophus meleagris* to reveal the toxins associated with sting. J. Proteom. 106, 17–29.

Li, R., Yu, H., Li, T., Li, P., 2020. Comprehensive proteome reveals the key lethal toxins in the venom of jellyfish *Nemopilema nomurai*. J. Proteome Res. 19, 2491–2500.

Li, Y.L., Dong, J., Wang, B., Li, Y.P., Yu, X.G., Fu, J., et al., 2016. Genetic characterization of different populations of *Rhopilema esculentum* based on the mitochondrial COI. sequence. J. Appl. Ecol. 27, 2340–2347.

Li, Y., Zhou, Z., Tian, M., Tian, Y., Dong, Y., Li, S., et al., 2017. Exploring single nucleotide polymorphism (SNP), microsatellite (SSR) and differentially expressed genes in the jellyfish (*Rhopilema esculentum*) by transcriptome sequencing. Mar. Genom. 34, 31–37.

Li, Y., Gao, L., Pan, Y., Tian, M., Li, Y., He, C., et al., 2020. Chromosome-level reference genome of the jellyfish *Rhopilema esculentum*. GigaScience 9, giaa036.

Li, Y., Bao, X., Li, Y., Zhou, Z., Fu, J., Gao, X., et al., 2022. Research on jellyfish diversity of marine aquaculture pond in Liaodong Bay using environmental DNA metabarcoding technique. Acta Ecol. Sin. 42, 5303–5313.

Lin, J., Feng, S., Wang, L., Qiu, Y., 2022. Complete mitochondrial genome sequence of *Acromitus flagellatus* and its phylogenetic relationship with related jellyfish species. Mitochondrial DNA B: Resour. 7, 1823–1824.

Linnaeus, C.v., 1758. Systema Naturae per regna tria naturae secundum classes, ordines, genera, species, cum characteibus, differentiis, synonymis, locis (tenth ed. Vol. Tomus I, Editio Decima, reformata). Stockholm: L. Salvii.

Ling, M.K., Yap, N.W.L., Iesa, I.B., Yip, Z.T., Huang, D., Quek, Z.B.R., 2023. Revisiting mitogenome evolution in Medusozoa with eight new mitochondrial genomes. Iscience 26, 108252.

Maas, O., 1903. Die Scyphomedusen der Siboga-Expedition. Siboga-Expeditie 11, 1–91.

Macri, S., 1778. Nuove Osservazioni intorno la storia naturale del Polmone Marino degli Antichi. Naples.

Maggio, T., Allegra, A., Bosch-Belmar, M., Cillari, T., Cuttitta, A., Falautano, M., et al., 2019. Molecular identity of the non-indigenous *Cassiopea* sp. from Palermo Harbour (central Mediterranean Sea). J. Mar. Biol. Assoc. 99, 1765–1773.

Maronna, M.M., Lawley, J.W., 2022. Current knowledge on genetic diversity of jellyfishes. In: Nandan, S.B., Thomas, L.C., Aneesh, B., Praved, P.H. (Eds.), Advances in DNA Barcoding and Molecular Taxonomy—Training Manual of International Workshop on DNA Barcoding and Molecular Taxonomy. Directorate of Public Relations and Publications, India, pp. 36–50.

Marques, A.C., Collins, A.G., 2004. Cladistic analysis of Medusozoa and cnidarian evolution. Inv. Biol. 123, 23–42.

Mayer, A.G., 1910. The medusae of the world. Vol. III. The Scyphomedusae. Carn. Inst. Wash. Publ. 109 III, 499–735.

Mellas, R.E., McIlroy, S.E., Fitt, W.K., Coffroth, M.A., 2014. Variation in symbiont uptake in the early ontogeny of the upside-down jellyfish, *Cassiopea* spp. J. Exp. Mar. Biol. Ecol. 459, 38–44.

Moon, J., Caron, J.B., Moysiuk, J., 2023. A macroscopic free-swimming medusa from the middle Cambrian Burgess Shale. Proc. R. Soc. B. 290, 20222490.

Morandini, A.C., 2024. Morphology of Rhizostomeae jellyfishes: what is known and what we advanced since the 1970s. Adv. Mar. Biol. 98.

Morandini, A.C., Stampar, S.N., Maronna, M.M., Silveira, F.L., 2017. All non-indigenous species were introduced recently? The case study of *Cassiopea* (Cnidaria: Scyphozoa) in Brazilian waters. J. Mar. Biol. Ass UK 97, 321–328.

Muffett, K., Miglietta, M.P., 2023. Demystifying *Cassiopea* species identity in the Florida Keys: *Cassiopea xamachana* and *Cassiopea andromeda* coexist in shallow waters. PLoS one 18, e0283441.

Mullis, K.B., Faloona, F.A., 1987. [21] Specific synthesis of DNA *in vitro* via a polymerase-catalyzed chain reaction. Methods Enzymol. 155, 335–350.

Neethling, S., Channing, A., Gershwin, L.A., Gibbons, M.J., 2011. A modern description of *Crambionella stuhlmanni* (Scyphozoa: Rhizostomeae) from St Lucia Estuary, South Africa. J. Exp. Mar. Biol. Ecol. 91, 357–367.

Nevárez-López, C., Hernández-Saavedra, N., Sánchez-Paz, A., Rojas-Posadas, D., Muhlia-Almazán, A., López-Martínez, J., 2020. Colour polymorphism and genetic structure in the cannonball jellyfish (*Stomolophus meleagris*, L. Agassiz, 1860) in the Gulf of California. Mar. Biol. Res. 16, 714–728.

Nishikawa, J., Ohtsuka, S., Mujiono, N., Lindsay, D.J., Miyamoto, H., Nishida, S., 2015. A new species of the commercially harvested jellyfish *Crambionella* (Scyphozoa) from central Java, Indonesia with remarks on the fisheries. J. Exp. Mar. Biol. Ecol. 95, 471–481.

Nong, W., Cao, J., Li, Y., Qu, Z., Sun, J., Swale, T., et al., 2020. Jellyfish genomes reveal distinct homeobox gene clusters and conservation of small RNA processing. Nat. Commun. 11, 3051.

Ohdera, A., Ames, C.L., Dikow, R.B., Kayal, E., Chiodin, M., Busby, B., et al., 2019. Box, stalked, and upside-down? Draft genomes from diverse jellyfish (Cnidaria, Acraspeda) lineages: *Alatina alata* (Cubozoa), *Calvadosia cruxmelitensis* (Staurozoa), and *Cassiopea xamachana* (Scyphozoa). GigaScience 8, giz069.

Omori, M., Nakano, E., 2001. Jellyfish fisheries in southeast Asia. Hydrobiologia 451, 19–26.

Ortman, B.D., Bucklin, A., Pages, F., Youngbluth, M., 2010. DNA barcoding the Medusozoa using mtCOI. Deep. Sea Res. Part. II Top. Stud. Oceanogr. 57, 2148–2156.

Owen, R., 1843. Lectures on the Comparative Anatomy and Physiology of the Invertebrate Animals. Longman, Brown, Green, and Longmans, London, pp. 1–392.

Pallas, P.S., 1774. Spicilegia Zoologica. Quibus novae imprimis et obscurae animalium species iconibus, descriptionibus atque comentariis illustrantur cura P.S. Pallas, fasciculus decimus. Berlin.

Parracho, T., Morais, Z., 2015. *Catostylus tagi*: partial rDNA sequencing and characterisation of nematocyte structures using two improvements in jellyfish sample preparation. J. Venom. Anim. Toxins Incl. Trop. Dis. 21, 40.

Péron, F., Lesueur, C.A., 1810. Tableau des caractères génériques et spécifiques de toutes les espèces de Méduses connues jusqu'à ce jour. Ann. Mus. Natl. Hist. Nat. Paris 14, 325–366.

Petersen, K.W., 1979. Development of coloniality in Hydrozoa. In: Larwood, G., Rosen, B.R. (Eds.), Biology and Systematics of Colonial Organisms. Academic Press, London, pp. 105–139.

Picciani, N., Kerlin, J.R., Sierra, N., Swafford, A.J., Ramirez, M.D., Roberts, N.G., et al., 2018. Prolific origination of eyes in Cnidaria with co-option of non-visual opsins. Curr. Biol. 28, 2413–2419.

Pourjomeh, F., Shokri, M.R., Rajabi-Maham, H., Rezai, H., Maghsoudlou, E., 2018. New records of the scyphozoan medusae (Cnidaria: Scyphozoa) in the north of Gulf of Oman, Iran. Mar. Biodivers. 48, 2193–2202.

Qiao, H., Liu, X., Zhang, X., Jiang, H., Wang, J., Zhang, L., 2013. Genetic diversity of populations and clones of *Rhopilema esculentum* in China based on AFLP analysis. Chin. J. Oceanol. Limnol. 31, 391–397.

Quoy, J.R.C., Gaimard, J.P., 1824. Voyage autour du monde, entrepis par ordre du Roi, sous les ministère et conformément aux instructions de S. Exc. M. Le Vicomte du Bouchage, secrétaire d`état au départment de la Marine, exécuté sur les corvettes de S. M. l'Uranie et la Physicienne, pendant les années 1817, 1818, 1819 et 1820 (Vol. Zoologie). Paris: Chez Pillet Aîné.

Ramšak, A., Stopar, K., Malej, A., 2012. Comparative phylogeography of meroplanktonic species, *Aurelia* spp. and *Rhizostoma pulmo* (Cnidaria: Scyphozoa) in European Seas. Hydrobiologia 690, 69–80.

Raposo, A., Alasqah, I., Alfheeaid, H.A., Alsharari, Z.D., Alturki, H.A., Raheem, D., 2022. Jellyfish as food: a narrative review. Foods 11, 2773.

Riyas, A., Kumar, A., Chandran, M., Jaleel, A., Kumar, A.B., 2021. The venom proteome of three common scyphozoan jellyfishes (*Chrysaora caliparea*, *Cyanea nozakii* and

Lychnorhiza malayensis) (Cnidaria: Scyphozoa) from the coastal waters of India. Toxicon 195, 93–103.

Rizman-Idid, M., Farrah-Azwa, A.B., Chong, V.C., 2016. Preliminary taxonomic survey and molecular documentation of jellyfish species (Cnidaria: Scyphozoa and Cubozoa) in Malaysia. Zool. Stud. 55, e35.

Rosales-Catalan, L., Estrada-González, M.C., Rivera-Pérez, C., Sanchez, M.A.R., Gamero-Mora, E., Morandini, A.C., et al., 2021. Genetic and morphological evidence of the presence of *Phyllorhiza punctata* in the southwestern Gulf of California (NE Pacific Ocean). Aquat. Invasions 16, 637–652.

Salvini-Plawen, L.V., 1978. On the origin and evolution of the lower Metazoa. J. Zoolog. Syst. Evol. Res. 16, 40–87.

Sanger, F., Nicklen, S., Coulson, A.R., 1977. DNA sequencing with chain-terminating inhibitors. PNAS 74, 5463–5467.

Santander, M.D., Maronna, M.M., Ryan, J.F., Andrade, S.C., 2022. The state of Medusozoa genomics: current evidence and future challenges. GigaScience 11, giac036.

Schiariti, A., Holst, S., Tiseo, G.R., Miyake, H., Morandini, A.C., 2024. Life cycles and reproduction of Rhizostomeae. Adv. Mar. Biol. 98.

Schuchert, P., 1993. Phylogenetic analysis of the Cnidaria. J. Zoolog. Syst. Evol. Res. 31, 161–173.

Sheeja, T.E., Kumar, I.P.V., Giridhari, A., Minoo, D., Rajesh, M.K., Babu, K.N., 2021. Amplified fragment length polymorphism: applications and recent developments. In: Besse, P. (Ed.), Molecular Plant Taxonomy. Methods in Molecular Biology 2222. Humana, New York, pp. 187–218.

Stampar, S.N., Gamero-Mora, E., Maronna, M.M., Fritscher, J.M., Oliveira, B.S.P., Sampaio, C.L.S., et al., 2020. The puzzling occurrence of the upside-down jellyfish *Cassiopea* (Cnidaria: Scyphozoa) along the Brazilian coast: a result of several invasion events? Zoologia 37, 1–10.

Stiasny, G., 1920. Die Scyphomedusen-Sammlung des Naturhistorischen Reichsmuseums in Leiden. III. Rhizostomeae. Zoologische Mededeelingen 5, 213–230.

Stiasny, G., 1921. Studien über Rhizostomeen mit besonderer berücksichtigung der Fauna des Malaiischen Archipels nebsteiner revision des Systems. Capita Zoologica 1, 1–179.

Straehler-Pohl, I., Jarms, G., 2022. Back to the roots, Part 2—Rhopaliophora (Scyphozoa, Cubozoa and Staurozoa) reborn based on early life cycle data. Plankton Benthos Res. 17, 105–126.

Sun, T., Dong, Z., Li, Y., 2019. *Versuriga anadyomene*, a newly recorded scyphozoan jellyfish (Scyphozoa: Rhizostomae) in Chinese waters. J. Oceanol. Limnol. 37, 266–272.

Swift, H.F., Gómez Daglio, L., Dawson, M.N., 2016. Three routes to crypsis: stasis, convergence, and parallelism in the *Mastigias* species complex (Scyphozoa, Rhizostomeae). Mol. Phylogenet. Evol. 99, 103–115.

Swift, H.F., Dawson, M.N., 2020. Demographic, environmental, and phenotypic change but genetic consistency in the jellyfish *Mastigias papua*. Biol. Bull. 239, 80–94.

Terrones, A., van der Bank, M., Moreno, J., Juan, A., 2022. DNA barcodes and microsatellites: How they complement for species identification in the complex genus *Tamarix* (Tamaricaceae). J. Syst. Evol. 60, 1140–1157.

Thé, J., Gamero-Mora, E., Da Silva, M.V.C., Morandini, A.C., Rossi, S., De Oliveira Soares, M., 2021. Non-indigenous upside-down jellyfish *Cassiopea* andromeda in shrimp farms (Brazil). Aquaculture 532, 735999.

Thiel, H., 1966. The evolution of the Scyphozoa, a review. In: Rees, W.J. (Ed.), The Cnidaria and Their Evolution. Academic Press, London, pp. 77–117.

Thiel, M.E., 1970. Über den zweifachen stammesgeschichtlichen ("biphyletischen") Ursprung der Rhizostomae (scyphomedusae) und ihre Aufteilung in die zwei neuen Ordnungen Cepheida und Rhizostomida. Abh. Verh. Naturwiss. Ver. Hambg. NF. 14, 145–168.

Thiel, M.E., 1976. Die Cassiopeida (Arcadomyariae MAAS, 1903, Kampylomyariae STIASNY, 1921) als dritte Ordnung der Rhizostomae (Scyphomedusae) und die Erhebung der letzteren zu der Überordnung Rhizostomoidea (Wurzelmundartige) Supraord. nov. Mitt. Hamb. Zool. Mus. Inst. 73, 1–16.

Thiel, M.E. (†), Reinicke, G.B., Holst, S., Morandini, A.C., Sötje, I., Straehler-Pohl, I., et al., 2024. Rhizostomeae—a remaining manuscript to complete the monograph on Scyphozoa by M.E. Thiel (published 1936–1977). Adv. Mar. Biol. 98 (electronic supplement), DOI:.

Tian, T., Chen, Z., Wang, M., Hu, Y., Wang, W., 2017. Inbreeding and genetic diversity analysis in a hatchery release population and clones of *Rhopilema esculentum* based on microsatellite markers. Chin. J. Oceanol. Limnol. 35, 580–586.

Tilesius, W.G., 1831. Beiträge zur Naturgeschichte der Medusen. I. Cassiopeae. Nova Acta Phys. Med. Acad. Caes. Leop. Carol. Nat. Curios 15, 247–288.

Uchida, T., 1926. The anatomy and development of a rhizostome medusa, *Mastigias papua* L. Agassiz, with observations on the phylogeny of Rhizostomae. J. Fac. Sci., Imp. Univ. Tokyo, Sect. IV 1, 45–95.

Ullah, M.S., Min, G.S., Dong, J., Yoon, W.D., Choi, J.K., 2015. First record of *Rhopilema esculentum* (Scyphozoa, Rhizostomae), edible jellyfish in Korea. Ocean. Polar Res. 37, 287–293.

Vanhöffen, E., 1888. Untersuchungen über semäostome und rhizostome Medusen. Bibl. Zool. 1, 5–52.

Verrill, A.E., 1865. Classifications of polyps (Extract condensed from a Synopsis of the Polypi of the North Pacific Exploring Expedition, under Captains Ringgold and Rodgers, U.S. N.). Proc. Essex Inst. 4, 145–152.

Von Montfort, G.M., Costello, J.H., Colin, S.P., Morandini, A.C., Migotto, A.E., Maronna, M.M., et al., 2023. Ontogenetic transitions, biomechanical trade-offs and macroevolution of scyphozoan medusae swimming patterns. Sci. Rep. 13, 9760.

Wang, G., Zhen, Y., Yu, Z., Shi, Y., Zhao, Q., Wang, J., et al., 2017. The physiological and molecular response of *Aurelia* sp. 1 under hypoxia. Sci. Rep. 7, 1558.

Wang, Y., Sun, S., 2017a. Complete mitochondrial genome of the jellyfish, *Nemopilema nomurai* (Cnidaria: Scyphozoa) and the phylogenetic relationship in the related species. Mitochondrial DNA B: Resour. 2, 165–166.

Wang, Y., Sun, S., 2017b. Complete mitochondrial genome of the jellyfish, *Rhopilema esculentum* Kishinouye 1891 (Cnidaria: Scyphozoa) and the phylogenetic relationship in the related species. Mitochondrial DNA B: Resour. 2, 167–168.

Wang, C., Wang, B., Wang, B., Wang, Q., Liu, G., Wang, T., et al., 2019. Unique diversity of sting-related toxins based on transcriptomic and proteomic analysis of the jellyfish *Cyanea capillata* and *Nemopilema nomurai* (Cnidaria: scyphozoa). J. Proteome Res. 18, 436–448.

Wang, J., Mi, T., Yu, Z., Wang, G., Wei, Q., Yang, J., et al., 2021. Species-specific detection and quantification of scyphomedusae in Jiaozhou Bay, China, using a quantitative real-time PCR assay. J. Oceanol. Limnol. 39, 1360–1372.

Werner, B., 1973. New investigations on systematics and evolution of the class Scyphozoa and the phylum Cnidaria. Publ. Seto Mar. Biol. Lab. 20, 35–61.

Xia, W., Jiang, H., Fan, W., Li, X., Liu, Y., Xu, L., Gou, X., 2023. Complete mitochondrial genome of *Mastigias papua* (Scyphozoa: Rhizostomeae: Mastigiidae) based on next-generation sequencing and phylogenetic analysis. Mitochondrial DNA Part B 8, 1294–1297.

Zhu, L., Yang, A.A., Zhou, C.Y., Yang, H., Luo, X.R., Chen, S.Q., et al., 2015. Characterization of 15 polymorphic microsatellite loci for the jellyfish *Rhopilema esculentum*. Conserv. Genet. Res. 7, 551–556.

CHAPTER SEVEN

Ecology of Rhizostomeae

Delphine Thibault[a], Zafrir Kuplik[b], Laura Prieto[c], Angelica Enrique-Navarro[c], Michael Brown[d], Shin Uye[e], Tom Doyle[f,g], Kylie Pitt[h], William Fitt[i], and Mark Gibbons[d,*]

[a]Aix Marseille University, Université de Toulon, CNRS, IRD, MIO, Marseille, France
[b]The Steinhardt Museum of Natural History, Tel Aviv University, Tel Aviv, Israel
[c]Department of Ecology and Coastal Management, Instituto de Ciencias Marinas de Andalucía (ICMAN), Consejo Superior de Investigaciones Científicas (CSIC), Puerto Real, Cádiz, Spain
[d]Department of Biodiversity and Conservation Biology, University of the Western Cape, Bellville, Republic of South Africa
[e]Graduate School of Integrated Sciences for Life, Hiroshima University, Higashi-Hiroshima, Japan
[f]School of Biological, Earth and Environmental Sciences, University College Cork, Cork, Ireland
[g]Science Foundation Ireland Research Centre for Energy, Climate and Marine, Environmental Research Centre, University College Cork, Cork, Ireland
[h]School of Environment and Science, Coastal and Marine Research Centre, Australian Rivers Institute, Griffith University, Australia
[i]Odum School of Ecology, University of Georgia, Athens, GA, United States
*Corresponding author. e-mail address: mgibbons@uwc.ac.za

Contents

Advances in Marine Biology, Volume 98
ISSN 0065-2881, https://doi.org/10.1016/bs.amb.2024.07.008

Abstract

Max Egon Thiel's chapter on the ecology of rhizostomes in his review up to 1970 covered a bewildering variety of topics, many of which are the focus of other chapters in this volume: their interactions with humans; their associations with other organisms; their venoms. Although he also discussed their habitats and habits, the effects of environmental conditions on distribution, and patterns in seasonality, he paid scant attention to blooms, he did not write about their role in the wider ecosystem, and he ignored alien introductions. It is clearly impossible to comprehensively update Thiel's review in a similar vein – we don't have the space – and so we have decided to focus our efforts here on either those topics that particularly fascinated him (seasonality), or those that he did not write about (alien introductions, their role in the ecosystem). Our narrative is based on case studies of well known taxa, from which we attempt to draw patterns of commonality, where appropriate. We conclude our account with a discussion of rhizostomes as Discomedusae, and whether we should be considering them separately from semaeostomes in terms of ecology.

1. Introduction

Max Egon Thiel's chapter on the ecology of rhizostomes in his review up to 1970 (published in Thiel et al., 2024) follows standard conventions of the time it was originally written and is akin to the works of John Mauchline on euphausiids (Mauchline and Fisher, 1969), mysids (Mauchline, 1980) and calanoid copepods (Mauchline, 1998). That is, separate sections of the text detail the distribution (in space and time) of each species recognised then, the relationships between medusae/polyps and their habitat and individual environmental factors, as well as the interactions with other species. This is done almost in a list-like fashion: species-by-species. Interestingly, Thiel has explicitly included humans amongst "*other species*", and their writing has sections on both stings (species, symptoms and treatments) and uses (food and remedies, the latter referencing their potential value as a laxative and in the cure of chilblains and gout) (Thiel et al., 2024). While the review is clearly thorough in its coverage, drawing together a diverse, and at times obscure, literature, it does not include any reference to the interactions between jellyfish and their prey. Indeed, nowhere in this chapter is there a mention of what they eat or of what their impact on food populations may be. Rather the authors include comments on diet in their chapter on Physiology.

The value of Thiel's work (in Thiel et al., 2024) is that it will provide a ready source of detailed species-specific information on a variety of topics related to the ecology of rhizostomes, and it will prove a handy font of early references. I do not believe that the authors synthesised the information in a way that improves our overall understanding of rhizostome ecology, but rather they provide the material that allows others to make and draw comparative conclusions. While this reflects Thiel's cautious approach as a scientist, it is undoubtedly too a reflection of lack of sufficient knowledge at the time the manuscript was originally penned.

The science of ecology, especially jellyfish ecology, has changed since Thiel crafted the first draft of his chapter, thanks in part to changes in the tools we have at our disposal to study the science. We take molecular techniques for granted (e.g. Bayha et al., 2010), as too we do electronic miniaturisation (e.g. Fossette et al., 2016), coupled ocean-atmosphere-biota models (e.g. El Rahi et al., 2020), satellite imagery (e.g. Johnson et al., 2005) and stable isotopes (e.g. Fleming et al., 2015). These are all methodological tools that scientists before Thiel did not have access to, and his synopsis of seasonal patterns, for example, is based on *"...more or less random observations made by individual researchers or on expeditions....[that were dependent].... on the accidental route, the season, the age[/size] of the animals... [sampled....and as a consequence] differences in the seasonal occurrence of the rhizostomes do not allow a uniform explanation..."*. The use of the various tools that we have available to us now, often in combination with community science initiatives (e.g. Marambio et al., 2021), which were largely unheard of at the time (especially in the context of jellyfish), has materially improved our understanding of *inter alia* seasonality in jellyfish populations.

But the emphasis of our ecological science has changed too, as we become increasingly focused on populations (e.g. Condon et al., 2013), fluctuations and drivers of change (e.g. Gibbons and Richardson, 2009), and of the role that jellyfish play within ecosystems (e.g. Doyle et al., 2014; Hays et al., 2018). This all reflects, of course, an imperative to understand and potentially ameliorate the impacts of human mediated change on the global environment, be it increased temperature (e.g. Domingues et al., 2008), reduced pH levels (e.g. Pörtner, 2008), declining concentrations of dissolved oxygen (e.g. Diaz and Rosenberg, 2008), resource exploitation (e.g. Pauly et al., 1998) or habitat modification (e.g. Heery et al., 2017). And changes in jellyfish populations (up or down) are seen as bellwethers of wider changes to marine ecosystems (e.g. Gibbons and Richardson, 2013). While most of the *Introductions* to recent papers on jellyfish will try to squeeze in (often unnecessary) references to jellyfish "blooms", regardless of their otherwise focus, Thiel devote but a single page of his review to the subject of "swarms" and swarm-formation (see Thiel et al., 2024). This latter account is entirely descriptive, and Thiel makes no attempt to explain the environmental drivers, though he does note that the majority of swarming species are temperate and not tropical, which is something that a recent review of blooming species (Fernández-Alías et al., 2021) emphasised.

Ecology is the study of the interactions between organisms and their environment and in the absence of direct experimentation it relies heavily on evidence derived from descriptive and comparative studies augmented with correlative statistics. Thiel's treatment of the interaction between species and their environment is almost entirely descriptive in his review (see Thiel et al., 2024). That said, one could argue that much of our current work on rhizostome ecology is descriptive and/or comparative and/or based on correlative statistics, though the number of experimental studies is increasing. While the majority of the latter have been conducted on polyps (e.g. Rato et al., 2021) or young medusae (e.g. Schäfer et al., 2021), and primarily relate to the impacts of temperature, salinity or pH on survival and growth (e.g. Algueró-Muñiz et al., 2016), efforts to understand interactions at the level of adult medusae are scarce (species of *Cassiopea* excepted: Rowe et al., 2022). In that sense too, our review here will be date-stamped by technology, and it is not unreasonable to assume that future laboratories will be able to examine the questions that we cannot answer here.

The most recent phylogeny of Discomedusae (Bayha et al., 2010) indicates that previously held ideas about the differences between Semaeostomeae and Rhizostomeae are, in part, artificial. The former is paraphyletic with respect to the latter and Ulmaridae would appear to be a sister taxon to Rhizostomeae (Bayha et al., 2010). Within the latter, Rhizostomida (=Daktyliophorae) also seems to be paraphyletic with respect to Cepheida (=Kolpophorae) (compare Gamero-Mora et al., 2024). While higher-level phylogeny may have an impact on some aspects of "rhizostome" biology, its role in shaping ecological responses to the environment is less well understood. As noted by Hamner and Dawson (2009) in the context of blooming, "*…Selection for convergent evolution in the sea is strong because mass occurrence has evolved multiple times in independent evolutionary lineages under similar circumstances…*". I guess at the end of the day it all comes down to details – detailed differences between those units that are shaped by interactions with the environment, though whether pH has the same material impact on feeding and growth of (e.g.) *Pelagia noctiluca* as it does of (e.g.) *Crambionella orsini* is unknown. A focus, therefore, on the ecology of Rhizostomeae may be similarly arbitrary, but we will conclude with a discussion of Rhizostomeae as Discomedusae, in the context of ecology.

In his chapter on the ecology of rhizostomes (see Thiel et al., 2024), Thiel attempts to review everything that was known about this subject at the time the draft was originally written in the 1960s and into the 1970s. To repeat such a task today would fill an unnecessary volume of its own, and so here we focus on the areas of ecology that Thiel clearly had a passion for, but which were based on the incomplete data of the day, or that he explored imperfectly. Thus, we pay attention to the drivers of population processes and seasonality, and on the role that rhizostomes play in the ecosystems they occupy.

The chapter before you has been assembled by scientists from across the globe. The team comprises individuals with their own interpretations of the published data/literature and although we have tried to harmonise and balance the text, we have nevertheless taken the liberty to allow individual opinions and voices to be heard. Unusually, therefore, you will see personal pronouns scattered throughout, for which we do not apologise.

2. Population fluctuations and seasonality

2.1 Introduction

All rhizostomes are metagenetic and display a life-cycle alternation between a sessile polyp stage that reproduces asexually and a free-living medusa stage that reproduces sexually (Schiariti et al. 2024). Typically, ephyrae are released by strobilae following a species-specific cue (Lucas et al., 2012) heralding the approach of environmental conditions that will maximise their survival. Ephyrae feed in the plankton and develop into medusae that will continue to feed, grow and increase in size before they in turn reproduce: information regarding the persistence of medusae thereafter is scant (e.g. Pitt et al., 2014). As Thiel highlights in his chapter, *"in general, a certain strobilation time can be assumed for each species, which determines the seasonal occurrence of the medusae in the plankton…[and] theoretically… there would be a regular alternation between the appearance of the larger specimens and the small individuals…."* (see Thiel et al., 2024). The authors go on to note, however, that the latter observation, reflecting growth of the medusae, may be only *"…partly the case…[as] they sometimes appear simultaneously or without any recognisable connection with the appearance of the small individuals. In three [of the documented] cases, the small and large medusae are found almost simultaneously throughout the year."* The latter three species are *Mastigias albipunctata* Stiasny, 1920, *Mastigias papua* (Lesson, 1830) and *Cassiopea xamachana* Bigelow, 1892 and while it is tempting to interpret this as a reflection of a lack of significant environmental periodicity in their habitats, in truth the majority of rhizostomes are almost *"…exclusively [found in] the shallow coastal waters of the tropical seas."*.

Thiel struggled to find explanations for many of the ecological observations made owing to the random nature of the data collected, and this was especially the case for population processes, as noted earlier (see Thiel et al., 2024). That said, in their conclusion to the section on seasonality Thiel wrote "*….it is not possible to determine the geographical distribution of rhizostomes only according to the locations of the large specimens fished in the open water. Rather, the small individuals must also be taken into account, which can connect individual finds that are far away from each other and can bring the distribution areas together.*". While written in the context of mapping geographic distribution, this content implies too that micro- and meso-scale differences in the spatial distribution of different age/size classes of medusae confounds conclusions in understanding population processes; processes that are in part at the whim of the prevailing oceanography.

This is exemplified if we consider the case of the pelagiid *Chrysaora fulgida* (Reynaud, 1830), which is found in the nearshore waters of the Benguela ecosystem (Gibbons et al., 2021). This review is about rhizostomes, but as I have little experience of the latter but some of the former, and this account is illustrative – please bear with us. Polyps of *C. fulgida* are presumed to occur on hard substrata in shallow water; ephyrae are released over a period of about five months during winter/spring in a latitudinally-dependent manner. As ephyrae grow and develop into medusae they are moved offshore by Ekman transport in surface waters and accumulate over the inner shelf thanks to the interaction of the shelf's topography and prevailing winds. There, they mature and reach maximum size before being returned onshore during periods of southerly wind stress. Sexual reproduction is then presumed to take place, and planulae settle in shallow waters. Sexual reproduction can take place at any time of year in this species, and although individuals may live for longer than 18 mo, it is unclear if the species dies after reproduction. At any one moment in time, then, the size of individuals observed in samples, will depend on where and, in part, when samples are collected. Observations at the population level are very strongly influenced by the oceanographic and atmospheric context, and are not just intrinsically biological.

Rather than review all the recent literature regarding population fluctuations of rhizostomes, we have decided to focus our efforts on those species that have been well-studied. In other words, our review is presented as a series of "case-studies", rather than a more conventional overview of all that is known. If readers are interested in isolated works on other species, I draw their attention to the comprehensive bibliography provided as an online supplement. Following a summary of what is known, penned by the experts of those species, we attempt to pull together their findings and make some tentative generalisations. Given that the Order comprises some 90-odd species, generalisations based on a small subset of species may be viewed as reckless, but hypotheses based on inductive reasoning form one of the cornerstones of the scientific method.

2.2 *Nemopilema nomurai*

Nemopilema nomurai (Kishinouye, 1922), or Nomura's jellyfish, also known as Echizen jellyfish, is perhaps the largest rhizostome, attaining a bell diameter of ca. 2 m and a wet weight of ca. 200 kg (Fig. 1, Shimomura, 1959; Yasuda, 2007). Prior to modern scientific study, this giant jellyfish was likely recognised in China as an edible jellyfish, given that another commonly eaten species, *Rhopilema esculentum* (Kishinouye, 1891), is often found concurrently with *N. nomurai* and was described by Zhang Hua (232–300 AD) in his book "Natural History" during the Tsin Dynasty (Wu, 1955). A large jellyfish that was presumably *N. nomurai* was first recorded in the southern Yellow Sea in a book by the Korean author Chong (1814) as "Haepaleo", which means jellyfish. In Japan, a jellyfish that was presumably *N. nomurai* was reported for the first time by William Griffish, an American teacher who lived in Fukui (formerly Echizen) Prefecture in 1871–1872 when he encountered this peculiar jellyfish at sea. He described them in his book as follows:

Fig. 1 A giant *Nemopilema nomurai* (maximum diameter: ca. 2 m) drifting off the coast of Ishikawa Prefecture, Japan, in October 2005. The diver is trying to put a pop-up satellite archival tag like a necklace around the oral arms. *Photo: Yomiuri Shimbun, with permission.*

Sometimes great jellyfish, two or three feet in diameter, sail gaily along near the shore, as proud as the long-handles-umbrella of a daimyo (meaning feudal lord in Japanese), and as brilliantly coloured as a Japanese parasol. Floating all around their bodies, like the streamers of a temple festival, or a court lady's ribbons, are their long tentacles or feelers. No peacock stretching his bannered tail could make a finer sight, or look prouder than these floating sun fishes, or bladders of living jelly. But alas for all things made of water! Let but a wave of unusual force, or a sudden gust of wind come, and this lump of pride lies collapsed and stranded on the shore, like a pancake upset into a turnover, in which batter and crust are hopelessly mixed together. When found fresh, men often come down to the shore and cut huge slices of blubber, as transparent as ice, they eat the solid water with their rice, in lieu of drink (Griffish, 1887, page 142–143).

Persistent occurrences of *N. nomurai* in Japanese coastal waters prompted local fishermen to use them as bait for red sea bream and filefish, and this traditional fishing method is still used today (see Brotz et al., 2024). However, scientists have seldom considered this jellyfish, primarily due to infrequent bloom events until the turn of the 21st century when massive blooms began occurring regularly. Now in recent decades, Chinese, Korean and Japanese researchers have made *N. nomurai* one of the most well-studied jellyfish species in the world. This article reviews these studies and provides up-to-date knowledge, including unpublished information, on the population fluctuations of *N. nomurai*.

2.2.1 Geographical distribution and expatriated migration to Japanese waters

The geographical origin of *N. nomurai* is the region consisting of the Bohai, Yellow and East China Seas. It is a large bay system flanked by the Korean Peninsula and the mainland of China (Fig. 2). A genetic analysis demonstrates that there are two distinct populations, one along the Korean coast and another along the Chinese coast (Gotoh et al., 2017). Although wild polyps have not yet been located, ephyrae and small medusae (<10 cm bell diameter) have been sighted and collected only in this region (Dong et al., 2018; Sun et al., 2015; Toyokawa et al., 2012). The robustness of planulae, polyps and ephyrae to low salinity (10–20) suggests that the polyps may exist in shallow coastal waters including estuaries (Dong et al., 2015; Feng et al., 2018a; Takao and Uye, 2018). Strobilation and subsequent release of ephyrae may begin during the spring in the southern area (e.g. the Changjiang River estuary) and during the early summer toward northern area (e.g. the Bohai Sea) (Dong et al., 2018; Feng et al., 2018b; Sun et al., 2015). Some medusae

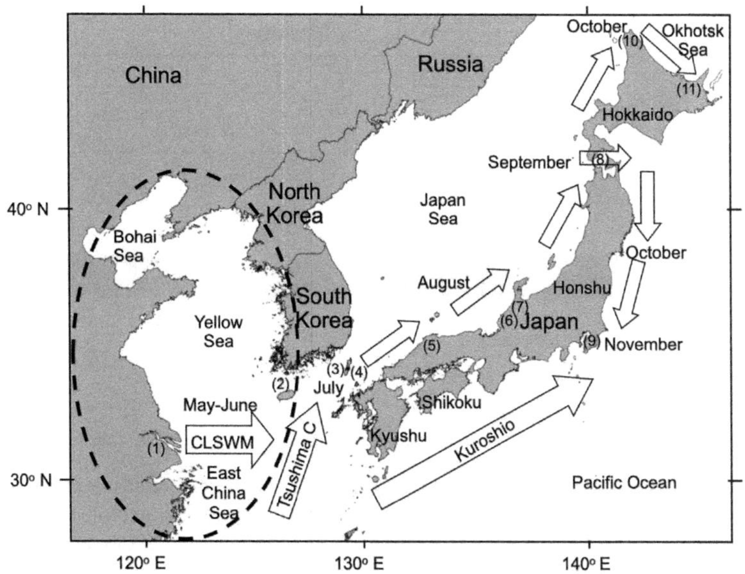

Fig. 2 A schematic of the advective transport of *Nemopilema nomurai* medusae from the polyp habitat and nursery ground (encircled by a dotted line) to Japanese waters by the Changjiang low salinity water mass (CLSWM) and the Tsushima Current. (1) Changjiang River, (2) Jeju Island, (3) Tsushima Strait, (4) Tsushima Island, (5) Shimane Prefecture, (6) Fukui Prefecture, (7) Ishikawa Prefecture, (8) Tsugaru Strait, (9) Boso Peninsula, (10) Cape Soya, and (11) Shiretoko Peninsula. Map lines delineate study areas and do not necessarily depict accepted national boundaries. *Modified from Uye, S., 2008. Blooms of the giant jellyfish Nemopilema nomurai: a threat to the fisheries sustainability of the East Asian Marginal Seas. Plankt. Bent. Res. 3 (Suppl.), 125–131.*

spend their lives in this area until they die in winter (Sun et al., 2015; Zhang et al., 2012), thereby maintaining their endemism (Fig. 3). However, the majority of medusae in the southern Yellow Sea and northern East China Sea are expatriated by water currents. Due to monsoonal rainfall in June and July in temperate East Asia, the Changjiang (Yangtze River) low salinity water mass (CLSWM) forms with its front extending close to Jeju Island, Korea. The medusae are entrained into the offshore-spreading CLSWM and then transported northward by the Tsushima Current to the Tsushima Strait between Japan and Korea (Chang and Isobe, 2003; Reizen and Isobe, 2006). The medusae continue to drift north-east by the Tsushima Current (often three-branched, Katoh, 1994), so that their distribution spreads over the southern half of the Japan Sea. In September, the frontal population reaches Hokkaido and they are divided into two groups. One group passes through the Tsugaru Strait between Honshu and Hokkaido to the Pacific Ocean and

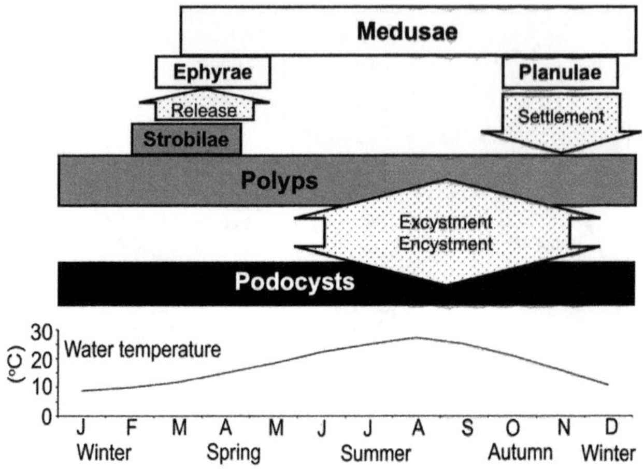

Fig. 3 A schematic of the seasonal life cycle of *Nemopilema nomurai* in the Changjiang River estuary and adjacent waters, based on physio-ecological properties determined in the laboratory and in the field (e.g. Jiaozhou Bay, China). The surface water temperature fluctuates seasonally from ca. 10 to 30 °C.

migrates as far south as the Boso Peninsula. Another group is transported further north along the coast of western Hokkaido to Cape Soya and enters the Sea of Okhotsk to migrate to Shiretoko Peninsula (Fig. 2). In 2005, in addition to the above transportation routes, some medusae were advected with the Kuroshio Current along the Pacific coast of Japan from the East China Sea to the Boso Peninsula. Accordingly, Japan was surrounded by *N. nomurai* during the fall of 2005. All expatriated medusae die in winter, and export enormous amounts of carbon to the sea floor (Yamamoto et al., 2008). The distance and time to travel from the Changjiang River estuary to the final destinations are ca. 2500 km and ca. 8 months, respectively.

2.2.2 Life cycle

2.2.2.1 Sexual reproduction

Nemopilema nomurai is a gonochoristic jellyfish having a ribbon-like ovary and testis at the bottom of a pleated, inter-radial gastric pouch, which protrudes beneath the bell (Fig. 4). In August and September, the immature gonads of young medusae (ca. 10 kg wet weight) are almost transparent. In October, some medusae (ca. 60 kg) begin to have coloured gonads, varying from milky white through pink to dark brown. In November and December, more medusae mature sexually, demonstrating that late fall and early winter are the main spawning seasons (Fig. 3, Iguchi et al., 2010; Kawahara et al., 2006; Ohtsu et al., 2007). The ovary of three females with 20, 39 and 56 kg

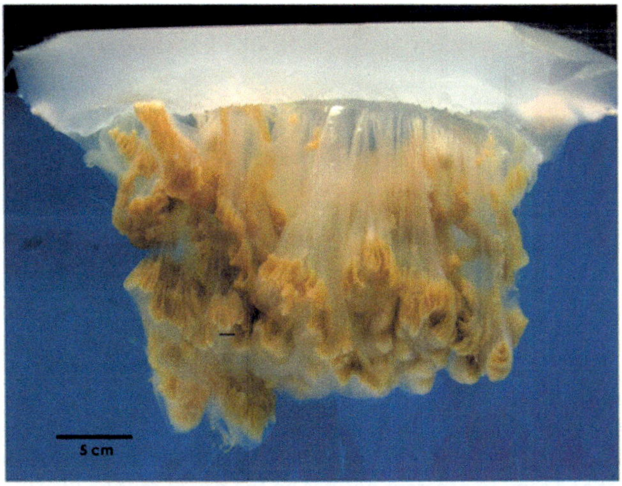

Fig. 4 View of *Nemopilema nomurai* ovary contained in the gastric pouch protruding beneath the bell. The gastric pouch is a large sac with numerous folds. The ovarian tissue, which becomes dark brown when matured, is located at the bottom of each fold. *From Ohtsu, K., Kawahara, M., Ikeda, H., et al., 2007. Experimental induction of gonadal maturation and spawning in the giant jellyfish Nemopilema nomurai (Scyphozoa: Rhizostomeae). Mar. Biol. 152, 667–676.*

wet weight caught in November 2009 in Shimane Prefecture carried 160 million, 690 million and 360 million oocytes, respectively (Uye et al., unpublished). Assuming a positive relationship to body weight, the fecundity of a fully grown female (ca. 200 kg) would attain more than one billion oocytes, perhaps the greatest in marine animals!

It is peculiar that even in November, physically intact and vigorously swimming medusae have exclusively immature gonads, and their gonadal maturation is induced when the medusae are fatally damaged. Ohtsu et al. (2007) and Ikeda et al. (2011a) have detailed the induction of gametogenesis after immature medusae were arrested in a net on the beach. On day 0, when a female medusa is caught, transparent and semitransparent oocytes less than 70 μm diameter are present in the ovary. On days 2–3, oocytes become larger and darker with extraordinarily large nuclei, or germinal vesicles, indicating that they are still in a primary oocyte stage with their maturation arrested at the prophase of the first meiotic division (Fig. 5). The maturation division, observed as breakdown of the germinal vesicles, is induced by light exposure, and spawning of eggs or mature secondary oocytes (ca. 110 μm diameter) from the ovary occurs ca. 90 min after light exposure (Fig. 6). For males, on day 0 the testis consists of small

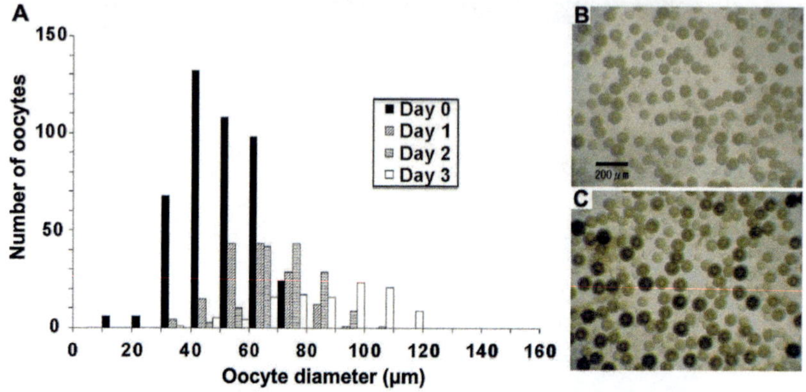

Fig. 5 Maturation of oocytes in *Nemopilema nomurai*. (A) Daily changes in the distribution of oocyte diameter in a female during captivity in a net on the beach. (B) Transparent and semi-transparent oocytes on the day of capture (day 0). (C) By day 2, oocytes become darker and larger, and have extraordinarily large nuclei. *(A) From Ikeda, H., Ohtsu, K., Uye, S., 2011a. Structural changes of gonads during artificially induced gametogenesis and spawning in the giant jellyfish Nemopilema nomurai (Scyphozoa: Rhizostomeae). J. Mar. Biol. Assoc. U.K. 91, 215–227; (B) From Ohtsu, K., Kawahara, M., Ikeda, H., et al., 2007. Experimental induction of gonadal maturation and spawning in the giant jellyfish Nemopilema nomurai (Scyphozoa: Rhizostomeae). Mar. Biol. 152, 667–676.*

ovoid sperm follicles (ca. 80 μm long axis) devoid of spermatozoa. By day 5 the follicles are enlarged (ca. 120 μm) and filled with spermatozoa. Light exposure also acts as a trigger for sperm release, which is complete within 20 min after illumination. As light plays an important role, dawn can function as a light trigger for spawning of both eggs and sperm (Fig. 7). Both are shed into the gastrovascular cavity and then released from the body through small orifices on the scapulets and oral arms. Egg fertilisation may take place not only in the females' gastrovascular cavity into which sperm are sucked from the surrounding water but also in the ambient water, suggesting that the proximity of mature females and males is important to increase the success of sexual reproduction. In November and December, when water temperature decreases < 20 °C, most medusae sink to deep, aphotic depths, and their spawning would be less successful: one or two orders of magnitudes lower compared to light-induced spawning (Ohtsu et al., 2007). However, for those trapped in the shallows, spawning would have high success.

Fertilised eggs develop and hatch into planula larvae about 24 h after fertilisation at 20 °C. Planulae are ca. 170 μm long and ca. 130 μm wide, with

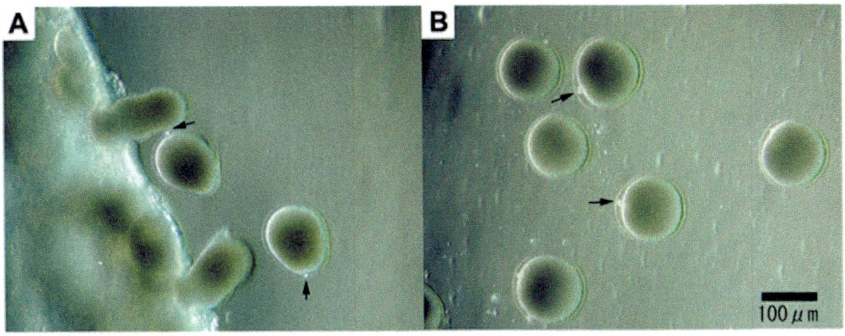

Fig. 6 Spawning of eggs or secondary oocytes from *Nemopilema nomurai*. (A) After light-induced breakdown of the germinal vesicles, the oocytes are squeezed out through pits on the ovarian epithelium. Spawned oocytes with a single polar body (with an arrow). (B) Completely matured secondary oocytes with two polar bodies (with an arrow). *From Ohtsu, K., Kawahara, M., Ikeda, H., et al., 2007. Experimental induction of gonadal maturation and spawning in the giant jellyfish Nemopilema nomurai (Scyphozoa: Rhizostomeae). Mar. Biol. 152, 667–676.*

10 μm long cilia or flagellae over the body surface. Planulae swim for several days to three weeks until they settle on hard substrates, either natural (e.g. pebble, bivalve shell, wood) or artificial (e.g. ceramic, glass, various plastics) (Kawahara et al., 2006; Takao and Uye, 2018) Schiariti et al., 2024.

2.2.2.2 Asexual reproduction

At 18 °C in the laboratory, it takes two days for settled planulae to develop into polyps with four tentacles and a calyx diameter of 250–300 μm. These polyps are capable of ingesting small and slow-swimming prey such as ciliates, own planulae, rotifers, copepod nauplii, planktonic larvae of various benthos. It is worth noting that conspecific planulae are suitable food for 4–8 tentacle polyps. It takes 6–10 days after settlement to develop to 8 tentacle polyps and a calyx diameter of 500 μm, which are capable of ingesting *Artemia* nauplii but excess feeding (≥2 nauplii polyp^{-1} d^{-1}) results in high polyp mortality. Within 10–20 days after settlement, the polyps develop to the adult stage with 16 tentacles and 800–1100 μm calyx diameter and begin asexual reproduction by producing podocysts. A stolon protrudes from the bottom of the calyx and attaches to the substrate 10–300 μm away from the base of the polyp. The polyp body mass gradually moves to the new attachment site, leaving a podocyst behind at the former position. The newly produced podocysts are whitish and ca. 300 μm in diameter. The duration of podocyst formation varies greatly from seven days to three months. A single productive founder polyp

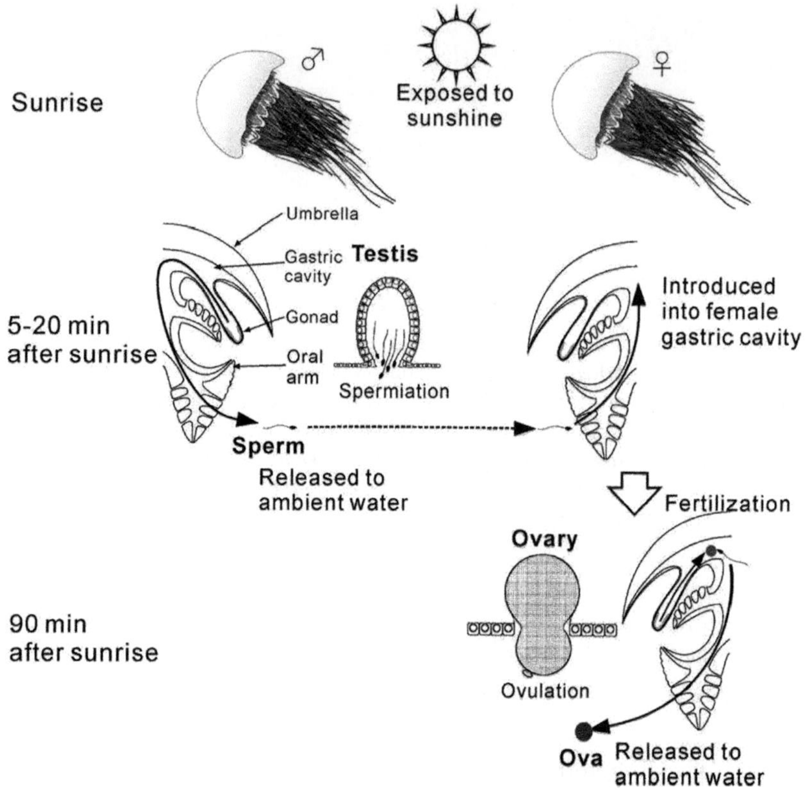

Fig. 7 A schematic of spermiation and ovulation of *Nemopilema nomurai* induced by exposure to natural sunlight. Spermiation takes place 5–20 min after sunrise, and sperm released into ambient water may be introduced into the female gastrovascular cavity. Ovulation takes place approximately 90 min after sunrise, and eggs may be fertilised in the gastrovascular cavity and released to ambient water. Fertilisation may also occur in the ambient water. *Based on findings by Ikeda, H., Ohtsu, K., Uye, S., 2011a. Structural changes of gonads during artificially induced gametogenesis and spawning in the giant jellyfish Nemopilema nomurai (Scyphozoa: Rhizostomeae). J. Mar. Biol. Assoc. U.K. 91, 215–227.*

produces 12 podocysts and six excysted polyps after six months (Fig. 8, Kawahara et al., 2006). The excystment into polyps occurs irregularly because of variable dormant periods of podocysts, described below.

The scant evidence collected to date suggests that in the laboratory, polyps maintained at ≥ 18 °C never strobilate, but those maintained at ≤ 15 °C do; it takes at least one month to start strobilation at 11 and 15 °C, and three months at 9 °C (Kawahara et al., 2013). These results suggest that a seasonal temperature decrease is a trigger that induces strobilation in wild polyps (Feng et al., 2015, 2018b). When strobilation begins, the calyx becomes elongated

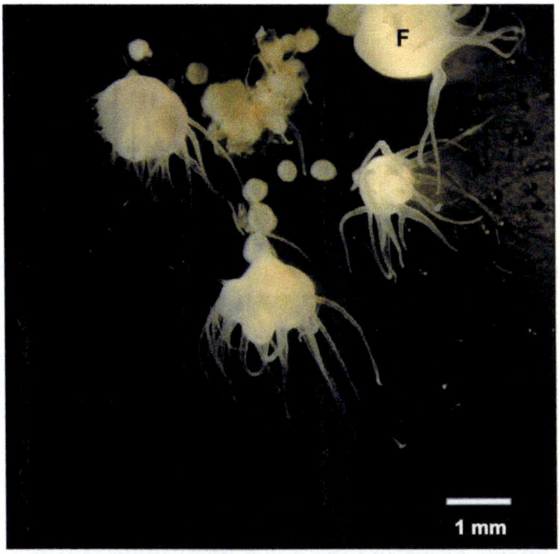

Fig. 8 A colony of 12 podocysts and 6 excysted polyps produced by a founder polyp (F) of *Nemopilema nomurai* after 6 months at 18 °C. *From Kawahara, M., Uye, S., Ohtsu, K., et al., 2006. Unusual population explosion of the giant jellyfish Nemopilemia nomurai (Scyphozoa: Rhizostomeae) in East Asian waters. Mar. Ecol. Prog. Ser. 307, 161–173.*

and segmented; after 1.5 days rhopalia with statoliths become apparent, and after 2 days ephyral lappets are elongated and pulsate rhythmically. Fully developed strobilae at 2.2–2.8 mm, measured from the base of the polyp to the top of the ephyral mouth, liberate ephyrae successively into the water. Under laboratory conditions, a strobila produces five ephyrae on average (Kawahara et al., 2006). Polyp colonies artificially established on plastic plates, which were submerged in Jiaozhou Bay, China, strobilated from 28 March to 9 June 2013, when temperature varied from 6.4 to 18.1 °C, and released two ephyrae per strobila on average (Feng et al., 2018b).

2.2.2.3 Podocyst dormancy

Dormancy is a characteristic feature in podocysts of *N. nomurai*, which consist of a dome-shaped chitinous capsule (200–500 μm diameter) enclosing a cell mass that has numerous nutrient granules containing protein, carbohydrates and lipids (Fig. 9, Ikeda et al., 2011b). Their metabolic activity is extremely low as evidenced by a scarcity of mitochondria, rough endoplasmic reticula, and Golgi complexes. They are capable of remaining dormant for more than 10 years at 19 °C in the laboratory (Ikeda and Uye, unpublished). Whether podocysts maintain dormancy or excystment into

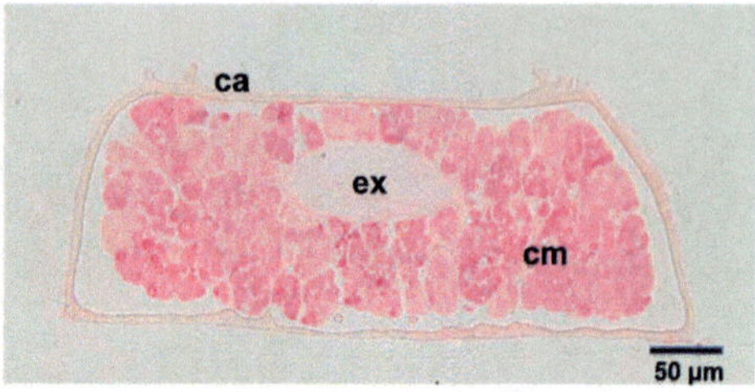

Fig. 9 Light microscopy of the sagittal section of a 4-month-old *Nemopilema nomurai* podocyst stained with hematoxylin and eosin. A chitinous capsule (ca) encapsulates a cell mass (cm) containing various nutrient reserves. At the centre of the cell mass is extracellular matrix (ex). *From Ikeda, H., Ohtsu, K., Uye, S., 2011b. Fine structure, histochemistry, and morphogenesis during excystment of the podocysts of the giant jellyfish Nemopilema nomurai (Scyphozoa, Rhizostomeae). Biol. Bull. 221, 248–260.*

polyps in Chinese coastal waters may affect the medusa population size in the following season, as schematically depicted in Fig. 10 (Kawahara et al., 2013). If few excystments occur and the polyp populations stay at a low level, the medusa population size would be small in the following year. On the other hand, if accumulated podocysts excyst extensively and create a large polyp population, an intense medusa bloom would be anticipated.

2.2.2.4 Development to mature medusae

Newly liberated ephyrae, 2.2–3.8 mm from lappet tip to lappet tip, typically ingest the same foods as polyp except for conspecific planulae and can grow if the temperature is at least 18 °C. However, they are unable to obtain sufficient food (i.e. *Artemia* nauplii) to sustain net growth at ≤ 11 °C. Therefore, the seasonal warming to ≥ca. 15 °C may enable and accelerate ephyral growth and development. At 22 °C and ca. 10 days after liberation in the laboratory, ephyrae develop to metephyrae, where the secondary lappets extend outwards to the primary lappets so that the bell margin looks polygonal, 8–14 mm wide from lappet tip to lappet tip. They carry eight oral arms, each of which branches into two wings at the tip. About 20 days after liberation, the central mouth closes and they advance to the medusa stage. A specific growth rate is 0.30 d^{-1} from 1-day-old ephyra (0.0035 g wet weight) to 20-day-old medusa (1.5 g) and decreases to 0.11 d^{-1} from 20-day-old to 48-day-old medusa (29 g) at 22–28 °C. Further rearing is impossible in small 50 L laboratory aquaria, perhaps because repeated contacts with the aquarium walls are fatal (Kawahara et al., 2006).

In the barrier-free sea, *N. nomurai* ingest micro- and mesozooplankton, dominated by copepods that are stung by nematocysts (Holst et al., 2024,

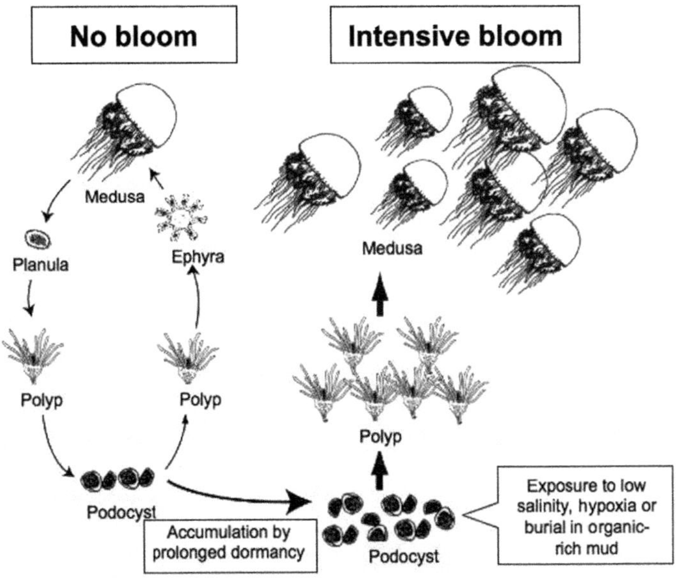

Fig. 10 A schematic of the possible distinction in podocyst abundance and behaviour (e.g. dormancy or excystment), which affects the bloom intensity of *Nemopilema nomurai* medusae in the following year. *From Kawahara, M., Ohtsu, K., Uye, S., 2013. Bloom or non-bloom in the giant jellyfish Nemopilema nomurai (Scyphozoa: Rhizostomeae): roles of dormant podocysts. J. Plank. Res. 35, 213–217.*

Nagata et al., 2024) and ingested through ca. 1 mm diameter mouthlets on the scapulets and oral arms. The zooplankton are assimilated in the gastric cavity of *N. nomurai* and the medusae grow to their giant size. In Shimane Prefecture, southwestern Japan Sea, we sampled a total of 341 medusae and measured their wet weights from July 2005 to January 2006. Mean wet weight increased from 3.4 kg in July to 108 kg in November (temperature range: 22–28 °C), with a specific growth rate of 0.03 d^{-1} (Fig. 11, Uye et al., unpublished). After November, medusae became senescent and disappeared by the end of January. In Japanese waters, we have never located ephyrae and small medusae of *N. nomurai*, which indicates that the expatriated medusae have not yet established outpost polyp populations in Japanese coastal waters.

2.2.3 Interannual fluctuations in population size

2.2.3.1 Bloom intensity and associated impact in fisheries

There are approximately 2000 set nets along the coast of Japan affected by *N. nomurai* blooms, the scale of which varies from small operations run by

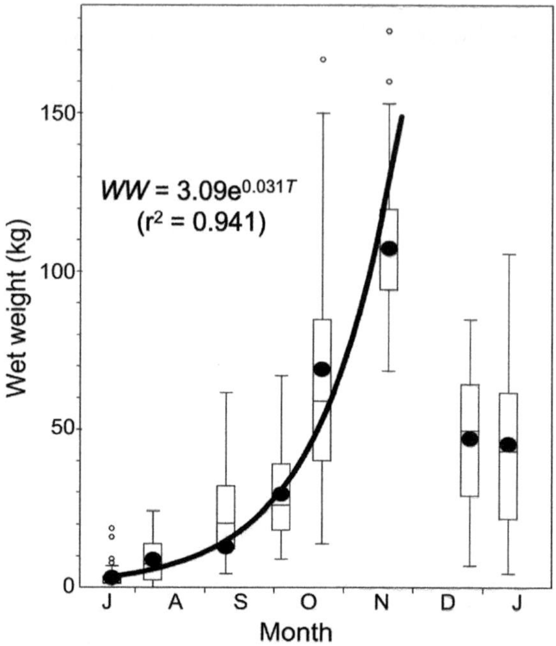

Fig. 11 Exponential wet weight (*WW*, kg) over time (*T*, days) for *Nemopilema nomurai* medusae collected in Shimane Prefecture from July to November 2005. After November, body weight decreases abruptly. The boxes extend from the top of the first quartile (lighter) to the top of the third quartile (heavier). The horizontal lines within the boxes denote the median, the whiskers extend to the limits of the range, and small open circles indicate outliers. Filled circles denote the mean (Uye et al., unpublished).

2–3 fishermen to large operations run by 20–30 fishermen. Among various types of fisheries that rely on nets, the set net fishery is most vulnerable to jellyfish blooms and hence suffers serious damage; the estimated monetary loss in Japanese fisheries was approximately 30 billion JPY (270 million USD) in 2005 (Uye, 2011). The annual economic damage in Korean fisheries was reported as 68–205 million USD from 2006 to 2010 (Kim et al., 2012). Since 2006, set net owners and cooperatives report the number of medusae entrapped in their nets each day to the Japan Fisheries Information Service Center (JAFIC, http://www.jafic.or.jp/kurage/). These reports consist of 9–427 reporters annually. We chose 30 relatively large-scale nets from Nagasaki (4), Shimane (3), Kyoto (7), Niigata (3), Aomori (4) and Iwate (9) Prefectures to calculate the annual medusa bycatch for each net and their grand mean, which represents the relative bloom intensity in Japanese waters (Fig. 12A, Uye et al., unpublished). In 2006, 2007 and 2009, the bloom was intense, ranging from 11,200 to 45,200 medusae net^{-1} year^{-1}, causing severe damage to their fisheries operations. Thereafter, the bloom was much less intense, with annual medusa bycatch ranging from 0.03 to 561 medusae net^{-1} year^{-1}.

Previous scientific reports together with informal information (e.g. newspaper articles) show that the intense bloom of *N. nomurai* occurred in 1920 (Kishinouye, 1922), 1957 (Shimomura, 1959), 1995 (Yasuda, 2007) and 2002–2005 (Uye, 2013). These indicate that an intense bloom historically occurred once per ca. 40 years before 1995, but in 2002,

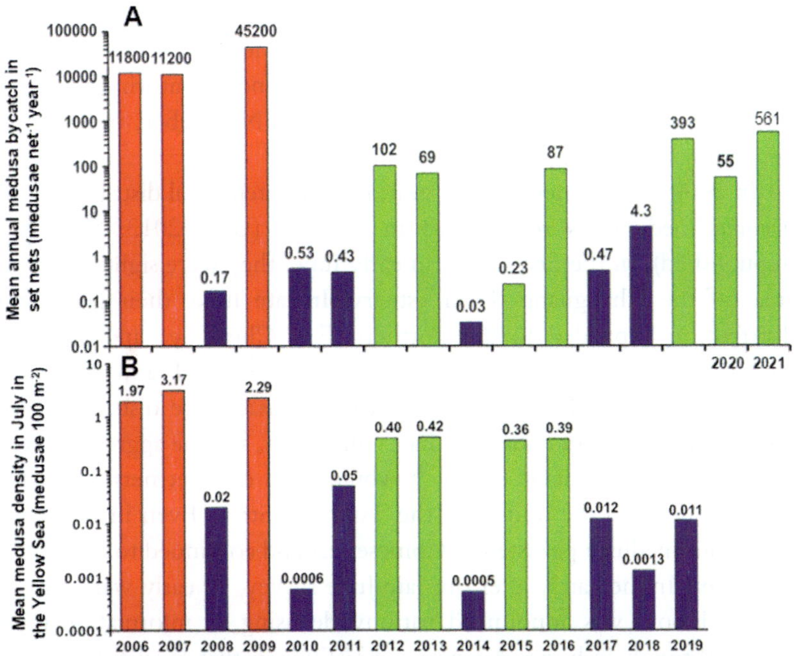

Fig. 12 (A) Mean annual bycatch of *Nemopilema nomurai* medusae per net for 30 Japanese set nets, with set net locations shown in the inset. (B) Mean medusa density in July in the Yellow Sea determined by monitoring using ships of opportunity. The bloom intensity is arbitrarily classified into three levels: intense (red columns), intermediate (green ones), and low (blue ones). Bycatch data from the 30 nets are available from the JAFIC (Japan Fisheries Information Service Centre) website. *From Uye, S., Brodeur, R.D. (eds.) (2017) Report of Working Group 26 on Jellyfish Blooms around the North Pacific Rim: Causes and Consequences. PICES Sci. Rep. No. 51, p. 221; Uye et al., unpublished.*

it occurred only 7 years after the previous intense bloom. The intense blooms were most frequent, as many as 7 times, during the first decade of this century. Then, the blooms have been less intense up to now.

2.2.3.2 Monitoring medusae using ships of opportunities

From 2006 to 2019, we developed an annual monitoring program using ships of opportunity in the Yellow Sea and East China Sea. After 2019, the COVID-19 pandemic prohibited us from boarding ships. The monitoring program was carried out from the ship deck, 15–20 m above sea level, during daytime by visually counting medusae near the surface. Counting occurred within 10 m or 15 m from the ship's side for 5 min intervals, and the ship's geographical position was determined by a hand-held GPS in 15 min intervals. Each year, juvenile medusae with bell diameter of 10–20 cm were first sighted in May or June, and monitoring was carried out once or twice per month until October, November, or December, depending on the year.

In 2009, a prominent bloom year, the spatiotemporal distribution of *N. nomurai* medusae is represented in Fig. 13 (Uye, 2013; Uye et al., unpublished). In the first survey in early June, they were sighted in a small area off the Changjiang River estuary. In late June, their geographical distribution expanded to the northern East China Sea and the central Yellow Sea at mean and maximum densities of 0.71 and 21 medusae 100 m^{-2}, respectively. In early July, the medusae were distributed across the Bohai, Yellow and northern East China Seas, and the aggregated frontal population (33 medusae 100 m^{-2}) would soon clog set nets at Tsushima Island, located in the middle of the Tsushima Strait. Even in late August, numerous medusae persisted in Chinese seas and continued to drift into the Japan Sea. Immediately after our late June survey, an early warning of the intense bloom was announced nationwide using an information transfer system through Fisheries Agency and JAFIC, so that local fishermen and fishery cooperatives could be made aware of the potential appearance of massive medusae in their fishing grounds.

Compiled data from our surveys from 2006 to 2019 show that the mean density of *N. nomurai* in the Yellow Sea was usually highest in July (Uye et al., unpublished), and the July density displayed remarkable year-to-year fluctuations (Fig. 12B, Uye and Brodeur, 2017; Uye et al., unpublished). In 2006, 2007 and 2009, the density was ≥ca. 2 medusae 100 m^{-2} (categorised as intense bloom years), whereas in 2008, 2010, 2011, 2014 and 2017–2019,

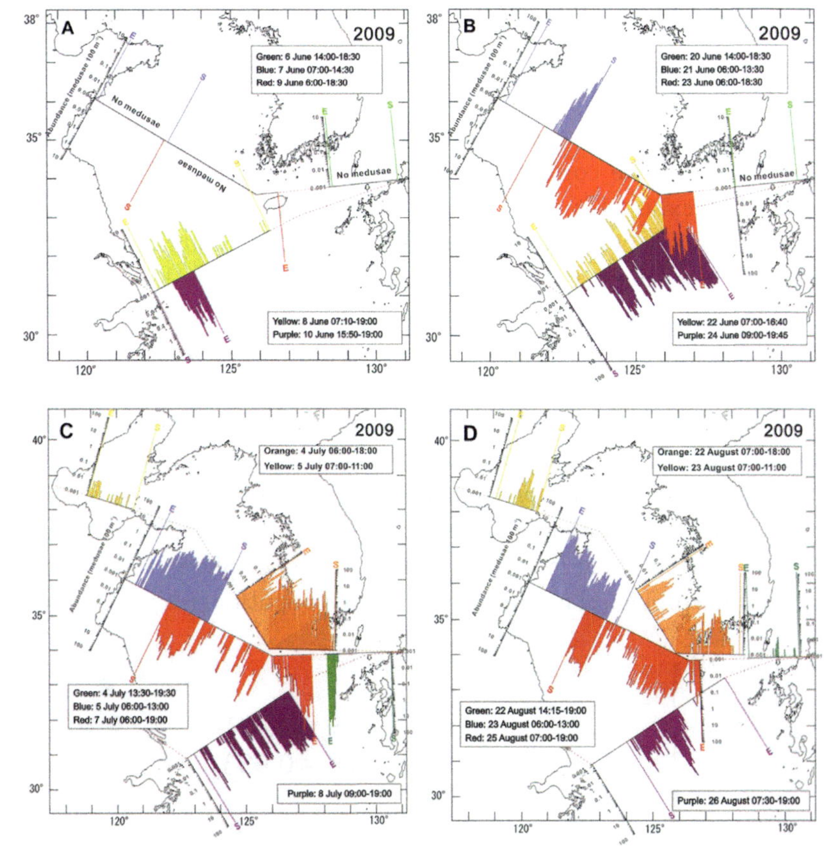

Fig. 13 Spatiotemporal distribution of *Nemopilema nomurai* medusae in 2009, the most intense bloom year, using ferries between Japan and China during (A) 6–9 June, (B) 20–24 June, (C) 4–8 July and (D) 22–26 August 2009. S and E denote the start and end points of daily sighting surveys, respectively. Each column represents the number of medusae counted during 5-minute intervals (Uye, 2010). *From Uye, S., 2010. Studies on functional roles of zooplankton in coastal marine ecosystem: toward restoring productive seas for global sustainability. Glob. sustainability. Oceanogr. Japan 19: 283–299. (in Japanese with English abstract); Uye et al., unpublished.*

it was ≤ 0.05 medusae 100 m^{-2} (low bloom years). In the remaining years, it ranged from 0.1 to 0.42 medusae 100 m^{-2} (intermediate bloom years).

2.2.3.3 Environmental factors causing year-to-year bloom intensity and using these data to forecast bloom intensity

Mean annual medusa bycatch in the 30 Japanese set nets (M, medusae net^{-1} year^{-1}) is plotted against mean medusa density in July in the Yellow Sea

(D, medusae $100\,m^{-2}$) in Fig. 14 (Uye et al., unpublished). Both data are variable, particularly in years of medusa densities < 0.1 medusae $100\,m^{-2}$ when mean annual bycatch is ca. 1.0 medusa net^{-1} $year^{-1}$ (excluding 2019). In years with densities ≥ 0.1 medusae $100\,m^{-2}$ (excluding 2015), a significant positive correlation expressed by $M = 1283D^{2.89}$ ($r^2 = 0.933$, $p < 0.01$) exists. This is a logical basis for our forecast of the bloom intensity in Japanese coastal waters by means of monitoring medusae in Chinese seas. Now, we can forecast the year-to-year bloom intensity roughly at three levels (intense, intermediate and low) in June–July, one to three months prior to the appearance of medusae in Japanese coastal waters.

The success of strobilation and the development of ephyrae to juvenile medusae are crucial to determine the magnitude of the medusa population, and may be affected by environmental conditions. Sea surface temperature (SST), U-wind, V-wind, precipitation, and shortwave radiation in Chinese seas acquired by weather satellites are available from Japan Meteorological Agency. In collaboration with physical oceanographers, we investigate which meteorological parameters explain the population size of *N. nomurai*

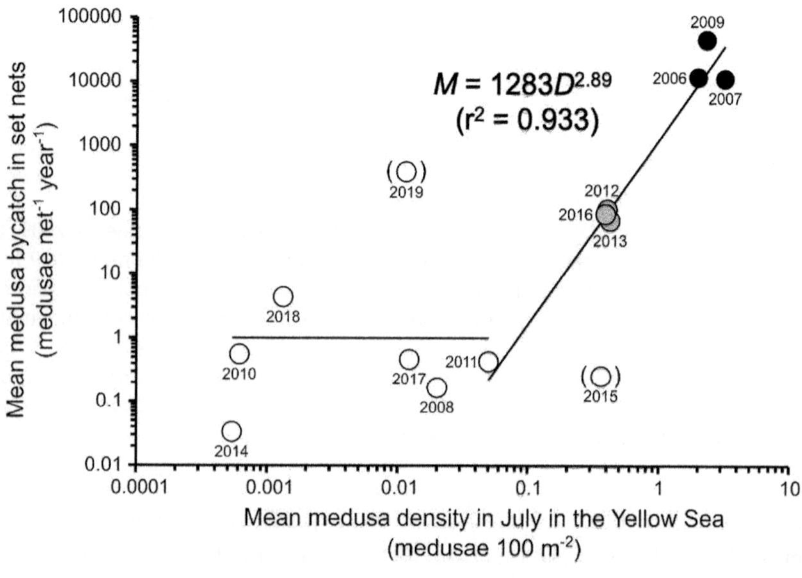

Fig. 14 Relationships between mean medusa density in July in the Yellow Sea (D, medusae $100\,m^{-2}$) and mean annual medusa bycatch in Japanese set nets (M, medusae net^{-1} $year^{-1}$). A significant positive relationship exists for years at densities ≥ 0.1 medusae $100\,m^{-2}$, excluding 2015, in July in the Yellow Sea. Grand mean of annual medusa bycatch in Japanese set nets is 1 medusa net^{-1} $year^{-1}$ for years at densities < 0.1 medusae $100\,m^{-2}$, excluding 2019 (Uye et al., unpublished).

in July in the Yellow Sea using multiple regression analysis. Using meteorological parameters from the most prominent polyp habitat in the Changjiang River estuary, which is represented by a $0.25 \times 0.25°$ grid at 122.125°E, 31.375°N, as explanatory variables and the mean medusa density in July in the Yellow Sea (D, medusae $100\,m^{-2}$) as objective variables, we obtained the best fit regression equation: $D = 0.56 \times$ (SST in February) $+ 0.37 \times$ (U-wind in May) $+ 0.20 \times$ (V-wind in June) ($r^2 = 0.81$, Uye et al., unpublished). The contribution of SST in February is most important, followed in decreasing order of importance by U-wind in May and V-wind in June. The biological significance of these parameters matches their statistical significance. In February when strobilation may begin (Fig. 3), warmer temperatures may accelerate strobilation, subsequent development to ephyrae, and feeding activity in ephyrae (Kawahara et al., 2006, 2013), which would lead to higher survivability and consequently to greater medusa population size. In May, when ephyrae develop to juvenile medusae, stronger U-wind (west wind) may increase transport from inshore to offshore Yellow Sea. In June, V-wind (south wind) may also increase offshore transport in the Yellow Sea. Currently, we can approximately forecast annual bloom intensity in February based on satellite-acquired SST data. More precise forecasts of bloom intensity occur in June–July, when the magnitude of medusa population in Chinese seas can be determined by in-situ monitoring surveys.

2.2.4 Conclusions

Repeated occurrences of intense blooms of the giant jellyfish, *N. nomurai*, in the East Asian marginal seas from 2000 to 2009 have prompted researchers to increase research into this jellyfish, which is now one of the most well-studied jellyfish species. The polyp habitat and geographical origin of *N. nomurai* is presumed to be in the Bohai, Yellow and East China Seas, and its endemism is maintained by the medusae that carry out their life cycle in the large bay flanked by the Korean Peninsula and the mainland of China. However, significant portions of medusae are expatriated by currents to Japanese waters after drifting for up to 8 months and 2500 km. Since 2006, annual relative bloom intensity can be assessed using mean annual medusa bycatch in 30 Japanese set nets. Bloom intensity was high and caused severe fisheries damages in 2006, 2007 and 2009, but bloom intensity has since decreased. Ships of opportunity have been used to monitor medusae in Chinese seas since 2006 (excluding 2020 and 2021); the mean medusa density in July in the Yellow Sea was high in 2006, 2007 and 2009, and had a significantly positive relationship with the mean annual medusa bycatch in Japanese set nets for years with densities ≥ 0.1 medusae 100 m^{-2}. In addition, the mean medusa density in July in the Yellow Sea was explained by meteorological parameters in the Changjiang River estuary; intense blooms occurred after high SST in February, followed by strong U-wind in May and strong V-wind in June. Now, it is possible to approximately forecast annual bloom intensity as early as February based on satellite-acquired SST, then provide a more precise forecast based on the results of *in situ* monitoring in June–July. The final forecast can be broadcasted to Japanese fishermen one to three months prior to the arrival of *N. nomurai* in their fishing grounds.

2.3 *Catostylus mosaicus*

2.3.1 *Introduction and geographical distribution*

Catostylus mosaicus (Quoy and Gaimard, 1824) (Fig. 15) is a large (up to ~35 cm bell diameter), bloom-forming rhizostome that is prevalent along Australia's east coast and in southeast Asia. The species is predominantly estuarine and occurs in coastal lagoons, embayments and the tidal reaches of

Fig. 15 *Catostylus mosaicus. Photo: Joanna Browne.*

rivers (Pitt and Kingsford, 2000a; Loveridge et al., 2021a). *Catostylus mosaicus* is an edible species (see Brotz et al. 2024) and the potential for it to be harvested commercially was the impetus for the early research into its population dynamics (Pitt and Kingsford, 2000a), life history (Pitt, 2000) and reproductive biology (Pitt and Kingsford, 2000b). The low value of the product relative to other readily available commercial taxa such as prawns, the high labour costs in Australia, and the need for the product to be processed immediately upon capture stalled the development of a fishery for this species in Australia.

The life cycle of *C. mosaicus* is typical of most rhizostomes (Schiariti et al., 2024). The medusae are gonochoristic (see Holst et al, 2024; Schiariti et al., 2024). Fertilisation is internal and females initially brood the larvae (see Holst et al, 2024; Schiariti et al., 2024). Polyps settle on a variety of natural (e.g. sandstone) and artificial (e.g. glass) surfaces in the laboratory (Pitt, 2000) but little is known about the distribution of polyps in the field; although they have been observed once on dead bivalve valves and on Perspex settlement plates deployed in a coastal lagoon (Pitt, unpublished). The initial description of the life cycle indicated that only one or two ephyrae formed during strobilation (Pitt, 2000) but since then cultures maintained in the laboratory have been observed undergoing polydisc strobilation with up to 10 ephyrae forming on a single polyp. *C. mosaicus* also prolifically produces podocysts, which can be induced to excyst in the laboratory by lowering the temperature by 2–3 °C (Lawley, unpublished).

Demographic (Pitt and Kingsford, 2000a) and phylogenetic (Dawson, 2005; Dawson et al., 2015) data both point to populations of *C. mosaicus* being relatively localised. Indeed, two sub-species have been identified in southern Australia with *C. mosaicus conservativus* and *Catostylus mosaicus mosaicus* occurring south and north of Cape Howe respectively (Dawson, 2005). Within each of these clades, substantial population structuring is also evident; indeed populations located in a permanently open coastal lagoon (Lake Illawarra) and embayment (Botany Bay) located only ~75 km apart are genetically distinct (Dawson et al., 2015). These findings align with the ecological observations of Pitt and Kingsford (2000a) who observed that the population dynamics and timing of recruitment were uncorrelated across six estuaries that span 800 km of the New South Wales coast. They concluded that local rather than large-scale environmental drivers had a greater influence on the population dynamics of this species.

2.3.2 Formation of blooms

Catostylus mosaicus is renowned for forming spectacular blooms (Fig. 16). The biomass of these blooms can be enormous; approximately 700 t km^{-2} of *C. mosaicus* was estimated to occur in a New South Wales coastal lagoon (Pitt and Kingsford, 2003). Blooms in the Port of Brisbane have disrupted shipping by clogging cooling water intakes and bow thrusters. In 2006, the USS *Ronald Reagan*, one of the world's largest nuclear-powered aircraft carriers departed the Port of Brisbane several days ahead of schedule to avoid the blooms of *C. mosaicus* which had been interfering with the vessel. Historical accounts indicate that blooms of *C. mosaicus* predated substantial anthropogenic development of the Australian coastline and are a natural part of the species' life cycle. The narrative of the Voyage of the HMS Rattlesnake (Macgillivray, 1852) observed in Port Phillip Bay, Victoria, that *"Vast numbers of a large sea-jelly (Rhizostoma mosaica) gave the water a milky appearance"* and Badham (1917) (referring to *Crambessa mosaicus* sensu *Catostylus mosaicus* in New South Wales) reported that *"This medusa is frequently found in large numbers, possibly brought together by currents and tidal action"*. Whether the frequency or intensity of blooms has changed, however, cannot be determined.

Two long-term data sets on the abundance of *C. mosaicus* exist but neither have been comprehensively analysed. The biomass of *C. mosaicus* was recorded in annual trawl surveys of Port Phillip Bay (Victoria) from

Fig. 16 Bloom of blue *Catostylus mosaicus* in the Port of Brisbane, Queensland (taken from the deck of RV Investigator) September 2019. *Photo: Kylie Pitt.*

1990 to 2010 by the Victorian Department of Primary Industries (Parry, unpublished data). Although there was a trend for the abundance of medusae to increase in the last few years of the survey, anecdotal observations indicated that populations declined after the trawling program concluded (Parry; personal communication). Since ~1999, the Queensland Department of Environment and Science (DES) has also recorded presence/absence of *C. mosaicus* at several hundred sites along the Queensland coast that are sampled monthly for water quality. Sample sites extend across 12 degrees of latitude, from the Daintree River (16.29°S) in tropical far north Queensland, to the Queensland/New South Wales border (28.16°S). A preliminary analysis of the data from Moreton Bay for the period 2000 – 2013 hinted that populations of *C. mosaicus* may exhibit cycles of 7–8 years periodicity (Joensen, 2014; Fig. 17) but a time-line that encompasses at least 3 complete cycles is usually required for confirmation of cyclical behaviour. A comprehensive analysis of this data set, which now extends to 24 years, is underway and this should rigorously test the population cycling hypothesis and identify potential drivers of jellyfish populations.

Analysis of a subset of the DES dataset from the Fitzroy River, in central Queensland revealed that extreme changes in salinity drive the population dynamics of *C. mosaicus* in that system (Loveridge et al., 2021a). The Fitzroy River is a macrotidal estuary and marine waters (salinities > 30) extend 53 km upstream from the river mouth until the riverine flow is

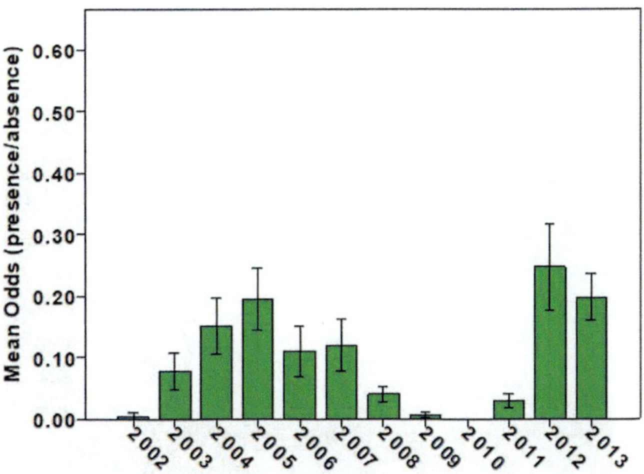

Fig. 17 Annual variation in the odds of occurrence of *Catostylus mosaicus* in Moreton Bay, Queensland (Joensen, 2014).

blocked by a barrage. *C. mosaicus* occurs throughout the tidal reach of the estuary and are often abundant immediately below the barrage. Floods that reduced the salinity to zero below the barrage occurred 16 times during the 20-year data set and high salinity waters took up to six months to re-establish. *C. mosaicus* disappeared from the sites below the barrage on all 16 times the estuary flooded. Laboratory experiments indicated that medusae may be able to survive transient freshwater events but that prolonged reduction in salinity (such as occurs after flooding) is lethal.

2.4 *Rhopilema nomadica*

2.4.1 Introduction and geographical distribution

Rhopilema nomadica Galil et al., 1990 has been the dominant scyphozoa species in the Levant since the 1980s. Reaching up to 100 cm in diameter (Kuplik, personal communication) and a wet weight of 40 kg (for an 80 cm individual; Lotan et al., 1992), this medusa is notoriously known for its painful sting. In just three decades, ever since a single specimen was first recorded in the Israeli coastal waters of eastern Mediterranean in 1977 (Galil et al., 1990), *R. nomadica* (Fig. 18) has earned itself a spot on the '100 worst invading species in the Mediterranean' list (Streftaris and Zenetos, 2006), and deservedly so: An immense population increase and annual swarming events in the region have become an annual phenomenon, mostly in the summer months of June-July. Following Israel, *R. nomadica* was later reported from

Fig. 18 *Rhopilema nomadica* cruising through the Israeli Mediterranean coastal waters. *Photo: Shevy Rothman.*

Lebanon (Lakkis and Zeidane, 1991), Syria (Lotan et al., 1994), Turkey (Kideys and Gücü, 1995), Greece and Malta (Siokou-Frangou et al., 2006; Deidun et al., 2011), Tunisia (Yahia et al., 2013), Egypt (Abu El-Regal and Temraz, 2016), and most recently Sardinia (Balistreri et al., 2017; Fig. 19). In assessing its potential to survive and reproduce in the temperatures of the western Mediterranean (12.2–26.2 °C), Dror and Angel (2023) concluded that the expansion of *R. nomadica* westward is probably just a matter of time; its polyps can survive the average minimum winter temperatures (12 °C), at least to some extent (Fig. 20), and then reproduce via podocysts and strobilate at 24 °C, well within the current maximum temperature of western waters. Considering the continuously warming seawater, its reproduction rate is expected to increase.

2.4.2 Seasonality

Since the 1980s, when swarming became apparent, there has been a noticeable change in the pattern of appearance of *R. nomadica* in the Israeli waters, mainly in the winter months. Lotan et al. (1994), who were the first to study the population of this species, showed that during 1989–1992 there was a summer aggregation that peaked in July-August and lower

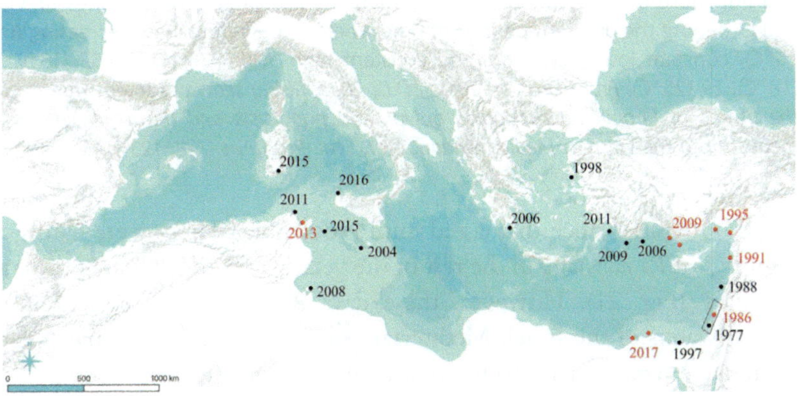

Fig. 19 Distribution and swarming of *Rhopilema nomadica* in the Mediterranean Sea. Extended from Yahia et al. (2013) and Balisteri et al. (2017). • = First record, • = First reported swarm. The two marks in the square along the coastline of Israel represent the 1st record and the 1st swarm of *R. nomadica* in the Mediterranean. Map lines delineate study areas and do not necessarily depict accepted national boundaries. *From Edelist, D., Guy-Haim, T., Kuplik, Z., et al., 2020. Phenological shift in swarming patterns of Rhopilema nomadica in the Eastern Mediterranean Sea. J. Plankton Res. 42, 211–219.*

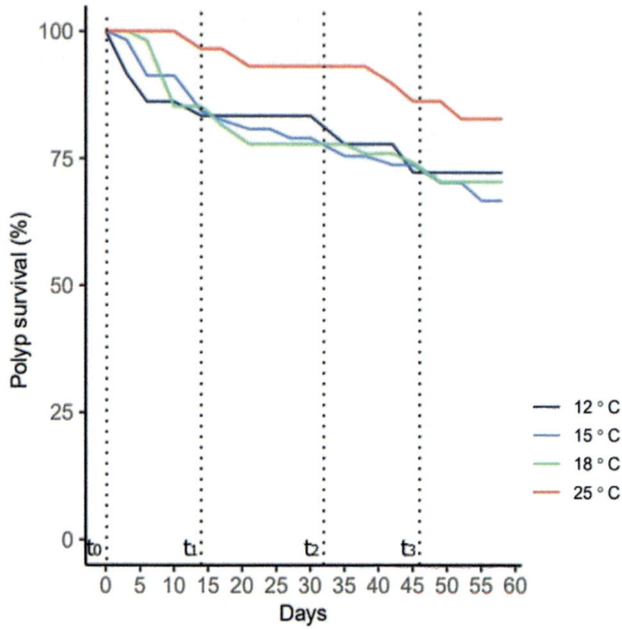

Fig. 20 Mean survival (%) of *Rhopilema nomadica* polyps and during 60 days of changing temperatures experiment (12, 15, 18, and 25 °C). Only polyps present at t0 and not those formed during the experiment were analysed. Dotted lines t0, t1, t2, and t3 denote the different periods of time where temperature was: Decreasing (from 25 °C), constant low, increasing, and constant high. $n_{(t0)}$ = 36 (12 °C), 57 (15 °C), 54 (18 °C), 29 (25 °C). *From Dror, H., Angel, D.L., 2023. Rising seawater temperatures affect the fitness of Rhopilema nomadica polyps and podocysts and the expansion of this medusa into the western Mediterranean. Mar. Ecol. Prog. Ser. 728, 123–143.*

jellyfish densities since that lasted until November. There were hardly any sightings of *R. nomadica* in the rest of the year, with only few in January and February (Fig. 21). However, these days *R. nomadica* is present in the coastal waters of the Israeli Mediterranean year-round, and winter swarms of medusae larger than 60 cm are a common sight (Edelist et al., 2020).

Rhopilema nomadica appears in the Israeli coastal waters almost exclusively as an adult medusa, rarely seen smaller than 5 cm in diameter (in multiple plankton tows conducted between 2011 and 2016, only few *R. nomadica* 2–3 cm in diameter were encountered but no ephyra was spotted). This is intriguing since most of the medusae that appear at the beginning of the summer (i.e. summer swarms) are in a developed reproductive state; large gonads that are apparent through the umbrella, contain advanced gametes (Fig. 22), which under laboratory conditions result in fertilisation and a large

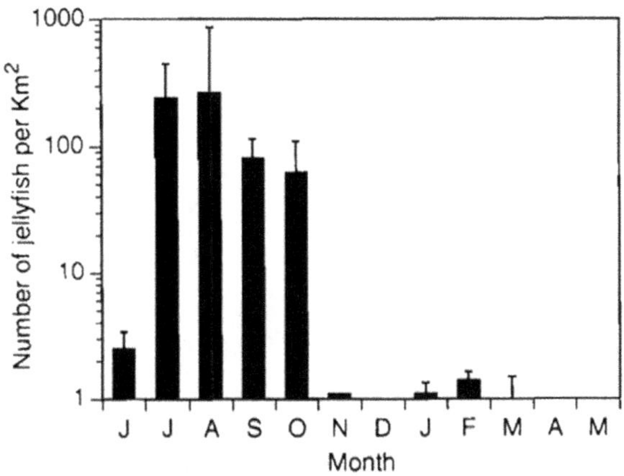

Fig. 21 Concentration of *Rhopilema nomadica* in Haifa Bay during June 1989 to May 1990. *From Lotan, A., Fine, M., Ben-Hillel, R., 1994. Synchronization of the life cycle and dispersal pattern of the tropical invader scyphomedusan Rhopilema nomadica is temperature dependent. Mar. Ecol. Prog. Ser. 109, 59–65.*

number of polyps and ephyrae. Taking in account the general south to north current direction along the Israeli coastline (Fig. 23), and it may well suggest that the offsprings of the summer *R. nomadica* drift northward, and that the origin of the sexually mature adult jellies is somewhere south of Israel, possibly at the south-east corner of the Mediterranean, perhaps in the nutrient rich waters of the Nile delta. The large number of ciliates that exist in that environment (Nour El-Din, 2001), should provide the polyps and the early stages of the medusa a more suitable and sufficient amount of food (Kamiyama, 2011, 2018) until they mature and reach the Israeli waters, when plankton is less abundant (i.e. in the summer months).

Interestingly, recent observations, confirmed by microscopic inspection, showed that winter *R. nomadica*, either individually or in swarms, also contained large and well developed gonads. Nevertheless, *R. nomadica* smaller than 5 cm are rarely seen in the winter months as well and the origin and the reason for the enormous size of the winter medusa are still to be investigated. One hypothesis is that these are the 'summer survivors' that continue roaming in the gyres of the eastern Mediterranean and manage to feed and increase size. Since *R. nomadica* is thought to be a sexually senescent species (i.e. sexual reproduction is followed by death) (Kuplik and Angel, 2020), these survivors should be medusae that didn't release gametes/reproduce during summertime and managed to continue drifting unharmed by waves and predators into the winter months. This hypothesis

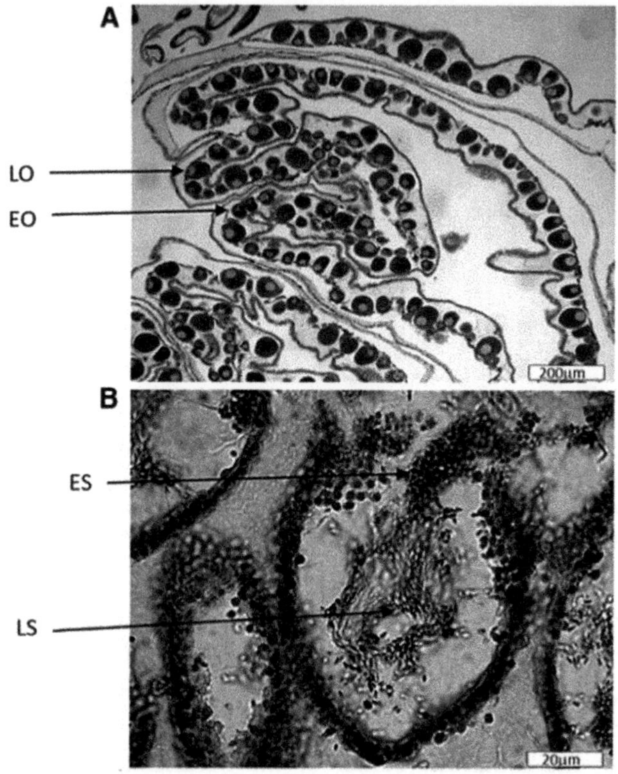

Fig. 22 Histological sections of *Rhopilema nomadica*'s gonads in advanced stages of sexual maturity. (A) female gonad; (B) male gonad. EO, Early oocyte; LO, Late oocyte; ES, Early spermatid; LS, Late spermatid. *Photo: Zafrir Kuplik.*

is supported by Strul (2022) who found shape similarities between statoliths of summer and large winter *R. nomadica*, contrary to statoliths of spring *R. nomadica*, suggesting that the large winter medusae are indeed 'old' summer medusae, remains of the summer swarms.

The second hypothesis derives from our recent findings where under laboratory conditions with ambient seawater temperature, two strobilation peaks occur: The first one at summer's tail-end (October) at a temperature of 28 °C, and the second one in the spring (April) at a temperature of 21 °C (Fig. 24) (Kuplik et al., unpublished). These strobilation events may produce more jellies and result in the enhanced medusae population as observed. Our observations are somewhat different from those of Lotan et al. (1994), who showed, also under laboratory conditions, that the optimal temperature for strobilation was 18–20 °C, suggesting that strobilation occurs in spring,

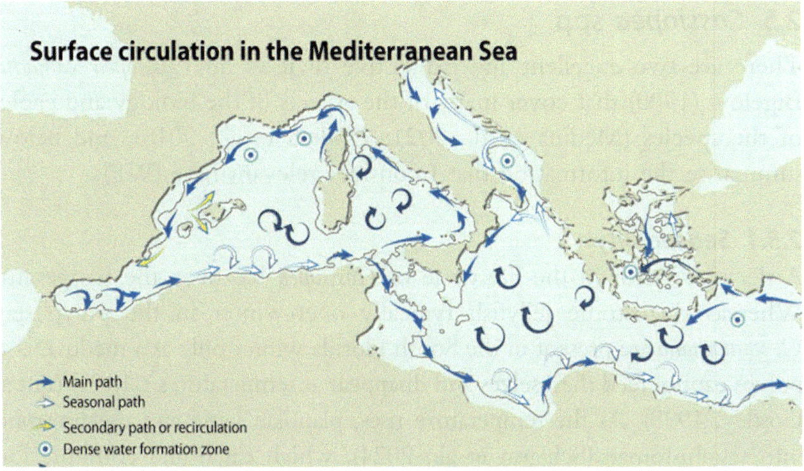

Fig. 23 Map of the Mediterranean Sea showing major features of surface circulation. Map lines delineate study areas and do not necessarily depict accepted national boundaries. *From https://www.grida.no/resources/5915.*

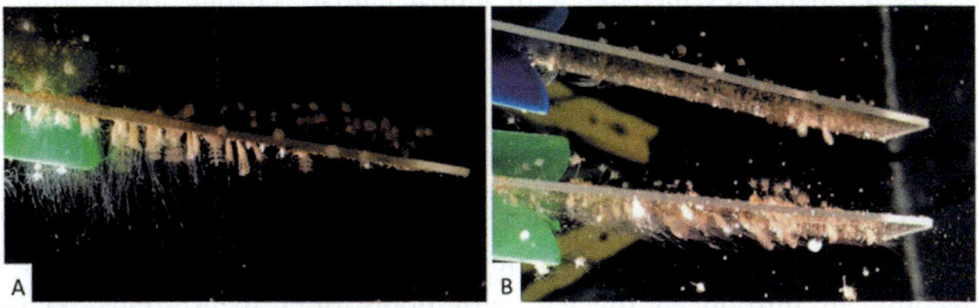

Fig. 24 Strobilation of *Rhopilema nomadica* polyps. (A) at 21 °C and (B) at 28 °C. *Photo: Zafrir Kuplik.*

whereas temperatures above 24 °C reduced strobilation rate. Based on these findings, Lotan et al. (1994) concluded that the low winter temperatures of the shallow waters act as a barrier that prohibits *R. nomadica* from expanding its range west of Greece. As mentioned above, this barrier 'has been broken' and *R. nomadica* has extended its range westwards, as far as Tunisia and Sardinia (Balistreri et al., 2017; Fig. 19), possibly due to the increase in sea water temperature (García-Monteiro et al., 2022; Guy-Haim et al., 2016; Yeruham et al., 2015). Since it is most likely that *R. nomadica* reproduces through all its distribution range, Edelist et al. (2020) hypothesise that the large winter medusa could be the offsprings of polyps located at the western edge of its expansion range, drifting eastward to the Levantine waters.

2.5 *Cassiopea* spp.

There are two excellent and exhaustive reviews on *Cassiopea xamachana* Bigelow (1900) that cover many of the aspects of the biology and ecology of the species (Medina et al., 2021; Ohdera et al., 2018), and below I summarise the information that I consider relevant here (WF).

2.5.1 Seasonality

A decisive factor in the life cycle of *Cassiopea* has been the temperature. Whereas rhizostome jellyfish typically over-winter in the polyp stage, *C. xamachana* are present in the South Florida winter only as a medusa, as the polyps cannot feed themselves and disappear at temperatures ≤ 18 °C (Fitt and Costley, 1998). As the temperature rises, planulae settle and metamorphose into scyphistomae (Schiariti et al., 2024), which catch and consume food. This could be a South Florida phenomenon, as it is unknown if scyphistomae can survive winter temperatures in lower latitudes of the Caribbean Sea. *C. xamachana* begins to strobilate when temperatures are ≥ 23 °C, thus completing the life cycle (Rahat and Adar, 1980).

As temperatures increase with global climate change, populations of *Cassiopea* appear to be expanding (Morandini et al., 2005, Morandini et al., 2017) with a longer season to strobilate (Richardson et al., 2009). *Cassiopea xamachana* can survive temperature up to 36 °C, albeit for short times during midday (Fitt and Costley, 1998), while Klein et al. (2016) found that *Cassiopea* from the Red Sea could tolerate 40 °C, so long as the temperatures decreased during the night. In addition, *Cassiopea* sp. has become an introduced species, with populations in Australia, Hawaii, Brazil, the Mediterranean and potentially the entire Caribbean (Çevik et al., 2006; Holland et al., 2004; Keable and Ahyong, 2016; Morandini et al., 2017; Schembri et al., 2010), possibly due in part to the higher temperatures of global climate change. Whether the exotic *Cassiopea*'s recent range expansions will harm the environment remains to be seen.

2.5.2 Growth

The polyps of *C. xamachana* grow much faster than those of *Cassiopea frondosa* (Pallas, 1774), and budding and settlement/metamorphosis occurs much more readily in *C. xamachana* (Fitt et al., 2021). That said, patterns of strobilation are similar between the two species and medusae of both species appear to grow at similar rates.

The growth of populations of medusae of *Cassiopea* was found to be patchy, but increasing during the last few decades (Niggl and Wild, 2010).

Populations have been documented as fluctuating rapidly (Mergner and Schuhmacher, 1981), reaching densities of 168 m^{-2} (Zarnoch et al., 2020). In the Florida Keys *C. xamachana* increases in abundance following strobilation in late May through September, and continues to grow throughout the summer and fall. Cold-fronts seasonally drop the temperature in South Florida, such that strobilation stops and populations of medusae decrease (Fitt and Costley, 1998).

2.5.3 Reproduction

Species of *Cassiopea* are gonochoristic brooders (see also Holst et al., 2024; Schiariti et al., 2024), with females identifiable through the presence of specialised central tentacles for holding developing eggs (Fig. 25). Larvae are released from females, and can survive over a month if kept without artificial or natural inducers of settlement. Cues for settlement and metamorphosis include peptides associated with bacterial biofilms associated with degrading mangrove leaves, and submerged sand and rock. Therefore, *Cassiopea* requires a separate genetic entity (e.g. species of bacteria) to complete the life cycle (see below). Asexual reproduction occurs in the polyps in the form of budding, which can occur at the rate of 2–3 buds per day in well-fed animals. Each bud can settle and metamorphose into a polyp (Schiariti et al., 2024).

Research efforts attempting to isolate a peptidic cue from *Cassiopea*'s natural substrate, degraded mangrove leaves, resulted in the extraction of several bioactive fractions, one of which contained a proline-rich 5.8 kDa peptide capable of inducing settlement and metamorphosis, although the

Fig. 25 *Cassiopea frondosa* (female, with central tentacles holding developing eggs/larvae) alongside *C. xamachama* (male). *Photo: Rich Aronson, with permission.*

precise amino acid sequence of this peptide or other natural cues have yet to be determined (Fleck and Fitt, 1999; Fleck et al., 1999). There are probably several peptides containing proline that function as natural cues (Fleck et al., 1999). Mangrove leaves have been shown to accumulate high levels of organic solutes such as proline, glycinebetaine, polyols and sugars in order to overcome the salinity-induced osmotic stress (Parida and Jha, 2010).

Noxious compounds leech out of mangrove leaves the first few days, followed by settlement of bacteria and fungus that begin degrading the leaves. Planula larvae of *Cassiopea* settle on leaves that have remained in seawater for over six months, or on submerged sand/rock, presumably on specific species of bacteria that make up some of the bacterial biofilm (Ohdera, unpublished).

Cassiopea requires colonisation of the polyp by an endosymbiont of the Symbiodineacea in order to strobilate (see also Holst et al., 2024; Nagata et al., 2024). Thus, *Cassiopea* depends on again another genetic entity to complete the life cycle (Gohar and Eisawy, 1960; Hofmann et al., 1996). Without infection of polyps they will not undergo strobilation, although a strobilation event in aposymbiotic polyps has been reported by Rahat and Adar (1980). Strobilation takes place ~3 weeks after initial infection of the polyp, depending on several factors including growth of the symbiont species, size of the polyp, temperature and feeding (Fitt, 1985; Fitt and Costley, 1998). While the precise cue(s) responsible for initiation of strobilation of *C. xamachana* is unknown, inhibition of photosynthesis with 3-(3,4-dichlorophenyl)-1,1- dimethylurea (DCMU) results in continuation of strobilation, albeit at lower rates (Hofmann and Kremer, 1981).

Indomethacin, a drug with structural similarities to a peptide involved in strobilation in *Aurelia,* induces strobilation of *C. xamachana* polyps to the medusa form (Cabrales-Arellano et al., 2017). This suggests that the evolution of symbiosis within members of the Rhizostomeae may have resulted in co-option of other pathways as a developmental trigger.

Cassiopea strobilates one jellyfish at a time; therefore the same polyp can regenerate tentacles and strobilate every 3–4 weeks until the water temperature decreases seasonally lower than 18 °C or feeding decreases leading to polyps too small to strobilate.

2.5.4 Life cycle

Cassiopea requires at least three things to stroblilate from a polyp to a medusa and complete the life cycle: temperatures above about 23 °C, a large enough size (about 1 mm diameter), and symbionts to infect and grow in the polyp. If the temperatures range from about 18–23 °C, the polyp could exist if enough food was around to catch, and the medusa flourish if sunny, but there would be no strobilation and the life cycle would not be complete. If the polyps are big enough and the water is warm, they depend on a different genetic entity (symbionts from the Symbiodineacea) to take up residence and grow in them to complete their life cycle.

The life cycle of *Cassiopea* apparently depends on yet another genetic entity for completion of the life cycle. The planula larva and buds from the polyps settle and metamorphose in response to specific bacterial species that presumably release peptide inducers (Hofmann and Brandt, 1987; Neumann, 1979). Bacterial biofilms are involved in decomposing mangrove leaves that could release peptide inducers (Fleck et al., 1999). A cholera toxin isolated from *Vibrio cholerae* was also inductive (Wolk et al., 1985), as well as cultures isolated from mangrove leaves containing *Pseudoalteromonas* sp. (Ohder et al., 2023).

2.6 *Rhizostoma octopus*

2.6.1 Overview

Rhizostoma octopus (Gmelin, 1791) or the barrel jellyfish (also known as dustbin lid jellyfish) is one of the largest rhizostomes, and can attain a bell diameter of ca. 90 cm and a wet weight of over 30 kg (Houghton et al., 2007) (Fig. 26). It is the most northerly distributed rhizostome with a population located as far north as Solway Firth and Clyde Sea area in Scotland (55°N latitude). The earliest definite record of *R. octopus* is from Cornwall in England from 1756 (Borlase, 1758). A local naturalist, William Borlase, described and made drawings of *R. octopus* that had washed up on local beaches between Penzance and Marazion. He wrote: *"It had eight legs, all dependent from a stem or stalk, this stem is dressed round with fourteen fangs; the back is round and convex, like that of rest; the whole transparent, colourless and cartilaginous… []. The margin is formed by little semi-circular fangs about half an inch in diameter, and as much distant from each other; transparent as the rest, but that at their very brim they are tipped with a very beautiful blue."*

More recently, Sir Frederick Russell wrote a comprehensive review (The Medusae of the British Isles) on the distribution, seasonality, morphology and general ecology of *R. octopus*, beginning with Borlase's earliest notes and ending with references to Thiel's research in Cuxhaven and the German Bight during the 1960s (Russell, 1970; Thiel et al., 2024).

Fig. 26 The barrel jellyfish *Rhizostoma octopus*, Beara Peninsula, Ireland. *Photo: Ciaran O Murchu, with permission.*

2.6.2 Geographic distribution

Rhizostoma octopus is the most northerly distributed rhizostome, being centred around the Irish Sea but located as far north as the Clyde Sea area in Scotland and as south as Arcachon and St. Jean de Luz in the Bay of Biscay (Houghton et al., 2006). Populations are also known from the southern North Sea off the Belgium, Dutch and German coasts such as the Elbe estuary (Holst et al., 2007; Russell, 1970; Thiel, 1966). Within this broad geographical area, *R. octopus* is mainly found in shallow bays (with riverine input) and as such has a punctuated distribution, with only a handful of areas known where populations regularly occur (Houghton et al., 2006; Lee et al., 2013; Van Walraven et al., 2015). Occasional records of individuals are known from many other localities, but these may represent 'sinks' where individuals are transported and stranded on beaches from 'source' populations (Lee et al., 2013).

A recent study using a combination of Lagrangian modelling and genetics found that different *R. octopus* populations are strongly differentiated from each other (Lee et al., 2013). Two locations out of five were relatively isolated from the others in terms of currents (i.e. in oceanographic simulations no particles from these sites dispersed to any other site). This isolation was further supported by genetic analyses which found two barriers within the study area to gene flow: one between the southerly located La Rochelle site and all the Irish Sea sites, and the other between one of the Irish Sea sites (Tremadoc Bay in Wales) and all the other Irish Sea sites (Lee et al., 2013). These analyses suggest that the distribution of *R. octopus* populations is tightly linked to shallow bays where populations of polyps successfully recruit and differentiate over time. Lee et al. (2013) also suggested that the present pattern of *R. octopus* distribution in the North East Atlantic and North Sea is the result of populations periodically expanding northwards and then contracting southwards from the Bay of Biscay area in response to the ice age cycles. This was further supported by Glynn et al. (2015) who investigated the genetic structure of *R. octopus* in conjunction with palaeodistribution modelling. They found that after the Last Glacial Maximum (~21 kya) *R. octopus* likely expanded northwards from a single, limited refugium located in the southern half of the Bay of Biscay area (Glynn et al., 2015).

2.6.3 Densities and interannual abundance

The highest reported density of *R. octopus* is from Pertuis Breton, near La Rochelle, where Fossette et al. (2015) observed 178 individuals during a 5-minute visual survey from a rigid inflatable boat over a linear distance of 4.84 km (and with 7 m survey width). This translates into a density of ~5 000 ind. per km^2, which would make it one of the highest reported biomasses (110,000 kg km^{-2}) of gelatinous zooplankton (Lilley et al., 2009). Estimates from aerial surveys in the Irish Sea and especially within Carmarthen Bay were much lower with maximum densities of 130 ind. km^{-2} (Elliott et al., 2017; Houghton et al., 2006). However, these estimates likely represent more realistic broad scale densities than the small scale visual surveys which may inflate numbers due to local patchiness.

Inter-annual abundance estimates for *R. octopus* are very rare. Within the Irish Sea, aerial surveys were carried out across the Irish Sea during 2003 to 2005 with population estimates gathered for three bays (Carmarthen and Tremadoc bays in Wales and Rosslare Harbour in Ireland) (Houghton et al., 2006). From 2008 until 2011, additional surveys were carried out in Carmarthen Bay. Examination of this *R. octopus* time series by Elliott et al. (2017) found that there is considerable interannual variation on jellyfish abundance with biomass varying between zero in

2005 and a maximum of 594.2 tonnes in 2009 (mean abundance ~220 tonnes). It is not known why there was no recruitment in 2005 nor what the principle drivers of such variable jellyfish biomass are.

Outside of Irish and UK waters, a notable 50 year time series of *R. octopus* abundance is known for the western Dutch Wadden Sea based on fisheries bycatch data (Van Walraven et al., 2015). These latter authors found that there was considerable variation in abundance over time with the highest mean catch of *R. octopus* occurring in the early to mid-1980s (Van Walraven et al., 2015). In most years, mean daily catches (index of annual abundance) are < 10 ind. day^{-1} with only six years with mean catches greater than this. The authors noted that highest annual abundances of *R. octopus* coincided with a period of high eutrophication period (1973–1991), however none of the covariates (inc. eutrophication, sea temperature and salinity) in their models were significant.

2.6.4 Seasonality and overwintering

The seasonality of *R. octopus* does not fit the classic jellyfish pattern of northern temperate waters with ephyrae released in January/February, medusae maturing March-May, medusae with planulae in May-July and then majority of medusae stranding during June and July (Lucas, 2001; Doyle et al., 2007; Houghton et al., 2007). For example, *R. octopus* medusae appear to strand all year round with many medium to large medusae stranding during winter months and in March and April (Doyle et al., 2007; Houghton et al., 2007). Houghton et al. (2007) interpreted these strandings as evidence that *R. octopus* overwinters so that they can capitalise on the spring phytoplankton bloom. This strategy may only be possible for rhizostomes, especially *R. octopus*, that have a robust umbrella and oral arms that can survive the winter months on the sea bed and avoid being broken up. To date, there are no direct observations of barrel jellyfish resting on the sea bed but there are reports of individuals being hauled up by fishermen. For example, Bastian et al. (2011) reported catches of *R. octopus* individuals in the Celtic Sea from the Irish Ground Fish Survey which operates in October to November. Historically, Russell (1970) reported that 100 s of individuals were hauled up in the Irish Sea in January 1966, but gives no details on location. Perhaps, more convincingly is the detailed research carried out by Fay and Husson (1983) in the Pertuis Charentais area of the Bay of Biscay. Their results clearly distinguished two cohorts of *R. octopus*. The first comprises juvenile medusae that appear fresh in July with bell diameters of 10 cm that grow rapidly to 40–50 cm by September, while the second is made up of individuals of 30–50 cm that appear in April and May and which have an aged appearance and some are damaged. This second cohort must have overwintered as they cannot have grown to 30–50 cm diameter in one or two months if released as ephyrae in January/February. Furthermore, Fay and Husson (1983) clearly indicate that strobilation for *R. octopus* must occur during the summer months as they document smaller individuals in July, as does Skinner (in Russell, 1970) who reported that many medusae of 16–35 mm diameter in the Solway Firth, Scotland, during July 1966. Within the Irish Sea, Doyle (personal observation) recorded very small *R. octopus* in Rosslare Harbour, Ireland, in July 2004. Evidence for strobilation in the wild has not been observed for *R. octopus* but detailed work carried out by Holst et al. (2007) has demonstrated that strobilation can occur with a temperature drop or increase, and so may happen in autumn or spring. However, considering that most observations of young medusae occur in June/July, these observations suggest that strobilaton occurs mostly in spring.

2.7 *Rhizostoma pulmo*

Rhizostoma pulmo (Macri, 1778) is a large (up to 60 cm umbrella diameter) rhizostome that is distributed along the coastline of the Mediterranean Sea and adjacent areas (Leoni et al., 2021b). The medusae are distinguished by having a whitish bell, and their marginal lappets and the tips and upper margins of their oral arms are imbued with a characteristic blue-violet colour (Fig. 27).

Fig. 27 *Rhizostoma pulmo* in the Mediterranean Sea. *Photo: Laura Prieto.*

Despite its wide distribution and problematic status (see below), there have been few longitudinal studies on population cycles. From a 3-year study in the northwest Mediterranean Sea, Fuentes et al. (2011) noted that both adult medusae and ephyrae in the Catalan Sea were largely restricted to the central, rather than the northern or southern, area. These authors characterised the central area as being subject to "high mesoscale activity due to the dynamic coupling of coastal and oceanic waters", whilst the northern and southern areas were "characterised by the Ebro River runoff associated with higher temperatures and lower salinities" in the latter, or the "dominance by [the] Liguro-Provençal-Catalan Current with low water temperatures and north winds" in the former. Despite the fact that Fuentes et al. (2011) suggest that *R. pulmo* can survive and reproduce in water experiencing a variety of temperature and salinity regimes, as evidenced by its wide distribution in the Mediterranean Sea (Leoni et al., 2021b), high/low temperatures and low salinities may indeed impact distribution. And although substratum type for scyphistomae may be important, it is clear that physical processes of "retention" and "aggregation" must play a part in influencing the distribution of pelagic medusae. Physical processes were invoked too by Reyes Suàrez et al. (2022) to explain (in part) the abnormally high abundances of *R. pulmo* in the northern Adriatic Sea during April 2021. Indeed, these authors suggested that "*...strong wind events, such as the bora wind for the Gulf of Trieste, which enhanced upwelling and mixing processes in the gulf, br[ought] the jellyfish from the deeper waters to the surface and cluster[ed] them along the coast...*".

The life cycle of *R. pulmo* has been fully redescribed recently by Fuentes et al. (2011), adding to the earlier work of Claus (1883, 1884), Paspaleff (1938) and others. In addition to polydisk strobilation, polyps can reproduce asexually by typical lateral or stolonal budding, and by the formation of podocysts, pedalocysts or stolonal podocysts (Fuentes et al., 2011; Schiariti et al. 2024). In general, scyphistoma can generate between 8–20 ephyrae (Claus, 1883, 1884; Fuentes et al., 2011; Purcell et al., 2012). In his review on the seasonal occurrence of rhizostomes, Thiel observed that the appearance of the medusae of *R. pulmo* in the Mediterranean plankton varied with location (Thiel et al., 2024). Thus, in the harbour at Cette sur Mer, small medusae first appeared in May and large medusae had disappeared by the end of October. In the Gulf of Naples, by contrast, animals were present all year around, albeit only as large animals in small numbers during winter but as swarms of smaller individuals in summer. Further east, in the northern Adriatic Sea, Thiel (reporting; Issel, 1922) noted that although this species was present for much of the year, there were periods when it was absent – and these periods of absence were not always in winter or consistent between years (Thiel et al., 2024).

More recent observations (Leoni et al., 2021b) largely substantiate Thiel's summary of seasonality (see Thiel et al., 2024). Fuentes et al. (2011) observed the species to first appear in the plankton of the Catalan Sea during May and had disappeared by the end of August/September. Over the three years of their study (2007–2009) in the Mar Menor (northwest Spain), Fuentes et al. (2011) observed the first ephyrae in plankton samples at the end of May, and the first appearance of mature medusae in August. The consistent presence of ephyrae indicated to Fuentes et al. (2011) that populations of *R. pulmo* in the Mar Menor are self-maintaining, though as elsewhere polyps have never been observed *in situ*. As a rule, the numbers of medusae decreased throughout the summer and they had mostly disappeared by January and March (Fuentes et al., 2011), though in some years, adult medusae *"did not disappear until [the] end of June of the following year, with some adults occurring with the new ephyrae"* (Fuentes et al., 2011). Effectively similar observations were made by Fernández-Alías et al. (2020) in the Mar Menor from data collected in 1997. These authors recorded a peak in abundance during mid-June, though interestingly, three more (lesser) peaks were noted over the period between late June and August and isolated individuals could be found overwintering. Further east, in the shallow Bages Sigean (southwestern France), Leoni et al. (2021a) too observed three peaks in the abundance of *R. pulmo*: a maximum in April and two lesser ones in June and July. Ephyrae occurred from April to June; juveniles from April to September and adults from May to October: no medusae were observed to overwinter. In the southern and eastern Adriatic Sea, Pestorić et al. (2021) noted that *R. pulmo* was most commonly seen during summer and autumn, but that like Thiel's earlier comments it was found throughout the year in the north (Thiel et al., 2024). Off the coast of Turkey, in the Levantine basin, mature medusae are observed mostly in late summer through the winter season, and it blooms in the Black Sea between September and November (Öztürk et al., 2018). Although the species is not common in the Gulf of Tunis, medusae are episodically present between May and October (Yahia et al., 2003).

In an effort to understand the effects of temperature on asexual reproduction of the polyps of Mediterranean jellyfish, Purcell et al. (2012) maintained those of *R. pulmo* in the laboratory at three temperatures (14 °C, 21 °C and 28 °C): all other environmental conditions were held constant (38–39 salinity; 12 h:12 h illumination cycle; fed once weekly with newly hatched *Artemia* sp. nauplii). Although polyps strobilated across this temperature range during the 108 days of study, they were slower to start strobilating at 14 °C than at the higher temperatures, and the period of strobilation was longer (Purcell et al., 2012). But, while more ephyrae were produced at 14 °C than at the other temperatures in the first strobilation, polyps were able to re-strobilate and the numbers of ephyrae released in the second event were generally higher than in the first. That said, Purcell et al. (2012) noted that the numbers of ephyrae per polyp were, on average, highest at 21 °C. Given the shorter time it took both for strobilation to be initiated and the duration of a strobilation event, Purcell et al. (2012) concluded that higher temperatures may be better than 14 °C for overall ephyra production.

The results presented above are interesting because they suggest that temperature change is not important in initiating strobilation, yet it is a change in the environment (temperature, salinity, day length etc.) that generally acts as a trigger for strobilation (Lucas et al., 2012). Although Purcell et al. (2012) did not stress its importance in the interpretation of their results, it is worth remembering that their laboratory cultures were maintained at ambient temperature prior to experimentation. This translates to approximately 13 °C (not stated), which means that all polyps were probably subject to a thermal change. And thermal change was indicated as driving the release of ephyrae by polyps of *R. pulmo* in Mar Menor by Fernández-Alías et al., 2021 (Fig. 28). These authors inferred the strobilation of polyps by tracking the appearance of the ephyrae and observed four separate pulses of ephyrae to be released between April and June, and in each case this was associated with an increase in temperature of ~0.4 °C day^{-1}.

Although a change in environment is to be expected with regards to strobilation, we must be careful not to over-generalise, because Fuentes et al. (2011) maintained scyphistomae in culture over a period of three years under constant environmental conditions (37–38 salinity; 21 °C temperature; 12 h:12 h illumination cycle; fed twice weekly with newly hatched *Artemia* sp. nauplii), and noted that they (only) strobilated spontaneously in spring. But while the jury may be out on the exact details cueing strobilation, it is clear that polyps can survive and strobilate over a range of temperatures > 10 °C (Purcell et al., 2012). Interestingly though, the thermal and salinity niches of ephyrae (which are narrower than those of juvenile or adult medusae), vary with location (Leoni et al., 2021a).

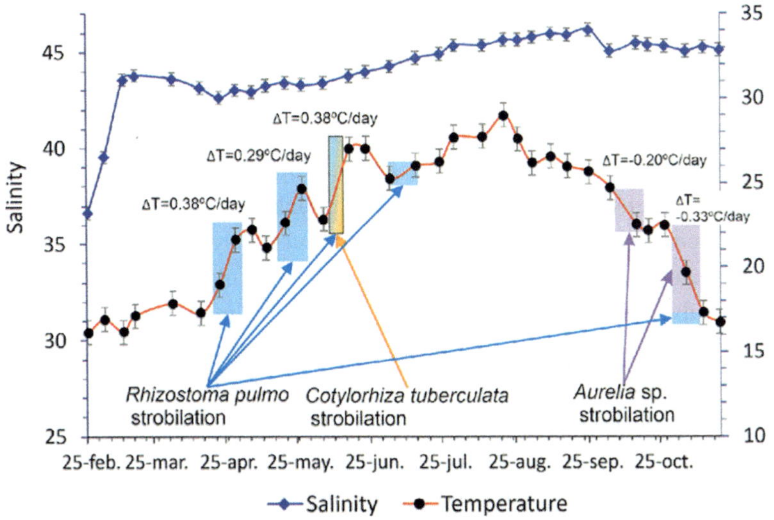

Fig. 28 Temperature and salinity records from Mar Menor lagoon through 1997 indicating the daily rate of increasing or decreasing temperature in the peaks of strobilation for each jellyfish species. Mean abundance of the variable within the lagoon ± standard error. *From Fernández-Alías, A., Marcos, C., Pérez-Ruzafa, A., 2021. Larger scyphozoan species dwelling in temperate, shallow waters show higher blooming potential. Mar. Pollut. Bull. 173, 113100.*

Thus, in the hypersaline Mar Menor, ephyrae are most likely to be found at temperatures between 15 °C and 24 °C, and at salinities between 33 and 37, whilst these same values in the Bages Sigean are between 13.5 °C and 27 °C, and 39 and 47, respectively (Leoni et al., 2021a). And herein lies the rub. *R. pulmo* is not "native" to either of these two systems, at least not in modern times, and the two lagoons were first "colonised" in the mid 1990s (Fuentes et al., 2011) and 2014 (Leoni et al., 2021a), respectively: they are but 670 km away (as the crow flies). Both are situated in the Western basin of the Mediterranean Sea, though the former falls within the Balearic sub-basin and the latter in the Gulf of Lyons. Both (probably) also fall within the same clade of *R. pulmo* identified by Ben Faleh et al. (2017) and so it is likely that both originated from a common ancestor. Yet, the thermal and salinity niches of the ephyrae of these two populations are very different implying a rapid local adaptation to prevailing environments.

The species is argued to have increased in abundance across much of the region since the late 1970s and early 1980s, and these blooms have caused a number of economic and health issues. Coastal areas that have been badly affected include the Adriatic Sea (Malej and Vukovic, 1984), the Ionian Sea

(Scalera-Liaci, 1991), the Black Sea (Mariottini and Pane, 2010), along the Ligurian Riviera (Carli, 1991) and a number of locations in Turkish waters (Bingel et al., 1991). It has also become a problem in semi-enclosed lagoons in the Mediterranean and Black Seas (Fuentes et al., 2011; Leoni et al., 2021a): Bages Sigean (France), Mar Menor (Spain) and Bizerte (Tunisia).

A comprehensive analysis of changes in populations of *R. pulmo* across the greater Mediterranean has recently been conducted by Leoni et al. (2021b), using semi-quantitative methods. Their results are informative. Firstly, they noted a substantial increase in the frequency and "intensity" of blooms in recent decades, which was explained partly by a general increase in temperature anomalies in the basin as a whole. Secondly, they observed that the start and end of the jellyfish season have shifted forward in the year over the last 10 years: spring temperature was correlated with the start and duration of the season while winter temperature was linked to the end of the season. Thirdly, they noted a positive correlation between latitude and bloom intensity, with northern locations (at a lower average temperature) showing less intense bloom events than those further south. And lastly, they observed substantial variability in the phenology of blooms by ecoregion, though they cautioned a too literal acceptance of all the results owing to patchiness in data quality for some areas. Regardless, the stronger patterns revealed that in the Adriatic Sea, blooms have begun to appear earlier in the year and there is a positive link between the duration of the bloom and winter temperatures. In the Western Mediterranean, the date of last occurrence has shifted from October (in 2008) to December (after 2017), and this is correlated with annual water temperatures. But in the Ionian Sea, by contrast, the month of the last occurrence has shifted from December (2008–2015), to September–October (2016–2018), which they linked to spring temperatures. Again, then there is strong evidence for local adaptations to regional environments.

Leoni et al. (2021b) have suggested that eutrophication *per se* is not driving changes in the appearance, magnitude or duration of blooms of *R. pulmo* in the Mediterranean Sea. And this despite the fact that populations may now be present, and indeed flourishing, in regions impacted by cultural eutrophication, and in which they have a strong ecological footprint (e.g. Perez-Ruzafa et al., 2002, 2019). That said, the relationship between the thermal environment and the structure and composition of the food environment is implicated by Leoni et al. (2021b) as impacting populations of *R. pulmo*, under some circumstances.

They argued that the enhanced duration of the medusae season might reflect various processes acting on the early life stages of the scyphomedusae, as well as a longer survival of adult stages (Leoni et al., 2021b). The greater length of the bloom period may be favoured by earlier blooms in those years with warmer spring and also to prolonged strobilation. Leoni et al. (2022) further hypothesised that the negative relationship between springtime temperatures noted in some areas and the start of the jellyfish season could be interpreted as meaning higher growth rates under warm springs, as evidenced by adults being observed earlier in the year. Bottom-up processes reflecting the effect of springtime temperatures on lower trophic levels was also postulated as leading to increases in the abundance of prey or changes to the composition of the prey field was also suggested. By contrast, warmer winters lead to an earlier collapse of jellyfish populations (particularly in the Levantine), which Leoni et al. (2021b) indicated could reflect the fact that prolonged stratification might exhaust nutrients in the surface waters, thereby limiting primary, and in turn secondary, production.

Although some of the arguments used to substantiate the conclusions reached by Leoni et al. (2021b) are based on experiments (e.g. Purcell et al., 2012), their key results are founded on correlative analyses and incomplete data, with all its flaws. Regardless, they provide a valuable framework for future work and a very useful insight into the possible mechanisms explaining seasonality and blooms in one of Thiel 's most covered subject animals (Thiel et al., 2024).

2.8 *Rhizostoma luteum*

Rhizostoma luteum (Quoy and Gaimard, 1827) is a large sized (up to 60 cm bell diameter) rhizostome (Fig. 29). The species was first described as *Orythia lutea* in 1827 from specimens found in the Strait of Gibraltar

Fig. 29 *Rhizostoma luteum* in the North-East Atlantic. *Photo: Laura Prieto.*

(Southern Iberian Peninsula). However, *R. luteum* was rediscovered and phylogenetically described in 2012 after being scientifically unnoticed and misidentified for more than 60 years (Prieto et al., 2013). At present, the species is recognised as the third species of the genus *Rhizostoma*, together with their congeners *R. pulmo* and *R. octopus*. The adult specimens can be morphologically identified by their milky-cream umbrella and their characteristic dark long purple-brown terminal clubs.

Rhizostoma luteum has a broad distribution along the eastern Atlantic Ocean, from the coast of Portugal (Stiasny, 1936, 1931) to South Africa (Kramp, 1959, 1955; Ranson, 1949), and in the southwestern Mediterranean Sea, including the Strait of Gibraltar and the Alboran Sea (Grenacher and Noll, 1876; Kienberger and Prieto, 2018; Prieto et al., 2013; Quoy and Gaimard, 1827). In 2013, Prieto et al. (2013) reported the presence of the species on the Atlantic coast of Morocco and along the south shore of the Iberian Peninsula, from the Gulf of Cádiz to the Alboran Sea. Later on it was again registered in the Atlantic coast of Morocco (Gibbons et al., 2021). Due to its wide distribution, *R. luteum* life stages must be adapted to a wide range of temperatures, with minimum sea surface temperature of 14 °C from Portugal to the Alboran Sea, to 25 °C for the Gulf of Guinea (Kienberger et al., 2018). Laboratory studies demonstrated that temperature shocks as those occurring in the field induced by coastal upwelling does not affect the survival of *R. luteum* scyphistomae (Kienberger et al., 2018), however, further experimental studies are needed.

Rhizostoma luteum has the typical metagenetic life cycle of scyphozoans (Kienberger et al., 2018). However, this species differs from their congeners *R. pulmo* and *R. octopus* by being a brood-carrying species, adult females are internally fertilised with the subsequent release of planulae. Then, larvae settle and develop into benthic polyps that reproduce asexually via podocyst formation by means of stolons or pedal disc (Kienberger et al., 2018). Strobilation in *R. luteum* is typically monodisc, unlike the other two *Rhizostoma* species that undergo polydisc strobilation (Kienberger et al., 2018).

2.9 Observations on invasive alien rhizostomes

It is widely considered that invasive alien species represent a critical threat to the biodiversity and functioning of terrestrial (Bajwa et al., 2016; Early et al., 2016) and freshwater ecosystems (Strayer, 2010), yet our appreciation of their impact in wider marine environments is less clear. Climate change (Worm and Lotze, 2021), including marine heatwaves (Smale et al., 2019), and the exploitation of living resources (Jennings and Kaiser, 1998; Mazaris et al., 2019) are all regarded as THE major threats to marine ecosystems and biodiversity, with alien invasive species rather having an impact mainly at the scale of local systems (e.g. but see Boudouresque et al., 2017). In a recent meta-analysis of the effects of exotic marine species on their environment, Anton et al. (2019) concluded that while alien predators (and primary producers) had a significant impact on assemblages, mobile species did not, and that in general the ecological effects of exotics were focused on communities and not on ecosystems as a whole.

At the time Thiel wrote his chapter on ecology (in the 1970s, see Thiel et al., 2024), alien and invasive marine species were poorly studied and their impacts on the environment only really guessed at. Indeed, data regarding the distribution of individual species was (and still is, we might add) far from complete and so extralimital records were regarded as just that. He notes *"If one considers the sporadic distribution of the individual species and the great distance of some localities of a species from each other, the thought suggests itself that these species could also occur in the intermediate areas, but have not yet been observed……provided they are really the same species"* (Thiel et al., 2024). In other words, unusual geographic records were considered as limital, and it is no surprise to note then that the account makes no mention of "alien", "exotic" or "invasive" species.

The most recent review of invasive marine jellyfish remains that of Bayha and Graham (2014), which builds on Graham and Bayha (2007). These authors noted that five species/taxa of Scyphozoa could be so regarded: *Rhopilema nomadica, Aurelia* spp. *Phyllorhiza* spp. *Cassiopea* spp. and *Mastigias* spp. Crypsis confuses our documentation and understanding of distribution and some of the latter four of the aforelisted genera display this in spades (Dawson and Jacobs, 2001; Swift et al., 2016); hence the "spp." epithet. To this list we can now add *Marivagia stellata* (Galil and Gershwin, 2010; Galil et al., 2010) as well as *Mawia benovici* (Piraino et al., 2014; Bayha et al., 2017), both of which were tentatively considered as aliens at the time of their first description. Interestingly, five of these seven taxa are Rhizostomeae. It is not our intention to repeat the reviews of Graham and Bayha (2007) or Bayha and Graham (2014) regarding this subject, but rather to make some general observations, as they pertain to rhizostomes.

But before we do that, it is perhaps worth exploring the "alien" nature of *R. nomadica* in the eastern Mediterranean. The species originally came to prominence in 1977, when it was observed for the "first" time in the

Mediterranean Sea off Israel. At that time, it was regarded as an unknown species. A name was assigned to the species some thirteen years later by Galil et al. (1990), with the holotype and paratypes being designated from specimens collected in 1939 and 1937, respectively, in the waters off Yemen (Kamaran) in the Red Sea. That description also included a large number of specimens collected off Israel in 1988 and 1989. Given that the first known specimens of this species had been recovered in the Red Sea, and it was not until some forty years later that it "appeared" in the Mediterranean Sea, Galil et al. (1990) were not wrong to suspect that the species was an alien.

However, there is an intriguing archival image of medusae from Haifa Bay taken in 1938 (Fig. 30), which strongly indicates that the species was already present in the eastern Mediterranean Sea at the same time the specimens used by Galil et al. (1990) to designate it were collected off Yemen. And this begins to question whether we were right to immediately assume the species was an alien. A photograph is not a specimen, of course (though we are quick to assign species names to photographs today), but there are no other species the snapped creatures could be. Confusion with the only possible contender, *R. pulmo,* is eliminated by the absence of terminal appendages, the lack of any pigment on the marginal lappets, and the overall shape of the oral arms.

Fig. 30 Yitzhak Rakotsh, Jellyfish in the Port. 19.10.1938. Silver print, 16.5 × 23 cm. From: Home Port, the story of Haifa Port, October 2002. Haifa Museums, Haifa City Museum. *Courtesy Haifa Port.*

Assuming that *R. nomadica* was already present in the waters of the Levant in 1938, why did it take so long to become problematic? Of course, one could argue that the original Israeli population went extinct and that the blooms witnessed in the 1970s were of newly introduced specimens. But it is more likely that the population persisted in the region as polyps and the medusae then emerged when conditions became favourable for them. Prolonged periods of population quiescence have been noted elsewhere in the Mediterranean Sea for *Drymonema dalmatinum* Haeckel, 1880 (Malej et al., 2014), and possibly also in the Gulf of Mexico for *Phyllorhiza punctata* von Lendenfeld, 1884 (Graham et al., 2003; Bolton and Graham, 2004). And recent observations in the laboratory support this cryptogenic species hypothesis (Kuplik, unpublished). Bear with us….

Cassiopea andromeda (Forskål, 1775) is a well-known Lessepsian invasive (Spanier, 1989), and yet the medusa stage is rarely seen *in-situ* – in the Levant. One day in 2013, however, an ephyra of *C. andromeda* appeared in an aquarium at the marine lab at Michmoret, Israel, housing an octopus that was being supplied with micro-filtered seawater on a flow-through basis (Fig. 31A). A search of the tank led to the discovery of a rock covered with the polyps of *C. andromeda*, which had been collected in the coastal waters 15 km north of the aquarium's location (Fig. 31B). Soon thereafter, a massive bloom of this species occurred in the tank (Fig. 31C) and a stable population of *C. andromeda* has subsequently persisted in the aquaria for the last ten years. But none are to be seen *in situ*. These observations suggest that because conditions in the laboratory favoured *C. andromeda*, the adult medusa stage could develop in the tanks, but that conditions in the field are not presently conducive to their survival. Changes in the food web, the disappearance of natural predators, or other changes in the environment can all serve as drivers for the proliferation and behaviour of scyphozoan

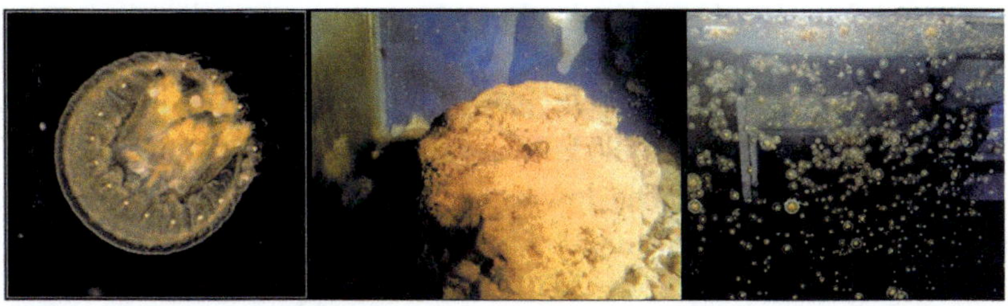

Fig. 31 *Cassipoea andromeda* as observed in the marine laboratory: (A) The 1st observed ephyra, (B) the 'polyp rock', (C) a bloom of *C. andromeda* in the aquarium. *Photo: Zafrir Kuplik.*

polyps (Lucas et al., 2012; Chi et al., 2019; Loveridge et al., 2021b; Schiariti et al., 2024). And when conditions are unfavourable, strobilation may be reduced or even inhibited and the species will remain cryptic until conditions improve (Lambert, 1935).

Arguing thus, *R. nomadica* may not be an alien to the eastern Mediterranean Sea, and it is possible instead that it is an anti-Lessepsian migrant! Referring to exotic species, Carlton (1996) wrote that *"the default in this view is to classify species without any obvious record of introduction as 'native'."* Only more detailed molecular work will resolve this question (see Gamero-Mora et al., 2024), but it is one we need to remember before we jump hastily to conclusions regarding nativity, because there is an ecological stigma attached to being an "alien".

Successful invasive alien species tend to be generalists that can tolerate a relatively wide range of climate conditions (Walther et al., 2009). And as a rule, jellyfish can; both as medusae and polyps. That said, Bayha and Graham (2014) note that *"…some species, such as P. punctata (Rippingale and Kelly, 1995), R. nomadica (Lotan et al., 1992), and M. papua (Dawson et al., 2001), have shown evidence of [relatively] narrow salinity and/or temperature tolerances, potentially restricting invasion possibilities to environments similar to their native ranges"*. In her review of non-indigenous species in the Mediterranean Sea, Galil (2008) suggested that most alien taxa there could be considered as thermophilic, with origins in the Indian Ocean, Red Sea and Indo-Pacific region. All of the invasive rhizostomes noted here are endemic to the latter parts of the world (accepting their arguments regarding the origins of *R. nomadica*) and all occur naturally in waters that are warmer than those experienced in their new ranges. That the first appearance of all but *Mastigias* spp. in the Mediterranean Sea is therefore in the Levantine Basin is perhaps no surprise.

These Erythrean or Lessepsian migrants have spread progressively westward from Israel, seemingly along the northern coastline in a manner that appears to follow patterns of surface circulation in the Mediterranean Sea (Fig. 23). In the case of *R. nomadica* and *P. punctata*, this expansion approximates a rate of ~80 $km.y^{-1}$ and between 70 and 95 $km.y^{-1}$, respectively (Table 1). The lower value in the latter range reflects the species' arrival off Sardinia in 2009 (Boero et al., 2009) and central Libya in 2015 (Rizgalla and Crocetta, 2020), while the upper value is based on its first appearance in Cadiz during 2018 (Enrique-Navarro and Prieto, 2020). All values are premised on the assumption that specimens originate from populations in the eastern Mediterranean and are transported according to

Table 1 Records of selected alien Scyphozoa in the Mediterranean Sea.

Species, Native range	Location	Date	Source
Marivagia stellata: Cepheida SW India	Israel	2006	Galil et al. (2010)
	Lebanon	2010	Bitar and Badreddine (2019)
	Syria	2015	Mamish and Durgham (2016)
Rhopilema nomadica Rhizostomatida NW Indian Ocean	Israel	1976	Galil et al. (1990)
	Lebanon	1988	Yahia et al. (2013)
	Syria	1991	Yahia et al. (2013)
	Turkey	1995	Kideys and Güccü (1995)
	Greece	2006	Sioukou-Frangou et al. (2006)
	Malta	2004	Deidun et al. (2011)
	Sardinia	2015	Balistreri et al. (2017)
	Egypt	<2015	Madkour et al. (2019)
	Tunisia	2008	Yahia et al. (2013)
Phyllorhiza spp. Cepheida SE Australia	Israel	1965	Galil et al. (1990)
	Greece	2005	Abed-Navandi and Kikinger (2007)

(continued)

Table 1 Records of selected alien Scyphozoa in the Mediterranean Sea. *(cont'd)*

Species, Native range	Location	Date	Source
	Italy (Sardinia)	2009	Boero et al. (2009)
	Tunisia	2012	Gueroun et al. (2015)
	Lebanon	2015	Dailianis et al. (2016)
	Syria	2011	Durgham (2011)
	Turkey	2010	Cevik et al. (2011)
	Malta	2016	Deidun et al. (2017)
	Libya	2015	Rizgalla and Crocetta (2020)
	Spain (Cadiz)	2018	Enrique-Navarro and Prieto (2020)

the currents shown in Fig. 23. While the westward spread is undoubtedly linked to recent warming in the region (López García 2015), it is undoubtedly, too, the result of natural and not human-assisted dispersal, following the likely sequential establishment of polyp beds *en-route*.

By contrast, *C. andromeda*, which was first recorded in the Suez Canal off Egypt in 1886 (Keller, 1888) only reached Tunisia in 2013 (Amor et al., 2016), via Cyprus (Maas, 1903), Greece (Schäfer, 1955) and Malta (Schembri et al., 2010). This equates to a dispersal rate of between 20 and 25 $km.y^{-1}$. Although Schembri et al. (2010) suggested its arrival in Malta was a result of anthropogenic intervention, the rate of westward spread (Egypt – Cyprus; Cyprus – Greece; Greece – Malta; Malta – Tunisia) appears relatively consistent through time, which hints at natural dispersal. Tempting though it is to interpret this slower dispersal (*cf. P. punctata, R. nomadica)* as a function of the benthic medusoid habitat, more work is clearly needed.

Although holopelagic ctenophores may be transported around the world in ballast tanks (e.g. Vinogradov et al., 1989), it is more likely that scyphozoans are shipped around the globe as fouling organisms on the hulls of vessels (Kramp, 1970; Bayha and Graham, 2014). By virtue of their benthic existence in shallow water, rhizostome polyps are likely exposed to a broad range of environmental conditions (temperature, oxygen, salinity) over the course of any one year in their natural habitat, alternating between periods of (possibly prolific) asexual population expansion when times are good and periods of stasis when things turn sour. Although some can also produce podocysts to weather out really unfavourable conditions, four of the five invasive rhizostomes belong to the family Cepheida, which do not produce podocysts (Arai, 2009).

And this introduces an interesting conundrum. Species of Cepheida reproduce asexually by generating planuloid swimming buds (Arai, 2009), which allows colonies to expand at a faster rate (Schiariti et al., 2014, 2024) than species such as *R. nomadica* that produce podocysts (Lotan et al., 1992). While the former mode of asexual reproduction may favour their success as invaders (for reasons that remain unclear), ephyra production is by monodisc strobilation (Helm, 2018). Like other species of Rhizostomatidae, *R. nomadica* displays polydisc strobilation (Schiariti et al., 2024) and Lotan et al., (1992) argue that its high reproductive potential (up to four strobilation events per year per strobila) allows populations of medusae to bloom. And in their review, Dawson and Hamner (2009) highlight the role of both of these features (podocysts and polydisc strobilation amongst others) as being commonly displayed across swarming taxa. Yet, *P. punctata* (and to a lesser extent species of *Cassiopea*), which in theory has characters that do not favour blooming, can and does bloom, especially in the northern Gulf of Mexico (Graham et al., 2003).

We suspect that this has less to do with the characteristics of the species, *per se*, than in the environment in which the species finds itself on introduction. In general, introduced species can become problematic in systems that are already compromised or that support relatively low levels of functional/species diversity (Bayha and Graham, 2014 and references therein). In the case of the Mediterranean Sea, diversity is highest in the west (especially northwest) and lowest in the east (Coll et al., 2010), and this pattern mirrors that of primary production (Colella et al., 2003). At face value then, the Levantine Basin is perhaps more at risk than elsewhere. But, the eastern Mediterranean also supports a comparatively greater number of fishing vessels than other areas (F.A.O., 2020) and was regarded by Tsikliras et al. (2015) as having its fisheries in "…a worst state…" than elsewhere. The latter point suggests perhaps that sufficient niche space was available for jellyfish to establish themselves in. While it is clear from the most recent State of the Environment and Development report for the Mediterranean (United Nations, 2020) that much of the region's coastal zone is impacted by humans, the eastern basin appears to be a particular focus of warming, urbanisation, and (potential) agricultural run-off. Unlike the Levantine Basin, the Gulf of Mexico is recognised as rich in species (Fautin et al., 2010) but, like the former, it too is impacted by a suite of human-mediated stressors (McKinney et al., 2021). And many of these stressors appear to be most severe in the northern subregion (Love et al., 2013), where *P. punctata* has bloomed, which again points to the value of jellyfish as bellwethers of environmental change.

2.10 Summary and thoughts on seasonality

All rhizostome jellyfish have metagenetic life cycles and strobilation marks the transition from benthic to planktonic life (Schiariti et al., 2024). Although it is clear that most species display seasonality, as typified by *N. nomurai* and *C. mosaicus*, it is not always pronounced. For example, *R. octopus* can be found stranded on beaches throughout the year, but large aggregations may only be visible during summer/autumn (Doyle et al., 2007). And there are, as ever, variations that are influenced by conditions in the surrounding environment, especially temperature (see above).

If we have been routinely monitoring the plankton for ephyrae, then we can make deductions about the cues that may lead to strobilation, *in situ*. If we haven't, we can culture polyps in the laboratory, and then conduct controlled experiments to determine the cues for strobilation (Fuentes et al., 2011; Holst et al., 2007; Kawahara et al., 2006; Kienberger et al., 2018). Although the cues determined for the laboratory kept polyps are not always applicable to the polyp populations *in situ*, their measurement allows us to determine which factors are responsible for the strobilation of wild polyps. Never have we managed to track the fate of rhizostome polyps, *in situ*, though polyps of *N. nomurai* and *R. esculentum* that were artificially established on plastic plates were monitored for strobilation by placing them *in situ* by Feng et al. (2017). Unfortunately, however, although both species did strobilate, long-term survival and subsequent monitoring was compromised by overgrowth from other sessile organisms (Feng et al., 2017).

Strobilation tends to take place in winter and spring in most temperate rhizostome species, indicating that seasonal chilling is a primary cue (Lucas et al., 2012, Schiariti et al., 2024). That said, for more tropical species, it may occur at warmer temperatures, as observed in *C. andromeda* (≥ca. 25 °C) from subtropical Gulf of Aqaba (Rahat and Adar, 1980). The cues for strobilation may not be fixed and it is worth noting that a change in the temperature required for strobilation to take place has been observed in the case of *R. nomadica* in the eastern Mediterranean Sea. Thus, where strobilation used to be noted once in spring (18–20 °C) in the late 1980s (Lotan et al., 1994), since the 2010s it is now observed twice: once in spring and once again in autumn (28 °C) and this has resulted in a phenological shift of medusa occurrence (Edelist et al., 2020).

If the timing of strobilation is clear and discrete, growth can be tracked following the release of ephyrae, as populations develop as recognisable cohorts and then decline (Leoni et al., 2021a). In temperate species, life cycles are generally less than a year, but some species can overwinter and live longer than a year (see below). For populations that occur within discrete areas, such as *C. mosaicus* off Australia, we are in a position to understand the process ourselves. This species is predominantly estuarine, and medusae are thought to be serial spawners, with regional populations comprising individuals in reproductive condition throughout the year. Patterns of recruitment to local populations vary across estuarine systems and there is little in the way of ephyral transport from one system to another, which would otherwise tend to synchronise recruitment. The high level of individual retention within separate estuaries effectively means that populations are self-seeding – though studies indicating exactly how different life-history stages capitalise on the vertically stratified current systems present in estuaries is unclear (as noted for copepods; Kimmerer et al., 1998). It is not unlikely that similar patterns are observed for other estuarine-dependent species. Certainly, ongoing work on a hitherto undescribed species of rhizostome along the south coast of South Africa reveals that within any (of five different) estuary (situated between Cape Agulhas and East London) there is < 5% pairwise sequence divergence (COI) between specimens, but there is ~12% pairwise divergence between estuaries. The latter is based on the analysis of a total of 468 bp for COI, of which 113 (24.14%) were invariant and 355 (75.86%) were parsimony informative.

But local recruitment is also implied for populations of species occurring over broader continental shelf environments (see e.g. Abboud et al., 2018). For example, pronounced genetic differences are seen between populations of *R. pulmo* in the Mediterranean Sea (Ben Faleh et al., 2017), and between populations of *R. octopus* in the Irish and Celtic Seas and the Bay of Biscay (Lee et al., 2013; Glynn et al., 2015). Although the former of these can be explained, in part, as a product of the separation of the eastern and western basins ~4.6 million years ago (Mizrahi et al., 2021), with accompanying current closure systems, the latter is a much more mesoscale feature (Lee et al., 2013). The latter authors suggest that patterns of surface circulation may limit the contact between some of the populations sampled, and that recurrent blooms in these locations reflect local recruitment. In other areas, however, their data suggest that populations are maintained by source-sink dynamics, because local polyp populations fail to establish themselves.

By contrast, in the case of taxa that have much more widespread distributions, the only way we can understand population processes is through collaboration, as in the case of *N. nomurai*. Prior to the turn of the twenty-first century, there were ca. 40-year gaps between major blooms of this species in the East Asian marginal seas, but this time period has shortened, and it now blooms more frequently. In 2006, Chinese, Korean and Japanese researchers established an international collaborative project in order to gain a comprehensive picture of the life cycle and geographical distribution of this ocean going rhizostome and have been working together since. Although this collaboration has begun to make it possible to forecast the year-to-year bloom intensity, thereby alleviating the damage to fisheries, it is still not possible to provide a clear answer to the very basic question about the culprit for the sudden population increase in the 2000s and subsequent decrease after the big bloom in 2009. To fully understand the mechanisms behind seasonal and interannual fluctuations of rhizostome populations, an integration of intensive regional and/or laboratory studies on the ecology and physiology of all life stages, particularly the polyp stage, and extensive, often international collaborative, studies on the geographical distributions of medusa population for as many years as possible is required.

Although *N. nomurai* shows no population structure across its range in the Yellow and Bohai Seas off the east coast of China (Dong et al., 2016), these are genetically distinct from populations off the west coast of the Korean Peninsula (Gotoh et al., 2017). This implies that polyp beds on the western and eastern side of the Yellow Sea, may be self-recruiting (or certainly self-maintained, but see Gotoh et al., 2017). Medusae populating waters off Japan and the eastern side of Korean Peninsula comprise a mix from both populations, and these are advected there in the Tsushima Current from a mixed pool previously aggregating in the low-salinity waters of the East China Sea. Given that *R. nomadica* has begun to shift the temperatures at which strobilation can take place, it will be interesting to see if *N. nomurai* can do likewise, and whether polyp populations can establish themselves off either the east coast of the Korean Peninsula or around Japan, where they are presently absent.

In the case of coastal fish populations, discrete geographic phases to their life cycle tend to reflect Bakun's (1990) triad of enrichment, retention and concentration. Thus, adults tend to reproduce at a particular time of year and in a particular place, spawning products are passively transported to nursery areas where they grow and then recruit, before individuals move once again through to spawning areas. This sequential pattern of movement is well documented in the case of short-lived anchovy and sardine populations across the world (e.g. Hutchings et al., 1998; Lett et al., 2007), and it permits these species to capitalise on spatial and temporal changes in the mesoscale environment that simultaneously allow them to maintain their position in natal systems and promote growth and survival. Although discrete stocks of non-anadromous species within regionally distributed populations may have separate spawning areas, mixing in common recruitment areas often ensures subsequent panmixia (e.g. Bakun and Weeks, 2006; de Moor et al., 2022). With the exception of some achelate decapods, which have phyllosoma larvae capable of long-distance migration (Sekiguchi and Inoue, 2002), the spatial patterns of distribution displayed by the different life-stages of fish contrast with those of most benthic invertebrates. In large part because fish have the ability to move horizontally against prevailing water currents, although most make use of local currents to limit energy expenditure (Bakun, 1998; Hutchings et al., 1998). In the case of benthic invertebrates, adults spawn *in habitus*, fertilised eggs are dispersed by local and regional currents, and larvae metamorphose and settle following behaviours that maximise their probabilities of return (e.g. DiBacco et al., 2011; Shanks and Brink, 2005; and references therein). Of course, timing is everything and spawning is attuned to the likely biological and physical environment that will favour survival, retention and settlement of offspring. Thus, the viability and persistence of many benthic invertebrate populations depends on the return of larvae to adult habitats, so that adult populations are maintained.

Jellyfish are neither fish nor are they members of the regular meroplankton, as Boero et al. (2008) have noted in their classic paper on "… irregularities rule…", and the maintenance of jellyfish populations is likely to be achieved through the "immortal" persistence of polyp stocks. And this may have nothing to do with replenishment or recruitment from subsequent life-stages. In one sense, once polyps have established themselves in a particular, generally favourable, area their populations are almost invulnerable – in the absence of anthropogenic interference. I say almost invulnerable because there are predators of polyps (see below) but how effective they are at controlling the population is unknown. The occasional recruitment of planulae seems to be "nice-to-have", but may not necessarily be vital to population persistence. In other words, as long as "some" planulae recruit to the polyp habitat "sometimes", medusae are free to do their "own thing" – disperse. Yet the high level of genetic structure observed in jellyfish populations over relatively short geographic distances (e.g. Abboud et al., 2018) implies that medusae are not very good at this – or at least that the products of sexual reproduction from dispersed medusae are not very effective at establishing separate populations of asexually reproducing polyps. Quite why this may be is at this stage unknown, and is surely deserving of further attention, but it might not be linked so much to the survival of planulae/polyps but to mismatches between an individual's strobilation cues and the "new" environment.

3. Feeding ecology and role in ecosystems

3.1 Introduction

The concept of the ecosystem emphasises the way that energy and materials move between living and non-living components in an interdependent and structured way. It is founded on the works of Elton (1927), Tansley (1935), Hutchinson (1940, 1948) and Lindemann (1942), amongst others, which similarly spawned an appreciation and understanding of concepts such as the biosphere and biogeochemistry. Modern approaches to the management of our environment are largely based on ensuring the integrity of ecosystems, their services and provisions, in the belief that in balanced ecosystems any impacts on one component will have knock-on effects to others. In the case of capture fisheries, for example, management agencies strive to implement the ecosystem approach to fisheries (EAF), which attempts to balance the needs/wants of humans with those of other ecosystem components (Garcia et al., 2003). But in order to determine the others' needs we require information on ecosystem structure, as well as on the biomass and production of ecosystem components and we need to understand flows and fluxes between said components. Easier said or written than done!

Jellyfish are naturally found in most coastal ecosystems around the world, being routinely or episodically conspicuous. They play a number of different roles within those ecosystems, and these are illustrated graphically in Fig. 32. As medusae, jellyfish are predators of a variety of pelagic (and migrating benthic) taxa (Flynn and Gibbons, 2007; Pitt et al., 2008) that range in size from single celled protistans (Leoni et al., 2022) to organisms almost as large as themselves – other jellyfish (Titelman et al., 2007); and some Cepheida contain photosynthetic zooxanthellae (Lampert, 2016; Holst et al., 2024; Nagata et al., 2024). The diets of jellyfish are broadly

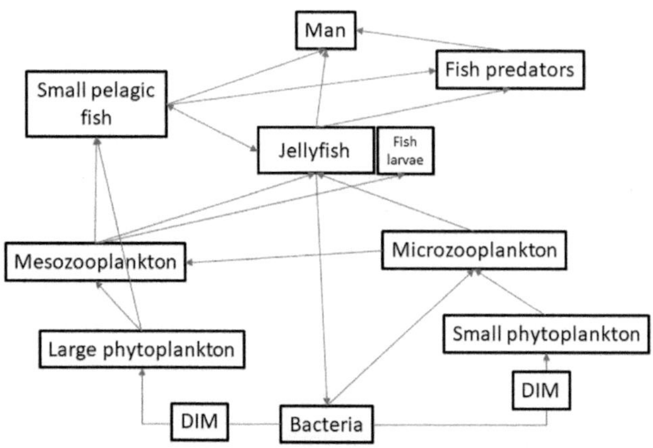

Fig. 32 Graphic showing the relationships between jellyfish and other ecosystem components.

similar to those of small pelagic fishes, and spatio-temporal overlap in populations of both taxa may lead (perhaps in the case of Semaeostomae, see below) to interspecific competition. Jellyfish are in turn eaten by a wide variety of different animals (e.g. Arai, 2005; Arkhipkin and Laptikhovsky, 2013; Harrison, 1984) and indeed some vertebrates (e.g. leatherback sea turtles) appear to be largely medusivorous (González Carman et al., 2013; Houghton et al., 2006). The (likely) constant release of mucous by jellyfish coupled with their feeding activities in the water column serve to fuel microbial and primary production there (Condon et al., 2011), whilst their sinking bodies end up as food for numerous benthic/demersal scavengers and decomposers (e.g. Sweetman et al., 2014). Alive, jellyfish may provide a "home" to diverse symbionts (e.g. Boco and Metillo, 2018), and they are widely recognised as giving refuge to young fishes (Griffin et al., 2019). As polyps, jellyfish feed on microcrustaceans and protozoans (Sun et al., 2021), with some again supporting photosynthetic zooxanthellae (Rahat and Adar, 1980). While they may be eaten by some benthic carnivores such as nudibranchs and decapods (Takao et al., 2014), inter-specific competition is likely to be for space and not food (though it has yet to be demonstrated). The aforegoing summary of the role that jellyfish may play is an abbreviation of significant research, primarily based on Semaeostomeae, and below we attempt to tease out the contributions to this understanding that has been derived from rhizostomes.

3.2 Diet and feeding

3.2.1 Autotrophy through symbiosis

Rhizostomes display a wide range of "diets" from autotrophy (via symbiosis) to true heterotrophy through mixotrophy. Facultative and obligate symbiosis with species of Dinophyceae of the genus *Symbiodinium* or *Cladocopium* have been reported for several medusae within the Classes Hydrozoa, Scyphozoa and Cubozoa. Among the Rhizostomeae, this feature appears to be most common among Kolpophorae, including Cepheidae, Mastigiidae, Cassiopeidae, Thysanostomatidae and Versurigidae (Djeghri et al., 2019 and references therein; Holst et al., 2024; Nagata et al., 2024). Medusae usually host only one phylotype of symbiont.

Unlike azooxanthellate jellyfish, symbiotic jellyfish rely strongly on the photosynthetic activity of their symbiont, thus taking part in the primary production of the ecosystem. Photosynthetic carbon fixation by the symbiont covers most or all the energetic requirements in the form of glycerol, sugars, organic acids, and amino acids of its host (Mammone et al., 2021; Verde and

McCloskey, 1998). But gross photosynthesis data for zooxanthellate medusae are scarce (Table 2) ranging from 3.7 to 14.8 (μg O_2 h^{-1} μg Chla $^{-1}$ h^{-1}). In the case of *Cassiopea* spp. primary production largely exceeds the respiration of the jellyfish.

Zooxanthellate jellyfish are often associated with oligotrophic environments where limiting nutrients affect the availability of heterotrophic food (Muscatine and Porter, 1977). Strong plasticity in photosynthesis capacity has been reported for *Cassiopea* allowing populations to thrive also in low irradiance and nutrients rich areas such as found in shady environments i.e. harbours, mangroves, seagrass beds (Niggl and Wild, 2010; Stoner et al., 2014; Verde and McCloskey, 1998). Adaptation to different light regimes can be explained by a potential shift in symbionts' density and cell size, larger cells increasing light gathering efficiency as well as higher chlorophyll *a* content (Verde and McCloskey, 1998).

As noted for phytoplankton, and macroalgae, photoinhibition can lead both pelagic (*M. papua*) and benthic (*Cassiopea* sp.) zooxanthellate jellyfish to adjust the concentration in mycosporine-like amino acids (MAAS: aromatic amino acids involved in the protection of aquatic organisms against solar radiation) of their symbionts as well as their preferred depth according to irradiance (Dawson and Hamner, 2003; Banaszak and Trench, 1995).

Autotrophy dominated nutrition by species of *Cassiopea* is certainly due to its benthic position, exposing its zooxanthellae-bearing photosynthetic oral arms to the sunlight, but also to its exumbrella lying against the sediment, and acting by its motion as a pump withdrawing nutrients through the benthic boundary layer (Jantzen et al., 2010) and augmenting, therefore, the vertical flow of water passing through its intricate body feature (Battista et al., 2022). *Cassiopea* spp. contribute to the bentho-pelagic coupling and all the more so as its population can be large. Benthic jellyfish have been shown to have negative impacts on seagrass cover and shoot densities (shading effects on primary producers, inhibiting gas exchange, preventing clonal development) (Stoner et al., 2014). Autotrophy has been reported mainly in monodisc strobilation species (Helm, 2018).

3.2.2 Heterotrophy

Out of the ninety-one species of Rhizostomeae (WoRMS, 2024), data on diet composition and feeding rates are only available for a few taxa: *Cassiopea* sp. *C. mosaicus, C. tuberculata* (Macri, 1778), *M. papua, N. nomurai, P. punctata, R. pulmo, R. nomadica,* and *Stomolophus meleagris* Agassiz, 1860 in

Table 2 Photosynthetic rates for zooxanthellate Rhizostomeae and Coronatae (*).

Species	Site	Max gross photosynthetic rates ($\mu g\ O_2\ h^{-1} \times 10^{-6}$ cells)	Max gross photosynthetic rates ($\mu g\ O_2\ h^{-1}\ \mu g\ Chla^{-1}\ h^{-1}$)	Gross production rates ($\mu mol\ O_2\ h^{-1}\ L^{-1}$)	References
Cassiopea andromeda	lab	—	3.7–4.2		Hofmann and Kremer (1981)
Cassiopea xamachana	Florida	13.6–15.8	6.3–11.1		Verde and McCloskey (1998)
Cassiopea sp.	Red Sea			144–288	Arossa et al. (2021)
Catostylus mosaicus	Australia			~40–55	West et al. (2009)
Mastigias sp.	Lake Palau	17.6–29.6	8.8–14.8		McCloskey et al. (1994)
Phylorhiza punctata	Australia			~25–50	West et al. (2009)
*Linuche unguiculata**	Bahamas	12.6	6.0		Kremer et al. (1990)

their natural environment except for *R. nomadica* for which data are only available for the Mediterranean Sea where the species is thought to be invasive (but see above).

3.2.2.1 Diet

The level of taxonomic effort put into describing the diet of rhizostomes appears to vary widely (Nagata et al., 2024). In some studies, large numbers of taxa (>40) have been reported [e.g. *C. mosaicus* (Browne and Kingsford, 2005); *Lychnorhiza lucerna* (Haeckel, 1880; Nagata and Morandini, 2018)], whilst in others there is little detailed information: Liu et al. (2011) described the diet of *Rhopilema esculenta* using only higher taxonomic levels (phylum, order). As a consequence, it is likely that we have strongly underestimated diet breadth.

Most published work presents direct measurements linked to the number of prey present on the oral arms or in the gut, though a few have inferred the diet from the structure and composition of the *in situ* zooplankton community (Shi et al., 2015).

Although filter-feeder jellyfish feed mainly on microplankton, they rely on a large panel of prey (Table 3) from diatoms, protozoans all the way to amphipods, and fish larvae. This size-dependent prey selectivity is probably related to the anatomy of their feeding structures (Larson, 1991; Lee et al., 2008, Liu et al., 2011) and to ontogenic processes (Álvarez-Tello et al., 2016; Kuplik and Angel, 2020; Nagata and Morandini, 2018; Wang et al., 2020). The morphology of rhizostomes affects their prey capture patterns, as the lack of tentacles and the enlarged "cauliflower" shape of the oral arms will impact the encounter rates between the feeding currents and the capture surface of these jellies (Bezio et al., 2018). Prey escape capacities will be higher when prey are caught in the feeding current of rhizostomes than of tentaculated semaeostomes i.e. for copepods contact efficiency of 32% and 45% respectively. Nevertheless, once captured, prey items are retained and consumed with the same efficiency (>95%) (Bezio et al., 2018). The "upside-down" jellyfish displayed the lowest contact efficiency, certainly linked with its benthic status and the narrowest exposed margins. Rhizostomes are mainly found in tropical or subtropical regions often oligotrophic and dominated by small size zooplankton, which could explain the evolution of their filter-feeding behaviour. The susceptibility of a prey item to be efficiently captured by a medusa depends therefore on a combination of several factors; its size, its swimming speed, the contact rates with the medusa and the size and speed of the medusa (see Nagata et al., 2024 for more

Table 3 Diet of Rhizostomeae.

	Lychnorhiza lucerna	*Stomolophus meleagris*				*Rhopilema nomadica*	*Rhopilema esculentum*	*Rhizostoma pulmo*	*Rhizostoma pulmo*	*Cotyloriza tuberculata*	*Catostylus mosaicus*			*Phyllorhiza punctata*	
Source	1	2	3	4	5	6	7	8	9	8	10	11	12	10	11
Diatoms	x			x	x	x	x	x	x	x			x		
Ceratium spp.	x								x						
Dinoflagellates									x						
Noctiluca									x		x		x		
Protozoans							x								
Ciliates				x		x	x								
Tintinnids	x				x			x	x				x		
Hydromedusae	x				x								x		
Actinula larvae															
Siphonophores	x														
Nematodes				x							x				
Platyhelminthes															

Polychaetes	x			x	x						x		x	x
Polychaete larvae	x								x	x		x		
Rotifers									x					
Cyphonaut larvae	x								x					
Phoronid larvae					x									
Brachiopod larvae					x									
Molluscan veligers	x	x	x	x		x	x	x	x	x		x	x	
Gastropod veligers	x		x	x	x	x	x		x	x		x	x	
Bivalve veligers	x	x	x	x	x	x	x		x	x		x	x	
Trocophore larvae									x					
Pteropods						x								
Insects												x	x	

(continued)

Table 3 Diet of Rhizostomeae. (*cont'd*)

	Lychnorhiza lucerna	*Stomolophus meleagris*				*Rhopilema nomadica*	*Rhopilema esculentum*	*Rhizostoma pulmo*	*Rhizostoma pulmo*	*Cotyloriza tuberculata*	*Catostylus mosaicus*			*Phyllorhiza punctata*	
Copepods	x	x	x	x		x	x	x	x		x	x		x	x
Calanoids	x				x	x			x		x		x	x	
Small calanoids	x														
Medium calanoids	x														
Large calanoids	x														
Copepodites	x						x		x		x		x	x	
Nauplii	x				x	x			x		x		x	x	
Unidentified eggs	x				x	x			x						
Temora turbinata	x														
Paracalanus spp.	x											x			x
Acartia sp.	x								x		x		x	x	
Pseudodiaptom-us acutus	x										x		x		

Acartia lilljeborgii	**x**								
Centropages sp.				**x**					
Temora stylifera	**x**						**x**		
Tortanus barbaratus							**x**		
Cyclopoids	**x**	**x**	**x**		**x**		**x**	**x**	
Oithona spp.	**x**				**x**	**x**	**x**	**x**	**x**
Poecilostomatoids	**x**								
Oncaea spp.	**x**						**x**		
Corycaeus spp.	**x**								
Harpacticoids	**x**	**x**	**x**						
Euterpina acutifrons	**x**								
Microsetella sp.	**x**								

(continued)

Table 3 Diet of Rhizostomeae. (*cont'd*)

	Lychnorhiza lucerna	***Stomolophus meleagris***				***Rhopilema nomadica***	***Rhopilema esculentum***	***Rhizostoma pulmo***	***Rhizostoma pulmo***	***Cotyloriza tuberculata***	***Catostylus mosaicus***			***Phyllorhiza punctata***	
Cladocerans	x	x			x				x				x		
Cirripede larvae		x	x	x	x				x				x		
Ostracods	x	x											x		
Amphipods	x	x		x	x						x		x	x	
Gammarids	x														
Isopods	x				x								x	x	
Decapods	x	x	x				x							x	
Brachyuran zoea	x	x		x	x				x		x		x		
Anomura adult				x											
Lucifer												x	x		x
Caridean shrimps					x								x		
Thalassinid larvae													x		

Shrimp larvae					x			x	
Chaetognaths	x			x	x			x	x
Larvaceans	x				x	x			
Ascidian larvae					x			x	
Fish eggs	x	x		x	x		x	x	x
Fish larvae	x		X	x					

References: (1) Nagata and Morandini (2018), (2) Padilla Serrato et al. (2013), (3) Alvarez-Tello et al. (2016), (4) Riascos et al. (2018), (5) Larson (1991), (6) Kuplik and Angel (2020), (7) Liu et al. (2011), (8) Perez-Ruzafa et al. (2002), (9) Dönmez and Bat, (2019), (10) Peach and Pitt (2005), (11) West et al. (2009), (12) Browne and Kingsford (2005).

information on feeding mechanisms). A few species can be considered as specialist predator as their diets highlight a positive selection on fish eggs, gastropods, bivalves and cirripeds; favoured prey were not the most abundant *in situ* (*Stomolophus meleagris*: Alvarez-Tello et al., 2016).

Meroplanktonic taxa such as gastropod and bivalve veligers appear frequently in the diet of rhizostome jellyfish. It is interesting to note that many *R. octopus* hotspots are also commercially important aquaculture areas. For example, both Rosslare Harbour (Irish Sea) and Pertuis Breton (La Rochelle, Bay of Biscay) are important areas for blue mussels (*Mytilus edulis*), and Carmarthen Bay is an important area for the cockle fishery (*Cerastoderma edule* (Linnaeus, 1758)) (Elliott et al., 2017). Considering that Leoni et al. (2022) found that *R. pulmo* can remove 8% of the ciliates daily standing stock, future studies should investigate the potential predatory impact *R. octopus* may have on mussel and cockle veliger larvae as Leoni et al. (2022) found that veliger larvae were one of the most consumed prey item by adults.

Chemical defences may be common amongst gelatinous zooplankton. *Stomolophus meleagris*, *C. andromeda*, and other species have been observed to release a sticky mucus when disturbed in a bucket but not when handled underwater. Mucus can contain both discharged and charged nematocysts (Holst et al., 2024) and have been shown to kill prey or sting humans (as personally experienced in the case of *Cassiopea* spp.). Clouds of nematocysts had a direct impact on the behaviour of potential predators (Shanks and Graham, 1988) and on associated fish (sheltering fish: juvenile planehead filefish *Stephanolepis hispida* (Linnaeus, 1767)) which swam away from the jellyfish; the crab *Libinia dubia* H. Milne Edwards, 1834 climbed as far as possible away from the mouth of the jelly over the bell. Associated fish have been shown to feed on prey paralysed by the jellyfish free released stinging cells rather than by the ones on the jelly itself. Nematocyst types and relative abundance in the epidermis of the oral arms do not seem to play a role in the capture of prey variations (Peach and Pitt, 2005; comparison of *P. punctata* and *C. mosaicus*) although cnidomes and absolute abundances of nematocysts are specific.

The diel vertical migration of prey has a strong impact on the diet composition with, in the case of *C. mosaicus,* emergent copepods, *Lucifer* sp. and mysids representing between 79 and 100% of the diet at night (Peach and Pitt, 2005; Pitt et al., 2008). Ontogenetic shifts in the diet composition of the jellyfish will exert, at times, pressures on the food web through both bottom up and top down controls (Leoni et al., 2022).

Increasing size of the medusae can lead to increase in number of prey (*R. pulmo,* Dönmez and Bat, 2019), or in the size of the prey but a decrease in diversity (*L. lucerna*: Nagata and Morandini, 2018). Decrease in diversity was also larger in individuals of *R. pulmo* (74 in young and 53 older/larger) (Leoni et al., 2022), but this seems to not be the rule as *S. meleagris* diet richness increased with the size of the umbrella from <3 taxa for specimen with umbrella ~10 mm to over 12 taxa for medusa with an umbrella >100 mm (Alvarez-Tello et al., 2016).

Spatial variations in diet are also reported for *S. meleagris* showing a dominance of fish eggs (58.6%) or veligers of bivalve (56%) in the Gulf of California and the Gulf of Mexico respectively (Alvarez-Tello et al., 2016; Larson, 1991). The size of a jellyfish's umbrella has therefore an impact on the diet of the organisms, as swimming capacity increases with size allowing the jellyfish to catch larger (Alvarez-Tello et al., 2016; Leoni et al., 2022) and faster prey (Dönmez and Bat, 2019; Kuplik and Angel, 2020). That said, it is very likely that *R. octopus* feeds in much the same way as *R. pulmo* in the Mediterranean, with smaller individuals feeding on copepods and then larger mature individuals feeding on much smaller prey such as ciliates (Leoni et al., 2022).

Zooxanthellate (*P. punctata*) and azooxanthellate (*C. mosaicus*) show mostly similar diets based on copepods (*Gladioferens* sp., *Paracalanus* sp., *Oithona* sp.), *Lucifer* sp. and polychaetes, though differences were reported with regards to gastropod and bivalve veligers (West et al., 2009). Peach and Pitt (2005) already reported these main differences between those two species, but they failed to highlight links between cnidome and diet composition, although symbiotic *P. punctata* had fewer nematocysts than the heterotrophic *C. mosaicus*. Algal endosymbionts and/or other nutritional sources have also been mentioned for *Phyllorhiza* in the Laguna Joyuda (García and Durbin, 1993) as daily rations on planktonic copepods appear low in relation to the observed growth rates of the jellyfish. Some azooxanthellate species, i.e. *R. pulmo* in Mar Menor display a strongly herbivorous diet with 62.4% represented by large diatoms (*Asterionella* spp. and *Coscinodiscus* spp.) with the population regularly located closest to the freshwater (and nutrient) inflows (Perez-Ruzafa et al., 2002).

Some species of rhizostome can also feed on benthic organisms in addition to their classical pelagic predatory behaviour (Riascos et al., 2018). Based on the ingestion rates of mesozooplankton by *L. lucerna* not reaching the minimum daily carbon requirement for this species, Skikne et al. (2009)

and Nagata and Morandini (2018) imply the potential use of dissolved and particulate organic matter as a food source (see also Nagata et al., 2024).

A good assessment of the diet composition relies on several factors. Net sampling is not appropriate as jellyfish will regurgitate part or all of their gut contents during sampling; scooping out jellies or collection by divers are recommended as these techniques will reduce potential loss in prey. Diet composition will rely also on the digestive status of the prey at the time of the observation, variable digestion time has been reported for different prey. And last but not least, diet composition description will depend on the taxonomic competences of the scientists. Shapes of the bell, the velar and marginal lappets, the lateral folds as well as the morphometry and functional features of the oral arms will have an impact on the feeding strategies of the jelly and will explain changes observed during the growth of the organisms (Jordano et al., 2022). Size of jellies seems therefore to be a key player in the composition of their diet, as young medusae have a higher capacity at catching faster prey i.e. copepods than larger jellies as they are more conspicuous in the ocean and their displacement more visible to their prey.

3.2.2.2 Ingestion rates

Little effort has been put, so far, towards feeding characterisation for this group of jellyfish as only ingestion rates have been measured for six species and clearance rates in only three (Table 4). The different protocols, the prey used (natural assemblages vs artemia nauplii), the environmental data provided and the way data are presented limit the comparison between the different species. All available information is presented in Table 4 (see also Nagata et al., 2024).

Gut contents is highly variable from 11.2 prey medusa^{-1} in *S. meleagris* (21.50 cm) in the Gulf of California, 6000 prey medusa^{-1} in the same species (10 cm) in Florida to values among the highest in Scyphomedusae with 12,138 prey reported in *L. lucerna* (Nagata and Morandini, 2018). On the other hand, *N. nomurai* (Wang et al., 2020) showed a clear increase in ingestion with increasing size with ~100 prey medusa^{-1} (~5 cm) to ~1600 prey medusa^{-1} (~40 cm). Clearance rates reported for rhizostomes appeared generally much higher than those found in Semaestomes (max 27.81 L h^{-1} for jellyfish bell diameter 12 cm) (Fig. 33), with values up to 3797 L h^{-1} in *P. punctata* (Graham et al., 2003) and 1697.7 L h^{-1} in *R. nomadica* (Kuplik 2018), with jellyfish much larger (WW 80,000, 6059 and 10,466 g respectively) and as high as 12,000 L h^{-1} and 60,000 L h^{-1} in *N. nomurai* (Uye, 2008; Wang et al., 2020). Clearance rates based on

Table 4 Number of prey per medusa, ingestion rate, and clearance rate of all published data on rhizostomes.

Species	Temperature (°C)	Prey density (ind L^{-1})	Bell Diameter (cm)	WW (g)	AFDW (µg)	Gut contents (prey $medusa^{-1}$)	Ingestion (prey d^{-1})	Clearance rate (L h^{-1})	Source
Lychnorhiza lucerna	18.8–21.5		5–30	14–2897			99–4345		1
Nemopilema nomurai				3000–80,000				4083–60,000	2
Rhizostoma pulmo			21.2–27.84	900.2–1800.8		77–208			3
									4
Ropilema nomadica	26.1–26.9	984	12–20	116–655			936–23,448	29.3–1697.7	5
	23–28	50–100	9–50	66–10,466			1056–16,176	1.5–15.2	6
	25.5–28.1	2536–14,566	16–42.20	500–5400		114–603			7
Stomolophus meleagris			1.50–10	9–325		700–6000	4000–65,000	41.23	8
									9
		124.66–347.18	21.50–87.80	5–261		11–428			10
Cotylorhiza tuberculata	20–30	400	0.35–0.75	0.003–0.033	140.18–486.58		122–783		11

(continued)

Table 4 Number of prey per medusa, ingestion rate, and clearance rate of all published data on rhizostomes. (*cont'd*)

Species	Temperature (°C)	Prey density (ind L^{-1})	Bell Diameter (cm)	WW (g)	AFDW (μg)	Gut contents (prey $medusa^{-1}$)	Ingestion (prey d^{-1})	Clearance rate (L h^{-1})	Source
Phyllorhiza punctata		24–121	4.8–19.2	10.17–1034.48	0.17–15.31		55,536–694, 704	23–201.4	12
		2294–3391	34–52	1938–6059	28.32–86.41	128–1651		63.8–3797.5	13
Aurelia	8	0.2–14.0	4–12			6–97		5.46–22.63	14

References: (1) Nagata and Morandini (2018), (2) Uye (2008), (3) Dönmez and Bat, 2019, (4) Leoni et al. (2022), (5) Kuplik (YEAR), (6) Kuplik et al. (2021), (7) Kuplik and Angel (2020), (8) Larson (1991), (9) Padilla Serrato et al. (2013), (10) Alvarez-Tello et al. (2016), (11) Astorga et al. (2012), (12) García and Durbin, (1993), (13) Graham et al. (2003), (14) Titelamn and Hansson (2006).

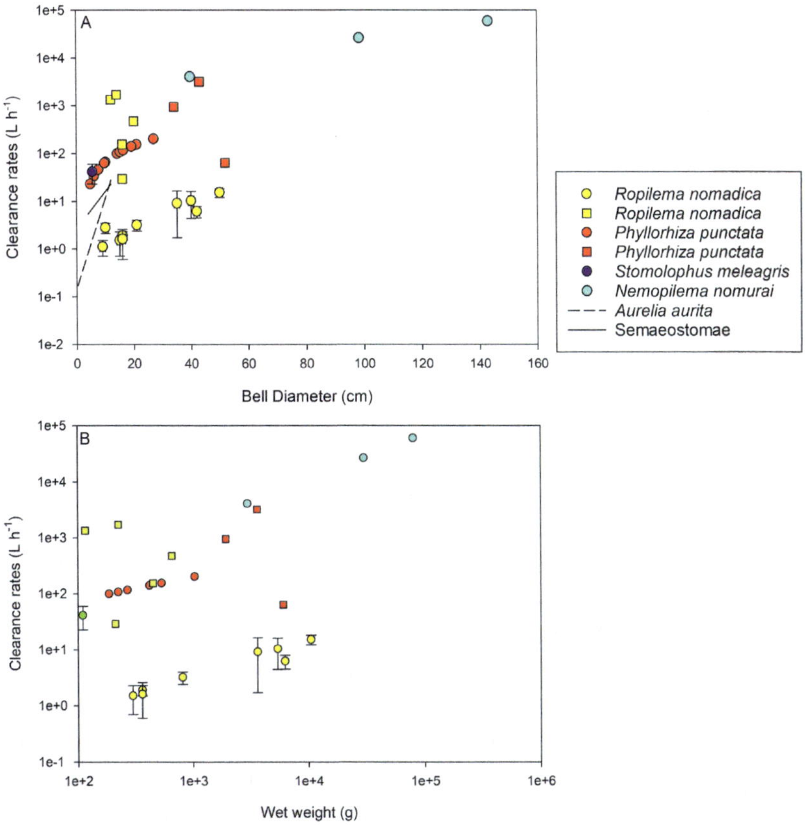

Fig. 33 Clearance rates (L h^{-1}) for several species of Rhizostoma as a function of (A) bell diameter and (B) wet weight (g). Kuplik et al., 2021 (yellow circle); Kuplik, 2017 (yellow square); García and Durbin, 1993 (red circle), Graham et al., 2003 (red square); Larson, 1991 (navy blue circle); Uye, 2008 (light blue circle). Values for *Aurelia aurita* (dotted line) and other Semastomae (full line) (Titelman and Hansson, 2006).

ambient zooplankton community (mostly *Acartia tonsa* Dana, 1849-1852) increased with the size of the medusae from 23 L h^{-1} for a 4.8 cm medusa to 201 L h^{-1} for a 26.9 cm medusa, but were independent of prey densities (Type I response) (García and Durbin, 1993). The same was reported by Kuplik et al. (2021) for *R. nomadica* on *Artemia* with increasing rate with increasing size (9 mm clearance and ingestion rates were 1.1 $\pm$ 0.4 L h^{-1} and 44 $\pm$ 15.5 ind h^{-1} respectively, increasing to 15.2 $\pm$ 3.1 L h^{-1} and 674 +/−425 ind h^{-1} for medusa of 50 mm in size). Integrity of the jellyfish will impact the observed rates as feeding capacity of the medusae will be reduced i.e. clearance rates (1.1 +/− 0.4 for 9 mm bell diameter to

$15.2 +/- 3.1\,L\,h^{-1}$ for a 50 mm bell diameter) values reported by Kuplik et al. (2021) are lower to those of García and Durbin (1993) as jellies oral arms were partially damaged.

Unfortunately, most studies lacked information on the temperature at which the experiments were run, impact of this parameter (Q10) cannot be fully investigated. When temperature is available, significant increase in ingestion rates is reported i.e. *C. tuberculata* from 419 ± 83.47 prey d^{-1} and 783.67 ± 83.58 prey d^{-1} at 20 and 25 °C (ephyrae showed rates between 2 and 4 times lower).

Daily ingestion of ephyrae represented between 290% and 850% of their dry weight. Based on mean carbon requirement, *N. nomurai* population can consume as little as 2.4% of the copepod production per day but as much as 62.9% during outbreaks occurring in small areas (Iguchi, 2017). In the Tsushima Strait, *N. nomurai* populations ingested 28% of the total micro- and mesozooplankton biomass per day (Uye 2014). Starvation in rhizostomes will lead to a retreat of the mouth parts and bell inversion, and potential shrinkage of the bell in ephyrae (Hatai, 1917; Meyer, 1914; Muffett et al., 2022; Nagata et al., 2024).

We have to keep in mind that diet and ingestion parameters based on gut contents suffer several biases leading usually to an underestimation of the pressure exerted by jellies on the plankton community. Items can be lost during collection of individuals if using net tows. Fast digestion rates (~1–2 h: Larson, 1991; Lee et al., 2008; Purcell, 2009) and digestibility of prey which can be more (hard bodied zooplankton) or less (soft bodied prey) easily done can also give a biased picture of the diet of jellies. Behaviours can also impact the diet (vertical distribution of jellies and prey). *Rhopilema nomadica* has been observed to target plankton patches within its otherwise oligotrophic environment. According to Reynolds (2014) a "novel kind of Lévy walk search pattern" can be associated with jellyfish search for food movement, an "optimised fast simulated annealing" helping them to locate highest concentration of prey within the water column or ability to easily locate prey olfactory trails.

Food availability may impact the size of rhizostome jellyfish. At 90 cm bell diameter, *R. octopus* is much larger than either *R. pulmo (60 cm)* or *R. luteum* (30 cm), and it is tempting to invoke Bergmann's Rule (Bergmann, 1848) to explain this (see also Fig. 34). In their meta-analysis of over 5000 bivalve species, Berke et al. (2012) failed to observe an overall trend of increasing size with latitude. That said, while mean sea surface temperature was indeed the strongest predictor of body size in some taxa,

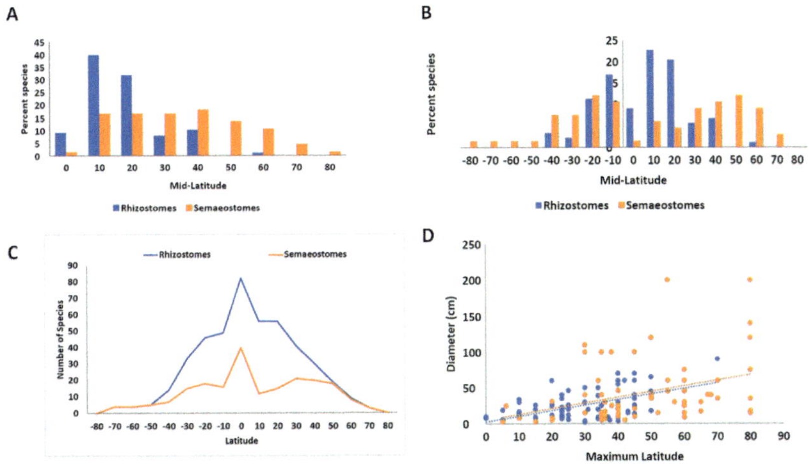

Fig. 34 Changes in the latitudinal diversity of Discomedusae. (A) Relationship between the number of species and the midpoint of their latitudinal range, irrespective of Hemisphere. (B) Relationship between the number of species and their midpoint of latitudinal range. (C) Relationship between species richness and latitude. (D) Relationship between the maximum latitude of distribution and medusa diameter. *All data from Jarms, G., Morandini, A.C., 2019. World Atlas of Jellyfish. Anhandlungen des Naturwissenschaftlichen Vereins, Hamburg, 815 pp.*

food availability was more important in others (Berke et al., 2012). *R. octopus* occurs in extremely productive temperate shelf seas where plankton abundance is, on average much higher than the Mediterranean Sea (where *R. pulmo* occurs), it may be that larger size is attainable due to available food. Of course, the fact that *R. octopus* overwinters and maximises food intake during the April/May spring bloom may also help individuals reach much larger sizes.

3.2.3 Mixotrophy

As a general assumption within any symbiotic relationships, respiration needs of the host are covered by the photosynthetic activities of the symbiont (Kikinger, 1992; Kremer et al., 1990; McCloskey et al., 1994; Verde and McCloskey, 1998) while nitrogen and phosphorus requirements still require prey predation (Kremer, 2005; Welsh et al., 2009). Symbionts will rely on light availability and nutrients excreted from the jellyfish as well as those present in the environment.

Symbiotic jellyfish are able to adapt to reduced light intensity by increasing [chlophyll *a*] (short-term response), and compensation by increasing heterotrophic feeding is not always observed depending on the level of obligatory association with zooxanthellae (Mortillaro et al., 2009). Eutrophication might favour zooxanthellate jellyfish populations (Fernández-Alías et al., 2020; Pérez-Ruzafa et al., 2002). Even though microbial dysbiosis disrupting the photosynthesis stability has been reported in several species of corals, the obligate photosymbiotic *C. xamachana* exhibits a strong resistance to dysbiosis in eutrophic environments (Röthig et al., 2021). Photosynthetic nutrition appears necessary for this species as bleached or light limited *C. xamachana* will not be able to maintain their population in the presence of high prey concentrations, although aposymbiotism for duration ~ 8 weeks and reinfection by free living symbionts has been reported. (Estes et al., 2003; Newkirk et al., 2020; Ohdera et al., 2018; Rädecker et al., 2017).

Isotopic signatures can bring additional information on the trophic functioning of jellies, as they can provide information on the contribution of the two main resources of food photosynthesis vs heterotrophy. Isotopic and elemental composition variation in zooxanthellate jellyfishes will reflect the contribution of the two types of nutrition. Photosynthesis of the symbiont will lead to a decreased $\delta^{15}N$, and an increased $\delta^{13}C$ and C/N ratio while heterotrophy led to increased $\delta^{15}N$, and decreased $\delta^{13}C$ and C/N ratio (Djeghri et al., 2020a).

Mixotrophic species will move along this nutritive spectrum depending on the species, populations, environments specificities, environmental conditions (and locations) and/or biotic interactions as well as life cycle (Bolton and Graham, 2004; Djeghri et al., 2021; McCloskey et al., 1994; Sugiura, 1969; Verde and McCloskey, 1998; Wilken et al., 2013). The cursor will move towards autotrophy as jellyfish population increases, decreasing the probability to catch prey as observed in *M. papua* (Djeghri et al., 2020b). Survival, population growth and potential outbreaks for zooxanthellate rhizostomeae will be largely influenced by either or both favourable photosynthesis conditions and prey abundance. *Symbiodinium in hospite* appeared to be also tightly limited by CO_2 derived from *Cassiopea* spp. metabolism (Rädecker et al., 2017).

Changes in food diet has been also observed between different life stages in *C. tuberculata* with higher ingestion rates but lower heterotrophic diet reported for young medusae than for ephyrae; switching therefore from a heterotrophic to a more autotrophic diet (Astorga et al., 2012). Zooxanthellated jellyfish will impact the availability of nutrients for other photosynthetic species by recycling internally the excretory products released by their host as well as using surrounding nutrients (West et al., 2009), and with a bottom-up effect on the zooplankton abundance.

3.3 Food for other organisms

For many years, jellyfish were considered a "trophic dead end" but recent studies using a variety of methods indicate that they likely play a far more prominent role in the marine food web than initially thought (Cardona et al., 2012). Despite the fact that jellyfish have a generally low energetic value (Doyle et al., 2007; Marques et al., 2016), studies using body mounted cameras indicate that predators often selectively feed on the more energy dense sections of the jellyfish, those often being the gonadal regions, tentacles or oral arms (as in Thiebot et al., 2016). And Hays et al. (2018) suggest that jellyfish may contribute far more to the energy budget of predators than previously thought due to their availability, low capture costs and rapid digestion.

Much of our appreciation of jellyfish to other organisms, as food, comes from the authoritative reviews of Arai (1988, 2005), Ates (2017) and Purcell and Arai (2001). Unfortunately, however, these efforts lack detailed information on the species of jellyfish consumed, which makes it difficult to put species into context. And further, even some of the more recent studies make inferences using data rather than direct observations, *per se*. For example, López-Martínez et al. (2018) suggest that *S. meleagris* may be consumed by the whale shark *Rhincodon typus* Smith, 1828, since the latter species has been observed associating with, and feeding among swarms of the former: but there is no empirical evidence that it is actually eaten. Similarly, Cardona et al. (2012) suggest that gelatinous organisms (including jellyfish) play a key role in supporting populations of top predators in the NW Mediterranean Sea, yet their conclusions are based on stable-isotope ratios and not hard observations. Table 5 provides a brief account of known rhizostome predators.

A retrospective analysis of the sightings database for *Dermochelys coriacea* (Vandelli, 1761) (>100 years of sightings) revealed that leatherbacks are more likely to be sighted in bays where you have *R. octopus* blooms than anywhere else, and suggests that these hotspots are consistent in space and time to drive long time foraging associations (Houghton et al., 2006). La Rochelle in the Bay of Biscay has the highest number of leatherback sea turtle sightings anywhere in the North East Atlantic and observations of them feeding on *R. octopus* are known from the 1970s (Duron, 1978). However, detailed observations of predator prey interactions are limited to few descriptive reports (Duron, 1978).

Explicit information on the predators of rhizostome polyps are also scant, though recent laboratory studies indicate that a variety of

Table 5 Direct observations of predation on living rhizostome jellyfish.

Species	Predator	Predator type	Location	Source
Mastigias papua	*Entacmaea medusivora*	Anthozoa	Palau	Fautin and Fitt (1991)
Cassiopea ornata			Kakaban Lake, Indonesia	Hoeksema et al. (2015)
Lynchnorhiza sp.	*Chelonia mydas*	Testudines	SW Atlantic	Carman et al. (2014)
Rhizostoma octopus	*Dermochelys coriacea*		Irish Sea	Houghton et al. (2006)
Stomolophus meleagris			North Carolina, USA	Grant and Ferrell (1993)
Rhizostoma octopus	*Fulmarus glacialis*	Procellariiformes	Irish Sea	McCanch and McCanch (1996)
Rhizostoma luteum	*Carcharodon carcharias*	Elasmobranchii	False Bay, South Africa	De Kock, pers. obs

nudibranchs and sea anemones will feed on the polyps, but not podocysts, of *N. nomurai* and *Rhopilema esculentum* (Tang et al., 2020). And polyps of the former species are known to also fall prey to gastropods and majid crabs (Kawahara et al., 2013).

3.4 Role in carbon pump and biogeochemistry

Although modelling studies suggest that jellyfish play an important part in nutrient cycling and that they represent a significant way to fast-track carbon transport to deep water (Luo et al., 2020; Wright et al., 2021), our understanding of their role in the global ecosystem is far from complete. I would also argue that our appreciation of their significance in local ecosystems is not that much better, and indeed I don't think that much has changed in this regard since Pauly et al. (2008) highlighted ways that we could more explicitly incorporate jellyfish into system models (MG). On the other hand, however, our knowledge about the processes involved in nutrient recycling are improving, largely as a result of some excellent experimental studies (but see Tinta et al., 2021).

Fig. 35 provides a schematic overview of the interactions between jellyfish and their environment, in the context of nutrient pools and cycling. Essentially, non-zooxanthellate jellyfish derive their C/N/P from zooplankton/ciliates; assimilated elements being incorporated into tissues

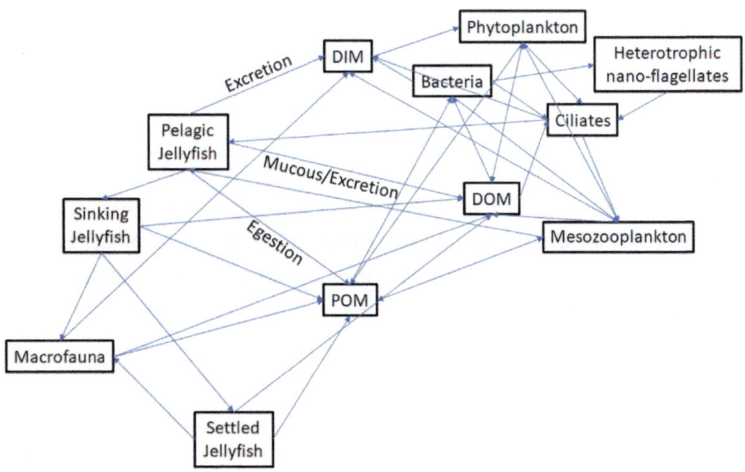

Fig. 35 Image showing the potential movement of dissolved (D) and particulate (P), organic (O) and inorganic (I), materials (M) through jellyfish and associated ecosystem components.

with a relatively low C:N ratio that reflects their high protein, low carbohydrate content (Lucas et al., 2011). The routine feeding activities of medusae will result in the production of particulate organic material (POM), whilst routine metabolism and excretion will generate dissolved organic material (DOM; amino acids, amines etc): individuals will also shed protein-rich mucous (Hubot et al., 2022) as POM and DOM (Niggl et al., 2010) as it has a low solubility (Hubot et al., 2022). The POM/DOM released by jellyfish is regarded as relatively labile (Blanchet et al., 2015; Condon et al., 2011) and is less refractory than that produced by other plankton (Sweetman et al., 2016). Although it has been suggested that jellyfish may also release dissolved inorganic materials (DIMs: NO_x, NH_4, PO_x), others have suggested that these are produced following the rapid bacterial remineralisation of DOM and POM (Pitt et al., 2009).

Although the bacterial communities associated with living jellyfish are quite different from those in the ambient environment (e.g. Early et al., 2016; Kos Kramar et al., 2019; Tinta et al., 2012; Titelman et al., 2006), they are very variable in composition and generalisations across taxa and landscapes are difficult to make. That said, the metabolism of ambient bacteria responds very quickly to the release of DOM by jellyfish, leading to elevated population growth and production as well as respiration (Condon et al., 2011; Blanchet et al., 2015). Communities rapidly change structure (Condon et al., 2011): those associated with jellyfish "thrive by oxidising" NH_4 to NO_2 and NO_3 (Hubot et al., 2021). The latter DIMs can be taken up by phytoplankton directly (Pitt et al., 2009), and Hubot et al. (2021) have recently suggested that jellyfish can supply 100% of local phytoplankton needs in some systems. Xiao et al. (2019) also noted that as jellyfish populations begin to develop, they can have a positive impact on the concentration of Chlophyll *a* because phytoplankton rapidly use the remineralised products of jellyfish exudates (especially P). While Hosia et al. (2015) did not detect any impact of jellyfish on overall primary production, they did notice that phytoplankton communities associated with jellyfish were different (greater numbers of diatoms) from those without.

Although pelagic medusae maintain themselves in the water column actively, their density alive is less than that when dead (Yamamoto et al., 2008), meaning they sink on death (Qu et al., 2015). Depending on the species and the size of the individual, the sinking carcass may be consumed and broken-up by other plankton and nekton *en route* to Davy Jones's Locker, and the extent of this fragmentation will influence the amount of time that this POM remains in suspension. And when in suspension, this

POM is subject to bacterial colonisation and remineralisation (Tinta et al., 2016, 2021), as outlined previously.

Jellyfish that arrive on the seabed, often *en masse* (Billett et al., 2006), may be consumed by mobile scavengers quite quickly (Sweetman et al., 2014; Yamamoto et al., 2008). The sudden impact of this windfall on other macrofauna is less clear though, because whilst epifauna may increase (Chelsky et al., 2016) more sessile infauna may decline in abundance (Chelsky et al., 2016; Marques et al., 2021). The latter observations have been attributed to the labile nature of jellyfish OM, which serves to fuel massive microbial rather than metazoan populations (Marques et al., 2016; Sweetman et al., 2016). Jellyfish decomposition is immediately associated with an increase in bacterial population sizes, changes in bacterial community composition and a decline in the concentration of dissolved oxygen (Frost et al., 2012; Guy-Haim et al., 2020). The latter may penetrate the sediment resulting in a change to the redox potential (Chelsky et al., 2016), with additional knock-on impacts to the infauna. It can also serve to lower the pH of bottom waters (Guy-Haim et al., 2020). Unlike at the surface, the accumulation of large quantities of DIM as a result of jellyfish decomposition may not stimulate primary production if nutrients are out of Redfield Ratio (Qu et al., 2015), while Blanchet et al. (2015) suggested that at the high concentrations observed, DIMs may actually prove toxic to plankton and benthos.

The aforegoing summary pulls together some of what we know about biogeochemical cycling in jellyfish and has been derived from studies across the Scyphozoa. There does not appear to be any significant difference in the elemental composition of semaeostome and rhizostome jellyfish (Lucas et al., 2011; Hubot et al., 2021), and so perhaps it is not unreasonable to suppose that the fate of a bloom of *Crambionella orsini* (Vanhöffen, 1888; Billett et al., 2006), *R. nomadica* (Guy-Haim et al., 2020), *N. nomurai* (Yamamoto et al., 2008) or *C. mosaicus* (Chelsky et al., 2015, 2016), will be the same as that of a bloom of *Chrysaora* spp. (Hubot et al., 2021). The latter reference was chosen deliberately because the response of the nitrifying bacterial community associated with the different species of *Chrysaora* investigated by Hubot et al. (2021) (*Chrysaora fulgida, C. hyscoscella* (Linnaeus, 1767), *C. pacifica* (Goette, 1886)) was not the same. But bacterial assemblages not only vary with jellyfish (e.g. Early et al., 2016), but also with water mass (Tinta et al., 2016), which means that the processes and impacts associated with biogeochemical remineralisation and cycling will vary on a case-by-case basis. Which is not helpful.

Cyanea capillata (Linnaeus, 1758) is generally considered to be the largest semaeostome medusa, while *N. nomurai* is thought to be the biggest rhizostome jellyfish. Using the length:weight equations supplied by Hay et al. (1990) and Zhang et al. (2012) for these two species (respectively), an individual *N. nomurai* of 200 cm diameter would weigh ~60% more than a *C. capillata* of equivalent size. The DW:WW ratios of rhizostome jellyfish (mean ratio=0.049, SE=0.0045, N = 5) are higher than those of semaeostome jellyfish (mean=0.0398, SE=0.0034, N = 23), based on data provided by Lucas et al. (2011) of whole animals (see also Nagata et al., 2024). They are certainly more "solid" than semaeostomes and they are likely to have a lower surface area to volume ratio too. This means that rhizostome jellyfish will probably sink quickly on death, with little in the way of degradation *en route* to the seabed (see also Billet et al., 2006), though of course what happens during an individual's senescence is anyone's guess. It is thus anticipated that the collapse of large rhizostome blooms would result in an initial flurry of scavenging by mobile epifauna and the release of (possibly) large quantities of POM and labile DOM, which would then fuel massive bacterial population growth and in turn lead to a localised reduction in the concentration of dissolved oxygen at the seabed accompanied by a decline in the pH and a change in local infaunal communities; a spike in the concentrations of DIMs will follow, which might (or might not) result in elevated phytoplankton growth, depending on water column depth and mixing processes.

Few biogeochemical studies of jellyfish have directed their attention at zooxanthellate taxa and those that have, have tended to study *Cassiopea*. Like jellyfish without symbionts, Pitt et al. (2009) suggested that *Cassiopea* derives N/P from carnivory but unlike jellyfish without symbionts, its C requirements are met by algal photosynthesis. Indeed, at saturating light intensities, Welsh et al. (2009) showed that photosynthesis was able to provide more than 100% of the animals' daily C needs, allowing for individual growth. The symbionts in turn get their N/P largely from *Cassiopea*, with little being released into the water column for use by phytoplankton. Welsh et al. (2009) demonstrated that jellyfish were slight sources of N/P for the water column at night, but that they were significant sinks for these DIMs during the day: these patterns reflecting diel changes in net oxygen production. The nocturnal release of NO_x by *Cassiopea*, in addition to NH_4, is suggested by Welsh et al. (2009) to reflect an associated microbiome comprising nitrifiers, in part. Jantzen et al. (2010) have shown that the zooxanthellae of *Cassiopea* derive their N/P not only from the surrounding water, but also from sediment pore-water. The routine flexing of the bell on the seabed draws pore-water to the sediment surface by pumping, which allows animals to harness the nutrients contained within. As Welsh et al. (2009) indicated, species like *Cassiopea* therefore generate spatial and temporal heterogeneity within a landscape (as Durieux et al., 2023).

4. Rhizostomeae as Discomedusae

How different is the ecology of rhizostome jellyfish from that of other Discomedusae, in the context of either seasonality and blooming, or their role in the ecosystem? Do they warrant separate consideration?

As a rule, rhizostome jellyfish are tropical and subtropical, whereas semaeostomes predominate in mid latitudes with a few species occurring as in polar seas (Fig. 34). In contrast, there are no high latitude rhizostomes, which may reflect the size structure of food webs present in such ecosystems (e.g. short food chains dominate high latitude seas vs long food chains in tropical seas) (Armengol et al., 2019; Murphy et al., 2007). The extreme seasonality of high latitude seas means that there is little time for long food chains to develop (Murphy et al., 2007) and as such, there may be limited scope for large bodied rhizostomes to capitalise on this production. In contrast, in temperate seas, large rhizostomes such as *R. octopus* may 'cheat' the short season (although longer than polar seas) by overwintering. In the tropics, rhizostomes can probably feed all year round. Indeed, this group is argued to have evolved in the Indo-Pacific (Kramp, 1970), being most diverse in the area of the Coral Triangle (data in Jarms and Morandini, 2019), which explains the asymmetry around the equator in Fig. 34. Caution should be exercised in the too literal interpretation of these figures because they lack information about recently erected (species of *Aurelia*: Lawley et al., 2021) and problematic taxa (species of *Pelagia*), because we are not yet familiar with their exact geographic delimitations. They also exclude those taxa in Jarms and Morandini (2019) that are not recognised as valid by WoRMS (2024). That said, diversity is clearly higher in tropical regions for both groups of jellyfish, as it is for many other taxa, and as exemplified by the excellent work of Gómez-Daglio and Dawson (2017). Of the 25 species of Discomedusae recovered by the latter authors in the tropical Eastern Pacific, twenty-two were new to science: nine semaeostomes and 13 rhizostomes.

Like semaeostome jellyfish, there is a tendency for rhizostome jellyfish to increase in size towards higher latitudes (Fig. 36), the latter plotted either as maximum ($r = 0.36$, $p < 0.005$, $N = 66$; $r = 0.49$, $p < 0.001$, $N = 86$; respectively) or mid distributional range ($r = 0.26$, $p < 0.05$; $r = 0.34$,

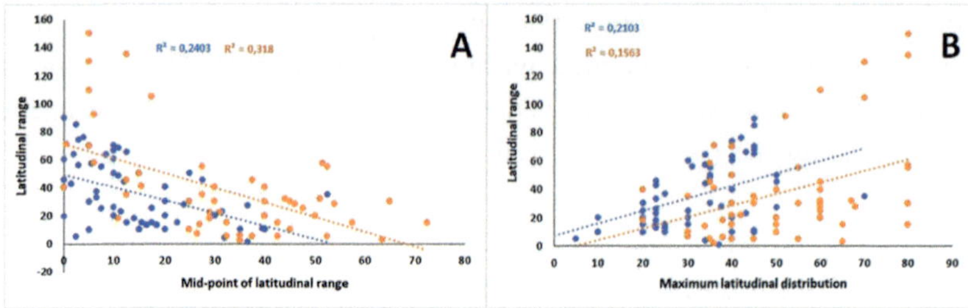

Fig. 36 Rapoport's Rule. Changes in the range size of rhizostome (blue) and semaeostome (orange) jellyfish by latitudinal mid-point (A) and maximum (B). *All data from Jarms, G., Morandini, A.C., 2019. World Atlas of Jellyfish. Anhandlungen des Naturwissenschaftlichen Vereins, Hamburg, 815 pp.*

$p < 0.005$; respectively). The data used to illustrate Bergmann's Rule were derived from Jarms and Morandini (2019) using taxa recognised as valid by WoRMS (2024), and are untransformed. Intriguingly, there is a relationship between latitudinal range and latitudinal distribution, with higher latitude species occupying significantly ($p < 0.05$) greater ranges than those from lower latitudes (Fig. 34: $N_{rhizostomes}=74$, $N_{semaeostomes}=56$). In other words, the data appear to conform to Rapoport's rule (Fortes and Absalão, 2004; *cf* Gaston and Chown, 1999; Rapoport, 1982; Stevens, 1989) for both scyphozoan groups. As noted previously, caution should be exercised in the interpretation of these data and note that those species with single distributional records have been ignored.

Rhizostomes are strictly coastal in their distribution, and they are often found in estuaries. There are no fully estuarine semaeostome jellyfish and there are no deep sea species of rhizostome jellyfish. The latter can probably be explained by the nature of the food environment, rather than by their apparently strict adherence to a metagenetic lifestyle. Though quite why there are no estuarine semaeostome species is intriguing. While it has been suggested that metagenetic taxa are restricted to relatively shallow water because of their polyp stages, perhaps especially in the case of those with symbiotic zooxanthellae (Djeghri et al., 2019), quite why rhizostome jellyfish are not as frequently encountered off the shelf as semaeostome medusae are, is intriguing. And those that are, have invariably been swept up in ocean currents and are being expatriated – as in the case of *N. nomurai* (see above). While it could be argued that (e.g.) *R. octopus* is abundant in the Celtic and Irish Seas (see above), it needs to be remembered that these are shelf seas with a bathymetry of, for the most part, less than 100 m depth.

Both their tropical origin and centre of diversity and their shallow coastal distribution may have a bearing on ecological differences with semaeostome jellyfish. Or do they? Populations of rhizostome jellyfish essentially display similar patterns in seasonal abundance as do coastal semaeostome jellyfish; likely because of commonalities in the underlying physiological and environmental processes. That said, a number of rhizostome taxa do appear to have components of the population that routinely overwinter as medusae (*R. octopus, R. nomadica*), though whether this reflects the relative aseasonal environment of tropical ancestors is unknown. Although one might expect that large animals would be more likely to overwinter (or survive overwintering) than smaller taxa (Fullerton et al., 2000; Geissinger et al., 2002; van Deurs et al., 2011), the fact that *N. nomurai* does not overwinter refutes this idea. Regardless, for those taxa that do overwinter, the strategy probably maximises growth potential since individuals are almost fully grown at the time of the first spring phytoplankton bloom, which allows them to capitalise early and attain a large size. Such may explain the large size attained by *R. octopus*, and contrasts with the behaviour of the even larger *C. capillata*, which clearly undergoes senescence in August and September (Brewer, 1989) with tentacles dropping off and large mass strandings washing up on beaches (Doyle et al., 2007).

By contrast, in an ecosystem context, there is some justification for considering rhizostome jellyfish as different from other Discomedusae. For a start, the anatomy of their feeding structures (Morandini, 2024) effectively means that they feed on slightly different, and smaller, prey items than semaeostomes. But it should be remembered that rhizostome jellyfish evolved in an area where long food chains dominated by small particles are the rule, not the exception, and so multiple small mouthlets may be a more efficient and effective way of eating such food than having a single, large and centralised mouth. And it is no coincidence that many species support symbiotic zooxanthellae because the greatest variety of other zooxanthellate cnidarians are similarly tropical (Veron et al., 2015). Another key difference is their robustness and as a result their greater ability to swim faster (Nagata et al., 2016, 2024), which may enable them to occur in areas where semaeostomes do not e.g. estuaries. A recent study by Malul et al. (2019) found that two different rhizostomes swim faster against the current than when swimming in the direction of the current. Such swimming ability may prevent them from stranding in shallow estuarine environments.

Jellyfish feature in the diet of a wide variety of predators (e.g. Purcell and Arai, 2001), especially as ephyrae and small medusae (Lamb et al., 2017), though with few exceptions (Hays et al., 2018; Houghton et al., 2006) it is unknown if they are routinely important components of the diet of any. It has been suggested that jellyfish may be significant competitors with fish, especially if there is a spatial and temporal overlap in distribution (Brodeur, 1998; Brodeur et al., 2008), owing to broad similarities in the diet – especially of small pelagic, filter-feeding species. And there is much data to support the idea that jellyfish can be important predators of fish eggs and larvae (Duffy et al., 1997); and that significant spatial and temporal overlap in distribution can influence the magnitude of finfish recruitment (Lynam et al., 2005; Möller, 1984).

But does this generalisation apply equally to semaeostome and rhizostome jellyfish? Perhaps not. As already noted, the diet of rhizostome jellyfish, as adults, largely comprises microzooplankton and meroplankton (see above), and as such they are able to thrive in estuarine and shallow waters where phytoplankton communities may not be dominated by diatoms during summer/autumn (e.g. Quinlan and Phlips, 2007). Rhizostome jellyfish do not appear to be major predators of large copepods or other crustacean zooplankton (see above), which means that unlike semaeostome jellyfish they are probably not strong competitors with finfish for food, because the two guild members are focusing on different components of the prey field. Although there is some evidence that some rhizostome jellyfish may feed on fish eggs and larvae (again, see above), no studies have as yet indicated that rhizostome jellyfish populations impact fish recruitment, as has been argued to be the case with some semaeostome jellyfish (Lynam et al., 2005; Möller, 1984). Which probably comes down, in part, to spatial and temporal mismatches in distribution. And while they could, in theory, influence the magnitude of bivalve recruitment, given their fondness for bivalve larvae, in truth the magnitude of spawn production by bivalves is such that only a very large and persistent bloom of jellyfish might have such an impact.

It has been suggested that cultural eutrophication may contribute to an increase in the abundance of jellyfish by increasing hypoxia and turbidity and modifying the N:P ratio (e.g. Purcell et al., 2007; Richardson et al., 2009). The latter can promote small-sized plankton which, given their dietary size preference, may favour rhizostomes. Although Haraldsson et al. (2012) have demonstrated that jellyfish may dominate the pelagic biomass of heavily eutrophic ecosystems, owing to their competitive advantage over fish in environments with limited light (jellyfish are tactile as opposed to visual predators), unambiguous, non-correlational evidence to support this modelled result is hard to find. Even with semaeostome species, which were the subject of Haraldsson et al. (2012) study, and thus more likely to compete with fish (see above). In their study of population fluctuations of *R. pulmo* in the Mediterranean Sea, Leoni et al. (2021a) considered that eutrophication played a much more local role than temperature in impacting inter- and intra-annual bloom dynamics.

That does not mean to say, however, that eutrophication may not lead to changes in the abundance of some species, as *Cassiopea* spp. appear to thrive in eutrophic waters. Mammone et al. (2021) have recently shown that the lower light levels associated with eutrophic waters does not impact the density of microalgal symbionts within medusa tissue, but that chlorophyll concentration does increase. On balance, these authors concluded that photosynthetic performance was high at both low and high light levels, and that no prior acclimation was necessary to achieve net carbon gains (Mammone et al., 2021). This suggests that eutrophication has no deleterious effect on *Cassiopea*. Exposure to elevated concentrations of dissolved organic carbon (DOC) can result in an elevated reliance on heterotophy by *Cassiopea*, as evidenced by an increase in respiration and decline in photosynthesis (Tilstra et al., 2022). But, as shown by Röthig et al. (2021) in experimental studies on the holobiont (medusa, zooxanthellae and associated bacterial community) of aposymbiotic *Cassiopea xamachana*, growth will be compromised over the medium term despite heterotrophy. Although DIC is rapidly taken up by zooxanthellae through photosynthesis, and then cycled through to medusae, both the bacterial communities associated with the jellyfish and the jellyfish itself appear to restrict access to DIN by the algal partner (Röthig et al., 2021). It has been suggested by these latter authors that this might allow *Cassiopea* to thrive in nutrient-rich waters, while scleractinian corals can not (D'Angelo and Wiedenmann, 2014), which helps to explain its presence in harbour environments in the Mediterranean Sea (see above). Their density means that they sink quickly, which impacts biogeochemistry – as outlined above.

In his review, Thiel wrote that some 20 species of rhizostome jellyfish had been observed in association with more than 25 species of fishes, and that few relationships were host-specific (Thiel et al., 2024). These observations were expanded upon by Purcell and Arai (2001), who summarised additional observations, and the list of associations has been added on to a number of other occasions (e.g. Kondo et al., 2014). Thiel (1970, 1978, 1979), building on earlier comments by Mansueti (1963) noted that the relationship between rhizostomes and fish may be different from that of semaeostomes, on the grounds of differences in anatomy and physiology

(Morandini, 2024; Nagata et al., 2024), noting that it is more likely to be symbiotic or commensal in the former and "parasitic" in the latter. In other words, Thiel (1970) suggests that rhizostomes likely play a greater role in the provision of refugia for juvenile fishes than do semaeostome jellyfish. Griffin et al. (2019) have recently reviewed the associations between these two groups of animals, and their data suggest that this ancient relationship is more important for demersal and reef fishes than it is for others. Although they indicate that semaeostome jellyfish are more frequently observed in association with fish than rhizostome jellyfish, the number of taxa involved is fewer (Griffin et al., 2019). Indeed, 65% of the Discomedusae that were observed in association with juvenile fishes by Griffin et al. (2019) were rhizostomes, which is greater than the approximately 50:50 ratio of semaeostomes to rhizostomes known to exist at present (WoRMS 2024). This suggests that the taxa of jellyfish associating with fishes is not a random subset of extant species. Griffin et al. (2019) suggested that for the behaviour to evolve, from the fish's point of view, the benefits of associating with jellyfish (feeding opportunities and protection from predators) must outweigh the immediate costs of association – namely the "risk of injury/death from jellyfish nematocyst stings", and it is likely that the costs are lower for rhizostomes than semaeostomes. But work is needed.

Semaeostome jellyfish have been linked to increases in phytoplankton through their consumption of herbivorous zooplankton (Møller and Riisgård, 2007): a top-down effect. By contrast, zooxanthellate species of rhizostome jellyfish may suppress phytoplankton populations through competition with them for nutrients (Pitt et al., 2009) and by allowing copepod populations to flourish (West et al., 2009): bottom-up and top-down effects, respectively. Xiao et al. (2019) noted a reduction in chlorophyll *a* concentrations at the peak of a bloom of the non-zooxanthellate *N. nomurai*, but in this case their data suggested it may have been due to bottom-up rather than top-down processes reflecting nutrient limitation.

From the perspective of population biology, rhizostomes do not appear to be materially different from semaeostomes. Or at least, scapulate rhizostomes do not appear to be materially different from semaeostomes – perhaps because, like most semeaostomes, they display polydisc strobilation (Schiariti et al., 2024) and are, for the most part, temperate in distribution. Information about population fluctuations for non-scapulate rhizostomes is restricted to the two species covered here, one with and one without symbiotic zooxanthellae. Quite how representative these are of the balance of inscapulate medusae is at present unknown; and so, like Thiel's contribution (see Thiel et al., 2024), we wait for more information. And while the macroecology of the two taxa may be similar (Figs. 34 and 36), it is very clear that rhizostomes (scapulate and inscapulate alike) play a very different role in the ecosystems they occupy than do semaeostomes, and these differences deserve increased attention in the future. Part of the reason for the need to study these animals in more detail reflects their value to humans as a resource, but their ability to feed on smaller plankton than do semaeostomes suggests that as global temperatures increase, and tropical organisms move poleward, they may come to play a more important role in our coastal marine ecosystems than at present. Although we may have employed electronic miniaturisation to greatest effect on rhizostomes (Hays et al., 2012), almost all of the other cutting-edge tools at our disposal (see above) have been directed primarily at semaeostomes and it is time to use these to greater effect.

Acknowledgements

S-IU would like to thank many of colleagues for their great contribution to the advancement of knowledge on *N. nomurai*. **MJG** would like to thank the National Research Foundation (NRF), PRIMALearn and the University of the Western Cape for ongoing support, and he is particularly grateful to the MLD (Mission Longue Duree) awarded by IRD for providing funds that allowed him, DT and ZK to throw ideas around and to craft the first full draft of the manuscript. **ZK** would like to thank the Maritime Aquaculture Department, Ramot-Yam School, Michmoret, Israel, and especially Rafi Yavetz and his staff, for their ongoing support. The Steinhardt Museum of Natural History is gratefully acknowledged for their financial support of his trip to work with MJG and DT at MIO, Marseille. **KP** would like to thank Griffith University and Sea World, Australia, for their ongoing support of the Griffith Sea Jellies Research Laboratory. **TD** would like to thank University College Cork and various funding agencies (INTERREG, IRC, SFI) for their support and colleagues who have contributed to the advancement of knowledge regarding *R. octopus*. **DT** would like to thank the Mediterranean Institute of Oceanography (MIO) and IRD for funding. **LP/AEN** would like to thank Govern des Illes Balears, SOCIB and CSIC (Spain) and the Vienna Zoo (Austria) for their ongoing support of the ICMAN-CSIC Jellyfish Laboratory. **MKB** would like to thank the University of the Western Cape (UWC) and the National Research Foundation (NRF) for their ongoing support.

References

Abboud, S.S., Gomez Daglio, L., Dawson, M.N., 2018. A global estimate of genetic and geographic differentiation in macromedusae—implications for identifying the causes of jellyfish blooms. Mar. Ecol. Progs. Ser. 591, 199–216.

Abed-Navandi, D., Kikinger, R., 2007. First record of the tropical scyphomedusa *Phyllorhiza punctata* von Lendenfeld. 1884 (Cnidaria: Rhizostomeae) Cent. Mediterranean Sea. Aquat. Inv. 2, 391–394.

Abu El-Regal, M., Temraz, T., 2016. Blooming of nomad jellyfish *Rhopilema nomadica* along the Egyptian Mediterranean coasts. Comm. Int. Explor. Sci. Mer. Mediterr. Rapp. 41, 489–490.

Algueró-Muñiz, M., Meunier, C.L., Holst, S., et al., 2016. Withstanding multiple stressors: ephyrae of the moon jellyfish (*Aurelia aurita*, Scyphozoa) in a high-temperature, high-CO_2 and low-oxygen environment. Mar. Biol. 163, 185.

Álvarez-Tello, F., López-Martínez, J., Lluch-Cota, D., 2016. Trophic spectrum and feeding pattern of cannonball jellyfish *Stomolophus meleagris* (Agassiz, 1862) from central Gulf of California. J. Mar. Biol. Assoc. U. K. 96, 1217–1227.

Amor, K.O.B., Rifi, M., Ghanem, R., et al., 2016. Update of alien fauna and new records from Tunisian marine waters. Mediterr. Mar. Sci. 17, 124–143.

Anton, A., Geraldi, N.R., Lovelock, C.E., et al., 2019. Global ecological impacts of marine exotic species. Nat. Ecol. Evol. 3, 787–800.

Arai, M.N., 2005. Predation on pelagic coelenterates: a review. J. Mar. Biol. Assoc. U. K. 85, 523–536.

Arai, M.N., 2009. The potential importance of podocysts to the formation of scyphozoan blooms: a review. Hydrobiologia 206, 241–246.

Arkhipkin, A., Laptikhovsky, V., 2013. From gelatinous to muscle food chain: rock cod *Patagonotothen ramsayi* recycles coelenterate and tunicate resources on the Patagonian Shelf. J. Fish. Biol. 83, 1210–1220.

Armengol, L., Calbet, A., Franchy, G., et al., 2019. Planktonic food web structure and trophic transfer efficiency along a productivity gradient in the tropical and subtropical Atlantic Ocean. Sci. Rep. 9, 2044.

Arossa, S., Barozzi, A., Callegari, M., et al., 2021. The internal microenvironment of the symbiotic jellyfish *Cassiopea* sp. from the Red Sea. Front. Mar. Sci. 8, 705915.

Astorga, D., Ruiz, J., Prieto, L., 2012. Ecological aspects of early life stages *of Cotylorhiza tuberculata* (Scyphozoa: Rhizostomae) affecting its pelagic population success. Hydrobiologia 690, 141–155.

Ates, R.M., 2017. Benthic scavengers and predators of jellyfish, material for a review. Plankton Benthos Res. 12, 71–77.

Badham, C., 1917. On a larval actinian parasitic in a rhizostome. Quart. J. Microscop. Sci. 62, 221–229.

Bajwa, A.A., Chauhan, B.S., Farooq, M., et al., 2016. What do we really know about alien plant invasion? A review of the invasion mechanism of one of the world's worst weeds. Planta 244, 39–57.

Bakun, A., 1998. Ocean triads and radical interdecadal variation: bane and boon to scientific fisheries management. In: Pitcher, T.J., Hart, P.J.B., Pauly, D. (Eds.), Reinventing Fisheries Management. Springer, Netherlands, pp. 331–358.

Bakun, A., Weeks, S.J., 2006. Adverse feedback sequences in exploited marine systems: Are deliberate interruptive actions warranted? Fish Fisheries 7, 316–333.

Balistreri, P., Spiga, A., Deidun, A., et al., 2017. Further spread of the venomous jellyfish *Rhopilema nomadica* Galil, Spannier & Ferguson, 1990 (Rhizostomeae, Rhizostomatidae) in the western Mediterranean. BioInvasions Rec. 6, 19–24.

Banaszak, A.T., Trench, R.K., 1995. Effects of ultraviolet (UV) radiation on marine microalgal-invertebrate symbioses. II. The synthesis of mycosporine-like amino acids in response to exposure to UV in *Anthopleura elegantissima* and *Cassiopeia xamachana*. J. Exp. Mar. Biol. Ecol. 194, 233–250.

Bastian, T., Haberlin, D., Purcell, J.E., et al., 2011. Large-scale sampling reveals the spatio-temporal distributions of the jellyfish *Aurelia aurita* and *Cyanea capillata* in the Irish Sea. Mar. Biol. 158, 2639–2652.

Battista, N., Gaddam, M.G., Hamlet, C.L., et al., 2022. The presence of a substrate strengthens the jet generated by upside-down jellyfish. Front. Mar. Sci. 9, 847061.

Bayha, K.M., Collins, A.G., Gaffney, P.M., 2017. Multigene phylogeny of the scyphozoan jellyfish family Pelagiidae reveals that the common U.S. Atlantic sea nettle comprises two distinct species (*Chrysaora quinquecirrha* and *C. chesapeakei*). PeerJ 5, e3863.

Bayha, K.M., Dawson, M.N., Collins, A.G., et al., 2010. Evolutionary relationships among scyphozoan jellyfish families based on complete taxon sampling and phylogenetic analysis of 18S and 28S ribosomal DNA. Int. Comp. Biol. 50, 436–455.

Bayha, K.M., Graham, W.M., 2014. Nonindigenous marine jellyfish: invasiveness, invasibility, and impacts. In: Pitt, K.A., Lucas, C.H. (Eds.), Jellyfish Blooms. Springer, Dordrecht, pp. 45–78.

Ben Faleh, A.R., Allaya, H., Armani, A., et al., 2017. Significant genetic differentiation among meroplanktonic barrel jellyfish *Rhizostoma pulmo* (Cnidaria: Scyphozoa) in the Mediterranean Sea. Afr. J. Mar. Sci. 39, 1–8.

Bergmann, C., 1848. Über die verhältnisse der wärmeökonomie der thiere zu ihrer grösse. Göttinger Studien 3, 595–708.

Berke, S.K., Jablonski, D., Krug, A.Z., et al., 2012. Beyond Bergmann's rule: size-latitude relationships in marine Bivalviaworld-wide. Global Ecol. Biogeog. 22, 173–183.

Bezio, N., Costello, J., Perry, E., et al., 2018. Effects of capture surface morphology on feeding success of scyphomedusae: a comparative study. Mar. Ecol. Prog. Ser. 596, 83–93.

Bigelow, R.P., 1900. Anat. Dev. *Cassiopea xamachana.* Mem. Boston Soc. Nat. Hist. 5, 751–785.

Billett, D.S.M., Bett, B., Jacobs, C.L., et al., 2006. Mass deposition of jellyfish in the deep Arabian Sea. Limnol. Oceanogr. 51, 2077–2083.

Bingel, F., Avsar, D., Gucu, A.C., 1991. Occurrence of jellyfish in Mersin Bay. Proceedings of the 2nd Workshop on Jellyfish Blooms in the Mediterranean Sea. UNEP Athens. Map Tech. Rep. Ser. 47, 65–71.

Bitar, G., Badreddine, A., 2019. First record of *Marivagia stellata* Galil and Gershwin, 2010 (Scyphozoa: Rhizostomeae:Cepheidae) from the Lebanese waters in the eastern Mediterranean sea. J. Black Sea/Medit. Environ 25, 172–177.

Blanchet, Pringault, M., Bouvy, O., et al., 2015. Changes in bacterial community metabolism and composition during the degradation of dissolved organic matter from the jellyfish *Aurelia aurita* in a Mediterranean coastal lagoon. Environ. Sci. Pollut. Res. 22, 13638–13653.

Boco, S.R., Metillo, E.B., 2018. Observations on the specific associations found between scyphomedusae and commensal fish and invertebrates in the Philippines. Symbiosis 75, 69–79.

Boero, F., Bouillon, J., Gravili, C., et al., 2008. Gelatinous plankton: Irregularities rule the world (sometimes). Mar. Ecol. Prog. Ser. 356, 299–310.

Boero, F., Putti, M., Trainito, E., et al., 2009. First records of *Mnemiopsis leidyi* (Ctenophora) from the Ligurian, Thyrrhenian and Ionian Seas (western Mediterranean) and first record of *Phyllorhiza punctata* (Cnidaria) from the western Mediterranean. Aquat. Invasion. 4, 675–680.

Bolton, T.F., Graham, W.M., 2004. Morphological variation among populations of an invasive jellyfish. Mar. Ecol. Prog. Ser. 278, 125–139.

Borlase, W., 1758. Natural History of Cornwall. Jackson, Oxford.

Boudouresque, C.F., Blanfuné, A., Fernandez, et al., 2017. Marine biodiversity – warming vs. biological invasions and overfishing in the Mediterranean Sea: Take care, 'One Train can hide another. M. O. J. Eco. Environ. Sci. 2, 00031.

Brewer, R.H., 1989. The annual pattern of feeding, growth, and sexual reproduction in *Cyanea* (Cnidaria: Scyphozoa) in the Niantic River Estuary, Connecticut. Biol. Bull. 176, 272–281.

Brodeur, R.D., 1998. *In situ* observations of the association between juvenile fishes and scyphomedusae in the Bering Sea. Mar. Ecol. Prog. Ser. 163, 11–20.

Brodeur, R.D., Suchman, C.L., Reese, D.C., et al., 2008. Spatial overlap and trophic interactions between pelagic fish and large jellyfish in the northern California Current. Mar. Biol. 154, 649–659.

Brotz, L., Angel, D.L., D'Ambra, I., Enrique-Navarro, A., Lauritano, C., Thibault, D., Prieto, L., 2024. Rhizostomes as a resource: The expanding exploitation of jellyfish by humans. Adv. Mar. Biol. 98, 511–548.

Browne, J.G., Kingsford, M.J., 2005. A commensal relationship between the scyphozoan medusae *Catostylus mosaicus* and the copepod *Paramacrochiron maximum*. Mar. Biol. 146, 1159–1168.

Cabrales-Arellano, P., Islas-Flores, T., Thogmé, P.E., et al., 2017. Indomethacin reproducibly induces metamorphosis in *Cassiopea xamachana* scyphistomae. PeerJ 5, e2979.

Cardona, L., Alvarez de Quevedo, I., Borrell, A., et al., 2012. Massive consumption of gelatinous plankton by mediterranean apex predators. PLoS One 7, e31329.

Carli, A., 1991. Macroplanktonic jellyfish in the Ligurian Sea (1984–1986). Monitoring and biological characteristics. Proc. 2nd Workshop Jellyfish Bloom. Mediterranean Sea. UNEP Athens. Map Tech. Rep. Ser. 47, 77–81.

Carlton, J.T., 1996. Biological invasions and cryptogenic species. Ecology 77, 1653–1655.

Carman, V.G., Botto, F., Gaitán, E., et al., 2014. A jellyfish diet for the herbivorous green turtle *Chelonia mydas* in the temperate SW Atlantic. Mar. Biol. 161, 339–349.

Çevik, C., Erkol, I.T., Toklu, B., 2006. A new record of an alien jellyfish from the Levantine coast of Turkey – *Cassiopea andromeda* (Forskål, 1775) [Cnidaria: Scyphozoa: Rhizostomea]. Aquat. Inv. 1, 196–197.

Cevik, C., Derici, O.B., Cevik, F., et al., 2011. First record of *Phyllorhiza punctata* von Lendenfeld, 1884 (Scyphozoa: Rhizostomeae: Mastigiidae) from Turkey. Aquat. Inv 6 (Suppl. 1), S27–S28.

Chang, P.H., Isobe, A., 2003. A numerical study on the Changjiang diluted water in the Yellow Sea and East China Seas. J. Geophys. Res. 108, 3299.

Chelsky, A., Pitt, K.A., Ferguson, A.J.P., et al., 2016. Decomposition of jellyfish carrion in situ: Short-term impacts on infauna, benthic nutrient fluxes and sediment redox conditions. Sci. Total Environ. 566/567, 929–937.

Chi, X., Mueller-Navarra, D.C., Hylander, S., et al., 2019. Food quality matters: Interplay among food quality, food quantity and temperature affecting life history traits of *Aurelia aurita* (Cnidaria: Scyphozoa) polyps. Sci. Total Environ. 656, 1280–1288.

Chong, Y.C., 1814. Jasaneodo. In: Lee, K.S. (Ed.), The History of Korea: 1970. Eu-Yoo Press, Seoul (in Korean).

Claus, C., 1883. Untersuchungen über die Organisation und Entwicklung der Medusen. Freytag.

Claus, C., 1884. Die Ephyren von *Cotylorhiza* und *Rhizostoma*. Zool. Inst. Wien. 5, 1–10.

Colella, S., D'Ortenzio, F., Marullo, S., et al., 2003. Primary production variability in the Mediterranean Sea from SeaWiFS data. Proc. SPIE – Int. Soc. Optical Eng. 5233. https://doi.org/10.1117/12.516791.

Coll, M., Piroddi, C., Steenbeek, J., et al., 2010. The biodiversity of the Mediterranean Sea: estimates, patterns, and threats. PLoS One 5, e11842.

Condon, R.H., Duarte, C.M., Pitt, K.A., et al., 2013. Recurrent jellyfish blooms are a consequence of global oscillations. Proc. Nat. Acad. Sci. 110, 1000–1005.

Condon, R.H., Steinberg, D.K., del Giorgio, P.A., et al., 2011. Jellyfish blooms result in a major microbial respiratory sink of carbon in marine systems. Proc. Natl. Acad. Sci. U. S. A. 108, 10225–10230.

D'Angelo, C., Wiedenmann, J., 2014. Impacts of nutrient enrichment on coral reefs: new perspectives and implications for coastal management and reef survival. Curr. Op. Environ. Sust. 7, 82–93.

Dailianis, T., Akyol, O., Babali, N., et al., 2016. New Mediterranean Biodiversity Records (July 2016). Mediter. Mar. Sci 17, 608–626.

Dana, J.D., 1849–1852. Crustacea. Part II. In: United States Exploring Expedition. During the years 1838–1842. Under the command of Charles Wilkes. U. S. N., C. Sherman Printer, Philadelphia. 14: 691–1618.

Dawson, M.N., 2005. Morphological and molecular resdescription of *Catostylus mosaicus conservativus* (Scyphozoa: Rhizostomeae: Catostylidae) from south-east Australia. J. Mar. Biol. Assoc. U. K. 85, 723–731.

Dawson, M.N., Cieciel, K., Decker, M.B., et al., 2015. Population-level perspectives on global change: genetic and demographic analyses indicate various scales, timing and causes of scyphozoan jellyfish blooms. Biol. Invasions 17, 851–867.

Dawson, M., Hamner, W.M., 2003. Geographic variation and behavioral evolution in marine plankton: the case of *Mastigias* (Scyphozoa, Rhizostomeae). Mar. Biol. 143, 1161–1174.

Dawson, M.N., Hamner, W.M., 2009. A character-based analysis of the evolution of jellyfish blooms: adaptation and exaptation. Hydrobiologia 616, 193–215.

Dawson, M.N., Jacobs, D., 2001. Molecular evidence for cryptic species of *Aurelia aurita* (Cnidaria, Scyphozoa). Biol. Bull. 200, 92–96.

Dawson, M.N., Martin, L.E., Penland, L.K., 2001. Jellyfish swarms, tourists, and the Christ-child. Hydrobiologia 451, 131–144.

de Moor, C., van der Lingen, C.D., Teske, P., 2022. A new hypothesis for South African sardine stock structure. University of Cape Town. Report. https://doi.org/10.25375/uct.21257664.v2.

Deidun, A., Arrigo, S., Piraino, S., 2011. The westernmost record of *Rhopilema nomadica* (Galil, 1990) in the Mediterranean – off the Maltese Islands. Aquat. Invasions. 6, S99–S103.

Deidun, A., Sciberras, J., Sciberras, A., et al., 2017. The first record of the white-spotted Australian jellyfish *Phyllorhiza punctata* von Lendenfeld, 1884 from Maltese waters (western Mediterranean) and from the Ionian coast of Italy. BioInvasions Rec 6, 119–124.

Diaz, R.J., Rosenberg, R., 2008. Spreading dead zones and consequences for marine ecosystems. Science 321, 926–929.

DiBacco, C., Fuchs, H.L., Pineda, J., et al., 2011. Swimming behavior and velocities of barnacle cyprids in a downwelling flume. Mar. Ecol. Prog. Ser. 433, 131–148.

Djeghri, N., Pondaven, P., Le Grand, F., et al., 2021. High trophic plasticity in the mixotrophic *Mastigias papua* - Symbiodiniaceae holobiont: implications for the ecology of zooxanthellate jellyfishes. Mar. Ecol. Prog. Ser. 666, 73–88.

Djeghri, N., Pondaven, P., Stibor, H., et al., 2019. Review of the diversity, traits, and ecology of zooxanthellate jellyfishes. Mar. Biol. 166, 147.

Djeghri, N., Pondaven, P., Stockenreiter, M., et al., 2020b. Isotopic and elemental compositions reveal density-dependent nutrition pathways in a population of mixotrophic jellyfish. Ecosphere 11 (11).

Djeghri, N., Stibor, H., Lebeau, O., et al., 2020a. $\delta^{13}C$, $\delta^{15}N$, and C:N ratios as nutrition indicators of zooxanthellate jellyfishes: insights from an experimental approach. J. Exp. Mar. Biol. Ecol. 522, 151257.

Domingues, C., Church, J., White, N., et al., 2008. Improved estimates of upper-ocean warming and multi-decadal sea-level rise. Nature 453, 1090–1093.

Dong, Z., Liu, Z., Liu, D., et al., 2016. Low genetic diversity and lack of genetic structure in the giant jellyfish *Nemopilema nomurai* in Chinese coastal waters. Mar. Biol. Res. 12, 769–775.

Dong, J., Sun, M., Purcell, J.E., et al., 2015. Effect of salinity and light intensity on somatic growth and podocyst production in polyps of the giant jellyfish *Nemopilema nomurai* (Scyphozoa: Rhiozostomeae). Hydrobiologia 754, 75–83.

Dong, J., Wang, B., Duan, Y., et al., 2018. Initial occurrence, ontogenic distribution-shifts and advection of *Nemopilema nomurai* (Scyphozoa: Rhizostomeae) in Liaodong Bay, China, from 2005–2015. Mar. Ecol. Prog. Ser. 591, 185–197.

Dönmez, M.A., Bat, L., 2019. Detection of feeding dietary *Rhizostoma pulmo* (Macri, 1778) in Samsun coasts of the Black Sea, Turkey. Ege J. Fish. Aquat. Sci. 36, 135–144.

Doyle, T.K., Hays, G.C., Harrod, C., et al., 2014. Ecological and societal benefits of jellyfish. In: Pitt, K.A., Lucas, C.H. (Eds.), Jellyfish Blooms. Springer, Dordrecht, pp. 105–127.

Doyle, T.K.J., Houghton, J.R., McDevitt, J., et al., 2007. The energy density of jellyfish: estimates from bomb-calorimetry and proximate-composition. J. Exp. Mar. Biol. Ecol. 343, 239–252.

Dror, H., Angel, D.L., 2023. Rising seawater temperatures affect the fitness of *Rhopilema nomadica* polyps and podocysts and the expansion of this medusa into the western Mediterranean. Mar. Ecol. Prog. Ser. 728, 123–143.

Duffy, J.T., Epifanio, C.E., Fuiman, L.A., 1997. Mortality rates imposed by three scyphozoans on red drum (*Sciaenops ocellatus* Linnaeus) larvae in field enclosures. J. Exp. Mar. Biol. Ecol. 212, 123–131.

Durieux, D.M., Scrogham, G.D., Fender, C., et al., 2023. Benthic jellyfish act as suction pumps to facilitate release of interstitial porewater. Sci. Rep. 13, 3770.

Durgham, H., 2011. First records of *Phyllorhiza punctata* von Lendenfeld, 1884 (Cnidaria: Rhizostomeae) from the Mediterranean coast of Syria. Int. J. Oceans Oceanogr 5, 153–155.

Duron, M., 1978. Contribution à l'étude de la biologie de *Dermochelys coriacea* (Linné) dans les Pertuis Charentais. Thèse Doctorale Biologie Animale. Université de. Bordx. Bordeaux 112.

Early, R., Bradley, B., Dukes, J., et al., 2016. Global threats from invasive alien species in the twenty-first century and national response capacities. Nat. Commun. 7, 12485.

Edelist, D., Guy-Haim, T., Kuplik, Z., et al., 2020. Phenological shift in swarming patterns of *Rhopilema nomadica* in the Eastern Mediterranean Sea. J. Plankton Res. 42, 211–219.

El Rahi, J., Weeber, M.P., El Serafy, G., 2020. Modelling the effect of behavior on the distribution of the jellyfish Mauve stinger (*Pelagia noctiluca*) in the Balearic Sea using an individual-based model. Ecol. Model. 433, 109230.

Elliott, A., Hobson, V., Tang, K.W., 2017. Balancing fishery and conservation: a case study of the barrel jellyfish *Rhizostoma octopus* in South Wales. ICES J. Mar. Sci. 74, 234–241.

Elton, C., 1927. Animal Ecology. Sedgewick & Jackson, London.

Enrique-Navarro, A., Prieto, L., 2020. First record of *Phyllorhiza punctata* (Cnidaria: Rhizostomae: Mastigiidae) in the Northeast Atlantic Ocean. Cah. Biol. Mar. 61, 349–353.

Estes, A.M., Kempf, S.C., Henry, R.P., 2003. Localization and quantification of carbonic anhydrase activity in the symbiotic scyphozoan *Cassiopea xamachana*. Biol. Bull. 204, 278–289.

F.A.O., 2020. The State of World Fisheries and Aquaculture 2020. Sustainability in Action. Rome.

Fautin, D., Dalton, P., Incze, L.S., et al., 2010. An overview of marine biodiversity in United States waters. PLoS One 5, e11914..

Fautin, D.G., Fitt, W.K., 1991. A jellyfish-eating sea anemone (Cnidaria, Actiniaria) from Palau: *Entacmaea medusivora* sp. nov. Hydrobiologia 216, 453–461.

Fay, S., Husson, B., 1983. Répartition, fréquence et régime alimentaire de la méduse *Rhizostoma pulmo*. Diplôme agronomie approfondie, École Nat. Sup. Agro., Rennes. pp. 67 pp..

Feng, S., Lin, J., Sun, S., et al., 2018a. Hyposalinity and incremental micro-zooplankton supply in early-developed *Nemopilema nomurai* polyp survival, growth, and podocyst reproduction. Mar. Ecol. Prog. Ser. 591, 117–128.

Feng, S., Wang, S.W., Sun, S., et al., 2018b. Strobilation of three scyphozoans (*Aurelia coelurea*, *Nemopilema nomurai*, and *Rhopilema esculentum*) in the field at Jiaozhou Bay, China. Mar. Ecol. Prog. Ser. 591, 141–153.

Feng, S., Wang, S.W., Zhang, G.T., et al., 2017. Selective suppression of in situ proliferation of scyphozoan polyps by biofouling. Mar. Pollut. Bull. 114, 1046–1056.

Feng, S., Zhang, F., Sun, S., et al., 2015. Effects of duration at low temperature on asexual reproduction in polyps of the scyphozoan *Nemopilema nomurai* (Scyphozoa: Rhizostomeae). Hydrobiologia 745, 97–111.
Fernández-Alías, A., Marcos, C., Pérez-Ruzafa, A., 2021. Larger scyphozoan species dwelling in temperate, shallow waters show higher blooming potential. Mar. Pollut. Bull. 173, 113100.
Fernández-Alías, A., Marcos, C., Quispe, J.I., et al., 2020. Population dynamics and growth in three scyphozoan jellyfishes, and their relationship with environmental conditions in a coastal lagoon. Estuar. Coast. Shelf. Sci. 243, 106901.
Fitt, W.K., 1985. Effect of different strains of the zooxanthella *Symbiodinium microadriaticum* on growth and survival of their coelenterate and molluscan hosts. Proc. Fifth Intl. Coral Reef. Symp. 6, 131–136.
Fitt, W.K., Costley, K., 1998. The role of temperature in survival of the polyp stage of the tropical rhizostome jellyfish *Cassiopea xamachana*. J. Exp. Mar. Biol. Ecol. 222, 79–91.
Fitt, W.K., Hofmann, D.K., Kemp, D.W., et al., 2021. Different physiology in the Jellyfish *Cassiopea xamachana* and *C. frondosa* in Florida Bay. Oceans 2, 811–821.
Fleck, J., Fitt, W.K., Hahn, M.G., 1999. A proline-rich peptide originating from decomposing mangrove leaves is one natural metamorphic cue of the tropical jellyfish *Cassiopea xamachana*. Mar. Ecol. Progr. Ser. 183, 115–124.
Fleck, J., Fitt, W.K., 1999. Degrading mangrove leaves of *Rhizophora mangle* provide a natural metamorphic cue for the upside-down jellyfish *Cassiopea xamachana*. J. Exp. Mar. Biol. Ecol. 234, 84–94.
Fleming, N.E.C., Harrod, C., Newton, J., et al., 2015. Not all jellyfish are equal: isotopic evidence for inter- and intraspecific variation in jellyfish trophic ecology. PeerJ 3, e1110.
Flynn, B.A., Gibbons, M.J., 2007. A note on the diet and feeding of *Chrysaora hysoscella* in Walvis Bay Lagoon, Namibia, during September 2003. Afri. J. Mar. Sci. 29, 303–307.
Forskål, P., 1775. Descriptiones Animalium, Avium, Amphibiorum, Piscium, Insectorum,Vermium; quae in Itinere Orientali Observavit Petrus Forskål. Post MortemAuctoris editit Carsten Niebuhr. Adjuncta est materia Medica Kahirina. Mölleri. Hafniae 19, 164 + xxxiv +.
Fortes, R.R., Absalão, R.S., 2004. The applicability of Rapoport's rule to the marine molluscs of the Americas. J. Biogeog. 31, 1909–1916.
Fossette, S., Gleiss, A.C., Chalumeau, J., et al., 2015. Current-oriented swimming by jellyfish and its role in bloom maintenance. Curr. Biol. 25, 342–347.
Fossette, S., Katija, K., Goldbogen, J.A., et al., 2016. How to tag a jellyfish? A methodological review and guidelines to successful jellyfish tagging. J. Plank. Res. 38, 1347–1363.
Frost, J.R., Jacoby, C.A., Frazer, T.K., et al., 2012. Pulse perturbations from bacterial decomposition of *Chrysaora quinquecirrha* (Scyphozoa: Pelagiidae). Hydrobiol 690, 247–256.
Fuentes, V., Straehler-Pohl, I., Atienza, D., et al., 2011. Life cycle of the jellyfish *Rhizostoma pulmo* (Scyphozoa: Rhizostomeae) and its distribution, seasonality and inter-annual variability along the Catalan coast and the Mar Menor (Spain, NW Mediterranean). Mar. Biol. 158, 2247–2266.
Fullerton, A.H., Garvey, J.E., Wright, R.A., et al., 2000. Overwinter growth and survival of largemouth bass: interactions among size, food, origin, and winter severity. Trans. Am. Fish. Soc. 129, 1–12.
Galil, B.S., 2008. Alien species in the Mediterranean Sea: which, when, where, why? Hydrobiologia 606, 105–116.
Galil, B.S., Gershwin, L.A., Douek, J., et al., 2010. *Marivagia stellata* gen. et sp. nov. (Scyphozoa: Rhizostomeae: Cepheidae), another alien jellyfish from the Mediterranean coast of Israel. Aquat. Inv. 5, 331–340.

Galil, B.S., Spanier, E., Ferguson, W.W., 1990. The scyphomedusae of the Mediterranean coast of Israel, including two lessepsian migrants new to the 1298 Mediterranean. Zool. Med. 64, 95–105.
Gamero-Mora, E., Lawley, J.W., Marona, M.M., Stampar, S.N., Morandini, A.C., 2024. Morphological and molecular data in the study of the evolution, population genetics and taxonomy of Rhizostomeae. Adv. Mar. Biol. 98, 361–396.
Garcia, S., Zerbi, A., Aliaume, C., et al., 2003. The ecosystem approach to fisheries: issues, terminology, principles, institution foundations, implementation and outlook. FAO Fisheries Technical Paper. No. 443. Rome, FAO. 81 pp.
García, J.R., Durbin, E., 1993. Zooplanktivorous predation by large scyphomedusae *Phyllorhiza punctata* (Cnidaria: Scyphozoa) in Laguna Joyuda. J. Exp. Mar. Biol. Ecol. 173, 71–93.
García-Monteiro, S., Sobrino, J.A., Julien, Y., et al., 2022. Surface Temperature trends in the Mediterranean Sea from MODIS data during years 2003–2019. Reg. Stud. Mar. Sci. 49, 102086.
Gaston, K.J., Chown, S.L., 1999. Why Rapoport's rule does not generalise. Oikos 84, 309–312.
Geissinger, E.A., Gregory, R.S., Laurel, B.J., et al., 2002. Food and initial size influence overwinter survival and condition of a juvenile marine fish (age-0 Atlantic cod). Can. J. Fish. Aquat. Sci. 78, 472–482.
Gibbons, M.J., Richardson, A.J., 2009. Patterns in jellyfish abundance in the North Atlantic. Hydrobiologia 616, 51–65.
Gibbons, M.J., Richardson, A.J., 2013. Beyond the jellyfish joyride and global oscillations: Advancing jellyfish research. J. Plankton Res. 35, 929–938.
Gibbons, M.J., Skrypzeck, H., Brodeur, R.D., et al., 2021. A comparative review of macromedusae in eastern boundary currents. Oceanogr. Mar. Biol. Ann. Rev. 59, 371–482.
Glynn, F., Houghton, J.D.R., Provan, J.I.M., 2015. Population genetic analyses reveal distinct geographical blooms of the jellyfish *Rhizostoma octopus* (Scyphozoa). Biol. J. Linn. Soc. 116, 582–592.
Gmelin, J.F., 1791. Vermes. In: Gmelin J.F. (Ed.) Caroli a Linnaei Systema Naturae per Regna Tria Naturae, Ed. 13. Tome 1(6). G.E. Beer, Lipsiae [Leipzig]. pp. 3021–3910. Systema Naturae. Linneaeus (ed.). Ed. 13. 1: pars. 6.
Goette, A., 1886. Verzeichniss der Medusen welche von Dr Sander, Stabsarzt auf S.M.S. "Prinz Adalbert" gesammelt wurden. Sitzungsberichte der preussischen Akademie der Wissenschaften 1886 (2), 831–837.
Gohar, H.A.F., Eisawy, M., 1960. The development of *Cassiopea andromeda* (Scyphomedusae). Publs Mar. Biol. Stat. Ghardaqa 11, 148–190.
Gomez Daglio, L., Dawson, M.N., 2017. Species richness of jellyfishes (Scyphozoa: Discomedusae) in the Tropical Eastern Pacific: Missed taxa, molecules, and morphology match in a biodiversity hotspot. Invert. Syst. 31, 635–663.
González Carman, V., Botto, F., Gaitán, E., et al., 2013. A jellyfish diet for the herbivorous green turtle *Chelonia mydas* in the temperate SW Atlantic. Mar. Biol. 161, 339–349.
Gotoh, O.R., Chang, S.H., Qu, X., Ito, C., Okuizumi, K., Yoon, W.D., et al., 2017. Distribution and dispersal of the giant jellyfish *Nemopilema nomurai* (Scyphozoa, Cnidaria) inferred from its genetic population structure. Biogeog 19, 93–100.
Graham, W.M., Bayha, K.M., 2007. Biological invasions by marine jellyfish. In: Nentwig, W. (Ed.), Biological Invasions. Ecological Studies 193. Springer, Berlin/Heidelberg, pp. 239–255.
Graham, W.M., Martin, D.L., Felder, D.L., Asper, V.L., Perry, H.M., 2003. Ecological and economic implications of a tropical jellyfish invader in the Gulf of Mexico. Biol. Inv. 5, 53–69.

Grant, G.S., Ferrell, D., 1993. Leatherback turtle, *Dermochelys coriacea* (Reptilia, Dermochelidae)—notes on near-shore feeding behavior and association with Cobia. Brimleyana 19, 77–81.

Grenacher, H., Noll, F.C., 1876. Beiträge zur Anatomie und Systematik der Rhizostomeen. Abhandlungen Senckenbergischen Naturforschenden Gesellschaft 10, 119–180.

Griffin, D.C., Harrod, C., Houghton, J.D.R., Capellini, I., 2019. Unravelling the macro-evolutionary ecology of fish–jellyfish associations: life in the 'gingerbread house. Proc. R. Soc. B 286, 20182325.

Griffish, W.E., 1887. Japanese Fairy World: Stories from the Wonder-Lore of Japan. Trübner & Co, London.

Gueroun, S.K.M., Kéfi-Daly, Y.O., Deidun, A., et al., 2015. First record and potential trophic impact of *Phyllorhiza punctata* (Cnidaria: Scyphozoa) along the north Tunisian coast (South Western Mediterranean Sea). It. J. Zool. 82, 95–100.

Guy-Haim, T., Rubin-Blum, M., Rahav, E., Belkin, N., Silverman, J., Sisma-Ventura, G., 2020. The effects of decomposing invasive jellyfish on biogeochemical fluxes and microbial dynamics in an ultra-oligotrophic sea. Biogeoscience 17, 5489–5511.

Guy-Haim, T., Silverman, J., Raddatz, S., Wahl, M., Israel, A., Rilov, G., 2016. The carbon turnover response to thermal stress of a dominant coralline alga on the fast warming Levant coast. Limnol. Oceanogr. 61, 1120–1133.

Haeckel, E., 1880. Das System der Acraspeden. 2te Hälfte des Systems der Medusen. AchtNachträge zur Vervollständigung des Systems. Denkschriften der Medicinisch-Naturwissenschaftlichen Gesellschaft zu Jena 2, 361–672.

Hamner, W.M., Dawson, M.N., 2009. A review and synthesis on the systematics and evolution of jellyfish blooms: advantageous aggregations and adaptive assemblages. Hydrobiologia 616, 161–191.

Haraldsson, M., Tönnesson, K., Tiselius, P., Thingstad, T.F., Aksnes, D.L., 2012. Analysis of the relationship between fish and jellyfish as a function of eutrophication and water clarity. Mar. Ecol. Prog. Ser. 471, 73–85.

Harrison, N.M., 1984. Predation on jellyfish and their associates by seabirds. Limnol. Oceanogr 29, 1335–1337.

Hatai, S., 1917. On the composition of the medusa, *Cassiopea xamachana* and the changes in it after starvation. Proc. Nat. Acad. Sci. 3, 22–24.

Hay, S.J., Hislop, J.R.G., Shanks, A.M., 1990. North Sea scyphomedusae; summer distribution, estimated biomass and significance particularly for 0-group gadoid fish. Neth. J. Sea Res. 25, 113–130.

Hays, G.C., Doyle, T.K., Houghton, J.D.R., 2018. A paradigm shift in the trophic importance of jellyfish? Trends Ecol. Evol. 33, 874–884.

Hays, G.C., Bastian, T., Doyle, T.K., et al., 2012. High activity and Lévy searches: jellyfish can search the water column like fish. Proc. R. Soc. B Biol. Sci 279, 465–473.

Heery, E.C., Bishop, M.J., Critchley, L.P., Bugnot, A.B., Airoldi, L., Mayer-Pinto, M., et al., 2017. Identifying the consequences of ocean sprawl for sedimentary habitats. J. Exp. Mar. Biol. Ecol. 492, 31–48.

Helm, R.R., 2018. Evolution and development of scyphozoan jellyfish. Biol. Rev. 93, 1228–1250.

Hoeksema, B.W., Tuti, Y., Becking, L.E., 2015. Mixed medusivory by the sea anemone *Entacmaea medusivora* (Anthozoa: Actiniaria) in Kakaban Lake, Indonesia. Mar. Biodiv. 45 (2), 141–142.

Hofmann, D.K., Brandt, U., 1987. Induction of metamorphosis in the symbiotic scyphozoan *Cassiopea andromeda*: Role of marine bacteria and of biochemicals. Symbiosis 4, 99–116.

Hofmann, D.K., Fitt, W.K., Fleck, J., 1996. Checkpoints in the life-cycle of *Cassiopea* spp: control of metagenesis and metamorphosis in a tropical jellyfish. Int. J. Dev. Biol. 40, 331–338.

Hofmann, D.K., Kremer, B.P., 1981. Carbon metabolism and strobilation in *Cassiopea andromedea* (Cnidaria: Scyphozoa): Significance of endosymbiotic dinoflagellates. Mar. Biol. 65, 25–33.

Holland, B.S., Dawson, M.N., Crow, G.L., et al., 2004. Global phylogeography of *Cassiopea* (Scyphozoa: Rhizostomeae): molecular evidence for cryptic species and multiple invasions of the Hawaiian Islands. Mar. Biol. 145, 1119–1128.

Holst, S., Sötje, I., Tiemann, H., et al., 2007. Life cycle of the rhizostome jellyfish *Rhizostoma octopus* (L.) (Scyphozoa, Rhizostomeae), with studies on cnidocysts and statoliths. Mar. Biol. 151 (5), 1695–1710.

Holst, S., Tiseo, G.R., Djeghri, N., Sötje, I., 2024. Approaches and findings in histological and micromorphological research on Rhizostomeae. Adv. Mar. Biol. 98, 99–192.

Hosia, A., Augustin, C.B., Dinasquet, J., et al., 2015. Autumnal bottom-up and top-down impacts of *Cyanea capillata*: a mesocosm study. J. Plank. Res. 37 (5), 1042–1055.

Houghton, J.D.R., Doyle, T.K., Davenport, J., et al., 2007. Stranding events provide indirect insights into the seasonality and persistence of jellyfish medusae. Hydrobiologia 589, 1–13.

Houghton, J.D., Doyle, T.K., Wilson, M.W., et al., 2006. Jellyfish aggregations and leatherback turtle foraging patterns in a temperate coastal environment. Ecology 87 (8), 1967–1972.

Hubot, N.D., Giering, S.L.C., Füssel, J., et al., 2021. Evidence of nitrification associated with globally distributed pelagic jellyfish. Limnol. Oceanogr. 66, 2159–2173.

Hubot, N., Giering, S.L.C., Lucas, C., 2022. Similarities between the biochemical composition of jellyfish body and mucus. J. Plank. Res. 44, 337–344.

Hutchings, L., Barange, M., Bloomer, S.F., et al., 1998. Multiple factors affecting south african anchovy recruitment in the spawning, transport and nursery areas. South. Afr. J. Mar. Sci. 19, 211–225.

Hutchinson, G.E., 1940. Bioecol. Ecol. 21, 267–268.

Hutchinson, G.E., 1948. Circular causal systems in ecology. Ann. N. Y. Acad. Sci. 40, 221–246.

Iguchi, N., 2017. Biomass, body elemental composition, and carbon requirement of *Nemopilema nomurai* (Scyphozoa: Rhizostomeae) in the southwestern Japan Sea. Plankton Benthos Res. 12 (2), 104–114.

Iguchi, N., Lee, H.E., Yoon, W.D., et al., 2010. Reproduction of the giant jellyfish, *Nemopilema nomurai* (Scyphozoa: Rhizostomeae), in 2006–2008 as peripherally-transported populations. Ocean. Sci. J. 45, 129–138.

Ikeda, H., Ohtsu, K., Uye, S., 2011a. Structural changes of gonads during artificially induced gametogenesis and spawning in the giant jellyfish *Nemopilema nomurai* (Scyphozoa: Rhizostomeae). J. Mar. Biol. Assoc. U. K. 91, 215–227.

Ikeda, H., Ohtsu, K., Uye, S., 2011b. Fine structure, histochemistry, and morphogenesis during excystment of the podocysts of the giant jellyfish *Nemopilema nomurai* (Scyphozoa, Rhizostomeae). Biol. Bull. 221, 248–260.

Issel, R., 1922. Nuove indagini sul plankton nelle acque di Rovigno. Rapporti Comitato Talassografico Italiano. Memoria 102, 1–36.

Jantzen, C., Wild, C., Rasheed, M., et al., 2010. Enhanced pore-water nutrient fluxes by the upside-down jellyfish *Cassiopea* sp. in a Red Sea coral reef. Mar. Ecol. Prog. Ser. 411, 117–125.

Jarms, G., Morandini, A.C., 2019. World Atlas of Jellyfish. Anhandlungen des Naturwissenschaftlichen Vereins. Hamburg 815 pp.

Jennings, S., Kaiser, M.J., 1998. The effects of fishing on marine ecosystems. Adv. Mar. Biol. 34, 201–352.

Joensen, B., 2014. Identifying the spatio-temporal drivers of population dynamics for the jellyfish *Catostylus mosaicus* (Scyphozoa: Rhizostomae) in Moreton Bay QLD. Honours Thesis. Griffith School of Environment. Griffith University, Australia.

Johnson, D.R., Perry, H.M., Graham, W.M., 2005. Using nowcast model currents to explore transport of non-indigenous jellyfish into the Gulf of Mexico. Mar. Ecol. Prog. Ser. 305, 139–146.

Jordano, M.D.A., Miyake, H., Nagata, R.M., et al., 2022. Patterns of morphological development in Scyphozoa ephyrae (Cnidaria, Medusozoa). Mar. Biodiv. 52, 31.

Kamiyama, T., 2011. Planktonic ciliates as a food source for the scyphozoan *Aurelia aurita* (sl): feeding activity and assimilation of the polyp stage. J. Exp. Mar. Biol. Ecol. 407, 207–215.

Kamiyama, T., 2018. Planktonic ciliates as food for the scyphozoan *Aurelia coerulea*: feeding and growth responses of ephyra and metephyra stages. Oceanogr 74, 53–63.

Katoh, O., 1994. Structure of the Tsushima Current in the southwestern Japan Sea. J. Oceanogr. 50, 317–338.

Kawahara, M., Ohtsu, K., Uye, S., 2013. Bloom or non-bloom in the giant jellyfish *Nemopilema nomurai* (Scyphozoa: Rhizostomeae): roles of dormant podocysts. J. Plank. Res. 35, 213–217.

Kawahara, M., Uye, S., Ohtsu, K., et al., 2006. Unusual population explosion of the giant jellyfish *Nemopilemia nomurai* (Scyphozoa: Rhizostomeae) in East Asian waters. Mar. Ecol. Prog. Ser. 307, 161–173.

Keable, S.J., Ahyong, S.T., 2016. First records of the invasive "upside-down jellyfish", *Cassiopea* (Cnidaria: Scyphozoa: Rhizostomeae: Cassiopeidae), from coastal lakes of New South Wales, Australia. Rec. Aust. Mus. 68; 23–30.

Keller, C., 1888. Die Wanderung der marinen Thierwelt im Suezcanal. Zool. Anz. 11, 389–395.

Kideys, A.E., Gücü, A.C., 1995. *Rhopilema nomadica*: a Lessepsian scyphomedusan new to the Mediterranean coast of Turkey. Isr. J. Zool. 41, 615–617.

Kienberger, K., Prieto, L., 2018. The jellyfish *Rhizostoma luteum* (Quoy & Gaimard, 1827): not such a rare species after all. Mar. Biodivers. 48, 1455–1462.

Kienberger, K., Riera-Buch, M., Schönemann, A.M., et al., 2018. First description of the life cycle of the jellyfish *Rhizostoma luteum* (Scyphozoa: Rhizostomeae). PLoS One 13. https://doi.org/10.1371/journal.pone.0202093.

Kikinger, R., 1992. *Cotylorhiza tuberculata* (Cnidaria:Scyphozoa)-Life history of a stationary population. Mar. Ecol. 13, 333–362.

Kim, D.H., Seo, J.N., Yoon, W.D., et al., 2012. Estimating the economic damage caused by jellyfish to fisheries in Korea. Fish. Sci. 78, 1147–1152.

Kimmerer, W.J., Burau, J.R., Bennett, W.A., 1998. Tidally oriented vertical migration and position maintenance of zooplankton in a temperate estuary. Limnol. Oceanogr. 43, 1697–1709.

Kishinouye, K., 1891. Bizen kurage. Zoological magazine (Dobutsugaku zasshi) 3, 53.

Kishinouye, K., 1922. Echizen kuarge (*Nemopilema nomurai*). Dobutsugaku Zasshi 34, 343–346.

Klein, S.G., Pitt, K.A., Carroll, A.R., 2016. Surviving but not thriving: inconsistent responses of zooxanthellate jellyfish polyps to ocean warming and future UV-B scenarios. Sci. Rep. 6, 28859.

Kondo, Y., Ohtsuka, S., Nishikawa, J., et al., 2014. Associations of fish juveniles with rhizostome jellyfishes in the Philippines, with taxonomic remarks on a commercially harvested species in Carigara Bay, Leyte Island. Plankton Benthos Res. 9, 51–56.

Kos Kramar, M., Tinta, T., Lučić, D., et al., 2019. Bacteria associated with moon jellyfish during bloom and post-bloom periods in the Gulf of Trieste (northern Adriatic). PLoS One 14 (1), e0198056.

Kramp, P.L., 1955. The medusae of the tropical west coast of Africa. Danish Science Press, Limited.

Kramp, P.L., 1959. Medusae, mainly from the west coast of Africa. Inst. R. des. Sci. Nat. Belgique 3, 1–33.

Kramp, P.L., 1970. Zoogeographical Stud. Rhizostomeae (Scyphozoa). Videnskabelige Meddelelser fra Dan. Naturhistorisk Foren. 133, 7–30.

Kremer, P., 2005. Ingestion and elemental budgets for *Linuche unguiculata*, a scyphomedusa with zooxanthellae. J. Mar. Biol. Assoc. UK. 85, 613–625.

Kremer, P., Costello, J., Kremer, J., et al., 1990. Significance of photosynthetic endosymbionts to the carbon budget of the scyphomedusa *Linuche unguiculata*. Limnol. Oceanog. 3535, 609–624.

Kuplik, Z., 2018. Proliferation of the scyphomedusa *Rhopilema nomadica* in the eastern Mediterranean: Efficiency in resource exploitation in an oligotrophic environment? PhD dissertation. University of Haifa, Israel.

Kuplik, Z., Angel, D.L., 2020. Diet composition and some observations on the feeding ecology of the rhizostome *Rhopilema nomadica* in Israeli coastal waters. J. Mar. Biol. Assoc. U. K. 100, 681–689.

Kuplik, Z., Kerem, D., Angel, D.L., 2021. Respiration rates, metabolic demands and feeding of ephyrae and young medusae of the Rhizostome *Rhopilema nomadica*. Diversity 13, 320.

Lakkis, S., Zeidane, R., 1991. Jellyfish swarm along the Lebanese coast (Abstract) Lebanese Association for the Advancement of Science 11th Science Meeting American University of Beirut.

Lamb, P.D., Hunter, E., Pinnegar, J.K., et al., 2017. Jellyfish on the menu: mtDNA assay reveals scyphozoan predation in the Irish Sea. R. Soc. open. sci. 4, 171421.

Lambert, F.J., 1935. Jellyfish. The difficulties of the study of their life history and other problems. Essex Naturalist 25, 70–86.

Lampert, K.P., 2016. *Cassiopea* and its zooxanthellae. In: Goffredo, S., Dubinsky, Z. (Eds.), The Cnidaria, Past, Present and Future. Springer, Cham, pp. 415–423.

Larson, R.J., 1991. Diet, prey selection and daily ration of *Stomolophus meleagris*, a filter-feeding scyphomedusa from the NE Gulf of Mexico. Estuar. Coast. Shelf Sci. 32 (5), 511–525.

Lawley, J.W., Gamero-Mora, E., Maronna, M.M., et al., 2021. The importance of molecular characters when morphological variability hinders diagnosability: systematics of the moon jellyfish genus *Aurelia* (Cnidaria: Scyphozoa). PeerJ 9, e11954.

Lee, P.L., Dawson, M.N., Neill, S.P., et al., 2013. Identification of genetically and oceanographically distinct blooms of jellyfish. J. Roy. Soc. Interface 10, 20120920.

Lee, H.E., Yoon, W.D., Lim, D., 2008. Description of feeding apparatus and mechanism in *Nemopilema nomurai* Kishinouye (Scyphozoa: Rhizostomeae). Ocean. Sci. J. 43, 61–65.

Leoni, V., Bonnet, D., Ramírez-Romero, E., et al., 2021a. Biogeography and phenology of the jellyfish *Rhizostoma pulmo* (Cnidaria: Scyphozoa) in southern European seas. Glob. Ecol. Biogeogr. 30, 622–639.

Leoni, V., Molinero, J.C., Crochemore, S., et al., 2022. Ontogenetic dietary shifts of the medusa *Rhizostoma pulmo* (Cnidaria: Scyphozoa). Hydrobiologia 849, 2933–2948.

Leoni, V., Molinero, J.C., Meffre, M., et al., 2021b. Variability of growth rates and thermohaline niches of *Rhizostoma pulmo*'s pelagic stages (Cnidaria: Scyphozoa). Mar. Biol. 168, 1–19.

Lesson, R.P., 1830-1831. Voyage autour du monde, exécuté par ordre duRoi, sur la corvette de Sa Majesté, La Coquille, pendant les années 1822, 1823,1824 et 1825. Zoologie 2, 1–471.

Lett, C., Penven, P., Ayón, P., Fréon, P., 2007. Enrichment, concentration and retention processes in relation to anchovy (*Engraulis ringens*) eggs and larvae distributions in the northern Humboldt upwelling ecosystem. Journal of Marine Systems 64 (1-4), 189–200.

Lilley, M.K.S., Houghton, J.D.R., Hays, G.C., 2009. Distribution, extent of inter-annual variability and diet of the bloom-forming jellyfish *Rhizostoma* in European waters. J. Mar. Biol. Assoc. U. K. 89, 39–48.

Lindemann, R.L., 1942. The trophic-dynamic aspect of Ecology. Ecology 23, 319–418.

Linnaeus, C., 1758. Systema Naturae per regna tria naturae, secundum classes, ordines, genera, species, cum characteribus, differentiis, synonymis, locis.Â *Editio decima, reformata [10th revised edition], vol. 1: 824 pp.*

Linnaeus, C., 1767. Systema naturae per regna tria naturae: secundum classes, ordines, genera, species, cum characteribus, differentiis, synonymis, locis. Ed. 12. 1., Regnum Animale. 1 & 2. Holmiae [Stockholm], Laurentii Salvii. pp. 1-532 [1766] pp. 533-1327 [1767].

Liu, C.Y., Wang, B., Li, Y.P., Dong, J., 2011. The feeding methods and feeding habit of jellyfish *Rhopilema esculentum* at different developmental phases. Fish. Sci. 8, 014.

López García, M.J., 2015. Recent warming in the Balearic Sea and Spanish Mediterranean coast. Towards an earlier and longer summer. Atmósfera 28 (3), 149–160.

López-Martínez, J., Porchas-Quijada, M., Álvarez-Tello, F.J., et al., 2018. Association of the whale shark *Rhincodon typus* with the cannonball jellyfish *Stomolophus meleagris*. J. Fish Biol. 93, 401–404.

Lotan, A., Ben-Hillel, R., Loya, Y., 1992. Life cycle of *Rhopilema nomadica*: a new immigrant scyphomedusan in the Mediterranean. Mar. Biol. 112, 237–242.

Lotan, A., Fine, M., Ben-Hillel, R., 1994. Synchronization of the life cycle and dispersal pattern of the tropical invader scyphomedusan *Rhopilema nomadica* is temperature dependent. Mar. Ecol. Prog. Ser. 109, 59–65.

Love, M., Baldera, A., Yeung, C., et al., 2013. The Gulf of Mexico Ecosystem: A Coastal and Marine Atlas. New Orleans, LA: Ocean Conservancy, Gulf Restoration Center.

Loveridge, A., Lucas, C.H., Pitt, K.A., 2021b. Shorter, warmer winters may inhibit production of ephyrae in a population of the moon jellyfish *Aurelia aurita*. Hydrobiologia 848 (3), 739–749.

Loveridge, A., Pitt, K.A., Lucas, C.H., et al., 2021a. Extreme changes in salinity drive population dynamics of *Catostylus mosaicus* in a modified estuary. Mar. Environ. Res. 168, 105306.

Lucas, C.H., 2001. Reproduction and life history strategies of the common jellyfish, *Aurelia aurita*, in relation to its ambient environment. In Jellyfish Blooms: Ecological and Societal Importance: Proceedings of the International Conference on Jellyfish Blooms, held in Gulf Shores, Alabama, 12–14 January 2000 (pp. 229–246). Springer Netherlands.

Lucas, C.H., Graham, W.M., Widmer, C., 2012. Jellyfish life histories: role of polyps in forming and maintaining scyphomedusa populations. Adv. Mar. Biol. 63, 133–196.

Lucas, C.H., Pitt, K.A., Purcell, J.E., et al., 2011. What's in a jellyfish? Proximate and elemental composition and biometric relationships for use in biogeochemical studies. Ecol. Arch. 92, E092–E144.

Luo, J.Y., Condon, R.H., Stock, C.A., Duarte, C.M., Lucas, C.H., Pitt, K.A., et al., 2020. Gelatinous zooplankton-mediated carbon flows in the global oceans: a data-driven modeling study. Glob. Biogeochem. Cycl. 34, e2020GB006704.

Lynam, C.P., Heath, M.R., Hay, S.J., Brierley, A.S., 2005. Evidence for impacts by jellyfish on North Sea herring recruitment. Mar. Ecol. Prog. Ser. 298, 157–167.

Maas, O., 1903. Die Scyphomedusen der Siboga expedition. Siboga Exped. 190 (11), 1–91.

Macgillivray, J., 1852. Narrative of the Voyage of HMS Rattlesnake 1 T & W Boon, London UK.

Madkour, F.F., Hanafy, M.H., Zaghloul, W.S., 2019. Record of aggregation of alien tropical schyphozoan *Rhopilema nomadica* Galil, 1990 in the Mediterranean Coast of Egypt. Int. Mar. Sci. J. 1 (2), 11.

Malej, A., Vodopivec, M., Lucic, D., et al., 2014. The lesser-known Medusa *Drymonema dalmatinum* Haeckel 1880 (Scyphozoa, Discomedusae) in the Adriatic Sea/LA Poco Nota Medusa *Drymonema dalmatinum* Haeckel 1880 (Scyphozoa, Discomedusae) Nel Mare Adriatico. In Annales: Series Historia Naturalis 24(2), 79. Scientific and Research Center of the Republic of Slovenia.

Malej, A., Vukovic, A., 1984. Some data on the occurrence and biology of the Scyphomedusa *Pelagia noctiluca* in the Gulf of Trieste, and the impact of jellyfish swarming on human activities. In: Workshop on Jellyfish Blooms in the Mediterranean. pp. 89–94.

Malul, D., Lotan, T., Makovsky, Y., et al., 2019. The Levantine jellyfish *Rhopilema nomadica* and *Rhizostoma pulmo* swim faster against the flow than with the flow. Sci. Rep. 9, 20337.

Mamish, S., Durgham, H., 2016. First record of the new alien sea jelly species *Marivagia stellata* Galil and Gershwin, 2010 off the Syrian coast. Mar. Biodivers. Rec. 9, 23.

Mammone, M., Ferrier-Pages, C., Lavorano, S., et al., 2021. High photosynthetic plasticity may reinforce invasiveness of upside-down zooxanthellate jellyfish in Mediterranean coastal waters. PLoS One 16, e0248814.

Mansueti, R., 1963. Symbiotic behaviour between small fishes and jellyfishes, with new data on that between the stromateid, *Peprilus alepidotus*, and the scyphomedusae, *Chrysaora quinquecirrha*. Copeia 1963, 40–80.

Marambio, M., Canepa, A., Lòpez, L., et al., 2021. Unfolding jellyfish bloom dynamics along the Mediterranean basin by transnational citizen science initiatives. Diversity 13, 274.

Mariottini, G.L., Pane, L., 2010. Mediterranean jellyfish venoms: a review on scyphomedusae. Mar. Drugs 8, 1122–1152.

Marques, R., Bouvier, C., Darnaude, A.M., et al., 2016. Jellyfish as an alternative source of food for opportunistic fishes. J. Exp. Mar. Biol. Ecol. 485, 1–7.

Marques, M., Rufino, M., Darnaude, A.M., et al., 2021. Jellyfish degradation in a shallow coastal Mediterranean lagoon. Estuar. Coast. Shelf Sci. 261, 107527.

Mauchline, J., 1980. The biology of mysids and euphausiids. Adv. Mar. Biol. 18, 3–369.

Mauchline, J., 1998. The biology of calanoid copepods. Adv. Mar. Biol. 33, 1–710.

Mauchline, J., Fisher, L.R., 1969. The biology of euphausiids. Adv. Mar. Biol. 7, 1–454.

Mazaris, A.D., Kallimanis, A., Gissi, E., et al., 2019. Threats to marine biodiversity in European protected areas. Sci. Total Environ. 677, 418–426.

McCanch, N.V., McCanch, M., 1996. Fulmars feeding on jellyfish. British Birds 89, 569.

McCloskey, L.R., Muscatine, L., Wilkerson, F.P., 1994. Daily photosynthesis, respiration, and carbon budgets in a tropical marine jellyfish (*Mastigias* sp.). Mar. Biol. 119, 13–22.

McKinney, L.D., Shepherd, J.G., Wilson, C.A., et al., 2021. The Gulf of Mexico: an overview. Oceanog 34, 30–43.

Medina, M., Ohdera, A., Bellantuono, A., et al., 2021. The upside-down Jellyfish *Cassiopea xamachana* as an emerging model system to study cnidarian-algal symbiosis. Ch. 11, In: Mar. Model. Org. Exp. Biol. 23 eBook ISBN 9781003217503.

Mergner, H., Schuhmacher, H., 1981. Quantitative analysis of the coral community of a fore reef area near Aqaba (Red Sea). Helgol. Meeresunters. 34, 337–354.

Meyer, A.G., 1914. The effects of temperature upon tropical marine animals. Pap. Tortugas Lab. 6, 1–24.

Mizrahi, G.A., Shemesh, E., Mizrachi, A., et al., 2021. Comparative genetics of scyphozoan species reveals the geological history and contemporary processes of the Mediterranean Sea. Ecol. Evol. 11, 10303–10319.

Möller, H., 1984. Reduction of a larval herring population by jellyfish predator. Science 224, 621–622.

Møller, L.F., Riisgård, H.U., 2007. Population dynamics, growth and predation impact of the common jellyfish *Aurelia aurita* and two hydromedusae, *Sarsia tubulosa*, and *Aequorea vitrina* in Limfjorden (Denmark). Mar. Ecol. Prog. Ser. 346, 153–165.

Morandini, A.C., Ascher, D., Stampar, S.N., 2005. Cubozoa e Scyphozoa (Cnidaria: Medusozoa) de águas costeiras do Brasil. Iheringia. Sér. Zool. 95, 281–294.
Morandini, A.C., Stampar, S.N., Maronna, M.M., et al., 2017. All non-indigenous species were introduced recently? The case study of *Cassiopea* (Cnidaria: Scyphozoa) in Brazilian waters. J. Mar. Biol. Assoc. U. K 97, 321–328.
Morandini, A.C., 2024. Morphology of Rhizostomeae jellyfishes: what is known and what we advanced since the 1970s. Adv. Mar. Biol. 98, 61–98.
Mortillaro, J.M., Pitt, K.A., Lee, S.Y., et al., 2009. Light intensity influences the production and translocation of fatty acids by zooxanthellae in the jellyfish *Cassiopea* sp. J. Exp. Mar. Biol. Ecol. 378 (1–2), 22–30.
Muffett, K.M., Aulgur, J., Miglietta, M.P., 2022. Impacts of light and food availability on early development of *Cassiopea* medusae. Front. Mar. Sci. 8, 783876.
Murphy, E.J., Watkins, J.L., Trathan, P.N., et al., 2007. Spatial and temporal operation of the Scotia Sea ecosystem: a review of large-scale links in a krill centred food web. Phil. Trans. Roy. Soc. Lond., Ser. B 362 (1477), 113–148.
Muscatine, L., Porter, J.W., 1977. Reef corals: mutualistic symbioses adapted to nutrient-poor environments. Biosciences 27, 454–460.
Nagata, R.M., Morandini, A.C., Colin, S.P., et al., 2016. Transitions in morphologies, fluid regimes, and feeding mechanisms during development of the medusa *Lychnorhiza lucerna*. Mar. Ecol. Prog. Ser. 557, 145–159.
Nagata, R.M., Morandini, A.C., 2018. Diet, prey selection, and individual feeding rates of the jellyfish *Lychnorhiza lucerna* (Scyphozoa, Rhizostomeae). Mar. Biol. 165, 187.
Neumann, R., 1979. Bacterial induction of settlement and metamorphosis in the planula larvae of *Cassiopea andromeda* (Cnidaria: Scyphozoa, Rhizostomeae). Mar. Ecol. Prog. Ser. 1, 21–28.
Newkirk, C.R., Frazer, T.K., Martindale, M.Q., et al., 2020. Adaptation to bleaching: are thermotolerant Symbiodiniaceae strains more successful than other strains under elevated temperatures in a model symbiotic cnidarian? Front. Microbiol. 11, 822.
Niggl, W., Naumann, M.S., Struck, U., et al., 2010. Organic matter release by the benthic upside-down jellyfish *Cassiopea* sp. fuels pelagic food webs in coral reefs. J. Exp. Mar. Biol. Ecol. 384, 99–106.
Niggl, W., Wild, C., 2010. Spatial distribution of the upside-down jellyfish *Cassiopea* sp. within fringing coral reef environments of the Northern Red Sea: implications for it is lifecycle. Helgol. Mar. Res. 64, 281–287.
Nour El-Din, N., 2001. Ecological investigation of the tintinnid community along the coastal waters of Alexandria, Egypt. Egypt. J. Aquat. Biol. Fish. 5, 165–177.
Ohder, A.H., Attarwala, K., Rubain, H., et al., 2023. Genomic insights of bacteria responsible for settlement and metamorphosis of *Cassiopea xamachana* larvae. mSphere 8 (3), 315–322.
Ohdera, A.H., Abrams, M.J., Ames, C.L., et al., 2018. Upside-down but headed in the right direction: review of the highly versatile *Cassiopea xamachana* system. Front. Ecol. Evol. 6, 35.
Ohtsu, K., Kawahara, M., Ikeda, H., et al., 2007. Experimental induction of gonadal maturation and spawning in the giant jellyfish *Nemopilema nomurai* (Scyphozoa: Rhizostomeae). Mar. Biol. 152, 667–676.
Öztürk, B., Topaloğlu, B., Sümen, S.G., et al., 2018. Jellyfish of the Black Sea and Eastern Mediterranean Waters. Turkish Mar. Res. Found. (TUDAV) Publ. 48, 75.
Padilla Serrato, J., Martinez, J., Acevedo-Cervantes, A., et al., 2013. Feeding of the scyphomedusa *Stomolophus meleagris* in the coastal lagoon Las Guásimas, northwest Mexico. Hidrobiológica 23, 218–1226.
Pallas, P.S., 1774. Spicilegia zoologica, quibus novae imprimis et obscurae animalium species iconibus, descriptionibus atque commentariis illustrantur cura P.S. Pallas. *[Book series, 14 volumes]*. Fasciculus 10, 1–41.

Parida, A.K., Jha, B., 2010. Salt tolerance mechanisms in mangroves: a review. Trees 24, 199–217.

Paspaleff, B.W., 1938. Über die Entwicklung von *Rhizostoma pulmo* Agass. Arb. Biol. Meeresst. Varna 7, 1–17.

Pauly, D., Christensen, V., Dalsgaard, J., et al., 1998. Fishing down marine food webs. Science 279, 860–863.

Pauly, D., Graham, W., Libralato, S., et al., 2008. Jellyfish in ecosystems, online databases, and ecosystem models. Hydrobiologia 616, 67–85.

Peach, M.B., Pitt, K.A., 2005. Morphology of the nematocysts of the medusae of two scyphozoans, *Catostylus mosaicus* and *Phyllorhiza punctata* (Rhizostomeae): implications for capture of prey. Invert. Biol. 124 (2), 98–108.

Pérez-Ruzafa, A., Gilabert, J., Gutiérrez, J.M., et al., 2002. Evidence of a planktonic food web response to changes in nutrient input dynamics in the Mar Menor coastal lagoon, Spain. Hydrobiologia 475, 359–369.

Pérez-Ruzafa, A., Pérez-Ruzafa, I., Newton, A., et al., 2019. Coastal Lagoons: Environmental Variability, Ecosystem Complexity, and Goods and Services Uniformity. In: Wolanski, E., Day, J.W., Elliott, M. (Eds.), Coasts and Estuaries. Elsevier, pp. 253–276.

Pestorić, B., Lučić, D., Bojanić, N., et al., 2021. Scyphomedusae and Ctenophora of the Eastern Adriatic: historical overview and new data. Divers 13 (5), 186.

Piraino, S., Aglieri, G., Martell, L., et al., 2014. *Pelagia benovici* sp. nov. (Cnidaria, Scyphozoa): a new jellyfish in the Mediterranean Sea. Zootaxa 3794, 455–468.

Pitt, K.A., 2000. Life history and settlement preferences of the edible jellyfish *Catostylus mosaicus* (Scyphozoa: Rhizostomeae). Mar. Biol. 136, 269–279.

Pitt, K.A., Budarf, A.C., Browne, J.G., et al., 2014. Bloom and bust: Why do blooms of jellyfish collapse? In: Pitt, K., Lucas, C. (Eds.), Jellyfish Blooms. Springer, Dordrecht.

Pitt, K.A., Clément, A.-L., Connolly, R.M., et al., 2008. Predation by jellyfish on large and emergent zooplankton: implications for benthic–pelagic coupling. Estuar. Coast. Shelf Sci. 76, 827–833.

Pitt, K.A., Connolly, R.M., Meziane, T., 2009. Stable isotope and fatty acid tracers in energy and nutrient studies of jellyfish: a review. Hydrobiologia 616, 119–132.

Pitt, K.A., Kingsford, M.J., 2000a. Geographic separation of stocks of the edible jellyfish *Catostylus mosaicus* (Rhizostomeae) in New South Wales, Australia. Mar. Ecol. Prog. Ser. 196, 143–155.

Pitt, K.A., Kingsford, M.J., 2000b. Reproductive biology of the edible jellyfish *Catostylus mosaicus* (Rhizostomeae). Mar. Biol. 137, 791–799.

Pitt, K.A., Kingsford, M.J., 2003. Temporal variation in the virgin biomass of the edible jellyfish, *Catostylus mosaicus* (Scyphozoa, Rhizostomeae). Fish. Res. 63, 303–313.

Pitt, K.A., Welsh, D.T., Condon, R.H., 2009. Influence of jellyfish blooms on carbon, nitrogen and phosphorus cycling and plankton production. Hydrobiologia 616, 133–149.

Pörtner, H., 2008. Ecosystem effects of ocean acidification in times of ocean warming: a physiologist's view. Mar. Ecol. Prog. Ser. 373, 203–217.

Prieto, L., Armani, A., Macías, D., 2013. Recent strandings of the giant jellyfish *Rhizostoma luteum* Quoy and Gaimard, 1827 (Cnidaria: Scyphozoa: Rhizostomeae) on the Atlantic and Mediterranean coasts. Mar. Biol. 160, 3241–3247.

Purcell, J.E., 2009. Extension of methods for jellyfish and ctenophore trophic ecology to large-scale research. Hydrobiologia 616, 23–50.

Purcell, J.E., Arai, M.N., 2001. Interactions of pelagic cnidarians and ctenophores with fish: a review. Hydrobiologia 451, 27–44.

Purcell, J.E., Atienza, D., Fuentes, V., et al., 2012. Temperature effects on asexual reproduction rates of scyphozoan species from the northwest Mediterranean Sea. Hydrobiologia 690, 169–180.

Purcell, J.E., Uye, S.-i, Lo, W.-T., 2007. Anthropogenic cause of jellyfish blooms and their direct consequences for humans: A review. Mar. Ecol. Prog. Ser. 350, 153–174.

Qu, C., Song, J.-M., Li, N., et al., 2015. Jellyfish (*Cyanea nozakii*) decomposition and its potential influence on marine environments studied via simulation experiments. Mar. Pollut. Bull. 97, 199–208.

Quinlan, E.L., Phlips, E.J., 2007. Phytoplankton assemblages across the marine to low-salinity transition zone in a blackwater dominated estuary. J. Plank. Res. 29, 401–416.

Quoy, J.R.C., Gaimard, J.P., 1824–1826. Zoologie. In: L. de Freycinet (ed.),. Voyage au tour du monde fait par ordre du roi, sur les corvettes de S. M: l'Uranie et la Physicienne pendant les annales 1817 à 1820. iv + 712 pp. [pp. 1-328, 1824; 329-616, 1825; 617-664, 1826].

Quoy, J.R.C., Gaimard, J.P., 1827. Observations zoologiques faites à bord de l'Astrolabe, en mai 1826, dans le Détroit de Gibraltar, reprint in: Annales Des Sciences Naturelles. pp. 5–239.

Rädecker, N., Pogoreutz, C., Wild, C., et al., 2017. Stimulated respiration and net photosynthesis in *Cassiopeia* sp. during glucose enrichment suggests *in hospite* CO_2 limitation of algal endosymbionts. Front. Mar. Sci. 4, 267.

Rahat, M., Adar, O., 1980. Effect of symbiotic zooxanthellae and temperature on budding and strobilation in *Cassiopeia andromeda* (Eschscholz). Biolog. Bull. 159, 394–401.

Ranson, G., 1949. Résultats scientifiques des croisières du navire-école belge "Mercator": II-Méduses. Musée R. d'histoire naturelle de. Belgique.

Rapoport, E.H., 1982. Areography: Geographical Strategies of Species. Pergamon Press, Oxford.

Rato, L.D., Pinto, C., Duarte, I.M., et al., 2021. Euryhalinity and thermal tolerance of *Phyllorhiza punctata* (Scyphozoa) scyphostomae: life history and physiological trade-offs. Mar. Biol. 168, 158.

Reizen, N., Isobe, A., 2006. Numerical tracer experiments representing behavior of the giant jellyfish, *Nemopilema nomurai*, in the Yellow and East China Seas. J. Oceanogr. 5, 425–436.

Reyes Suàrez, N.C., Tirelli, V., Ursella, L., et al., 2022. Multi-platform study of the extreme bloom of the barrel jellyfish *Rhizostoma pulmo* (Cnidaria: Scyphozoa) in the northernmost gulf of the Mediterranean Sea (Gulf of Trieste) in April 2021. Ocean. Sci. 18, 1321–1337. https://doi.org/10.5194/egusphere-2022-393.

Reynaud, A.A.M., 1830. Medusa (Rhyzostoma) fulgida. In: Lesson, R.P. (Ed.), Centurie Zoologique, ou choix d'animaux rares, nouveaux ou imparfaitement connus. F.G. Levrault, Paris, pp. 79–80.

Reynolds, A.M., 2014. Signatures of active and passive optimized Lévy searching in jellyfish. J. R. Soc. Interface 11, 20140665.

Riascos, J.M., Aguirre, W., Hopfe, C., et al., 2018. Floating nurseries? Scyphozoan jellyfish, their food, and their rich symbiotic fauna in a tropical estuary. PeerJ 6, e5057. https://doi.org/10.7717/peerj.5057.

Richardson, A.J., Bakun, A., Hays, G.C., et al., 2009. The jellyfish joyride: causes, consequences and management responses to a more gelatinous future. Trends Ecol. Evol. 24, 312–322.

Rippingale, R.J., Kelly, S.J., 1995. Reproduction and survival of *Phyllorhiza punctata* (Cnidaria: Rhizostomeae) in a seasonally fluctuating salinity regime in Western Australia. Mar. Fresh. Res. 46, 1145–1151.

Rizgalla, J., Crocetta, F., 2020. First record of *Phyllorhiza punctata* von Lendenfeld, 1884 (Cnidaria: Scyphozoa: Rhizostomeae) in Libya through social media data mining. BioInv. Rec. 9 (3), 490–495.

Röthig, T., Puntin, G., Wong, J.C.Y., et al., 2021. Holobiont nitrogen control and its potential for eutrophication resistance in an obligate photosymbiotic jellyfish. Microbiome 9, 127.

Rowe, C.E., Keable, S.J., Ahyong, S.T., et al., 2022. Physiological responses of the upside-down jellyfish, *Cassiopea* (Cnidaria: Scyphozoa: Cassiopeidae) to temperature and implications for their range expansion along the east coast of Australia. J. Exp. Mar. Biol. Ecol. 554, 151765.

Russell, F.S., 1970. The medusae of the British Isles. II. Pelagic Scyphozoa with a supplement to the first volume on hydromedusae. Cambridge University Press, Cambridge, pp. 281.

Scalera-Liaci, L., 1991. Jellyfish in the Mediterranean sea. First results. MAP Tech. Rep. Ser. 47, 163–166.

Schäfer, W., 1955. Eine Qualle aus dem Indischen Ozean in der Agais. Natur. Volk 85, 241–245.

Schäfer, S., Gueroun, S.K.M., Andrade, C., et al., 2021. Combined effects of temperature and salinity on polyps and ephyrae of *Aurelia solida* (Cnidaria: Scyphozoa). Diversity 13, 573.

Schembri, P.J., Deideun, A., Vella, P.J., 2010. First record of *Cassiopea andromeda* (Scyphozoa: Rhizostomeae: Cassiopeidae) from the central Mediterranean Sea. Mar. Biodiv. Rec. 3, e6.

Schiariti, A., Morandini, A.C., Jarms, G., et al., 2014. Asexual reproduction strategies and blooming potential in Scyphozoa. Mar. Ecol. Prog. Ser. 510, 241–253.

Schiariti, A., Holst, S., Tiseo, G.R., Miyake, H., Morandini, A.C., 2024. Life cycles and reproduction of Rhizostomeae. Adv. Mar. Biol. 98, 193–254.

Sekiguchi, H., Inoue, N., 2002. Recent advances in larval recruitment processes of scyllarid and palinurid lobsters in Japanese waters. J. Oceanog. 58, 747–757.

Shanks, A.L., Brink, L., 2005. Upwelling, downwelling, and cross-shelf transport of bivalve larvae: test of a hypothesis. Mar. Ecol. Prog. Ser. 302, 1–12.

Shanks, A.L., Graham, W.M., 1988. Chemical defense in a scyphomedusa. Mar. Ecol. Prog. Ser. 45, 81–86.

Shi, Y.-Q., Sun, S., Zhayung, G.-T., et al., 2015. Distribution pattern of zooplankton functional groups in the Yellow Sea in June: a possible cause for geographical separation of giant jellyfish species. Hydrobiologia 754, 43–58.

Shimomura, T., 1959. On the unprecedented flourishing of 'Echizen Kurage', *Stomolophus nomurai* (Kishinouye), in the Tsushima current regions in autumn, 1958. Bull. Jap. Sea Reg. Fish. Res. Lab. 7, 85–107.

Siokou-Frangou, I., Sarantakos, K., Christou, E.D., 2006. First record of the scyphomedusa *Rhopilema nomadica* Galil. 1990 (Cnidaria: Scyphozoa: Rhizostomeae) Greece. Aquat. Inv. 1, 194–195.

Siokou-Frangou, I., Sarantakos, K., Christou, E.D., 2006. First record of the scyphomedusa *Rhopilema nomadica* Galil, 1990 (Cnidaria: Scyphozoa: Rhizostomeae) in Greece. Aquatic Invasions 1, 194–195.

Skikne, S.A., Sherlock, R.E., Robison, B.H., 2009. Uptake of dissolved organic matter by ephyrae of two species of scyphomedusae. J. Plank. Res. 31, 1563–1570.

Smale, D.A., Wernberg, T., Oliver, E.C.J., et al., 2019. Marine heatwaves threaten global biodiversity and the provision of ecosystem services. Nat. Clim. Chang. 9, 306–312.

Spanier, E., 1989. Swarming of jellyfishes along the Mediterranean coast of Israel. Israel J. Zool. 36, 55–56.

Stevens, G.C., 1989. The latitudinal gradient in geographical range: how so many species coexist in the tropics. Am. Nat. 133, 240–256.

Stiasny, G., 1920. Die Scyphomedusen-Sammlung des Naturhistorischen Reichsmuseums in Leiden. III. Rhizostomeae. Zool. Meded. 5, 213–230.

Stiasny, G., 1931. Die rhizostomeen-sammlung des British museum (natural history) in London. Zool. Meded. 14, 137–178.

Stiasny, G., 1936. *Rhizostoma luteum* (Quoy und Gaimard), im Tejo vor Lissabon nachgewiesen. Arch. Mus. Bocage 7, 1–6.

Stoner, E.W., Yeager, L.A., Sweatman, J.L., et al., 2014. Modification of a seagrass community by benthic jellyfish blooms and nutrient enrichment. J. Exp. Mar. Biol. Ecol. 461, 185–192.

Strayer, D.L., 2010. Alien species in fresh waters: ecological effects, interactions with other stressors, and prospects for the future. Fresh. Biol. 55, 152–174.

Streftaris, N., Zenetos, A., 2006. Alien marine species in the Mediterranean – the 100 'Worst Invasives' and their impact. Mediterr. Mar. Sci. 7, 87.

Strul, D., 2022. The use of statoliths and statocysts in the scyphomedusa *Rhopilema nomadica* as a tool for monitoring jellyfish swarms and for estimating the age of the jellyfish (MA thesis). Department of Maritime Civilizations, University of Haifa, Israel.

Sugiura, Y., 1969. On the life-history of Rhizostome medusae V. On the relation between zooxanthellae and the strobilation of *Cephea cephea*. Bull. Mar. Biol. Stn. Asamushi 8, 227–233.

Sun, T., Wang, L., Zhao, J., Dong, Z., 2021. Application of DNA metabarcoding to characterize the diet of the moon jellyfish *Aurelia coerulea* polyps and ephyrae. Acta Oceanol. Sin. 40, 160–167.

Sun, S., Zhang, F., Li, C., et al., 2015. Breeding places, population dynamics, and distribution of the giant jellyfish *Nemopilema nomurai* (Scyphozoa: Rhizostomeae) in the Yellow Sea and the East China Sea. Hydrobiologia 754, 59–74.

Sweetman, A.K., Chelsky, A., Pitt, K.A., et al., 2016. Jellyfish decomposition at the seafloor rapidly alters biogeochemical cycling and carbon flow through benthic food-webs. Limnol. Oceanogr. 61, 1449–1461.

Sweetman, A.K., Smith, C.R., Dale, T., et al., 2014. Rapid scavenging of jellyfish carcasses reveals the importance of gelatinous material to deep-sea food webs. Proc. Roy. Soc. B Biol. Sci. 281, 1–8.

Swift, H.F., Gómez Daglio, L., Dawson, M.N., 2016. Three routes to crypsis: stasis, convergence, and parallelism in the *Mastigias* species complex (Scyphozoa, Rhizostomeae). Mol. Phylo. Evol. 99, 103–115.

Takao, M., Okawachi, H., Uye, S., 2014. Natural predators of polyps of *Aurelia aurita* s.l. (Cnidaria: Scyphozoa: Semaeostomeae) and their predation rates. Plankton Benthos Res 9, 105–113.

Takao, M., Uye, S., 2018. Effects of low salinity on the physiological ecology of planulae and polyps of scyphozoans in the East Asian marginal seas: potential impacts of monsoon rainfall on medusa population size. Hydrobiologia 815, 165–176.

Tang, C., Sun, S., Zhang, F., 2020. Intraguild predation by polyps of three scyphozoan jellyfish: *Nemopilema nomurai*, *Aurelia coerulea*, and *Rhopilema esculentum*. J. Oceanol. Limnol. 38, 1755–1761.

Tansley, A.G., 1935. The use and abuse of vegetational concepts and terms. Ecology 16, 284–307.

Thiebot, J.B., Ito, K., Raclot, T., et al., 2016. On the significance of Antarctic jellyfish as food for Adélie penguins, as revealed by video loggers. Mar. Biol. 163 (5), 108.

Thiel, M.E., 1966. Untersuchungen über die Herkunft, das Auftreten, das Wachstum und die Fortpflanzung von Rhizostoma octopus L. Ag. im Elbmündungsgebiet. Abh Verhnaturw Ver Hamburg NF 10, 59–88.

Thiel, M.E., 1970. Das Zusammenleben von Jung- und Kleinfischen mit Rhizostomeen (Scyphomedusae). Ber. Dtsch. Wiss. Komm. Meeresforsch. 21, 444–473.

Thiel, M.E., 1978. Das Zusammenleben von Jung- und Kleinfischen mit Semaeostomen (Scyphomedusae). Mitt. Hamb. Zool. Mus. Inst. 75, 19–47.

Thiel, H., 1979. Assoziationen von Quallen und Fischen. Nat. Mus. 109, 353–360.

Thiel, M.E. (†), Reinicke, G.B., Holst, S., Morandini, A.C., Sötje, I., Straehler-Pohl, I., Wiesenthal, A.A., Thiel, H., 2024. Rhizostomeae—a remaining manuscript to complete the Monograph series on Scyphozoa by M. E. Thiel (published 1936–1977). Advances in Marine Biology, 98 (electronic supplement).

Tilstra, A., El-Khaled, Y.C., Meier, S., et al., 2022. Invasive upside-down jellyfish tolerate organic eutrophication and warming. Bull. Mar. Sci. 98, 381–392.

Tinta, T., Klun, K., Herndl, G.J., 2021. The importance of jellyfish–microbe interactions for biogeochemical cycles in the ocean. Limnol. Oceanogr. 66, 2011–2032.

Tinta, T., Kogovšek, T., Malej, A., et al., 2012. Jellyfish modulate bacterial dynamic and community structure. PLoS ONE 7, e39274.

Tinta, T., Kogovšek, T., Turk, V., et al., 2016. Microbial transformation of jellyfish organic matter affects the nitrogen cycle in the marine water column – a Black Sea case study. J. Exp. Mar. Biol. Ecol. 475, 19–30.

Titelman, J., Gandon, L., Goarant, A., et al., 2007. Intraguild predatory interactions between the jellyfish *Cyanea capillata* and *Aurelia aurita*. Mar. Biol. 152, 745–756.

Titelman, J., Hansson, L.J., 2006. Feeding rates of the jellyfish *Aurelia aurita* on fish larvae. Mar. Biol. 149, 297–306.

Titelman, J., Riemann, L., Sørnes, T.A., et al., 2006. Turnover of dead jellyfish: stimulation and retardation of microbial activity. Mar. Ecol. Prog. Ser. 325, 43–58.

Toyokawa, M., Shibata, M., Cheng, J.H., et al., 2012. First record of wild ephyrae of the giant jellyfish *Nemopilema nomurai*. Fish. Sci. 78, 1213–1218.

Tsikliras, A.C., Dinouli, A., Tsiros, V.-Z., et al., 2015. The Mediterranean and Black Sea fisheries at risk from overexploitation. PLoS One 10 (3), e0121188.

United Nations Environment Programme/Mediterranean Action Plan and Plan Bleu (2020). State of the Environment and Development in the Mediterranean. Nairobi.

Uye, S., 2008. Blooms of the giant jellyfish *Nemopilema nomurai*: a threat to the fisheries sustainability of the East Asian Marginal Seas. Plankt. Bent. Res. 3 (Suppl.), 125–131.

Uye, S., 2010. Studies on functional roles of zooplankton in coastal marine ecosystem: toward restoring productive seas for global sustainability. Oceanogr. Japan 19, 283–299 (in Japanese with English abstract).

Uye, S., 2011. Human forcing of the copepod-fish-jellyfish triangular trophic relationship. Hydrobiol 666, 71–83.

Uye, S., 2013. The giant jellyfish *Nemopilema nomurai* in East Asian Marginal Seas. In: Pitt, K.A., Lucas, C.H. (Eds.), Jellyfish Blooms. Springer, Dordrecht, pp. 185–205.

Uye, S. 2014. Recurrent jellyfish blooms in the East Asian seas: how to adapt and manage for fisheries sustainability. Annual Science Conference 2014, A Coruña, Spain. CM 2014/3064 A:01.

Uye, S., Brodeur, R.D. (eds.) (2017) Report of Working Group 26 on Jellyfish Blooms around the North Pacific Rim: Causes and Consequences. PICES Sci. Rep. No. 51, p. 221.

van Deurs, M., Hartvig, M., Steffensen, J.F., 2011. Critical threshold size for overwintering sandeels (*Ammodytes marinus*). Mar. Biol. 158, 2755–2764.

Van Walraven, L., Langenberg, V.T., Dapper, R., et al., 2015. Long-term patterns in 50 years of scyphomedusae catches in the western Dutch Wadden Sea in relation to climate change and eutrophication. J. Plankton Res. 37, 151–167.

Vanhöffen, E., 1888. Untersuchungen über semäostome und rhizostome Medusen. Bib. Zool 1, 1–52.

Verde, E., McCloskey, L., 1998. Production, respiration, and photophysiology of the mangrove jellyfish *Cassiopea xamachana* symbiotic with zooxanthellae: effect of jellyfish size and season. Mar. Ecol. Prog. Ser. 168, 147–162.

Veron, J., Stafford-Smith, M., DeVantier, L., et al., 2015. Overview of distribution patterns of zooxanthellate Scleractinia. Front. Mar. Sci. 1, 81.

Vinogradov, M.E., Shushkina, E.A., Musaeva, E.I., et al., 1989. Ctenophore *Mnemiopsis leidyi* (A Agassiz) (Ctenophora: Lobata) – new settlers in the Black Sea. Oceanology 29, 293–298.

Walther, G.-R., Roques, A., Hulme, P.E., et al., 2009. Alien species in a warmer world: risks and opportunities. Trends Ecol. Evol. 24, 686–693.

Wang, L., Guo, M., Li, T., et al., 2020. Small jellyfish as a supplementary autumnal food source for juvenile chaetognaths in Sanya Bay, China. J. Mar. Sci. Engin. 8 (12), 956.
Welsh, D.T., Dunn, R.J.K., Meziane, T., 2009. Oxygen and nutrient dynamics of the upside down jellyfish (*Cassiopea* sp.) and its influence on benthic nutrient exchanges and primary production. Hydrobiologia 635, 351–362.
West, E.J., Pitt, K.A., Welsh, D.T., et al., 2009. Top-down and bottom-up influences of jellyfish on primary productivity and planktonic assemblages. Limnol. Oceanog. 54, 2058–2071.
Wilken, S., Huisman, J., Naus-Wiezer, S., et al., 2013. Mixotrophic organisms become more heterotrophic with rising temperature. Ecol. Lett. 16, 225–233.
Wolk, M., Rahat, M., Fitt, W.K., et al., 1985. Cholera toxin and thyrotropin can replace natural inducers required for the metamorphosis of larvae and buds of the scyphozoan *Cassiopea andromeda*. Wilhelm. Roux's Arc. Devel. Biol. 194, 487–490.
Worm, B., Lotze, H.K., 2021. Chapter 21: Marine biodiversity and climate change. Letcher, T (ed.), climate and global change: observed impacts on planet earth, third ed. Elsevier, pp. 445–464.
WoRMS Editorial Board, 2024. World Register of Marine Species. Available from https://www.marinespecies.org at VLIZ. Accessed 2024-08-12. https://doi.org/10.14284/170.
Wright, R.M., Le Quéré, C., Buitenhuis, E., et al., 2021. Role of jellyfish in the plankton ecosystem revealed using a global ocean biogeochemical model. Biogeosci 18, 1291–1320.
Wu, B.L., 1955. *Rhopilema esculentum*. Biol. Newslet. 4, 35–40.
Xiao, W., Zeng, Y., Liu, X., et al., 2019. The impact of giant jellyfish *Nemopilema nomurai* blooms on plankton communities in a temperate marginal sea. Mar. Pollut. Bul. 149, 110507.
Yahia, M.N.D., Goy, J., Yahia-Kéfi, O.D., 2003. Distribution et écologie des méduses (Cnidaria) du golfe de Tunis (Méditerranée sud occidentale). Oceanol. Acta 26, 645–655.
Yahia, M.N.D., Yahia, O.K.D., Gueroun, S.K.M., et al., 2013. The invasive tropical scyphozoan *Rhopilema nomadica* Galil, 1990 reaches the Tunisian coast of the Mediterranean Sea. BioInv. Rec. 2, 319–323.
Yamamoto, J., Hirose, M., Ohtani, T., et al., 2008. Transportation of organic matter to the sea floor by carrion falls of the giant jellyfish *Nemopilema nomurai* in the Sea of Japan. Mar. Biol. 153, 311–317.
Yasuda, T., 2007. The Giant Jellyfish and the Moon Jellyfish. Seizando-shoten, Tokyo. (in Japanese).
Yeruham, E., Rilov, G., Shpigel, M., et al., 2015. Collapse of the echinoid *Paracentrotus lividus* populations in the Eastern Mediterranean – result of climate change? Sci. Rep. 5, 1–6.
Zarnoch, C.B., Hossain, N., Fusco, E., et al., 2020. Size and density of upside-down jellyfish, *Cassiopea* sp. and their Impact on benthic fluxes in a Caribbean Lagoon. Mar. Environ. Res. 154, 104845.
Zhang, F., Sun, S., Jin, X., Li, C., 2012. Associations of large jellyfish distributions with temperature and salinity in the Yellow Sea and East China Sea. Hydrobiologia 690, 81–96.

CHAPTER EIGHT

Rhizostomes as a resource: The expanding exploitation of jellyfish by humans

Lucas Brotz[a,b,*], Dror L. Angel[c], Isabella D'Ambra[d,e], Angélica Enrique-Navarro[f], Chiara Lauritano[g], Delphine Thibault[h], and Laura Prieto[f]

[a]Sea Around Us, Institute for the Oceans and Fisheries, University of British Columbia, Vancouver, BC, Canada
[b]SeaLifeBase, Quantitative Aquatics, Khush Hall, International Rice Research Institute (IRRI), Los Baños, Philippines
[c]Department of Maritime Civilizations & Recanati Institute for Maritime Studies, University of Haifa, Mt. Carmel, Haifa, Israel
[d]Integrative Marine Ecology Department, Stazione Zoologica Anton Dohrn, Villa Comunale, Naples, Italy
[e]National Center for Future Biodiversity (NFBC), Palermo, Italy
[f]Department of Ecology and Coastal Management, Instituto de Ciencias Marinas de Andalucía (ICMAN), Consejo Superior de Investigaciones Científicas (CSIC), Puerto Real Cádiz, Spain
[g]Ecosustainable Marine Biotechnology Department, Stazione Zoologica Anton Dohrn, Napoli, Italy
[h]Aix Marseille University, Université de Toulon, CNRS, IRD, MIO, Marseille, France
*Corresponding author. e-mail address: lucasbrotz@gmail.com

Contents

Advances in Marine Biology, Volume 98
ISSN 0065-2881, https://doi.org/10.1016/bs.amb.2024.08.001

Abstract

While jellyfish are often considered to be a nuisance, their value to ecosystems and for human exploitation is shifting this perception. People have been eating jellyfish for millennia. In recent decades, the scale of jellyfish fisheries has expanded dramatically, with annual catches in the hundreds of thousands of tonnes. The overwhelming majority of jellyfish species targeted for human consumption are from the order Rhizostomeae, which can also be fed to livestock and certain species in mariculture operations. The use of rhizostome jellyfish is expanding beyond food applications, such as pharmaceuticals and cosmetics, especially for collagen and other bioactive compounds. Jellyfish collagen is high in antioxidants, can act as an immunostimulator, and has applications for tissue engineering and medical implements. Jellyfish venom extracts exhibit high biological activities, including those that are antihypertensive, antimicrobial, and anticancer. Jellyfish can also be used as fertilizers and insecticides, and jellyfish mucus appears to have potential as a filter for nanoparticles and microplastics, suggesting possible applications in wastewater treatment. Most of these applications are still in developmental stages, and beyond their use as food, jellyfish are not targeted at commercial scale, apart from collagen extraction. As research advances, exploitation of jellyfish is expected to continue expanding. Given the lack of knowledge and understanding regarding jellyfish fisheries and their management, caution should be exhibited to avoid overfishing.

1. Introduction

Rhizostome jellyfish have been used by humans for thousands of years (Tian et al., 2016). Historically, this use has almost exclusively been for direct consumption as food, and catches of jellyfish now eclipse those of many other popular seafoods, such as lobsters and clams (see www.seaaroundus.org). In his review on Rhizostomeae up to 1970, Thiel discusses the consumption of rhizostomes by humans in both Europe and Asia, and even describes several potential health benefits, including as a laxative and a remedy for gallstones (Thiel et al., 2024). Presently, jellyfish are also used as feed inputs for livestock and mariculture. As demand for edible rhizostomes continues to increase, there are a variety of other applications where the use of jellyfish appears to have promising potential (Fig. 1).

Most proteins in jellyfish are organized into collagen, which is in high demand for pharmaceutical and cosmetic use. Indeed, collagen has been regarded as one of the most useful and promising biomaterials (Lee et al., 2001).

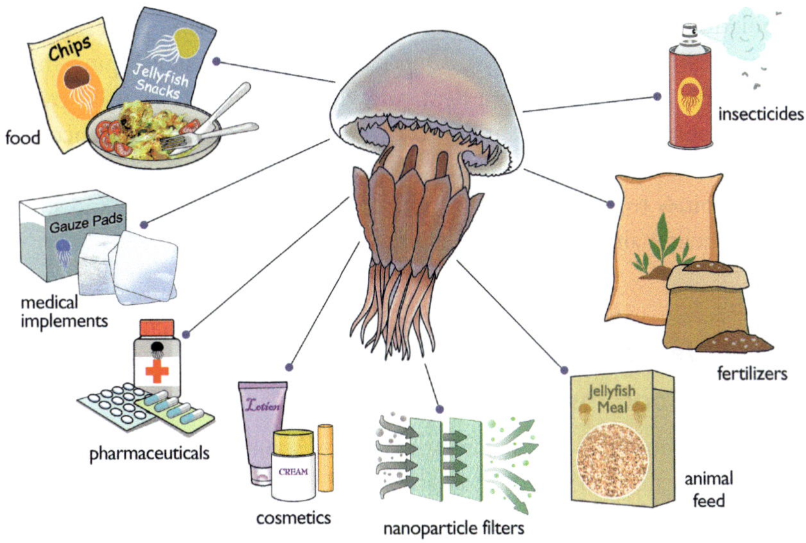

Fig. 1 Potential uses for rhizostome jellyfish.

More traditional sources of collagen have potential health concerns, and exploitation of marine collagen (including jellyfish-derived collagen) has emerged as an attractive biocompatible alternative. Jellyfish collagen is a promising source of antioxidants, can act as an anticoagulant potentially useful in treating hypertension, and can have immunostimulatory effects. In addition to pharmaceuticals and cosmetics, collagen has several biomedical applications, such as tissue engineering, whereby tissue scaffolds can play an important role in cartilage and bone regeneration. Collagen can also aid as a vehicle for drug delivery and has potential for use in medical implements due to its impressive absorptive capacity.

Venom extracts from jellyfish also have a wide variety of pharmacological applications, exhibiting numerous biological activities including those that are anti-inflammatory, anti-hypertensive, antimicrobial, analgesic, antioxidant, anticoagulant, immunomodulatory, and anticancer (Nisa et al., 2021). More extensive information on Rhizostomeae venoms is presented in this volume by Nagata et al. (2024), although some of the bioprospecting applications are discussed below.

In a number of locations worldwide, jellyfish blooms appear to be increasing in frequency or intensity in recent decades (Brotz et al., 2012). With the increased use of jellyfish by humans, it has been suggested that expanding jellyfish exploitation even further will help to control their populations; however, this line of thinking is problematic for a number of reasons (Gibbons et al., 2016). One possibility is that developing markets for jellyfish can result in more demand, which is then met by implementing management measures intended to protect stocks or even enhance them, such as hatchery programs (e.g., Dong et al., 2009). Nonetheless, new opportunities to exploit jellyfish will likely continue as finfish and shellfish stocks decline and research on jellyfish expands.

2. Jellyfish as food

2.1 Food for humans

Consumption of jellyfish occurs primarily in Asian cultures; however, there are attempts afoot to create alternative food products to appeal to a broader audience (e.g., Pedersen et al., 2017; Raposo et al., 2018; Villalba-Urquidy et al., 2022; Youssef et al., 2019). The research team led by Antonella Leone in Italy has spearheaded research into the nutritional aspects and culinary potential of medusae for the European palate, despite the Western neophobic view of jellyfish as food (Leone et al., 2015, 2019, 2022). This also included the investigation of alternative processing methodologies (see below), as well as safety and quality assessments based on microbiological, chemical, and biochemical analyses. Much of the development for culinary recipes has been done using rhizostomes *Rhizostoma pulmo* (Macri, 1778) and occasionally *Cotylorhiza tuberculata* (Macri, 1778) (Leone, 2020).

Fisheries for jellyfish have also expanded worldwide due to increased demand in Asia and as more traditional fish stocks collapse (Brotz et al., 2017; Fahrenbruch, 2018). Interestingly, most jellyfish species preferred for human consumption are rhizostomes. Approximately 35 species of jellyfish have been reported being consumed by humans (Brotz et al., 2017), including at least 29 rhizostomes (Table 1). With respect to jellyfish catches, which globally can exceed 800,000 tonnes annually (Brotz, 2016a), the composition is nearly all rhizostomes, as many of the non-rhizostome species reported to be consumed are in localized regions or are from unique reports. This is because rhizostomes produce the desired crispy and crunchy texture that is characteristic of jellyfish prepared according to traditional processing techniques. Although there is much variation, the general traditional preparation method involves soaking the jellyfish in various mixtures or slurries of salt and alum (usually potassium aluminum sulfate – $KAl[SO_4]_2 \cdot 12H_2O$), a process that is sometimes referred to as "pickling." The

Table 1 Species of Rhizostomeae eaten by humans.

Family	Species	Location	References
Cassiopeidae	*Cassiopea ndrosia*	Philippines	Omori and Nakano (2001)
Catostylidae	*Acromitus hardenbergi*	Malaysia; Indonesia; Thailand	Kitamura and Omori (2010)
	Catostylus mosaicus	Australia	Victoria (2006), Raoult and Gaston (2018)
	Catostylus perezi	Pakistan	Gul and Morandini (2013), Muhammed and Sultana (2008)
	Catostylus tagi	Portugal	Raposo et al. (2018)
	Crambione mastigophora	Indonesia	Kitamura and Omori (2010), Omori and Nakano (2001)
	Crambionella annandalei	Myanmar; India	Behera et al. (2018), Kitamura and Omori (2010)
	Crambionella orsini	India; Sri Lanka	Karunarathne et al. (2024)
	Crambionella helmbiru	Indonesia	Nishikawa et al. (2015)
	Crambionella stuhlmanni	India	Kuthalingam et al. (1989), Mohan et al. (2011)
Cepheidae	*Cephea cephea*	Thailand	Omori and Nakano (2001)
	Cotylorhiza tuberculata	Italy	Brotz et al. (2017), Leone (2020), Leone et al. (2019)

(continued)

Table 1 Species of Rhizostomeae eaten by humans. (*cont'd*)

Family	Species	Location	References
Lobonematidae	*Lobonema smithi*	China; India; Philippines	Hong (2002), Kingsford et al. (2000), Murugan and Durgekar (2008)
	Lobonemoides gracilis	China; Philippines; Sri Lanka	Hong (2002), Kitamura and Omori (2010), Omori (1981), Karunarathne et al. (2024)
	Lobonemoides robustus	Indonesia; Myanmar; Vietnam; Thailand; Philippines	Kitamura and Omori (2010)
Lychnorhizidae	*Lychnorhiza lucerna*	Argentina	Brotz et al. (2017)
Mastigiidae	*Mastigias* sp.	Thailand	Sloan and Gunn (1985)
	Phyllorhiza punctata	Australia	Coleman et al. (1990), Kailola et al. (1993)
Rhizostomatidae	*Nemopilema nomurai*	China; Japan; Korea	Li et al. (2014), Morikawa (1984), Omori (1978)
	Rhizostoma octopus	United Kingdom	Elliott et al. (2016)
	Rhizostoma pulmo	Turkey; Italy	Leone (2020), Leone et al. (2019), Ozer and Celikkale (2001)
	Rhizostoma sp.	India	Chidambaram (1984)
	Rhopilema asamushi	China; Japan	Hsieh and Rudloe (1994)

	Rhopilema esculentum	China; India; Indonesia; Japan; Korea; Malaysia; Thailand; Russia; Vietnam	Kingsford et al. (2000), Morikawa (1984), Nishikawa et al. (2008), Omori (1978), Omori and Kitamura (2004), Panda and Madhu (2009), Sloan (1986), Yakovlev et al. (2005)
	Rhopilema hispidum	China; Indonesia; Japan; Malaysia; Pakistan; Thailand; Vietnam	Gul and Morandini (2015), Kingsford et al. (2000), Kitamura and Omori (2010), Muhammed and Sultana (2008), Omori and Kitamura (2004)
	Rhopilema nomadica	Turkey	Kingsford et al. (2000)
	Rhopilema verrilli	USA	Kingsford et al. (2000), Rudloe (1992)
Rhizostomatidae?	(suspected unique sp.)	Indonesia	Kitamura and Omori (2010), Omori and Nakano (2001)
Stomolophidae	*Stomolophus meleagris*	USA; Mexico; Nicaragua; Ecuador; Honduras	Brotz et al. (2017), Fahrenbruch (2018), Hsieh et al. (2001), López-Martinez and Álvarez-Tello (2013)
	Stomolophus sp. 1	Mexico	Brotz et al. (2021), Getino-Mamet et al. (2019), Gómez Daglio and Dawson (2017)
	Stomolophus sp. 2	Mexico	Brotz et al. (2021), Getino-Mamet et al. (2019), Gómez Daglio and Dawson (2017)

Source: Adapted from Brotz, L., Schiariti, A., López-Martínez, J., Álvarez-Tello, J., Hsieh, Y.-H.P., Jones, R.P., et al., 2017. Jellyfish fisheries in the Americas: origin, state of the art, and perspectives on new fishing grounds. Rev. Fish. Biol. Fish. 27, 1–29.

procedure is usually stepwise, with the jellyfish being transferred to various salt/alum mixtures; the whole process taking days to weeks. Pickling reduces the water content and lowers the pH, inhibiting microbial growth and thus extending the product's shelf life. Processed jellyfish is a semi-dried, almost rubbery product that is typically 15–20% of the original wet weight (Brotz, 2016b).

Preserved jellyfish are served in a variety of ways. Usually, the product is soaked for hours or overnight, often with multiple water changes. This is done to remove the salt and alum, and partially rehydrate the product. The jellyfish is then typically sliced into thin strips, blanched, and served either on its own (with a sauce such as sesame oil, soy sauce, or vinegar) or as an ingredient in more elaborate dishes. In China, jellyfish are savored as a delicacy, and are often served at special occasions such as weddings, holidays, and other celebrations (Hsieh and Rudloe, 1994). There are a variety of purported health benefits to eating jellyfish, although most of the assertions have not been tested by Western science. Recently, Lv et al. (2022) demonstrated that collagen hydrolysate derived from *Nemopilema nomurai* Kishinouye, 1922 produced several benefits when fed to mice on a high-fat diet, suggesting that jellyfish collagen could potentially be used for the prevention and treatment of diet-induced obesity. There are also health concerns about the detectable amounts of aluminum in the product due to the alum used in processing (Ogimoto et al., 2012; Zhang et al., 2016). Methods are being developed that allow processing of jellyfish without the use of alum (e.g., Bleve et al., 2021; Esina et al., 2023; Leone et al., 2019; Ramires et al., 2022a, 2022b; Yuferova, 2017) or the removal of aluminum after processing (e.g., Li et al., 2023); however, they have not yet been widely adopted. Gelatin derived from rhizostome jellyfish also has the potential to be used in a variety of food applications, such as gelling, thickening, or binding agents (Chiarelli et al., 2021).

Stocks of *Rhopilema esculentum* Kishinouye, 1891 are also grown in aquaculture operations in China for human food. This includes hatchery programs, as well as culturing in closed nets inshore and in large saltwater ponds where they are grown to market size. The scale of these operations is considerable, with jellyfish pond aquaculture reaching 8000 ha two decades ago in Liaoning Province alone (Guan et al., 2004 in You et al., 2007). Hatchery programs, or so-called "reseeding," whereby juvenile jellyfish are reared in the laboratory and then released into the wild in an attempt to rebuild stocks, began in the 1980s and expanded in the 1990s and 2000s, with millions or occasionally billions of individual jellyfish released annually (Dong et al., 2009; You et al., 2007).

2.2 Food for animals consumed by humans

Jellyfish may be used as a feedstock input for aquaculture (i.e., mariculture). For example, Liu et al. (2015) demonstrated the rearing of juvenile silver pomfret *Pampus argenteus* (Euphrasen, 1788) using a diet supplemented with *R. esculentum*. Wakabayashi et al. (2012) did feeding experiments with larval slipper lobsters *Ibacus novemdentatus* Gibbes, 1850 using jellyfish, including rhizostomes *Mastigias papua* (Lesson, 1830) and *N. nomurai*. In addition, there are several reports and trials that were successful using non-rhizostome jellyfish as feed, so it remains possible that rhizostomes could successfully be used in similar applications (e.g., Goldstein and Nelson, 2011; Miyajima et al., 2011a, 2011b; Miyajima-Taga et al., 2014; Wakabayashi et al., 2012, 2016a, 2016b). Jellyfish also appear to be useful in multi-trophic aquaculture systems whereby other farmed animals, such as shrimp, clams, and sea cucumbers benefit from the particulate organic matter generated by jellyfish, as well as helping to alleviate benthic nutrient loading (Duarte et al., 2022; Guo et al., 2014, 2017; Ren et al., 2014).

Jellyfish are also used as bait for fishing. In Japan, parts of *N. nomurai* are sometimes used for fishing sea bream (likely *Pagrus* spp.) (Omori and Kitamura, 2004). In India, jellyfish have been used as bait in fish traps for at least decades (Prabhu, 1954; Thomas, 1969). For example, fishers in the Gulf of Mannar and Palk Bay report that traps that include jellyfish as bait appear to increase catches of spinefoot fish (*Siganus spp.*) (Varghese et al., 2008). Adjacent trap fisheries in Sri Lanka use rhizostomes *Acromitus flagellatus* and *Lychnorhiza malayensis* as bait (Karunarathne and de Croos, 2021).

Although documentation of using jellyfish as a major feedstock input for animal agriculture is rare, one unconfirmed report indicates that an average of 100,000 tonnes of jellyfish are being processed for chicken meal annually (CIESM, 2010). Hsieh and Rudloe (1994) outline several studies whereby both chickens and pigs increased their growth rates while being fed a diet supplemented with a paste derived from jellyfish, indicating the potential of jellyfish as a feedstock additive.

3. Bioactive compounds

3.1 Collagen

Overall, Rhizostomeae contain higher amounts of proteins compared to Semaeostomeae (D'Ambra and Merquiol, 2022; Merquiol et al., 2019). Like Semaeostomeae, most proteins in rhizostome jellyfish are organized into collagen. Collagen (from the ancient Greek words *kolla*, glue, and *genos*, common descent) is a structural protein of the connective tissue and constitutes the most abundant protein in the animal kingdom, characterized by a high content of amino acids (Hickman et al., 2006). The protein structure of collagen is made up of three polypeptide chains twisted together to form a triple helix, with two identical chains (α-1) and the third (α-2) being different from the other two in its primary structure and sequence. Each collagen chain has about 1050 amino acids with the triplet Gly-X-Y, where Gly stands for glycine and X and Y can be any amino acid, frequently proline and hydroxyproline (Ricard-Blum et al., 2005). Glycine is essential to pack the three α chains into hexagonal structures which then form elongated fibrils. There are more than 20 types of collagen (Ferreira et al., 2012) resulting from the complex structural and hierarchical organization of collagen and the combination of amino acids. Collagen Type I is a basic component of bones, dermis, tendons, ligaments, and cornea tissues. Type II is present in cartilages, vitreous body, and nucleus pulposus. Type III constitutes skin, vessel walls, and reticular fibers of most tissues, including lungs, liver, and spleen. Type IV is part of the basement membranes, which form the supporting structure on which epithelial and endothelial cells grow and surround muscle, fat, and peripheral nerves. Type V is often associated with Type I and is part of cornea tissues.

At present, most collagen used for biomedical and pharmacological applications is obtained from terrestrial mammals, especially from bovine and porcine sources, as they have high amounts of this compound (Lin et al., 2011). However, the use of porcine and bovine-derived collagen remains controversial and is becoming restricted, as collagen from bovine sources may induce health problems in humans (Cho et al., 2005). The use of bovine-derived products dramatically decreased during the spreading of bovine spongiform encephalopathy (BSE) in Europe at the beginning of the 2000s. In addition to BSE, other diseases such as transmissible spongiform encephalopathy and foot-and-mouth disease resulted in the restriction of bovine-derived products, which are a potential vehicle of transmission. Consequently, strict regulations have been implemented within the European Union (CE n. 999/2001 and UE n. 142/2011). Additionally, religious beliefs and alimentary restrictions may prevent some populations from using mammalian products. Finally, cattle and pigs bred in captivity under unhealthy conditions have raised the concern of animal rights movements (Eriksson et al., 2013).

In response to these issues, marine animals are being investigated as new sources of collagen. Recent analyses of collagen content of marine organisms have elucidated that a remarkable amount of collagen may be obtained from fish (skin and bones) as well as sponges, and are considered a good alternative to collagen derived from bovine and porcine sources (Lin et al., 2011; Silva et al., 2014). In this context, jellyfish have emerged within marine organisms as an attractive alternative source of next-generation collagen (Widdowson et al., 2018). Jellyfish-derived collagen shares several features with its vertebrate counterpart (e.g., bovine), showing optimal absorption and biocompatibility, making it highly attractive for biomedical applications (Addad et al., 2011; Widdowson et al., 2018). This new collagen source provides a safer alternative for humans, avoiding transmission risks of diseases from bovines and potential viral vectors.

As mentioned, jellyfish proteins are mainly organized in the collagen complex, which accounts for up to 40–60% of the dry mass of scyphozoans (Addad et al., 2011; Khong et al., 2016; Leone et al., 2015; Lucas et al., 2011). Additionally, it has been demonstrated that some rhizostome jellyfish have greater protein content (collagen) than other marine organisms, including sponges (Merquiol et al., 2019). The determination of collagen content is challenged by different factors, mainly the type (dry or wet) mass and the extraction protocol used, which make a remarkable difference considering that jellyfish, particularly scyphomedusae, are >90% water (Arai, 1997). As a consequence, determinations of collagen content made using dry mass can be an order of magnitude greater than those made using wet mass (Table 2). As for extraction methodologies, the most popular protocols are based either on acid or pepsin, with pepsin solubilization being the most effective to maximize the yield of collagen, based on a direct comparison of the two methods (Barzideh et al., 2014; Leone et al., 2015).

As reported in previous reviews, *Rhopilema asamushi* Uchida, 1927 and *Stomolophus meleagris* Agassiz, 1860 possess the highest content of collagen within the scyphomedusae, both Semaeostomeae and Rhizostomeae, where it has been investigated (Table 2).

Collagen from several rhizostome jellyfish has been isolated and characterized (Addad et al., 2011; Calejo et al., 2009; Cheng et al., 2017; Khong et al., 2016; Nagai et al., 2000). The collagen extracted from the abundant Mediterranean jellyfish *R. pulmo* is compatible and has a biological effect on human cells similar to that of mammalian (bovine-derived) Type I collagen (Addad et al., 2011). Collagen from rhizostome jellyfish may be recognized by human integrins and heparin-sulfate receptors; thus, human cells may adhere to jelly-derived collagen as they do to mammalian collagen. *R. esculentum*, abundant in the East China Sea and Japan (Jarms and Morandini, 2019), has

Table 2 Collagen content (percentage of dry (DM) or wet mass (WM)) in different species of scyphomedusae extracted using two different protocols based either on acid or pepsin solubilization. B, bell; M, mesoglea; ND, no data; OA, oral arms; W, whole.

		Collagen content				
		Pepsin		Acid		
Species	**Tissue**	**(% DM)**	**(% WM)**	**(% DM)**	**(% WM)**	**Reference**
Cassiopea andromeda	W	2.2–6.0	ND	ND	ND	De Rinaldis et al. (2021)
Catostylus tagi	B	2.7	ND	ND	ND	Calejo et al. (2009)
	B	ND	4.5	ND	ND	Addad et al. (2011)
Cotylorhiza tuberculata	OA	ND	19.4	ND	ND	Addad et al. (2011)
	B	<10	ND	ND	ND	Addad et al. (2011)
	B	ND	8.3–31.5	ND	ND	Addad et al. (2011)
Rhizostoma pulmo	OA	ND	26–90	ND	ND	Addad et al. (2011)
	B	<10	ND	ND	ND	Addad et al. (2011)
Rhopilema asamushi	M	35.2	ND	ND	ND	Nagai et al. (2000)
Rhopilema esculentum	M	ND	0.28	ND	0.12	Cheng et al. (2017)
Stomolophus meleagris	M	46.4	ND	ND	ND	Nagai et al. (1999)
Nemopilema nomurai	M	2.2	ND	ND	ND	Miura and Kimura (1985)

collagen that appears highly biocompatible with humans (Nagai et al., 2000; Song et al., 2006). In 2015, 4.3 tonnes of *Rhizostoma octopus* (Gmelin, 1791) were caught in Carmarthen Bay in the United Kingdom as part of an emerging fishery for jellyfish collagen (Elliott et al., 2016).

3.2 Antioxidants, anticancers, and antihypertensives

Antioxidants, or free-radical scavengers, are compounds that can prevent or slow cell damage caused by free radicals. Rhizostome jellyfish collagen is a source of a great number of antioxidants (De Domenico et al., 2019; Leone et al., 2015; Zhuang et al., 2009b). Recently, hydrolyzed collagen peptides obtained from *R. pulmo* showed higher antioxidant activity (Leone et al., 2015) and a protective effect against oxidative stress in human keratinocytes treated with hydrogen peroxide, than collagen derived from vertebrates, along with no cytotoxic effect (De Domenico et al., 2019).

In addition to *R. pulmo*, collagen peptides from *R. esculentum* exhibited radical scavenging and antifatigue activities (Ding et al., 2011; Yu et al., 2005b, 2006). Collagen hydrolysate from the same species can chelate copper ions, inhibiting the intracellular tyrosinase activity and consequently, display superoxide anion-scavenging and melanogenesis inhibitory activities (Zhuang et al., 2009a). Both collagen and its hydrolysate exhibited melanogenesis inhibitory activities and may act as UV radiation protectors (Zhuang et al., 2009b). *R. esculentum* collagen may reduce the abnormal changes of antioxidative indicators, being a potential novel antiphotoaging agent from natural resources (Zhuang et al., 2009a, 2009b). Similarly, collagen from *S. meleagris* presents anti-melanogenesis activity, likely because collagen peptides act at glutathione levels decreasing melanin content and inhibiting tyrosinase activity (Zhuang and Sun, 2012).

Within Mediterranean Rhizostomeae, remarkable antioxidant activity of collagen peptides was found for *R. pulmo*, the zooxanthellate jellyfish *C. tuberculata* (Leone et al., 2015), and *Rhizostoma luteum* (Quoy and Gaimard, 1827) (Prieto et al., 2018). The antioxidant activity of these species, with a high abundance of collagen, peptides, and other bioactive molecules, makes them a largely untapped source of natural compounds of nutraceutical, cosmeceutical, and pharmacological interest. On the other hand, the zooxanthellate *Cassiopea andromeda* (Forskål, 1775), which is an alien species in the Mediterranean Sea (Thibault et al., 2024), constitutes a potential source of bioactive compounds (antioxidants) different from collagen (De Domenico et al., 2023). The endosymbiotic dinoflagellates (see Holst et al., 2024), in addition to supplying photosynthates to their host (see Nagata et al., 2024), are also known to produce bioactive compounds such as long-chain unsaturated fatty acids, polyphenols, and pigments, including carotenoids, with antioxidant properties and other beneficial biological activities. Therefore, the holobiont (jellyfish-dinoflagellate) might be considered a promising natural source of multiple bioactive compounds produced through mixotrophic metabolism, which are of interest for a wide range of biotechnological applications. However, *Catostylus tagi* (Haeckel, 1869), which appears to enter the Mediterranean through the Strait of Gibraltar, showed low levels of antioxidant capacity (Morais et al., 2009).

As mentioned, extracts from *C. tuberculata* showed antioxidant capacity, and three fractions were also able to induce a dose-dependent decrease of breast cancer (MCF-7) cell viability (Leone et al., 2013). To study the action mechanism, the possible activity of *C. tuberculata* fractions was evaluated on the gap junction intercellular communication (GJIC) functionality. A time- and dose-dependent enhancement of GJIC in MCF-7 cells was found (Leone et al., 2013). The antiproliferative activity was less pronounced on epidermal keratinocyte (HEKa) human cell cultures, where the correlation with GJIC functionality was found only during the first 2–4 h of exposure to the jellyfish fractions (Leone et al., 2013).

Other bioactivities are reported for *N. nomurai* venom, which was found to have a chymotrypsin-like serine protease with fibrinolytic activity with a therapeutic potential for thrombotic diseases (Bae et al., 2017) as well as anticancer activity against human hepatocellular carcinoma (HepG2) cells (Choudhary et al., 2018). In particular, the *N. nomurai* venom anticancer activity was time- and concentration- dependent already at about 1 μg/mL. Additionally, *N. nomurai* venom has been recently tested on human HaCaT keratinocytes to shed light on the skin symptoms after jellyfish envenomation (Choudhary et al., 2021). Treatment with *N. nomurai* venom induced a concentration-dependent reduction in HaCaT cell viability, a decrease in the expression of proteins involved in cell survival and development (such as plasminogen, vinculin, emilin-1, basonuclin2 and focal adhesion kinase 1) and the increased expression of stress or immune response-related proteins (such as Toll-like receptor 4, aminopeptidase N, heat shock protein 105 kDa and ephrin type-A receptor 1) identifying putative players during skin damage and key targets for possible therapeutic strategies (Choudhary et al., 2021).

Recently, Riccio et al. (2022) tested the effects of the methanolic extract and enriched fractions, obtained by solid-phase extraction fractionation, from *C. tuberculata* on various human tumor cell lines in order to evaluate its potential antiproliferative activity. Data showed that raw extract from *C. tuberculata* oral arms was active against melanoma cells (A2058) only at 100 μg/mL. Regarding fractions, fraction C from *C. tuberculata* oral arms was the most active, at both 10 and 100 μg/mL, to reduce melanoma cell viability and showed a dose-dependent activity. LC/MS based dereplication analyses showed that the active fraction mainly contained fatty acids and derivatives. The authors also used PCR arrays in order to investigate the mechanism of action in A2058 cells treated in the presence of *C. tuberculata* oral arms fraction C identifying the up-regulation of apoptotic peptidase activating factor 1 and the down-regulation of junctophilin 3, tumor necrosis factor and DENN/MADD domain containing 4A.

The cytotoxic and antitumor efficacy of *R. octopus* venom extract was tested against different tumor cell lines and Ehrlich ascites carcinoma-bearing mice. Venom extract reduced the viability of human hepatocellular (HepG-2) and breast cancer (MCF-7) cell lines, and tumor volume and tumor cell count decreased in mice bearing tumors, along with an overall improvement of liver and kidney functions (Salama et al., 2022).

In recent years, hypertension, a condition characterized by high blood pressure, has become one of the most common lifestyle-related diseases (Zhang et al., 2009; Zhuang et al., 2012a). The angiotensin-I-converting enzyme (ACE) plays a key role in regulating blood pressure (Skeggs et al., 1956). This enzyme raises blood pressure by converting angiotensin-I (inactive decapeptide) into the potent vasoconstrictor angiotensin-II (octapeptide) and inactivating the vasodilating nonapeptide (Li et al., 2004). Thus, inhibition of ACE suppresses blood pressure elevation.

ACE inhibitory peptides were isolated from *R. esculentum* hydrolysates (Liu et al., 2012, 2013, 2016). The antihypertensive effect of collagen peptides from *R. esculentum* collagen was demonstrated on renovascular hypertension in rats. When rats were orally administered those peptides over a month-long period, a reduction in their blood pressure and an inhibitory effect on angiotensin II concentration was observed (Zhuang et al., 2012b). Additionally, a peptide solution derived from *N. nomurai* showed high antihypertensive activity associated with high ACE inhibitory activity (Morinaga et al., 2010).

3.3 Q-mucin

Mucins, particularly Q-mucin extracted from scyphomedusae, appear to be compounds with promising applications in biotechnology, especially in biomedicines. Mucins belong to a family of glycoconjugates, defined as glycosylated proteins with high molecular mass, and are produced by epithelial tissues in most animals. Because of their ability to form gels, they play a key role in most gel-like secretions, from lubrication to cell signaling to the formation of chemical barriers (Marin et al., 2008). Mucins often serve as inhibitors or control mineralization, such as nacre formation in mollusks (Marin et al., 2000), calcification in echinoderms (Boskey, 2003), and bone formation in vertebrates (Midura and Hascall, 1996). They also bind to pathogens as part of the immune system. However, over-expression of mucin proteins, especially MUC1, are associated with many types of cancer (Brockhausen and Melamed, 2021; Niv, 2008).

Most mucins, particularly those found in mammals, are membrane-bound due to the presence of a hydrophobic membrane-spanning domain that leads to the formation of gel-like mucins or to build up network structures commonly found in saliva or mucosal barriers (Thornton et al., 2008). Conversely, the production of mucin in scyphomedusae occurs in the mesoglea (Ushida et al., 2022). Considering the high yield of mucins obtained from jellyfish compared to secretions from other organisms, protocols to extract Q-mucin from scyphomedusae are likely to continue to improve (Ushida et al., 2024).

Masuda et al. (2007) were the first to extract a glycoprotein belonging to the mucin family from five species of scyphozoan jellyfish, including the rhizostomes *N. nomurai* and *R. esculentum*. The protein possessed a high molecular mass, and its main chain structure contained a unique sequence of eight amino acids, of which two glycosylated ones were repeated twice. The authors named this compound *qniumucin* (*Q-mucin*) and determined that it was highly abundant in jellyfish and similar to the human mucin MUC5AC.

Within biomedical applications, Q-mucin appears to be very effective in the treatment of cartilage degeneration. Qniumucin was used to treat articular cartilage degeneration through intra-articular injections using a model of osteoarthritis (rabbits). Qniumucin extracted from the semaeostome *Aurelia aurita* (Linnaeus, 1758) and the rhizostome *N. nomurai* inhibited articular cartilage degeneration in a model of osteoarthritis (Ohta et al., 2009).

The extraction of qniumucin from scyphozoan jellyfish has been carried out using different methods and protocols. With the perspective of an increased use of this compound for biotechnological applications, Ushida et al. (2022) recently proposed a refined extraction protocol to maximize the yield of Q-mucin and allow the extraction from frozen scyphozoans.

To enhance the extraction protocol, the authors added a chelate reagent (e.g., EDTA) to inactivate endogenous proteases that could induce the decomposition of amino acids. The addition of ethanol allowed the precipitation of amino acid polymers, within which Q-mucin dissociated from the others and was easily soluble in deionized water. Since the remaining mesoglea at the end of the treatment appeared to contain a large amount of the precursor of Q-mucin, the authors suggested that future studies should focus on the development of innovative protocols to effectively separate polymers precipitated after the addition of ethanol in order to increase the yield of extracted Q-mucin.

3.4 Immunostimulators

Immunomodulators are molecules that affect the immune system by inducing their suppression or stimulation. Immunostimulation by natural compounds has become a promising way to prevent and cure diseases. The collagen extracts of the giant jellyfish *N. nomurai*, which is commonly consumed in Japan and China, had immunostimulatory effects on human cultured cells (Suganthi and Bragadeeswaran, 2012). The collagen extracted from this species stimulates the transcription and translation activity to enhance the production of immunoglobulin and cytokine (Nishimoto et al., 2008). The immunostimulatory effect was previously observed on human hybridoma and blood cells in vivo, without associated allergic complications (Morishige et al., 2011). Another study showed that collagen from this species enhanced the production of immunoglobulin (Ig) M of human hybridoma and accelerated IgM and IgG production of peripheral blood lymphocytes (Sugahara et al., 2006).

4. Technomedical applications

4.1 Tissue engineering

Tissue scaffolds or matrices are structures to support cell growth composed of biocompatible biomaterials used for cell attachment and migration in tissue engineering. The use of collagen extracted from rhizostome jellyfish for tissue scaffolds has increased over the last decade due to the demonstrated biocompatibility of human and jellyfish collagen (Song et al., 2006).

Collagen has also been combined with other bioactive molecules to build hybrid matrices (Jeong et al., 2007; Lee et al., 2007). The combination of collagen from *N. nomurai* and hyaluronic acid to build 3-dimensional and highly porous scaffolds (Lee et al., 2007), allows binding with bioactive molecules (DNA, peptides, growth factors, and drugs) and was used for cartilage regeneration. Porous scaffolds using refibrillized collagen of *R. esculentum* were also proposed for potential application in cartilage regeneration (Hoyer et al., 2014; Pustlauk et al., 2017). In addition, tubular scaffolds made up of collagen extracted from *S. meleagris* and polylactic-*co*-glycolic acid fibrils were used to build vascular grafts (Jeong et al., 2007). The implementation of hybrid scaffolds enhanced vascular endothelial cell development by improving mechanical strength and retaining the differentiated cell phenotypes. Finally, a recent study developed biphasic scaffolds combining fibrillized jellyfish collagen from *R. esculentum*, alginate, and mineralized salmon collagen (Bernhardt et al., 2018) for potential application in the repair and regeneration of osteochondral defects.

Collagen from *R. pulmo* was used to develop a therapeutic collagen-based biomaterial for articular cartilage repair (Pugliano et al., 2017). The combination of a jelly-derived collagen scaffold with human stem cells and the TGF-β3 growth factor resulted in chondrogenesis and cartilage differentiation and repair. Widdowson et al. (2018) also used *R. pulmo* collagen to build implantable scaffolds by tissue engineering that were successfully implanted into a rat model. Collagen from the same species has been suggested as a safe and biocompatible biomaterial for osteoarthritis repair (Ahmed et al., 2021) as well as for bone regeneration (Flaig et al., 2020).

Given the similarity to collagen, Nudelman et al. (2019) used Q-mucin from *Rhopilema nomadica* Galil, Spanier, and Ferguson, 1990 and *N. nomurai* made of nanometric fibers to produce scaffolds. Nanometric fibers allowed control of the mechanical, morphological, and chemical properties of the scaffolds, which showed a high degree of biocompatibility and biodegradability. Assays of cell proliferation indicated that the scaffolds were suitable for cardiac cell growth. By including silver within the nanofibers, antibacterial activity was promoted, with the formation of antibacterial mats. Pre-clinical trials on porcine wound models indicated a complete healing of wounds.

4.2 Drug delivery

The collagen derived from *R. pulmo* can be used as a cell matrix to culture human-induced pluripotent stem cell-derived microglial-like cell lines (iMGL) (Mearns-Spragg et al., 2020). iMGL culture possesses the characteristics required for microglia, which play a key role within the central nervous system. Microglia ensure synaptic plasticity, immune activity, neurogenesis, and homeostasis. Thus, collagen from this rhizostome jellyfish offers a highly biocompatible alternative to mammalian collagen and other extracellular matrices, as well as potential applications in the study of neural transmission and new treatments for neural degeneration diseases, such as Alzheimer's disease. In line with this study, collagen of the rhizostome *C. tagi* was isolated, biochemically characterized (Calejo et al., 2009), and has been proposed to be used as a drug delivery biomaterial (Calejo et al., 2012). A microparticulate protein delivery system was developed for the controlled release of therapeutic proteins (Calejo et al., 2012).

4.3 Medical implements

Collagen is an effective hemostatic for the prevention of extensive blood loss and promotes tissue regeneration after wounds. These properties are prompting the use of jellyfish-derived collagen for medical applications (Gajbhiye and Wairkar, 2022). Collagen Type I isolated from *R. esculentum* mesoglea has been proposed to produce biomaterials to control massive blood loss (Cheng et al., 2017). Sponges derived from jellyfish collagen exhibited higher water absorption than medical gauze and constituted a suitable surface for the adhesion and proliferation of hemocytes and platelets. Thus, rhizostome jellyfish collagen may become a hemostatic material for wound-healing applications.

Attempting to capitalize on the absorptive quality of dehydrated jellyfish, the Israeli start-up Cine'al has developed their proprietary "Hydromash" by homogenizing medusa biomass (likely from *R. nomadica*) which can then purportedly be used to manufacture a variety of products such as diapers, tampons, paper towels, and bandages. In addition to producing super-absorbent polymers, the company claims that their products also biodegrade rapidly; a distinct advantage considering the massive problems associated with solid wastes. Despite making headlines nearly a decade ago (e.g., Main, 2014), Cine'al does not appear to have brought any products to market as of yet. Nevertheless, the process has been published (Steinberger et al., 2019), and the resultant hydrogel has been used by others (see Section 5.3.).

In a study by Derkus et al. (2016), researchers utilized collagen extracted from *R. pulmo* to fabricate a highly promising aptasensor with a detection limit of 6.25 nM for thrombin. Thrombin is a key protease involved in the blood coagulation process that plays a critical role in chelating fibrinogen and forming fibrin clots (Huntington, 2005). This aptasensor holds potential applications in clinical settings for analyzing thrombin levels in both blood and cerebrospinal fluid samples obtained from patients with various pathologies. Among these pathologies, hypertension, a condition characterized by high blood pressure, is known to contribute to the formation of blood clots in arteries, thereby increasing the risk of strokes by reducing or blocking blood flow to the brain. Furthermore, metabolic levels of thrombin have been linked to both hemorrhage and thrombosis, as well as neurological diseases. The cross-linking of jellyfish collagen to the designed amine thrombin aptasensor using glutaraldehyde enhances the anticoagulant properties of collagen, providing a dual benefit in mitigating thrombotic risks.

5. Biotechnological applications

5.1 Cosmetics

Collagen-like proteins can help improve skin elasticity and reduce the appearance of fine lines and wrinkles by preventing moisture loss from the skin and binding water vapor to the skin. Additionally, collagen extracts are rich in minerals and amino acids that can nourish and hydrate the skin. Jellyfish extracts are sometimes used in cosmetic products for their purported moisturizing and anti-aging benefits, as jellyfish collagen is similar to mammalian collagen in amino acid composition. Arunwuttipong and Khunngam (2023) found that women treated with a jellyfish extract cream derived from *R. esculentum* had reduced wrinkles, as well as improved skin elasticity and hydration compared with a placebo. Gelatin from jellyfish tissue did not show rheological properties superior to that of mammalian tissue, as had been shown in fish. However, Cho et al. (2014) showed that it could be used in various food and cosmetic products. Skin aging is usually linked to collagen Type I reduction which may be partially compensated for by consumption of collagen peptides (Felician et al., 2018). The ratio of collagen Type III to Type I has been shown to increase in mice skin under UV radiation exposure. Again, damage can be reduced by using collagen peptides to improve moisture absorption and retention properties (Felician et al., 2018). In their study based on collagen extracts from *N. nomurai*, Kim et al. (2016) showed the interesting potential for jellyfish extracts as a moisturizer which could be used as an additive in cosmetics.

Several companies are currently advertising for skin care products based on jellyfish extracts (either collagen or venom) as part of the ingredients in creams, masks, or complements (e.g., see Table 3). Others advertise products

Table 3 Sample skin care products containing jellyfish. ND, no data.

Product name	Price	Link
Bisutang Deep Sea Jellyfish Mask	€4.99 for 25 g	https://www.aliexpress.com/i/1005001833731874.html
JM Solution Active Jellyfish Vital Skin Care	€100 for set	https://caldstore.com/products/jm-solution-active-jellyfish-vital-skin-care-set-prime-korean-cosmetic
Diet Esthetic Jellyfish Venom Essence	ND	https://www.glamot.com/p/25353/diet-esthetic-jellyfish-venom-essence
KollaJell	$39.95 for 60 capsules	https://www.certifiednutra.com/kolla_jell.php
Immortality Day Cream	€42 for 50 mL	https://soulcenter.co.za/shop/apothecary/immortality-day-cream/
Masstige jellyfish collagen day cream	€4.55 for 50 mL	https://bestkosmex.com.ua/ua/p970715216-krem-dlya-litsa.html#
Royal beauty jellyfish cream	$59 for 30 mL	https://www.thaibestsellers.com/product/royal-beauty-jellyfish-cream/
Skin Therapy Eternal Jellyfish Complex DNA Miracle Crème	€18 for 30 mL	https://www.amazon.com.au/Skin-Therapy-Eternal-Miracle-Creme/dp/B075K3PB5C
ByKleo – Ultrahydrating Medusa Cream	€21 for 50 mL	https://www.bykleo.com/producto/crema-facial-medusa/
Schnaphil Whitening Moisturizing Repairing Scar Jellyfish Pure Extract	€11 for 10 mL	https://www.crea-diem.com/schnaphil-whitening-moisturzing-repairing-scar-jellyfish-pure-extract.html

based on jellyfish; however, they do not contain any medusozoan ingredients, such as one inspired by the "immortal" jellyfish (*Turritopsis* sp.), projected to reach annual sales in the millions of U.S. dollars (Manso, 2022).

5.2 Nanoparticle filters

Jellyfish may also have potential for use as an efficient means to capture small particles. Patwa et al. (2015) found that mucus obtained from non-rhizostome semaeostome medusae (*A. aurita* and *Pelagia noctiluca* (Forsskål, 1775)) and the ctenophore *Mnemiopsis leidyi* A. Agassiz, 1865 were all successful in the capture of gold and quantum dots nanoparticles (NPs) that are typically released from nanotechnology industries. They proposed that the binding between these synthetic NPs and the mucus glycoproteins involved hydrogen bonding, ionic interactions, and dehydration of polar groups. Ha et al. (2020) achieved similar results using *A. aurita* extract to capture silver citrate NPs. The EU GoJelly project took the NP-capture idea one step further and tested mucus obtained from a number of different jellyfish, including *R. nomadica*, as a means to capture microplastics for potential use in wastewater treatment plants (Freeman et al., 2020). A protocol was devised to test the binding efficiency of mucus to micro- and nanoplastics, as well as to further study the binding mechanism (Ben-David et al., 2023). Scyphozoan mucus was found to be more effective at removing nanoparticles than many commonly used coagulants and flocculants.

5.3 Fertilizers

Jellyfish medusae may also have value as fertilizers for trees and food crops. Emadodin et al. (2020) reviewed the potential of this concept, concluding that the outlook is promising, however, more research is needed. Using jellyfish as fertilizer can add nutrients to the soil and provide increased water retention, while avoiding many of the negative effects associated with more conventional chemical fertilizers. However, the salt concentration needs to be taken into account, as too much salt can negatively affect the growth of many seedlings and plants. Samaraweera and Dissanayake (2022) treated okra *Abelmoschus esculentus* Moench, 1794 plants with fertilizers prepared from four different jellyfish species, including two rhizostomes: *Lychnorhiza malayensis* Stiasny, 1920 and *Marivagia stellata* Galil and Gershwin, 2010. Treatments with jellyfish performed as well as commercially available compost fertilizer for the first seven weeks of the ten-week experiment, and much better than a negative control (no fertilizer).

N. nomurai that had been processed into a powder was added to the soil of cypress trees *Chamaecyparis obtusa* Siebold and Zuccarini that were planted on an eroded slope. Those trees treated with the jellyfish fertilizer showed increased growth and decreased withering compared with untreated plots (Ezaki et al., 2011; Kim et al., 2012). Chun et al. (2011) also found increased growth for several tree species treated with *N. nomurai*, although the seedling survival of some species declined. Seo et al. (2014) examined the effects of using jellyfish fertilizer on the growth of seedlings in a fire-ravaged forest area in Korea. Both black pine *Pinus thunbergii* Parlatore, 1868 and pin oak *Quercus palustris* Rugel ex Candolle, 1864 exhibited increased survival and growth with the addition of jellyfish fertilizer. Kim et al. (2018) reported positive results using jellyfish fertilizer (derived from *N. nomurai* and the non-rhizostome *A. aurita*) for seedlings of black pine *P. thunbergii* planted as part of a coastal disaster prevention forest. Seedlings treated with jellyfish fertilizer had increased survival and growth compared with controls.

The tropical tree *Shorea macrophylla* (de Vriese, 1963) is endemic to Borneo and is important both ecologically and socio-economically. Perumal et al. (2019) demonstrated that morphological quality attributes of *S. macrophylla* seedlings were significantly higher for those treated with a jellyfish-derived fertilizer compared to chemical and controlled-release fertilizers. The jellyfish species in question is suspected to be *N. nomurai* as the fertilizer used was from Marutomo Corporation in Japan (Watanabe et al., 2015).

In order to mitigate the undesired effect of high salt content, Hossain et al. (2013) used desalinated, dried jellyfish “chips” as fertilizer. When mixed into the soil before transplanting, the jellyfish chips resulted in increased growth of rice plants and helped to effectively control weeds. Watanabe et al. (2015) similarly demonstrated the inhibitory effects of jellyfish fertilizer derived from *N. nomurai* on weeds during rice cultivation. Other experiments have also been successful using non-rhizostome jellyfish as fertilizer (e.g., Borchert et al., 2021; Fukushi et al., 2004). Hydrogels can also be used to deliver additional nutrients to the soil, as was demonstrated for copper by Gavriely et al. (2022), using a hydrogel derived from *R. nomadica*.

Soaking seeds in mixtures of jellyfish also appears to have beneficial effects. Hussein and Saleh (2014) soaked wheat *Triticum aestivum* Linnaeus, 1753 and lentil *Lens culinaris* Medikus, 1787 seeds in liquefied jellyfish and found improvements in germination rate and seedling height, weight, and number of leaves. Hussein et al. (2015) similarly found enhanced germination parameters for several herb and food crop species by soaking the seeds in liquefied jellyfish; however, the jellyfish species used remains unclear as it is likely incorrectly reported as *Cyanea capillata* (Linnaeus, 1758) from the Mediterranean Sea.

5.4 Insecticides

Insecticidal activity has also been found using rhizostome venom extract. Proteinous venom extracted from *R. esculentum* showed insecticidal properties against the pest pear lace bug *Stephanitis pyri* (Fabricius, 1775) (Felician et al., 2019; Yu et al., 2005a). Venom from *N. nomurai* showed activity against the cotton bollworm *Helicoverpa armigera* (Hubner 1808), which can infest and damage important crops such as cotton, maize, sorghum, sunflower, tomato, okra, and a range of legumes worldwide (Yu et al., 2014). *N. nomurai* venom was also effective against the citrus spider mite *Panonychus citri* (McGregor 1916), one of the primary pests affecting citrus crops, especially oranges (Yu et al., 2021). Hussein et al. (2015) reported effective treatments against the peach fruit fly *Bactrocera zonata* (Saunders, 1842) and the saw-toothed grain beetle *Oryzaephilus surinamensis* (Linnaeus, 1758); however, the species of jellyfish used is unclear, as the stated *C. capillata* from the Mediterranean Sea is suspected to be incorrect.

5.5 Environmental monitoring

The effects and concentrations of different contaminants in marine environments have been studied in some jellyfish, but only in two rhizostome species, *N. nomurai* (Hanaoka et al., 2001) and *Cassiopea* sp. (Epstein et al., 2016; Templeman and Kingsford, 2010). Hanaoka et al. (2001) found that water-soluble arsenic content was found in the umbrella, oral arms, and tentacles at concentrations of 0.183, 0.111, and 0.664 μg/g wet weight, respectively. In the case of *Cassiopea* sp., it has the capacity to bioaccumulate several trace elements such as aluminum, arsenic, barium, cadmium, chromium, copper, iron, manganese, and zinc, potentially serving as a biomonitor in coastal marine systems (Epstein et al., 2016; Templeman and Kingsford, 2010).

Morabito et al. (2014) studied the effects of different heavy metals (zinc, cadmium, cobalt, and lanthanum) on the nematocysts discharge (see Holst et al., 2024) response in the non-rhizostome *P. noctiluca* and found that this jellyfish could be used as a model for ecotoxicological research. Recently, Macali and Bergami (2020) demonstrated that the same species can be used as a bioindicator of plastic pollution. Further work needs to be done to explore if these findings are also applicable to rhizostome species of jellyfish.

6. Concluding remarks

The recent developments using rhizostomes in bioactive, medical, and biotechnological applications support the notion that jellyfish can be a valuable resource and should not be perceived solely as mere pests. Of course, rhizostome jellyfish have been highly valued in many cultures as food for centuries, and demand for them continues to increase. Evidently, before jellyfish can be used, they must first be caught. Fishing for jellyfish is most often done using dip-nets from small boats, although a variety of other gears have been employed including set nets, drift nets, gill nets, hooks, purse seines, weirs, beach seines, and trawl nets. With contemporary catches approaching nearly one million tonnes annually (Brotz, 2016a), surprisingly scant research has been devoted to the impacts of jellyfish fisheries or their management (see Kingsford et al., 2000). Indeed, many of the methods developed for finfish fisheries analysis can be adapted for jellyfish (Brotz and Pauly, 2017). Given that we know so little of jellyfish population dynamics, combined with our increasing knowledge of their importance to ecosystems (Doyle et al., 2014), it should behoove regulators to exhibit caution with respect to regulating jellyfish fisheries (Gibbons, 2023).

There are also other ways to obtain jellyfish aside from targeted fishing. Jellyfish accumulate at the intake screens of coastal power plants and desalination plants, at times causing major problems and expense (Purcell et al., 2007; Richardson et al., 2009). Using such jellyfish may be a more cost-effective method of collection, especially for jellyfish bound for uses other than human food where markets prefer intact and undamaged individuals (Edelist et al., 2021). Fisheries targeting finfish and other seafoods may also unintentionally catch jellyfish as bycatch (Bosch-Belmar et al., 2020), which are normally discarded, but could be used for many of the applications discussed above (D'Ambra and Merquiol, 2022). Nevertheless, even with these approaches, the still unknown consequences of removing jellyfish from ecosystems should not be ignored.

Importantly, we must rapidly scale up our understanding of jellyfish as well as their impacts on society, both positive and negative (Graham et al., 2014).

Jellyfish have historically been understudied, and we still lack fundamental understanding about many aspects of their ecology (see Thibault et al., 2024), including how the two bipartite phases of their life history are related to one another with respect to population dynamics (see Schiariti et al, 2024). Developing deeper insights into their ecology is critical as we find new ways to exploit them.

Acknowledgments

The *Sea Around Us* is a research initiative supported by the Beagle Charitable Foundation, David and Lucille Packard Foundation, Marisla Foundation, Oak Foundation, and Summit Foundation, among others. A.E.-N. was supported by the project PIE-MEDUSAS-202330E57 (CSIC), Spain.

References

Addad, S., Exposito, J.-Y., Faye, C., Ricard-Blum, S., Lethias, C., 2011. Isolation, characterization and biological evaluation of jellyfish collagen for use in biomedical applications. Mar. Drugs 9, 967–983.

Ahmed, Z., Powell, L.C., Matin, N., Mearns-Spragg, A., Thornton, C.A., Khan, I.M., et al., 2021. Jellyfish collagen: a biocompatible collagen source for 3D scaffold fabrication and enhanced chondrogenicity. Mar. Drugs 19.

Arai, M.N., 1997. A Functional Biology of Scyphozoa. Chapman & Hall,, London.

Arunwuttipong, S., Khunngam, P., 2023. Jellyfish extract cream, effects of jellyfish extract cream on wrinkle, skin elasticity and hydration of facial skin. Lat. Am. J. Pharm. 42, 1164–1170.

Bae, S.K., Lee, H., Heo, Y., Pyo, M.J., Choudhary, I., Han, C.H., et al., 2017. In vitro characterization of jellyfish venom fibrin(ogen)olytic enzymes from *Nemopilema nomurai*. J. Venomous Anim. Toxins Incl. Trop. Dis. 23, 35.

Barzideh, Z., Latiff, A.A., Gan, C.-Y., Abedin, M.Z., Alias, A.K., 2014. ACE inhibitory and antioxidant activities of collagen hydrolysates from the ribbon jellyfish (*Chrysaora* sp.). Food Technol. Biotechnol. 52, 495–504.

Behera, P., Jishnudev, M., Rao, P., Ghosh, S., 2018. An emerging fishery for the jellyfish *Crambionella annandalei* along the coast of Andhra Pradesh. Mar. Fish. Inf. Serv. Tech. Ext. Ser. 236, 24–25.

Ben-David, E.A., Habibi, M., Haddad, E., Sammar, M., Angel, D.L., Dror, H., et al., 2023. Mechanism of nanoplastics capture by jellyfish mucin and its potential as a sustainable water treatment technology. Sci. Total. Environ. 869.

Bernhardt, A., Paul, B., Gelinsky, M., 2018. Biphasic scaffolds from marine collagens for regeneration of osteochondral defects. Mar. Drugs 16.

Bleve, G., Ramires, F.A., De Domenico, S., Leone, A., 2021. An alum-free jellyfish treatment for food applications. Front. Nutr. 8.

Borchert, F., Emadodin, I., Kluss, C., Rotter, A., Reinsch, T., 2021. Grass growth and N_2O emissions from soil after application of jellyfish in coastal areas. Front. Mar. Sci. 8.

Bosch-Belmar, M., Milisenda, G., Basso, L., Doyle, T.K., Leone, A., Piraino, S., 2020. Jellyfish impacts on marine aquaculture and fisheries. Rev. Fish. Sci. Aquac. 29, 242–259.

Boskey, A.L., 2003. Biomineralization: an overview. Connect. Tissue Res. 44, 5–9.

Brockhausen, I., Melamed, J., 2021. Mucins as anti-cancer targets: perspectives of the glycobiologist. Glycoconjugate J. 38, 459–474.

Brotz, L., 2016a. Jellyfish fisheries – a global assessment. In: Pauly, D., Zeller, D. (Eds.), Global Atlas of Marine Fisheries: A Critical Appraisal of Catches and Ecosystem Impacts. Island Press, Washington, DC, USA, pp. 110–124.

Brotz, L., Cheung, W.W.L., Kleisner, K., Pakhomov, E., Pauly, D., 2012. Increasing jellyfish populations: trends in Large Marine Ecosystems. Hydrobiologia 690, 3–20.

Brotz, L., Cisneros-Montemayor, A.M., Cisneros-Mata, M.A., 2021. The race for jellyfish: winners and losers in Mexico's Gulf of California. Mar. Policy 134.

Brotz, L., Pauly, D., 2017. Studying jellyfish fisheries: toward accurate national catch reports and appropriate methods for stock assessments. In: Mariottini, G.L. (Ed.), Jellyfish: Ecology, Distribution Patterns and Human Interactions. Nova Publishers, Hauppauge, NY, USA, pp. 313–329.

Brotz, L., Schiariti, A., López-Martínez, J., Álvarez-Tello, J., Hsieh, Y.-H.P., Jones, R.P., et al., 2017. Jellyfish fisheries in the Americas: origin, state of the art, and perspectives on new fishing grounds. Rev. Fish. Biol. Fish. 27, 1–29.

Brotz, L., 2016b. Jellyfish fisheries of the world. University of British Columbia, Vancouver, Canada, PhD dissertation, pp. ix + 170.

Calejo, M.T., Almeida, A.J., Fernandes, A.I., 2012. Exploring a new jellyfish collagen in the production of microparticles for protein delivery. J. Microencapsulation 29, 520–531.

Calejo, M.T., Morais, Z.B., Fernandes, A.I., 2009. Isolation and biochemical characterisation of a novel collagen from *Catostylus tagi*. J. Biomater. Sci. Polym. Ed. 20, 2073–2087.

Cheng, X.C., Shao, Z.Y., Li, C.B., Yu, L.J., Raja, M.A., Liu, C.G., 2017. Isolation, characterization and evaluation of collagen from jellyfish *Rhopilema esculentum* Kishinouye for use in hemostatic applications. PLoS One 12.

Chiarelli, P.G., Pegg, R.B., Kumar, G.D., Solval, K.M., 2021. Exploring the feasibility of developing novel gelatin powders from salted, dried cannonball jellyfish (*Stomolophus meleagris*). Food Biosci. 44.

Chidambaram, L., 1984. Export oriented processing of Indian jelly fish (*Muttai chori*, Tamil) by Indonesian Method at Pondicherry Region, Technical and Extension Series. Marine Fisheries Information Service, Central Marine Fisheries Research Institute. Indian Council of Agricultural Research,, Cochin, India, pp. 11–13.

Cho, S., Ahn, J.-R., Koo, J.-S., Kim, S.-B., 2014. Physicochemical properties of gelatin from jellyfish *Rhopilema hispidum*. Fish. Aquat. Sci. 17, 299–304.

Cho, S.M., Gu, Y.S., Kim, S.B., 2005. Extracting optimization and physical properties of yellowfin tuna (*Thunnus albacares*) skin gelatin compared to mammalian gelatins. Food Hydrocolloids 19, 221–229.

Choudhary, I., Hwang, D., Chae, J., Yoon, W., Kang, C., Kim, E., 2021. Proteomic changes during the dermal toxicity induced by *Nemopilema nomurai* jellyfish venom in HaCaT human keratinocyte. Toxins 13, 311.

Choudhary, I., Lee, H., Pyo, M.J., Heo, Y., Chae, J., Yum, S.S., et al., 2018. Proteomic investigation to identify anticancer targets of *Nemopilema nomurai* jellyfish venom in human hepatocarcinoma HepG2 cells. Toxins 10, 194.

Chun, K.W., Damdinsuren, E., Yeom, K., Ezaki, T., 2011. Effect of jellyfish fertilizer application on seedling growth. J. Jpn. Soc. Reveget. Technol. 37, 155–158.

CIESM, 2010. New partnerships for blue biotechnology development – innovative solutions from the sea. In: Briand, F. (Ed.), Report on CIESM International Workshop. The Mediterranean Science Commission, Monaco, pp. 44 11–12 November 2010.

Coleman, A., Micin, S., Mulvay, P., Rippingale, R., 1990. The brown jellyfish (*Phyllorhiza punctata*) in the Swan-Canning Estuary. Waterways Information No. 2 Swan River Trust, Perth, Australia, pp. 4.

D'Ambra, I., Merquiol, L., 2022. Jellyfish from fisheries by-catches as a sustainable source of high-value compounds with biotechnological applications. Mar. Drugs 20. https://www.mdpi.com/1660-3397/20/4/266

De Domenico, S., De Rinaldis, G., Mammone, M., Bosch-Belmar, M., Piraino, S., Leone, A., 2023. The zooxanthellate jellyfish holobiont *Cassiopea andromeda*, a source of soluble bioactive compounds. Mar. Drugs 21, 272.

De Domenico, S., De Rinaldis, G., Paulmery, M., Piraino, S., Leone, A., 2019. Barrel jellyfish (*Rhizostoma pulmo*) as source of antioxidant peptides. Mar. Drugs 17.

De Rinaldis, G., Leone, A., De Domenico, S., Bosch-Belmar, M., Slizyte, R., Milisenda, G., et al., 2021. Biochemical characterization of *Cassiopea andromeda* (Forsskal, 1775), another Red Sea jellyfish in the western Mediterranean Sea. Mar. Drugs 19.

Derkus, B., Arslan, Y.E., Bayrac, A.T., Kantarcioglu, I., Emregul, K.C., Emregul, E., 2016. Development of a novel aptasensor using jellyfish collagen as matrix and thrombin detection in blood samples obtained from patients with various neurodisease. Sens. Actuators B: Chem. 228, 725–736.

Ding, J.-F., Li, Y.-Y., Xu, J.-J., Su, X.-R., Gao, X., Yue, F.-P., 2011. Study on effect of jellyfish collagen hydrolysate on anti-fatigue and anti-oxidation. Food Hydrocolloids 25, 1350–1353.

Dong, J., Jiang, L.X., Tan, K.F., Liu, H.Y., Purcell, J.E., Li, P.J., et al., 2009. Stock enhancement of the edible jellyfish (*Rhopilema esculentum*) in Liaodong Bay, China: a review. Hydrobiologia 616, 113–118.

Doyle, T.K., Hays, G.C., Harrod, C., Houghton, J.D.R., 2014. Ecological and societal benefits of jellyfish. In: Lucas, C.H., Pitt, K.A. (Eds.), Jellyfish Blooms. Springer, Dordrecht, the Netherlands, pp. 105–127.

Duarte, I.M., Marques, S.C., Leandro, S.M., Calado, R., 2022. An overview of jellyfish aquaculture: for food, feed, pharma and fun. Rev. Aquac. 14, 265–287.

Edelist, D., Angel, D.L., Canning-Clode, J., Gueroun, S.K.M., Aberle, N., Javidpour, J., et al., 2021. Jellyfishing in Europe: current status, knowledge gaps, and future directions towards a sustainable practice. Sustainability 13.

Elliott, A., Hobson, V., Tang, K.W., 2016. Balancing fishery and conservation: a case study of the barrel jellyfish *Rhizostoma octopus* in South Wales. ICES J. Mar. Sci 74, 234-241. https://academic.oup.com/icesjms/article/74/1/234/2669573

Emadodin, I., Reinsch, T., Rotter, A., Orlando-Bonaca, M., Taube, F., Javidpour, J., 2020. A perspective on the potential of using marine organic fertilizers for the sustainable management of coastal ecosystem services. Environ. Sustain. 3, 105–115.

Epstein, H.E., Templeman, M.A., Kingsford, M.J., 2016. Fine-scale detection of pollutants by a benthic marine jellyfish. Mar. Pollut. Bull. 107, 340–346.

Eriksson, A., Burcharth, J., Rosenberg, J., 2013. Animal derived products may conflict with religious patients' beliefs. BMC Med. Ethics 14.

Esina, L.M., Belyakova, I.A., Ushakova, Z.E., Shtenina, D.V., 2023. Development of the technology for salted products derived from barrel jellyfish *Rhizostoma pulmo* (Macri, 1778). Aquat. Bioresour. Environ. 6, 107–120.

Ezaki, T., Kohno, S., Tetsuro, K., Chun, K.W., Damdinsuren, E., 2011. The effect of jellyfish fertilizer on tree growth in hillside erosion control works. J. Jpn. Soc. Reveget. Technol. 37, 151–154.

Fahrenbruch, M.L., 2018. Like scooping money out of the sea': Chinese luxury seafood consumption, roving bandits, and the boom-and-bust jellyfish fishery on the Miskitu Coast of Nicaragua. J. Lat. Am. Geogr. 17, 209–238.

Felician, F.F., Xia, C.L., Qi, W.Y., Xu, H.M., 2018. Collagen from marine biological sources and medical applications. Chem. Biodivers. 15.

Felician, F.F., Yu, R.H., Li, M.Z., Li, C.J., Chen, H.Q., Jiang, Y., et al., 2019. The wound healing potential of collagen peptides derived from the jellyfish *Rhopilema esculentum*. Chin. J. Traumatol. 22, 12–20.

Ferreira, A.M., Gentile, P., Chiono, V., Ciardelli, G., 2012. Collagen for bone tissue regeneration. Acta Biomater. 8, 3191–3200.

Flaig, I., Radenkovic, M., Najman, S., Prohl, A., Jung, O., Barbeck, M., 2020. In vivo analysis of the biocompatibility and immune response of jellyfish collagen scaffolds and its suitability for bone regeneration. Int. J. Mol. Sci. 21.

Freeman, S., Booth, A.M., Sabbah, I., Tiller, R., Dierking, J., Klun, K., et al., 2020. Between source and sea: the role of wastewater treatment in reducing marine microplastics. J. Environ. Manag. 266.
Fukushi, K., Ishio, N., Tsujimoto, J., Yokota, K., Hamatake, T., Sogabe, H., et al., 2004. Preliminary study on the potential usefulness of jellyfish as fertilizer. Bull. Soc. Sea Water Sci. Jpn. 58, 209–217.
Gajbhiye, S., Wairkar, S., 2022. Collagen fabricated delivery systems for wound healing: a new roadmap. Biomater. Adv. 142.
Gavriely, S., Gulakhmedova, T., Yecheskel, Y., Rubin, A.E., Xing, B.S., Richter, S., et al., 2022. Slow release of copper from jellyfish-based hydrogels for soil enrichment. NanoImpact 27.
Getino-Mamet, L.N., Gómez Daglio, L., García De-Léon, J.G., 2019. High genetic differentiation in the edible cannonball jellyfish (cnidaria: Scyphozoa: *Stomolophus* spp.) from the Gulf of California, Mexico. Fish. Res. 219, 105328.
Gibbons, M.J., Boero, F., Brotz, L., 2016. We should not assume that fishing jellyfish will solve our jellyfish problem. ICES J. Mar. Sci. 73, 1012–1018.
Gibbons, M.J., 2023. Jellyfish, people and the United Nations' Sustainable Development Goals. 7th International Jellyfish Blooms Symposium, Kerala, India.
Goldstein, J.S., Nelson, B., 2011. Application of a gelatinous zooplankton tank for the mass production of larval Caribbean spiny lobster, *Panulirus argus*. Aquat. Living Resour. 24, 45–51.
Gómez Daglio, L., Dawson, M.N., 2017. Species richness of jellyfishes (Scyphozoa: Discomedusae) in the Tropical Eastern Pacific: missed taxa, molecules, and morphology match in a biodiversity hotspot. Invertebr. Syst. 31, 635–663.
Graham, W.M., Gelcich, S., Robinson, K.L., Duarte, C.M., Brotz, L., Purcell, J.E., et al., 2014. Linking human well-being and jellyfish: ecosystem services, impacts, and societal responses. Front. Ecol. Environ. 12, 515–523.
Guan, S., Zhang, P.-G., Liu, C.-Y., 2004. Pond cultivation state, existing problem and successful example of Rhopilema esculenta of Liaoning Province (in Chinese). Fish. Sci. 23, 30–31.
Gul, S., Morandini, A.C., 2013. New records of scyphomedusae from Pakistan coast: *Catostylus perezi* and *Pelagia* cf. *noctiluca*. Mar. Biodivers. Rec. 6, 1–6.
Gul, S., Morandini, A.C., 2015. First record of the jellyfish *Rhopilema hispidum* from the coast of Pakistan. Mar. Biodivers. Rec. 8, 1–4.
Guo, K., Zhao, W., Jiang, Z.Q., Dong, S.L., 2017. A study of organic carbon, nitrogen and phosphorus budget in jellyfish-shellfish-fish-prawn polyculture ponds. Aquac. Res. 48, 68–76.
Guo, K., Zhao, W., Wang, S., Dong, S.L., 2014. Study of food web structure and trophic level in the sea ponds of an optimized culture model (jellyfish-shellfish-fish-prawn). Aquac. Int. 22, 1783–1791.
Ha, J., Kim, E., Lee, B.G., Yeo, M.K., 2020. Capture and toxicity assessment of Ag citrate nanoparticles using jellyfish extract. Mol. Cell. Toxicol. 16, 431–439.
Hanaoka, K., Ohno, H., Wada, N., Ueno, S., Goessler, W., Kuehnelt, D., et al., 2001. Occurrence of organo-arsenicals in jellyfishes and their mucus. Chemosphere 44, 743–749.
Hickman, C., Ober, W., Garrison, C., 2006. Principios Integrales de Zoología. McGraw-Hill-Interamericana.
Holst, S., Tiseo, G.R., Djeghri, N., Sötje, I., 2024. Approaches and findings in histological and micromorphological research on Rhizostomeae since 1970. Adv. Mar. Biol. 98, 99–192.
Hong, H., 2002. The jellyfish and *Rhopilema esculentum*. Biol. Bull. 37, 13–16.
Hossain, S.T., Sugimoto, H., Asagi, N., Araki, T., Ueno, H., Morokuma, M., et al., 2013. The use of desalinated-dried jellyfish and rice bran for controlling weeds and rice yield. J. Org. Syst. 8, 28–37.

Hoyer, B., Bernhardt, A., Lode, A., Heinemann, S., Sewing, J., Klinger, M., et al., 2014. Jellyfish collagen scaffolds for cartilage tissue engineering. Acta Biomater. 10, 883–892.

Hsieh, Y.H.P., Leong, F.M., Rudloe, J., 2001. Jellyfish as food. Hydrobiologia 451, 11–17.

Hsieh, Y.H.P., Rudloe, J., 1994. Potential of utilizing jellyfish as food in Western countries. Trends Food Sci. Technol. 5, 225–229.

Huntington, J.A., 2005. Molecular recognition mechanisms of thrombin. J. Thromb. Haemost. 3, 1861–1872.

Hussein, O.S., Saleh, O.I., 2014. Effect of soaking in jellyfish on some parameters of wheat and lentil seedlings. J. Environ. Sci. Toxicol. Food Technol. 8, 32–39.

Hussein, O.S., Sayed, R.M., Saleh, O.I., 2015. Uses of jellyfish in pre sowing seeds treatment and pest control. Am. J. Exp. Agric. 5, 60–69.

Jarms, G., Morandini, A.C., 2019. World Atlas of Jellyfish. Dölling und Galitz Verlag, Germany, pp. 816.

Jeong, S.I., Kim, S.Y., Cho, S.K., Chong, M.S., Kim, K.S., Kim, H., et al., 2007. Tissue-engineered vascular grafts composed of marine collagen and PLGA fibers using pulsatile perfusion bioreactors. Biomaterials 28, 1115–1122.

Kailola, P.J., Williams, M.J., Stewart, P.C., Reichelt, R.E., McNee, A., Grieve, C., 1993. Australian Fisheries Resources. Bureau of Resource Science and Fisheries Research and Development Corporation, Canberra, Australia.

Karunarathne, K.D., Amaya, H.K.T., Digamadulla, D.S., de Croos, M.D.S.T., 2024. Redescription of the commercially harvested jellyfish, *Crambionella orsini and Lobonemoides gracilis* (Cnidaria: Scyphozoa) from Sri Lanka, Mar. Biol. Res. https://www.tandfonline.com/doi/abs/10.1080/17451000.2024.2371322

Karunarathne, K.D., de Croos, M.D.S.T., 2021. Jellyfish species used as live baits in traditional trap fishery of Sri Lanka: *Acromitus flagellatus* and *Lychnorhiza malayensis* (Cnidaria: Scyphozoa). J. Indian. Soc. Coastal Agric. Res. 39, 172–182. https://epubs.icar.org.in/ejournal/index.php/JISCAR/article/view/110892

Khong, N.M.H., Yusoff, F.M., Jamilah, B., Basri, M., Maznah, I., Chan, K.W., et al., 2016. Nutritional composition and total collagen content of three commercially important edible jellyfish. Food Chem. 196, 953–960.

Kim, D.W., Baek, T.S., Kim, Y.J., Choi, S.K., Lee, D.W., 2016. Moisturizing effect of jellyfish collagen extract. J. Soc. Cosmet. Sci. Korea 42, 153–162.

Kim, S.-W., Ezaki, T., Lee, Y.-T., Teramoto, Y., Chun, K.-W., 2018. Evaluating the effect of jellyfish chips on the survival and growth of *Pinus thunbergii* seedlings planted in a coastal area of Ehime Prefecture. Jpn. J. For. Environ. Sci. 34, 196–198.

Kim, Y.-R., Kim, S.-W., Damdinsuren, E., Ezaki, T., Chun, K.W., 2012. The effect of jellyfish (*Nemopilema nomurai*) fertilizer on tree growth in hillside erosion control works. J. For. Sci. 28, 227–231.

Kingsford, M.J., Pitt, K.A., Gillanders, B.M., 2000. Management of jellyfish fisheries, with special reference to the Order Rhizostomeae. Oceanogr. Mar. Biol. 38, 85–156.

Kitamura, M., Omori, M., 2010. Synopsis of edible jellyfishes collected from Southeast Asia, with notes on jellyfish fisheries. Plankton Benthos Res 5, 106–118.

Kuthalingam, M.D.K., James, D.B., Sarvesan, R., Devadoss, P., Manivasagam, S., Thirumilu, P., 1989. A note on the processing of the jelly fish at Alambaraikuppam near Mahabalipuram. Marine Fisheries Information Service, Technical and Extension Series (98). Central Marine Fisheries Research Institute, Indian Council of Agricultural Research, Cochin, India, pp. 8–10.

Lee, S.J., Kim, S.Y., Lee, Y.M., 2007. Preparation of porous collagen/hyaluronic acid hybrid scaffolds for biomimetic functionalization through biochemical binding affinity. J. Biomed. Mater. Res. Part. B Appl. Biomater. 82B, 506–518.

Lee, C.H., Singla, A., Lee, Y., 2001. Biomedical applications of collagen. Int. J. Pharm. 221, 1–22.

Leone, A., 2020. European Jellyfish Cookbook. Consiglio Nazionale delle Ricerche – Istituto di Scienze delle Produzioni Alimentari, Lecce, Italy.

Leone, A., De Domenico, S., Medoro, C., Cianciabella, M., Daniele, G.M., Predieri, S., 2022. Development of sensory lexicon for edible jellyfish. Agriculture 12.

Leone, A., Lecci, R., Durante, M., Piraino, S., 2013. Extract from the zooxanthellate jellyfish *Cotylorhiza tuberculata* Modulates gap junction intercellular communication in human cell cultures. Mar. Drugs 11, 1728–1762.

Leone, A., Lecci, R.M., Durante, M., Meli, F., Piraino, S., 2015. The bright side of gelatinous blooms: nutraceutical value and antioxidant properties of three Mediterranean jellyfish (Scyphozoa). Mar. Drugs 13, 4654–4681.

Leone, A., Lecci, R.M., Milisenda, G., Piraino, S., 2019. Mediterranean jellyfish as novel food: effects of thermal processing on antioxidant, phenolic, and protein contents. Eur. Food Res. Technol. 245, 1611–1627.

Li, G.H., Le, G.W., Shi, Y.H., Shrestha, S., 2004. Angiotensin I-converting enzyme inhibitory peptides derived from food proteins and their physiological and pharmacological effects. Nutr. Res. 24, 469–486.

Li, J., Ling, J., Cheng, J., 2014. On utilization of two edible macro-jellyfish and evaluation of the biomass of *Nemopilema nomurai* in China Sea. Mar. Fish. 36, 202–207.

Li, C.S., Sun, W.Q., Liu, S.C., Pan, C., Wang, D., Feng, Y., et al., 2023. Efficient aluminum removal and microorganism inhibition in ready-to-eat jellyfish by slightly acidic electrolyzed water. Food Biosci. 53.

Lin, Z., Solomon, K.L., Zhang, X.L., Pavlos, N.J., Abel, T., Willers, C., et al., 2011. In vitro evaluation of natural marine sponge collagen as a scaffold for bone tissue engineering. Int. J. Biol. Sci. 7, 968–977.

Liu, C.-s., Chen, S.-q., Zhuang, Z.-m., Yan, J.-p., Liu, C.-l., Cui, H.-t., 2015. Potential of utilizing jellyfish as food in culturing *Pampus argenteus* juveniles. Hydrobiologia 754, 189–200.

Liu, X., Zhang, M.S., Jia, A.R., Zhang, Y.G., Zhu, H.M., Zhang, C., et al., 2013. Purification and characterization of angiotensin I converting enzyme inhibitory peptides from jellyfish *Rhopilema esculentum*. Food Res. Int. 50, 339–343.

Liu, X., Zhang, M.S., Shi, Y.P., Qiao, R.J., Tang, W., Sun, Z.L., 2016. Production of the angiotensin I converting enzyme inhibitory peptides and isolation of four novel peptides from jellyfish (*Rhopilema esculentum*) protein hydrolysate. J. Sci. Food Agric. 96, 3240–3248.

Liu, X., Zhang, M.S., Zhang, C., Liu, C.H., 2012. Angiotensin converting enzyme (ACE) inhibitory, antihypertensive and antihyperlipidaemic activities of protein hydrolysates from *Rhopilema esculentum*. Food Chem. 134, 2134–2140.

López-Martinez, J., Álvarez-Tello, J., 2013. The jellyfish fishery in Mexico. Agric. Sci. 4, 57–61.

Lucas, C.H., Pitt, K.A., Purcell, J.E., Lebrato, M., Condon, R.H., 2011. What's in a jellyfish? Proximate and elemental composition and biometric relationships for use in biogeochemical studies. Ecology 92, 1704.

Lv, Z., Zhang, C.Y., Song, W., Chen, Q.S., Wang, Y.H., 2022. Jellyfish collagen hydrolysate alleviates inflammation and oxidative stress and improves gut microbe composition in high-fat diet-fed mice. Mediators Inflamm. 2022, 5628702.

Macali, A., Bergami, E., 2020. Jellyfish as innovative bioindicator for plastic pollution. Ecol. Indic. 115.

Main, D., 2014. Company plans to make paper towels and diapers out of jellyfish. Popular Science. https://www.popsci.com/article/science/company-plans-make-diapers-out-jellyfish/, April 11, 2014.

Manso, J., 2022. How jellyfish inspired Tina Craig's newest product. Yahoo!Finance. https://finance.yahoo.com/news/jellyfish-inspired-tina-craig-newest-150041022.html. (accessed 13. 07.23.).

Marin, F., Corstjens, P., de Gaulejac, B., de Vrind-De Jong, E., Westbroek, P., 2000. Mucins and molluscan calcification. Molecular characterization of mucoperlin, a novel mucin-like protein from the nacreous shell layer of the fan mussel *Pinna nobilis* (Bivalvia, pteriomorphia). J. Biol. Chem. 275, 20667–20675.

Marin, F., Luquet, G., Marie, B., Medakovic, D., 2008. Molluscan shell proteins: primary structure, origin, and evolution. Curr. Top. Dev. Biol. 80, 209–276.

Masuda, A., Baba, T., Dohmae, N., Yamamura, M., Wada, H., Ushida, K., 2007. Mucin (qniumucin), a glycoprotein from jellyfish, and determination of its main chain structure. J. Nat. Products 70, 1089–1092.

Mearns-Spragg, A., Tilman, J., Tams, D., Barnes, A., 2020. The biological evaluation of jellyfish collagen as a new research tool for the growth and culture of iPSC derived microglia. Front. Mar. Sci. 7.

Merquiol, L., Romano, G., Ianora, A., D'Ambra, I., 2019. Biotechnological applications of scyphomedusae. Mar. Drugs 17, 604.

Midura, R.J., Hascall, V.C., 1996. Bone sialoprotein – a mucin in disguise? Glycobiology 6, 677–681.

Miura, S., Kimura, S., 1985. Jellyfish mesoglea collagen – characterization of molecules as alpha-1-alpha-2-alpha-3 heterotrimers. J. Biol. Chem. 260, 5352–5356.

Miyajima, Y., Masuda, R., Yamashita, Y., 2011. Feeding preference of threadsail filefish *Stephanolepis cirrhifer* on moon jellyfish and lobworm in the laboratory. Plankton Benthos Res. 6, 12–17.

Miyajima, Y., Masuda, R., Kurihara, A., Kamata, R., Yamashita, Y., Takeuchi, T., 2011. Juveniles of threadsail filefish, *Stephanolepis cirrhifer*, can survive and grow by feeding on moon jellyfish *Aurelia aurita*. Fish. Sci. 77, 41–48.

Miyajima-Taga, Y., Masuda, R., Kurihara, A., Yamashita, Y., Takeuchi, T., 2014. Feeding moon jellyfish improves the tilting behavior of hatchery-reared red sea bream juveniles. Nippon Suisan Gakkaishi 80, 934–945.

Mohan, S., Rajapackiam, S., Rajan, S., 2011. Unusual Heavy Landings of Jellyfish *Crambionella stuhlmanni* and Processing Methods at Pulicat Landing Centre, Chennai. Marine Fisheries Information Service Technical and Extension Series. Marine Fisheries Information Service (208), Technical and Extension Series Central Marine Fisheries Research Institute, Indian Council of Agricultural Research, Cochin, India, pp. 27–29.

Morabito, R., Dossena, S., La Spada, G., Marino, A., 2014. Heavy metals affect nematocysts discharge response and biological activity of crude venom in the jellyfish *Pelagia noctiluca*. Cell. Physiol. Biochem. 34, 244–254.

Morais, Z.B., Pintao, A.M., Costa, I.M., Calejo, M.T., Bandarra, N.M., Abreu, P., 2009. Composition and in vitro antioxidant effects of jellyfish *Catostylus tagi* from Sado Estuary (SW Portugal). J. Aquat. Food Product. Technol. 18, 90–107.

Morikawa, T., 1984. Jellyfish. FAO Infofish Mark. Dig. 1, 37–39.

Morinaga, Y., Iwai, K., Tomita, H., Takaya, Y., Naraoka, T., Matsue, H., 2010. Chemical nature of a new antihypertensive peptide derived from jellyfish. Food Sci. Technol. Res. 16, 333–340.

Morishige, H., Sugahara, T., Nishimoto, S., Muranaka, A., Ohno, F., Shiraishi, R., et al., 2011. Immunostimulatory effects of collagen from jellyfish in vivo. Cytotechnology 63, 481–492.

Muhammed, F., Sultana, R., 2008. New record of edible jellyfish, *Rhizostoma pulmo* from Pakistani waters. Mar. Biodivers. Rec. 1, 1–3.

Murugan, A., Durgekar, R., 2008. Beyond the Tsunami: Status of Fisheries in Tamil Nadu, India: A Snapshot of Present and Long-Term Trends. United Nations Development Programme. Ashoka Trust for Research in Ecology and the Environment, Bangalore, India.

Nagai, T., Ogawa, T., Nakamura, T., Ito, T., Nakagawa, H., Fujiki, K., et al., 1999. Collagen of edible jellyfish exumbrella. J. Sci. Food Agric. 79, 855–858.

Nagai, T., Worawattanamateekul, W., Suzuki, N., Nakamura, T., Ito, T., Fujiki, K., et al., 2000. Isolation and characterization of collagen from rhizostomous jellyfish (*Rhopilema asamushi*). Food Chem. 70, 205–208.

Nagata, R.M., D'Ambra, I., Lauritano, C., von Montfort, G.M., Djeghri, N., Jordano, M.A., et al., 2024. Physiology and functional biology of Rhizostomeae jellyfish. Adv. Mar. Biol. 98, 255–360.

Nisa, S.A., Vinu, D., Krupakar, P., Govindaraju, K., Sharma, D., Vivek, R., 2021. Jellyfish venom proteins and their pharmacological potentials: a review. Int. J. Biol. Macromol. 176, 424–436.

Nishikawa, J., Ohtsuka, S., Mulyadi, Mujiono, N., Lindsay, D.J., Miyamoto, H., et al., 2015. A new species of the commercially harvested jellyfish *Crambionella* from central Java, Indonesia with remarks on the fisheries. J. Mar. Biol. Assoc. U.K. 95, 471–481.

Nishikawa, J., Thu, N.T., Ha, T.M., Thu, P.T., 2008. Jellyfish fisheries in northern Vietnam. Plankton Benthos Res. 3, 227–234.

Nishimoto, S., Goto, Y., Morishige, H., Shiraishi, R., Doi, M., Akiyama, K., et al., 2008. Mode of action of the immunostimulatory effect of collagen from jellyfish. Biosci. Biotechnol. Biochem. 72, 2806–2814.

Niv, Y., 2008. MUC1 and colorectal cancer pathophysiology considerations. World J. Gastroenterol. 14, 2139–2141.

Nudelman, R., Alhmoud, H., Delalat, B., Fleicher, S., Fine, E., Guliakhmedova, T., et al., 2019. Jellyfish-based smart wound dressing devices containing in situ synthesized antibacterial nanoparticles. Adv. Funct. Mater. 29, 1902783.

Ogimoto, M., Suzuki, K., Kabashima, J., Nakazato, M., Uematsu, Y., 2012. Aluminium content in foods with aluminium-containing food additives. Food Hyg. Saf. Sci. 53, 57–62.

Ohta, N., Sato, M., Ushida, K., Kokubo, M., Baba, T., Taniguchi, K., et al., 2009. Jellyfish mucin may have potential disease-modifying effects on osteoarthritis. BMC Biotechnol. 9, 98.

Omori, M., 1978. Zooplankton fisheries of the world: a review. Mar. Biol. 48, 199–205.

Omori, M., 1981. Edible jellyfish in the far east waters: a brief review of the biology and fishery. Bull. Plankton Soc. Jpn. 28, 1–111.

Omori, M., Kitamura, M., 2004. Taxonomic review of three Japanese species of edible jellyfish. Plankton Biol. Ecol. 51, 36–51.

Omori, M., Nakano, E., 2001. Jellyfish fisheries in southeast Asia. Hydrobiologia 451, 19–26.

Ozer, N.P., Celikkale, M.S., 2001. Utilization possibilities of jellyfish *Rhizostoma pulmo*, as a food in the Black Sea. J. Food Sci. Technol. 38, 175–178.

Panda, S.K., Madhu, V.R., 2009. Studies on the preponderance of jellyfish in coastal waters of Veraval. Fish. Technol. 46, 99–106.

Patwa, A., Thiery, A., Lombard, F., Lilley, M.K.S., Boisset, C., Bramard, J.F., et al., 2015. Accumulation of nanoparticles in "jellyfish" mucus: a bio-inspired route to decontamination of nano-waste. Sci. Rep. 5, 11387.

Pedersen, M.T., Brewer, J.R., Duelund, L., Hansen, P.L., 2017. On the gastrophysics of jellyfish preparation. Int. J. Gastronomy Food Sci. 9, 34–38.

Perumal, M., Wasli, M.E., Ho, S.Y., 2019. Influences of inorganic and organic fertilizers to morphological quality attributes of *Shorea macrophylla* seedlings in a tropical nursery. Biodiversitas 20, 2110–2118.

Prabhu, M.S., 1954. The perch fishery by special traps in the area around Mandapam in the Gulf of Mannar and Palk Bay. Indian J. Fish. 1, 94–129.

Prieto, L., Enrique-Navarro, A., Volsi, R.L., Ortega, M.J., 2018. The large jellyfish *Rhizostoma luteum* as sustainable a resource for antioxidant properties, nutraceutical value and biomedical applications. Mar. Drugs 16.

Pugliano, M., Vanbellinghen, X., Schwinté, P., Benkirane-Jessel, N., Keller, L., 2017. Combined jellyfish collagen Type II, human stem cells and Tgf-β3 as a therapeutic implant for cartilage repair. J. Stem Cell Res. Ther. 7.

Purcell, J.E., Uye, S., Lo, W.T., 2007. Anthropogenic causes of jellyfish blooms and their direct consequences for humans: a review. Mar. Ecol. Prog. Ser. 350, 153–174.

Pustlauk, W., Paul, B., Brueggemeier, S., Gelinsky, M., Bernhardt, A., 2017. Modulation of chondrogenic differentiation of human mesenchymal stem cells in jellyfish collagen scaffolds by cell density and culture medium. J. Tissue Eng. Regenerat. Med. 11, 1710–1722.

Ramires, F.A., Bleve, G., De Domenico, S., Leone, A., 2022. Combination of solid state and submerged fermentation strategies to produce a new jellyfish-based food. Foods 11 (24), 3974.

Ramires, F.A., De Domenico, S., Migoni, D., Fanizzi, F.P., Angel, D.L., Slizyte, R., et al., 2022. Optimization of a calcium-based treatment method for jellyfish to design food for the future. Foods 11 (17), 2697.

Raoult, V., Gaston, T.F., 2018. Rapid biomass and size-frequency estimates of edible jellyfish populations using drones. Fish. Res. 207, 160–164.

Raposo, A., Coimbra, A., Amaral, L., Gonçalves, A., Morais, Z., 2018. Eating jellyfish: safety, chemical and sensory properties. J. Sci. Food Agric. 98, 3973–3981.

Ren, Y.C., Dong, S.L., Wang, X.B., Gao, Q.F., Jiang, S.H., 2014. Beneficial co-culture of jellyfish *Rhopilema esculenta* (Kishinouye) and sea cucumber *Apostichopus japonicus* (Selenka): implications for pelagic-benthic coupling. Aquac. Res. 45, 177–187.

Ricard-Blum, S., Ruggiero, F., van der Rest, M., 2005. The collagen superfamily. In: Brinckmann, J., Notbohm, H., Müller, P.K. (Eds.), Collagen. Springer, pp. 35–84.

Riccio, G., Martinez, K.A., Martin, J., Reyes, F., D'Ambra, I., Lauritano, C., 2022. Jellyfish as an alternative source of bioactive antiproliferative compounds. Mar. Drugs 20.

Richardson, A.J., Bakun, A., Hays, G.C., Gibbons, M.J., 2009. The jellyfish joyride: causes, consequences and management responses to a more gelatinous future. Trends Ecol. Evol. 24, 312–322.

Rudloe, J., 1992. Jellyfish: A New Fishery for the Florida Panhandle. Report prepared for The Apalachee Regional Planning Council, Blounstown, Florida. U.S. Department of Commerce, Economic Development Administration, EDA Project No. 04-06-03801, p. 35.

Salama, W.M., El-Naggar, S.A., Attia, H.M., Hamed, E.H., 2022. The cytotoxic effect of rhizostome jellyfish venom on cancer cell lines and its antitumor activity in Ehrlich ascites-bearing mice. Egypt. J. Zool. 78, 15–30.

Samaraweera, V.D., Dissanayake, D.C.T., 2022. Use of jellyfish as a potential organic fertilizer and its effects on the growth of okra, *Abelmoschus esculentus*. Ceylon J. Sci. 51, 299–306.

Schiariti, A., Holst, S., Tiseo, G.R., Miyake, H., Morandini, A.C., 2024. Life cycles and reproduction of Rhizostomeae. Adv. Mar. Biol. 98, 193–254.

Seo, J.-I., Kim, Y.-R., Otsuki, K., Kim, S.-W., Yeom, K.-J., Yi, J.-S., et al., 2014. Contributions of jellyfish fertilizer to survival and growth of seedlings planted in a recently burned forest, Republic of Korea. J. Fac. Agric. Kyushu Univ. 59, 181–189.

Silva, T.H., Moreira-Silva, J., Marques, A.L.P., Domingues, A., Bayon, Y., Reis, R.L., 2014. Marine origin collagens and its potential applications. Mar. Drugs 12, 5881–5901.

Skeggs, L.T., Kahn, J.R., Shumway, N.P., 1956. The preparation and function of the hypertensin-converting enzyme. J. Exp. Med. 103, 295–299.

Sloan, N.A., 1986. World jellyfish and tunicate fisheries, and the Northeast Pacific echinoderm fishery. Can. Spec. Publ. Fish. Aquat. Sci. 92, 23–33.

Sloan, N.A., Gunn, C.R., 1985. Fishing, processing, and marketing of the jellyfish, *Aurelia aurita*, from southern British Columbia. Can. Ind. Rep. Fish. Aquat. Sci. 157, 29 + v.

Song, E., Kim, S.Y., Chun, T., Byun, H.J., Lee, Y.M., 2006. Collagen scaffolds derived from a marine source and their biocompatibility. Biomaterials 27, 2951–2961.

Steinberger, L.R., Gulakhmedova, T., Barkay, Z., Gozin, M., Richter, S., 2019. Jellyfish-based plastic. Adv. Sustain. Syst. 3.

Sugahara, T., Ueno, M., Goto, Y., Shiraishi, R., Doi, M., Akiyama, K., et al., 2006. Immunostimulation effect of jellyfish collagen. Biosci. Biotechnol. Biochem. 70, 2131–2137.

Suganthi, K., Bragadeeswaran, S., 2012. Antimicrobial and immunomodulatory activities of jellyfish (*Chrysaora quinquecirrha*) venom. In: Sabu, A., Augustine, A. (Eds.), Prospects in Bioscience: Addressing the Issues. Springer, pp. 283–292.

Templeman, M.A., Kingsford, M.J., 2010. Trace element accumulation in *Cassiopea* sp. (Scyphozoa) from urban marine environments in Australia. Mar. Environ. Res. 69, 63–72.

Thibault, D., Kuplik, Z., Prieto, L., Brown, M.K., Uye, S.-I., Doyle, T.K., Pitt, K., Fitt, W.K., Gibbons, M.J. 2024. Ecology of Rhizostomeae. Adv. Mar. Biol. 98, 397–510.

Thiel, M.E.(†), Reinicke, G.B., Holst, S., Morandini, A.C., Sötje, I., Straehler-Pohl, I., Wiesenthal, A.A., Thiel, H., 2024. Rhizostomeae-a remaining manuscript to complete the Monograph series on Scyphozoa by M. E. Thiel (published 1936–1977). Advances in Marine Biology 98 (electronic supplement) at (*Work in progress, to be published soon*).

Thomas, P.A., 1969. Goat fishes (Family Mullidae) of the Indian Seas. Marine Biological Association of India Memoirs III.

Thornton, D.J., Rousseau, K., McGuckin, M.A., 2008. Structure and function of the polymeric mucins in airways mucus. Annu. Rev. Physiol. 459–486.

Tian, T., Chen, Z., Wang, M., Hu, Y., Wang, W., 2016. Inbreeding and genetic diversity analysis in a hatchery release population and clones of *Rhopilema esculentum* based on microsatellite markers. Chin. J. Oceanol. Limnol.

Ushida, K., Inui, H., Kaneko, T., Tanaka, S., Mochizuki, A., Kaise, S., et al., 2024. Preparation of jellyfish mucin. In: Kameyama, A. (Ed.), Mucins: Methods and Protocols. Methods in Molecular Biology 2763 Springer Nature.

Ushida, K., Sato, R., Momma, T., Tanaka, S., Kaneko, T., Morishita, H., 2022. Jellyfish mucin (qniumucin) extracted with a modified protocol indicated its existence as a constituent of the extracellular matrix. Biochim. Biophys. Acta Gen. Subj. 1866, 130189.

Varghese, M., Kasinathan, C., Gandhi, A., 2008. Trap Fishing in the Gulf of Mannar and Palk Bay. Marine Fisheries Information Service (197), Technical and Extension Series Central Marine Fisheries Research Institute. Indian Council of Agricultural Research,, Cochin, India, pp. 7–8.

Victoria, F., 2006. Statement of Management Arrangements for the Victorian Developmental Jellyfish Fishery (*Catostylus mosaicus*), Statement prepared for the Australian Department of Heritage and Environment. Fisheries Victoria Division of the Department of Primary Industries,, Victoria, Australia, pp. 15.

Villalba-Urquidy, B.D., Velazquez-Valdez, L.P., Bracamontes-Picos, S.J., Del Toro-Sánchez, C.L., Chan-Higuera, J.E., Ezquerra-Brauer, J.M., 2022. Conversion of dry-salted cannonball jellyfish (*Stomolophus meleagris*) umbrella and oral arms to cornmeal snacks and gelatin with antioxidant properties. Fishes 7.

Wakabayashi, K., Nagai, S., Tanaka, Y., 2016. The complete larval development of *Ibacus ciliatus* from hatching to the nisto and juvenile stages using jellyfish as the sole diet. Aquaculture 450, 102–107.

Wakabayashi, K., Sato, R., Ishii, H., Akiba, T., Nogata, Y., Tanaka, Y., 2012. Culture of phyllosomas of *Ibacus novemdentatus* in a closed recirculating system using jellyfish as food. Aquaculture 330, 162–166.

Wakabayashi, K., Sato, H., Yoshie-Stark, Y., Ogushi, M., Tanaka, Y., 2016. Differences in the biochemical compositions of two dietary jellyfish species and their effects on the growth and survival of *Ibacus novemdentatus* phyllosomas. Aquac. Nutr. 22, 25–33.

Watanabe, Y., Ochi, Y., Sugimoto, H., Kato-Noguchi, H., 2015. Weed inhibitory activity of Nomura's jellyfish. Environ. Control. Biol. 53, 165–167.

Widdowson, J.P., Picton, A.J., Vince, V., Wright, C.J., Mearns-Spragg, A., 2018. In vivo comparison of jellyfish and bovine collagen sponges as prototype medical devices. J. Biomed. Mater. Res. Part. B Appl. Biomater. 106, 1524–1533.

Yakovlev, Y.M., Borodin, P.A., Osipov, E.V., 2005. The fishery of jellyfish *Rhopilema* in Peter the Great Bay. Rybnoe Khozyaistvo 5, 72–75.

You, K., Ma, C.H., Gao, H.W., Li, F.Q., Zhang, M.Z., Qiu, Y.T., et al., 2007. Research on the jellyfish (*Rhopilema esculentum*) and associated aquaculture techniques in China: current status. Aquac. Int. 15, 479–488.

Youssef, J., Keller, S., Spence, C., 2019. Making sustainable foods (such as jellyfish) delicious. Int. J. Gastronomy Food Sci. 16.

Yu, H.H., Li, R.F., Dong, X.L., Xing, R.G., Liu, S., Li, P.C., 2014. Efficacy of venom from tentacle of jellyfish *Nemopilema nomurai* against the cotton bollworm *Helicoverpa armigera*. Biomed. Res. Int.

Yu, H.H., Li, R.F., Wang, X.Q., Yue, Y., Liu, S., Xing, R.O.G., et al., 2021. Field experiment effect on citrus spider mite *Panonychus citri* of venom from jellyfish *Nemopilema nomurai*: the potential use of jellyfish in agriculture. Toxins 13.

Yu, H.H., Liu, X.G., Dong, X.L., Li, C.P., Xing, R.E., Liu, S., et al., 2005. Insecticidal activity of proteinous venom from tentacle of jellyfish *Rhopilema esculentum* Kishinouye. Bioorg. Med. Chem. Lett. 15, 4949–4952.

Yu, H.H., Liu, X.G., Xing, R.E., Liu, S., Li, C., Li, P.C., 2005. Radical scavenging activity of protein from tentacles of jellyfish *Rhopilema esculentum*. Bioorg. Med. Chem. Lett. 15, 2659–2664.

Yu, H.H., Liu, X.G., Xing, R.E., Liu, S., Guo, Z.Y., Wang, P.B., et al., 2006. In vitro determination of antioxidant activity of proteins from jellyfish *Rhopilema esculentum*. Food Chem. 95, 123–130.

Yuferova, A.A., 2017. The impact of different drying modes of scyphozoan jellyfish Rhopilema esculentum and Aurelia aurita on the protein and carbohydrate components in their composition and the possibility of their use as dried prepared food. J. Food Process. Eng. 40.

Zhang, C.H., Cao, W.H., Hong, P.Z., Ji, H.W., Qin, X.M., He, J.F., 2009. Angiotensin I-converting enzyme inhibitory activity of *Acetes chinensis* peptic hydrolysate and its antihypertensive effect in spontaneously hypertensive rats. Int. J. Food Sci. Technol. 44, 2042–2048.

Zhang, H., Zhang, J.Y., Wang, H.L., Luo, P.J., Zhang, J.B., 2016. The revision of aluminum-containing food additive provisions in China. Biomed. Environ. Sci. 29, 461–466.

Zhuang, Y.L., Hou, H., Zhao, X., Zhang, Z.H., Li, B.F., 2009. Effects of collagen and collagen hydrolysate from jellyfish (*Rhopilema esculentum*) on mice skin photoaging induced by UV irradiation. J. Food Sci. 74, H183–H188.

Zhuang, Y., Sun, L., Li, B., 2012. Production of the angiotensin-I-converting enzyme (ACE)-inhibitory peptide from hydrolysates of jellyfish (*Rhopilema esculentum*) collagen. Food Bioprocess. Technol. 5, 1622–1629.

Zhuang, Y., Sun, L., Zhang, Y., Liu, G., 2012. Antihypertensive effect of long-term oral administration of jellyfish (*Rhopilema esculentum*) collagen peptides on renovascular hypertension. Mar. Drugs 10, 417–426.

Zhuang, Y.L., Sun, L.P., Zhao, X., Wang, J.F., Hou, H., Li, B.F., 2009. Antioxidant and melanogenesis-inhibitory activities of collagen peptide from jellyfish (*Rhopilema esculentum*). J. Sci. Food Agric. 89, 1722–1727.

Zhuang, Y.L., Sun, L.P., 2012. Anti-melanogenic activities of collagen peptides from jellyfish (*Stomolophus meleagris*) In: International Conference on Materials for Environment Protection Energy Applications (MEPEA), Kuala Lumpur, Malays, pp. 505–512.

9780443296468